T0328619

Differential Equations with *Mathematica*

FOURTH EDITION

Differential Equations with *Mathematica*

FOURTH EDITION

Martha L. Abell

Georgia Southern University, Statesboro, USA

James P. Braselton

Georgia Southern University, Statesboro, USA

AMSTERDAM • BOSTON • HEIDELBERG • LONDON
NEW YORK • OXFORD • PARIS • SAN DIEGO
SAN FRANCISCO • SINGAPORE • SYDNEY • TOKYO
Academic Press is an imprint of Elsevier

Academic Press is an imprint of Elsevier
125 London Wall, London EC2Y 5AS, United Kingdom
525 B Street, Suite 1800, San Diego, CA 92101-4495, United States
50 Hampshire Street, 5th Floor, Cambridge, MA 02139, United States
The Boulevard, Langford Lane, Kidlington, Oxford OX5 1GB, United Kingdom

Notices
Knowledge and best practice in this field are constantly changing. As new research and experience
broaden our understanding, changes in research methods, professional practices, or medical
treatment may become necessary.

Practitioners and researchers must always rely on their own experience and knowledge in evaluating
and using any information, methods, compounds, or experiments described herein. In using such
information or methods they should be mindful of their own safety and the safety of others, including
parties for whom they have a professional responsibility.

To the fullest extent of the law, neither the Publisher nor the authors, contributors, or editors, assume
any liability for any injury and/or damage to persons or property as a matter of products liability,
negligence or otherwise, or from any use or operation of any methods, products, instructions, or ideas
contained in the material herein.

Library of Congress Cataloging-in-Publication Data
A catalog record for this book is available from the Library of Congress

British Library Cataloguing-in-Publication Data
A catalogue record for this book is available from the British Library

ISBN: 978-0-12-804776-7

For information on all Academic Press publications
visit our website at https://www.elsevier.com/

Working together
to grow libraries in
developing countries

www.elsevier.com • www.bookaid.org

Publisher: Nikki Levy
Acquisition Editor: Graham Nisbet
Editorial Project Manager: Susan Ikeda
Production Project Manager: Mohanapriyan Rajendran
Cover Designer: Maria Inês Cruz

Typeset by SPi Global, India

Contents

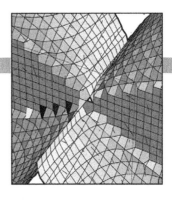

Preface . . . . **xiii**

1 Introduction to Differential Equations **1**

  1.1 Definitions and Concepts 2

  1.2 Solutions of Differential Equations 7

  1.3 Initial and Boundary-Value Problems 19

  1.4 Direction Fields 28

      1.4.1 Creating Interactive Applications 41

2 First-Order Ordinary Differential Equations **45**

  2.1 Theory of First-Order Equations: A Brief Discussion 45

  2.2 Separation of Variables 52

      Application: Kidney Dialysis 62

  2.3 Homogeneous Equations 66

      Application: Models of Pursuit 72

  2.4 Exact Equations 76

  2.5 Linear Equations 82

      2.5.1 Integrating Factor Approach 83

      2.5.2 Variation of Parameters and the Method of Undetermined Coefficients . . 96

      Application: Antibiotic Production 100

2.6   Numerical Approximations of Solutions to First-Order Equations . 103

    2.6.1   Built-In Methods 103

    Application: Modeling the Spread of a Disease 108

    2.6.2   Other Numerical Methods 114

    Euler's Method 115

    Improved Euler's Method 120

    The Runge-Kutta Method 124

3   Applications of First-Order Equations **133**

3.1   Orthogonal Trajectories 133

    Application: Oblique Trajectories 142

3.2   Population Growth and Decay 145

    3.2.1   The Malthus Model 145

    3.2.2   The Logistic Equation 152

    Application: Harvesting 163

    Application: The Logistic Difference Equation 166

3.3   Newton's Law of Cooling 171

3.4   Free-Falling Bodies 177

4   Higher-Order Differential Equations **187**

4.1   Preliminary Definitions and Notation 187

    4.1.1   Introduction 187

    4.1.2   The nth-Order Ordinary Linear Differential Equation 193

    4.1.3   Fundamental Set of Solutions 200

    4.1.4   Existence of a Fundamental Set of Solutions 205

    4.1.5   Reduction of Order 207

4.2   Solving Homogeneous Equations With Constant Coefficients 210

    4.2.1   Second-Order Equations 211

    4.2.2   Higher-Order Equations 215

4.3   Introduction to Solving Nonhomogeneous Equations 223

4.4   Nonhomogeneous Equations With Constant Coefficients:
The Method of Undetermined Coefficients 229

    4.4.1   Second-Order Equations 231

    4.4.2   Higher-Order Equations 248

4.5   Nonhomogeneous Equations With Constant Coefficients:
Variation of Parameters 255

    4.5.1   Second-Order Equations 255

    4.5.2   Higher-Order Nonhomogeneous Equations 259

4.6 Cauchy-Euler Equations 262

    4.6.1 Second-Order Cauchy-Euler Equations 263

    4.6.2 Higher-Order Cauchy-Euler Equations 267

    4.6.3 Variation of Parameters 272

4.7 Series Solutions 275

    4.7.1 Power Series Solutions About Ordinary Points 275

    4.7.2 Series Solutions About Regular Singular Points 287

    4.7.3 Method of Frobenius 289

    Application: Zeros of the Bessel Functions of the First Kind 302

    Application: The Wave Equation on a Circular Plate 304

4.8 Nonlinear Equations 308

5 Applications of Higher-Order Differential Equations **329**

5.1 Harmonic Motion 329

    5.1.1 Simple Harmonic Motion 329

    5.1.2 Damped Motion 339

    5.1.3 Forced Motion 352

    5.1.4 Soft Springs 369

    5.1.5 Hard Springs 372

    5.1.6 Aging Springs 374

    Application: Hearing Beats and Resonance 376

5.2 The Pendulum Problem 377

5.3 Other Applications 390

    5.3.1 L-R-C Circuits 390

    5.3.2 Deflection of a Beam 394

    5.3.3 Bodé Plots 397

    5.3.4 The Catenary 402

6 Systems of Ordinary Differential Equations **415**

6.1 Review of Matrix Algebra and Calculus 415

    6.1.1 Defining Nested Lists, Matrices, and Vectors 415

    6.1.2 Extracting Elements of Matrices 421

    6.1.3 Basic Computations With Matrices 423

    6.1.4 Systems of Linear Equations 426

    6.1.5 Eigenvalues and Eigenvectors 429

    6.1.6 Matrix Calculus 434

6.2   Systems of Equations: Preliminary Definitions and Theory   435
 6.2.1   *Preliminary Theory*   439
 6.2.2   *Linear Systems*   451

6.3   Homogeneous Linear Systems With Constant Coefficients   461
 6.3.1   *Distinct Real Eigenvalues*   462
 6.3.2   *Complex Conjugate Eigenvalues*   468
 6.3.3   *Alternate Method for Solving Initial-Value Problems*   477
 6.3.4   *Repeated Eigenvalues*   480

6.4   Nonhomogeneous First-Order Systems: Undetermined
 Coefficients, Variation of Parameters, and the Matrix Exponential .   488
 6.4.1   *Undetermined Coefficients*   489
 6.4.2   *Variation of Parameters*   493
 6.4.3   *The Matrix Exponential*   499

6.5   Numerical Methods   507
 6.5.1   *Built-In Methods*   507
 Application: Controlling the Spread of a Disease   514
 6.5.2   *Euler's Method*   525
 6.5.3   *Runge-Kutta Method*   531

6.6   Nonlinear Systems, Linearization, and Classification
 of Equilibrium Points   535
 6.6.1   *Real Distinct Eigenvalues*   535
 6.6.2   *Repeated Eigenvalues*   542
 6.6.3   *Complex Conjugate Eigenvalues*   546
 6.6.4   *Nonlinear Systems*   550
 Classification of Equilibrium Points   551

7 Applications of Systems of Ordinary Differential Equations **565**

7.1   Mechanical and Electrical Problems With First-Order
 Linear Systems   565
 7.1.1   *L-R-C Circuits With Loops*   565
 7.1.2   *L-R-C Circuit With One Loop*   566
 7.1.3   *L-R-C Circuit With Two Loops*   569
 7.1.4   *Spring-Mass Systems*   572

7.2   Diffusion and Population Problems With First-Order
 Linear Systems   574
 7.2.1   *Diffusion Through a Membrane*   574
 7.2.2   *Diffusion Through a Double-Walled Membrane*   577
 7.2.3   *Population Problems*   581

7.3   Applications That Lead to Nonlinear Systems   585
    7.3.1   *Biological Systems: Predator-Prey Interactions, The Lotka-Volterra*
        System, and Food Chains in the Chemostat   586
    7.3.2   *Physical Systems: Variable Damping*   603
    7.3.3   *Differential Geometry: Curvature*   609

8   Laplace Transform Methods **613**

8.1   The Laplace Transform   613
    8.1.1   *Definition of the Laplace Transform*   613
    8.1.2   *Exponential Order, Jump Discontinuities and*
        Piecewise-Continuous Functions   617
    8.1.3   *Properties of the Laplace Transform*   620
8.2   The Inverse Laplace Transform   626
    8.2.1   *Definition of the Inverse Laplace Transform*   626
    8.2.2   *Laplace Transform of an Integral*   633
8.3   Solving Initial-Value Problems With the Laplace Transform   635
8.4   Laplace Transforms of Step and Periodic Functions   643
    8.4.1   *Piecewise-Defined Functions: The Unit Step Function*   643
    8.4.2   *Solving Initial-Value Problems With Piecewise-Continuous*
        Forcing Functions   648
    8.4.3   *Periodic Functions*   652
    8.4.4   *Impulse Functions: The Delta Function*   662
8.5   The Convolution Theorem   668
    8.5.1   *The Convolution Theorem*   668
    8.5.2   *Integral and Integrodifferential Equations*   670
8.6   Applications of Laplace Transforms, Part I   673
    8.6.1   *Spring-Mass Systems Revisited*   673
    8.6.2   *L-R-C Circuits Revisited*   678
    8.6.3   *Population Problems Revisited*   685
    Application: The Tautochrone   686
8.7   Laplace Transform Methods for Systems   690
8.8   Applications of Laplace Transforms, Part II   704
    8.8.1   *Coupled Spring-Mass Systems*   704
    8.8.2   *The Double Pendulum*   709
    Application: Free Vibration of a Three-Story Building   715

9 Eigenvalue Problems and Fourier Series **721**

9.1   Boundary-Value Problems, Eigenvalue Problems,
Sturm-Liouville Problems 721
    9.1.1   Boundary-Value Problems 721
    9.1.2   Eigenvalue Problems 724
    9.1.3   Sturm-Liouville Problems 729

9.2   Fourier Sine Series and Cosine Series 731
    9.2.1   Fourier Sine Series 731
    9.2.2   Fourier Cosine Series 739

9.3   Fourier Series 743
    9.3.1   Fourier Series 743
    9.3.2   Even, Odd, and Periodic Extensions 754
    9.3.3   Differentiation and Integration of Fourier Series 760
    9.3.4   Parseval's Equality 765

9.4   Generalized Fourier Series 767

10 Partial Differential Equations **781**

10.1 Introduction to Partial Differential Equations and
Separation of Variables 781
    10.1.1 Introduction 781
    10.1.2 Separation of Variables 783

10.2 The One-Dimensional Heat Equation 785
    10.2.1 The Heat Equation With Homogeneous Boundary Conditions 786
    10.2.2 Nonhomogeneous Boundary Conditions 790
    10.2.3 Insulated Boundary 794

10.3 The One-Dimensional Wave Equation 798
    10.3.1 The Wave Equation 798
    10.3.2 D'Alembert's Solution 805

10.4 Problems in Two Dimensions: Laplace's Equation 809
    10.4.1 Laplace's Equation 809

10.5 Two-Dimensional Problems in a Circular Region 816
    10.5.1 Laplace's Equation in a Circular Region 817
    10.5.2 The Wave Equation in a Circular Region 821

Appendix: Getting Started **835**

Introduction to Mathematica 835
    A Note Regarding Different Versions of Mathematica 836

Getting Started With Mathematica 837
Five Basic Rules of Mathematica Syntax 843

Getting Help From Mathematica 844
Mathematica Help 849

The Mathematica Menu **853**

Bibliography **855**

Index **857**

Preface

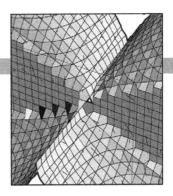

Differential Equations with Mathematica is an appropriate reference for all users of Mathematica who encounter differential equations in their profession, in particular, for beginning users like students, instructors, engineers, business people, and other professionals using Mathematica to solve and visualize solutions to differential equations. *Differential Equations with Mathematica* is a valuable supplement for students and instructors at engineering schools that use Mathematica.

Differential Equations with Mathematica introduces basic commands and includes typical examples of applications of them. A study of differential equations relies on concepts from calculus and linear algebra so the text also includes discussions of relevant commands useful in those areas. In many cases, seeing a solution graphically is most meaningful so *Differential Equations with Mathematica* relies heavily on Mathematica's outstanding graphics capabilities.

Taking advantage of Version 10 of Mathematica, *Differential Equations with Mathematica*, Fourth Edition, introduces the fundamental concepts of Mathematica to solve (analytically, numerically, and/or graphically) differential equations of interest to students, instructors, and scientists. Other features to help make *Differential Equations with Mathematica*, Fourth Edition, as easy to use and as useful as possible include the following.

1. **Version 10 Compatibility.** All examples illustrated in *Differential Equations with Mathematica*, Fourth Edition, were completed using Version 10 of Mathematica. Although most computations can continue to be carried out with earlier versions of Mathematica, we have taken advantage of the new features in Version 10 as much as possible.

2. **Applications.** Applications, many of which are documented by references, from a variety of fields, especially biology, physics, and engineering, are included throughout the text.

3. **Detailed Table of Contents.** The table of contents includes all chapter, section, and subsection headings. Along with the comprehensive index, we hope that users will be able to locate information quickly and easily.

4. **Comprehensive Index.** In the index, mathematical examples and applications are listed by topic, or name, as well as commands along with frequently used options: particular mathematical examples as well as examples illustrating how to use frequently used commands are easy to locate. In addition, commands in the index are cross-referenced with frequently used options. Functions available in the various packages are cross-referenced both by package and alphabetically.

Finally, we must express our appreciate to those who assisted in this project. We would like to express appreciation to our editors at Elsevier/Academic Press for providing a pleasant environment in which to work. We especially thank those close to us: Imogene Abell, Lori Braselton, and David Ewing for enduring with us the pressures of meeting a deadline and for graciously accepting our demanding work schedules. We certainly could not have completed this task without their care and understanding.

Martha L. Abell
(E-Mail: martha@georgiasouthern.edu)
James P. Braselton
(E-Mail: jbraselton@georgiasouthern.edu)
Statesboro, Georgia
August, 2016

Introduction to Differential Equations

1

The purpose of *Differential Equations with Mathematica*, Fourth Edition, is twofold. First, we introduce and discuss the topics covered in typical undergraduate and beginning graduate courses in ordinary and partial differential equations including topics such as Laplace transforms, Fourier series, eigenvalue problems, and boundary-value problems. Second, we illustrate how Mathematica is used to enhance the study of differential equations not only by eliminating the computational difficulties, but also by overcoming the visual limitations associated with the explicit solutions to differential equations. In each chapter, we first present the material in a manner similar to most differential equations texts and then illustrate how Mathematica can be used to solve some typical problems. For example, in Chapter 2, we introduce the topic of first-order equations. First, we show how to solve certain types of problems by hand and then show how Mathematica can be used to assist in the same solution procedures. Finally, we illustrate how Mathematica commands like DSolve and NDSolve can be used to solve some frequently encountered equations exactly and/or numerically. In Chapter 3 we discuss some applications of first-order equations. Since we are experienced and understand the methods of solution covered in Chapter 2, we make use of DSolve and similar commands to obtain solutions. In doing so, we are able to emphasize the applications themselves as opposed to becoming bogged down in calculations.

The advantages of using Mathematica in the study of differential equations are numerous, but perhaps some of the most useful include that of being able to produce the graphics associated with solutions of differential equations as well as generating very accurate numerical solutions of equations that can only be solved numerically. This is in the discussion of applications because many physical situations are modeled with differential equations. For example, we will

Differential Equations with Mathematica. http://dx.doi.org/10.1016/B978-0-12-804776-7.00001-2

see that the motion of a pendulum can be modeled by a differential equation. When we solve the problem of the motion of a pendulum, we use technology to actually watch the pendulum move. The same is true for the motion of a mass attached to the end of a spring as well as many other problems. In having this ability, the study of differential equations becomes much more meaningful as well as interesting.

If you are a beginning Mathematica user and, especially, new to Version 10, the **Appendix** contains an introduction to Mathematica, including discussions about entering and evaluating commands, and taking advantage of Mathematica's extensive help facilities.

Although Chapter 1 is short in length, Chapter 1 introduces examples that will be investigated in subsequent chapters. Also, the vocabulary introduced in Chapter 1 will be used throughout the text. Consequently, even though, to a large extent, it may be read quickly, subsequent chapters will take advantage of the terminology and techniques discussed here.

1.1 Definitions and Concepts

We begin our study of differential equations by explaining what a differential equation is.

Definition 1 (Differential Equation). *A **differential equation** is an equation that contains the derivative or differentials of one or more dependent variables with respect to one or more independent variables. If the equation contains only ordinary derivatives (of one or more dependent variables) with respect to a single independent variable, the equation is called an **ordinary differential equation**.*

EXAMPLE 1.1.1:   Discuss properties of the following differential equations.

1. $dy/dx = x^2/\left(y^2 \cos y\right)$ and $dy/dx + du/dx = u + x^2 y$
2. $(y-1)dx + x \cos y \, dy = 1$
3. $xy'' + xy' + \left(x^2 - n^2\right) y = 0$
4. $\begin{cases} \dfrac{dx}{dt} = (a - by)x \\[2mm] \dfrac{dy}{dt} = (-m + nx)y. \end{cases}$

SOLUTION: The equations $dy/dx = x^2/\left(y^2 \cos y\right)$ and $dy/dx + du/dx = u + x^2 y$ are examples of *ordinary differential equations* because they have only one independent variable (specified as x because of the dy/dx notation).

The equation $(y-1)dx + x \cos y\, dy = 1$ is an *ordinary differential equation* written in *differential form*. If y is the dependent variable, the equation is nonlinear in y because of the product of the y terms or because of the $\cos y$ term. A differential equation is nonlinear if it is *nonlinear in terms of the dependent variable*.

Using *prime notation*, a *solution* of the *ordinary differential equation* $xy'' + xy' + \left(x^2 - n^2\right)y = 0$, which is called **Bessel's equation**, is a function $y = y(x)$ with the property that $x\, d^2y/dx^2 + x\, dy/dx + \left(x^2 - n^2\right)y$ simplifies to identically the 0 function.

On the other hand,

Remember that if $y = y(t)$, t is the independent variable and y is the dependent variable.

$$\begin{cases} \dfrac{dx}{dt} = (a - by)x \\ \dfrac{dy}{dt} = (-m + nx)y \end{cases} \tag{1.1}$$

where a, b, m, and n are positive constants, is a *system* of two ordinary differential equations, called the **predator-prey equations**. A *solution* consists of two functions $x = x(t)$ and $y = y(t)$ that satisfy **both** equations. Predator-prey models can exhibit *very* interesting behavior as we will see when we study systems of differential equations and nonlinear systems of differential equations in more detail later.

See texts like Giordano, Weir, and Fox's *A First Course in Mathematical Modeling*, [12], and similar texts for detailed descriptions of predator-prey models.

∎

Note that a system of differential equations can consist of more than two equations. For example, the basic equations that describe the competition between two organisms, with population densities x_1 and x_2, respectively, in a chemostat are

See Smith and Waltman's *The Theory of the Chemostat*, [24], for a detailed discussion of chemostat models.

$$\begin{cases} S' = 1 - S - \dfrac{m_1 S}{a_1 + S}x_1 - \dfrac{m_2 S}{a_2 + S}x_2 \\[2mm] x_1' = x_1\left(\dfrac{m_1 S}{a_1 + S} - 1\right) \\[2mm] x_2' = x_2\left(\dfrac{m_2 S}{a_2 + S} - 1\right) \end{cases} \tag{1.2}$$

where $'$ denotes differentiation with respect to t; $S = S(t)$, $x_1 = x_1(t)$, and $x_2 = x_2(t)$. For equation (1.2), we remark that S denotes the concentration of the nutrient available to the competitors with population densities x_1 and x_2. We investigate chemostat models in more detail in Chapter 7.

If an equation contains partial derivatives of one or more dependent variables with respect to one or more independent variables, then the equation is called a **partial differential equation**.

EXAMPLE 1.1.2: Discuss properties of the following partial differential equations.

1. $u\dfrac{\partial u}{\partial t} = \dfrac{\partial u}{\partial x}$

2. $uu_x + u = u_{yy}$

SOLUTION: Because the equations involve partial derivatives of an unknown function, equations like $u\dfrac{\partial u}{\partial t} = \dfrac{\partial u}{\partial x}$ and $uu_x + u = u_{yy}$ are partial differential equations. For **Laplace's equation**, $\dfrac{\partial^2 u}{\partial x^2} + \dfrac{\partial^2 u}{\partial y^2} = 0$ a *solution* would be a function $u = u(x, y)$ such that $u_{xx} + u_{yy}$ is identically the 0 function. A *solution* $u = u(x, t)$ of the **wave equation** is a function satisfying $\dfrac{\partial^2 u}{\partial t^2} = \dfrac{\partial^2 u}{\partial x^2}$.

The partial differential equation $\dfrac{\partial u}{\partial t} = \dfrac{\partial^2 u}{\partial x^2}$ is known as the **heat equation**.

As with systems of ordinary differential equations, systems of partial differential equations can be considered. With exceptions, their study is beyond the scope of this text.

■

Generally, given a differential equation, our goal in this course will most often be to construct a *solution* (or a numerical approximation of the *solution*). The approach to solving an equation depends on various features of the equation. The first level of classification, distinguishing between *ordinary* and *partial* differential equations, was discussed above. Generally, equations with higher *order* are more difficult to solve than those with lower *order*.

Definition 2 (Order). *The **order** of a differential equation is the order of the highest-order derivative appearing in the equation.*

EXAMPLE 1.1.3: Determine the order of each of the following differential equations: (a) $dy/dx = x^2/(y^2 \cos y)$; (b) $u_{xx} + u_{yy} = 0$; (c) $(dy/dx)^4 = y + x$; and (d) $y^3 + dy/dx = 1$.

SOLUTION: (a) The order of this equation is one because the only derivative it includes is a first-order derivative, dy/dx; (b) This equation is classified as second-order because the highest-order derivatives, both u_{xx}, representing $\partial^2 u/\partial x^2$, and u_{yy}, representing $\partial^2 u/\partial y^2$, are of order two. Hence, **Laplace's equation**, $u_{xx} + u_{yy} = 0$, is a second-order partial differential equation; (c) This is a first-order equation because the highest-order derivative is the first derivative. Raising that derivative to the fourth power does not affect the order of the equation. The expressions

$$\left(\frac{dy}{dx}\right)^4 \quad \text{and} \quad \frac{d^4 y}{dx^4}$$

do not represent the same quantities: $(dy/dx)^4$ represents the derivative of y with respect to x raised to the fourth power; d^4y/dx^4 represents the fourth derivative of y with respect to x. (d) Again, we have a first-order equation, because the highest-order derivative is the first derivative.

■

Linear differential equations are defined in a manner similar to algebraic linear equations that are introduced in algebra and pre-calculus courses.

Definition 3 (Linear Differential Equation). *An ordinary differential equation (of order n) is **linear** if it is of the form*

$$a_n(x)\frac{d^n y}{dx^n} + a_{n-1}(x)\frac{d^{n-1}y}{dx^{n-1}} + \cdots + a_2(x)\frac{d^2 y}{dx^2} + a_1(x)\frac{dy}{dx} + a_0(x)y = f(x), \quad (1.3)$$

where the functions $a_i(x)$, $i = 0, 1, \ldots, n$, and $f(x)$ are given and $a_n(x)$ is not the zero function.

For the linear differential equation (1.3), $f(x)$ is called the **forcing function**.

If an equation is in the form given by equation (1.3), it is said to be in **general form**. Dividing by a_n so that the lead coefficient is 1 gives us the **standard form** of the equation:

$$\frac{d^n y}{dx^n} + a_{n-1}(x)\frac{d^{n-1}y}{dx^{n-1}} + \cdots + a_2(x)\frac{d^2 y}{dx^2} + a_1(x)\frac{dy}{dx} + a_0(x)y = f(x). \quad (1.4)$$

If the equation does not meet the requirements, definitions, or can be written in the forms described by equation (1.3) or (1.4), the equation is **nonlinear**.

A similar classification is followed for partial differential equations. In this case, the coefficients in a linear partial differential equation are functions of the independent variables.

If $f(x)$ is identically equal to the zero function, the linear equations (1.3) or (1.4) are said to be **homogeneous**.

EXAMPLE 1.1.4: Determine which of the following differential equations are linear: (a) $dy/dx = x^3$; (b) $d^2u/dx^2 + u = x^x$; (c) $(y-1)dx+x\cos y \, dy = 0$; (d) $y^{(3)}+yy' = x$; (e) $y'+x^2y = x$; (f) $x''+\sin x = 0$; (g) $u_{xx} + yu_y = 0$; and (h) $u_{xx} + u\,u_y = 0$.

SOLUTION: (a) This equation is linear, because the nonlinear term x^3 is the function $f(x)$ of the independent variable in equation (1.3). (b) This equation is also linear. Using u as the dependent variable name does not affect the linearity. (c) Solving for dy/dx we have $dy/dx = (1-y)/(x\cos y)$. Because the right-hand side of this equation includes a nonlinear function of y, the equation is *nonlinear* in y. However, solving for dx/dy, we see that

$$\frac{dx}{dy} = \frac{\cos y}{1-y}x \quad \text{or} \quad \frac{dx}{dy} - \frac{\cos y}{1-y}x = 0.$$

This equation is linear in the variable x, if we take the dependent variable to be x and the independent variable to be y in this equation. (d) The coefficient of the y' term is y and, thus, depends on y. Hence, this equation is nonlinear. (e) This equation is linear. The term x^2 is the coefficient function $a_0(x) = x^2$ of y. (f) This equation, known as the **pendulum equation** because it models the motion of a pendulum, is nonlinear because it involves a nonlinear function of x, the dependent variable in this case. (t is assumed to be the independent variable.) For this equation, the nonlinear function of x is $\sin x$. (g) This partial differential equation is linear because the coefficient of u_y is a function of one of the independent variables. (h) In this case, there is a product of u and one of its derivatives. Therefore, the equation is nonlinear.

■

In the same manner that we consider systems of equations in algebra, we can also consider systems of differential equations. For example, if x and y represent functions of t, we will learn to solve the **system of linear equations**

$$\begin{cases} dx/dt = ax + by \\ dy/dt = cx + dy \end{cases},$$

where a, b, c, and d represent constants and differentiation is with respect to t in Chapter 8. On the other hand, systems (1.1) and (1.2) involve products of the dependent variables (x and y; S, x_1, and x_2, respectively) so are **nonlinear systems of ordinary differential equations**.

We will see that linear and nonlinear systems of differential equations arise naturally in many physical situations that are modeled with more than one equation and involve more than one dependent variable.

1.2 Solutions of Differential Equations

When faced with a differential equation, our goal is frequently, but not always, to determine explicit and/or numerical *solutions* to the equation.

Definition 4 (Solution). *A **solution** to the nth-order ordinary differential equation*

$$F\left(x, y, y', y'', \ldots, y^{(n)}\right) = 0 \tag{1.5}$$

on the interval $a < x < b$ is a function $\phi(x)$ that is continuous on the interval $a < x < b$ and has all the derivatives present in the differential equation such that

$$F\left(x, \phi, \phi', \phi'', \ldots, \phi^{(n)}\right) = 0$$

on $a < x < b$.

In subsequent chapters, we will discuss methods for solving differential equations. Here, in order to understand what is meant to be a solution, we either give both the equation and a solution and then verify the solution or use Mathematica to solve equations directly.

EXAMPLE 1.2.1: Verify that the given function is a solution to the corresponding differential equation: (a) $dy/dx = 3y$, $y(x) = e^{3x}$; (b) $u'' + 16u = 0$, $u(x) = \cos 4x$; and (c) $y'' + 2y' + y = 0$, $y(x) = xe^{-x}$.

Solution: (a) Differentiating y we have $dy/dx = 3e^{3x}$ so that substitution yields

$$\frac{dy}{dx} = 3y \quad \text{or} \quad 3e^{3x} = 3e^{3x}.$$

(b) Two derivatives are required in this case: $u' = -4\sin 4x$ and $u'' = -16\cos 4x$. Therefore,

$$u'' + 16u = -16\cos 4x + 16\cos 4x = 0.$$

If you are a beginning Mathematica user, see the **Appendix** for help getting started with Mathematica. When a semi-colon (;) is included at the end of a command, the resulting output is not displayed.

(c) In this case, we illustrate how to use Mathematica. After defining y,

```
y[x_] = xExp[-x];
```

we use $' = d/dx$ to compute $y' = e^{-x} - xe^{-x}$, naming the resulting output dy.

```
dy = y'[x]
```

$$e^{-x} - e^{-x}x$$

Similarly, we use $'\,'$ to compute $y'' = 2e^{-x} + xe^{-x}$, naming the resulting output d2y.

```
d2y = y''[x]
```

$$-2e^{-x} + e^{-x}x$$

Finally, we compute $y'' + 2y' + y = 2e^{-x} + 2\left(e^{-x} - xe^{-x}\right) + xe^{-x} = 0$. The result is not automatically simplified so we use Simplify to simplify the output.

```
y''[x] + 2y'[x] + y[x]
```

$$-2e^{-x} + 2e^{-x}x + 2\left(e^{-x} - e^{-x}x\right)$$

```
d2y + 2dy + y[x]
```

$$-2e^{-x} + 2e^{-x}x + 2\left(e^{-x} - e^{-x}x\right)$$

```
Simplify[y''[x] + 2y'[x] + y[x]]
```

$$0$$

```
y''[x] + 2y'[x] + y[x]//Simplify
```

$$0$$

We graph this solution with Plot. Entering

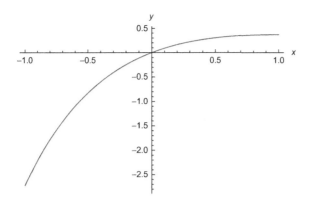

Figure 1-1   Plot of $y(x) = xe^{-x}$ on the interval $[-1, 1]$

```
Plot[y[x],{x,-1,1},PlotStyle->CMYKColor[0,0.89,0.94,0.28],
    AxesStyle → Black, AxesLabel → {x, y}]
```

graphs $y(x) = xe^{-x}$ on the interval $[-1, 1]$. See Figure 1-1. In the `Plot` command we include several options such as `PlotStyle-> CMYKColor [0, 0.89, 0.94, 0.28]`, which indicates a color choice. In the hard copy of this text, you will see a shade of gray. On the other hand, in the electronic version, you will see a deep red. Omitting the command will cause Mathematica to return to its default, which is currently a dark blue. Similarly, `AxesStyle->Black` is included because in printed versions of manuscripts, the default gray for axes is sometimes difficult to see. Our personal preference is to label axes when possible, hence the option `AxesLabel->{x,y}` is included to simply label the axes. For a basic graph, you would not need to include any of these options. Refer to the **Appendix** for more details regarding getting started with Mathematica and frequently used options with frequently used commands.

■

In the previous example, the solution is given as a function of the independent variable. In these cases, the solution is said to be explicit. In solving some differential equations, however, we can only find an equation involving x and y that the solution satisfies. In this case, the solution is said to be **implicit**.

EXAMPLE 1.2.2: Verify that the given implicit function satisfies the differential equation.

$$\text{Function: } 2x^2 + y^2 - 2xy + 5y = 0$$

$$\text{Differential Equation: } \frac{dy}{dx} = \frac{2y - 4x - 5}{2y - 2x}$$

SOLUTION: We use implicit differentiation to compute the derivative of the equation $2x^2 + y^2 - 2xy + 5y = 0$:

Assuming that $y = y(x)$, $\frac{dy}{dx} = y'$.

$$4x + 2y\frac{dy}{dx} - 2x\frac{dy}{dx} - 2y + 5 = 0$$

$$(2y - 2x)\frac{dy}{dx} = 4x - 5$$

$$\frac{dy}{dx} = \frac{2y - 4x - 5}{2y - 2x}.$$

Hence, the given implicit solution satisfies the differential equation.

We also illustrate how to use Mathematica to differentiate the equation $2x^2 + y^2 - 2xy + 5y = 0$ with respect to x. After clearing all prior definitions of x, y, and eq, if any, with Clear, we define eq to be the equation $2x^2 + y^2 - 2xy + 5y = 0$. Note how we use a double equals sign (==) to separate the left- and right-hand sides of the equation.

```
Clear[x, y]

eq = 2x^2 + y^2 − 2xy + 5x == 0
```

$$5x + 2x^2 - 2xy + y^2 == 0$$

Next, we use Dt to differentiate eq with respect to x, naming the resulting output step1. The symbol Dt[y,x] appearing in the result represents dy/dx; step1 represents the equation $4x + 2y\,y' - 2xy' - 2y + 5 = 0$.

```
step1 = Dt[eq, x]
```

$$5 + 4x - 2y - 2x\text{Dt}[y, x] + 2y\text{Dt}[y, x] == 0$$

Finally, we obtain $y' = dy/dx$ by solving the equation step1 for Dt[y,x] with Solve.

> step2 = Solve[step1, Dt[y, x]]

$$\left\{\left\{Dt[y, x] \rightarrow \frac{5 + 4x - 2y}{2(x - y)}\right\}\right\}$$

Generally, to graph an equation of the form $f(x, y) = C$, where C is a constant, we use the ContourPlot command which is used to graph level curves of surfaces: the graph of $f(x, y) = C$ is the same as the graph of the level curve of $z = f(x, y)$ corresponding to $z = C$. Thus, the graph of the equation $2x^2 + y^2 - 2xy + 5y = 0$ is the same as the graph of the level curve of $z = f(x, y) = 2x^2 + y^2 - 2xy + 5y$ corresponding to 0. Note how $2x^2 + y^2 - 2xy + 5y$, the left-hand side of the equation eq, is extracted from eq with Part ([[...]]): $2x^2 + y^2 - 2xy + 5y$ is the first part of eq.

> eq[[1]]

$$5x + 2x^2 - 2xy + y^2$$

Thus, entering

> ContourPlot[Evaluate[eq[[1]]], {x, −7, 2}, {y, −7, 2},
>
> Frame → False, Axes → Automatic, AxesStyle → Black,
>
> AxesOrigin → {0, 0}, PlotPoints → 100, Contours → {0},
>
> ContourShading → False]

graphs the equation $2x^2 + y^2 - 2xy + 5y = 0$ as shown in Figure 1-2 for $-7 \leq x \leq 2$ and $-7 \leq y \leq 2$ (the option Contours->{0} instructs Mathematica to graph only the level curve of corresponding to 0). The option ContourShading->False specifies to not shade the regions between contours, Frame->False specifies that a frame is to not be placed around the resulting graphics object, Axes->Automatic specifies that axes are to be placed on the resulting graphics object while the option AxesOrigin->{0,0} specifies that they intersect at the point $(0, 0)$ and the option AxesStyle->Black specifies that they be drawn in black so that they are easily seen when printing. The option PlotPoints->100 instructs Mathematica to increase the number of sample points to 100 (the default is 15), helping assure that the resulting graphics object appears smooth.

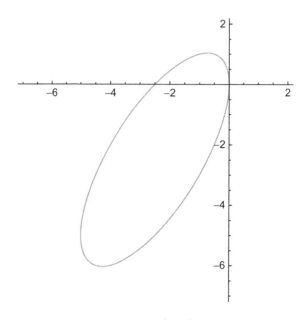

Figure 1-2   Graph of $2x^2 + y^2 - 2xy + 5y = 0$

■

For details, see Graff's *Wave Motion in Elastic Solids*, [13].

EXAMPLE 1.2.3: On a rectangular membrane, the solution of the **wave equation**,

$$\frac{\partial^2 w}{\partial x^2} + \frac{\partial^2 w}{\partial y^2} = \frac{1}{c_0}\frac{\partial^2 w}{\partial t^2} \tag{1.6}$$

takes the form

$$w = \sum_{n=1}^{\infty}\sum_{m=1}^{\infty}\left(A_{mn}\cos\omega_{mn}t + B_{mn}\sin\omega_{mn}t\right)W_{mn},$$

where A_{mn} and B_{mn} are constants, $\omega_{mn}^2 = c_0^2\left(\xi_n^2 + \zeta_m^2\right)$, $\xi_n = n\pi/a$, $\zeta_m = m\pi/b$, and the **normal modes**, W_{nm}, are given by

$$W_{nm}(x,y) = \sin\xi_n x \sin\zeta_m y.$$

(a) Verify that $f_c(x,y,t) = \cos\omega_{mn}t\, W_{mn}(x,y)$ and $f_s(x,y,t) = \sin\omega_{mn}t\, W_{mn}(x,y)$ satisfy the wave equation (1.6) on a rectangular membrane.
(b) Plot the first few normal modes of the membrane.

SOLUTION: (a) After defining $\omega_{mn}^2 = c_0^2 (\xi_n^2 + \zeta_m^2)$, $\xi_n = n\pi/a$, $\zeta_m = m\pi/b$, we define $f_c(x, y, t) = \cos \omega_{mn} t W_{mn}(x, y)$ and $f_s(x, y, t) = \sin \omega_{mn} t W_{mn}(x, y)$.

```
ω[m_, n_] = PicoSqrt[n^2/a^2 + m^2/b^2];
```

```
ξ1[n_] = nPi/a;
```

```
ξ2[m_] = mPi/b;
```

```
fc[x_, y_, t_] = Cos[ω[m, n]t]Sin[ξ1[n]x]Sin[ξ2[m]y];
```

```
fs[x_, y_, t_] = Sin[ω[m, n]t]Sin[ξ1[n]x]Sin[ξ2[m]y];
```

To verify that $f_c(x, y, t)$ satisfies equation (1.6), we compute

```
fcxx = D[fc[x, y, t], {x, 2}]
```

$$-\frac{n^2\pi^2 \text{Cos}\left[c_0\sqrt{\frac{m^2}{b^2} + \frac{n^2}{a^2}}\,\pi t\right] \text{Sin}\left[\frac{n\pi x}{a}\right]\text{Sin}\left[\frac{m\pi y}{b}\right]}{a^2}$$

```
fcyy = D[fc[x, y, t], {y, 2}]
```

$$-\frac{m^2\pi^2 \text{Cos}\left[c_0\sqrt{\frac{m^2}{b^2} + \frac{n^2}{a^2}}\,\pi t\right] \text{Sin}\left[\frac{n\pi x}{a}\right]\text{Sin}\left[\frac{m\pi y}{b}\right]}{b^2}$$

```
fctt = D[fc[x, y, t], {t, 2}]
```

$$-c_0^2\left(\frac{m^2}{b^2} + \frac{n^2}{a^2}\right)\pi^2 \text{Cos}\left[c_0\sqrt{\frac{m^2}{b^2} + \frac{n^2}{a^2}}\,\pi t\right] \text{Sin}\left[\frac{n\pi x}{a}\right]\text{Sin}\left[\frac{m\pi y}{b}\right]$$

`fcxx + fcyy - 1/c0 fctt` is simplified with `Simplify`; the result is 0 so $f_c(x, y, t)$ satisfies equation (1.6).

```
Simplify[fcxx + fcyy - 1/co^2fctt]//Simplify
```

```
0
```

On the other hand, to verify that $f_s(x, y, t)$ satisfies equation (1.6), we compute and simplify

$$\frac{\partial^2 f_s}{\partial x^2} + \frac{\partial^2 f_s}{\partial y^2} - \frac{1}{c_0}\frac{\partial^2 f_s}{\partial t^2}$$

in a single step. The result is identically equal to 0 so $f_s(x, y, t)$ also satisfies equation (1.6).

```
Simplify[D[fs[x, y, t], {x, 2}] + D[fs[x, y, t], {y, 2}]

  -1/co^2D[fs[x, y, t], {t, 2}]]
```

0

(b) To graph the normal modes, we choose $a = b = 1$. We then use
   Table and Plot3D to plot $W_{nm}(x, y)$, $0 \le x \le 1$ and $0 \le y \le 1$, for
   $n = 1, 2, 3$, and 4, $m = 1, 2$, and 3. The resulting array of graphics is
   displayed as a graphics array using Show and GraphicsArray. In
   Figure 1-3, the first row corresponds to $n = 1$, the second to $n = 2$,
   and so on; the first column corresponds to $m = 1$, the second to
   $m = 3$, and so on.

```
tp = Table[Plot3D[Sin[nPix]Sin[mPiy], {x, 0, 1}, {y, 0, 1},

   BoxRatios → {1, 1, 1}, PlotPoints → 45], {n, 1, 4}, {m, 1, 3}];

Show[GraphicsGrid[tp]]
```

■

As indicated in the previous example, without added initial or boundary
conditions many differential equations may have more than one solution, a unique
solution, or no solutions. We further illustrate this property in the following
examples where we use the Mathematica command DSolve to solve the indicated
equations. Generally, the command

```
DSolve[F[x,y[x],y'[x],...,D[y[x],{x,n}]]==0,y[x],x]
```

attempts to solve the differential equation (1.5) for y. Detailed help regarding DSolve is obtained by entering ?DSolve or by going to **Help** under the Mathematica menu

and then selecting **Help...** Once Mathematica opens the **Help Browser**, you
can either type DSolve and select **Go To** or select **Algebraic Computation**
followed by **Equation Solving** and DSolve to obtain a description of the DSolve

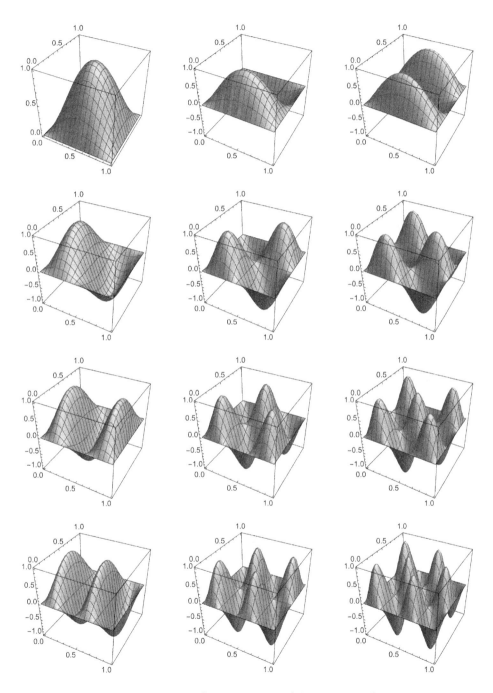

Figure 1-3 W_{nm} for $n = 1, 2, 3,$ and 4, $m = 1, 2,$ and 3

command, a discussion of its various options, and several examples, as illustrated in the following screen shot.

EXAMPLE 1.2.4: Verify that the differential equation $dy/dx = -y\cos x$ has infinitely many solutions.

SOLUTION: We use `DSolve` to solve this first-order linear equation and name the resulting list `sol`. We see that the solution is given in terms of a *replacement rule* and interpret the result to mean that if `C[1]` is any number, a solution to the equation is $y = $ `C[1]`$\mathrm{E}^{-\mathrm{Sin}[x]}$. In traditional mathematical notation, we could write that $y = Ce^{-\sin x}$ is a solution of $dy/dx = -y\cos x$ for any number C.

> The formula for the solution is extracted from `sol` with `sol[[1,1,2]]`. or, if you are using Version 10, by selecting and copying.

```
Clear[y]

sol = DSolve[y′[x] == Cos[x]y[x], y[x], x]
```

$$\left\{\left\{y[x] \to e^{\mathrm{Sin}[x]}C[1]\right\}\right\}$$

Observe that the output in `sol` is a *list*. Parts of lists are extracted with `Part`, `[[...]]`. Thus, `sol[[1]]` returns the first part of `sol`, `sol[[1,1]]` returns the first part of the first part of `sol`, and

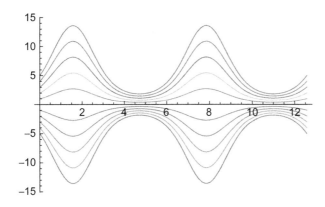

Figure 1-4 $y = Ce^{-\sin x}$ for various values of C

`sol[[1,1,2]]` returns the second part of the first part of the first part of `sol`.

> `sol[[1]]`
>
> `sol[[1, 1]]`
>
> `sol[[1, 1, 2]]`

$$\left\{ y[x] \rightarrow e^{\text{Sin}[x]} C[1] \right\}$$

$$y[x] \rightarrow e^{\text{Sin}[x]} C[1]$$

$$e^{\text{Sin}[x]} C[1]$$

To graph the solution for various values of C, we use `->` `ReplaceAll` to replace the values of `C[1]` by i for $i = -5, -4, \ldots, 4, 5$ and name the resulting list `toplot`. We then use `Plot` to graph the functions in the resulting list and show the results in Figure 1-4.

> `toplot = Table[sol[[1, 1, 2]]/. C[1] → i, {i, −5, 5}];`
>
> `Plot[toplot, {x, 0, 4Pi}]`

■

EXAMPLE 1.2.5: Verify that $y'' + y = 0$ has infinitely many solutions.

SOLUTION: We use `DSolve` to solve this second-order linear equation and name the resulting list `sol`. We interpret the result to mean that if C[1] and C[2] are any numbers, a solution to the equation is $y = $ C[1]Cos[x] + C[2]Sin[x]. In traditional mathematical notation, we write that $y = C_1 \cos x + C_2 \sin x$ is a solution of $y'' + y = 0$ for any constant values of C_1 and C_2.

```
Clear[y, sol]

sol = DSolve[y''[x] + y[x] == 0, y[x], x]
```

$\{\{y[x] \to C[1]Cos[x] + C[2]Sin[x]\}\}$

In particular, this result indicates that $y = C \cos x$ is a solution of $y'' + y = 0$ for any value of C (set $C_2 = 0$) and that $y = C \sin x$ is a solution of $y'' + y = 0$ for any value of C (set $C_1 = 0$). Some of the members of the family of solutions graphed with Plot. First, we use Table to generate a set of eleven functions obtained by replacing C in $y = C \cos x$ by -2.5, -2, -1.5, ..., 1.5, 2, and 2.5, naming the resulting set toplot1 and then a set of eleven functions obtained by replacing C in $y = C \sin x$ by $-2.5, -2, -1.5, \ldots, 1.5, 2,$ and 2.5, naming the resulting set toplot2.

```
toplot1 = Table[sol[[1, 1, 2]]/.{C[1] → i, C[2] → 0},

{i, −2.5, 2.5, .5}];

toplot2 = Table[sol[[1, 1, 2]]/.{C[2] → i, C[1] → 0},

{i, −2.5, 2.5, .5}];
```

Then, the set of functions toplot1 and toplot2 are graphed with Plot for $0 \le x \le 4\pi$. Neither graph is displayed as it is generated because we include a semicolon at the end of each Plot command. Instead, we show the graphs side-by-side using Show together with GraphicsRow in Figure 1-5.

```
q1 = Plot[toplot1, {x, 0, 4Pi}, AxesLabel → {x, y},

PlotLabel → "Plot 1"];
```

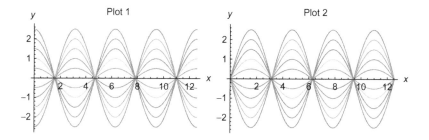

Figure 1-5   Plots of $y = C \cos x$ and $y = C \sin x$ for various values of C

```
q2 = Plot[toplot2, {x, 0, 4Pi}, AxesLabel → {x, y},

  PlotLabel → "Plot 2"];

Show[GraphicsRow[{q1, q2}]]
```

∎

1.3 Initial and Boundary-Value Problems

In many applications, we are not only given a differential equation to solve but we are given one or more conditions that must be satisfied by the solution(s) as well. For example, suppose that we want to find an antiderivative of the function $f(x) = 3x^2 - 4x$. Then, we solve the differential equation $dy/dx = 3x^2 - 4x$ by integrating.

$$\frac{dy}{dx} = 3x^2 - 4x \Longrightarrow y = \int \left(3x^2 - 4x\right) dx \Longrightarrow y = x^3 - 2x^2 + C.$$

```
step1 = Integrate[3x^2 - 4x, x] + c
```

$c - 2x^2 + x^3$

Because the solution involves an arbitrary constant and all solutions to the equation can be obtained from it, we call this a **general solution**. On the other hand, if we want to find a solution that passes through the point $(1, 4)$, we must find a solution that satisfies the *auxiliary condition* $y(1) = 4$. Substitution into $y = x^3 - 2x^2 + c$ yields $y(1) = 1^3 - 2 \cdot 1^2 + c = 4 \Longrightarrow c = 5$. Therefore, *the* member of the family of solutions $y = x^3 - 2x^2 + c$ that satisfies $y(1) = 4$ is $y = x^3 - 2x^2 + 5$. The following commands illustrate how to graph some members of the family of solutions by substituting various values of c into the general solution. We also graph the solution to the problem

$$\begin{cases} y' = 3x^2 - 4x \\ y(1) = 4 \end{cases}.$$

First, we use `Table` to generate a table of functions $x^3 - 2x^2 + c$ for $c = -20$, $-18, \ldots, 18, 20$, naming the resulting set of functions `toplot`. Note that we use c to avoid conflict with the built-in symbol C. The set of functions `toplot` is not displayed (for length reasons) because a semi-colon (;) is included at the end of the command.

```
toplot = Table[step1, {c, −20, 20, 2}];
```

To graph the functions contained in `toplot` we use `Table`. Then, we graph `toplot` with `Plot` in Figure 1-6(a).

```
q1 = Plot[toplot, {x, −2, 3}, AxesLabel → {x, y}, PlotLabel →

    "Various Plots", PlotRange → {−25, 25}];
```

The solution that satisfies the initial condition is graphed in Figure 1-6(b).

```
q2 = Plot[x^3 − 2x^2 + 5, {x, −2, 3}, AxesLabel → {x, y},

    PlotLabel → "A Particular Solution",

    PlotRange → {−25, 25}, PlotStyle → Black];
```

```
Show[GraphicsRow[{q1, q2}]]
```

Notice that this first-order equation requires one auxiliary condition to eliminate the unknown coefficient in the general solution. Frequently, the independent variable in a problem is t, which usually represents time. Therefore, we call the *auxiliary condition* of a first-order equation an **initial condition**, because it indicates the initial-value (at $t = t_0$) of the dependent variable. Problems that involve an initial condition are called **initial-value problems**.

If `DSolve` cannot find an exact solution to an initial-value problem or if numerical results are desired, the command

```
NDSolve[{y'[x]==f[x,y[x]],y[x0]==y0},y[x],{x,a,b}]
```

attempts to find a numerical solution to the initial-value problem $\{y' = f(x, y), y(x_0) = y_0\}$ valid for $a \le x \le b$. We use the **Help Browser** to obtain information about `NDSolve` and its options as well as several examples illustrating its use.

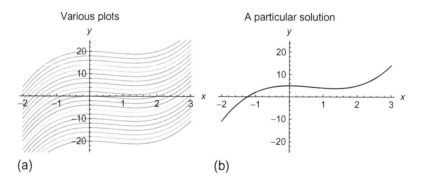

(a)   (b)

Figure 1-6   (a) Plot of $y = x^3 − 2x^2 + C$ for various values of C. (b) Plot of *the* solution that satisfies $y(1) = 4$

Notice that the syntax of the NDSolve command is almost identical to that of the DSolve command, except what we must specify an interval $[a, b]$ on which we want the numerical solution to be valid.

EXAMPLE 1.3.1: Graph the solution to the initial-value problem $\begin{cases} y' = \sin\left(x^2\right) \\ y(0) = 0 \end{cases}$ on the interval $[0, 10]$. Evaluate $y(5)$.

SOLUTION: In this case, we see that DSolve is able to solve the initial-value problem although the result is given in terms of the FresnelS function.

$$\text{exactsol} = \text{DSolve}[\{y'[x] == \text{Sin}[x\text{\textasciicircum}2], y[0] == 0\}, y[x], x]$$

$$\left\{\left\{y[x] \rightarrow \sqrt{\frac{\pi}{2}}\,\text{FresnelS}\left[\sqrt{\frac{2}{\pi}}x\right]\right\}\right\}$$

Here is Mathematica's description of the FresnelS function.

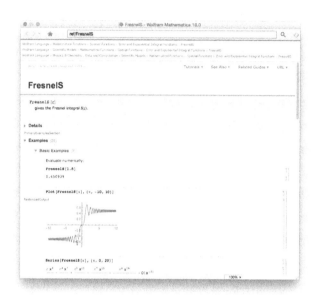

Using `NDSolve`, we obtain a numerical solution to the initial-value problem valid on the interval $[0, 10]$:

$$\texttt{numsol = NDSolve[\{y'[x] == Sin[x\^2], y[0] == 0\}, y[x], \{x, 0, 10\}]}$$

$$\{\{y[x] \rightarrow \texttt{InterpolatingFunction[][}x]\}\}$$

which we graph with `Plot` in Figure 1-7.

$$\texttt{Plot[y[x]/.numsol,\{x,0,10\},PlotRange} \rightarrow \texttt{All, PlotStyle} \rightarrow \texttt{Black,}$$

$$\texttt{AxesStyle} \rightarrow \texttt{Black, AxesLabel} \rightarrow \texttt{\{x, y\}]}$$

The value of $y(5)$ is found with `ReplaceAll`, \..

$$\texttt{numsol/.x} \rightarrow \texttt{5}$$

$$\{\{y[5] \rightarrow \texttt{0.527917}\}\}$$

and indicates that $y(5) \approx 0.5279$.

■

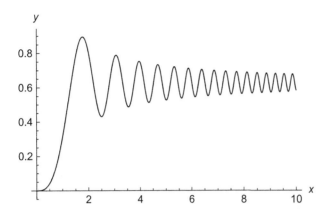

Figure 1-7 Plot of the solution of $y' = \sin\left(x^2\right)$, $y(0) = 0$ for $0 \le x \le 10$

Because first-order equations involve a single auxiliary condition, which is usually referred to as an initial condition, we use the following examples to distinguish between **initial-value** and **boundary-value** problems which involve higher-order equations.

EXAMPLE 1.3.2: Consider the second-order differential equation $x'' + x = 0$, which models the motion of a mass with $m = 1$ attached to the end of a spring with spring constant $k = 1$, where $x(t)$ represents the displacement of the mass from the equilibrium position $x = 0$ at time t. A general solution of this differential equation is found to be $x(t) = A\cos t + B\sin t$, where A and B are arbitrary constants, with DSolve.

```
Clear[x]

gensol = DSolve[x"[t] + x[t] == 0, x[t], t]
```

$$\{\{x[t] \rightarrow C[1]Cos[t] + C[2]Sin[t]\}\}$$

Because this is a second-order equation, we need two auxiliary conditions to determine the two unknown constants. Suppose that the initial displacement of the mass is $x(0) = 0$ and the initial velocity is $x'(0) = 1$. This is an **initial-value problem** because we have two auxiliary conditions given at the same value of t, namely $t = 0$. Use these initial conditions to determine the solution of this problem.

SOLUTION: Because we need the first derivative of the general solution, we calculate $x'(t) = B \cos t - A \sin t$. Substitution yields $x(0) = A = 0$ and $x'(0) = B = 1$. Hence, the solution is $x(t) = \sin t$. DSolve can solve this initial-value problem as well.

```
Clear[x]

toplot2 = DSolve[{x''[t] + x[t] == 0,x[0] == 0,x'[0] == 1},x[t],t]
```

$\{\{x[t] \to \text{Sin}[t]\}\}$

We plot various solutions to the differential equation by generating a list of functions in toplot1 obtained by replacing C[1] and C[2] in gensol with specific numbers.

```
toplot1 = Table[gensol[[1, 1, 2]]/.{C[1] → i,C[2] → j},{i,−2,2},

{j, −2, 2}];
```

We then plot the functions in toplot1 with Plot in q1

```
q1 = Plot[toplot1, {t, 0, 4Pi}, AxesStyle → Black,

    AxesLabel → {x, y}, PlotLabel → "Plot 1"];
```

and the solution to the initial value problem in q2.

```
q2 = Plot[toplot2[[1, 1, 2]], {t, 0, 4Pi}, AxesStyle → Black,

    AxesLabel → {x, y}, PlotLabel → "Plot 2"];
```

We then show the two graphs side-by-side using Show together with GraphicsRow in Figure 1-8

```
Show[GraphicsRow[{q1, q2}]]
```

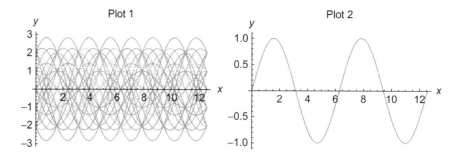

Figure 1-8   In Plot 1, various solutions of the second-order differential equation. In Plot 2, the specific function that satisfies the initial conditions $x(0) = 0$ and $x'(0) = 1$

■

EXAMPLE 1.3.3: The shape of a bendable beam of length 1 unit that is subjected to a compressive force at one end is described by the graph of the solution $y(x)$ of the differential equation $\dfrac{d^2y}{dx^2} + \dfrac{\pi^2}{4}y = 0, 0 < x < 1$. If the height of the beam above the x-axis is known at the endpoints $x = 0$ and $x = 1$, then we have a **boundary-value problem**. Use the boundary conditions $y(0) = 1$ and $y(1) = 2$ to find the shape of the beam.

SOLUTION: First, we use DSolve to find a general solution to the equation. The result indicates that a general solution is $y(x) = A\cos\left(\frac{\pi}{2}x\right) + B\sin\left(\frac{\pi}{2}x\right)$.

Later, we will see that under reasonable conditions, initial-value problems have unique solutions. On the other hand, boundary-value problems may have no solutions, infinitely many solutions or a unique solution.

```
DSolve[y"[x] + Pi^2/4y[x] == 0, y[x], x]
```

$$\left\{\left\{y[x] \to C[1]\mathrm{Cos}\left[\frac{\pi x}{2}\right] + C[2]\mathrm{Sin}\left[\frac{\pi x}{2}\right]\right\}\right\}$$

Applying the condition $y(0) = 0$ to the general solution yields

$$y(0) = A\cos 0 + B\sin 0 = A = 0.$$

Similarly, $y(1) = 2$ indicates that

$$y(1) = B\sin\frac{\pi}{2} = B = 2,$$

so the solution to the boundary value problem is $y(x) = 2\sin\left(\frac{\pi}{2}x\right)$, $0 < x < 1$. DSolve is also able to solve this boundary value problem.

```
sol = DSolve[{y"[x] + Pi^2/4y[x] == 0, y[0] == 0, y[1] == 2},
y[x], x]
```

$$\left\{\left\{y[x] \to 2\mathrm{Sin}\left[\frac{\pi x}{2}\right]\right\}\right\}$$

This function that describes the shape of the beam is graphed with Plot in Figure 1-9.

```
Plot[y[x]/.sol, {x, 0, 1}, PlotStyle → Black, AxesStyle → Black,
AxesLabel → {x, y}]
```

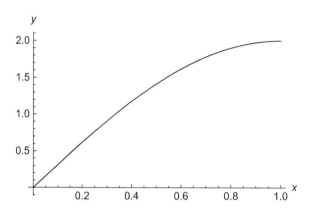

Figure 1-9   A plot of a solution to the beam equation

■

We will see that it is usually impossible to find exact solutions of higher-order nonlinear initial-value problems. In those cases, we can often use NDSolve to generate an accurate approximation of the solution.

Also see Example 1.4.4.   **EXAMPLE 1.3.4 (Rayleigh's Equation):** **Rayleigh's equation** is the nonlinear equation

$$\frac{d^2x}{dt^2} + \left[\frac{1}{3}\left(\frac{dx}{dt}\right)^2 - 1\right]\frac{dx}{dt} + x = 0 \tag{1.7}$$

and arises in the study of the motion of a violin string. Graph the solution to Rayleigh's equation on the interval $[0, 15]$ if (a) $x(0) = 1$, $x'(0) = 0$; (b) $x(0) = 0.1, x'(0) = 0$; and (c) $x(0) = 0, x'(0) = 1.9$.

SOLUTION: First, observer that DSolve cannot solve the differential equation.

DSolve[x″[t] + (1/3x′[t]^2 − 1)x′[t] + x[t] == 0, x[t], t]

$$\text{DSolve}\left[x[t] + x'[t]\left(-1 + \frac{1}{3}x'[t]^2\right) + x''[t] == 0, x[t], t\right]$$

In each case, we use NDSolve to approximate the solution to the initial-value problem, naming the results numsol1, numsol2, and numsol3, respectively.

numsol1 = NDSolve[{x″[t] + (1/3x′[t]^2 − 1)x′[t] + x[t] == 0,

  x[0] == 1, x′[0] == 0}, x[t], {t, 0, 15}]

{{x[t] → InterpolatingFunction[][t]}}

numsol2 = NDSolve[{x″[t] + (1/3x′[t]^2 − 1)x′[t] + x[t] == 0,

  x[0] == .1, x′[0] == 0}, x[t], {t, 0, 15}]

{{x[t] → InterpolatingFunction[][t]}}

numsol3 = NDSolve[{x″[t] + (1/3x′[t]^2 − 1)x′[t] + x[t] == 0,

  x[0] == 0, x′[0] == 1.9}, x[t], {t, 0, 15}]

{{x[t] → InterpolatingFunction[][t]}}

All three solutions are graphed together on the interval $[0, 15]$ with Plot in Figure 1-10. Notice that the solution to (c) appears to be periodic.

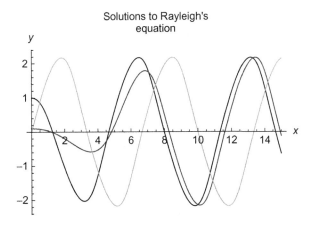

Figure 1-10 Plot of three solutions of Rayleigh's equation (equation (1.7))

```
Plot[Evaluate[x[t]/.{numsol1, numsol2, numsol3}],

  {t, 0, 15}, PlotStyle → {GrayLevel[0], GrayLevel[.3],

  GrayLevel[.6]}, PlotRange → All, AxesLabel → {x, y},

  AxesStyle → Black, PlotLabel → "Solutions to Rayleigh's

  Equation"]
```

■

1.4 Direction Fields

The geometrical interpretation of solutions to first-order differential equations of the form $dy/dx = f(x, y)$ is important to the basic understanding of problems of this type. Suppose that a solution to this equation is a function $y = \psi(x)$, so a solution is the graph of the function ψ. Therefore, if (x, y) is a point on this graph, the slope of the tangent line is given by $f(x, y)$. A set of short line segments representing the tangent lines can be constructed for a large number of points. This collection of line segments is known as the **direction field** of the differential equation and provides a great deal of information concerning the behavior of the family of solutions. This is due to the fact that by determining the slope of the tangent line for a large number of points in the plane, the shape of the graphs of the solutions can be seen without actually having a formula for them. The direction field for a differential equation provides a geometric interpretation about the behavior of the solutions of the equation. Throughout this text, we will frequently display graphs of various solutions to a differential equation along with a graph of the direction field. Direction fields are generated with the StreamPlot and StreamDensityPlot functions.

To generate a basic direction field, use StreamPlot.

```
StreamPlot[{1,f[x,y]},{x,x0,x1},{y,y0,y1}]
```

graphs a basic direction field associated with $dy/dx = f(x, y)$ for $x_0 \leq x \leq x_1$ and $y_0 \leq y \leq y_1$. We use the **Help Browser** to obtain detailed information about these commands.

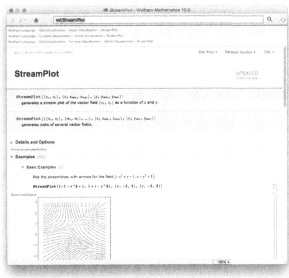

To obtain optimal results with `StreamPlot` or `StreamDensityPlot`, we find that it is useful to be able to refer to the extensive options that these two functions have. Use the appropriate help window shown above to obtain more information about the various options that allow you to customize your graphics.

```
Options[StreamPlot]

{AlignmentPoint →Center, AspectRatio →1, Axes →False, AxesLabel →None,
 AxesOrigin →Automatic, AxesStyle →{}, Background →None, BaselinePosition →Automatic,
 BaseStyle →{}, BoundaryStyle →None, BoxRatios →Automatic, ColorFunction →None,
 ColorFunctionScaling →True, ColorOutput →Automatic, ContentSelectable →Automatic,
 CoordinatesToolOptions →Automatic, DisplayFunction :→$DisplayFunction, Epilog →{},
 Evaluated →Automatic, EvaluationMonitor →None, FormatType :→TraditionalForm,
 Frame →True, FrameLabel →None, FrameStyle →{}, FrameTicks →Automatic,
 FrameTicksStyle →{}, GridLines →None, GridLinesStyle →{}, ImageMargins →0.,
 ImagePadding →All, ImageSize →Automatic, ImageSizeRaw →Automatic, LabelStyle →{},
 LightingAngle →None, MaxRecursion →Automatic, Mesh →None, MeshFunctions →Automatic,
 MeshShading →None, MeshStyle →Automatic, Method →Automatic, NormalsFunction →Automatic,
 PerformanceGoal :→$PerformanceGoal, PlotLabel →None, PlotLegends →None,
 PlotRange →{Full, Full}, PlotRangeClipping →True, PlotRangePadding →Automatic,
 PlotRegion →Automatic, PlotTheme :→$PlotTheme, PreserveImageOptions →Automatic,
 Prolog →{}, RegionFunction →{True &}, RotateLabel →True, StreamColorFunction →None,
 StreamColorFunctionScaling →True, StreamPoints →Automatic, StreamScale →Automatic,
 StreamStyle →Automatic, TargetUnits →Automatic, Ticks →Automatic, TicksStyle →{},
 VectorColorFunction →None, VectorColorFunctionScaling →True, VectorPoints →None,
 VectorScale →Automatic, VectorStyle →Automatic, WorkingPrecision →MachinePrecision}
```

EXAMPLE 1.4.1: Graph the direction field associated with the differential equation $dy/dx = e^{-x} - 2y$.

SOLUTION: Entering

```
p1 = StreamPlot[{1, Exp[-x] - 2y}, {x, -1/2, 1}, {y, -3/4, 3/4},
      Frame → False, Axes → Automatic, AxesOrigin → {0, 0},
   AxesLabel → {x,y}, StreamStyle → Black, PlotLabel → "(a)"];

p2 = StreamDensityPlot[{1, Exp[-x] - 2y}, {x, -1/2, 1}, {y, -3/4,
      3/4}, Frame → False, Axes → Automatic, AxesOrigin → {0, 0},
      AxesLabel → {x, y}, PlotLabel → "(b)"];

p3 = StreamPlot[{1, Exp[-x] - 2y}, {x, -1/2, 1}, {y, -3/4, 3/4},
      Frame → False, Axes → Automatic, AxesOrigin → {0, 0},
   AxesLabel → {x,y}, StreamStyle → Black, StreamScale → None,
      PlotLabel → "(c)"];

Show[GraphicsRow[{p1, p2, p3}]]
```

graphs various direction fields of $dy/dx = e^{-x} - 2y$ using `StreamPlot` and `StreamDensityPlot`. We show the three plots together using `Show` together with `GraphicsRow` in Figure 1-11.

Generally, Mathematica is able to find a general solution of first-order linear equations like this with `DSolve`.

There are many ways to visualize the solutions to the problem. For example, a general solution of this first-order linear equation is found to be $y = e^{-x} + Ce^{-2x}$ with `DSolve`.

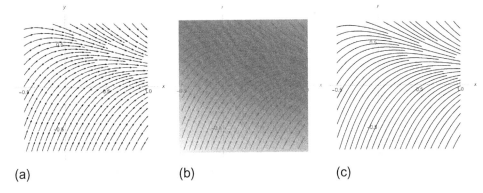

(a)   (b)   (c)

Figure 1-11   Various views of the direction field of $dy/dx = e^{-x} - 2y$

```
Clear[x, y, diffeq]
gensol = DSolve[y'[x] == Exp[-x] - 2y[x], y[x], x]
```

$$\{\{y[x] \to e^{-x} + e^{-2x}C[1]\}\}$$

```
gensol[[1]]
gensol[[1, 1]]
gensol[[1, 1, 2]]
```

$$\{y[x] \to e^{-x} + e^{-2x}C[1]\}$$

$$y[x] \to e^{-x} + e^{-2x}C[1]$$

$$e^{-x} + e^{-2x}C[1]$$

first clears all prior definitions of x, y, and diffeq, if any, and then solves the equation $dy/dx = e^{-x} - 2y$ for $y = y(x)$, naming the resulting output gensol. In the DSolve command, the first argument (y'[x] == e^{-x} − 2y[x]) represents the equation $dy/dx = e^{-x} - 2y$, the second argument (y[x]) instructs Mathematica that we are solving for $y = y(x)$, and the third argument (x) instructs Mathematica that the independent variable is x. Note that gensol is a nested list. The first part of gensol, extracted with gensol[[1]], is the list $\{y(x) \to e^{-x} + e^{-2x}C[1]\}$ the first part of this list, extracted with gensol[[1,1]], is the list y(x) → e^{-x} + e^{-2x}C[1]; and the first part of this list, extracted with gensol[[1,1,1]], is $y(x)$ while the second part of this list (which represents the formula for the solution), extracted with gensol[[1,1,2]], is $y = e^{-x} + Ce^{-2x}$. Of course, if you are using Version 10 (or later), you can extract these results by selecting, copying, and pasting the results to the desired location in your Mathematica notebook.

Note that in the formula for the solution the built-in symbol C is used to denote arbitrary constants. Here C[1] represents C in the solution $y = e^{-x} + Ce^{-2x}$.

To graph the solution for various values of the arbitrary constant, we use Table and ReplaceAll (/.) to replace C[1] in the formula for the general solution obtained in gensol by i for $i = -3, -2.8, -2.6,$ $\dots, 2.60, 2.80,$ and 3, naming the resulting set of functions toplot. The list toplot is not displayed because a semi-colon (;) is included at the end of the Table command.

> `toplot = Table[gensol[[1, 1, 2]]/. C[1] → i, {i, −3, 3, .2}];`

We then use Plot to graph the set of functions toplot for $-1/2 \le x \le 1$ in Figure 1-11, naming the resulting graphics object p4. The option PlotRange->{-3/4,3/4} instructs Mathematica that the range of y-values displayed corresponds to the interval $[-3/4, 3/4]$, AspectRatio->1 specifies that the ratio of the lengths of x and y-axes in the resulting graph is to be 1, AxesStyle->Black specifies that the axes are to be black (which helps them to be seen easier when being printed), Notice that we can predict the behavior of the solutions of this equation by observing the direction field, as we confirm with the following Show together with GraphicsRow commands in Figure 1-12.

> `p4 = Plot[toplot, {x, −1/2, 1}, PlotRange → {−3/4, 3/4},`
>
>   `AspectRatio → 1, AxesStyle → Black, AxesLabel → {x, y}];`
>
> `p5 = Show[p1, p4, PlotLabel → None];`
>
> `Show[GraphicsRow[{p4, p5}]]`

This is the primary purpose of direction fields: most differential equations cannot be solved by the elementary methods covered in an introductory text on differential equations but the numerically generated direction fields can help us understand the behavior of the solutions of the differential equation or system of two differential equations.

∎

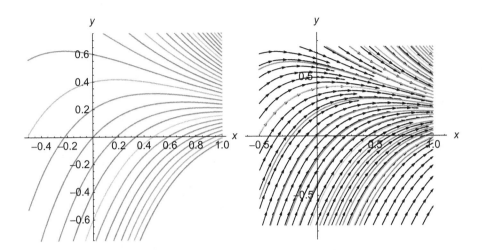

Figure 1-12   Direction field associated with $dy/dx = e^{-x} - 2y$ together with plots of several solutions

Mathematica allows us to graph solutions to equations and associated direction fields that would be nearly impossible by traditional methods.

EXAMPLE 1.4.2: Graph the direction field associated with the differential equation

$$\frac{dy}{dx} = \frac{\cos y - y \cos x}{x \sin y + \sin x - 1}.$$

SOLUTION: As in the previous example, we illustrate using `StreamPlot` and `StreamDensityPlot` to graph the direction field for the equation. The results are shown side-by-side using `Show` together with `GraphicsRow` in Figure 1-13

```
p1 = StreamPlot[{1, (Cos[y] - yCos[x])/(xSin[y] + Sin[x] - 1)},
    {x, 0, 4Pi}, {y, 0, 4Pi},
    Frame → False, Axes → Automatic, AxesOrigin → {0, 0},
    AxesLabel → {x,y}, StreamStyle → Black, PlotLabel → "(a)"];

p2 = StreamDensityPlot[{1, (Cos[y] - yCos[x])/
    (xSin[y] + Sin[x] - 1)}, {x, 0, 4Pi}, {y, 0, 4Pi},
```

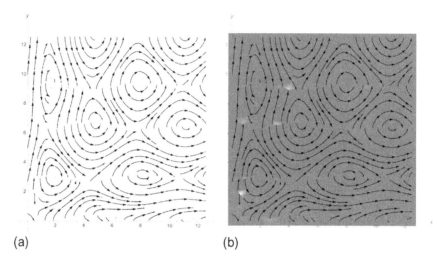

(a)   (b)

Figure 1-13   Direction field associated with $dy/dx = (\cos y - y \cos x) / (x \sin y + \sin x - 1)$

```
Frame → False, Axes → Automatic, AxesOrigin → {0, 0},

AxesLabel → {x,y}, StreamStyle → Black, PlotLabel → "(b)"];

Show[GraphicsRow[{p1, p2}]]
```

Next, we use `DSolve` to find a general solution of the equation, naming the result `gensol`. Observe that the result is an implicit solution given in terms of a `Solve` command.

```
Clear[x, y]

gensol = DSolve[(xSin[y[x]] + Sin[x] − 1)y′[x] ==

Cos[y[x]] − y[x]Cos[x], y[x], x]
```

$\text{Solve}[-x\text{Cos}[y[x]] - y[x] + \text{Sin}[x]y[x] == C[1], y[x]]$

The result indicates that a general solution is $y \sin x - x \cos y - y = C$.

We use `Part` (`[[...]]`) to extract the formula for the solution from `gensol` in the same way as in previous examples.

```
toplot = gensol[[1, 1]]/.{y[x] → y}
```

$-y - x\text{Cos}[y] + y\text{Sin}[x]$

To graph solutions (integral curves) for various values of C, we note that the graph of the equation $y \sin x - x \cos y - y = C$ for various values

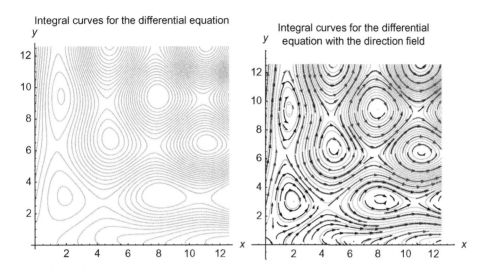

Figure 1-14   Various solutions of $dy/dx = (\cos y - y\cos x)/(x\sin y + \sin x - 1)$ together with the direction field

of C is the same as the graph of the level curves of $z = f(x, y) = y\sin x - x\cos y - y$ for various values of z. The integral curves for the equation are then graphed with `ContourPlot` in p3 and shown together with the direction field in p5. We show the integrals curves and the integral curves together with the direction field in Figure 1-14.

Depending upon your printer, Mathematica's default shade of gray for coloring axes can sometimes be difficult to see when printed. Therefore, we almost always include the option `AxesStyle->Black` to be sure that the axes are displayed in black and, consequently, when printed are easy to see.

```
p3 = ContourPlot[toplot, {x,0,4Pi}, {y,0,4Pi}, ContourStyle
    → Gray, ContourShading → False, Frame → False, Axes
    → Automatic, AxesStyle → Black, AxesLabel → {x, y},
    Contours → 30, PlotLabel → "Integral Curves for the
    Differential Equation"];
```

Use the `PlotLabel` option to create a caption for your graphics. We usually like to label our axes so include options like `AxesLabel->{x,y}` or `AxesLabel->{t,y}` to label the axes according to the convention being used in the problem.

```
p5 =

Show[p1, p3,

PlotLabel → "Integral Curves for the Differential Equation;

     with the Direction Field"];

Show[GraphicsRow[{p3, p5}]]
```

∎

Mathematica is particularly useful in graphing the direction field associated with a system of equations.

```
StreamPlot[{f[x,y],g[x,y]},{x,x0,x1},{y,y0,y1}]
```

graphs the direction field associated with the system $\begin{cases} x' = f(x,y) \\ y' = g(x,y) \end{cases}$ for $x_0 \leq x \leq x_1$ and $y_0 \leq y \leq y_1$. You can generate a direction field together with a density plot using $\texttt{StreamDensityPlot}$.

EXAMPLE 1.4.3 (Competing Species): Under certain assumptions the system of equations

$$\begin{cases} \dfrac{dx}{dt} = x\,(a - b_1 x - b_2 y) \\[2mm] \dfrac{dy}{dt} = y\,(c - d_1 x - d_2 y) \end{cases}, \tag{1.8}$$

where a, b_1, b_2, c, d_1, and d_2 represent positive constants, can be used to model the population of two species, represented by $x(t)$ and $y(t)$, competing for a common food supply. Graph the direction field associated with the system if (a) $a = 1$, $b_1 = 2$, $b_2 = 1$, $c = 1$, $d_1 = 0.75$, and $d_2 = 2$; and (b) $a = 1$, $b_1 = 1$, $b_2 = 1$, $c = 0.67$, $d_1 = 0.75$, and $d_2 = 1$.

SOLUTION: After identifying $f(x, y) = x\,(a - b_1 x - b_2 y)$ and $g(x, y) = y\,(c - d_1 x - d_2 y)$, we define f and g. Observe that in Mathematica, the way we defined f and g might be interpreted as subscripted definitions. For example, the definition of f might be interpreted to be $f_{a,b_1,b_2,c,d_1,d_2}(x, y)$ and that of g to be interpreted as $g_{a,b_1,b_2,c,d_1,d_2}(x, y)$

```
Clear[f, g, x, y]

f[a_, b1_, b2_, c_, d1_, d2_][x_, y_] = x(a − b1x − b2y);

g[a_, b1_, b2_, c_, d1_, d2_][x_, y_] = y(c − d1x − d2y);
```

Then, for (a) we use $a = 1$, $b_1 = 2$, $b_2 = 1$, $c = 1$, $d_1 = 0.75$, and $d_2 = 2$, and graph the direction field associated with the system for $0 \leq x \leq 1$ and $0 \leq y \leq 1$ with StreamPlot in Figure 1-15(a). In this case, we see that the both species appear to approach some equilibrium population. In fact, later we will see that this equilibrium population is obtained by solving the system of equations $\begin{cases} a - b_1 x - b_2 y = 0 \\ c - d_1 x - d_2 y = 0 \end{cases}$ for x and y.

```
p1 = StreamPlot[{f[1,2,1,0.67,0.75,2][x,y],g[1,2,1,1,0.75,

      2][x,y]},{x,0,1},{y,0,1}, Frame → False,Axes → Automatic,

      AxesOrigin → {0, 0}, AxesStyle → Black, StreamStyle →

      Black, PlotLabel → "(a)"];
```

For (b), we re-enter the StreamPlot command using the different parameter values. See Figure 1-15(b).

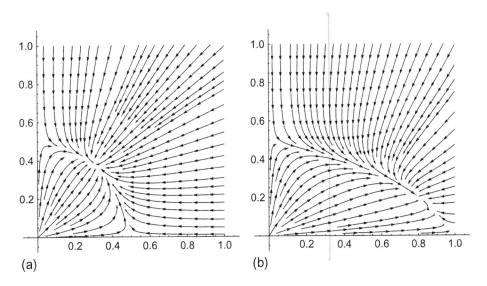

(a)   (b)

Figure 1-15 Direction fields associated with the competing species system using the parameter values in (a) on the left and using the parameter values for (b) on the right

```
p2 = StreamPlot[{f[1,1,1,1,0.75,2][x,y],g[1,2,1,1,0.75,

2][x,y]}, {x,0,1}, {y,0,1}, Frame → False,Axes → Automatic,

  AxesOrigin → {0, 0}, AxesStyle → Black, StreamStyle →

  Black, PlotLabel → "(b)"];
```

```
Show[GraphicsRow[{p1, p2}]]
```

In the figure, we see that it appears as though the species with population given by $y(t)$ eventually dies out while the species with population given by $x(t)$ eventually dominates and approaches some equilibrium population. We will see that this is true and the equilibrium population of the species with population given by $x(t)$ will be found by computing the limit as $t \to \infty$ of the solution to the differential equation $dx/dt = ax - b_1 x^2$. To find the equilibrium solutions we solve the system $f(x,y) = 0$, $g(x,y) = 0$ for x and y.

For (a), we illustrate how Solve generally give the same results numerically.

```
Solve[{f[1, 2, 1, 0.67, 0.75, 2][x, y] == 0,

g[1, 2, 1, 1, 0.75, 2][x, y] == 0}, {x, y}]
```

```
{{x → 0., y → 0.5}, {x → 0.307692, y → 0.384615}, {x → 0.5,

y → 0.}, {x → 0., y → 0.}}
```

```
NSolve[{f[1, 2, 1, 0.67, 0.75, 2][x, y] == 0,

g[1, 2, 1, 1, 0.75, 2][x, y] == 0}, {x, y}]
```

```
{{x → 0.5, y → 0.}, {x → 0., y → 0.5}, {x → 0.307692,

y → 0.384615}, {x → 0., y → 0.}}
```

The result indicates that the equilibrium points are $(0, .5)$ and $(0.307692, 0.384615)$. From the direction field, we see that $(0.5, 0)$ is *unstable* because solutions near this point are heading away from the point. On the other hand, $(0.8, 0.2)$ is *stable* because solutions starting near the point move closer to it.

For (b), we follow the same approach. In this case, we only use NSolve.

```
NSolve[{f[1, 1, 1, 1, 0.75, 2][x, y] == 0, g[1, 2, 1, 1, 0.75, 2]

[x, y] == 0}, {x, y}]
```

$\{\{x \rightarrow 1., y \rightarrow 0.\}, \{x \rightarrow 0.8, y \rightarrow 0.2\}, \{x \rightarrow 0., y \rightarrow 0.5\}, \{x \rightarrow 0., y \rightarrow 0.\}\}$

Observe that $(0.8, 0.2)$ is *stable* (nearby solutions tend to the equilibrium point) while $(0, 0.5)$ is unstable (nearby solutions tend away from the equilibrium point).

■

Often, we can generate the direction field of a higher-order equation by rewriting it as a system of first-order equations.

EXAMPLE 1.4.4 (Rayleigh's Equation): Write **Rayleigh's equation,** (1.7), as a system of two first-order equations. Graph the direction field associated with the resulting system on the rectangle $[-4, 4] \times [-4, 4]$.

Also see Example 1.3.4.

SOLUTION: We write Rayleigh's equation as a system by letting $y = x'$. Then Rayleigh's equation, (1.7), becomes

$$y' = x'' = -\left[\frac{1}{3}(x')^2 - 1\right]x' - x = -\left(\frac{1}{3}y^2 - 1\right)y - x$$

so Rayleigh's equation is equivalent to the system

$$\begin{cases} x' = y \\ y' = -\left(\frac{1}{3}y^2 - 1\right)y - x \end{cases}.$$

The direction field associated with this system is then graphed with `StreamPlot` in Figure 1-16(a).

```
p1 = StreamPlot[{y, -(1/3y^2 - 1)y - x},
    {x, -4, 4}, {y, -4, 4}, Frame → False, Axes → Automatic,
    AxesOrigin → {0, 0},
    AxesStyle → Black, StreamStyle → Gray, AxesLabel → {x,y},
    PlotLabel → "(a)"];
```

Next, we define a function numsol$\{$x0, y0$\}$]. Given an ordered pair (x_0, y_0), numsol [$\{$x0, y0$\}$] uses NDSolve to generate a solution of Rayleigh's equation that satisfies the initial conditions $x(0) = x_0$,

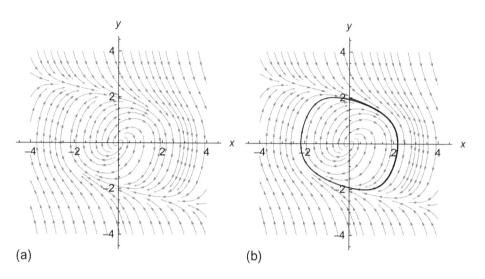

(a)   (b)

Figure 1-16   Direction field associated with the Rayleigh system on the left. On the right, the direction field together with a graph of the solution that satisfies $x(0) = 0$, $y(0) = 2$

$y(0) = y_0$ valid for $0 \leq t \leq 20$. In Figure 1-16(b) we show the graph of this solution together with the direction field.

```
Clear[numsol];

numsol[{x0_, y0_}]:=NDSolve[{x'[t] == y[t],

    y'[t] == -(1/3y[t]^2 - 1)y[t] - x[t], x[0] == x0,

    y[0] == y0}, {x[t], y[t]}, {t, 0, 20}]

q1 = numsol[{0, 2}]
```

$\{\{x[t] \to \text{InterpolatingFunction}[\,][t], y[t] \to \text{InterpolatingFunction}[\,][t]\}\}$

```
p2a = ParametricPlot[{x[t], y[t]}/.q1, {t, 0, 15},

    PlotRange → {{-4, 4}, {-4, 4}}, AspectRatio → 1,

    PlotStyle → {{Thickness[.01], Black}}];

p2 = Show[p1, p2a, PlotLabel → "(b)"];

Show[GraphicsRow[{p1, p2}]]
```

In the direction field, we see that solutions appear to tend to a closed curve, C. We can accurately approximate C. First, we use NDSolve to approximate the solution to the equation. We start by creating a list of initial conditions in `initconds`.

```
initconds = {{1, 0}, {0.1, 0}, {0, 1.9}, {−4, 4}, {−3, 0},
    {0, 3}, {3, 0}, {0, −3.5}, {−3.5, 3.5}, {−3.5, −3.5}, {3.5, 4},
    {−2, −4}, {2, 4}};
```

Next, we use `Map` to apply `numsol` to each of the initial conditions in `initconds` and name the resulting list of numerical solutions `sols`.

```
Clear[sols]

sols = Map[numsol, initconds];
```

We then use `Table` together with `ParametricPlot` to graph each of the numerical solutions in `sols`.

```
p3a = Table[ParametricPlot[Evaluate[{x[t], y[t]}/.sols[[i]]],
    {t, 0, 15}, PlotRange → {{−4, 4}, {−4, 4}}, PlotStyle →
    {{Thickness[0.0075], Black}}], {i, 1, Length[sols]}];

p3 = Show[p3a];

Show[p1, p3, PlotLabel → "Direction Field with Various
Solutions to Rayleigh's Equation"]
```

The solution plots together with the direction field are shown in Figure 1-17. In the graph, we see that the graph of the solutions appear to converge to a closed periodic orbit, C. The graphs of the other solutions also appear to tend to C, which is called a *limit cycle*.

■

1.4.1 Creating Interactive Applications

One of the more interesting developments in recent releases of Mathematica is the `Manipulate` function. With `Manipulate`, you can design programs that allow you to experiment with changing parameter values and more. With some practice, objects that you create with `Manipulate` can be exported to be essentially stand-alone applications. In some case, Wolfram's *CDF Player* may be needed to take advantage of all the features of your `Manipulate` object.

Basic information about `Manipulate` is obtained with `?Manipulate`.

Direction field with various solutions to Rayleigh's equation

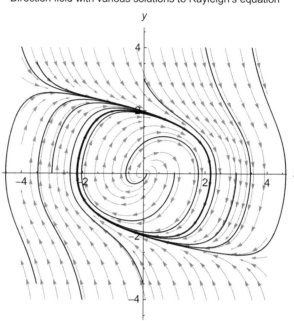

Figure 1-17   Solutions of Rayleigh's equation tend to a limit cycle

More advanced features are various options are obtained with the **Help Browser**

```
? Manipulate

Manipulate[expr, {u, u_min, u_max}] generates a version
      of expr with controls added to allow interactive manipulation of the value of u.
Manipulate[expr, {u, u_min, u_max, du}] allows the value of u to vary between u_min and u_max in steps du.
Manipulate[expr, {{u, u_init}, u_min, u_max, ...}] takes the initial value of u to be u_init.
Manipulate[expr, {{u, u_init, u_lbl}, ...}] labels the controls for u with u_lbl.
Manipulate[expr, {u, {u_1, u_2, ...}}] allows u to take on discrete values u_1, u_2, ....
Manipulate[expr, {u, ...}, {v, ...}, ...] provides controls to manipulate each of the u, v, ....
Manipulate[expr, c_u → {u, ...}, c_v → {v, ...}, ...] links the controls to the specified controllers on an external device.  ≫
```

Using Rayleigh's equation as an example, we use Manipulate to create an example of showing how solutions to Rayleigh's equation change.

```
Manipulate[

  p1 = StreamPlot[{y, -(1/3y^2 - 1)y - x},

{x, -4, 4}, {y, -4, 4}, Frame → False, Axes → Automatic, AxesOrigin

  → {0, 0}, AxesStyle → Black, StreamStyle → Gray, AxesLabel
```

```
        → {x, y}]; p2 = NDSolve[{x'[t] == y[t],

        y'[t] == -(1/3y[t]^2 - 1)y[t] - x[t], x[0] == x0,

        y[0] == y0}, {x[t], y[t]}, {t, 0, 20}];

    p3 = ParametricPlot[Evaluate[{x[t], y[t]}/.p2], {t, 0, 15},

PlotRange → {{-4, 4}, {-4, 4}}, PlotStyle → {{Thickness[0.0075],

    Black}}]; Show[p1, p3, PlotLabel → "Direction Field with a

    Solution to Rayleigh's Equation",

        PlotRange → {{-4, 4}, {-4, 4}}, AspectRatio → 1],

    {{x0, 2.2}, -4, 4}, {{y0, 0}, -4, 4}]
```

You can use the sliders to adjust x_0 and y_0 and you will be able to see how the solution to the initial value problem changes. Manipulate can be a useful interactive tool (Figure 1-18).

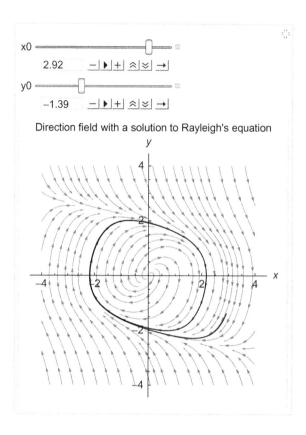

Figure 1-18   Use the sliders to change the initial conditions. To see all options click on the plus button

First-Order Ordinary Differential Equations

2

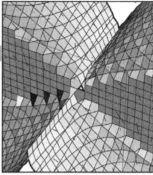

We will devote a considerable amount of time in this text to developing explicit, implicit, numerical, and graphical solutions of differential equations. In this chapter we introduce frequently encountered forms of first-order ordinary differential equations and methods to construct explicit, numerical, and graphical solutions of them. Several of the equations along with the methods of solution discussed here will be used in subsequent chapters of the text.

2.1 Theory of First-Order Equations: A Brief Discussion

In order to understand the types of first-order initial-value problems that have a unique solution, the Picard-Lindelöf Existence and Uniqueness theorem is stated.

Theorem 1 (Existence and Uniqueness). *Consider the initial-value problem*

$$\begin{cases} dy/dx = f(x, y) \\ y(x_0) = y_0 \end{cases}. \tag{2.1}$$

If f and $\partial f / \partial y$ are continuous functions on the rectangular region R,

$$R = \{(x, y) | a < x < b, c < y < d\},$$

containing the point (x_0, y_0), there exists an interval $|x - x_0| < h$ centered at x_0 on which there exists one and only one solution to the differential equation that satisfies the initial condition.

Charles Emile Picard
(July 24, 1856, Paris, France-December 11, 1941, Paris, France) According to O'Connor, "Picard and his wife had three children, a daughter and two sons, who were all killed in World War I. His grandsons were wounded and captured in World War II."
In regards to Picard's teaching, the famous French mathematician Jacques Salomon Hadamard (1865–1963) wrote in Picard's obituary "A striking feature of Picard's scientific personality was the perfection of his teaching, one of the most marvelous, if not the most marvelous that I have known."
See texts like [6], [7], or [3].

Differential Equations with Mathematica. http://dx.doi.org/10.1016/B978-0-12-804776-7.00002-4

Often, we can use the command

```
DSolve[{y'[x]==f[x,y[x]],y[x0]==y0},y[x],x]
```

to solve the initial-value problem (2.1).

When exact solutions are not possible or not desired, try to use `NDSolve` to generate a numerical solution. The command

```
NDSolve[{y'[x]==f[x,y[x]],y[x0]==y0},y[x],{x,x0,x1}]
```

attempts to generate a numerical solution of $dy/dx = f(x, y)$, $y(x_0) = y_0$ valid for $x_0 \le x \le x_1$.

EXAMPLE 2.1.1: Solve the initial-value problem

$$\begin{cases} dy/dx = x/y \\ y(0) = 0 \end{cases}.$$

Does this result contradict the Existence and Uniqueness Theorem?

SOLUTION: We begin by using `StreamPlot` to graph the direction field associated with the equation in Figure 2-1(a).

p1 = StreamPlot[{1, x/y}, {x, −5, 5}, {y, −5, 5}, StreamStyle

→ Gray, Frame → False, Axes → Automatic, AxesOrigin

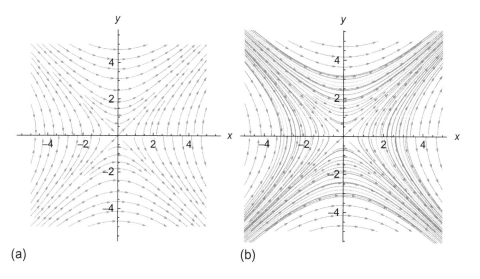

(a)   (b)

Figure 2-1   (a) Direction field for $dy/dx = x/y$. (b) Various solutions of $dy/dx = x/y$ with the direction field

```
→ {0, 0}, AxesLabel → {x, y}, AxesStyle → Black, PlotLabel

→ "(a)"];
```

This equation is solved with `DSolve` to determine the family of solutions $y = -\sqrt{x^2 + C}$ and $y = \sqrt{x^2 + C}$.

```
gensol = DSolve[y'[x] == x/y[x], y[x], x]
```

$$\left\{ \left\{ y[x] \rightarrow -\sqrt{x^2 + 2C[1]} \right\}, \left\{ y[x] \rightarrow \sqrt{x^2 + 2C[1]} \right\} \right\}$$

```
gensol[[1, 1, 2]]
```

$$-\sqrt{x^2 + 2C[1]}$$

```
gensol[[2, 1, 2]]
```

$$\sqrt{x^2 + 2C[1]}$$

We extract the formulas for the solutions using `Part` (`[[...]]`). In this case, there are two formulas.

```
gensol[[1, 1, 2]]
```

$$-\sqrt{x^2 + 2C[1]}$$

```
gensol[[2, 1, 2]]
```

$$\sqrt{x^2 + 2C[1]}$$

To graph the solutions for various values of the arbitrary constant, `Table` is used to create two lists of solutions to the differential equation. Generally, `Table[f[x], {x,a,b,n}]` creates a list of $f(x)$ for x-values from a to b in steps of n. These are then graphed with `Plot` and shown together with the direction field in Figure 2-1(b).

```
toplot1 = Table[gensol[[1, 1, 2]]/.C[1] → i, {i, −5, 5, 10/19}];

toplot2 = Table[gensol[[2, 1, 2]]/.C[1] → i, {i, −5, 5, 10/19}];

p2a = Plot[{toplot1, toplot2}, {x, −5, 5}];

p2 = Show[p2a, p1, Frame → False, Axes → Automatic,

AxesOrigin → {0, 0},

AxesLabel → {x, y}, AxesStyle → Black, PlotRange →

{{−5, 5}, {−5, 5}}, AspectRatio → 1, PlotLabel → "(b)"];

Show[GraphicsRow[{p1, p2}]]
```

Application of the initial condition yields $0^2 - 0^2 = C$, so $C = 0$. Therefore, solutions that pass through $(0, 0)$, satisfy $y^2 - x^2 = 0$, so there are four solutions, $y = x$, $y = -x$, $y = |x|$, and $y = -|x|$ that satisfy the differential equation and the initial condition.

```
partsol = DSolve[{y'[x] == x/y[x], y[0] == 0}, y[x], x]
```

$$\left\{ \left\{ y[x] \to -\sqrt{x^2} \right\}, \left\{ y[x] \to \sqrt{x^2} \right\} \right\}$$

The solution is graphed in Figure 2-2.

```
Plot[{Abs[x], -Abs[x]}, {x, -5, 5}, PlotRange → {{-5, 5},

    {-5, 5}}, AspectRatio → 1, PlotStyle → Black, AxesStyle

    → Black, AxesLabel → {x, y}]
```

Although more than one solution satisfies this initial-value problem, the Existence and Uniqueness Theorem is *not* contradicted because

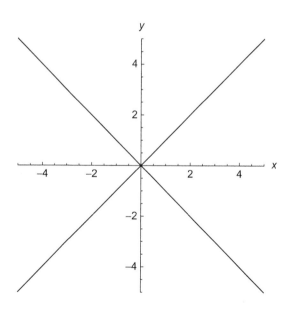

Figure 2-2   Solutions of $dy/dx = x/y$, $y(0) = 0$

the function $f(x, y) = x/y$ is not continuous at the point $(0,0)$; the requirements of the theorem are not met.

■

EXAMPLE 2.1.2: Verify that the initial-value problem $\{dy/dx = y,\ y(0) = 1\}$ has a unique solution.

SOLUTION: In this case, $f(x, y) = y$, $x_0 = 0$, and $y_0 = 1$. Hence, both f and $\partial f/\partial y = 1$ are continuous on all rectangular regions containing the point $(x_0, y_0) = (0, 1)$. Therefore by the Existence and Uniqueness Theorem, there exists a unique solution to the differential equation that satisfies the initial condition $y(0) = 1$.

We can verify this by solving the initial-value problem. The unique solution is $y = e^x$, which is computed with DSolve and then graphed with Plot in Figure 2-3(a). Notice that the graph passes through the point $(0, 1)$, as required by the initial condition. We show the graph together with the direction field in Figure 2-3(b).

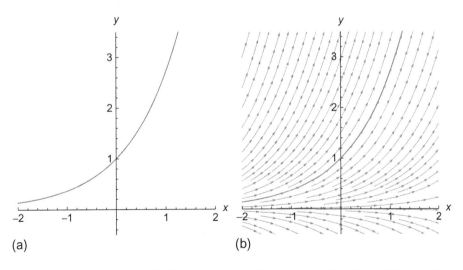

(a)   (b)

Figure 2-3   (a) Plot of $y = e^x$. (b) The solution together with the direction field for $dy/dx = y$

```
sol = DSolve[{y'[x] == y[x], y[0] == 1}, y[x], x]
```

$$\{\{y[x] \to e^x\}\}$$

```
p1 = Plot[sol[[1, 1, 2]], {x, -2, 2}, PlotStyle->CMYKColor
    [0, 0.89, 0.94, 0.28], PlotRange → {{-2, 2}, {-.5, 3.5}},
    AspectRatio → 1, AxesStyle → Black, AxesLabel → {x, y},
    PlotLabel → "(a)"];

p2a = StreamPlot[{1, y}, {x, -2, 2}, {y, -.5, 3.5},
    StreamStyle → Gray, Frame → False, Axes → Automatic,
    AxesOrigin→ {0,0}, AxesLabel → {x,y}, AxesStyle→ Black];
    p2 = Show[p1, p2a, PlotLabel → "(b)"];

Show[GraphicsRow[{p1, p2}]]
```

■

EXAMPLE 2.1.3: Show that the initial-value problem

$$\begin{cases} x\dfrac{dy}{dx} - y = x^2 \cos x \\ y(0) = 0 \end{cases}$$

has infinitely many solutions.

SOLUTION: Writing $xy' - y = x^2 \cos x$ in the form $y' = f(x, y)$ results in

$$\frac{dy}{dx} = \frac{x^2 \cos x + y}{x}$$

and because $f(x, y) = \left(x^2 \cos x + y\right)/x$ is not continuous on an interval containing $x = 0$, the Existence and Uniqueness theorem does not guarantee the existence or uniqueness of a solution. In fact, using DSolve we see that a general solution of the equation is $y = x \sin x + Cx$ and for every value of C, $y(0) = 0$.

```
sol = DSolve[{xy'[x] - y[x] == x^2Cos[x], y[0] == 0},
y[x], x]
```

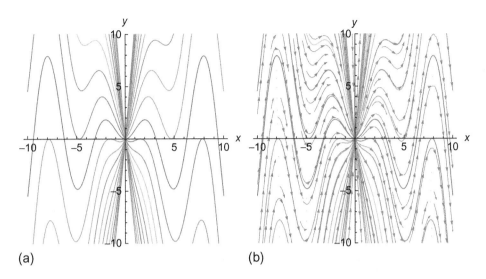

Figure 2-4   (a) Every solution satisfies $y(0) = 0$. (b) Solutions with the direction field

$$\{\{y[x] \to xC[1] + x\text{Sin}[x]\}\}$$

We confirm this graphically by graphing several solutions. First, we use Table to define toplot to be a set of functions obtained by replacing the arbitrary constant in $y(x)$ by $-10, -9, \ldots, 9, 10$ (Figure 2-4).

```
toplot = Table[sol[[1, 1, 2]]/. C[1] → i, {i, −10, 10}];

p1 = Plot[toplot, {x, −10, 10}, PlotRange → {−10, 10},

    AxesLabel → {x, y}, PlotLabel → "(a)", AspectRatio → 1,

    AxesStyle → Black];

p2a = StreamPlot[{1, (x^2Cos[x] + y)/x}, {x, −10, 10},

{y, −10, 10}, StreamStyle → Gray,

    Frame → False, Axes → Automatic, AxesOrigin → {0, 0},

    AxesLabel → {x, y}, AxesStyle → Black];

p2 = Show[p1, p2a, PlotLabel → "(b)"];

Show[GraphicsRow[{p1, p2}]]
```

■

2.2 Separation of Variables

Definition 5 (Separable Differential Equation). *A differential equation that can be written in the form $g(y)y' = f(x)$ or $g(y)\,dy = f(x)\,dx$ is called a* **separable differential equation**.

Separable differential equations are solved by collecting all the terms involving y on one side of the equation, all the terms involving x on the other side of the equations and integrating:

$$g(y)\,dy = f(x)\,dx \implies \int g(y)\,dy = \int f(x)\,dx + C,$$

where C is a constant.

EXAMPLE 2.2.1: Show that the equation

$$\frac{dy}{dx} = \frac{2\sqrt{y} - 2y}{x}$$

is separable, and solve by separation of variables.

SOLUTION: As with previous examples, we start by graphing the direction field with `StreamPlot` in Figure 2-5(a).

```
p1 = StreamPlot[{1, (2Sqrt[y] − 2y)/x}, {x, −1, 1}, {y, 0, 2},

    PlotRange → {{−1, 1}, {0, 2}},

    AxesLabel → {x, y}, PlotLabel → "(a)", AspectRatio → 1,

    AxesStyle → Black, StreamStyle → Black, Frame → False,

    Axes, → Automatic, AxesOrigin → {0, 0}];
```

The equation $y' = \left(2\sqrt{y} - 2y\right)/x$ is separable because it can be written in the form

$$\frac{1}{2\sqrt{y} - 2y}\,dy = \frac{1}{x}\,dx.$$

To solve the equation, we integrate both sides and simplify. Observe that we can write this equation as

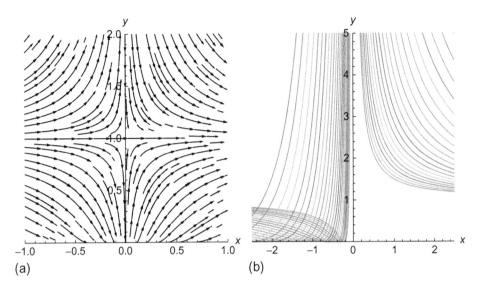

Figure 2-5 (a) Direction field. (b) Various solutions of $y' = \left(2\sqrt{y} - 2y\right)/x$

$$\int \frac{1}{2\sqrt{y}} \frac{1}{1 - \sqrt{y}} dy = \frac{1}{x} dx + C.$$

To evaluate the integral on the left-hand side, let $u = 1 - \sqrt{y}$ so $-du = \frac{1}{2\sqrt{y}} dy$. We then obtain

$$-\int \frac{1}{u} du = \int \frac{1}{x} dx + C_1$$

so that $-\ln|u| = \ln|x| + C_1$. Recall that $-\ln|u| = \ln|u|^{-1}$, so we have

$$\ln \frac{1}{|u|} = \ln|x| + C_1.$$

Using Mathematica, we use Integrate.

```
step1 = Integrate[1/(2Sqrt[y] - 2y), y]
```

$$-\text{Log}\left[1 - \sqrt{y}\right]$$

The integral on the right-hand side of the equation is computed in the same way.

Note that Log [x] represents the **natural logarithm function**, $y = \ln x$.

```
step2 = Integrate[1/x, x]
```

Log[x]

Simplification yields

$$\frac{1}{|u|} = e^{\ln|x|+C_1} = C_2|x|,$$

where $C_2 = e^{C_1}$. Resubstituting we find that

$$\frac{1}{|1 - \sqrt{y}|} = C_2|x| \quad \text{or} \quad x = \frac{C_3}{1 - \sqrt{y}}.$$

Solving for y shows us that

$$\sqrt{y} - 1 = \frac{C_3}{x}$$

$$\sqrt{y} = \frac{x + C_3}{x}$$

$$y = \left(\frac{x + C_3}{x}\right)^2$$

is a general solution of the equation $y' = (2\sqrt{y} - 2y)/x$. We obtain the same results with Mathematica,

We use constant to represent the arbitrary constant C to avoid ambiguity with the built-in symbol C.

```
step3 = Solve[step1 == step2 + constant, y]
```

$$\left\{\left\{y \to \frac{e^{-2\text{constant}}\left(-1 + e^{\text{constant}}x\right)^2}{x^2}\right\}\right\}$$

where $e^{-\text{constant}}$ represents the arbitrary constant in the solution. We obtain an equivalent result with DSolve. Entering

```
Clear[x, y]
```

```
gensol = DSolve[y'[x] == (2Sqrt[y[x]] - 2y[x])/x, y[x], x]
```

$$\left\{\left\{y[x] \to \frac{\left(e^{\frac{C[1]}{2}} + x\right)^2}{x^2}\right\}\right\}$$

finds a general solution of the equation which is equivalent to the one we obtained by hand and names the result gensol. The formula for the solution, which is the second part of the first part of the first

part of gensol, is extracted from gensol with gensol [[1,1,2]].
Alternatively, if you are using Version 10, you can select, copy, and
paste the result to any location in the notebook.

To graph the solution for various values of C[1], which represents
the arbitrary constant in the formula for the solution, we use Table
together with ReplaceAll (/.) to generate a set of functions obtained
by replacing C[1] in the formula for the solution by i for $i = -3$,
-2.75, ..., 2.75, and 3, naming the resulting set of functions
toplot.

```
toplot = Table[gensol[[1, 1, 2]]/. C[1] → i,

    {i, −3, 3, .25}];
```

We then graph the set of functions toplot with Plot in
Figure 2-5(b).

```
p2 = Plot[toplot, {x, −2.5, 2.5}, PlotRange → {{−2.5, 2.5},

    {0, 5}}, AxesLabel → {x, y}, PlotLabel → "(b)",

    AspectRatio → 1, AxesStyle → Black, Frame → False,

    Axes → Automatic, AxesOrigin → {0, 0}];

Show[GraphicsRow[{p1, p2}]]
```

■

An initial-value problem involving a separable equation is solved through the
following steps.

1. Find a general solution of the differential equation using separation of
   variables.
2. Use the initial condition to determine the unknown constant in the general
   solution.

EXAMPLE 2.2.2: Solve (a) $y\cos x\,dx - \left(1 + y^2\right) dy = 0$ and (b) the
initial-value problem $\left\{y\cos x\,dx - \left(1 + y^2\right) dy = 0,\ y(0) = 1\right\}$.

SOLUTION: As in the previous examples, we begin by graphing the
direction field in Figure 2-7(a) with StreamPlot.

```
p1 = StreamPlot[{1, yCos[x]/(1 + y^2)}, {x, -Pi, 2Pi},

  {y, 0, 3Pi}, AxesLabel → {x, y}, PlotLabel → "(a)",

  AspectRatio → 1, AxesStyle → Black, StreamStyle → Black,

  Frame → False, Axes → Automatic, AxesOrigin → {0, 0}];
```

(a) Note that this equation can be rewritten as $dy/dx = (y \cos x) / (1 + y^2)$. We first use DSolve to solve the equation.

```
gensol1 = DSolve[y'[x]==y[x]Cos[x]/(1 + y[x]^2), y[x], x]
```

$$\left\{\left\{y[x] \rightarrow -\sqrt{\text{ProductLog}\left[e^{2C[1]+2\sin[x]}\right]}\right\},\right.$$

$$\left.\left\{y[x] \rightarrow \sqrt{\text{ProductLog}\left[e^{2C[1]+2\sin[x]}\right]}\right\}\right\}$$

In this case, we see that DSolve is able to solve the nonlinear equation, although the result contains the ProductLog function. Given z, the **Product Log function** returns the principal value of w that satisfies $z = we^w$. See Figure 2-6. A more familiar form of the solution is found

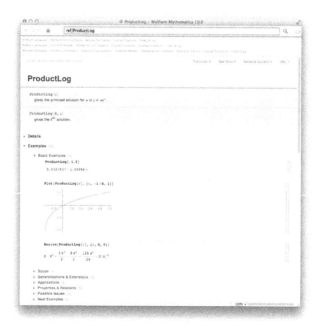

Figure 2-6   Mathematica's help for ProductLog

using traditional techniques. Separating variables and integrating gives us

$$\frac{1+y^2}{y}\,dy = \cos x\,dx$$

$$\left(\frac{1}{y}+y\right)dy = \cos x\,dx$$

$$\ln|y| + \frac{1}{2}y^2 = \sin x + C.$$

We can also use Mathematica to implement the steps necessary to solve the equation by hand. To solve the equation, we must integrate both the left- and right-hand sides which we do with Integrate, naming the resulting output lhs and rhs, respectively.

```
lhs = Integrate[(1 + y^2)/y, y]
```

```
rhs = Integrate[Cos[x], x]
```

$$\frac{y^2}{2} + \text{Log}[y]$$

Sin[x]

Therefore, a general solution to the equation is $\ln|y| + \frac{1}{2}y^2 = \sin x + C$. We now use ContourPlot to graph $\ln|y| + \frac{1}{2}y^2 = \sin x + C$ in Figure 2-7(b) for various values of C by observing that the level curves of $f(x,y) = \ln|y| + \frac{1}{2}y^2 - \sin x$ correspond to the graph of $\ln|y| + \frac{1}{2}y^2 = \sin x + C$ for various values of C.

```
p2a = ContourPlot[lhs - rhs, {x, -Pi, 2Pi}, {y, 0, 3Pi},

    AxesLabel → {x, y}, AspectRatio → 1, ContourShading →

    False, Contours → 50, ContourStyle → {{Thickness[.001],

    Black}}, AxesStyle → Black, Frame → False, Axes →

    Automatic, AxesOrigin → {0, 0}];

p2 = Show[p1, p2a, AxesLabel → {x, y}, PlotLabel → "(b)",

    AspectRatio → 1, AxesStyle → Black, Frame → False,

    Axes → Automatic, AxesOrigin → {0, 0}];
```

By substituting $y(0) = 1$ into this equation, we find that $C = 1/2$, so the implicit solution is given by $\ln|y| + \frac{1}{2}y^2 = \sin x + 1/2$.

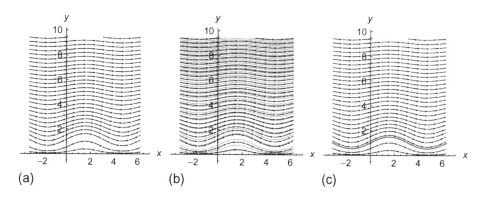

Figure 2-7   (a) Direction field for the equation. (b) Plot of $\ln |y| + \frac{1}{2}y^2 = \sin x + C$ for various values of C. (c) The solution that satisfies $y(0) = 1$ is highlighted

We can also use DSolve to solve the initial value problem as well. The solution is then graphed in Figure 2-7(c) with Plot.

```
sol2 = DSolve[{y'[x]==y[x]Cos[x]/(1 + y[x]^2), y[0] == 1},

y[x], x]
```

$$\left\{\left\{y[x] \rightarrow \sqrt{\texttt{ProductLog}\left[e^{1+2\text{Sin}[x]}\right]}\right\}\right\}$$

```
p3a = Plot[y[x]/.sol2, {x, -Pi, 2Pi}, AxesStyle → Black,

   AspectRatio → 1, PlotRange → {{-Pi, 2Pi}, {0, 3Pi}},

   PlotStyle → {{Thickness[.01], CMYKColor[0, 0.89, 0.94,

   0.28]}}];

p3 = Show[p1, p3a, AxesLabel → {x, y}, PlotLabel → "(c)",

   AspectRatio → 1, AxesStyle → Black, Frame → False,

   Axes → Automatic, AxesOrigin → {0, 0}];

Show[GraphicsRow[{p1, p2, p3}]]
```

■

EXAMPLE 2.2.3: Solve each of the following equations. (a) $y' - y^2 \sin t = 0$, (b) $y' = \alpha y \left(1 - \frac{1}{K}y\right)$, $K, \alpha > 0$ constant.

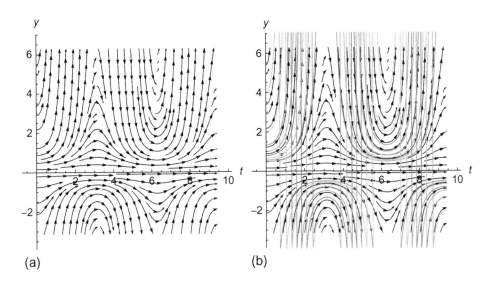

(a) (b)

Figure 2-8 (a) Direction field for the equation. (b) Several solutions of $y' - y^2 \sin t = 0$ with the direction field

SOLUTION: (a) The direction field is shown in Figure 2-8(a).

```
p1 = StreamPlot[{1, y^2Sin[t]}, {t, 0, 3Pi}, {y, -Pi, 2Pi},
   AxesLabel → {t, y}, PlotLabel → "(a)", AspectRatio → 1,
   AxesStyle → Black, StreamStyle → Black, Frame → False,
   Axes → Automatic, AxesOrigin → {0, 0}];
```

The equation is separable:

$$\frac{1}{y^2} dy = \sin t \, dt$$

$$\int \frac{1}{y^2} dy = \int \sin t \, dt$$

$$-\frac{1}{y} = -\cos t + C$$

$$y = \frac{1}{\cos t + C}.$$

We check our result with DSolve.

```
gensola = DSolve[y'[t] - y[t]^2Sin[t] == 0, y[t], t]
```

$$\left\{ \left\{ y[t] \to \frac{1}{-C[1] + \text{Cos}[t]} \right\} \right\}$$

Observe that the result is given as a list. The formula for the solution is the second part of the first part of the first part of `sola`.

```
gensola[[1, 1, 2]]
```

$$\frac{1}{-C[1] + \text{Cos}[t]}$$

We then graph the solution for various values of C with `Plot` in Figure 2-8(b).

```
toplot = Table[gensola[[1, 1, 2]]/.C[1] → i, {i, −1, 1, .25}];

p2a = Plot[toplot, {t, 0, 3Pi}];

p2 = Show[p1, p2a, AxesLabel → {t, y}, PlotLabel → "(b)",

AspectRatio → 1,

  AxesStyle → Black, Frame → False, Axes → Automatic,

  AxesOrigin → {0, 0}];

Show[GraphicsRow[{p1, p2}]]
```

expression /. x->y
replaces all occurrences of x
in expression by y.
Table[a[k],{k,n,m}]
generates the list a_n, a_{n+1},
..., a_{m-1}, a_m.
To graph the list of functions
list for $a \le x \le b$, enter
Plot[list,{x,a,b}].

(b) After separating variables, we use partial fractions to integrate.

$$y' = \alpha y \left(1 - \frac{1}{K}y\right)$$

$$\frac{1}{\alpha y \left(1 - \frac{1}{K}y\right)} dy = dt$$

$$\frac{1}{\alpha} \left(\frac{1}{y} + \frac{1}{K - y}\right) = dt$$

$$\frac{1}{\alpha} (\ln |y| - \ln |K - y|) = C_1 t$$

$$\frac{y}{K - y} = C e^{\alpha t}$$

$$y = \frac{CK e^{\alpha t}}{C e^{\alpha t} - 1}.$$

We check the calculations with Mathematica. First, we use `Apart` to find the partial fraction decomposition of $\dfrac{1}{\alpha y \left(1 - \frac{1}{K}y\right)}$.

```
step1 = Apart[1/(αy(1 − 1/ky)), y]
```

$$\frac{1}{y\alpha} - \frac{1}{(-k + y)\alpha}$$

Then, we use `Integrate` to check the integration.

> `step2 = Integrate[step1, y]`

$$k\left(\frac{\text{Log}[y]}{k\alpha} - \frac{\text{Log}[-k+y]}{k\alpha}\right)$$

Last, we use `Solve` to solve $\frac{1}{\alpha}(\ln|y| - \ln|K - y|) = ct$ for y.

> `step3 = Solve[step2 == constant, y]`

$$\left\{\left\{y \to \frac{e^{\text{constant}\,\alpha}\,k}{-1 + e^{\text{constant}\,\alpha}}\right\}\right\}$$

We can use `DSolve` to find a general solution of the equation

> `gensol = DSolve[y'[t] == αy[t](1 − 1/ky[t]), y[t], t]`

$$\left\{\left\{y[t] \to \frac{e^{t\alpha + kC[1]}\,k}{-1 + e^{t\alpha + kC[1]}}\right\}\right\}$$

as well as find the solution that satisfies the initial condition $y(0) = y_0$.

> `initsol = DSolve[{y'[t] == αy[t](1 − 1/ky[t]), y[0] == y0},`
>
> `y[t], t]`

$$\left\{\left\{y[t] \to \frac{e^{t\alpha}\,k\,y0}{k - y0 + e^{t\alpha}y0}\right\}\right\}$$

The equation $y' = \alpha y\left(1 - \frac{1}{K}y\right)$ is called the **Logistic equation** (or **Verhulst equation**) and is used to model the size of a population that is not allowed to grow in an unbounded manner. Assuming that $y(0) > 0$, then all solutions of the equation have the property that $\lim_{t \to \infty} y(t) = K$.

To see this, we set $\alpha = K = 1$ and use `StreamPlot` to graph the direction field associated with the equation in Figure 2-9(a).

Logistic growth is discussed in more detail in Section 3.2.2.

> `p1 = StreamPlot[{1, y(1 − y)}, {t, 0, 3}, {y, 0, 3}, AxesLabel`
>
> `→ {t, y}, PlotLabel → "(a)", AspectRatio → 1,`
>
> `AxesStyle → Black, StreamStyle → Black, Frame → False,`
>
> `Axes, → Automatic, AxesOrigin → {0, 0}];`

The property is more easily seen when we graph various solutions along with the direction field as done next in Figure 2-9(b).

> `step2 = initsol[[1, 1, 2]]/.{k → 1, α → 1}`

$$\frac{e^t\,y0}{1 - y0 + e^t\,y0}$$

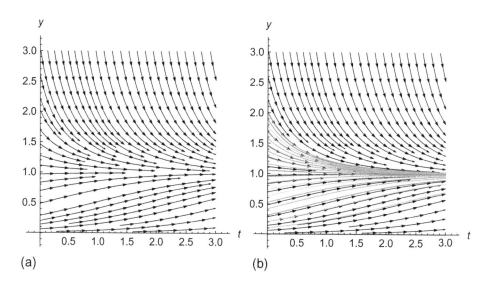

(a)   (b)

Figure 2-9   (a) A typical direction field for the Logistic equation. (b) A typical direction field for the Logistic equation along with several solutions

```
toplot = Table[step2/.y0 → i, {i, 0, 2, 2/19}];

p2a = Plot[toplot, {t, 0, 3}, AxesLabel → {t, y},

    AspectRatio → 1, AxesStyle → Black, Frame → False,

    Axes → Automatic, AxesOrigin → {0, 0}];

p2 = Show[p1, p2a, PlotLabel → "(b)"];

Show[GraphicsRow[{p1, p2}]]
```

∎

Sources: D. N. Burghes and M. S. Borrie, *Modeling with Differential Equations*, Ellis Horwood Limited, pp. 41–45. Joyce M. Black and Esther Matassarin-Jacobs, *Luckman and Sorensen's Medical-Surgical Nursing: A Psychophysiologic Approach*, Fourth Edition, W. B. Saunders Company (1993), pp. 1509–1519, 1775–1808.

Application: Kidney Dialysis

The primary purpose of the kidney is to remove waste products, like urea, creatinine, and excess fluid, from blood. When kidneys are not working properly, wastes accumulate in the blood; when toxic levels are reached, death is certain. The leading causes of chronic kidney failure in the United States are hypertension (high blood pressure) and diabetes mellitus. In fact, one-quarter of all patients requiring **kidney dialysis** have diabetes. Fortunately, **kidney dialysis** removes waste products from the blood of patients with improperly working kidneys. During the hemodialysis process, the patient's blood is pumped through a

dialyser, usually at a rate of 1–3 deciliters per minute. The patient's blood is separated from the "cleaning fluid" by a semi-permeable membrane, which permits wastes (but not blood cells) to diffuse to the cleaning fluid; the cleaning fluid contains some substances beneficial to the body which diffuse to the blood. The "cleaning fluid," called the **dialysate**, is flowing in the *opposite* direction as the blood, usually at a rate of 2–6 deciliters per minute. Waste products from the blood diffuse to the dialysate through the membrane at a rate proportional to the difference in concentration of the waste products in the blood and dialysate. If we let $u(x)$ represent the concentration of wastes in blood, $v(x)$ represent the concentration of wastes in the dialysate, where x is the distance along the dialyser, Q_D represent the flow rate of the dialysate through the machine, and Q_B represent the flow rate of the blood through the machine, then

$$\begin{cases} Q_B u' = -k(u - v) \\ -Q_D v' = k(u - v) \end{cases},$$

where k is the proportionality constant.

If we let L denote the length of the dialyser and the initial concentration of wastes in the blood is $u(0) = u_0$ while the initial concentration of wastes in the dialysate is $v(L) = 0$, then we must solve the initial-value problem

$$\begin{cases} Q_B u' = -k(u - v) \\ -Q_D v' = k(u - v) \\ u(0) = u_0, \ v(L) = 0 \end{cases}.$$

Solving the first equation for u' and the second equation for $-v'$, we obtain the equivalent system

$$\begin{cases} u' = -\dfrac{k}{Q_B}(u - v) \\ -v' = \dfrac{k}{Q_D}(u - v) \\ u(0) = u_0, \ v(L) = 0 \end{cases}.$$

Adding these two equations results in a separable (and linear) equation in $u - v$,

$$u' - v' = -\frac{k}{Q_B}(u - v) + \frac{k}{Q_D}(u - v)$$

$$(u - v)' = -\left(\frac{k}{Q_B} - \frac{k}{Q_D}\right)(u - v).$$

Let $\alpha = k/Q_B - k/Q_D$ and $y = u - v$. Then we must solve the separable equation $y' = -\alpha y$, which is done with DSolve, naming the resulting output step1. We then name y the result obtained in step1 by extracting the formula for y[x] from step1 with Part ([[...]]) and replacing C[1] by c with ReplaceAll (/.).

```
Clear[x, y]

step1 = DSolve[y'[x] == -αy[x], y[x], x]
```

$$\{\{y[x] \to e^{-x\alpha} C[1]\}\}$$

```
y = step1[[1, 1, 2]]/.C[1] → c
```

$$ce^{-x\alpha}$$

Using the facts that $u' = -\frac{k}{Q_B}(u - v)$ and $v = u - y$, we are able to use DSolve to find $u(x)$.

```
step2 = DSolve[{u'[x] == -k/Q_s cExp[-αx], u[0] == u0},

u[x], x]
```

$$\left\{\left\{u[x] \to \frac{e^{-x\alpha}\left(ck - ce^{x\alpha}k + e^{x\alpha}u0\alpha Q_S\right)}{\alpha Q_S}\right\}\right\}$$

Note that we use cap1 to represent L.

Because $y = u - v$, $v = u - y$. Consequently, because $v(L) = 0$ we are able to compute c.

```
leftside = step2[[1, 1, 2]] - y/.x->cap1
```

$$-ce^{-cap1\alpha} + \frac{e^{-cap1\alpha}\left(ck - ce^{cap1\alpha}k + e^{cap1\alpha}u0\alpha Q_S\right)}{\alpha Q_S}$$

```
cval = Solve[leftside == 0, c]
```

$$\left\{\left\{c \to \frac{e^{cap1\alpha}u0\alpha Q_S}{-k + e^{cap1\alpha}k + \alpha Q_S}\right\}\right\}$$

and determine u and v. Next, we substitute the value of C into the formula for u and v.

```
u = Simplify[step2[[1, 1, 2]]/.cval[[1]]]
```

$$\frac{u0\left(\left(-1 + e^{(cap1-x)\alpha}\right)k + \alpha Q_S\right)}{\left(-1 + e^{cap1\alpha}\right)k + \alpha Q_S}$$

```
v = Simplify[u - y/.cval[[1]]]
```

$$\frac{e^{-x\alpha}\left(e^{cap1\alpha} - e^{x\alpha}\right)u0(k - \alpha Q_S)}{\left(-1 + e^{cap1\alpha}\right)k + \alpha Q_S}$$

For example, in healthy adults, typical urea nitrogen levels are 11–23 milligrams per deciliter, while serum creatinine levels range from 0.6 to 1.2 milligrams per deciliter and the total volume of blood is 4–5 L.

Suppose that hemodialysis is performed on a patient with urea nitrogen level of 34 mg/dL and serum creatinine level of 1.8 using a dialyser with $k = 2.25$ and $L = 1$. If the flow rate of blood, Q_B, is 2 dL/minute while the flow rate of the dialysate, Q_D, is 4 dL/minute, will the level of wastes in the patient's blood reach normal levels after dialysis is performed?

After defining the appropriate constants, we evaluate u and v

$\alpha = k/Q_s - k/Q_o$;

$k = 2.25$;

$cap1 = 1$;

$u0 = 34 + 1.8$;

$Q_s = 2$; $Q_o = 4$;

$u//\text{Simplify}$

$v//\text{Simplify}$

$-14.2623 + 50.0623 e^{-0.5625x}$

$-14.2623 + 25.0312 e^{-0.5625x}$

and then graph u and v on the interval $[0,1]$ with Plot in Figure 2-10. Remember that the dialysate is moving in the direction opposite the blood. Thus, we see from the graphs that as levels of waste in the blood decrease, levels of waste in the

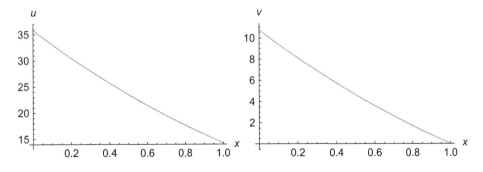

Figure 2-10   Remember that the dialysate moves in the opposite direction as the blood

dialysate increase and at the end of the dialysis procedure, levels of waste in the blood are within normal ranges.

```
p1 = Plot[u, {x, 0, 1}, AxesLabel → {x, "u"}, AxesStyle → Black];

p2 = Plot[v, {x, 0, 1}, AxesLabel → {x, "v"}, AxesStyle → Black];

Show[GraphicsRow[{p1, p2}]]
```

Typically, hemodialysis is performed 3–4 hours at a time 3 or 4 times per week. In some cases, a kidney transplant can free patients from the restrictions of dialysis. Of course, transplants have other risks not necessarily faced by those on dialysis; the number of available kidneys also affects the number of transplants performed. For example, in 1991 over 130,000 patients were on dialysis while only 7000 kidney transplants had been performed.

2.3 Homogeneous Equations

Definition 6 (Homogeneous Differential Equation). *A differential equation that can be written in the form*

$$M(x, y)\, dx + N(x, y)\, dy = 0,$$

where

$$M(tx, ty) = t^n M(x, y) \quad and \quad N(tx, ty) = t^n N(x, y)$$

is called a **homogeneous differential equation of degree** *n.*

It is a good exercise to show that an equation is homogeneous if we can write it in either of the forms $dy/dx = F(y/x)$ or $dy/dx = G(x/y)$.

EXAMPLE 2.3.1: Show that the equation $(x^2 + xy)\, dx - y^2\, dy = 0$ is homogeneous.

SOLUTION: Let $M(x, y) = x^2 + xy$ and $N(x, y) = -y^2$. Because $M(tx, ty) = (tx)^2 + (tx)(ty) = t^2 (x^2 + xy) = t^2 M(x, y)$ and $N(tx, ty) = -t^2 y^2 = t^2 N(x, y)$, the equation $(x^2 + xy)\, dx - y^2\, dy = 0$ is homogeneous of degree two.

■

Homogeneous equations can be reduced to separable equations by either of the substitutions

$$y = ux \quad \text{or} \quad x = vy.$$

Generally, use the substitution $y = ux$ if $N(x, y)$ is less complicated than $M(x, y)$ and use $x = vy$ if $M(x, y)$ is less complicated than $N(x, y)$. If a difficult integration problem is encountered after a substitution is made, try the other substitution to see if it yields an easier problem.

EXAMPLE 2.3.2: Solve the equation $\left(x^2 - y^2\right) dx + xy \, dy = 0$.

SOLUTION: For this example, $M(x, y) = x^2 - y^2$ and $N(x, y) = xy$. Then, $M(tx, ty) = t^2 M(x, y)$ and $N(tx, ty) = t^2 N(x, y)$, which means that $\left(x^2 - y^2\right) dx + xy \, dy = 0$ is a homogeneous equation of degree two. Assume $x = vy$. Then, $dx = v \, dy + y \, dv$ and substituting into the equation and simplifying yields

$$0 = \left(x^2 - y^2\right) dx + xy \, dy$$
$$0 = \left(v^2 y^2 - y^2\right) (v \, dy + y \, dv) + vy \cdot y \, dy$$
$$0 = \left(v^2 - 1\right) (v \, dy + y \, dv) + v \, dy$$
$$0 = v^3 \, dy + y \left(v^2 - 1\right) dv.$$

We solve this equation by rewriting it in the form

$$\frac{1}{y} dy = \frac{1 - v^2}{v^3} dv = \left(\frac{1}{v^3} - \frac{1}{v}\right) dv$$

and integrating. This yields

$$\ln |y| = -\frac{1}{2v^2} - \ln |v| + C_1,$$

which can be simplified as

$$\ln |vy| = -\frac{1}{2v^2} + C_1, \quad \text{so} \quad vy = Ce^{-1/(2v^2)}, \quad \text{where } C = \pm e^{C_1}.$$

Because $x = vy$, $v = x/y$, and resubstituting into the above equation yields

$$x = Ce^{-y^2/(2x^2)}$$

as a general solution of the equation $(x^2 - y^2)\,dx + xy\,dy = 0$. We see that DSolve is able to solve the equation as well.

```
Clear[x, y]

gensol = DSolve[x^2 - y[x]^2 + xy[x]y'[x] == 0, y[x], x]
```

$$\left\{\left\{y[x] \to -x\sqrt{C[1] - 2\text{Log}[x]}\right\}, \left\{y[x] \to x\sqrt{C[1] - 2\text{Log}[x]}\right\}\right\}$$

The result means that a general solution of the equation is $y^2 = x^2\,(C - 2\ln|x|)$. We can graph this implicit solution for various values of C by solving this equation for C

```
f = Solve[y==gensol[[1, 1, 2]], C[1]]
```

$$\left\{\left\{C[1] \to \frac{y^2 + 2x^2\,\text{Log}[x]}{x^2}\right\}\right\}$$

```
f[[1, 1, 2]]
```

$$\frac{y^2 + 2x^2\,\text{Log}[x]}{x^2}$$

and then noting that graphs of the equation $y^2 = x^2\,(C - 2\ln|x|)$ for various values of C are the same as the graphs of the level curves of the function $f(x, y) = (y^2 + 2x^2 \ln|x|) / (2x^2)$.

The ContourPlot command graphs several level curves $z = f(x, y)$, C a constant, of the function $z = f(x, y)$. We may instruct Mathematica to graph the level curves of $z = f(x, y)$ for particular values of C by including the Contours option. For example, the level curves of $f(x, y) = (y^2 + 2x^2 \ln|x|) / (2x^2)$ that intersect the x-axis at $x = 1$, $3/2$, 2, ..., $19/2$, and 10 are the contours with values obtained by replacing each occurrence of y in $f(x, y)$ by 0 and x by 1, $3/2$, 2, ..., $19/2$, and 10 which we do now with Table and ReplaceAll (/.), naming the resulting set of ten numbers contourvals.

```
contourvalues = Table[f[[1, 1, 2]]/.{x → i, y → 0},

    {i, 1, 10, .5}];


cp1 = ContourPlot[f[[1, 1, 2]], {x, 0.01, 10}, {y, −5, 5},

    PlotPoints → 150, Frame → False, Contours →

    contourvalues, Axes → Automatic, ContourStyle →
```

```
{{Gray, Thickness[.01]}}, ContourShading → False,

AxesOrigin → {0, 0}, AxesStyle → Black, AxesLabel →

{x, y}, PlotLabel → "(a)"];
```

graphs several level curves of $z = f(x, y)$ for $0.01 \leq x \leq 10$ and $-5 \leq y \leq 5$ and names the resulting graphics object cp1. cp1 is not displayed because we include a semi-colon (;) at the end of the ContourPlot command. The option Contours->contourvals instructs Mathematica to draw contours with values given in the list of numbers contourvals. We use StreamPlot to graph the direction field associated with the equation on the same rectangle, $[0.01, 10 \times [-5, 5]$

> We avoid $x = 0$ because $f(x, y)$ is undefined if $x = 0$.

```
dirf = StreamPlot[{1, (y^2 − x^2)/(xy)}, {x, 0.01, 10},

    {y, −5, 5}, Frame → False, StreamStyle → Black];
```

and then display cp1 (Figure 2-11(a)) and the direction field together with Show together with GraphicsRow in Figure 2-11(b).

```
p2 = Show[cp1, dirf, PlotLabel → "(b)"];

Show[GraphicsRow[{cp1, p2}]]
```

■

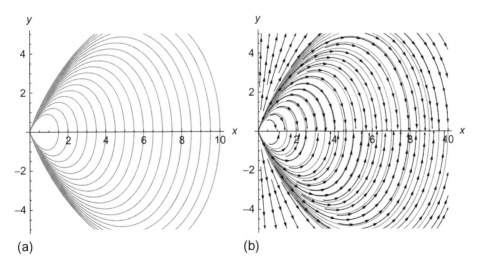

(a)   (b)

Figure 2-11 (a) Various solutions for the homogeneous equation $\left(x^2 - y^2\right)$ $dx + xy\,dy = 0$. (b) Various solutions and direction field for the homogeneous equation $\left(x^2 - y^2\right) dx + xy\,dy = 0$

EXAMPLE 2.3.3: Solve $\left(y^2 + 2xy\right) dx - x^2 dy = 0$.

SOLUTION: In this case, letting $F(t) = t^2 + 2t$, we note that $dy/dx = F(y/x) = (y/x)^2 + 2(y/x)$ so the equation is homogeneous.

Let $y = ux$. Then, $dy = u\, dx + x\, du$. Substituting into $\left(y^2 + 2xy\right) dx - x^2 dy = 0$ and separating gives us

$$\left(y^2 + 2xy\right) dx - x^2 dy = 0$$

$$\left(u^2 x^2 + 2ux^2\right) dx - x^2 (u\, dx + x\, du) = 0$$

$$\left(u^2 + 2u\right) dx - (u\, dx + x\, du) = 0$$

$$\left(u^2 + u\right) dx = -x\, du$$

$$\frac{1}{u(u+1)}\, du = -\frac{1}{x}\, dx.$$

Integrating the left- and right-hand sides of this equation with Integrate,

```
Integrate[1/(u(u + 1)), u]
```

Log[u] − Log[1 + u]

```
Integrate[1/x, x]
```

Log[x]

exponentiating, resubstituting $u = y/x$, and solving for y gives us

$$\ln|u| - \ln|u + 1| = -\ln|x| + C$$

$$\frac{u}{u+1} = \frac{C}{x}$$

$$\frac{\dfrac{y}{x}}{\dfrac{y}{x} + 1} = \frac{C}{x}$$

$$y = \frac{Cx}{x - C}.$$

```
Solve[(y/x)/(y/x + 1) == cx, y]
```

$$\left\{\left\{y \to -\frac{cx^2}{-1 + cx}\right\}\right\}$$

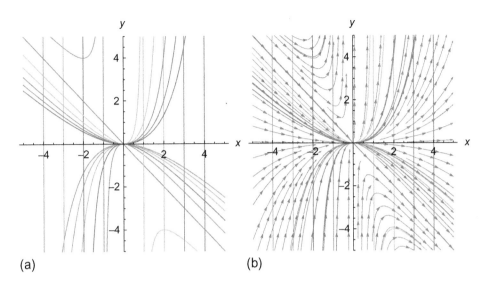

(a)   (b)

Figure 2-12 (a) Graphs of several solutions of $\left(y^2 + 2xy\right) dx - x^2 dy = 0$. (b) Graphs of several solutions together with the direction field

We confirm this result with DSolve and then graph several solutions with Plot in Figure 2-12(a).

```
sol = DSolve[y[x]^2 + 2xy[x] − x^2y′[x] == 0, y[x], x]
```

$$\left\{\left\{y[x] \rightarrow -\frac{x^2}{x - C[1]}\right\}\right\}$$

```
Solve[y[x]^2 + 2xy[x] − x^2y′[x] == 0, y′[x]]
```

$$\left\{\left\{y′[x] \rightarrow \frac{2xy[x] + y[x]^2}{x^2}\right\}\right\}$$

```
toplot = Table[sol[[1, 1, 2]]/.C[1] → i, {i, −5, 5}];

p1 = Plot[toplot, {x, −5, 5}, PlotRange → {−5, 5},

    AxesStyle → Black, AxesLabel → {x, y}, PlotLabel → "(a)",

    AspectRatio → 1];
```

We use StreamPlot to graph the direction field and then display the direction field together with the solutions in Figure 2-12(b).

```
p2a = StreamPlot[{1, (2xy + y^2)/x^2}, {x, −5, 5},

    {y, −5, 5}, StreamStyle → Gray];

p2 = Show[p1, p2a, PlotLabel → "(b)", AspectRatio → 1];

Show[GraphicsRow[{p1, p2}]]
```

In Figure 2-12, observe that the vertical lines correspond to discontinuities in the solution and are not graphs of the solutions to the equations. *Vertical lines are never a portion or part of the graph of a real-valued function of a single variable.* With careful use of the `Exclusion` option, the vertical lines can be deleted from the plots.

■

Application: Models of Pursuit

Sources: A particularly interesting and fun-to-read discussion of flight paths and models of pursuit can be found in *Differential Equations: A Modeling Perspective* by Robert L. Borrelli and Courtney S. Coleman and published by John Wiley & Sons.

Suppose that one object pursues another whose motion is known by a predetermined strategy. For example, suppose that an airplane is positioned at $B(1000,0)$ to fly to another airport $A(0,0)$ that is 1000 miles directly west of its position B, as illustrated in the following figure. Assume that the airplane aims toward A at all times. If the wind goes from south to north at a constant speed, w, and the airplane's speed in still air is b, determine conditions on b so that the airplane eventually arrives at A and describe its path.

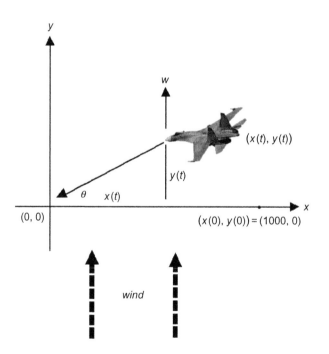

As described, the speed of the airplane, b, must be greater than the speed of the wind, w: $b > w$, in order for the plane to arrive at A. Observe that dx/dt describes the airplane's velocity in the x direction:

$$\frac{dx}{dt} = -b\cos\theta = \frac{-bx}{\sqrt{x^2+y^2}},$$

because from right-triangle trigonometry we know that $\cos\theta = \text{adjacent}/$ hypotenuse $= x/\sqrt{x^2+y^2}$. Similarly,

$$\frac{dy}{dt} = -b\sin\theta + w = \frac{-by}{\sqrt{x^2+y^2}} + w,$$

so

$$\frac{dy}{dx} = \frac{dy/dt}{dx/dt} = \frac{\dfrac{-by}{\sqrt{x^2+y^2}} + w}{\dfrac{-bx}{\sqrt{x^2+y^2}}} = \frac{by - w\sqrt{x^2+y^2}}{bx}.$$

This is a homogeneous equation (of degree one) because it can be written in the form $dy/dx = F(y/x)$:

$$\frac{dy}{dx} = \frac{by - w\sqrt{x^2+y^2}}{bx} = \frac{y}{x} - \frac{w}{b}\sqrt{1 + \left(\frac{y}{x}\right)^2}.$$

Therefore, we must solve the initial-value problem

$$\begin{cases} \dfrac{dy}{dx} = \dfrac{by - w\sqrt{x^2+y^2}}{bx} \\ y(1000) = 0 \end{cases}.$$

In this case, we see DSolve is both able to find a general solution of the equation

```
Clear[x, y, w, b]
DSolve[y'[x] == (by[x] − wSqrt[x^2 + y[x]^2])/(bx),
    y[x], x]
```

$$\left\{\left\{y[x] \rightarrow x\text{Sinh}\left[\frac{bC[1] - w\text{Log}[x]}{b}\right]\right\}\right\}$$

as well as solve the initial-value problem.

```
Clear[x, y, w, b]
DSolve[{y'[x] == (by[x] − wSqrt[x^2 + y[x]^2])/(bx), y[1000] == 0},
    y[x], x]
```

$$\left\{\left\{y[x] \rightarrow x\text{Sinh}\left[\frac{w\text{Log}[1000] - w\text{Log}[x]}{b}\right]\right\}\right\}$$

Alternatively, letting $y = ux$, differentiating to obtain $dy = u\,dx + x\,du$, and substituting into the homogeneous equation results in the separable equation

$$\frac{dy}{dx} = \frac{du}{dx}x + u = \frac{bux - w\sqrt{x^2 + u^2 x^2}}{bx}$$

$$\frac{du}{dx}x + u = u - \frac{w}{b}\sqrt{1 + u^2}$$

$$\frac{1}{\sqrt{1 + u^2}}du = -\frac{w}{b}\frac{1}{x}dx$$

```
y = ux;

eqn = Dt[y] == (by - wSqrt[x^2 + y^2])/(bx);

step1 = PowerExpand[Simplify[eqn]]
```

$$\frac{\sqrt{1 + u^2}\,w}{b} + xDt[u] + uDt[x] == u$$

Integrating the left-hand side of this equation yields $\int \dfrac{1}{\sqrt{1 + u^2}}du = \ln\left|u + \sqrt{1 + u^2}\right| + C_1$

```
leftint = Integrate[1/Sqrt[1 + u^2], u]

ArcSinh[u]

leftint = TrigToExp[leftint]
```

$$\text{Log}\left[u + \sqrt{1 + u^2}\right]$$

and integrating the right results in $-\dfrac{w}{b}\int\dfrac{1}{x}dx = -\dfrac{w}{b}\ln|x| + C_2$. Note that absolute value bars are not necessary because x and y are both positive and, hence, u is nonnegative. Thus, $\ln\left(u + \sqrt{1 + u^2}\right) = -\dfrac{w}{b}\ln x + C$.

```
rightint = Integrate[-w/(bx), x] + c
```

$$c - \frac{w\text{Log}[x]}{b}$$

Because $y(1000) = 0$, $C = \frac{w}{b}\ln 1000$

```
cval = Solve[leftint == rightint/.{x → 1000, u → 0}, c]
```

$$\left\{\left\{c \to \frac{w\text{Log}[1000]}{b}\right\}\right\}$$

and $\ln\left(u + \sqrt{1 + u^2}\right) = -\dfrac{w}{b}\ln x + \dfrac{w}{b}\ln 1000$.

```
step2 = leftint == rightint/.cval[[1]]
```

$$\mathrm{Log}\left[u + \sqrt{1 + u^2}\right] == \frac{w\mathrm{Log}[1000]}{b} - \frac{w\mathrm{Log}[x]}{b}$$

Solving for u gives us

$$\ln\left(u + \sqrt{1 + u^2}\right) = \ln\left(\frac{x}{1000}\right)^{-w/b}$$

$$u + \sqrt{1 + u^2} = \left(\frac{x}{1000}\right)^{-w/b}$$

$$\sqrt{1 + u^2} = \left(\frac{x}{1000}\right)^{-w/b} - u$$

$$1 + u^2 = \left(\frac{x}{1000}\right)^{-2w/b} - 2u\left(\frac{x}{1000}\right)^{-w/b} + u^2$$

$$2u\left(\frac{x}{1000}\right)^{-w/b} = \left(\frac{x}{1000}\right)^{-2w/b} - 1$$

$$u = \frac{1}{2}\left[\left(\frac{x}{1000}\right)^{-w/b} - \left(\frac{x}{1000}\right)^{w/b}\right].$$

```
step3 = Solve[step2, u]
```

$$\left\{\left\{u \rightarrow \frac{1}{2}\left(10^{\frac{3w}{b}}x^{-\frac{w}{b}} - 10^{-\frac{3w}{b}}x^{\frac{w}{b}}\right)\right\}\right\}$$

We solve for y by resubstituting $u = y/x$ and multiplying by x:

$$\frac{y}{x} = \frac{1}{2}\left[\left(\frac{x}{1000}\right)^{-w/b} - \left(\frac{x}{1000}\right)^{w/b}\right]$$

$$y = \frac{1}{2}x\left[\left(\frac{x}{1000}\right)^{-w/b} - \left(\frac{x}{1000}\right)^{w/b}\right]$$

$$y = \frac{1}{2x}\left[\left(x\left(10^{-3}\right)\right)^{-w/b} - \left(x\left(10^{-3}\right)\right)^{w/b}\right].$$

```
Clear[y]
```

```
y[x_] = xstep3[[1, 1, 2]]
```

$$\frac{1}{2}x\left(10^{\frac{3w}{b}}x^{-\frac{w}{b}} - 10^{-\frac{3w}{b}}x^{\frac{w}{b}}\right)$$

We graph y for various values of w/b by setting $b = 1$ and then using `Table` to generate the value of y for $w = 0.25, 0.50, \ldots, 2.0$. These functions are then graphed with `Plot` in Figure 2-13. Notice that the airplane never arrives at A if $w/b \geq 1$.

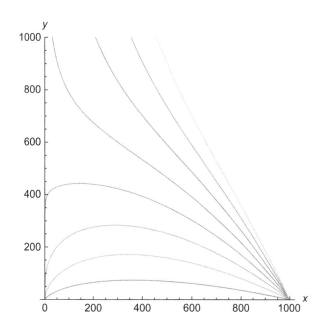

Figure 2-13   If $w/b \geq 1$, the airplane never reaches its destination

```
b = 1;

toplot = Table[y[x], {w, 0.2, 2.0, .25}];

Plot[toplot, {x, 0, 1000}, PlotRange → {0, 1000},
    AxesLabel → {x, y}, AxesStyle → Black,
    AspectRatio → 1]
```

2.4 Exact Equations

Definition 7 (Exact Differential Equation). *A differential equation that can be written in the form*

$$M(x, y)\, dx + N(x, y)\, dy = 0$$

where

$$M(x, y)\, dx + N(x, y)\, dy = \frac{\partial f}{\partial x}(x, y)\, dx + \frac{\partial f}{\partial y}(x, y)\, dy$$

for some function $z = f(x, y)$ *is called an* **exact differential equation**.

We can show that the differential equation $M(x, y)\, dx + N(x, y)\, dy = 0$ is exact if and only if $\partial M/\partial y = \partial N/\partial x$.

EXAMPLE 2.4.1: Show that the equation $2xy^3\, dx + \left(1 + 3x^2y^2\right) dy = 0$ is exact and that the equation $x^2 y\, dx + 5xy^2\, dy = 0$ is not exact.

SOLUTION: Because

$$\frac{\partial}{\partial y}\left(2xy^3\right) = 6xy^2 = \frac{\partial}{\partial x}\left(1 + 3x^2y^2\right),$$

the equation $2xy^3\, dx + \left(1 + 3x^2y^2\right) dy = 0$ is an exact equation. On the other hand, the equation $x^2 y\, dx + 5xy^2\, dy = 0$ is not exact because

$$\frac{\partial}{\partial y}\left(x^2 y\right) = x^2 \neq 5y^2 = \frac{\partial}{\partial x}\left(5xy^2\right).$$

(However, the equation $x^2 y\, dx + 5xy^2\, dy = 0$ is separable.)

∎

If an equation is exact, we can find a function $z = f(x, y)$ such that $M(x, y) = \frac{\partial f}{\partial x}(x, y)$ and $N(x, y) = \frac{\partial f}{\partial y}(x, y)$.

1. Assume that $M(x, y) = \frac{\partial f}{\partial x}(x, y)$ and $N(x, y) = \frac{\partial f}{\partial y}(x, y)$.
2. Integrate $M(x, y)$ with respect to x. (Add an arbitrary function of y, $g(y)$.)
3. Differentiate the result in Step 2 with respect to y and set the result equal to $N(x, y)$. Solve for $g'(y)$.
4. Integrate $g'(y)$ with respect to y to obtain an expression for $g(y)$. (There is no need to include an arbitrary constant.)
5. Substitute $g(y)$ into the result obtained in Step 2 for $f(x, y)$.
6. A general solution is $f(x, y) = C$ where C is a constant.
7. If given an initial-value problem, apply the initial condition to determine C.

Remark. A similar algorithm can be stated so that in Step 2 $N(x, y)$ is integrated with respect to y.

EXAMPLE 2.4.2: Solve $2x \sin y\, dx + \left(x^2 \cos y - 1\right) dy = 0$ subject to $y(0) = 1/2$.

SOLUTION: The equation $2x \sin y \, dx + (x^2 \cos y - 1) \, dy = 0$ is exact because

$$\frac{\partial}{\partial y} (2x \sin y) = 2x \cos y = \frac{\partial}{\partial x} \left(x^2 \cos y - 1 \right).$$

Let $z = f(x, y)$ be a function with $\partial f / \partial x = 2x \sin y$ and $\partial f / \partial y = x^2 \cos y - 1$. Then, integrating $\partial f / \partial x$ with respect to x yields

$$f(x, y) = \int 2x \sin y \, dx = x^2 \sin y + g(x).$$

Notice that the arbitrary function $g = g(y)$ of y serves as a "constant" of integration with respect to x. Because we have $\partial f / \partial y = x^2 \cos y - 1$ from the differential equation, and

$$\frac{\partial f}{\partial y}(x, y) = x^2 \cos y + g'(y)$$

from differentiation of $f(x, y)$ with respect to y, $g'(y) = -1$. Integrating $g'(y)$ with respect to y gives us $g(y) = -y$. Therefore, $f(x, y) = x^2 \sin y - y$, so a general solution of the exact equation is $x^2 \sin y - y = C$, where C is a constant. Because our solution requires that $y(0) = 1/2$, we must find the solution in the family of solutions that passes through the point $(0, 1/2)$. Substituting these values of x and y into the general solution, we obtain $0^2 \cdot \sin(1/2) - 1/2 = C$ so that $C = -1/2$. Therefore, the desired solution is $x^2 \sin y - y = -1/2$. We are able to use DSolve to solve the initial-value problem implicitly as well.

Notice that we do not have to include the constant in calculating $g(y)$ because we combine it with the constant in the general solution.

```
Clear[x, y]

partsol = DSolve[{2xSin[y[x]] + (x^2Cos[y[x]] − 1)y'[x] == 0,

   y[0] == 1/2}, y[x], x]
```

$\mathrm{Solve}\left[x^2 \mathrm{Sin}[y[x]] - y[x] == -\frac{1}{2}, \ y[x] \right]$

```
partsol[[1]]
```

$x^2 \mathrm{Sin}[y[x]] - y[x] == -\dfrac{1}{2}$

```
partsol[[1, 1]]
```

$x^2 \mathrm{Sin}[y[x]] - y[x]$

To graph the function partsol[[1,1,2]], we use ContourPlot to graph the equation for $-10 \leq x \leq 10$ and $-10 \leq y \leq 10$ in

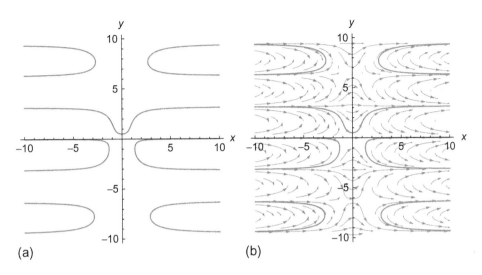

Figure 2-14   Plot of $x^2 \sin y - y = -1/2$

Figure 2-14(a). We include the option AxesOrigin->{0,0} to specify that the axes intersect at the point $(0,0)$. We use StreamPlot to graph the direction field and show the direction field together with several solutions in Figure 2-14(b).

```
p1 = ContourPlot[Evaluate[partsol[[1, 1]]/.y[x] → y],

    {x, −10, 10}, {y, −10, 10}, Frame → False, ContourShading

    → False, Contours → {−1/2}, ContourStyle → {{Black,

    Thickness[.01]}}, Axes → Automatic, AxesOrigin → {0, 0},

    AxesLabel → {x, y}, PlotLabel → "(a)", AxesStyle

    → Black];

p2a = StreamPlot[{1, −2xSin[y]/(x^2Cos[y] − 1)},

    {x, −10, 10}, {y, −10, 10}, StreamStyle → Gray];

p2 = Show[p1, p2a, PlotLabel → "(b)", AspectRatio → 1];

Show[GraphicsRow[{p1, p2}]]
```

∎

The following example illustrates how we can use Mathematica to assist us in carrying out the necessary steps encountered when solving an exact equation.

EXAMPLE 2.4.3: Solve

$(2x - y^2 \sin(xy)) \, dx + (\cos(xy) - xy \sin(xy)) \, dy = 0.$

SOLUTION: We begin by identifying $M(x, y) = 2x - y^2 \sin(xy)$ and $N(x, y) = \cos(xy) - xy \sin(xy)$. We then define capm, corresponding to M, and capn, corresponding to N. We then see that the equation is exact because $\partial M / \partial y = \partial N / \partial x$.

```
capm[x_, y_] = 2x - y^2Sin[xy];

capn[x_, y_] = Cos[xy] - xySin[xy];

D[capm[x, y], y] == D[capn[x, y], x]
```

```
True
```

Next, we compute $\int M(x, y) \, dx$ and add an arbitrary function of y, g[y], to the result.

```
f = Integrate[capm[x, y], x] + g[y]
```

$x^2 + y\text{Cos}[xy] + g[y]$

Differentiating f with respect to y gives us

```
D[f, y] == capn[x, y]
```

and because we must have that $\partial f / \partial y = N(x, y)$, we obtain the equation

$\text{Cos}[xy] - xy\text{Sin}[xy] + g'[y] == \text{Cos}[xy] - xy\text{Sin}[xy]$

which we solve for $g'(y)$ with Solve.

```
Solve[D[f, y] == capn[x, y], g'[y]]
```

$\{\{g'[y] \to 0\}\}$

```
f = f /. g[y] → 0
```

Thus, $g(y)$ is a (real-valued) constant and a general solution of the equation is $x^2 + y \cos(xy) = C$. We can graph this general solution for various values of C by observing that the level curves of the function $z = x^2 + y\cos(xy)$ correspond to the graphs of the equation $x^2 + y\cos(xy) = C$ for various values of C.

$$x^2 + y\text{Cos}[xy]$$

We now use `ContourPlot` to graph several level curves of $z = x^2 + y\cos(xy)$ on the rectangle $[0, 3\pi] \times [0, 3\pi]$ in Figure 2-15. In this case, the option `Frame->False` instructs Mathematica to not place a frame around the resulting graphics object, the option `Axes->Automatic` specifies that axes are to be placed on the graph, `AxesOrigin->{0,0}` specifies that the axes are to intersect at the point $(0, 0)$, `AxesStyle->GrayLevel[.5]` specifies that the axes are to be drawn in a light gray, `ContourShading->False` specifies that the region between contours is to not be shaded and the option `PlotPoints->150` helps assure that the resulting contours appear smooth. `Contours->n` instructs Mathematica to graph n contours. If you prefer that the contours correspond to specific function values us `Contours->{list of function values}`. Use `ContourStyle` to specify the style of your contours, such as their thickness and color.

```
cp1 = ContourPlot[f, {x, 0, 3Pi}, {y, 0, 3Pi},

    Frame → False, Axes → Automatic, AxesOrigin → {0, 0},
```

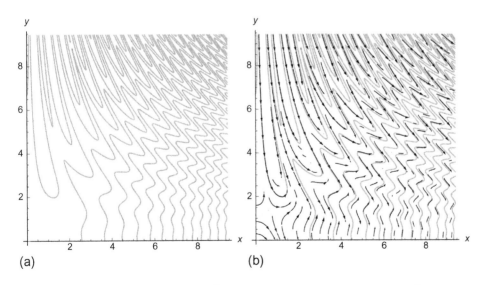

(a)   (b)

Figure 2-15   (a) Level curves of $z = x^2 + y\cos(xy)$. (b) Level curves of $z = x^2 + y\cos(xy)$ together with the direction field

```
AxesLabel → {x, y}, ContourShading → False,

Contours → 15, AxesStyle → Black,

ContourStyle → {{CMYKColor[0, 0.89, 0.94, 0.28],

Thickness[.0075]}}, PlotPoints → 150,

PlotLabel → "(a)"];

p2 = StreamPlot[{1, −capm[x, y]/capn[x, y]},

  {x, 0, 3Pi}, {y, 0, 3Pi}, StreamStyle → Black];

p3 = Show[cp1, p2, PlotLabel → "(b)"];

Show[GraphicsRow[{cp1, p3}]]
```

We see that DSolve is able to find an implicit solution of the equation after we rewrite it in the form $\left(2x - y^2 \sin(xy)\right) + (\cos(xy) - xy\sin(xy))\, y' = 0$.

```
gensol = DSolve[capm[x, y[x]] + capn[x, y[x]]y'[x] == 0,

y[x], x]
```

$\mathsf{Solve}\left[x^2 + \mathsf{Cos}[xy[x]]y[x] == C[1],\ y[x]\right]$

```
step2 = gensol[[1, 1]]
```

$x^2 + \mathsf{Cos}[xy[x]]y[x]$

```
implicitplot = step2/.y[x] → y
```

$x^2 + y\,\mathsf{Cos}[xy]$

∎

2.5 Linear Equations

Definition 8 (First-Order Linear Equation). *A differential equation of the form*

$$a_1(x)\frac{dy}{dx} + a_0(x)y = f(x), \tag{2.2}$$

*where $a_1(x)$ is not identically the zero function, is a first-order **linear differential equation**.*

Assuming that $a_1(x)$ is not identically the zero function, dividing equation (2.2) by $a_1(x)$ gives us the **standard form** of the first-order linear equation:

$$\frac{dy}{dx} + p(x)y = q(x). \tag{2.3}$$

If $q(x)$ is identically the zero function, we say that the equation is **homogeneous**. The **corresponding homogeneous equation** of equation (2.3) is

$$\frac{dy}{dx} + p(x)y = 0. \tag{2.4}$$

2.5.1 Integrating Factor Approach

Multiplying equation (2.3) by $e^{\int p(x)\,dx}$ yields

$$e^{\int p(x)\,dx}\frac{dy}{dx} + e^{\int p(x)\,dx}p(x)y = e^{\int p(x)\,dx}q(x).$$

By the product rule and the Fundamental Theorem of Calculus,

$$\frac{d}{dx}\left(e^{\int p(x)\,dx}y\right) = e^{\int p(x)\,dx}\frac{dy}{dx} + e^{\int p(x)\,dx}p(x)y$$

so equation (2.3) becomes

$$\frac{d}{dx}\left(e^{\int p(x)\,dx}y\right) = e^{\int p(x)\,dx}q(x).$$

Integrating and dividing by $e^{\int p(x)\,dx}$ yields a general solution of $y' + p(x)y = q(x)$:

$$e^{\int p(x)\,dx}y = \int e^{\int p(x)\,dx}q(x)\,dx$$

$$y = \frac{1}{e^{\int p(x)\,dx}}\int e^{\int p(x)\,dx}q(x)\,dx = e^{-\int p(x)\,dx}\int e^{\int p(x)\,dx}q(x)\,dx.$$

The term $\mu(x) = e^{\int p(x)\,dx}$ is called an **integrating factor** for the linear equation (2.3).

Thus, first-order linear equations can always be solved, although the resulting integrals may be difficult or impossible to evaluate exactly.

As we see with the following command, DSolve is always able to solve first-order linear differential equations, although the result might contain unevaluated integrals.

```
Clear[x, y, p, q]
```

```
DSolve[y'[x] + p[x]y[x] == q[x], y[x], x]
```

$$\left\{\left\{y[x] \rightarrow e^{\int_1^x -p[K[1]]\,dK[1]}C[1] + e^{\int_1^x -p[K[1]]\,dK[1]}\int_1^x e^{-\int_1^{K[2]} -p[K[1]]\,dK[1]}\right.\right.$$
$$\left.\left. q[K[2]]\,dK[2]\right\}\right\}$$

EXAMPLE 2.5.1: Solve $x\,dy/dx + y = x\cos x$, $x > 0$.

SOLUTION: First, we place the equation in the form used in the derivation above. Dividing the equation by x yields

$$\frac{dy}{dx} + \frac{1}{x}y = \cos x, \tag{2.5}$$

where $p(x) = 1/x$ and $q(x) = \cos x$. Then, an integrating factor is

$$\mu(x) = e^{\int \frac{1}{x}dx} = e^{\ln|x|} = x, \text{ for } x > 0,$$

and multiplying equation (2.5) by the integrating factor gives us

$$\frac{d}{dx}(xy) = x\frac{dy}{dx} + y = x\cos x.$$

Integrating once we have

$$xy = \int x\cos x\,dx.$$

Using the integration by parts formula, $\int u\,dv = uv - \int v\,du$, with $u = x$ and $dv = \cos x\,dx$, we obtain $du = dx$ and $v = \sin x$ so

$$xy = \int x\cos x\,dx = x\sin x - \int \sin x\,dx = x\sin x + \cos x + C.$$

Therefore, a general solution of the equation $x\,dy/dx + y = x\cos x$ for $x > 0$ is $y = (x\sin x + \cos x + C)/x$. We see that DSolve is also successful in finding a general solution of the equation.

If we want to solve the equation for $x < 0$, then we would have $e^{\int \frac{1}{x}dx} = e^{\ln|x|} = -x$ for $x < 0$.

```
Clear[x, y]
```

```
gensol = DSolve[xy'[x] + y[x] == xCos[x], y[x], x]
```

$$\left\{\left\{y[x] \rightarrow \frac{C[1]}{x} + \frac{\cos[x] + x\sin[x]}{x}\right\}\right\}$$

As we have seen in previous examples, we can graph the solution for various values of the arbitrary constant by generating a set of functions obtained by replacing the arbitrary constant with numbers using `Table` and `ReplaceAll` (`/.`).

```
toplot = Table[gensol[[1, 1, 2]]/.C[1] → i, {i, −4, 4}];
```

In this case, Mathematica generates several error messages, which are not displayed here because the solution is undefined if $x = 0$. Nevertheless, the resulting graph shown in Figure 2-16(a) is displayed correctly.

```
q1 = Plot[toplot, {x, 0, 4Pi}, PlotRange → {−2Pi, 2Pi},

    AspectRatio → 1, AxesStyle → Black, AxesLabel → {x, y},

    PlotLabel → "(a)"];
```

To graph the direction field, we use `StreamPlot`. First, we write the equation in the form $dy/dx = f(x, y)$.

```
step2 = Solve[xy′[x] + y[x] == xCos[x], y′[x]]/.y[x] → y
```

$$\left\{\left\{y'[x] \rightarrow \frac{-y + x\text{Cos}[x]}{x}\right\}\right\}$$

We then use `StreamPlot` to graph the direction field and display the result in Figure 2-16(b).

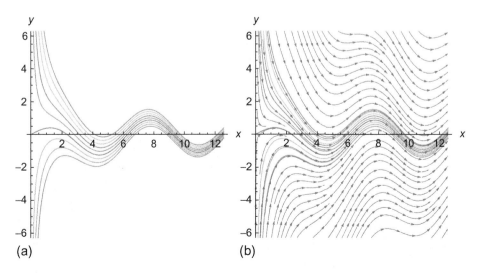

(a)   (b)

Figure 2-16   (a) Various solutions of $x\,dy/dx + y = x\cos x$, $x > 0$. (b) Various solutions with the direction field

```
q2a = StreamPlot[{1, step2[[1, 1, 2]]}, {x, 0, 4Pi}, {y, -2Pi, 2Pi},
    PlotRange → {-2Pi, 2Pi}, AspectRatio → 1,
    AxesStyle → Black, AxesLabel → {x, y}, PlotLabel → "(b)"];

q2 = Show[{q1, q2a}, PlotRange→ {{0, 4Pi}, {-2Pi, 2Pi}}, PlotLabel
    → "(b)"];

Show[GraphicsRow[{q1, q2}]]
```

■

    As with other types of equations, we solve initial-value problems by first finding a general solution of the equation and then applying the initial condition to determine the value of the constant.

EXAMPLE 2.5.2: Solve the initial-value problem $\begin{cases} dy/dx + 5x^4 y = x^4 \\ y(0) = -7 \end{cases}$.

SOLUTION: As we have seen in many previous examples, DSolve can be used to find a general solution of the equation and the solution to the initial-value problem, as done in gensol and partsol, respectively.

```
Clear[x, y, gensol]
gensol = DSolve[y'[x] + 5x^4y[x] == x^4, y[x], x]
```

$$\left\{ \left\{ y[x] \to \frac{1}{5} + e^{-x^5} C[1] \right\} \right\}$$

```
partsol = DSolve[{y'[x] + 5x^4y[x] == x^4, y[0] == -7}, y[x], x]
```

$$\left\{ \left\{ y[x] \to \frac{1}{5} e^{-x^5} \left(-36 + e^{x^5} \right) \right\} \right\}$$

    We now graph various solutions to the differential equation with Table followed by Plot in Figure 2-17(a).

```
toplot1 = Table[gensol[[1, 1, 2]]/. C[1] → c, {c, -20, 20, 2}];
p1 = Plot[toplot1, {x, -1, 2}, PlotStyle → Gray,
```

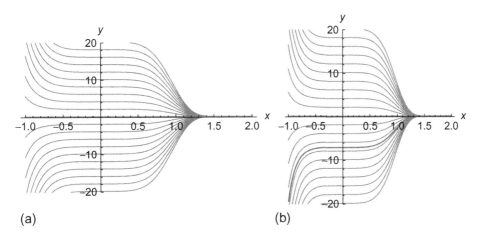

Figure 2-17 (a) Various solutions of $dy/dx + 5x^4y = x^4$. (b) The solution of $dy/dx + 5x^4y = x^4$ that satisfies $y(0) = -7$

```
AxesStyle → Black, AxesLabel → {x, y}, PlotRange → {-20, 20},

PlotLabel → "(a)", AspectRatio → 1];
```

Next we graph the solution to the initial-value problem obtained in partsol with Plot in Figure 2-17(b).

```
p2a = Plot[partsol[[1, 1, 2]], {x, -1, 2},

    PlotStyle → {{Thickness[.01], CMYKColor[0, 0.89, 0.94, 0.28]}},

    AxesStyle → Black, AxesLabel → {x, y}, PlotRange → {-20, 20},

    PlotLabel → "(b)", AspectRatio → 1];

p2 = Show[p2a, p1];
```

```
Show[GraphicsRow[{p1, p2}]]
```

We can also use Mathematica to carry out the steps necessary to solve first-order linear equations. We begin by identifying the integrating factor $e^{\int 5x^4\,dx} = e^{x^5}$, computed as follows with Integrate.

```
intfac = Exp[Integrate[5x^4, x]]
```

e^{x^5}

Therefore, the equation can be written as

$$\frac{d}{dx}\left(e^{x^5}y\right) = x^4 e^{x^5}$$

so that integration of both sides of the equation yields

$$e^{x^5} y = \frac{1}{5} e^{x^5} + C.$$

```
rightside = Integrate[intfacx^4, x]
```

$$\frac{e^{x^5}}{5}$$

Hence, a general solution is $y = \frac{1}{5} + Ce^{-x^5}$. Note that we compute y by using Solve to solve the equation $e^{x^5} y = \frac{1}{5} e^{x^5} + C$ for y.

```
step1 = Solve[Exp[x^5]y == rightside + c, y]
```

$$\left\{ \left\{ y \to \frac{1}{5} e^{-x^5} \left(5c + e^{x^5} \right) \right\} \right\}$$

We find the unknown constant C by substituting the initial condition $y(0) = -7$ into the general solution and solving for C.

```
findc = Solve[-7==step1[[1, 1, 2]]/.x → 0]
```

$$\left\{ \left\{ c \to -\frac{36}{5} \right\} \right\}$$

Therefore, the solution to the initial-value problem is $y = \frac{1}{5} - \frac{36}{5} e^{-x^5}$.

```
step1[[1, 1, 2]]/.findc[[1]]
```

$$\frac{1}{5} e^{-x^5} \left(-36 + e^{x^5} \right)$$

■

We can use DSolve to solve a first-order linear equation even if the coefficient functions are discontinuous or piecewise-defined. In such situations, it is often useful to take advantage of the *unit step function*. The **unit step function**, $\mathcal{U}(t)$, is defined by

$$\mathcal{U}(t) = \begin{cases} 1, t \geq 0 \\ 0, t < 0 \end{cases}.$$

The Mathematica command UnitStep[t] returns $\mathcal{U}(t)$.

EXAMPLE 2.5.3 (Drug Concentration): If a drug is introduced into the bloodstream in dosages $D(t)$ and is removed at a rate proportional to the concentration, the concentration $C(t)$ at time t is given by

$$\begin{cases} dC/dt = D(t) - kC \\ C(0) = 0 \end{cases},$$

where $k > 0$ is the constant of proportionality.

Suppose that over a 24-hour period, a drug is introduced into the bloodstream at a rate of $24/t_0$ for exactly t_0 hours and then stopped so that $D_{t_0}(t) = \begin{cases} 24/t_0, \ 0 \le t \le t_0 \\ 0, \ t > t_0 \end{cases}$. Calculate and then graph $C(t)$ on the interval $[0, 30]$ if $k = 0.05, 0.10, 0.15, 0.20,$ and 0.25 for $t_0 = 4, 8, 12, 16,$ and 25. How does increasing t_0 affect the concentration of the drug in the bloodstream? Then consider the effect of increasing k.

See J. D. Murray's *Mathematical Biology*, Springer-Verlag, 1990, pp. 645–649.

SOLUTION: To compute $C(t)$, we must keep in mind that $D_{t_0}(t)$ is a piecewise defined function. In terms of the unit step function, $\mathcal{U}(t)$,

$$D_{t_0}(t) = \frac{24}{t_0}\mathcal{U}(t_0 - t)$$

Note that we use lower-case letters to avoid any ambiguity with built-in objects like C and D.

```
d[t_, t0_] = 24/t0UnitStep[t0 - t];
```

For example, entering d(t,4) returns $D_4(t) = \begin{cases} 6, \ 0 \le t \le 4 \\ 0, \ t > 4 \end{cases}$.

```
d[t, 4]
```

```
6UnitStep[4 - t]
```

Given k and t_0, the function sol returns the solution to the initial-value problem $\begin{cases} dC/dt = D_{t_0}(t) - kC \\ C(0) = 0 \end{cases}$.

```
Clear[sol, k, c, t, t0]
```

```
sol[k_, t0_]:=DSolve[{c'[t] == d[t, t0] - kc[t], c[0] == 0}, c[t], t]
```

```
[[1, 1, 2]]
```

Then, for $k = 0.05$ we solve the initial-value problem $\begin{cases} dC/dt = D_{t_0}(t) - kC \\ C(0) = 0 \end{cases}$ for $t = 4, 8, 12, 16,$ and 20 by applying sol to the list $\{4, 8, 12, 16, 20\}$ using Map. These solutions are graphed with Plot in Figure 2-18.

Map[f,list] computes $f(x)$ for each element x of list.

```
toplot05 = Map[sol[0.05, #]&, {4, 8, 12, 16, 20}];
```

```
p1 = Plot[toplot05, {t, 0, 30}, PlotRange → {0, 30},

   AspectRatio → 1, AxesStyle → Black,

   AxesLabel → {t, c}]
```

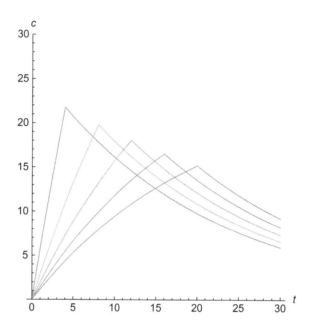

Figure 2-18   As t_0 increases, the maximum concentration of the drug decreases

Similar steps are repeated for $k = 0.10, 0.15, 0.20,$ and 0.25 by defining the function toplot. Given k, toplot[k] solves the initial-value problem $\begin{cases} dC/dt = D_{t_0}(t) - kC \\ C(0) = 0 \end{cases}$ for $t_0 = 4, 8, 12, 16,$ and 20.

```
Clear[toplot, sols]

toplot[k_]:=Map[sol[k, #]&, {4, 8, 12, 16, 20}];
```

We then apply toplot to the list $\{0.1, 0.15, 0.20, 0.25\}$ naming the resulting list of lists of functions sols.

```
sols = Map[toplot, {0.1, 0.15, 0.2, 0.25}];
```

Each list of functions in sols is then graphed with Plot by applying the pure function

```
Plot[Evaluate[#], {t, 0, 30}, PlotRange -> {0, 30},
AspectRatio -> 1, AxesStyle -> Black,
AxesLabel->{t,c}] &
```

to each element of sols with Map.

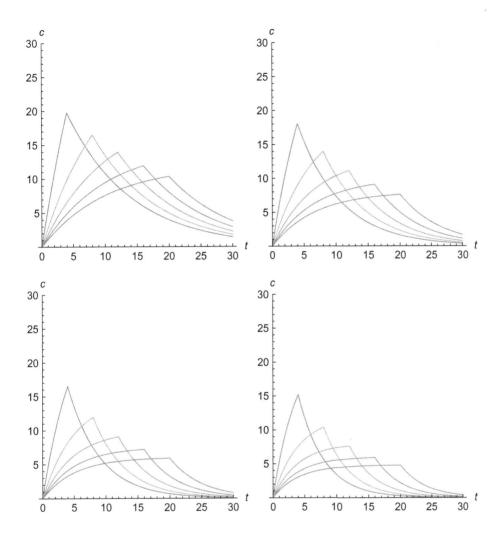

Figure 2-19 $C(t)$ for various values of t_0 and k

```
toshow = Map[Plot[#, {t, 0, 30}, PlotRange → {0, 30},
    AspectRatio → 1, AxesStyle → Black,
    AxesLabel → {t, c}]&, sols];
```

Finally, all four graphs are shown together as a graphics array using Show and GraphicsGrid in Figure 2-19.

```
Show[GraphicsGrid[Partition[toshow, 2]]]
```

From the graphs, we see that as t_0 is increased, the maximum concentration level decreases and occurs at later times, while increasing k increases the rate at which the drug is removed from the bloodstream.

Another way to see how t_0 and k affect the concentration of the drug is to use Manipulate. To avoid some computational errors, we use NDSolve rather than DSolve in the Manipulate function. In Figure 2-20, use the sliders to experiment with different t_0 and k values.

```
Manipulate[Plot[Evaluate[NDSolve[{c'[t] ==

    d[t, t0] − kc[t], c[0] == 0},

    c[t], {t, 0, 30}]][[1, 1, 2]]], {t, 0, 30},

    PlotRange → {{0, 30}, {0, 30}},
```

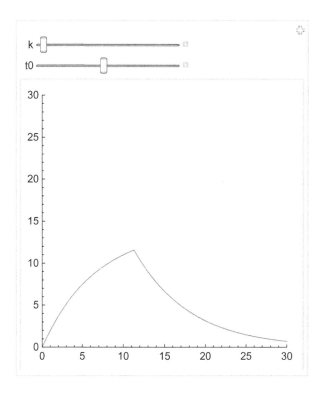

Figure 2-20   Use the sliders to change k and t_0 to see how the solution to the initial value problem changes. To see the specific values k and t_0, click on the + button to expand the sliders

> ```
> AspectRatio → 1, AxesStyle → Black],
> ```
>
> ```
> {{k, 1}, 0, 25}, {{t0, 12}, 0, 24}]
> ```

∎

If the integration cannot be carried out, the solution can often by approximated numerically by taking advantage of numerical integration techniques. Generally,

```
NDSolve[{y'[t]==f[t,y[t]],y[t0]=y0},y[t],{t,tmin,tmax}]
```

attempts to numerically solve $dy/dt = f(t, y)$, $y(t_0) = y_0$ for $t_{min} \leq t \leq t_{max}$. Observe that the syntax for `NDSolve` is nearly identical to that of the `DSolve` command except that we must specify an interval on which the solution is to be accurate (Figure 2-21).

EXAMPLE 2.5.4: Graph the solution to the initial-value problem $y' - y \sin(2\pi x) = 1$, $y(0) = 1$ on the interval $[0, 2\pi]$.

Figure 2-21   Mathematica's help for `NDSolve`

SOLUTION: Note that DSolve is successful in finding the solution to the initial-value problem even though the result contains unevaluated integrals.

> `Clear[x, y, partsol]`

> `partsol = DSolve[{y'[x] − Sin[2Pix]y[x] == 1,`

> `y[0] == 1}, y[x], x]`

$$\left\{\left\{y[x] \to e^{-\frac{\text{Cos}[2\pi x]}{2\pi}} \left(e^{\frac{1}{2}/\pi} + \text{BesselI}\left[0, \frac{1}{2\pi}\right] + \int_1^x e^{\frac{\text{Cos}[2\pi K[1]]}{2\pi}} dK[1]\right)\right\}\right\}$$

We can evaluate the result for particular numbers. For example, entering

> `partsol[[1, 1, 2]]/.x → 1`

$$e^{-\frac{1}{2}/\pi} \left(e^{\frac{1}{2}/\pi} + \text{BesselI}\left[0, \frac{1}{2\pi}\right]\right)$$

returns the value of the solution to the initial-value problem if $x = 1$. This result is a bit complicated to understand so we use N to obtain a numerical approximation.

> `N[%]`

    1.85827

Before graphing the solution to the initial value problem, we use StreamPlot to graph the direction field for the equation in Figure 2-22(a).

> `p1 = StreamPlot[{1, 1 + Sin[2Pix]y}, {x, 0, 8}, {y, 0, 8},`

> `Axes → Automatic, Frame → False, AxesStyle → Black,`

> `AxesLabel → {x, y}, PlotLabel → "(a)"];`

To graph the solution on the interval $[0, 2\pi]$, we use NDSolve to generate a numerical solution to the initial-value problem valid for $0 \le x \le 2\pi$. As stated previously, the command

> `NDSolve[{deq,ics},fun,{var,varmin,varmax}]`

returns a numerical solution fun (which is a function of the variable var) of the differential equation deq that satisfies the initial conditions ics valid on the interval [varmin,varmax]. In some cases, the

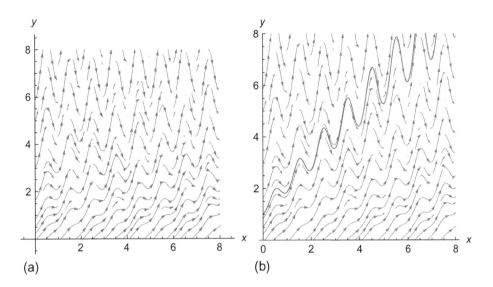

Figure 2-22 (a) Plot of the direction field. (b) Plot of a numerical solution to an initial value problem together with the direction field

interval on which the solution returned by NDSolve is smaller than the interval requested.

We see that the syntax for the NDSolve command is nearly the same as the syntax of the DSolve command although we must specify an interval on which we want the approximation to be valid. In this case, including $\{x, 0, 2\pi\}$ in the NDSolve command instructs Mathematica to (try to) make the resulting numerical solution valid for $0 \le x \le 2\pi$.

Note that the number of initial conditions in ics must equal the order of the differential equation deq.

```
numsol = NDSolve[{y′[x] − Sin[2Pix]y[x] == 1, y[0] == 1}, y[x],

{x, 0, 8}]
```

```
{{y[x] → InterpolatingFunction[][x]}}
```

The resulting output is an InterpolatingFunction which represents an approximate function obtained through interpolation. We can evaluate the result for particular values of x as long as $0 \le x \le 2\pi$. For example, entering

```
numsol/.x → 1
```

```
{{y[1] → 1.85827}}
```

approximates the value of the solution to the initial-value problem if $x = 1$. In this case, the result means that $y(1) \approx 1.85828$. We can graph the result returned by NDSolve in the same way as we graph results returned by DSolve: entering

```
p2a = Plot[numsol[[1, 1, 2]], {x, 0, 8},
    PlotStyle → {{Thickness[.01], CMYKColor[0, 0.89, 0.94,
    0.28]}}, AxesStyle → Black, AxesLabel → {x, y}, PlotRange
    → {0, 8}, PlotLabel → "(b)", AspectRatio → 1];
p2 = Show[p2a, p1];

Show[GraphicsRow[{p1, p2}]]
```

graphs the solution to the initial-value problem on the interval $[0, 2\pi]$
as shown in Figure 2-22(b). Note that we obtain the same graph by
entering `Plot[y[x] /. numsol, {x, 0, 2Pi}]`.

■

2.5.2 Variation of Parameters and the Method of Undetermined Coefficients

Observe that equation (2.4) is separable:

$$\frac{dy}{dx} + p(x)y = 0$$

$$\frac{1}{y}dy = -p(x)\,dx$$

$$\ln|y| = -\int p(x)\,dx + C$$

$$y = Ce^{-\int p(x)\,dx}.$$

Notice that any constant multiple of a solution to a linear homogeneous equation
is also a solution. Now suppose that y is any solution of equation (2.3) and y_p is a
particular solution of equation (2.3). Then,

A **particular solution** is a
specific function that is a
solution to the equation that
does not contain any
arbitrary constants.

$$\left(y - y_p\right)' + p(x)\left(y - y_p\right) = y' + p(x)y - \left(y_p' + p(x)y_p\right)$$
$$= q(x) - q(x) = 0.$$

Thus, $y - y_p$ is a solution to the corresponding homogeneous equation of equation (2.3). Hence,

$$y - y_p = Ce^{-\int p(x)\,dx}$$

$$y = Ce^{-\int p(x)\,dx} + y_p$$

$$y = y_h + y_p,$$

where $y_h = Ce^{-\int p(x)\,dx}$. That is, a general solution of equation (2.3) is

$$y = y_h + y_p,$$

where y_p is a particular solution to the nonhomogeneous equation and y_h is a general solution to the corresponding homogeneous equation. Thus, to solve equation (2.3), we need to first find a general solution to the corresponding homogeneous equation, y_h, which we can accomplish through separation of variables, and then find a particular solution, y_p, to the nonhomogeneous equation.

If y_h is a solution to the corresponding homogeneous equation of equation (2.3) then for any constant C, Cy_h is also a solution to the corresponding homogeneous equation. Hence, it is impossible to find a particular solution to equation (2.3) of this form. Instead, we search for a particular solution of the form $y_p = u(x)y_h$, where $u(x)$ is *not* a constant function. Assuming that a particular solution, y_p, to equation (2.3) has the form $y_p = u(x)y_h$, differentiating gives us

$$y_p{}' = u'y_h + uy_h{}'$$

and substituting into equation (2.3) results in

$$y_p{}' + p(x)y_p = u'y_h + uy_h{}' + p(x)uy_h = q(x).$$

Because $uy_h{}' + p(x)uy_h = u\left[y_h{}' + p(x)y_h\right] = u \cdot 0 = 0$, we obtain

y_h is a solution to the corresponding homogeneous equation so $y_h{}' + p(x)y_h = 0$.

$$u'y_h = q(x)$$

$$u' = \frac{1}{y_h}q(x)$$

$$u' = e^{\int p(x)\,dx}q(x)$$

$$u = \int e^{\int p(x)\,dx}q(x)\,dt$$

so

$$y_p = u(x)\,y_h = Ce^{-\int p(x)\,dx}\int e^{\int p(x)\,dx}q(x)\,dx.$$

Because we include an arbitrary constant of integration when evaluating $\int e^{\int p(x)\,dx}q(x)\,dx$, it follows that we can write a general solution of equation (2.3) as

$$y = e^{-\int p(x)\, dx} \int e^{\int p(x)\, dx} q(t)\, dt. \tag{2.6}$$

Exponential growth is discussed in more detail in Section 3.2.1.

EXAMPLE 2.5.5 (Exponential Growth): Let $y = y(t)$ denote the size of a population at time t. If y grows at a rate proportional to the amount present, y satisfies

$$\frac{dy}{dt} = \alpha y, \tag{2.7}$$

where α is the **growth constant**. If $y(0) = y_0$, using equation (2.6) results in $y = y_0 e^{\alpha t}$. We use DSolve to confirm this result.

```
Clear[t, y]

DSolve[{y'[t] == αy[t], y[0] == y0}, y[t], t]
```

$$\{\{y[t] \rightarrow e^{t\alpha} y0\}\}$$

$dy/dt = k\,(y - y_s)$ models *Newton's Law of Cooling*: the rate at which the temperature, $y(t)$, changes in a heating/cooling body is proportional to the difference between the temperature of the body and the constant temperature, y_s, of the surroundings. Newton's Law of Cooling is discussed in more detail in Section 3.3.

EXAMPLE 2.5.6: Solve each of the following equations: (a) $dy/dt = k\,(y - y_s)$, $y(0) = y_0$, k and y_s constant and (b) $y' - 2ty = t$.

SOLUTION: (a) By hand, we rewrite the equation and obtain

$$\frac{dy}{dt} - ky = -ky_s.$$

A general solution of the corresponding homogeneous equation

$$\frac{dy}{dt} - ky = 0$$

is $y_h = e^{kt}$. Because k and $-ky_s$ are constants, we suppose that a particular solution of the nonhomogeneous equation, y_p, has the form $y_p = A$, where A is a constant.

Assuming that $y_p = A$, we have $y_p' = 0$ and substitution into the nonhomogeneous equation gives us

This will turn out to be a lucky guess. If there is not a solution of this form, we would not find one of this form.

$$\frac{dy_p}{dt} - ky_p = -KA = -ky_s \quad \text{so} \quad A = y_s.$$

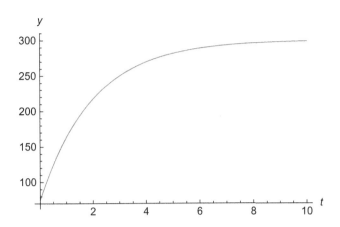

Figure 2-23 The temperature of the body approaches the temperature of its surroundings

Thus, a general solution is $y = y_h + y_p = Ce^{kt} + y_s$. Applying the initial condition $y(0) = y_0$ results in $y = y_s + (y_0 - y_s)e^{kt}$.

We obtain the same result with DSolve. We graph the solution satisfying $y(0) = 75$ assuming that $k = -1/2$ and $y_s = 300$ in Figure 2-23. Notice that $y(t) \to y_s$ as $t \to \infty$.

```
sola = DSolve[{y'[t] == k(y[t] - ys),

  y[0] == y0}, y[t], t]
{{y[t] → e^kt y0 + ys - e^kt ys}}

tp = sola[[1, 1, 2]]/.{k → -1/2,

  ys → 300, y0 → 75};
p1 = Plot[tp, {t, 0, 10}, AxesLabel → {t, y},

  AxesStyle → Black]
```

(b) The equation is in standard form and we identify $p(t) = -2t$. Then, the integrating factor is $\mu(t) = e^{\int p(t)\, dt} = e^{-t^2}$. Multiplying the equation by the integrating factor, $\mu(t)$, results in

$$e^{-t^2}(y' - 2ty) = te^{-t^2} \quad \text{or} \quad \frac{d}{dt}\left(ye^{-t^2}\right) = te^{-t^2}.$$

Integrating gives us

$$ye^{-t^2} = -\frac{1}{2}e^{-t^2} + C \quad \text{or} \quad y = -\frac{1}{2} + Ce^{t^2}.$$

We confirm the result with DSolve.

```
solb = DSolve[y'[t] - 2 t y[t] == t, y[t], t]
```

$$\left\{\left\{y[t] \rightarrow -\frac{1}{2} + e^{t^2} C[1]\right\}\right\}$$

∎

Application: Antibiotic Production

Source: Kevin H. Dykstra and Henry Y. Wang, "Changes in the Protein Profile of *Streptomyces Griseus* during a Cycloheximide Fermentation," *Biochemical Engineering V*, Annals of the New York Academy of Sciences, Volume 56, New York Academy of Sciences (1987), pp. 511–522.

When you are injured or sick, your doctor may prescribe antibiotics to prevent or cure infections. In the journal article "Changes in the Protein Profile of *Streptomyces Griseus* during a Cycloheximide Fermentation" we see that production of the antibiotic cycloheximide by Streptomyces is typical of antibiotic production. During the production of cycloheximide, the mass of *Streptomyces* grows relatively quickly and produces little cycloheximide. After approximately 24 hours, the mass of *Streptomyces* remains relatively constant and cycloheximide accumulates. However, once the level of cycloheximide reaches a certain level, extracellular cycloheximide is degraded (**feedback inhibited**). One approach to alleviating this problem to maximize cycloheximide production is to continuously remove extracellular cycloheximide. The rate of growth of *Streptomyces* can be described by the separable equation

$$\frac{dX}{dt} = \mu_{\max}\left(1 - \frac{1}{X_{\max}}X\right)X,$$

where X represents the mass concentration in g/L, $\mu_{\max}$ is the maximum specific growth rate, and $X_{\max}$ represents the maximum mass concentration. We now solve

Note that this equation can be converted to a linear equation with the substitution $y = X^{-1}$.

the initial-value problem $\begin{cases} dX/dt = \mu_{\max}\left(1 - \frac{1}{X_{\max}}X\right)X \\ X(0) = 1 \end{cases}$ with DSolve, naming the

result sol1.

```
Clear[x]

sol1 = DSolve[{x'[t] == μ(1 - x[t]/xmax)x[t],

    x[0] == 1}, x[t], t]
```

$$\left\{\left\{x[t] \rightarrow \frac{e^{t\mu}\text{xmax}}{-1 + e^{t\mu} + \text{xmax}}\right\}\right\}$$

Experimental results have shown that $\mu_{\max} = 0.3\,\text{hr}^{-1}$ and $X_{\max} = 10\,\text{g/L}$. For these values, we use `Plot` to graph $X(t)$ on the interval $[0, 24]$ in Figure 2-24(a). Then, we use `Table` and `TableForm` to determine the mass concentration at the end of 4, 8, 12, 16, 20, and 24 hours.

```
μ = 0.3; xmax = 10.;

p1 = Plot[x[t]/.sol1, {t, 0, 24}, PlotStyle → Black,

    AxesStyle → Black, AxesLabel → {t, x},

    PlotLabel → "(a)"];

TableForm[Table[{t, sol1[[1, 1, 2]]}, {t, 4, 24, 4.}]]
```

4.	2.69487
8.	5.50521
12.	8.02624
16.	9.3104
20.	9.78178
24.	9.93326

The rate of accumulation of cycloheximide is the difference between the rate of synthesis and the rate of degradation:

$$\frac{dP}{dt} = R_s - R_d.$$

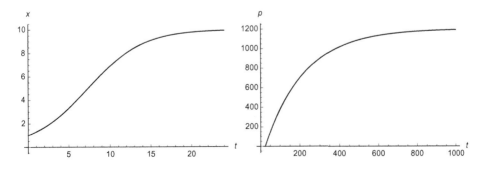

Figure 2-24 (a) Plot of the mass concentration, $x(t)$. (b) Accumulation of the antibiotic

It is known that $R_d = K_d P$, where $K_d = 5 \times 10^{-3}\,\mathrm{h}^{-1}$, so $dP/dt = R_s - R_d$ is equivalent to $dP/dt = R_s - K_d P$. Furthermore,

$$R_s = Q_{po} EX \left(1 + \frac{P}{K_I}\right)^{-1},$$

where Q_{po} represents the specific enzyme activity with value $Q_{po} \approx 0.6\,\mathrm{g\,CH/g}$, protein $\cdot$ h and K_I represents the inhibition constant. E represents the intracellular concentration of an enzyme which we will assume is constant. For large values of K_I and t, $X(t) \approx 10$ and $(1 + P/K_I)^{-1} \approx 1$. Thus, $R_s \approx 10 Q_{po} E$ so

$$\frac{dP}{dt} = 10 Q_{po} E - K_d P.$$

After defining K_d and Q_{po}, we solve the initial-value problem $\begin{cases} dP/dt = 10 Q_{po} E - K_d P \\ p(24) = 0 \end{cases}$ and then graph $\dfrac{1}{E} P(t)$ on the interval $[0, 24]$ in Figure 2-24(b).

```
Clear[p]

kd = 5/1000;

qp0 = 0.6;

sol2 = DSolve[{p'[t] == 10qp0 - kdp[t],

  p[24] == 0}, p[t], t]
```

$$\{\{p[t] \to 1200 . e^{-0.005t} (-1.1275 + 1 . e^{0.005t})\}\}$$

```
p2 = Plot[p[t]/.sol2, {t, 24, 1000}, PlotStyle → Black,

  AxesStyle → Black, AxesLabel → {t, p},

  AxesOrigin → {0, 0},

  PlotLabel → "(b)"];

Show[GraphicsRow[{p1, p2}]]
```

From the graph, we see that the total accumulation of the antibiotic approaches a limiting value, which in this case is 1200.

2.6 Numerical Approximations of Solutions to First-Order Equations

2.6.1 Built-In Methods

Numerical approximations of solutions to differential equations can be obtained with NDSolve, which is particularly useful when working with nonlinear equations for which DSolve alone is unable to find an explicit or implicit solution. The command

NDSolve[{y'[t]==f[t,y[t]],y[t0]==y0},y[t],{t,a,b}]

attempts to generate a numerical solution of

$$\begin{cases} \dfrac{dy}{dt} = f(t, y) \\ y(t_0) = y_0 \end{cases}$$

valid for $a \le t \le b$. In some cases, the interval on which the solution returned by NDSolve is smaller than the interval requested. You can obtain basic information regarding NDSolve by entering ?NDSolve or detailed information by accessing Mathematica's on-line help facility by selecting **Help** from the Mathematica help menu.

EXAMPLE 2.6.1: Consider

$$\frac{dy}{dt} = \left(t^2 - y^2\right)\sin y, \ y(0) = -1.$$

(a) Determine $y(1)$. (b) Graph $y(t)$ for $-1 \le t \le 10$.

SOLUTION: We first use StreamPlot to plot the direction field for $y' = (t^2 - y^2)\sin y$ in Figure 2-25(a).

```
f[t_, y_] = (t^2 - y^2)Sin[y];

p1 = StreamPlot[{1, f[t, y]}, {t, 0, 10}, {y, 0, 10},

    AspectRatio → 1, StreamStyle → Black, AxesOrigin → {0, 0},
```

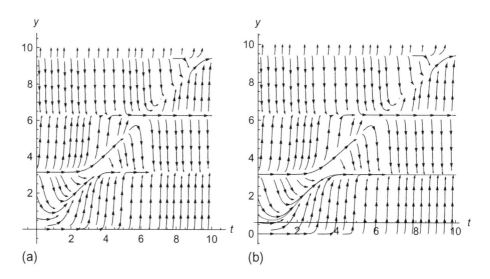

Figure 2-25   (a) The direction field for the equation. (b) Graph of the solution to $y' = \left(t^2 - y^2\right)\sin y$, $y(0) = -1$

                  `Frame → False, Axes → Automatic, AxesStyle → Black,`

                  `PlotLabel → "(a)", AxesLabel → {t, y}];`

We obtain a numerical solution valid for $0 \le t \le 1000$ using the `NDSolve` function.

                  `sole = NDSolve[{y'[t] == f[t,y[t]],y[0] == 1},y[t],{t,0,1000}]`

                  `{{y[t] → InterpolatingFunction[][t]}}`

Entering `sole /.t->1` evaluates the numerical solution if $t = 1$.

                  `sole/.x → 1`

                  `{{y[1] → -0.766}}`

The result means that $y(1) \approx -.766$. We use the `Plot` command to graph the solution for $0 \le t \le 10$ in Figure 2-25.

                  `p2a = Plot[y[t]/.sole, {t, 0, 10}, PlotStyle →`

                  `{{Thickness[.01], CMYKColor[0, 0.89, 0.94, 0.28]}}];`

                  `p2 = Show[p2a, p1, AxesStyle → Black, AxesLabel → {t, y},`

                  `PlotRange → {{0, 10},{0, 10}}, PlotLabel → "(b)",`

                  `AspectRatio → 1];`

                  `Show[GraphicsRow[{p1, p2}]]`

■

EXAMPLE 2.6.2: Graph the solution to the initial-value problem

$$\begin{cases} dy/dx = \sin(2x - y) \\ y(0) = 0.5 \end{cases},$$

on the interval [0, 15]. What is the value of $y(1)$?

SOLUTION: As with the previous examples, we use `StreamPlot` to generate a plot of the direction field shown in Figure 2-26(a).

```
Clear[x, y]

f[x_, y_] = Sin[2x - y];

p1 = StreamPlot[{1, f[x, y]}, {x, 0, 10}, {y, 0, 10},

    AspectRatio → 1, StreamStyle → Black, AxesOrigin → {0, 0},

    Frame → False, Axes → Automatic, AxesStyle → Black,

    PlotLabel → "(a)", AxesLabel → {x, y}];
```

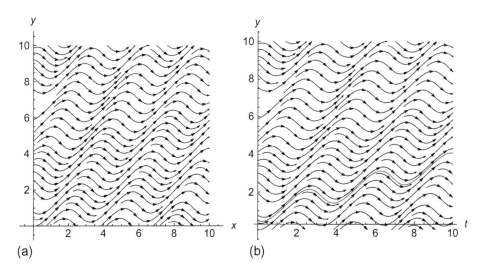

(a)   (b)

Figure 2-26   (a) Plot of the direction field. (b) Graph of the solution to $y' = \sin(2x - y)$, $y(0) = 0.5$

We use NDSolve to approximate the solution to the initial-value problem, naming the resulting output numsol. The resulting InterpolatingFunction is a procedure that represents an approximate function obtained through interpolation.

numsol = NDSolve[{$y'[x]$ == $f[x, y[x]]$, $y[0]$ == .5}, $y[x]$,

{x, 0, 15}]

{{$y[x] \rightarrow$ InterpolatingFunction[][x]}}

numsol/.$x \rightarrow$ 1

{{$y[1] \rightarrow$ 0.875895}}

returns a list corresponding to the value of $y(x)$ if $x = 1$. We interpret the result to mean that $y(1) \approx 0.875895$. We then graph the solution returned by NDSolve using Plot in the same way that we graph solutions returned by DSolve. As you probably expect, entering Plot[numsol[[1,1,2]]], {x,0,15}] produces the same graph as the one shown in Figure 2-26(b) generated by the following Plot command.

p2a = Plot[$y[x]$/.numsol, {x, 0, 10},

   PlotStyle $\rightarrow$ {{Thickness[.01], CMYKColor[0, 0.89, 0.94,

   0.28]}}];

p2 = Show[p2a, p1, AxesStyle $\rightarrow$ Black, AxesLabel $\rightarrow$ {x, y},

   PlotRange$\rightarrow$ {{0, 10}, {0, 10}}, PlotLabel$\rightarrow$ "(b)",

   AspectRatio$\rightarrow$ 1];

Show[GraphicsRow[{p1, p2}]]

A different way to graph solutions that satisfy different initial conditions is to define a function as we do here. Given i, sol[i] returns a numerical solution to the initial-value problem $y' = \sin(2x-y)$, $y(0) = i$.

Clear[x, y, i, sol]

sol[i_]:=NDSolve[{$y'[x]$ == Sin[$2x - y[x]$], $y[0]$ == i},

   $y[x]$, {x, 0, 15}]

For example, to use sol, we first use Table to define inits to be the list of numbers $i/2$ for $i = 1, 2, \ldots, 5$ and then use Map to apply sol to the list of numbers inits. The command

interpfunctions=Map[sol,inits]

computes `sol[i]` for each value of *i* in `inits`. The result is a nested list consisting of `InterpolatingFunction`'s.

We graph the set of `InterpolatingFunction`'s with `Plot` in the same way as we graph other sets of functions. See Figure 2-27(b).

```
inits = Table[i/2, {i, 1, 10}];
```

```
interpfunctions = Map[sol, inits];
```

Last, we show these graphs together with the direction field associated with the equation in Figure 2-27(b).

```
p3 = Plot[Evaluate[y[x]/.interpfunctions],

    {x, 0, 10}, PlotRange → {{0, 10}, {0, 10}},

    AspectRatio → 1,

    PlotStyle → {{CMYKColor[0, 0.89, 0.94, 0.28]}},

    AxesLabel → {x, y}, PlotLabel → "(a)"];
```

```
p4 = Show[p3, p1, PlotLabel → "(b)"];
```

```
Show[GraphicsRow[{p3, p4}]]
```

■

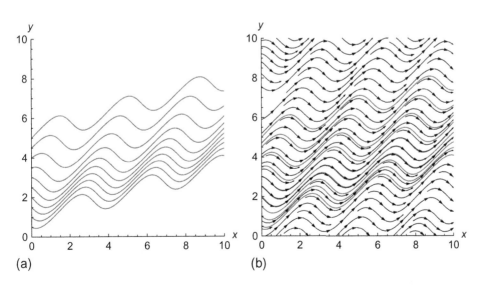

(a)   (b)

Figure 2-27   Various solutions of $y' = \sin(2x - y)$

Source: Herbert W. Hethcote, "Three Basic Epidemiological Models," in *Applied Mathematical Ecology*, edited by Simon A. Levin, Thomas G. Hallan, and Louis J. Gross, New York, Springer-Verlag (1989), pp. 119–143.

Application: Modeling the Spread of a Disease

Suppose that a disease is spreading among a population of size N. In some diseases, like chickenpox, once an individual has had the disease, the individual becomes immune to the disease. In other diseases, like most venereal diseases, once an individual has had the disease and recovers from the disease, the individual does not become immune to the disease; subsequent encounters can lead to recurrences of the infection.

Let $S(t)$ denote the percent of the population susceptible to a disease at time t, $I(t)$ the percent of the population infected with the disease, and $R(t)$ the percent of the population unable to contract the disease. For example, $R(t)$ could represent the percent of persons who have had a particular disease, recovered, and have subsequently become immune to the disease. In order to model the spread of various diseases, we begin by making several assumptions and introducing some notation.

1. Susceptible and infected individuals die at a rate proportional to the number of susceptible and infected individuals with proportionality constant μ called the **daily death removal rate**; the number $1/\mu$ is the **average lifetime** or **life expectancy**.
2. The constant λ represents the **daily contact rate**: on average, an infected person will spread the disease to λ people per day.
3. Individuals recover from the disease at a rate proportional to the number infected with the disease with proportionality constant γ. The constant γ is called the **daily recovery removal rate**; the **average period of infectivity** is $1/\gamma$.
4. The **contact number** $\sigma = \lambda/(\gamma + \mu)$ represents the average number of contacts an infected person has with both susceptible and infected persons.

If a person becomes susceptible to a disease after recovering from it (like gonorrhea, meningitis, and streptococcal sore throat), then the percent of persons susceptible to becoming infected with the disease, $S(t)$, and the percent of people in the population infected with the disease, $I(t)$, can be modeled by the system of differential equations

$$\begin{cases} \dfrac{dS}{dt} = -\lambda IS + \gamma I + \mu - \mu S \\[2mm] \dfrac{dI}{dt} = \lambda IS - \gamma I - \mu I \\[2mm] S(0) = S_0, \ I(0) = I_0, \ S(t) + I(t) = 1 \end{cases} \qquad (2.8)$$

This model is called an **SIS model** (susceptible-infected-susceptible model) because once an individual has recovered from the disease, the individual again becomes susceptible to the disease.

We can write $dI/dt = \lambda IS - \gamma I - \mu I$ as $dI/dt = \lambda I(1 - I) - \gamma I - \mu I$ because $S(t) = 1 - I(t)$. Therefore, we need to solve the initial-value problem

$$\begin{cases} \dfrac{dI}{dt} = [\lambda - (\gamma + \mu)]I - \lambda I^2 \\ I(0) = I_0 \end{cases}. \tag{2.9}$$

In the following, we use $\mathtt{i}$ to represent I, thus avoiding conflict with the built-in constant $\mathtt{I} = \sqrt{-1}$. After defining $\mathtt{eq}$, we use $\mathtt{DSolve}$ to find the solution to the initial-value problem.

```
eq = i'[t] + (γ + μ − λ)i[t]== − λi[t]²;
```

```
sol = DSolve[{eq, i[0] == i0}, i[t], t]//FullSimplify
```

$$\left\{\left\{i[t] \to \frac{e^{\frac{(i\pi + t\lambda)(\gamma + \mu)}{\gamma - \lambda + \mu}} i0^{\frac{\gamma + \mu}{\gamma - \lambda + \mu}}(-\gamma + \lambda - \mu)}{e^{\frac{(i\pi + t\lambda)(\gamma + \mu)}{\gamma - \lambda + \mu}} i0^{\frac{\gamma + \mu}{\gamma - \lambda + \mu}}\lambda - e^{t\gamma + t\mu + \frac{t\lambda^2}{\gamma - \lambda + \mu} + \frac{\lambda(i\pi + \text{Log}[i0] - \text{Log}[-\gamma + \lambda - i0\lambda - \mu])}{\gamma - \lambda + \mu}}(-\gamma + \lambda - i0\lambda - \mu)} \frac{\gamma + \mu}{\gamma - \lambda + \mu}}\right\}\right\}$$

We can use this result to see how a disease might spread through a population. For example, we compute the solution to the initial-value problem, which is extracted from $\mathtt{sol}$ with $\mathtt{sol[[1,1,2]]}$, if $\lambda = 0.50$, $\gamma = 0.75$, and $\mu = 0.65$. In this case, we see that the contact number is $\sigma = \lambda/(\gamma + \mu) \approx 0.357143$.

```
λ = 0.5;
```

```
γ = 0.75;
```

```
μ = 0.65;
```

$$\sigma = \frac{\lambda}{\gamma + \mu}$$
```
sol[[1, 1, 2]]
```

```
0.357143
```

$$-\frac{0.9e^{1.55556(i\pi+0.5t)}i0^{1.55556}}{-e^{1.67778t+0.555556(i\pi-\text{Log}[-0.9-0.5i0]+\text{Log}[i0])}(-0.9 - 0.5i0)^{1.55556} + 0.5e^{1.55556(i\pi+0.5t)}i0^{1.55556}}$$

Next, we use $\mathtt{Table}$ to substitute various initial conditions into $\mathtt{sol[[1,1,2]]}$, naming the resulting set of nine functions $\mathtt{toplot1}$. We then graph the functions in $\mathtt{toplot1}$ for $0 \le t \le 5$ in Figure 2-28(a). Apparently, regardless of the initial percent of the population infected, under these conditions, the disease is eventually removed from the population. This makes sense because the contact number is less than one.

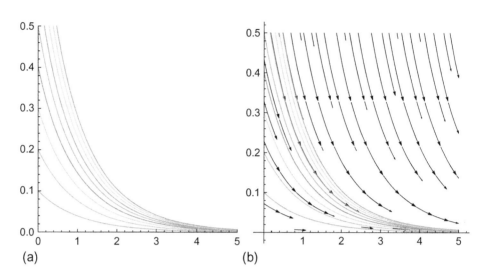

(a)   (b)

Figure 2-28   (a) The disease is removed from the population. (b) Illustrating the removal of the disease using the direction field

```
toplot1 = Table[sol[[1, 1, 2]], {i0, 0.1, 0.9, 0.1}];

p1 = Plot[Evaluate[toplot1], {t, 0, 5}, AspectRatio → 1,

    PlotLabel → "(a)", PlotRange → {{0, 5}, {0, 0.5}}];
```

After writing the equation in the form $dI/dt = f(t, I)$ in step1,

```
step1 = Solve[eq, i'[t]]
```

$$\{\{i'[t] \to -0.9 i[t] - 0.5 i[t]^2\}\}$$

```
toplot2 = step1[[1, 1, 2]] /. i[t] → i
```

$$-0.9 i - 0.5 i^2$$

we use Streamplot to graph the direction field shown in Figure 2-28(b).

```
p2a = StreamPlot[{1, toplot2}, {t, 0, 5}, {i, 0, .5},

    Frame → False, Axes → Automatic, AxesStyle → Black,

    StreamStyle → Black, StreamPoints → Fine,

    PlotLabel → "(b)", AspectRatio → 1];
```

```
p2 = Show[p2a, p1, PlotRange → {{0, 5}, {0, .5}}];
```

```
Show[GraphicsRow[{p1, p2}]]
```

On the other hand, if $\lambda = 1.5$, $\gamma = 0.75$, and $\mu = 0.65$, we see that the contact number is $\sigma = \lambda/(\gamma + \mu) \approx 1.07143$.

```
λ = 1.5;
```

```
γ = 0.75;
```

```
μ = 0.65;
```

$$\sigma = \frac{\lambda}{\gamma + \mu}$$

```
sol[[1, 1, 2]]
```

```
1.07143
```

$$\frac{0.1e^{-14.(i\pi+1.5t)}}{\left(-\frac{e^{-21.1t-15.(i\pi-Log[0.1-1.5i0]+Log[i0])}}{(0.1-1.5i0)^{14.}} + \frac{1.5e^{-14.(i\pi+1.5t)}}{i0^{14.}}\right)i0^{14.}}$$

Proceeding as before, we graph the solution using different initial conditions in Figure 2-29. In this case, we see that no matter what percent of the population is initially infected, a certain percent of the population is always infected. This makes sense because the contact number is greater than one. In fact, it is a theorem that

$$\lim_{t\to\infty} I(t) = \begin{cases} 1 - 1/\sigma, & \text{if } \sigma > 1 \\ 0, & \text{if } \sigma \le 1 \end{cases}.$$

```
toplot2 = Table[sol[[1, 1, 2]], {i0, 0.1, 0.9, 0.1}];
```

```
p1 = Plot[toplot2, {t, 0, 5}];
```

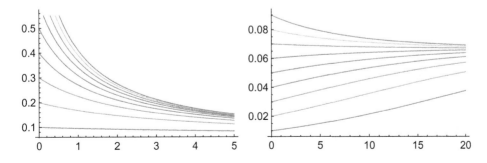

Figure 2-29   The disease persists

```
toplot3 = Table[sol[[1, 1, 2]], {i0, 0.01, 0.09, 0.01}];

p2 = Plot[toplot3, {t, 0, 20}];

Show[GraphicsRow[{p1, p2}]]
```

The incidence of some diseases, such as measles, rubella, and gonorrhea, oscillate seasonally. To model these diseases, we may wish to replace the constant contact rate λ, by a periodic function $\lambda(t)$. For example, to graph the solution to the SIS model for various initial conditions if (a) $\lambda(t) = 3 - 2.5 \sin 6t$, $\gamma = 2$, and $\mu = 1$ and (b) $\lambda(t) = 3 - 2.5 \sin 6t$, $\gamma = 1$, and $\mu = 1$ we proceed as follows. For (a), we begin by defining λ, γ, and μ, and eq.

```
Clear[λ, i, t, γ, μ]

λ[t_] = 3 - 2.5Sin[6t];

γ = 2;

μ = 1;

eq = i'[t]==(λ[t] - (γ + μ))i[t] - λ[t]i[t]²
```

$$i'[t] == -i[t]^2(3 - 2.5\text{Sin}[6t]) - 2.5i[t]\text{Sin}[6t]$$

We will graph the solutions satisfying the initial conditions $I(0) = I_0$ for $I_0 = 0.1, 0.2, \ldots, 0.9$. We begin by defining `graph`. Given `i0`, `graph[i0]` graphs the solution to the initial-value problem

$$\begin{cases} \dfrac{dI}{dt} = [\lambda(t) - (\gamma + \mu)]I - \lambda(t)I^2 \\ I(0) = I_0 \end{cases}$$

on the interval $[0, 10]$. Next, we use `Table` to define the list of numbers `inits`, corresponding to the initial conditions, and then use `Map` to apply the function `graph` to the list of numbers `inits`. We see that the result is a list of nine graphics objects that we name `toshow`.

```
graph[i0_]:=Module[{numsol},

    numsol = NDSolve[{eq, i[0]==i0}, i[t],

    {t, 0, 10}];

    Plot[i[t]/.numsol, {t, 0, 10}, PlotRange → {{0, 10}, {0, 1}},

    AspectRatio → 1]]

inits = Table[i, {i, 0.1, 0.9, 0.1}];

toshow = Map[graph, inits];
```

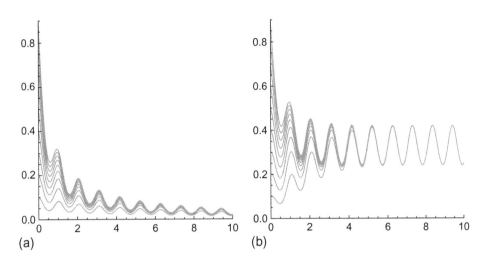

Figure 2-30   The disease persists periodically in the population

Finally, we use Show to view the list of nine graphs toshow in Figure 2-30(a). For (b), we proceed in the same manner as in (a). See Figure 2-30(b).

```
p1 = Show[toshow, PlotRange → {{0, 10}, All}, PlotLabel → "(a)"];
```

```
Clear[λ, i, t, γ, μ]
```

```
λ[t_] = 3 - 2.5Sin[6t];
```

```
γ = 1;
```

```
μ = 1;
```

```
eq = i'[t]==(λ[t] - (γ + μ))i[t] - λ[t]i[t]²
```

$$i'[t] == i[t](1 - 2.5\mathrm{Sin}[6t]) - i[t]^2(3 - 2.5\mathrm{Sin}[6t])$$

```
graph[i0_]:=Module[{numsol},
```

```
numsol = NDSolve[{eq, i[0]==i0}, i[t], {t, 0, 10}];
```

```
   Plot[i[t]/.numsol, {t, 0, 10}, PlotRange → {{0, 10}, {0, 1}},
```

```
   AspectRatio → 1]]
```

```
inits = Table[i, {i, 0.1, 0.9, 0.1}];
```

```
toshow = graph/@inits;
```

Figure 2-31   `NDSolve` can implement a wide range of numerical methods when solving differential equations

```
p2 = Show[toshow, PlotRange → {{0, 10}, All}, PlotLabel → "(b)"];

Show[GraphicsRow[{p1, p2}]]
```

2.6.2 Other Numerical Methods

In other cases, you may wish to implement your own numerical algorithms to approximate solutions of differential equations. We briefly discuss three familiar methods (Euler's method, the improved Euler's method, and the Runge-Kutta method) and illustrate how to implement these algorithms using Mathematica. Details regarding these and other algorithms, including discussions of the error involved in implementing them, can be found in most numerical analysis texts or other references like the Zwillinger's *Handbook of Differential Equations*, [29]. Also, note that `NDSolve` has numerous options for implementing different numerical methods when numerically solving differential equations. Use the **Help** menu and explore the `NDSolve` **Options** for a comprehensive discussion of `NDSolve`'s capabilities. See Figure 2-31.

Euler's Method

In many cases, we cannot obtain an explicit formula for the solution to an initial-value problem of the form

$$\begin{cases} \dfrac{dy}{dx} = f(x, y) \\ y(x_0) = y_0 \end{cases}$$

but we can approximate the solution using a numerical method like **Euler's method**, which is based on tangent line approximations. Let h represent a small change, or **step size**, in the independent variable x. Then, we approximate the value of y at the sequence of x-values, $x_1, x_2, \ldots, x_n$, where

$$x_1 = x_0 + h$$
$$x_2 = x_1 + h = x_0 + 2h$$
$$x_3 = x_2 + h = x_0 + 3h$$
$$\vdots$$
$$x_n = x_{n-1} + h = x_0 + nh.$$

The slope of the tangent line to the graph of y at each value of x is found with the differential equation $y' = dy/dx = f(x, y)$. For example, at $x = x_0$, the slope of the tangent line is $f(x_0, y(x_0)) = f(x_0, y_0)$. Therefore, the tangent line to the graph of y is

$$y - y_0 = f(x_0, y_0)(x - x_0) \quad \text{or} \quad y = f(x_0, y_0)(x - x_0) + y_0.$$

Using this line to find the value of y, which we call y_1, at x_1 then yields

$$y_1 = f(x_0, y_0)(x_1 - x_0) + y_0 = hf(x_0, y_0) + y_0.$$

Therefore, we obtain the approximate value of y at x_1. Next, we use the point (x_1, y_1) to estimate the value of y when $x = x_2$. Using a similar procedure, we approximate the tangent line at $x = x_1$ with

$$y - y_1 = f(x_1, y_1)(x - x_1) \quad \text{or} \quad y = f(x_1, y_1)(x - x_1) + y_1.$$

Then, at $x = x_2$,

$$y_2 = f(x_1, y_1)(x_2 - x_1) + y_1 = hf(x_1, y_1) + y_1.$$

Continuing with this procedure, we see that at $x = x_n$,

$$y_n = hf(x_{n-1}, y_{n-1}) + y_{n-1}. \tag{2.10}$$

Using this formula, we obtain a sequence of points of the form (x_n, y_n), $n = 1, 2, \ldots$ where y_n is the approximate value of $y(x_n)$.

```
f[x_, y_] =?;

h =?;

x0 =?;

y0 =?;

xe[n_] = x0 + nh;

ye[n_] := ye[n] = hf[xe[n - 1], ye[n - 1]] + ye[n - 1];

ye[0] = y0;
```

EXAMPLE 2.6.3: Use Euler's method with (a) $h = 0.1$ and (b) $h = 0.05$ to approximate the solution of $y' = xy$, $y(0) = 1$ on $0 \le x \le 1$. Also, determine the exact solution and compare the results.

SOLUTION: Because we will be considering this initial-value problem in subsequent examples, we first determine the exact solution with `DSolve` and graph the result with `Plot`, naming the graph `p1`.

```
exactsol = DSolve[{y'[x]==xy[x], y[0]==1}, y[x], x]
```

$$\left\{ \left\{ y[x] \to e^{\frac{x^2}{2}} \right\} \right\}$$

```
p1 = Plot[E^{\frac{x^2}{2}}, {x, 0, 1}, PlotStyle → GrayLevel[0.4]];
```

To implement Euler's method (2.10), we note that $f(x, y) = xy$, $x_0 = 0$, and $y_0 = 1$. (a) With $h = 0.1$, we have the formula

$$y_n = hf(x_{n-1}, y_{n-1}) + y_{n-1} = 0.1x_{n-1}y_{n-1} + y_{n-1}.$$

For $x_1 = x_0 + h = 0.1$, we have

$$y_1 = 0.1x_0y_0 + y_0 = 0.1 \cdot 0 \cdot 1 + 1 = 1.$$

Similarly, for $x_2 = x_0 + 2h = 0.2$,

$$y_2 = 0.1x_1y_1 + y_1 = 0.1 \cdot 0.1 \cdot 1 + 1 = 1.01.$$

In the following, we define f, h, x, and y to calculate y_n given by equation (2.10). We define ye using the form

$$ye [n\_] := ye [n] = \ldots$$

so that Mathematica "remembers" the values of ye computed, and thus, when computing ye [n], Mathematica need not recompute ye [n-1] if ye [n-1] has previously been computed.

```
f[x_, y_] = xy;

h = 0.10;

x₀ = 0;

y₀ = 1;

xₑ[n_] = x₀ + nh;

yₑ[n_]:=yₑ[n] = h * f[xₑ[n − 1], yₑ[n − 1]] + yₑ[n − 1];

yₑ[0] = y₀;
```

Next, we use Table to calculate the set of ordered pairs (x_n, y_n) for $n = 0, 1, 2, \ldots, 9, 10$, naming the result first, and then TableForm to view first in traditional row-and-column form.

```
first = Table[{xₑ[n], yₑ[n]}, {n, 0, 10}];

TableForm[first]
```

```
0.    1

0.1   1.

0.2   1.0025

0.3   1.00751

0.4   1.01507

0.5   1.02522
```

```
0.6   1.03803

0.7   1.05361

0.8   1.07204

0.9   1.09348

1.    1.11809
```

To compare these results to the exact solution, we use `ListPlot` to graph the list of ordered pairs first in `t2` and display `t2` together with `p1` with `Show` in Figure 2-32.

lp = Map[Point, first]

{Point[{0., 1}], Point[{0.1, 1.}], Point[{0.2, 1.0025}],

Point[{0.3, 1.00751}], Point[{0.4, 1.01507}], Point[{0.5,

1.02522}], Point[{0.6, 1.03803}], Point[{0.7, 1.05361}]

Point[{0.8, 1.07204}], Point[{0.9, 1.09348}],

Point[{1., 1.11809}]}

Alternatively, we can produce Figure 2-32(a) using `ListPlot` together with the option `PlotStyle->PointSize[.02]` with the following command.

t2 = Graphics[{PointSize[0.02], lp}];

q1 = Show[p1, t2, PlotLabel → "(a)"];

(b) For $h = 0.05$, we use

$$y_n = hf(x_{n-1}, y_{n-1}) + y_{n-1} = 0.05x_{n-1}y_{n-1} + y_{n-1}$$

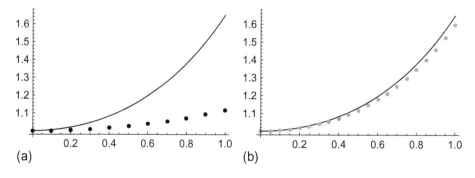

(a)   (b)

Figure 2-32   (a) Comparison of Euler's method to the exact solution using $h = 0.1$. (b) Comparison of Euler's method to the exact solution using $h = 0.05$

to obtain an approximation. In the same manner as in (a), we define f, h, x, and y to calculate y_n given by equation (2.10). Then, we use `Table` to calculate the set of ordered pairs for $n = 0, 1, 2, \ldots, 19, 20$, naming the result `second`, followed by `TableForm` to view `second` in traditional row-and-column form.

```
Remove[x, y, f]
f[x_, y_] = xy;
h = 0.05;
x₀ = 0;
y₀ = 1;

xₑ[n_] = x₀ + nh;
yₑ[n_]:=yₑ[n] = h * f[xₑ[n - 1], yₑ[n - 1]] + yₑ[n - 1];
yₑ[0] = y₀;

second = Table[{xₑ[n], yₑ[n]}, {n, 0, 20}];
TableForm[second]
```

0.	1
0.05	1.
0.1	1.0025
0.15	1.00751
0.2	1.01507
0.25	1.02522
0.3	1.03803
0.35	1.05361
0.4	1.07204
0.45	1.09348
0.5	1.11809
0.55	1.14604
0.6	1.17756

0.65	1.21288
0.7	1.2523
0.75	1.29613
0.8	1.34474
0.85	1.39853
0.9	1.45796
0.95	1.52357
1.	1.59594

We graph the approximation obtained with $h = 0.05$ together with the graph of $y = e^{x^2/2}$ in Figure 2-32(b). Notice that the approximation is more accurate when h is decreased.

```
t3 = ListPlot[second, PlotStyle->PointSize[0.02]];

q2 = Show[p1, t3, PlotLabel → "(b)"];

Show[GraphicsRow[{q1, q2}]]
```

■

Improved Euler's Method

Euler's method can be improved by using an average slope over each interval. Using the tangent line approximation of the curve through (x_0, y_0), $y = f(x_0, y_0)(x - x_0) + y_0$, we find the approximate value of y at $x = x_1$ which we now call $y_1{}^*$:

$$y_1{}^* = hf(x_0, y_0) + y_0.$$

With the differential equation $y' = f(x, y)$, we find that the approximate slope of the tangent line at $x = x_1$ is $f(x, y_1{}^*)$. Then, the average of the two slopes, $f(x_0, y_0)$ and $f(x_1, y_1{}^*)$, is $\frac{1}{2}(f(x_0, y_0) + f(x_1, y_1{}^*))$, and an equation of the line through (x_0, y_0) with slope $\frac{1}{2}(f(x_0, y_0) + f(x_1, y_1{}^*))$ is

$$y = \frac{1}{2} \left(f(x_0, y_0) + f(x_1, y_1^*) \right) (x - x_0) + y_0.$$

Therefore, at $x = x_1$, we find the approximate value of f with

$$y_1 = \frac{1}{2} \left(f(x_0, y_0) + f(x_1, y_1^*) \right) (x_1 - x_0) + y_0 = \frac{1}{2} h \left(f(x_0, y_0) + f(x_1, y_1^*) \right) + y_0.$$

Continuing in this manner, the approximation at each step of the **improved Euler's method** depends on the following two calculations:

$$y_n^* = hf(x_{n-1}, y_{n-1}) + y_{n-1}$$

$$y_n = \frac{1}{2} h \left(f(x_{n-1}, y_{n-1}) + f(x_n, y_n^*) \right) + y_{n-1}. \tag{2.11}$$

```
f[x_,y_]=?;

h=?;

x0=?;

y0=?;

xᵢ[n_]= x0 + nh;

yᵢ[n_]:=

yᵢ[n] = 1/2 h (f[xᵢ[n - 1], yᵢ[n - 1]] + f[xᵢ[n], hf[xᵢ[n - 1], yᵢ[n - 1]]

 +yᵢ[n - 1]]) + yᵢ[n - 1]; yᵢ[0] = y0;
```

EXAMPLE 2.6.4: Use the improved Euler's method to approximate the solution of $y' = xy$, $y(0) = 1$ on $0 \le x \le 1$ for $h = 0.1$. Also, compare the results to the exact solution.

SOLUTION: In this case, $f(x, y) = xy$, $x_0 = 0$, and $y_0 = 1$ so equation (2.11) becomes

$$y_n^* = hx_{n-1}y_{n-1} + y_{n-1}$$

$$y_n = \frac{1}{2} h \left(x_{n-1}y_{n-1} + x_n y_n^* \right) + y_{n-1}$$

for $n = 1, 2, \ldots, 10$. For example, if $n = 1$, we have

$$y_1{}^* = hx_0 y_0 + y_0 = 0.1 \cdot 0 \cdot 1 + 1 = 1$$

and

$$y_1 = \frac{1}{2} h \left(x_0 y_0 + x_1 y_1{}^* \right) + y_0 = \frac{1}{2} \cdot 0.1 \cdot (0 \cdot 1 + 0.1 \cdot 1) + 1 = 1.005.$$

Similarly,

$$y_2{}^* = hx_1 y_1 + y_1 = 0.1 \cdot 0.1 \cdot 1.005 + 1.005 = 1.01505$$

and

$$
\begin{aligned}
y_2 &= \frac{1}{2} h \left(x_1 y_1 + x_2 y_2{}^* \right) + y_1 \\
&= \frac{1}{2} \cdot 0.1 \cdot (0.1 \cdot 1.005 + 0.2 \cdot 1.01505) + 1.005 = 1.0201755.
\end{aligned}
$$

In the same way as in the previous example, we define f, x, h, and y. We define yi using the form

$$\texttt{yi[n\_]:=yi[n]=} \ldots,$$

so that Mathematica "remembers" the values of ystar and y computed. Thus, to compute yi[n], Mathematica need not recompute yi[n-1] if yi[n-1] has previously been computed.

```
Remove[f, x, y]

f[x_, y_] = xy;

h = 0.1;

x₀ = 0;

y₀ = 1;

xᵢ[n_] = x₀ + nh;

yᵢ[n_]:=
yᵢ[n] = N[ (1/2) h (f [xᵢ[n − 1], yᵢ[n − 1]]
        +f [xᵢ[n], hf [xᵢ[n − 1], yᵢ[n − 1]] + yᵢ[n − 1]])
        +yᵢ[n − 1]] ;

yᵢ[0] = y₀;
```

We then compute (x_n, y_n) for $n = 0, 1, \ldots, 10$ and name the resulting list of ordered pairs `third`.

```
third = Table[{x_i[n], y_i[n]}, {n, 0, 10}];
```

```
TableForm[third]
```

0.	1
0.1	1.005
0.2	1.02018
0.3	1.04599
0.4	1.08322
0.5	1.13305
0.6	1.19707
0.7	1.27739
0.8	1.37677
0.9	1.49876
1.	1.64788

We graph the approximation obtained using the improved Euler's method together with the graph of the exact solution in Figure 2-33. From the results, we see that the approximation using the improved Euler's method results in a slight improvement from that obtained in the previous example.

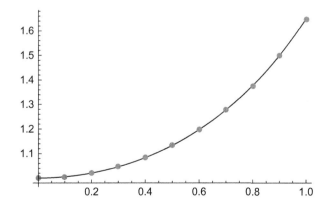

Figure 2-33 Comparison of the improved Euler's method to the exact solution using $h = 0.1$

```
t4 = ListPlot[third, PlotStyle->PointSize[0.02]];

Show[p1, t4]
```

∎

The Runge-Kutta Method

In an attempt to improve on the approximation obtained with Euler's method as well as avoid the analytic differentiation of the function $f(x, y)$ to obtain y'', y''', ..., the *Runge-Kutta method* is introduced. Let us begin with the *Runge-Kutta method of order two*. Suppose that we know the value of y at x_n. We now use the point (x_n, y_n) to approximate the value of y at a nearby value $x = x_n + h$ by assuming that

$$y_{n+1} = y_n + Ak_1 + Bk_2,$$

where

$$k_1 = hf(x_n, y_n) \quad \text{and} \quad k_2 = hf(x_n + ah, y_n + bk_1).$$

We can use the Taylor series expansion of y to obtain another representation of $y_{n+1} = y(x_n + h)$ as follows:

$$y(x_n + h) = y(x_n) + hy'(x_n) + \frac{h^2}{2!}y''(x_n) + \cdots = y_n + hy'(x_n) + \frac{h^2}{2!}y''(x_n) + \cdots$$

Now, because

$$y_{n+1} = y_n + Ak_1 + Bk_2 = y_n + Ahf(x_n, y_n) + Bhf(x_n + ah, y_n + bhf(x_n, y_m)),$$

we wish to determine values of A, B, a, and b such that these two representations of y_{n+1} agree. Notice that if we let $A = 1$ and $B = 0$, then the relationships match up to order h. However, we can choose these parameters more wisely so that agreement occurs up through terms of order h^2. This is accomplished by considering the Taylor series expansion of a function $z = F(x, y)$ of two variables about (x_0, y_0) which is given by

$$F(x_0, y_0) + \frac{\partial F}{\partial x}(x_0, y_0)(x - x_0) + \frac{\partial F}{\partial y}(x_0, y_0)(y - y_0) + \cdots$$

In our case, we have

$$f(x_n + ah, y_n + bhf(x_n, y_m)) = f(x_n, y_n) + ah\frac{\partial f}{\partial x}(x_n, y_n)$$

$$+ bhf(x_n, y_n)\frac{\partial f}{\partial y}(x_n, y_n)(y - y_0) + O(h^2).$$

The power series is then substituted into the following expression and simplified to yield:

$$y_{n+1} = y_n + Ahf(x_n, y_n) + Bhf(x_n + ah, y_n + bhf(x_n, y_m))$$

$$= y_n + (A + B)hf(x_n, y_n) + aBh^2 \frac{\partial f}{\partial x}(x_n, y_n) + bBh^2 f(x_n, y_n) \frac{\partial f}{\partial x}(x_n, y_n) + O\left(h^3\right).$$

Comparing this expression to the following power series obtained directly from the Taylor series of y,

$$y(x_n + h) = y(x_n) + hf(x_n, y_n) + \frac{1}{2}h^2 \frac{\partial f}{\partial x}(x_n, y_n) + \frac{1}{2}h^2 \frac{\partial f}{\partial y}(x_n, y_n) + O\left(h^3\right)$$

or

$$y_{n+1} = y_n + hf(x_n, y_n) + \frac{1}{2}h^2 \frac{\partial f}{\partial x}(x_n, y_n) + \frac{1}{2}h^2 \frac{\partial f}{\partial y}(x_n, y_n) + O\left(h^3\right),$$

we see that $A, B, a,$ and b must satisfy the following system of nonlinear equations:

$$A + B = 1, \quad aA = \frac{1}{2}, \quad \text{and} \quad bB = \frac{1}{2}.$$

Therefore, choosing $a = b = 1$, the Runge-Kutta method of order two uses the equation:

$$y_{n+1} = y(x_n + h) = y_n + \frac{1}{2}hf(x_n, y_n) + \frac{1}{2}hf(x_n + h, y_n + hf(x_n, y_n))$$
$$= y_n + \frac{1}{2}(k_1 + k_2), \tag{2.12}$$

where $k_1 = hf(x_n, y_n)$ and $k_2 = hf(x_n + h, y_n + k_1)$.

```
f[x_,y_]=?;

h=?;

x0 =?;

y0 =?;

xr[n_] = x0 + nh;

yr[n_]:=

yr[n] = yr[n - 1] + 1/2 hf[xr[n - 1], yr[n - 1]] +
1/2 hf[xr[n - 1] + h, yr[n - 1] + hf[xr[n - 1], yr[n - 1]]]

yr[0] = y0;
```

EXAMPLE 2.6.5: Use the Runge-Kutta method of order two with $h = 0.1$ to approximate the solution of the initial-value problem $y' = xy$, $y(0) = 1$ on $0 \le x \le 1$.

SOLUTION: As with the previous examples, $f(x, y) = xy$, $x_0 = 0$, and $y_0 = 1$. Therefore, on each step we use the three equations

$$k_1 = hf(x_n, y_n) = 0.1x_ny_n,$$
$$k_2 = hf(x_n + h, y_n + k_1) = 0.1(x_n + 0.1)(y_n + k_1),$$

and

$$y_{n+1} = y_n + \frac{1}{2}(k_1 + k_2).$$

For example, if $n = 0$, then

$$k_1 = 0.1x_0y_0 = 0.1 \cdot 0 \cdot 1 = 0,$$
$$k_2 = 0.1(x_0 + 0.1)(y_0 + k_1) = 0.1 \cdot 0.1 \cdot 1 = 0.01,$$

and

$$y_1 = y_0 + \frac{1}{2}(k_1 + k_2) = 1 + \frac{1}{2} \cdot 0.01 = 1.005.$$

Therefore, the Runge-Kutta method of order two approximates that the value of y at $x = 0.1$ is 1.005.

In the same manner as in the previous two examples, we define a function yr to implement the Runge-Kutta method of order two and use Table to generate a set of approximations for $n = 0, 1, \ldots, 10$.

```
Remove[f, x, y]

f[x_, y_] = xy;

h = 0.1;

x0 = 0;

y0 = 1;

xr[n_] = x0 + nh;

yr[n_]:=
  yr[n] = yr[n - 1] + 1/2 hf[xr[n - 1], yr[n - 1]] +
```

$$\frac{1}{2}hf\left[x_r[n-1]+h,\ y_r[n-1]+hf\left[x_r[n-1],\ y_r[n-1]\right]\right]$$

$$y_r[0]=y_0;$$

```
rktable1 = Table[{x_r[i], y_r[i]}, {i, 0, 10}];
```

```
TableForm[rktable1]
```

0.	1
0.1	1.005
0.2	1.02018
0.3	1.04599
0.4	1.08322
0.5	1.13305
0.6	1.19707
0.7	1.27739
0.8	1.37677
0.9	1.49876
1.	1.64788

We then use ListPlot to graph the set of points determined in rktable1. The graphs in p1 and p2 are shown together with Show in Figure 2-34.

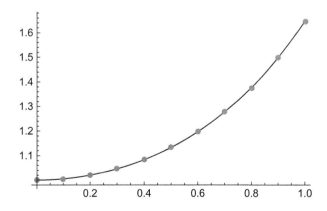

Figure 2-34 Comparison of the Runge-Kutta method of order two to the exact solution using $h = 0.1$

```
p2 = ListPlot[rktable1, PlotStyle->PointSize[0.02]];

Show[p1, p2]
```

■

The terms of the power series expansions used in the derivation of the Runge-Kutta method of order two can be made to match up to order four. These computations are rather complicated, so they will not be discussed here. However, after much work, the **fourth-order Runge-Kutta method** approximation at each step is found to be made with

$$y_{n+1} = y_n + \frac{1}{6}h\left(k_1 + 2k_2 + 2k_3 + k_4\right), \ n = 0, 1, 2, \ldots$$

where

$$k_1 = f\left(x_n, y_n\right)$$

$$k_2 = f\left(x_n + \frac{1}{2}h, y_n + \frac{1}{2}hk_1\right) \tag{2.13}$$

$$k_3 = f\left(x_n + \frac{1}{2}h, y_n + \frac{1}{2}hk_2\right)$$

and

$$k_4 = f\left(x_{n+1}, y_n + hk_3\right).$$

```
f[x_, y_] =?;

h =?;

x₀ =?;

y₀ =?;

xᵣ[n_] = x₀ + nh;

yᵣ[n_]:=yᵣ[n] = yᵣ[n - 1] + 1/6 h (k₁[n - 1] + 2k₂[n - 1] + 2k₃[n - 1] + k₄[n - 1]);

yᵣ[0] = y₀;

k₁[n_]:=k₁[n] = f[xᵣ[n], yᵣ[n]];

k₂[n_]:=k₂[n] = f[xᵣ[n] + h/2, yᵣ[n] + 1/2 hk₁[n]];

k₃[n_]:=k₃[n] = f[xᵣ[n] + h/2, yᵣ[n] + 1/2 hk₂[n]];

k₄[n_]:=k₄[n] = f[xᵣ[n + 1], yᵣ[n] + hk₃[n]]
```

EXAMPLE 2.6.6: Use the fourth-order Runge-Kutta method with $h = 0.1$ to approximate the solution of the problem $y' = xy$, $y(0) = 1$ on $0 \le x \le 1$.

SOLUTION: With $f(x, y) = xy$, $x_0 = 0$, and $y_0 = 1$, using equation (2.13), the formulas are

$$y_{n+1} = y_n + \frac{0.1}{6} (k_1 + 2k_2 + 2k_3 + k_4), \, n = 0, 1, 2, \ldots$$

where

$$k_1 = f(x_n, y_n) = x_n y_n$$

$$k_2 = f\left(x_n + \frac{1}{2}h, y_n + \frac{1}{2}hk_1\right) = \left(x_n + \frac{1}{2} \cdot 0.1\right)\left(y_n + \frac{1}{2} \cdot 0.1k_1\right)$$

$$k_3 = f\left(x_n + \frac{1}{2}h, y_n + \frac{1}{2}hk_2\right) = \left(x_n + \frac{1}{2} \cdot 0.1\right)\left(y_n + \frac{1}{2} \cdot 0.1k_2\right)$$

and

$$k_4 = f(x_{n+1}, y_n + hk_3) = x_{n+1}(y_n + 0.1k_3).$$

For $n = 0$, we have

$$k_1 = x_0 y_0 = 0 \cdot 1 = 0$$

$$k_2 = \left(x_0 + \frac{1}{2} \cdot 0.1\right)\left(y_0 + \frac{1}{2} \cdot 0.1k_1\right) = 0.05 \cdot 1 = 0.05$$

$$k_3 = \left(x_0 + \frac{1}{2} \cdot 0.1\right)\left(y_0 + \frac{1}{2} \cdot 0.1k_2\right) = 0.05 \cdot (1 + 0.0025) = 0.050125$$

and

$$k_4 = x_1(y_0 + 0.1k_3) = 0.1 \cdot (1 + 0.0050125) = 0.10050125.$$

Therefore,

$$y_1 = y_0 + \frac{0.1}{6}(k_1 + 2k_2 + 2k_3 + k_4) = 1.005012521.$$

We list the results for the Runge-Kutta method of order four and compare these results to the exact solution in Figure 2-35. Notice that this method yields the most accurate approximation of the methods used to this point.

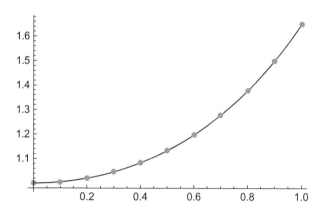

Figure 2-35　Comparison of the fourth-order Runge-Kutta method to the exact solution using $h = 0.1$

```
Remove[f, x, y]

f[x_, y_] = xy;

h = 0.1;

x0 = 0;

y0 = 1;

xr[n_] = x0 + nh;

yr[n_]:=yr[n] = yr[n - 1] + 1/6 h (k1[n - 1] + 2 k2[n - 1] + 2 k3[n - 1]
+k4[n - 1]) ;

yr[0] = y0 ;

k1[n_]:=k1[n] = f[xr[n], yr[n]] ;

k2[n_]:=k2[n] = f[xr[n] + h/2, yr[n] + 1/2 hk1[n]] ;

k3[n_]:=k3[n] = f[xr[n] + h/2, yr[n] + 1/2 hk2[n]] ;

k4[n_]:=k4[n] = f[xr[n + 1], yr[n] + hk3[n]]

rktable2 = Table[{xr[i], yr[i]}, {i, 0, 10}];

TableForm[rktable2]
```

```
0.    1

0.1   1.00501
```

```
0.2   1.0202

0.3   1.04603

0.4   1.08329

0.5   1.13315

0.6   1.19722

0.7   1.27762

0.8   1.37713

0.9   1.4993

1.    1.64872
```

```
p3 = ListPlot[rktable2, PlotStyle->PointSize[0.02]];

Show[p1, p3]
```

■

Applications of First-Order Equations

When the space shuttle was launched from the Kennedy Space Center, its escape velocity could be determined by solving a first-order ordinary differential equation. The same can be said for finding the flow of electromagnetic forces, the temperature of a cup of coffee, the population of a species as well as numerous other applications. In this chapter, we show how these problems can be expressed as first-order equations. We will focus our attention on setting up the problems and explaining the meaning of the subsequent solutions because the techniques for solving these problems were discussed in Chapter 2.

3.1 Orthogonal Trajectories

We begin our discussion with *orthogonal trajectories*, a topic that is encountered in the study of electromagnetic fields and heat flow frequently in your physics and thermodynamics classes.

Differential Equations with Mathematica. http://dx.doi.org/10.1016/B978-0-12-804776-7.00003-6

Definition 9 (Orthogonal Curves). *Two lines, ℓ_1 and ℓ_2, with slopes m_1 and m_2, respectively, are **orthogonal** (or **perpendicular**) if their slopes satisfy the relationship $m_1 = -1/m_2$. Two curves, C_1 and C_2, are **orthogonal** (or **perpendicular**) at a point if their respective tangent lines to the curves at that point are perpendicular.*

EXAMPLE 3.1.1: Use the definition of orthogonality to verify that the curves given by $y = x$ and $y = \sqrt{1 - x^2}$ are orthogonal at the point $\left(\sqrt{2}/2, \sqrt{2}/2\right)$.

SOLUTION: First note that the point $\left(\sqrt{2}/2, \sqrt{2}/2\right)$ lies on both the graph of $y = x$ and $y = \sqrt{1 - x^2}$. The derivatives of the functions are given by $y' = 1$ and $y' = -x/\sqrt{1 - x^2}$, respectively.

```
Clear[x, y]

y[1][x_] = x;

y[2][x_] = Sqrt[1 - x^2];

y[1]'[x]
y[2]'[x]
```

```
1
```

$$-\frac{x}{\sqrt{1 - x^2}}$$

```
y[2]'[Sqrt[2]/2]
```

```
-1
```

Thus, the curves are orthogonal at the point $\left(\sqrt{2}/2, \sqrt{2}/2\right) = (1/\sqrt{2}, 1/\sqrt{2})$ because the slopes of the lines tangent to the graphs of $y = x$ and $y = \sqrt{1 - x^2}$ at the point $\left(\sqrt{2}/2, \sqrt{2}/2\right)$ are negative reciprocals. We graph these two curves along with the tangent line to $y = \sqrt{1 - x^2}$ at $\left(\sqrt{2}/2, \sqrt{2}/2\right)$ in Figure 3-1 to illustrate that the two are orthogonal. The option `AspectRatio->1` specifies that the ratio of lengths of the x-axis to the y-axis in the resulting graphics object be 1. That is, the graphic is plotted to scale.

```
Plot[{y[1][x], y[2][x], -x + Sqrt[2]}, {x, -3/2, 3/2},

    PlotRange → {{-3/2, 3/2}, {-1, 2}},

    AspectRatio → 1, AxesStyle → Black]
```

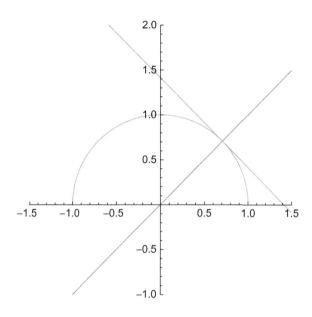

Figure 3-1 The curves are orthogonal at the point $\left(\sqrt{2}/2, \sqrt{2}/2\right)$

■

The next step in our discussion of orthogonal curves is to determine the set of orthogonal curves to a given family of curves. We refer to this set of orthogonal curves as the **family of orthogonal trajectories**. Suppose that a family of curves is defined as $F(x, y) = C$ and that the slope of the tangent line at any point on these curves is $dy/dx = f(x, y)$. Then, the slope of the tangent line on the orthogonal trajectory is $dy/dx = -1/f(x, y)$ so the family of orthogonal trajectories is found by solving the first-order equation $dy/dx = -1/f(x, y)$.

EXAMPLE 3.1.2: Determine the family of orthogonal trajectories to the family of curves $y = cx^2$.

SOLUTION: First, we must find the slope of the tangent line at any point on the parabola $y = cx^2$. Differentiating with respect to x results in $dy/dx = 2cx$. However from $y = cx^2$, we have that $c = y/x^2$. Substitution into $dy/dx = 2cx$ then yields $dy/dx = 2 \cdot y/x^2 \cdot x = 2y/x$ on the parabolas. Hence, we must solve $dy/dx = -x/(2y)$ to determine the orthogonal trajectories. This equation is separable, so we write it as $2y\,dy = x\,dx$, and then integrating both sides gives us $2y^2 + x^2 = k$, where k is a constant, which we recognize as a family of ellipses. Note that an equivalent result is obtained with `DSolve`.

```
sol = DSolve[y'[x] == -x/(2y[x]), y[x], x]
```

$$\left\{\left\{y[x] \to -\frac{\sqrt{-x^2 + 4C[1]}}{\sqrt{2}}\right\}, \left\{y[x] \to \frac{\sqrt{-x^2 + 4C[1]}}{\sqrt{2}}\right\}\right\}$$

We graph several members of the family of parabolas $y = cx^2$, the family of ellipses $2y^2 + x^2 = k$, and the two families of curves together. First, we define `parabs` to be the list of functions obtained by replacing c in $y = cx^2$ by nine equally spaced values of c between $-3/2$ and $3/2$.

```
step1 = Solve[sol[[1, 1, 2]] == y[x], C[1]]
```

$$\left\{\left\{C[1] \to \frac{1}{4}\left(x^2 + 2y[x]^2\right)\right\}\right\}$$

```
step2 = step1[[1, 1, 2]]/.y[x] → y
```

$$\frac{1}{4}\left(x^2 + 2y^2\right)$$

```
parabs = Table[cx^2, {c, -3/2, 3/2, 1/8}];
```

Next, we graph the list of functions `parabs` for $-3 \le x \le 3$ with `Plot` and name the result `p1`. The graphs in `p1` are not displayed because a semi-colon is included at the end of the `Plot` command. We graph several ellipses $2y^2 + x^2 = k$ by using `ContourPlot` to graph several level curves of $f(x, y) = y^2 + \frac{1}{2}x^2$ and name the result `p2`. Including the option `PlotPoints->130` helps assure that the ellipses appear smooth in the result. Including the option `ContourShading->False` specifies that the region between contours is not shaded. As with `p1`, `p2` is not displayed. Finally, `p1` and `p2` are displayed together with `Show` in Figure 3-2. Notice that these two families appear orthogonal, confirming the results we obtained.

```
p1 = Plot[parabs, {x, -3, 3}];

p2 = ContourPlot[step2, {x, -3, 3}, {y, -3, 3},
    Contours → 30, ContourShading → False,
    PlotPoints → 130];

Show[p1, p2, PlotRange → {{-3, 3}, {-3, 3}},
    AspectRatio → 1, AxesLabel → {x, y},
    AxesStyle → Black]
```

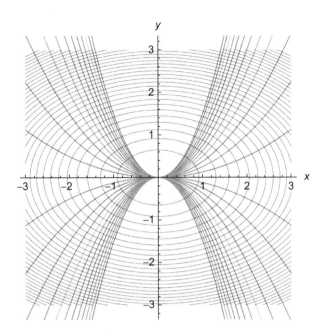

Figure 3-2 The two sets of curves are orthogonal to each other

∎

EXAMPLE 3.1.3 (Temperature): Let $T(x, y)$ represent the temperature at the point (x, y). The curves given by $T(x, y) = c$ (where c is constant) are called **isotherms**. The orthogonal trajectories are curves along which heat will flow. Determine the isotherms if the curves of heat flow are given by $y^2 + 2xy - x^2 = c$.

SOLUTION: We begin by finding the slope of the tangent line at each point on the heat flow curves $y^2 + 2xy - x^2 = c$ using implicit differentiation.

```
eq1 = y[x]^2 + 2xy[x] - x^2 == c;
```

```
step1 = Dt[eq1, x]
```

$$-2x + 2y[x] + 2xy'[x] + 2y[x]y'[x] == \text{Dt}[c, x]$$

Because c represents a constant, $d/dx\,(c) = 0$. We interpret `step2` to be equivalent to the equation $2yy' + 2y + 2xy' - 2x = 0$, where $y' = dy/dx$.

```
step2 = step1/.Dt[c, x] → 0
```

$$-2x + 2y[x] + 2xy'[x] + 2y[x]y'[x] == 0$$

We calculate $y' = dy/dx$ by solving `step2` for `y'[x]` with `Solve` and name the result `imderiv`.

```
imderiv = Solve[step2, y'[x]]
```

$$\left\{\left\{y'[x] \to \frac{x - y[x]}{x + y[x]}\right\}\right\}$$

Thus, $dy/dx = (x-y)/(x+y)$ so the orthogonal trajectories satisfy the differential equation $dy/dx = -(x+y)/(x-y)$.

This equation is also homogeneous of degree one.

Writing this equation in differential form as $(x+y)dx + (x-y)dy = 0$, we see that this equation is exact because $\partial/\partial y\,(x+y) = 1 = \partial/\partial x\,(x-y)$. Thus, we solve the equation by integrating $x+y$ with respect to x to yield $f(x,y) = \frac{1}{2}x^2 + xy + g(y)$. Differentiating f with respect to y then gives us $f_y(x,y) = x + g'(y)$. Then, because the equation is exact, $x + g'(y) = x - y$. Therefore, $g'(y) = -y$ which implies that $g(y) = -\frac{1}{2}y^2$. This means that the family of orthogonal trajectories (isotherms) is given by $\frac{1}{2}x^2 + xy - \frac{1}{2}y^2 = k$.

Note that `DSolve` is able to solve this differential equation.

```
step3 = DSolve[y'[x] == imderiv[[1, 1, 2]], y[x], x]//Simplify
```

$$\left\{\left\{y[x] \to -x - \sqrt{e^{2C[1]} + 2x^2}\right\}, \left\{y[x] \to -x + \sqrt{e^{2C[1]} + 2x^2}\right\}\right\}$$

To graph $y^2 + 2xy - x^2 = c$ and $\frac{1}{2}x^2 + xy - \frac{1}{2}y^2 = k$ for various values of c and k to see that the curves are orthogonal, we use `ContourPlot`. First, we graph several level curves of $y^2 + 2xy - x^2 = c$ on the rectangle $[-4, 4] \times [-4, 4]$ and name the result `cp1`. The option `Contours->40` instructs Mathematica to graph forty contours instead of the default of ten.

```
cp1 = ContourPlot[y^2 + 2xy − x^2, {x, −4, 4},

    {y, −4, 4}, ContourShading → False, Axes → Automatic,

    Contours → 40, PlotPoints → 120, AxesOrigin → {0, 0},

    Frame → False, AxesLabel → {x, y}, PlotLabel → "(a)"];
```

Next we graph several level curves of $\frac{1}{2}x^2 + xy - \frac{1}{2}y^2 = k$ on the same rectangle and name the result `cp2`. In this case, the option

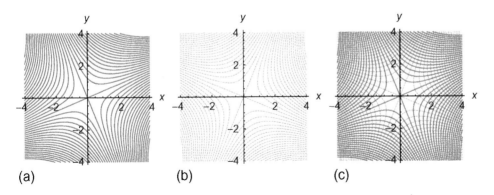

Figure 3-3   (a) and (b) Several members of each family of curves. (c) The two sets of curves are orthogonal to each other

```
ContourStyle->{{GrayLevel[0.4],Dashing[{0.01}]}}
```

specifies that the contours are to be dashed in a medium gray.

```
cp2 = ContourPlot[xy + x^2/2 - y^2/2, {x, -4, 4},
    {y, -4, 4}, ContourShading → False, Axes → Automatic,
    Contours → 40, PlotPoints → 120, AxesOrigin → {0, 0},
    ContourStyle → {{Gray, Dashing[{0.01}]}},
    Frame → False, AxesLabel → {x, y}, PlotLabel → "(b)"];
```

The graphs are then displayed side-by-side using Show and GraphicsRow in Figure 3-3(a) and (b) and together using Show in Figure 3-3(c).

```
Show[GraphicsRow[{cp1, cp2, Show[cp1, cp2, PlotLabel →
    "(c)"]}]]
```

∎

EXAMPLE 3.1.4: Determine the orthogonal trajectories of the family of curves given by $y^2 - 2cx = c^2$. Graph several members of both families of curves on the same set of axes.

SOLUTION: After defining eq to be the equation $y^2 - 2cx = c^2$, we implicitly differentiate.

```
Clear[x, y, c]

eq = y[x]^2 - 2cx == c^2;

step1 = Dt[eq]
```

$$-2x\,\text{Dt}[c] - 2c\,\text{Dt}[x] + 2\text{Dt}[x]y[x]y'[x] == 2c\,\text{Dt}[c]$$

As in the previous examples, we interpret Dt [x] to be 1, Dt [c] to be 0, and Dt [y] to represent dy/dx.

```
step2 = step1/.{Dt[c] → 0, Dt[x] → 1}
```

$$-2c + 2y[x]y'[x] == 0$$

The equation $y^2 - 2cx = c^2$ is a quadratic in c. Solving for c,

```
cval = Solve[eq, c]
```

$$\left\{\left\{c \to -x - \sqrt{x^2 + y[x]^2}\right\}, \left\{c \to -x + \sqrt{x^2 + y[x]^2}\right\}\right\}$$

we choose to substitute the first value into the equation $y\,dy/dx = c$.

```
imderiv = Solve[step2, y'[x]]/.c → cval[[1, 1, 2]]
```

$$\left\{\left\{y'[x] \to \frac{-x - \sqrt{x^2 + y[x]^2}}{y[x]}\right\}\right\}$$

```
imderiv[[1, 1, 2]]
```

$$\frac{-x - \sqrt{x^2 + y[x]^2}}{y[x]}$$

Then, we must solve $\dfrac{dy}{dx} = \dfrac{y}{x + \sqrt{x^2 + 2y^2}}$.

```
de = y'[x] == -1/imderiv[[1, 1, 2]]
```

$$y'[x] == -\frac{y[x]}{-x - \sqrt{x^2 + y[x]^2}}$$

Note that Mathematica is able to solve this equation. The formula for the first part of the solution is extracted from cval with cval[[1,1,2]] and the second part is extracted with cval[[2,1,2]].

```
DSolve[de, y[x], x]//Simplify
```

$$\left\{\left\{y[x] \to -e^{\frac{C[1]}{2}}\sqrt{e^{C[1]} + 2x}\right\}, \left\{y[x] \to e^{\frac{C[1]}{2}}\sqrt{e^{C[1]} + 2x}\right\}\right\}$$

```
cval[[1, 1, 2]]

cval[[2, 1, 2]]
```

$$-x - \sqrt{x^2 + y[x]^2}$$

$$-x + \sqrt{x^2 + y[x]^2}$$

Thus, $y^2 = 4C^2 + 4Cx$ and replacing $2C$ with C yields $y^2 = C^2 + 2Cx$ or $y^2 - 2Cx = c^2$, which means that this family of curves is self-orthogonal. We confirm that the family is self-orthogonal with ContourPlot in Figure 3-4.

```
cp1 = ContourPlot[cval[[1, 1, 2]], {x, −10, 10},

  {y[x], −10, 10}, ContourShading → False, Frame → False,

  Axes → Automatic, AxesOrigin → {0, 0}, Contours → 40,

  ContourStyle → Black,

  PlotPoints → 120, AxesStyle → Black, AxesLabel → {x, y},

  PlotLabel → "(a)"];

cp2 = ContourPlot[cval[[2, 1, 2]], {x, −10, 10},

  {y[x], −10, 10}, ContourShading → False, Frame → False,

  Axes → Automatic, AxesOrigin → {0, 0}, Contours → 40,

  ContourStyle → {{Gray, Dashing[{0.01}]}},
```

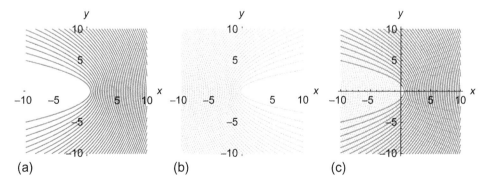

(a) (b) (c)

Figure 3-4   (a) and (b) The plots are symmetric about the y-axis. (c) The family of curves is self-orthogonal

```
PlotPoints → 120, AxesStyle → Black, AxesLabel → {x, y},

PlotLabel → "(b)"];

Show[GraphicsRow[{cp1, cp2, Show[cp1, cp2, PlotLabel → "(c)"]}]]
```

■

Application: Oblique Trajectories

If we are given a family of curves that satisfies the differential equation $dy/dx = f(x, y)$ and we want to find a family of curves that intersects this family at a constant angle θ, we must solve the differential equation

$$\frac{dy}{dx} = \frac{f(x, y) \pm \tan \theta}{1 \mp f(x, y) \tan \theta}.$$

For example, to find a family of curves that intersects the family of curves $x^2 + y^2 = c^2$ at an angle of $\pi/6$, we first implicitly differentiate the equation to obtain

$$2x + 2y \frac{dy}{dx} = 0 \implies \frac{dy}{dx} = -\frac{x}{y} = f(x, y).$$

Because $\tan \theta = \tan \pi/6 = 1/\sqrt{3}$, we solve

$$\frac{dy}{dx} = \frac{-x/y + 1/\sqrt{3}}{1 - (-x/y)(1/\sqrt{3})} = \frac{-x\sqrt{3} + y}{y\sqrt{3} + x},$$

which is a first-order homogeneous equation. With the substitution $x = vy$, we obtain the separable equation

$$\frac{1 - v\sqrt{3}}{1 + v^2} dv = \frac{\sqrt{3}}{y} dy.$$

Integrating yields

$$-\frac{\sqrt{3}}{2} \ln \left(1 + v^2 \right) + \tan^{-1} v = \sqrt{3} \ln |y| + k_1$$

so

$$-\frac{\sqrt{3}}{2} \ln \left(1 + \frac{x^2}{y^2} \right) + \tan^{-1} \frac{x}{y} = \sqrt{3} \ln |y| + k_1.$$

Mathematica finds an equivalent implicit solution.

```
sol1 = DSolve[y'[x] == (-Sqrt[3]x + y[x])/(Sqrt[3]y[x] + x), y[x], x]
```

$$\text{Solve}\left[\text{ArcTan}\left[\frac{y[x]}{x}\right] + \frac{1}{2}\sqrt{3}\text{Log}\left[1 + \frac{y[x]^2}{x^2}\right] == C[1] - \sqrt{3}\text{Log}[x], y[x]\right]$$

```
sol1a = Solve[sol1[[1]], C[1]]/.y[x] → y
```

$$\left\{\left\{C[1] \to \frac{1}{2}\left(2\text{ArcTan}\left[\frac{y}{x}\right] + 2\sqrt{3}\text{Log}[x] + \sqrt{3}\text{Log}\left[1 + \frac{y^2}{x^2}\right]\right)\right\}\right\}$$

Similarly, for

$$\frac{dy}{dx} = \frac{-x/y - 1/\sqrt{3}}{1 + (-x/y)(1/\sqrt{3})} = \frac{-x\sqrt{3} - y}{y\sqrt{3} - x},$$

we obtain

$$\frac{1 + v\sqrt{3}}{1 + v^2}dv = -\frac{\sqrt{3}}{y}dy.$$

So that the trajectories are

$$\frac{\sqrt{3}}{2}\ln\left(1 + \frac{x^2}{y^2}\right) + \tan^{-1}\frac{x}{y} = -\sqrt{3}\ln|y| + k_1.$$

```
sol2 = DSolve[y'[x] == (-Sqrt[3]x - y[x])/(Sqrt[3]y[x] - x), y[x], x]
```

$$\text{Solve}\left[-\text{ArcTan}\left[\frac{y[x]}{x}\right] + \frac{1}{2}\sqrt{3}\text{Log}\left[1 + \frac{y[x]^2}{x^2}\right] == C[1] - \sqrt{3}\text{Log}[x], y[x]\right]$$

```
sol2a = Solve[sol2[[1]], C[1]]/.y[x] → y
```

$$\left\{\left\{C[1] \to \frac{1}{2}\left(-2\text{ArcTan}\left[\frac{y}{x}\right] + 2\sqrt{3}\text{Log}[x] + \sqrt{3}\text{Log}\left[1 + \frac{y^2}{x^2}\right]\right)\right\}\right\}$$

To confirm the result graphically, we graph several members of each family of curves in Figure 3-5(b) and (c).

```
cp1 = ContourPlot[x^2 + y^2, {x, -10, 10}, {y, -10, 10},

    Frame → False, Contours → 20, ContourStyle → {{Gray, Dashing[{0.01}]}},

    Axes → Automatic, ContourShading → False,

    PlotPoints → 120, AxesOrigin → {0, 0},

    AxesLabel → {x, y}, PlotLabel → "(a)"];
```

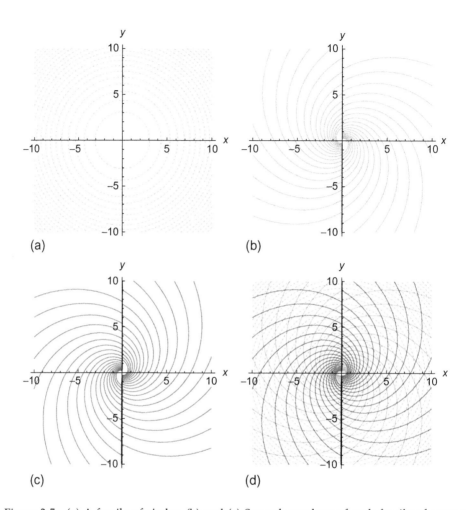

Figure 3-5   (a) A family of circles. (b) and (c) Several members of each family of curves. (d) The curves intersect at an angle of $\pi/6$

$$cp2 = \text{ContourPlot}\left[\frac{1}{2}\left(2\text{ArcTan}\left[\frac{y}{x}\right] + 2\sqrt{3}\text{Log}[\text{Abs}[x]] + \sqrt{3}\text{Log}\left[1 + \frac{y^2}{x^2}\right]\right),\right.$$

$$\{x, -10, 10\}, \{y, -10, 10\},$$

Frame $\rightarrow$ False, Contours $\rightarrow$ 20, ContourStyle $\rightarrow$ {{Gray}},

Axes $\rightarrow$ Automatic, ContourShading $\rightarrow$ False,

PlotPoints $\rightarrow$ 120, AxesOrigin $\rightarrow$ {0, 0},

AxesLabel $\rightarrow$ {x, y}, PlotLabel $\rightarrow$ "(b)"];

```
cp3 = ContourPlot[1/2(-2ArcTan[y/x] + 2√3Log[Abs[x]] + √3Log[1 + y²/x²]),

  {x, -10, 10}, {y, -10, 10},

  Frame → False, Contours → 20, ContourStyle → {{Black}},

  Axes → Automatic, ContourShading → False,

  PlotPoints → 120, AxesOrigin → {0, 0},

  AxesLabel → {x, y}, PlotLabel → "(c)"];

Show[GraphicsGrid[{{cp1, cp2},

  {cp3, Show[cp1, cp2, cp3, PlotLabel → "(d)"]}}]]
```

and then show the curves together in Figure 3-5(d) to see that they intersect at an angle of $\pi/6$.

3.2 Population Growth and Decay

Many interesting problems involving population can be solved through the use of first-order differential equations. These include the determination of the number of cells in a bacteria culture, the number of citizens in a country, and the amount of radioactive substance remaining in a fossil. We begin our discussion by solving a population problem.

3.2.1 The Malthus Model

Suppose that the rate at which a population of size $y(t)$ at time t changes is proportional to the population, $y(t)$, at time t. Mathematically, this statement is represented as the first-order initial-value problem

$$\begin{cases} dy/dt = ky \\ y(0) = y_0 \end{cases}, \tag{3.1}$$

where y_0 is the initial population. If $k > 0$, then the population increases (growth) while the population decreases (decay) if $k < 0$. Problems of this nature arise in such fields as cell population growth in biology as well as radioactive decay in physics. Equation (3.1) is known as the Malthus model due to the work of the English clergyman and economist Thomas R. Malthus. We solve the Malthus model for all values of k and y_0 which enables us to refer to the solution in other problems without solving the differential equation again. Rewriting $dy/dt = ky$ in the form $dy/y = k\,dt$, we see that this is a separable differential equation. Integrating and simplifying results in:

$$\int \frac{1}{y}dy = \int k\,dt$$
$$\ln|y| = kt + C_1$$
$$y = Ce^{kt}, \text{ where } C = e^{C_1}.$$

Notice that because y represents population, $y \geq 0$ and, therefore, $|y| = y$. To find C, we apply the initial condition obtaining $y_0 = y(0) = Ce^{k \cdot 0} = C$. Thus, the solution to the initial-value problem (3.1) is

$$y = y_0 e^{kt}. \tag{3.2}$$

We obtain the same result with DSolve.

```
Clear[x, y, t]

DSolve[{y'[t] == ky[t], y[0] == y0}, y[t], t]
```

$$\{\{y[t] \to e^{kt} y0\}\}$$

EXAMPLE 3.2.1 (Radioactive Decay): Forms of a given element with different numbers of neutrons are called **nuclides**. Some nuclides are not stable. For example, potassium-40 (^{40}K) naturally decays to reach argon-40 (^{40}Ar). This decay which occurs in some nuclides was first observed, but not understood, by Henri Becquerel (1852–1908) in 1896. Marie Curie, however, began studying this decay in 1898, named it **radioactivity**, and discovered the radioactive substances polonium and radium. Marie Curie (1867–1934), along with her husband, Pierre Curie (1859–1906), and Henri Becquerel, received the Nobel Prize in Physics in 1903 for their work on radioactivity. Marie Curie subsequently received the Nobel Prize in Chemistry in 1910 for discovering polonium and radium.

Given a sample of ^{40}Ar of sufficient size, after 1.2×10^9 years approximately half of the sample will have decayed to ^{40}Ar. The half-life of a nuclide is the time for half the nuclei in a given sample to decay. We see that the rate of decay of a nuclide is proportional to the amount present because the half-life of a given nuclide is constant and independent of the sample size.

Marie and Pierre Curie discover radium on December 21, 1898.

If the half-life of Polonium ^{209}Po is 100 years, determine the percentage of the original amount of ^{209}Po that remains after 50 years.

SOLUTION: Let y_0 represent the original amount of ^{209}Po that is present. Then the amount present after t years is $y(t) = y_0e^{kt}$. Because $y(100) = y_0/2$ and $y(100) = y_0e^{100k}$, we solve $y_0e^{100k} = y_0/2$ for e^k:

$$e^{100k} = \frac{1}{2} \quad \text{or} \quad e^k = \left(\frac{1}{2}\right)^{1/100}$$

so

$$y(t) = y_0e^{kt} = y_0\left(\frac{1}{2}\right)^{t/100}.$$

```
k = -Log[2]/100;

y[t_] = y0Exp[kt];

Simplify[y[t]]
```

$2^{-t/100}$y0

In order to determine the percentage of y_0 that remains, we evaluate $y(50) = y_0(1/2)^{50/100} = y_0/\sqrt{2} \approx 0.7071y_0$.

```
y[50]
```

$$\frac{y0}{\sqrt{2}}$$

```
N[y[50]]
```

0.707107y0

Therefore, 70.71% of the original amount of ^{209}Po remains after 50 years.

∎

In the previous example, we see that we can determine the percentage of y_0 that remains even though we do not know the value of y_0. Hence, instead of letting $y(t)$

represent the amount of the substance present after time t, we can let it represent the fraction (or percent) of y_0 that remains after time t. In doing this, we use the initial condition $y(0) = 1$ to indicate that 100% of y_0 is present at $t = 0$.

EXAMPLE 3.2.2: The wood of an Egyptian sarcophagus (burial case) is found to contain 63% of the carbon-14 found in a present day sample. What is the age of the sarcophagus?

SOLUTION: The half-life of carbon-14 is 5730 years. Let $y(t)$ be the percent of carbon-14 in the sample after t years. Then, $y(0) = 1$. Because $y(t) = y_0 e^{kt}$, $y(5730) = e^{5730k}$. Solving for k yields:

$$\ln\left(e^{5730k}\right) = \ln\left(\frac{1}{2}\right)$$

$$5730k = \ln\left(\frac{1}{2}\right)$$

$$k = \frac{\ln\left(\frac{1}{2}\right)}{5730} = -\frac{\ln 2}{5730}.$$

Thus, $y(t) = e^{kt} = e^{-\frac{\ln 2}{5730}t} = 2^{-t/5730}$.

```
Clear[k, y]

k = -Log[2]/5730;

y[t_] = Exp[kt]
```

$2^{-t/5730}$

In this problem, we must find the value of t for which $y(t) = 0.63$. Solving the equation $2^{-t/5730} = 0.63 = 63/100$ results in:

$$2^{-t/5730} = \frac{63}{100}$$

$$\ln\left(2^{-t/5730}\right) = \ln\frac{63}{100}$$

$$-\frac{t}{5730}\ln 2 = \ln\frac{63}{100}$$

$$t = -\frac{5730\ln\frac{63}{100}}{\ln 2} \approx 3819.48.$$

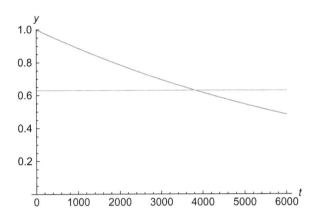

Figure 3-6 The age of the sarcophagus is the *t*-coordinate of the point of intersection of $y(t)$ and $y = 0.63$

We conclude that the sarcophagus is approximately 3819 years old.

An alternative way to approximate the age of the sarcophagus is to first graph $y(t)$ and the line $y = 0.63$ with Plot as shown in Figure 3-6. The age of the sarcophagus is the *t*-coordinate of the point of intersection of $y(t)$ and $y = 0.63$.

```
Plot[{y[t], 0.63}, {t, 0, 6000},
    PlotRange → {0, 1}, AxesStyle → Black,
    AxesLabel → {t, y}]
```

We see that the *t*-coordinate of the point of intersection is approximately $t \approx 3770$. A more accurate approximation is obtained with FindRoot. The command FindRoot [equation, {variable, first guess}] attempts to find a solution to the equation equation in the variable variable "near" the number firstguess. Thus, entering

```
FindRoot[y[t] == 0.63, {t, 4000}]
```

```
{t → 3819.48}
```

returns an approximation of the solution to the equation $y(t) = 0.63$ using an initial approximation of 3770. Alternatively, NSolve [equation(s), variables] attempts to find numerical solutions to equation(s).

```
NSolve[y[t] == 0.63]
```

```
{{t → 3819.48}}
```

■

To observe some of the limitations of the Malthus model, we next consider a population problem in which the rate of growth of the population does not exclusively depend on the population present.

EXAMPLE 3.2.3: The population of the United States was recorded as 5.3 million in 1800. Use the Malthus model to approximate the population for years after 1800 if k was experimentally determined to be 0.03. Compare these results to the actual population. Is this a good approximation for years after 1800?

SOLUTION: In this example, $k = 0.03$ and $y_0 = 5.3$ and our model for the population of the United States at time t (where t is the number of years from 1800) is $y(t) = 5.3e^{0.03t}$.

```
Clear[k, y, t]

pop[t_, k_, y0_] = y0Exp[kt];

pop[t, 0.03, 5.3]
```

$5.3e^{0.03t}$

In order to compare this model with the actual population of the United States, census figures for the population of the United States for various years are listed in Table 3-1 along with the corresponding value of $y(t)$.

Table 3-1   Population of the United States for various years

Year (t)	Actual Population (in millions)	Value of $y(t) =$ $5.3e^{0.03t}$	Year (t)	Actual Population (in millions)	Value of $y(t) =$ $5.3e^{0.03t}$
1800 (0)	5.30	5.30	1870 (70)	38.56	43.28
1810 (10)	7.24	7.15	1880 (80)	50.19	58.42
1820 (20)	9.64	9.66	1890 (90)	62.98	78.86
1830 (30)	12.68	13.04	1900 (100)	76.21	106.45
1840 (40)	17.06	17.60	1910 (110)	92.23	143.70
1850 (50)	23.19	23.75	1920 (120)	106.02	193.97
1860 (60)	31.44	32.06	1930 (130)	123.20	261.83

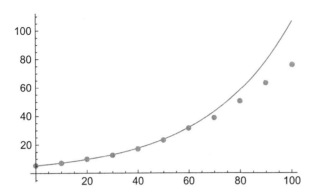

Figure 3-7   Over time, the Malthus model does not provide a good model of the growth of the population of the United States

   Although the model appears to closely approximate the data for several years after 1800, the accuracy of the approximation diminishes over time. This is because the population of the United States does not exclusively increase at a rate proportional to the population. Hence, another model which better approximates the population taking other factors into account is needed. The graph of $y(t) = 5.3e^{0.03t}$ is shown along with the data points in Figure 3-7 to show how the approximation becomes less accurate as t increases.

```
realpop = {{0, 5.3}, {10, 7.24}, {20, 9.64},
    {30, 12.68}, {40, 17.06}, {50, 23.19},
    {60, 31.44}, {70, 38.56}, {80, 50.19},
    {90, 62.98}, {100, 76.21}, {110, 92.23},
    {120, 106.02}, {130, 123.2}};

toshow = ListPlot[realpop, PlotStyle->PointSize[0.02]];

popplot = Plot[pop[t, 0.03, 5.3], {t, 0, 100},
    PlotStyle->GrayLevel[.3],
    DisplayFunction->Identity];

Show[popplot, toshow]
```

■

3.2.2 The Logistic Equation

The **logistic equation** (or **Verhulst equation**) is the equation

$$\frac{dy}{dt} = (r - ay(t))\, y(t), \tag{3.3}$$

where r and a are constants. Equation (3.3) was first introduced by the Belgian mathematician Pierre Verhulst to study population growth. The logistic equation differs from the Malthus model in that the term $r - ay(t)$ is not constant. This equation can be written as $dy/dt = (r - ay)y = ry - ay^2$ where the term $-y^2$ represents an inhibitive factor. Under these assumptions the population is neither allowed to grow out of control nor grow or decay constantly as it was with the Malthus model.

The logistic equation is separable, and, thus, can be solved by separation of variables. We solve equation (3.3) subject to the condition $y(0) = y_0$.

Separating variables and using partial fractions to integrate with respect to y, we have

$$\frac{1}{(r - ay)y}\, dy = dt$$

$$\left(\frac{a}{r}\frac{1}{r - ay} + \frac{1}{r}\frac{1}{y} \right) dy = dt$$

$$\left(a\frac{1}{r - ay} + \frac{1}{y} \right) dy = r\, dt$$

$$-\ln|r - ay| + \ln|y| = rt + C.$$

Using the properties of logarithms to solve this equation for y yields

$$\ln\left| \frac{y}{r - ay} \right| = rt + C$$

$$\frac{y}{r - ay} = e^{rt+C} = Ke^{rt}, \text{ where } K = e^{C}$$

$$y = r\left(\frac{1}{K}e^{-rt} + a \right)^{-1}.$$

```
DSolve[y'[t] == (r - a y[t]) y[t], y[t], t]
```

$$\left\{ \left\{ y[t] \to \frac{e^{rt+rC[1]}r}{1 + ae^{rt+rC[1]}} \right\} \right\}$$

Applying the initial condition $y(0) = y_0$ and solving for K, we find that

$$K = \frac{y_0}{r - ay_0}.$$

After substituting this value into the general solution and simplifying, the solution of equation (3.3) that satisfies the initial condition $y(0) = y_0$ can be written as

$$y = \frac{ry_0}{ay_0 + (r - ay_0)\,e^{-rt}}. \tag{3.4}$$

Notice that if $r > 0$, $\lim_{t\to\infty} y(t) = r/a$ because $\lim_{t\to\infty} e^{-rt} = 0$. This makes the solution to the logistic equation different from that of the Malthus model in that the solution to the logistic equation approaches a finite nonzero limit as $t \to \infty$ while that of the Malthus model approaches either infinity or zero as $t \to \infty$.

```
DSolve[{y'[t] == (r - ay[t])y[t], y[0] == y0}, y[t], t]
```

$$\left\{\left\{y[t] \to \frac{e^{rt}\,ry0}{r - ay0 + ae^{rt}y0}\right\}\right\}$$

Some prefer to write the Logistic equation as $\dfrac{dy}{dt} = ay\left(1 - \dfrac{1}{K}y\right)$, where a and K are positive constants. In this formulation of the model, a is usually called the **growth rate** and K is called the **carrying capacity**. The general solution of $\dfrac{dy}{dt} = ay\left(1 - \dfrac{1}{K}y\right)$ is often written in the form

$$y = \frac{1}{\dfrac{1}{K} + Ce^{-at}}.$$

Writing the solution in this form helps us see that

$$\lim_{t\to\infty} y = \lim_{t\to\infty} \frac{1}{\dfrac{1}{K} + Ce^{-at}} = K$$

because if $a > 0$, $\lim_{t\to\infty} Ce^{-at} = 0$.

```
DSolve[y'[t] == a(1 - 1/ky[t])y[t], y[t], t]
```

$$\left\{\left\{y[t] \to \frac{e^{at+kC[1]}k}{-1 + e^{at+kC[1]}}\right\}\right\}$$

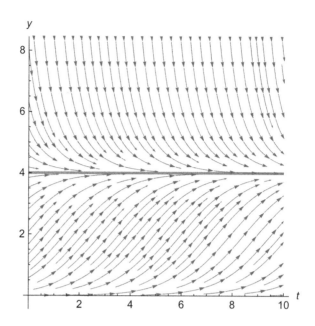

Figure 3-8   All nontrivial solutions of the Logistic equation tend to the carrying capacity as $t \to \infty$

DSolve[{y'[t] == a(1 − 1/ky[t])y[t], y[0] == y0}, y[t], t]

$$\left\{\left\{y[t] \to \frac{e^{at}ky0}{k - y0 + e^{at}y0}\right\}\right\}$$

To see that all nontrivial solutions tend to the carrying capacity as $t \to \infty$, we graph a direction field. Figure 3-8 graphs the direction field for $\dfrac{dy}{dt} = ay\left(1 - \dfrac{1}{K}y\right)$ if $a = 1$ and $K = 4$. In the figure, observe that all nontrivial solutions tend to $y = K = 4$.

```
p1 = Plot[4, {t, 0, 10}, PlotStyle →

    {{Thickness[.01], CMYKColor[0.64, 0, 0.95, 0.40]}}];

p2 = StreamPlot[{1, (1 − 1/4y)y}, {t, 0, 10}, {y, 0, 10},

    Frame → False, Axes → Automatic, AxesStyle → Black,

    AxesLabel → {t, y}, StreamStyle->CMYKColor[0, 0.89, 0.94, 0.28],

    StreamPoints → Fine];

Show[p1, p2, AspectRatio → 1, AxesLabel → {t, y},

    AxesStyle → Black]
```

EXAMPLE 3.2.4: Use the logistic equation to approximate the population of the United States using $r = 0.03$, $a = 0.0001$, and $y_0 = 5.3$. Compare this result with the actual census values shown in Table 3-1. Use the model obtained to predict the population of the United States in the year 2000.

SOLUTION: We substitute the indicated values of r, a, and y_0 into equation (3.4) to obtain the approximation of the population of the United States at time t, where t represents the number of years since 1800,

$$y(t) = \frac{0.159}{0.00053 + 0.02947e^{-0.03t}}.$$

```
pop[t_] =
```
$$\frac{0.159}{0.00053 + 0.0294\text{Exp}[-0.03t]}$$

$$\frac{0.159}{0.00053 + 0.0294e^{-0.03t}}$$

```
realpop = Union[realpop, {{140, 132.16}, {150, 151.33},

    {160, 179.32}, {170, 203.3}, {180, 226.54},

    {190, 248.71}}];
```

We compare the approximation of the population of the United States given by the approximation with the actual population obtained from census figures. Note that this model appears to more closely approximate the population over a longer period of time than the Malthus model which was considered in the previous examples as we can see in the graph shown in Figure 3-9.

```
toshow = ListPlot[realpop, PlotStyle->PointSize[0.02]];

popplot = Plot[pop[t], {t, 0, 200}, PlotStyle → GrayLevel[0.3]];

Show[popplot, toshow]

pop[200]

263.736

pop[2010]

300.
```

Be sure that you have defined
`realpop` as in
Example 3.2.3 before
entering the following
commands.

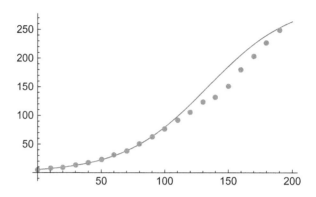

Figure 3-9   Over time, the logistic equation appears to provide a better model of the population of the United States than the Malthus model

To predict the population of the United States in the year 2000 with this model, we evaluate $y(200)$. Thus, we predict that the population will be approximately 263.66 million in the year 2000. Note that projections of the population of the United States in the year 2000 made by the Bureau of the Census range from 259.57 million to 278.23 million.

pop[200]

263.736

pop[2010]

300.

■

EXAMPLE 3.2.5 (Logistic Equation With Predation):   Incorporating predation into the **logistic equation**, $y' = \alpha y \left(1 - \dfrac{1}{K} y \right)$, results in

$$\frac{dy}{dt} = \alpha y \left(1 - \frac{1}{K} y \right) - P(y),$$

where $P(y)$ is a function of y describing the rate of predation. A typical choice for P is $P(y) = ay^2/(b^2 + y^2)$ because $P(0) = 0$ and P is bounded above: $\lim_{t \to \infty} P(y) < \infty$.

Remark. Of course, if $\lim_{t\to\infty} y(t) = Y$, then $\lim_{t\to\infty} P(y) = aY^2/(b^2 + Y^2)$. Generally, however, $\lim_{t\to\infty} P(y) \neq a$ because $\lim_{t\to\infty} y(t) \leq K \neq \infty$, for some $K \geq 0$, in the predation situation.

If $\alpha = 1$, $a = 5$ and $b = 2$, graph the direction field associated with the equation as well as various solutions if (a) $K = 19$ and (b) $K = 20$.

SOLUTION: (a) We define eqn[{k,y0}] to be

$$\frac{dy}{dt} = y\left(1 - \frac{1}{K}y\right) - \frac{5y^2}{4 + y^2}.$$

```
eqn[{k_, y0_}]:=NDSolve[{y'[t]==y[t](1 - 1/ky[t]) - 5y[t]^2/
(4 + y[t]^2), y[0] == y0}, y[t], {t, 0, 10}];
```

(a) and then the direction field along with the solutions that satisfy $y(0) = .5$, $y(0) = .2$, and $y(0) = 4$ in Figure 3-10(b).

```
pvf19 = StreamPlot[{1, y(1 - 1/19y) - 5y^2/(4 + y^2)},
{t, 0, 10}, {y, 0, 6}, Axes → Automatic, Frame → False,
AxesStyle → Black, AxesOrigin → {0, 0}, StreamStyle →
Gray,
AxesLabel → {t, y}, PlotLabel → "(a)"];
```

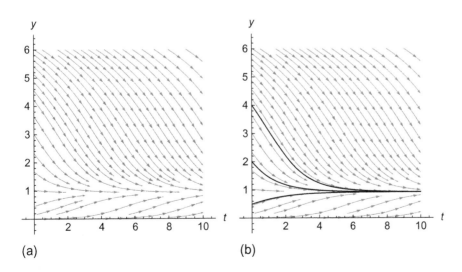

(a)   (b)

Figure 3-10   (a) Direction field and (b) direction field with three solutions

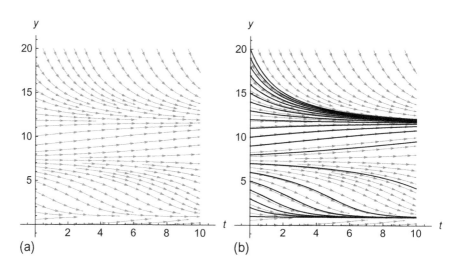

Figure 3-11   (a) Direction field. (b) Direction field with several solutions

```
sols = Map[eqn, {{19, .5}, {19, 2}, {19, 4}}];

p2 = Plot[y[t]/.sols, {t, 0, 10}, PlotRange → {{0, 10}, {0, 6}},
    AxesStyle→ Black, PlotStyle→ Black, AxesLabel → {t, y}];

Show[GraphicsRow[{pvf19,Show[pvf19,p2, PlotLabel→ "(b)"]}]]
```

In the plot, notice that all nontrivial solutions appear to approach an equilibrium solution. We determine the equilibrium solution by solving $y' = 0$

```
Solve[y(1 − 1/19y) − 5y^2/(4 + y^2) == 0, y]//N
```

$\{\{y \;\to\; 0.\}, \{y \;\to\; 0.923351\}, \{y \;\to\; 9.03832 + 0.785875i\}, \{y \to 9.03832 − 0.785875i\}\}$

to see that it is $y \approx 0.923$.

(b) We carry out similar steps for (b). First, we graph the direction field with StreamPlot in Figure 3-11.

```
pvf20 = StreamPlot[{1, y(1 − 1/20y) − 5y^2/(4 + y^2)},
    {t, 0, 10}, {y, 0, 20}, Axes → Automatic, Frame → False,
    AxesStyle→ Black, AxesOrigin→ {0,0}, StreamStyle→ Gray,
    AxesLabel → {t, y}, PlotLabel → "(a)"];
```

```
sols2 = Table[eqn[{20., i}], {i, 1, 20}];
```

$p3 = Plot[y[t]/.sols2, \{t, 0, 10\}, PlotRange \rightarrow \{\{0, 10\}, \{0, 20\}\},$
$\quad AxesStyle \rightarrow Black, PlotStyle \rightarrow Black, AxesLabel \rightarrow \{t, y\}];$

$Show[GraphicsRow[\{pvf20, Show[pvf20, p3, PlotLabel \rightarrow "(b)"]\}]]$

In Figure 3-11, notice that there are three nontrivial equilibrium solutions that are found by solving $y' = 0$. In this example, $y \approx .926$ and $y \approx 11.687$ are stable while $y \approx 7.386$ is unstable.

■

EXAMPLE 3.2.6 (Growth in the Chemostat): The *scaled* equations for the growth of a population in a chemostat are

$$\begin{cases} \dfrac{dS}{dt} = 1 - S - \dfrac{mS}{a+S}x \\[2mm] \dfrac{dx}{dt} = x\left(\dfrac{mS}{a+S} - 1\right) \\[2mm] S(0) \geq 0, \quad x(0) > 0 \end{cases} \qquad (3.5)$$

where $S(t)$ denotes the concentration of the nutrient at time t for the organism with concentration $x(t)$ at time t.

See Smith and Waltman's, *The Theory of the Chemostat: Dynamics of Microbial Competition*, [24], for a detailed discussion of various chemostat models.

SOLUTION: Letting $\Sigma = 1 - S - x$, we see that $\Sigma' = -\Sigma$.

```
seq = 1 - s - msx/(a + s);

xeq = x(ms/(a + s) - 1);

sigma = 1 - s - x;

seq + xeq - sigma//Simplify

    0
```

and system (3.5) can be written as

$$\begin{cases} \dfrac{d\Sigma}{dt} = -\Sigma \\[2mm] \dfrac{dx}{dt} = x\left(\dfrac{m(1 - \Sigma - x)}{a + (1 - \Sigma - x)} - 1\right) \\[2mm] \Sigma(0) > 0, x(0) > 0 \end{cases} \qquad (3.6)$$

Because $\Sigma(t) = \Sigma(0)e^{-t}$, $\lim_{t\to\infty} \Sigma(t) = 0$ so system (3.6) can be rewritten as the single first-order equation

$$\frac{dx}{dt} = x\left(\frac{m(1-x)}{a+(1-x)} - 1\right) \quad \text{or} \quad \frac{dx}{dt} = x\left[\frac{m(1-x)}{1+a-x} - 1\right], \quad 0 \le x \le 1, \quad (3.7)$$

where $x(0) > 0$. The rest points (or equilibrium points) of equation (3.7) are found by solving

$$x\left[\frac{m(1-x)}{1+a-x} - 1\right] = 0$$

for x.

```
s1 = Solve[1 - s - x == 0, s]
```

$$\{\{s \to 1 - x\}\}$$

```
xeq/.s1[[1]]
```

$$\left(-1 + \frac{m(1-x)}{1+a-x}\right)x$$

$$\text{Solve}\left[\left(-1 + \frac{m(1-x)}{1+a-x}\right)x == 0, x\right]$$

$$\left\{\{x \to 0\}, \left\{x \to \frac{-1-a+m}{-1+m}\right\}\right\}$$

For $m \ne 1$, observe that the nonzero rest point can be written as $x = 1 - \lambda$, where $\lambda = a/(m-1)$ is called the **break-even** concentration.

$$\text{Apart}\left[\frac{-1-a+m}{-1+m}\right]$$

$$1 - \frac{a}{-1+m}$$

We use Plot to graph $y = 1 - \lambda$ in Figure 3-12(a).

```
p1 = Plot[1 - λ, {λ, 0, 1}, AspectRatio → Automatic,
    AxesLabel → {, ""}, PlotLabel → "(a)"];
```

On the other hand, in Figure 3-12(b) we use ContourPlot to generate a plot of the level curve of $f(a, m) = (m - a - 1)/(m - 1)$ corresponding to 0. Points (a, m) in the white region are points where $1 - \lambda$ is positive.

```
p2 = ContourPlot[ (-1 - a + m)/(-1 + m) , {a, 0, 10}, {m, 0, 10}, Contours
    → {0}, PlotPoints → 300,
```

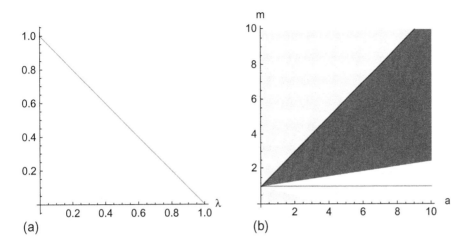

Figure 3-12 (a) Plot of $y = 1 - \lambda$. (b) The blue (dark gray in print versions) region corresponds to points (a, m) where $1 - \lambda$ is negative

```
        AxesLabel → {a, m}, Frame → False, Axes → Automatic,

        PlotLabel → "(b)"];

    Show[GraphicsRow[{p1, p2}]]
```

To see how a, m, and $x(0) = x_0$ affect the solutions of equation (3.7), we define the function x. The command x[m, a] [x0] plots the solution of equation (3.7) satisfying $x(0) = x_0$ for $0 \le t \le 10$. You can include Plot options, opts, with x[m, a] [x0, opts].

$$\frac{-1 - a + m}{-1 + m} / . \{a \to 2, m \to 6\}$$

$$\frac{3}{5}$$

$$\frac{-1 - a + m}{-1 + m} / . \{a \to 8, m \to 2\}$$

-7

```
    Clear[x]

    x[m_, a_][x0_, opts___] := Module[{λ},

    λ = a/(m - 1);

    numsol = NDSolve[{x'[t] == x[t](m - 1)/(1 + a - x[t])

        (1 - λ - x[t]), x[0] == x0}

        x[t], {t, 0, 100}]; Plot[Evaluate[x[t]/.numsol], {t, 0, 10}, opts,
```

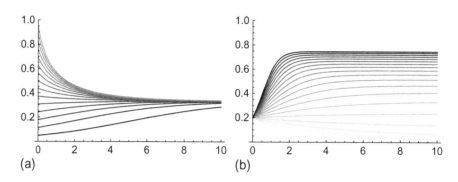

(a)   (b)

Figure 3-13   (a) If $m = 4$ and $a = 2$, all solutions of equation (3.7) approach the equilibrium solution $x = 1/3$. (b) If m is sufficiently small, the organism becomes extinct

```
            DisplayFunction → Identity, PlotRange → {0, 1}]

        ]
```

For example,

```
        g1 = Table[x[4, 2][x0, PlotStyle → GrayLevel[0 + x0/2]],

            {x0, .05, .95, .9/14}];
```

$$\frac{-1 - a + m}{-1 + m} / . \{m \to 4, a \to 2\}$$

$$\frac{1}{3}$$

```
        g1b = Show[g1, PlotLabel → "(a)"];
```

plots solutions to equation (3.7) if $m = 4$ and $a = 2$ for various initial conditions. The results are shown in Figure 3-13(a). In this case, we see that all solutions approach the equilibrium solution $x = 1/3$. On the other hand, entering

```
        g2 = Table[x[m, 2][.2, PlotStyle → GrayLevel[(9 − m)/8]],

            {m, 1.01, 9, 8.99/24}];

        g2b = Show[g2, PlotLabel → "(b)"];

        Show[GraphicsRow[{g1b, g2b}]]
```

solves equation (3.7) if $a = 2$ and $x_0 = 0.2$ for various values of m. The results are shown in Figure 3-13(b).

■

Application: Harvesting

If we wish to take a constant harvest rate, H (like hunting, fishing, or disease) into consideration, then we might instead modify the logistic equation (3.3) and use the equation

Source: David A. Sanchez, "Populations and Harvesting," *Mathematical Modeling: Classroom Notes in Applied Mathematics*, Murray S. Klamkin, Editor, SIAM (1987), pp. 311–313.

$$\frac{dP}{dt} = rP - aP^2 - H$$

to model the population under consideration. Notice that Mathematica can find a solution to this equation if $r^2 - 4aH < 0$.

```
gensol = DSolve[p'[t]==rp[t] - ap[t]^2 - h, p[t], t]
```

$$\left\{ \left\{ p[t] \to \frac{r + \sqrt{4ah - r^2}\,\mathrm{Tan}\left[\frac{1}{2}\left(-\sqrt{4ah - r^2}\,t + \sqrt{4ah - r^2}\,C[1]\right)\right]}{2a} \right\} \right\}$$

If H does not depend on P, the equilibrium solutions are found by solving $rP - aP^2 - H = 0$ for P.

```
eqsols = Solve[rp - ap^2 - h == 0, p]
```

$$\left\{ \left\{ p \to -\frac{-r + \sqrt{-4ah + r^2}}{2a} \right\}, \left\{ p \to \frac{r + \sqrt{-4ah + r^2}}{2a} \right\} \right\}$$

```
r^2 - 4ah/.{r → 7/10, a → 1/10, h → 1}
```

Observe that there are two distinct equilibrium solutions if $r^2 - 4aH > 0$. There is one equilibrium solution if $r^2 - 4aH = 0$, and none if $r^2 - 4aH < 0$. Suppose that for a certain species it is found that $r = 7/10$, $a = 1/10$, and $H = 1$. Then, the model becomes $dP/dt = \frac{7}{10}P - \frac{1}{10}P^2 - 1$ with equilibrium solutions $P = 2$ and $P = 5$.

$$\frac{9}{100}$$

For these parameter values, the solution obtained in `gensol` is not valid because $r^2 - 4aH = 9/100 > 0$.

```
eqsols/.{r → 7/10, a → 1/10, h → 1}
```

$$\{\{p \to 2\}, \{p \to 5\}\}$$

Mathematica can solve the initial-value problem $dP/dt = \frac{7}{10}P - \frac{1}{10}P^2 - 1$, $P(0) = P_0$.

```
Solve[rp - ap^2 - h == 0, p]
```

$$\left\{ \left\{ p \to -\frac{-r + \sqrt{-4ah + r^2}}{2a} \right\}, \left\{ p \to \frac{r + \sqrt{-4ah + r^2}}{2a} \right\} \right\}$$

```
Solve[r^2 - 4ah == 0, h]
```

$$\left\{ \left\{ h \to \frac{r^2}{4a} \right\} \right\}$$

exactsol=DSolve[{p'[t]==7/10p[t] − 1/10p[t]^2 − 1, p[0]==p0}, p[t], t]

$$\left\{ \left\{ p[t] \rightarrow \frac{10 - 10e^{3t/10} - 2p0 + 5e^{3t/10}p0}{5 - 2e^{3t/10} - p0 + e^{3t/10}p0} \right\} \right\}$$

Proceeding numerically, we define the function numgraph. Given i, numgraph[i] attempts to graph the numerical solution to $dP/dt = \frac{7}{10}P - \frac{1}{10}P^2 - 1$, $P(0) = P_0$ on the rectangle $[-10, 10] \times [-10, 10]$. Any options specified are passed through to the Plot command.

```
Clear[p, numgraph];

numgraph[i_ , opts___]:=

  Module[{numsol}, numsol = NDSolve[{p'[t]==7/10p[t] − 1/10p[t]^2 − 1,

  p[0] == i}, p[t], {t, 0, 10}];

  Plot[p[t]/.numsol, {t, 0, 10}, opts, DisplayFunction → Identity,

  PlotStyle → {{GrayLevel[0], Thickness[0.01]}},

  PlotRange → {{0, 10}, {0, 10}}]

]
```

For example, entering

```
numgraph[11, PlotRange → All]
```

displays the graph of the solution to $dP/dt = \frac{7}{10}P - \frac{1}{10}P^2 - 1$, $P(0) = 11$ for $0 \leq t \leq 10$, which is shown in Figure 3-14. We then use Map to graph the solution of $dP/dt = \frac{7}{10}P - \frac{1}{10}P^2 - 1$, $P(0) = i$, $i = 1, 1/2, \ldots, 10$, naming the resulting set of graphs toshow. Notice that Mathematica generates several error messages, not

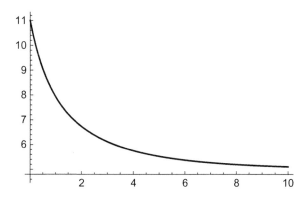

Figure 3-14   A solution to the logistic equation with harvesting

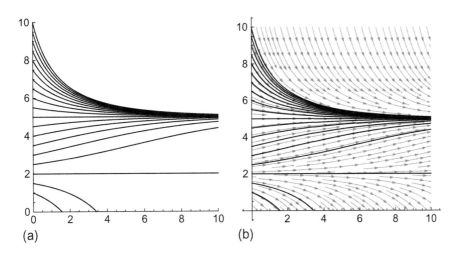

Figure 3-15   Solutions to the logistic equation with harvesting with the associated direction field

all of which are shown here, because solutions satisfying $P(0) = P_0$ for $P_0 < 2$ become unbounded very quickly.

```
toshow = Map[numgraph, Table[i, {i, 1, 10, 1/2}]];

p1 = Show[toshow, AspectRatio → 1, PlotLabel → "(a)"]'
```

The unbounded behavior of the solutions is particularly evident when we display the graphs in `toshow` together with the direction field associated with the equation in Figure 3-15.

```
p2a = StreamPlot[{1, 7/10y − 1/10y^2 − 1}, {x, 0, 10}, {y, 0, 10}, Frame

→ False, Axes → Automatic,

  AxesOrigin → {0, 0}, StreamStyle → GrayLevel[0.5], StreamPoints

  → Fine];

p2 = Show[p2a, p1, PlotRange → {{0, 10}, {0, 10}}, AspectRatio → 1,

PlotLabel → "(b)"];

Show[GraphicsRow[{p1, p2}]]
```

Thus, if $P_0 < 2$ and harvesting is allowed to continue, the species becomes extinct. If $P_0 > 2$, the population of the species has an equilibrium population of $P = 5$.

You can determine the extinction times with `FindRoot` or `NSolve`. Here we use `FindRoot` (Figure 3-16).

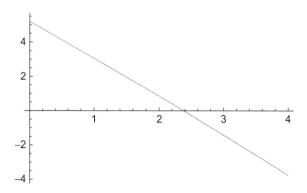

Figure 3-16   Use the plot to estimate the extinction times. Then use `FindRoot` to obtain more accurate approximations

```
Clear[p1]
```

$$p1[t\_] = \frac{r + \sqrt{4ah - r^2}\,\mathrm{Tan}\left[\dfrac{\sqrt{4ah - r^2}(-at + C[1])}{2a}\right]}{2a} /.\{r \to 0.03, a \to$$

$$0.0001, h \to 2.26, C[1] \to -0.150185\}$$

```
5000.(0.03 + 0.002Tan[10.(−0.150185 − 0.0001t)])
```

```
Plot[p1[t], {t, 0, 4}]
```

```
FindRoot[p1[t]==0, {t, 2.5}]
```

```
{t → 2.37816}
```

Sources: See texts like Boyce and DiPrima's *Elementary Differential Equations and Boundary-Value Problems*, [5], and Edwards and Penney's *Differential Equations and Boundary Value Problems: Computing and Modeling*, [10] for elementary but precise discussions.

Application: The Logistic Difference Equation

Given x_0, the **Logistic difference equation** is

$$x_{n+1} = rx_n(1 - x_n). \tag{3.8}$$

Assume that $x_0 = 0.5$.

Given r, we use Mathematica to define the function $x_r(n)$ using the form

```
x[r_][n_]:=x[r][n]=...
```

so that Mathematica "remembers" the values of $x_r(n)$ computed. In doing so, Mathematica need not recompute $x_r(n-1)$ to compute $x_r(n)$ if $x_r(n-1)$ has already been computed.

```
Clear[x]
x[r_][0] = 0.5;
x[r_][n_] := x[r][n] = rx[r][n - 1](1 - x[r][n - 1]);
```

For example,

```
t4 = Table[x[3.83][n], {n, 1, 50}]
```

```
{0.9575, 0.155857, 0.503896, 0.957442, 0.156061, 0.504433,

0.957425, 0.156121, 0.504592, 0.957419, 0.15614, 0.504642,

0.957417, 0.156146, 0.504658, 0.957417, 0.156148, 0.504664,

0.957417, 0.156149, 0.504666, 0.957417, 0.156149, 0.504666,

0.957417, 0.156149, 0.504666, 0.957417, 0.156149, 0.504666,

0.957417, 0.156149, 0.504666, 0.957417, 0.156149, 0.504666,

0.957417, 0.156149, 0.504666, 0.957417, 0.156149, 0.504666,

0.957417, 0.156149, 0.504666, 0.957417, 0.156149, 0.504666,

0.957417, 0.156149}
```

computes x_n for $n = 1, 2, \ldots, 50$ if $r = 3.83$ in equation (3.8) and

```
p1 = ListLinePlot[t4, PlotLabel → "(a)", AxesLabel → {n, x}];
```

plots the resulting list and connects successive points with line segments as shown in Figure 3-17(a). Use `ListLinePlot` to graph points that are successively connected with line segments and use `ListPlot` to graph lists of points.

To investigate how the behavior of equation (3.8) changes as r changes, we enter

```
t1 = Table[{r, x[r][n]}, {r, 2.8, 4, 1.2/249}, {n, 101, 300}];
```

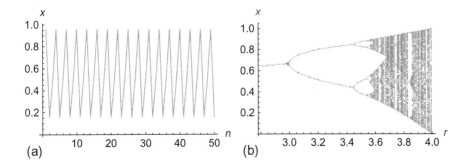

(a)

(b)

Figure 3-17 (a) A 3-cycle. (b) The "Pitchfork Diagram"

which computes a nested list: for 250 equally spaced values of r between 2.8 and 4.0, the list consisting of $(r, x_r(n))$ for $n = 101, \dots, 300$ is returned. The nested list is converted to a list of ordered pairs with `Flatten`

```
toshow = Flatten[t1, 1];
```

and then plotted with `ListPlot` in Figure 3-17(b).

```
p2 = ListPlot[t1, AxesLabel → {r, x}, PlotLabel → "(b)"];
```

```
Show[GraphicsRow[{p1, p2}]]
```

However, if you immediately request `x[r][M]` for a large value of `M` without computing `x[r][n]` for `n` less than `M`, Mathematica has a difficult time working backwards. We finally abort the calculation.

```
x[3.9][1000]
```

```
0.835629
```

```
x[3.9][10000]
```

```
Hold[3.9x[3.9][8978 − 1](1 − x[3.9][8978 − 1])]
```

When using a recursive definition like the one illustrated and to be on the safe side, a calculation like this should be carried out after the first 9999 terms are computed. For situations like this, we prefer using `Nest`. For repeated compositions of a function with itself, `Nest[f,x,n]` computes the composition

$$\underbrace{(f \circ f \circ f \circ \cdots f)}_{n \text{ times}}(x) = \underbrace{(f\,(f\,(f\,\cdots)))}_{n \text{ times}}(x) = f^n(x).$$

In terms of a composition, computing x_n in equation (3.8) is equivalent to composing $f(x) = rx(1 - x)$ with itself n times.

```
Clear[f, x]
```

```
f[r_][x_] = rx(1 − x)//N;
```

Thus, entering

```
Table[Nest[f[2.5], 0.5, n], {n, 1, 50}]
```

```
{0.625, 0.585938, 0.606537, 0.596625, 0.601659, 0.599164,
0.600416, 0.599791, 0.600104, 0.599948, 0.600026, 0.599987,
0.600007, 0.599997, 0.600002, 0.599999, 0.6, 0.6, 0.6, 0.6, 0.6,
```

```
0.6, 0.6, 0.6, 0.6, 0.6, 0.6, 0.6, 0.6, 0.6, 0.6, 0.6, 0.6,

0.6, 0.6, 0.6, 0.6, 0.6, 0.6, 0.6, 0.6, 0.6, 0.6, 0.6, 0.6,

0.6, 0.6, 0.6, 0.6, 0.6}
```

computes x_n for $n = 1, 2, \ldots, 50$ if $r = 2.5$ while entering

```
Table[Nest[f[3.5], .5, n], {n, 1, 50}]
```

```
{0.875, 0.382813, 0.826935, 0.500898, 0.874997, 0.38282, 0.826941,

0.500884, 0.874997, 0.38282, 0.826941, 0.500884, 0.874997, 0.38282,

0.826941, 0.500884, 0.874997, 0.38282, 0.826941, 0.500884, 0.874997,

0.38282, 0.826941, 0.500884, 0.874997, 0.38282, 0.826941, 0.500884,

0.874997, 0.38282, 0.826941, 0.500884, 0.874997, 0.38282, 0.826941,

0.500884, 0.874997, 0.38282, 0.826941, 0.500884, 0.874997, 0.38282,

0.826941, 0.500884, 0.874997, 0.38282, 0.826941, 0.500884,

0.874997, 0.38282}
```

computes x_n for $n = 1, 2, \ldots, 50$ if $r = 3.5$.

You can see how solutions depend on r in a variety of ways. For example, entering the following command computes x_n for $n = 1000, \ldots 1050$ for 9 equally spaced values of r between 3.2 and 3.9.

```
t1 = Table[{n, Nest[f[r], 0.5, n]},

   {r, 3.2, 3.9, .7/8}, {n, 1000, 1050}];
```

Each list is plotted with `ListLinePlot` so that successive points are connected with line segments.

```
t1b = Map[ListLinePlot[#, PlotRange → {0, 1}]&, t1];
```

All nine graphs are shown together as an array using `Show` and `GraphicsGrid` in Figure 3-18.

```
Show[GraphicsGrid[Partition[t1b, 3]]]
```

We regenerate the "Pitchfork diagram" with the following commands in Figure 3-19(a).

```
t1 = Table[{r, Nest[f[r], .5, n]},

   {r, 2.8, 4, 1.2/249}, {n, 101, 300}];
```

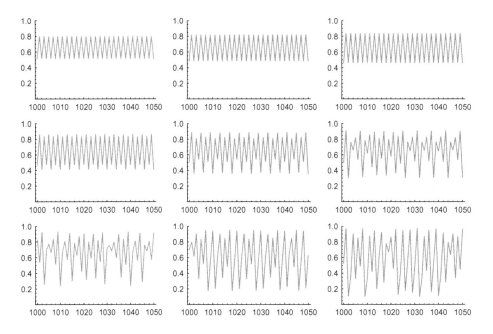

Figure 3-18   As *r* increases, we see *chaos*

```
toshow = Flatten[t1, 1];

p1 = ListPlot[toshow, AxesLabel → {r, x},

    PlotStyle → PointSize[0.004], PlotLabel → "(a)"];
```

You can zoom in on areas of interest, as well. In Figure 3-19(b), we restrict *r* to $3.7 \le r \le 4.0$.

```
t3 = Table[{r, Nest[f[r], .5, n]},

    {r, 3.7, 4, 0.4/349}, {n, 101, 300}];

toshow = Flatten[t3, 1];

p2 = ListPlot[toshow, AxesLabel → {r, x},

    PlotRange → {{3.7, 4.}, {0, 1}},

    PlotStyle → PointSize[0.004], PlotLabel → "(b)"];

Show[GraphicsRow[{p1, p2}]]
```

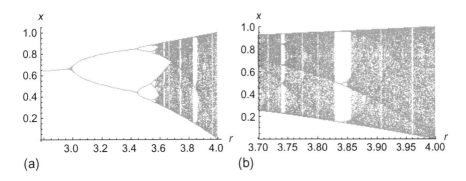

x

Figure 3-19   (a) A copy of the "Pitchfork diagram." (b) The "Pitchfork diagram" for $3.7 \le r \le 4.0$

3.3 Newton's Law of Cooling

First-order linear differential equations can be used to solve a variety of problems that involve temperature. For example, a medical examiner can find the time of death in a homicide case, a chemist can determine the time required for a plastic mixture to cool to a hardening temperature, and an engineer can design the cooling and heating system of a manufacturing facility. Although distinct, each of these problems depend on a basic principle, *Newton's Law of Cooling*, that is used to develop the associated differential equation.

> **Newton's Law of Cooling:** The rate at which the temperature $T(t)$ changes in acooling body is proportional to the difference between the temperature of the body and the constant temperature T_s of the surrounding medium.

Newton's Law of Cooling is modeled with the first-order initial-value problem

$$
\begin{cases}
\dfrac{dT}{dt} = k\,(T - T_s) \\[2mm]
T(0) = T_0
\end{cases}
, \tag{3.9}
$$

Equation (3.9) is also linear.
See Example 2.5.5 where we
solved equation (3.9) by
viewing it as a linear
equation.

where T_0 is the initial temperature of the body and k is the constant of proportionality. If T_s is constant, equation (3.9) is separable and separating variables gives us

$$\frac{1}{T - T_s} dT = k \, dt \quad \text{so} \quad \ln |T - T_s| = kt + C_1.$$

Using the properties of the natural logarithm and simplifying yields $T(t) = Ce^{kt} + T_s$, where $C = \pm e^{C_1}$. Applying the initial condition implies that $T_0 = C + T_s$, so $C = T_0 - T_s$. Therefore, the solution of equation (3.9) is

$$T(t) = (T_0 - T_s) e^{kt} + T_s. \tag{3.10}$$

```
Clear[t, capt]

del = DSolve[{capt'[t]==k(capt[t] - capts), capt[0] == t0}, capt[t], t]
```

$$\{\{\text{capt}[t] \to \text{capts} - \text{capts}e^{kt} + e^{kt}t0\}\}$$

Recall that if $k < 0$, $\lim_{t \to \infty} e^{kt} = 0$. Therefore, $\lim_{t \to \infty} T(t) = T_s$, so the temperature of the body approaches that of its surroundings.

EXAMPLE 3.3.1: A pie is removed from a 350°F oven and placed to cool in a room with temperature 75°F. In 15 minutes, the pie has a temperature of 150°F. Determine the time required to cool the pie to a temperature of 80°F so that it may be eaten.

SOLUTION: In this example, $T_0 = 350$ and $T_s = 75$. Substituting these values into equation (3.10), we obtain $T(t) = (350 - 75)e^{kt} + 75 = 2755e^{kt} + 75$.

```
step1 = capts + E^{kt}(-capts + t0)/.{t0 → 350, capts → 75}
```

$$75 + 275e^{kt}$$

To solve the problem we must find k or e^k. Because we also know that $T(15) = 150$, $T(15) = 275e^{15k} = 75$. Solving this equation for k or e^k gives us:

$$275e^{kt} = 75$$

$$e^{15k} = \frac{3}{11}$$

$$275e^{kt} = 75$$

$$\ln\left(e^{15k}\right) = \ln\left(\frac{3}{11}\right)$$

$$e^{15k} = \frac{3}{11}$$

$$15k = \ln\left(\frac{3}{11}\right) \quad \text{or} \quad e^{k} = \left(\frac{3}{11}\right)^{1/15}$$

$$k = \frac{\ln\left(\frac{3}{11}\right)}{15} \qquad e^{k} = \left(\frac{11}{3}\right)^{-1/15}.$$

$$k = -\frac{\ln\left(\frac{11}{3}\right)}{15}$$

Thus, $T(t) = 275e^{-t\,\ln(11/3)/15} + 75 = 275\left(\frac{11}{3}\right)^{-t/15} + 75.$

```
step2 = step1/.k → 1/15 Log[3/11]
```

$$75 + 25 \; 3^{t/15} 11^{1-\frac{t}{15}}$$

To find the value of t for which $T(t) = 80$, we solve the equation $275\left(\frac{11}{3}\right)^{-t/15} + 75 = 80$ for t:

$$275\left(\frac{11}{3}\right)^{-t/15} = 5$$

$$\left(\frac{11}{3}\right)^{-t/15} = \frac{1}{55}$$

$$\ln\left(\frac{11}{3}\right)^{-t/15} = \ln\left(\frac{1}{55}\right) = -\ln 55$$

$$-\frac{t}{15}\ln\left(\frac{11}{3}\right) = -\ln 55$$

$$t = \frac{15\,\ln 55}{\ln\left(\frac{11}{3}\right)} \approx 46.264.$$

Alternatively, we can graph the solution together with the line $y = 80$ as shown in Figure 3-20

```
Plot[{step2, 80}, {t, 0, 90}, PlotStyle → {GrayLevel[0], Dashing

[{0.01}]}, PlotRange → {0, 350}]
```

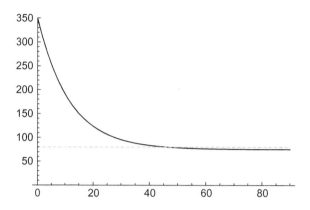

Figure 3-20   The value of t where $y = T(t)$ and $y = 80$ is the solution to the problem

and then use `FindRoot` to approximate the time at which the temperature of the pie reaches 80°F.

```
FindRoot[step2==80,{t, 45}]
```

$\{t \to 46.264\}$

Thus, the pie will be ready to eat after approximately 46 minutes.

An interesting question associated with cooling problems is to determine if the pie reaches room temperature. From the formula, $T(t) = 275 \left(\dfrac{11}{3}\right)^{-t/15} + 75$, we see that the component $275 \left(\dfrac{11}{3}\right)^{-t/15} > 0$, so $T(t) = 275 \left(\dfrac{11}{3}\right)^{-t/15} + 75 > 75$. Therefore, the pie never actually reaches room temperature according to our model. However, we see from the graph and from the values in the following table that its temperature approaches 75°F as t increases.

```
Table[{t, step2//N}, {t, 60, 100, 10}]//TableForm
```

60	76.5214
70	75.6398
80	75.2691
90	75.1132
100	75.0476

■

If the temperature of the surroundings, T_s, is not constant the situation may be more complicated. For example, consider the problem of heating and cooling a building. Over the span of a twenty-four hour day, the outside temperature, T_s, varies so the problem of determining the temperature inside the building becomes more complicated. Assuming that the building has no heating or air conditioning system, the differential equation that needs to be solved to find the temperature $u(t)$ at time t inside the building is

$$\frac{du}{dt} = k \left(C(t) - u(t) \right), \tag{3.11}$$

where $C(t)$ is a function that describes the outside temperature and $k > 0$ is a constant that depends on the insulation of the building. According to this equation, if $C(t) > u(t)$, then $du/dt > 0$, which implies that u increases. On the other hand, if $C(t) < u(t)$, then $du/dt < 0$ which means that u decreases.

EXAMPLE 3.3.2: (a) Suppose that during the month of April in Atlanta, Georgia, the outside temperature in degrees F is given by $C(t) = 70 - 10\cos(\pi t/12)$, $0 \leq t \leq 24$. Determine the temperature in a building that has an initial temperature of $60°F$ if $k = 1/4$. (b) Compare this to the temperature in June when the outside temperature is $C(t) = 80 - 10\cos(\pi t/12)$ and the initial temperature is $70°F$.

The first choice of $C(t)$ has average value of $70°$F; the second choice has an average value of $80°$F.

SOLUTION: (a) The initial-value problem that we must solve is (b)

$$\begin{cases} \dfrac{du}{dt} = k \left[70 - 10\cos\left(\dfrac{\pi}{12}t\right) - u \right] \\ u(0) = 60 \end{cases}.$$

The differential equation can be solved if we write it as $du/dt + ku = k\left[70 - 10\cos(\pi t/12) \right]$ and then use an integrating factor. This gives us

$$\frac{d}{dt}\left(e^{kt}u \right) = ke^{kt}\left[70 - 10\cos(\pi t/12) \right],$$

so we must integrate both sides of the equation. Of course, solving the equation is most easily carried out through the use of DSolve.

```
sol1 = DSolve[{u'[t]==1/4 (70 - 10Cos[πt/12] - u[t]) , u[0]==60},
   u[t], t]//Simplify
```

$$\left\{\left\{ u[t] \rightarrow \frac{10\left(63 + 7\pi^2 - e^{-t/4}\pi^2 - 9\mathrm{Cos}\left[\frac{\pi t}{12}\right] - 3\pi\mathrm{Sin}\left[\frac{\pi t}{12}\right]\right)}{9 + \pi^2} \right\}\right\}$$

We then use Plot to graph the solution for $0 \le t \le 24$ in Figure 3-21.

p1 = Plot[u[t]/.sol1, {t, 0, 24}, PlotLabel → "(a)",

    AxesLabel → {t, u}];

Note that the temperature reaches its maximum (approximately 77^0) near $t \approx 15.5$ hours which corresponds to 3:30 pm A more accurate estimate is obtained with FindRoot by setting the first derivative of the solution equal to zero and solving for t.

    FindRoot$\left[\text{Evaluate}\left[\partial_t(\text{sol1[[1, 1, 2]])==0}\right], \{t, 15\}\right]$

    $\{t \rightarrow 15.1506\}$

(b) This problem is solved in the same manner as the previous case.

sol2 = DSolve$\left[\left\{u'[t]==\frac{1}{4}\left(80 - 10\text{Cos}\left[\frac{\pi t}{12}\right] - u[t]\right), u[0]==70\right\}, \right.$
$u[t], t\Big]$//Simplify

$$\left\{\left\{ u[t] \rightarrow \frac{10\left(72 + 8\pi^2 - e^{-t/4}\pi^2 - 9\mathrm{Cos}\left[\frac{\pi t}{12}\right] - 3\pi\mathrm{Sin}\left[\frac{\pi t}{12}\right]\right)}{9 + \pi^2} \right\}\right\}$$

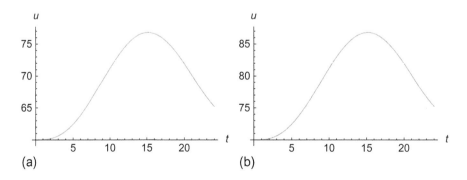

Figure 3-21   (a) The temperature in a hypothetical building over a period of 24 hours. (b) The plot is almost identical to the plot obtained in (a) except with higher associated temperatures

The solution is graphed with `Plot` in Figure 3-21(b). From the graph, we see that the maximum temperature appears to occur near $t \approx 15$ hours.

```
p2 = Plot[u[t]/.sol2,{t, 0, 24}, PlotLabel → "(b)",
    AxesLabel → {t, u}];
```

Again, a more accurate value is obtained with `FindRoot` by setting the first derivative of the solution equal to zero and solving for t. This calculation yields 15.15 hours, the same as that in (a).

```
FindRoot[Evaluate[∂ₜ(sol2[[1, 1, 2]])==0], {t, 15}]
```

```
{t → 15.1506}
```

```
Show[GraphicsRow[{p1, p2}]]
```

■

3.4 Free-Falling Bodies

The motion of objects can be determined through the solution of first-order initial-value problems. We begin by explaining some of the theory that is needed to set up the differential equation that models the situation.

> **Newton's Second Law of Motion:** The rate at which the momentum of a body changes with respect to time is equal to the resultant force acting on the body.

Because the body's momentum is defined as the product of its mass and velocity, this statement is modeled as

$$\frac{d}{dt}(mv) = F,$$

where m and v represent the body's mass and velocity, respectively, and F is the sum of the forces (the resultant force) acting on the body. Because m is constant, differentiation leads to the well-known equation

$$m\frac{dv}{dt} = F.$$

If the body is subjected only to the force due to gravity, then its velocity is determined by solving the differential equation

$$m\frac{dv}{dt} = mg \quad \text{or} \quad \frac{dv}{dt} = g,$$

where $g = 32 \, \text{ft/s}^2$ (English system) and $g = 9.8 \, \text{m/s}^2$ (international system). This differential equation is applicable only when the resistive force due to the medium (such as air resistance) is ignored. If this offsetting resistance is considered, we must discuss all of the forces acting on the object. Mathematically, we write the equation as

$$m\frac{dv}{dt} = \sum \left(\text{forces acting on the object}\right),$$

where the direction of motion is taken to be the positive direction. Because air resistance acts against the object as it falls and g acts in the same direction of the motion, we state the differential equation in the form

$$m\frac{dv}{dt} = mg + (-F_R) \quad \text{or} \quad m\frac{dv}{dt} = mg - F_R,$$

where F_R represents this resistive force. Note that down is assumed to be the positive direction. The resistive force is typically proportional to the body's velocity, v, or the square of its velocity, v^2. Hence, the differential equation is linear or nonlinear based on the resistance of the medium taken into account.

EXAMPLE 3.4.1: Determine the velocity and displacement functions of an object with $m = 1$, where 1 slug $= \text{lb} \, \text{s}^2/\text{ft}$, that is thrown downward with an initial velocity of 2 ft/s from a height of 1000 feet. Assume that the object is subjected to air resistance that is equivalent to the instantaneous velocity of the object. Also, determine the time at which the object strikes the ground and its velocity when it strikes the ground.

SOLUTION: First, we set up the initial-value problem to determine the velocity of the object. Because the air resistance is equivalent to the instantaneous velocity, we have $F_R = v$. The formula $m \, dv/dt = mg - F_R$

then gives us $dv/dt = 32 - v$. Of course, we must impose the initial velocity $v(0) = 2$. Therefore, the initial-value problem is

$$\begin{cases} dv/dt = 32 - v \\ v(0) = 2 \end{cases},$$

which is both separable and first-order linear. We solve it as a linear first-order equation and so we multiply both sides of the equation by the integrating factor e^t, which results in $d/dt\,(e^t v) = 32t$. Integrating both sides gives us $e^t v = 32 e^t + C$, so $v = 32 + Ce^{-t}$. Applying the initial velocity, we have $v(0) = 32 + C = 0$. Therefore, the velocity of the object is $v = 32 - 30e^{-t}$. We obtain the same result with DSolve, naming the resulting output step1.

```
Clear[v, t]

step1 = DSolve[{v'[t]==32 - v[t], v[0]==2} , v[t], t]
```

$$\{\{v[t] \rightarrow 2e^{-t}(-15 + 16e^t)\}\}$$

To determine the position, or distance traveled at time t, $s(t)$, we solve the first-order equation $ds/dt = 32 - e^{-t}$ with initial displacement $s(0) = 0$. Notice that we use the initial displacement as a reference and let $s = s(t)$ represent the distance traveled from this reference point.

```
step2 = DSolve[{s'[t]==32 - 30E^-t, s[0]==0} , s[t], t]
```

$$\{\{s[t] \rightarrow 2e^{-t}(15 - 15e^t + 16e^t t)\}\}$$

Thus, the displacement of the object at time t is given by $s = 32t + 30e^{-t} - 30$.

Because we are taking $s(0) = 0$ as our starting point, the object strikes the ground when $s(t) = 1000$. Therefore, we must solve $s = 32t + 30e^{-t} - 30 = 1000$. The roots of this equation can be approximated with FindRoot. We begin by graphing the function $s = s(t)$ and the line $s = 1000$ with Plot in Figure 3-22.

```
Plot[{-30 + 30E^-t + 32t, 1000} , {t, 0, 70}, PlotStyle →

{GrayLevel[0], GrayLevel[0.5]}]
```

From the graph of this function, we see that $s(t) = 1000$ near $t \approx 35$. To obtain a better approximation, we use FindRoot.

```
t00 = FindRoot[-30 + 30E^-t + 32t==1000, {t, 35}]
```

$$\{t \rightarrow 32.1875\}$$

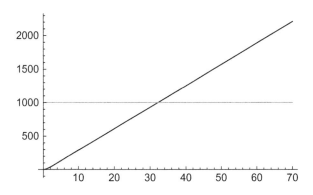

Figure 3-22   Plots of $s = s(t)$ and the line $s = 1000$

so the object strikes the ground after approximately 32.1875 seconds.

The velocity at the point of impact is found to be 32.0 ft/s by evaluating the derivative, $s'(t) = v(t) = 32 - 30e^{-t}$, at the time at which the object strikes the ground, $t \approx 32.1875$.

$32 - 30E^{-t}/.(t00[[1]])$

$32.$

■

EXAMPLE 3.4.2: Determine a solution (for the velocity and the displacement) of the differential equation that models the motion of an object of mass m when directed upward with an initial velocity of v_0 from an initial displacement y_0 assuming that the air resistance equals cv, where c is constant.

SOLUTION: Because the motion of the object is upward, mg and F_R act against the upward motion of the object; mg and F_R are in the negative direction. Therefore, the initial-value problem that must be solved in this case is the linear problem,

$$\begin{cases} \dfrac{dv}{dv} = -g - \dfrac{c}{m}v \\ v(0) = v_0 \end{cases},$$

which we solve with DSolve, naming the resulting output sol.

```
Clear[v, t, s]
```

$$sol = DSolve\left[\left\{v'[t] == -g - \frac{cv[t]}{m}, v[0] == v0\right\}, v[t], t\right]$$

$$\left\{\left\{v[t] \rightarrow -\frac{e^{-\frac{ct}{m}}\left(-gm + e^{\frac{ct}{m}}gm - cv0\right)}{c}\right\}\right\}$$

```
Clear[v, t, s]
```

$$sol = DSolve\left[\left\{v'[t] == -g - \frac{cv[t]}{m}, v[0] == v0\right\}, v[t], t\right]$$

$$\left\{\left\{v[t] \rightarrow -\frac{e^{-\frac{ct}{m}}\left(-gm + e^{\frac{ct}{m}}gm - cv0\right)}{c}\right\}\right\}$$

Next, we use sol to define velocity. This function can be used to investigate numerous situations without re-solving the differential equation each time.

$$velocity[m\_, c\_, g\_, v0\_, t\_] = \frac{-gm + cE^{-\frac{ct}{m}}\left(\frac{gm}{c} + v0\right)}{c};$$

For example, the velocity function for the case with $m = 128$ slugs, $c = 1/160$, $g = 32$ ft/s^2, and $v_0 = 48$ ft/s is $v(t) = 88e^{-4t/5} - 40$.

$$velocity\left[\frac{1}{128}, \frac{1}{160}, 32, 48, t\right] //Expand$$

$$-40 + 88e^{-4t/5}$$

The displacement function $s(t)$ that represents the distance above the ground at time t is determined by integrating the velocity function. This is accomplished here with DSolve using the initial displacement y_0. As with the previous case, the output is named pos so that the displacement formula may be used to define the function position.

$$pos = DSolve\left[\left\{y'[t] == velocity[m, c, g, v0, t], y[0] == y0\right\}, y[t], t\right]$$

$$
\left\{\left\{\left\{ y[t] \rightarrow \frac{e^{-\frac{ct}{m}}\left(-gm^2 + e^{\frac{ct}{m}}gm^2 - ce^{\frac{ct}{m}}gmt - cmv0 + ce^{\frac{ct}{m}}mv0 + c^2e^{\frac{ct}{m}}y0\right)}{c^2} \right\}\right\}\right\}
$$

```
position[m_, c_, g_, v0_, y0_, t_] =
```
$$
-\frac{1}{c^2}\left(E^{-\frac{ct}{m}}gm^2 + cgmt + cE^{-\frac{ct}{m}}mv0 - c^2\left(\frac{gm^2}{c^2} + \frac{mv0}{c} + y0\right)\right);
$$

The displacement and velocity functions are plotted in the following
using the parameters $m = 128$ slugs , $c = 1/160$, $g = 32\,\text{ft/s}^2$, and $v_0 = 48\,\text{ft/s}$ as well as $y_0 = 0$.

The time at which the object reaches its maximum height occurs
when the derivative of the displacement is equal to zero. From the
graph in Figure 3-23(a) we see that $s'(t) = v(t) = 0$ when $t \approx 1$.

```
p1 = Plot[{velocity[ 1/128 , 1/160 , 32, 48, t],
    position[ 1/128 , 1/160 , 32, 48, 0, t]}, {t,0,2},
```

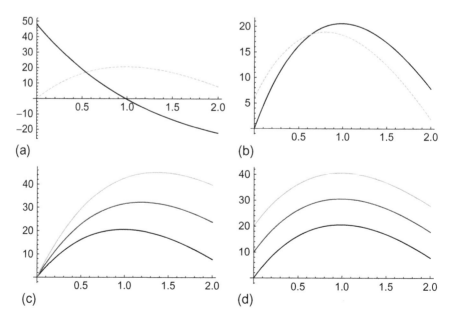

(a)

(b)

(c)

(d)

Figure 3-23   (a) The maximum height of the object occurs when its velocity is 0. (b) Varying
v_0 and y_0. (c) Varying v_0. (d) Varying y_0

```
PlotStyle→{GrayLevel[0], Dashing[{0.01}]},
PlotLabel → "(a)"];
```

A more accurate approximation, $t \approx 0.985572$, is obtained using `Solve` together with `N`

$$\texttt{root = N}\left[\texttt{Solve}\left[\partial_t\texttt{position}\left[\frac{1}{128}, \frac{1}{160}, 32, 48, 0, t\right]\texttt{==0}, t\right]\right]$$

```
{{t   →   ConditionalExpression[1.25(0.788457 + (0. +
6.28319i)C[1]), C[1] ∈ Integers]}}
```

with `FindRoot`.

$$\texttt{FindRoot}\left[\texttt{velocity}\left[\frac{1}{128}, \frac{1}{160}, 32, 48, t\right]\texttt{==0}, \{t, 1\}\right]$$

```
{t → 0.985572}
```

We now compare the effect that varying the initial velocity and displacement has on the displacement function. Suppose that we use the same values used earlier for m, c, and g. However, we let $v_0 = 48$ in one function and $v_0 = 36$ in the other. We also let $y_0 = 0$ and $y_0 = 6$ in these two functions, respectively. See Figure 3-23(b).

$$\texttt{p2 = Plot}\left[\left\{\texttt{position}\left[\frac{1}{128}, \frac{1}{160}, 32, 48, 0, t\right],\right.\right.$$
$$\left.\left.\texttt{position}\left[\frac{1}{128}, \frac{1}{160}, 32, 36, 6, t\right]\right\}, \{t, 0, 2\}, \texttt{PlotStyle} \rightarrow\right.$$
```
{GrayLevel[0], Dashing[{0.01}]}, PlotLabel → "(b)"];
```

Figure 3-23(c) demonstrates the effect that varying the initial velocity only has on the displacement function. The values of v_0 used are 48, 64, and 80. The darkest curve corresponds to $v_0 = 48$. Notice that as the initial velocity is increased the maximum height attained by the object is increased as well.

$$\texttt{p3 = Plot}\left[\left\{\texttt{position}\left[\frac{1}{128}, \frac{1}{160}, 32, 48, 0, t\right], \texttt{position}\right.\right.$$
$$\left.\left[\frac{1}{128}, \frac{1}{160}, 32, 64, 0, t\right], \texttt{position}\left[\frac{1}{128}, \frac{1}{160}, 32, 80, 0, t\right]\right\},$$
```
{t, 0, 2},  PlotStyle → {GrayLevel[0],  GrayLevel[0.3],
GrayLevel[0.6]},  PlotLabel → "(c)"];
```

Figure 3-23 indicates the effect that varying the initial displacement and holding all other values constant has on the displacement function. We use values of 0, 10, and 20 for y_0. Notice that the value of the initial displacement vertically translates the displacement function.

```
p4 = Plot[{position[1/128, 1/160, 32, 48, 0, t],
    position[1/128, 1/160, 32, 48, 10, t],
    position[1/128, 1/160, 32, 48, 20, t]}, {t, 0, 2},
    PlotStyle → {GrayLevel[0], GrayLevel[0.3], GrayLevel[0.6]},
    PlotLabel → "(d)"];

Show[GraphicsGrid[{{p1, p2}, {p3, p4}}]]
```

■

We now combine several of the topics discussed in this section to solve the following problem.

EXAMPLE 3.4.3: An object of mass $m = 1$ slug is dropped from a height of 50 feet above the surface of a small pond. While the object is in the air, the force due to air resistance is v. However, when the object is in the pond, it is subjected to a buoyancy force equivalent to $6v$. Determine how much time is required for the object to reach a depth of 25 feet in the pond.

SOLUTION: This problem must be broken into two parts: an initial-value problem for the object above the pond, and an initial-value problem for the object below the surface of the pond. The initial-value problem above the pond's surface is found to be

$$\begin{cases} dv/dt = 32 - v \\ v(0) = 0 \end{cases}.$$

However, to define the initial-value problem to find the velocity of the object beneath the ponds surface, the velocity of the object when it reaches the surface must be known. Hence, the velocity of the object above the surface must be determined by solving the initial-value problem above. The equation $dv/dt = 32 - v$ is separable and solved with DSolve in step1.

```
Clear[s, y, t, v]

step1 = DSolve[{v'[t]==32 - v[t], v[0]==0}, v[t], t]
```

$$\{\{v[t] \rightarrow 32e^{-t}(-1 + e^t)\}\}$$

In order to find the velocity when the object hits the pond's surface we must know the time at which the distance traveled by the object (or the displacement of the object) is 50. Thus, we must find the displacement function which is done by integrating the velocity function obtaining $s(t) = 32e^{-t} + 32t - 32$.

```
step2 = DSolve[{y'[t]==32 - 32E^-t, y[0]==0}, y[t], t]
```

$$\{\{y[t] \to 32e^{-t}(1 - e^t + e^t t)\}\}$$

The displacement function is graphed with `Plot` in Figure 3-24(a). The value of t at which the object has traveled 50 feet is needed. This time appears to be approximately 2.5 seconds.

```
p1=Plot[{step2[[1,1,2]], 50}, {t,0, 5}, PlotStyle→

    {GrayLevel[0], Dashing[{0.01}]}, PlotLabel → "(a)"];
```

A more accurate value of the time at which the object hits the surface is found using `FindRoot`. In this case, we obtain $t \approx 2.47864$. The velocity at this time is then determined by substitution into the velocity function resulting in $v(2.47864) \approx 29.3166$. Note that this value is the initial velocity of the object when it hits the surface of the pond.

```
t1 = FindRoot[-32 + 32E^-t + 32t==50, {t, 2.5}]
```

$$\{t \to 2.47864\}$$

```
v1 = 32 - 32E^-t/.(t1[[1]])
```

```
29.3166
```

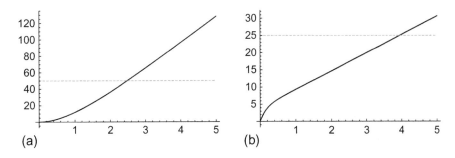

(a)   (b)

Figure 3-24   (a) The object has traveled 50 feet when $t \approx 2.5$. (b) After approximately 4 seconds, the object is 25 feet below the surface of the pond

Thus, the initial-value problem that determines the velocity of the object beneath the surface of the pond is given by

$$\begin{cases} dv/dt = 32 - 6v \\ v(0) = 29.3166 \end{cases}.$$

The solution of this initial-value problem is $v(t) = \frac{16}{3} + 23.9833e^{-t}$ and integrating to obtain the displacement function (the initial displacement is 0) we obtain $s(t) = 3.99722 - 3.99722e^{-6t} + \frac{16}{3}t$. These steps are carried out in step3 and step4.

```
step3 = DSolve [{v'[t]==32-6v[t], v[0]==v1} , v[t], t] //
Simplify
```

$$\{\{v[t] \rightarrow 5.33333 + 23.9832e^{-6t}\}\}$$

```
step4 = DSolve [{y'[t] == step3[[1, 1, 2]], y[0]==0} , y[t], t]
```

$$\{\{y[t] \rightarrow 5.33333e^{-6\cdot t}(-0.749476 + 0.749476e^{6\cdot t} + 1.e^{6\cdot t}t)\}\}$$

This displacement function is then plotted in Figure 3-24(b) to determine when the object is 25 feet beneath the surface of the pond. This time appears to be near 4 seconds.

```
p2 = Plot[{y[t]/.step4, 25}, {t, 0, 5}, PlotStyle →
{GrayLevel[0], Dashing[{0.01}]}, PlotLabel → "(b)"];

Show[GraphicsRow[{p1, p2}]]
```

A more accurate approximation of the time at which the object is 25 feet beneath the pond's surface is obtained with FindRoot. In this case, we obtain $t \approx 3.93802$. Finally, the time required for the object to reach the pond's surface is added to the time needed for it to travel 25 feet beneath the surface to see that approximately 6.41667 seconds are required for the object to travel from a height of 50 feet above the pond to a depth of 25 feet below the surface.

```
t2 = FindRoot[(step4[[1, 1, 2]])==25, {t, 4}]
```

$$\{t \rightarrow 3.93802\}$$

```
t1[[1, 2]] + t2[[1, 2]]
```

```
6.41667
```

∎

Higher-Order Differential Equations

$$4$$

In Chapters 2 and 3 we saw that first-order differential equations can be used to model a variety of physical situations. However, many physical situations need to be modeled by higher-order differential equations. In this chapter, we discuss several methods for solving higher-order linear differential equations.

4.1 Preliminary Definitions and Notation

4.1.1 Introduction

In the same way as in previous chapters, we can frequently use `DSolve` to generate exact solutions of higher-order equations and `NDSolve` to generate numerical solutions to higher-order initial-value problems.

EXAMPLE 4.1.1 (Van-der-Pol Equation): The **Van-der-Pol equation**, which arises in the study of nonlinear damping, is the nonlinear second-order equation

$$\frac{d^2x}{dt^2} + \mu\left(x^2 - 1\right)\frac{dx}{dt} + x = 0. \tag{4.1}$$

(a) If $x(0) = 1$ and $x'(0) = 0$, graph the solution to the Van-der-Pol equation (4.1) on the interval $[0, 15]$ for $\mu = 1/32, 1/16, 1/8, 1/4, 1/2, 1,$

Differential Equations with Mathematica. http://dx.doi.org/10.1016/B978-0-12-804776-7.00004-8

3/2, 2, 3, 5, 7, and 9. (b) Compare the graphs of these solutions to the graph of the solution to the initial-value problem

$$\begin{cases} x'' + x = 0 \\ x(0) = 1,\ x'(0) = 0 \end{cases}.$$

SOLUTION: We begin by defining the function `vanderpol`. Given μ, `vanderpol[`μ`]` numerically solves the initial-value problem

$$\begin{cases} x'' + \mu \left(x^2 - 1 \right) x' + x = 0 \\ x(0) = 1,\ x'(0) = 0 \end{cases}. \tag{4.2}$$

```
vanderpol[μ_]:=

    NDSolve[{x"[t] + μ(x[t]^2 − 1)x'[t] + x[t] == 0,

    x[0] == 1, x'[0] == 0}, x[t], {t, 0, 15}]
```

For example, entering

```
numsol1 = vanderpol[1]
```

$\{\{x[t] \to \text{InterpolatingFunction}[][t]\}\}$

```
p1 = Plot[x[t]/.numsol1, {t, 0, 15}, PlotRange → {−3.5, 3.5},

    AspectRatio → 1, PlotStyle → Black, AxesStyle → Black,

    AxesLabel → {t, x}, PlotLabel → "(a)"]
```

returns a numerical solution to the initial-value problem (4.2) if $\mu = 1$ and then graphs the result on $[0, 15]$, as shown in Figure 4-1(a). Entering

```
numsol1/.t → 1
```

$\{\{x[1] \to 0.497616\}\}$

shows us that if $\mu = 1$, $x(1) \approx 0.497615$ and entering

```
p2 = Plot[Evaluate[D[numsol1[[1, 1, 2]], t]], {t, 0, 15},

    PlotRange → {−3.5, 3.5},

    AspectRatio → 1, PlotStyle → Black, AxesStyle → Black,

    AxesLabel → {t, x}, PlotLabel → "(b)"]

Show[GraphicsRow[{p1, p2}]]
```

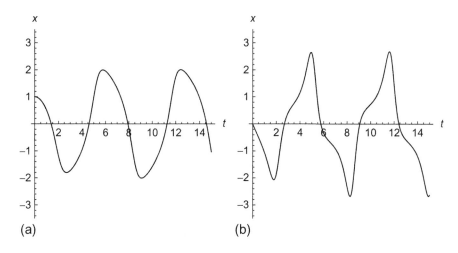

Figure 4-1 (a) Plot of $x(t)$ if $\mu = 1$. (b) Plot of $x'(t)$ if $\mu = 1$

graphs the derivative of the numerical solution on the interval $[0, 15]$ as shown in Figure 4-1(b).

Because we will be graphing the solution for many values of μ, we now define the function solgraph. Given μ, solgraph$[\mu]$ graphs the solution to the initial-value problem (4.2) on $[0, 15]$. Note that the resulting graph is not displayed because the option DisplayFunction->Identity is included in the Plot command. Any options included in the solgraph command are passed to the Plot command but are "optional" as illustrated in the example.

```
Remove[solgraph]

solgraph[μ_, opts___]:=Module[{numsol},

    numsol = vanderpol[μ];

    Plot[x[t]/.numsol, {t, 0, 15}, opts, PlotStyle → Black,

    PlotRange → {-3.5, 3.5}, Ticks → {{0, 15}, {-3, 3}},

    AspectRatio → 1]]
```

For example, entering

```
solgraph[11]
```

displays the graph of the solution to equation (4.2) on the interval $[0, 15]$ if $\mu = 11$ shown in Figure 4-2. Thus, entering

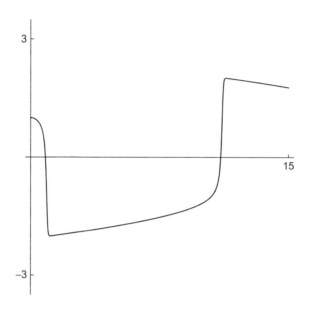

Figure 4-2   Plot of $x(t)$ if $\mu = 11$

```
muvals = {1/32, 1/16, 1/8, 1/4, 1/2, 1,

3/2, 2, 3, 5, 7, 9};

graphs = Map[solgraph, muvals];
```

graphs the solution to the initial-value problem on the interval for $\mu = 1/32, 1/16, 1/8, 1/4, 1/2, 1, 3/2, 2, 3, 5, 7,$ and 9.

After partitioning this list of graphics into three-element subsets in toshow with Partition, the resulting array of graphics is displayed with Show and GraphicsGrid in Figure 4-3.

```
toshow = Partition[graphs, 3];
```

```
Show[GraphicsGrid[toshow]]
```

We find the solution to $\begin{cases} x'' + x = 0 \\ x(0) = 1,\, x'(0) = 0 \end{cases}$ with DSolve. The graph of $y = \cos t$ looks most like the first graph in toshow, corresponding to $\mu = 1/32$.

```
exactsol = DSolve[{x''[t] + x[t]==0, x[0]==1, x'[0]==0}, x[t], t]
```

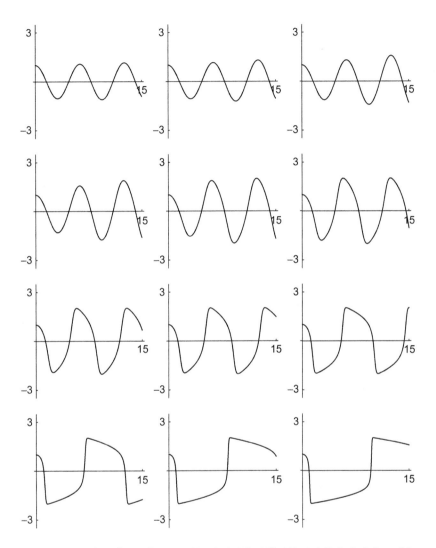

Figure 4-3   Plot of $x(t)$ if $\mu = 1/32, 1/16, 1/8, 1/4, 1/2, 1, 3/2, 2, 3, 5, 7$, and 9

$\{\{x[t] \rightarrow \text{Cos}[t]\}\}$

Last, we show the two graphs together to see how similar they are in Figure 4-4.

```
sol2 = vanderpol[1/32]
```

$\{\{x[t] \rightarrow \text{InterpolatingFunction}[][t]\}\}$

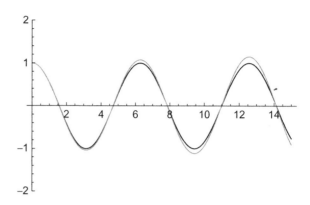

Figure 4-4   Plots of $x(t)$ if $\mu = 1/32$ (in gray) and $y = \cos t$ (in black)

```
Plot[Evaluate[x[t]/.{exactsol, sol2}], {t, 0, 15},
  PlotStyle → {GrayLevel[0], GrayLevel[0.5]},
  PlotRange → {-2, 2}]
```

The example is well-suited to experiment with using Manipulate.

```
Manipulate[
p1 = NDSolve[{x″[t] + μ(x[t]^2 − 1)x′[t] + x[t] == 0,
  x[0] == x0, x′[0] == x0p}, x[t], {t, 0, 50}];
  Plot[{x[t]}/.p1, {t, 0, 25}, PlotRange → {-12.5, 12.5},
  AspectRatio → 1, PlotStyle → {Black}, AxesStyle → Black,
  AxesLabel → {t, x}], {{μ, 2.}, 0, 10}, {{x0, 0}, 0, 10},
  {{x0p, 1}, -5, 10}]
```

Figure 4-5 illustrates how using Manipulate can help explore the behavior of solutions that satisfy various initial conditions.

∎

The example illustrates an important difference between linear and nonlinear equations. Exact solutions of linear equations with constant coefficients can often be found. Nonlinear equations can often be approximated by linear equations. Thus, we concentrate our study on linear differential equations.

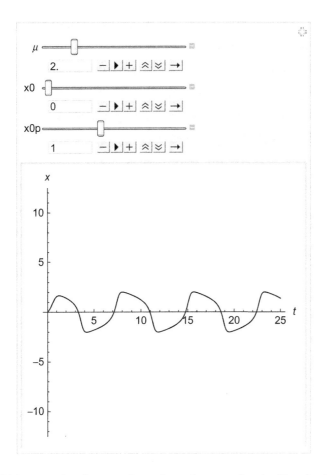

Figure 4-5   Using `Manipulate` to investigate the second-order Van-der-Pol equation

4.1.2 The *n*th-Order Ordinary Linear Differential Equation

In order to develop the methods needed to solve higher-order differential equations, we must state several important definitions and theorems. We begin by introducing the types of higher-order equations that we will be solving in this chapter by restating the following definition that was given in Chapter 1.

Definition 10 (Linear Differential Equation).   *An ordinary differential equation (of order n) is **linear** if it is of the form*

$$a_n(x)\frac{d^n y}{dx^n} + a_{n-1}(x)\frac{d^{n-1}y}{dx^{n-1}} + \cdots + a_2(x)\frac{d^2 y}{dx^2} + a_1(x)\frac{dy}{dx} + a_0(x)y = f(x), \qquad (4.3)$$

For the linear differential equation (4.3), $f(x)$ is called the **forcing function**.

where the functions $a_i(x)$, $i = 0, 1, \ldots, n$, and $f(x)$ are given and $a_n(x)$ is not the zero function.

Dividing equation (4.3) (the nth order linear equation in **general form**) by $a_n(x)$ gives us the nth order linear equation in **standard form**:

$$\frac{d^n y}{dx^n} + a_{n-1}(x)\frac{d^{n-1}y}{dx^{n-1}} + \cdots + a_2(x)\frac{d^2 y}{dx^2} + a_1(x)\frac{dy}{dx} + a_0(x)y = f(x), \qquad (4.4)$$

If $f(x)$ is identically the zero function, equations (4.3) and (4.4) are said to be **homogeneous**; if $f(x)$ is not the zero function, equations (4.3) and (4.4) are said to be **nonhomogeneous**; and if the functions $a_i(x)$, $i = 1, 2, \ldots, n$ are constants, equation (4.3) is said to have **constant coefficients**. An nth-order equation accompanied by the conditions

$$y(x_0) = y_0, \quad y'(x_0) = y_0', \ldots, y^{(n-1)}(x_0) = y_0^{(n-1)}$$

where $y_0, y_0', \ldots, y_0^{(n-1)}$ are constants is called an **nth-order initial-value problem**. For equation (4.3), the **corresponding homogeneous equation** is

$$a_n(x)\frac{d^n y}{dx^n} + a_{n-1}(x)\frac{d^{n-1}y}{dx^{n-1}} + \cdots + a_2(x)\frac{d^2 y}{dx^2} + a_1(x)\frac{dy}{dx} + a_0(x)y = 0. \qquad (4.5)$$

The following theorem gives sufficient conditions for the existence of a unique solution of the nth-order initial-value problem.

Theorem 2 (Existence and Uniqueness). *If $a_n(x), a_{n-1}(x), \ldots, a_1(x), a_0(x)$ and $f(x)$ are continuous throughout an interval I and $a_n(x) \neq 0$ for all x in the interval I, then for every x_0 in I there is a unique solution to the initial-value problem*

$$\begin{cases} a_n(x)\dfrac{d^n y}{dx^n} + a_{n-1}(x)\dfrac{d^{n-1}y}{dx^{n-1}} + \cdots + a_1(x)\dfrac{dy}{dx} + a_0(x)y = f(x) \\ y(x_0) = y_0,\ y'(x_0) = y_0',\ \ldots, y^{(n-1)}(x_0) = y_0^{(n-1)} \end{cases}, \qquad (4.6)$$

on I where $y_0, y_0', \ldots, y_0^{(n-1)}$ represent arbitrary constants.

Now that we have conditions that indicate the existence of solutions, we become familiar with the properties of the functions that form the solution. We will see that solutions to nth-order ordinary linear differential equations require n solutions with the following property.

Definition 11 (Linearly Dependent and Linearly Independent Functions). *Let*

$$S = \{f_1(x), f_2(x), \ldots, f_n(x)\}$$

be a set of n functions. S is **linearly dependent** *on an interval I if there are constants c_1, $c_2, \ldots, c_n$, not all zero, so that*

$$c_1 f_1(x) + c_2 f_2(x) + \cdots + c_n f_n(x) = 0$$

for every value of x in the interval I. S is **linearly independent** *if S is not linearly dependent.*

It is a good exercise to use the definition of linear dependence to show that a set of two functions is linearly dependent if and only if the two functions are constant multiples of each other.

Definition 12 (Wronskian). *Let $S = \{f_1(x), f_2(x), \ldots, f_n(x)\}$ be a set of n functions for which each is differentiable at least $n - 1$ times. The* **Wronskian** *of S, W(S), denoted by*

$$W(S) = W(\{f_1(x), f_2(x), \ldots, f_n(x)\})$$

is the determinant

$$W(S) = \begin{vmatrix} f_1(x) & f_2(x) & \cdots & f_n(x) \\ f_1'(x) & f_2'(x) & \cdots & f_n'(x) \\ \vdots & \vdots & \vdots & \vdots \\ f_1^{(n-1)}(x) & f_2^{(n-1)}(x) & \cdots & f_n^{(n-1)}(x) \end{vmatrix}. \tag{4.7}$$

EXAMPLE 4.1.2: Compute the Wronskian for each of the following sets of functions: (a) $S = \{\sin x, \cos x\}$ and (b) $S = \{\cos 2x, \sin 2x, \sin x \cos x\}$.

SOLUTION: The 2×2 determinant $\begin{vmatrix} a_{11} & a_{12} \\ a_{21} & a_{22} \end{vmatrix}$ is computed with $a_{11}a_{22} - a_{12}a_{21}$. Thus, for (a) we have

$$W(S) = \begin{vmatrix} \sin x & \cos x \\ \dfrac{d}{dx}(\sin x) & \dfrac{d}{dx}(\cos x) \end{vmatrix} = \begin{vmatrix} \sin x & \cos x \\ \cos x & -\sin x \end{vmatrix} = -\sin^2 x - \cos^2 x = -1.$$

For (b), we need to compute the determinant

$$
\begin{vmatrix}
\cos 2x & \sin 2x & \sin x \cos x \\
\dfrac{d}{dx}(\cos 2x) & \dfrac{d}{dx}(\sin 2x) & \dfrac{d}{dx}(\sin x \cos x) \\
\dfrac{d^2}{dx^2}(\cos 2x) & \dfrac{d^2}{dx^2}(\sin 2x) & \dfrac{d^2}{dx^2}(\sin x \cos x)
\end{vmatrix}.
$$

The 3×3 determinant $\begin{vmatrix} a_{11} & a_{12} & a_{13} \\ a_{21} & a_{22} & a_{23} \\ a_{31} & a_{32} & a_{33} \end{vmatrix}$ can be computed in several equivalent ways. For example,

$$
\begin{vmatrix} a_{11} & a_{12} & a_{13} \\ a_{21} & a_{22} & a_{23} \\ a_{31} & a_{32} & a_{33} \end{vmatrix} = a_{11} \begin{vmatrix} a_{22} & a_{23} \\ a_{32} & a_{33} \end{vmatrix} - a_{12} \begin{vmatrix} a_{21} & a_{23} \\ a_{31} & a_{33} \end{vmatrix} + a_{13} \begin{vmatrix} a_{21} & a_{22} \\ a_{31} & a_{32} \end{vmatrix}.
$$

To carry out the sequence of steps to compute the Wronskian, we take advantage of the Det command, which computes the determinant of a square matrix.

First, we define caps to be the set of functions $S = \{\cos 2x,\ \sin 2x,\ \sin x \cos x\}$.

    caps = {Cos[2x], Sin[2x], Sin[x]Cos[x]}

    {Cos[2x], Sin[2x], Cos[x]Sin[x]}

Next, we use D to compute the list

$$
\left\{ \frac{d}{dx}(\cos 2x),\ \frac{d}{dx}(\sin 2x),\ \frac{d}{dx}(\sin x \cos x) \right\}.
$$

Note that D automatically computes the derivative (with respect to x) of each function in caps.

    row2 = D[caps, x]

    {-2Sin[2x], 2Cos[2x], Cos[x]² - Sin[x]²}

$$
\left\{ \frac{d^2}{dx^2}(\cos 2x),\ \frac{d^2}{dx^2}(\sin 2x),\ \frac{d^2}{dx^2}(\sin x \cos x) \right\}.
$$

Observe that entering `row3=D[caps,{x,2}]` yields the same result.

row3 = *D*[caps, {*x*, 2}]

{−4Cos[2*x*], −4Sin[2*x*], −4Cos[*x*]Sin[*x*]}

row3 = *D*[row2, *x*]

{−4Cos[2*x*], −4Sin[2*x*], −4Cos[*x*]Sin[*x*]}

Finally, we use `Det` to see that the determinant

{caps, row2, row3}//MatrixForm

$$\begin{pmatrix} \text{Cos}[2x] & \text{Sin}[2x] & \text{Cos}[x]\text{Sin}[x] \\ -2\text{Sin}[2x] & 2\text{Cos}[2x] & \text{Cos}[x]^2 - \text{Sin}[x]^2 \\ -4\text{Cos}[2x] & -4\text{Sin}[2x] & -4\text{Cos}[x]\text{Sin}[x] \end{pmatrix}$$

Det[{caps, row2, row3}]

0

is zero.

∎

In Example 4.1.2, we see that in (a) the Wronskian is not 0 while in (b) the Wronskian is 0. Moreover, the sets of functions in (a) is linearly independent because $y = \sin x$ and $y = \cos x$ are not multiples of each other while the set of functions in (b) is linearly dependent: $\sin 2x = 2 \sin x \cos x$. In fact, we will see that we can often use the Wronskian to determine if a set of functions is linearly dependent or linearly independent.

Use the command `Wronskian[{y1[x],y2[x],...,yn[x]},x]` to compute the Wronskian of the set of functions $S = \{f_1(x), f_2(x), \ldots, f_n(x)\}$. Mathematica's help for `Wronskian` is shown in Figure 4-6.

We illustrate the use of `Wronskian` for the set of functions

$$\text{caps} = S = \left\{ \frac{1}{\sqrt{x}} \sin 4x, \frac{1}{\sqrt{x}} \cos 4x \right\}.$$

Wronskian[{Sin[4*x*]/Sqrt[*x*], Cos[4*x*]/Sqrt[*x*]}, *x*]

$-\dfrac{4}{x}$

Figure 4-6   Mathematica's help for the `Wronskian` function

Because the Wronskian for these two functions is not 0 and they are both solutions of $4x^2y'' + 4xy' + (64x^2 - 1)y = 0$, $x > 0$, as verified with the following commands,

```
y1[x_] = Sin[4x]/Sqrt[x];

y2[x_] = Cos[4x]/Sqrt[x];

4x^2y''[x] + 4xy'[x] + (64x^2 - 1)y[x]//.
    {{y[x] → y1[x], y'[x] → y1'[x], y''[x] → y1''[x]},
    {y[x] → y2[x], y'[x] → y2'[x], y''[x] → y2''[x]}}//
    Simplify

{0, 0}
```

we can conclude they are linearly independent by the following theorem.

Theorem 3. *Let $S = \{f_1(x), f_2(x), \ldots, f_m(x)\}$ be a set of m solutions of equation (4.5) on an interval I. S is linearly independent if and only if $W(S) \neq 0$ for at least one value of x in the interval I.*

EXAMPLE 4.1.3: Use the Wronskian to classify each of the following sets of functions as linearly independent or linearly dependent: (a) $S = \{1 - 2\sin^2 x, \cos 2x\}$ and (b) $S = \{e^x, xe^x, x^2 e^x\}$.

SOLUTION: (a) Note that both functions in S are solutions of $y'' + 4y = 0$. Here, we must compute the determinant of the 2×2 matrix

$$\begin{vmatrix} 1 - 2\sin^2 x & \cos 2x \\ \dfrac{d}{dx}\left(1 - 2\sin^2 x\right) & \dfrac{d}{dx}(\cos 2x) \end{vmatrix}.$$

We use `Wronskian` to compute the determinant

```
Wronskian[{1 − 2Sin[x]^2, Cos[2x]}, x]//Simplify
```

> 0

and see that the result is 0. Therefore, the set of functions $S = \{1 - 2\sin^2 x, \cos 2x\}$ is linearly dependent. This makes sense because these functions are multiples of each other: $\cos 2x = \frac{1}{2}\left(1 - 2\sin^2 x\right)$.

(b) Note that all three functions in S are solutions of $y''' - 3y'' + 3y' - y = 0$. Here, we must compute the determinant

$$\begin{vmatrix} e^x & xe^x & x^2 e^x \\ \dfrac{d}{dx}(e^x) & \dfrac{d}{dx}(xe^x) & \dfrac{d}{dx}\left(x^2 e^x\right) \\ \dfrac{d^2}{dx^2}(e^x) & \dfrac{d^2}{dx^2}(xe^x) & \dfrac{d^2}{dx^2}\left(x^2 e^x\right) \end{vmatrix}.$$

```
Wronskian[{Exp[x], xExp[x], x^2Exp[x]}, x]
```

> $2e^{3x}$

We conclude that S is linearly independent because the Wronskian of S is not identically zero.

■

4.1.3 Fundamental Set of Solutions

A **nontrivial solution** is a solution that is not identically the zero function.

Obtaining a collection of n linearly independent solutions to the nth-order linear homogeneous differential equation (4.5) is of great importance in solving it.

Definition 13 (Fundamental Set of Solutions). *A set S of n linearly independent nontrivial solutions of the nth-order linear homogeneous equation (4.5) is called a* **fundamental set of solutions** *of the equation.*

EXAMPLE 4.1.4: Show that $S = \{e^{-5x}, e^{-x}\}$ is a fundamental set of solutions of the equation $y'' + 6y' + 5y = 0$.

SOLUTION: Because

$$\frac{d^2}{dx^2}\left(e^{-5x}\right) + 6\frac{d}{dx}\left(e^{-5x}\right) + 5e^{-5x} = 25e^{-5x} - 30e^{-5x} + 5e^{-5x} = 0$$

and

$$\frac{d^2}{dx^2}\left(e^{-x}\right) + 6\frac{d}{dx}\left(e^{-x}\right) + 5e^{-x} = e^{-x} - 6e^{-x} + 5e^{-x} = 0$$

each function is a solution of the differential equation. It follows that S is linearly independent because

$$W(S) = \begin{vmatrix} e^{-5x} & e^{-x} \\ -5e^{-5x} & -e^{-x} \end{vmatrix} = -e^{-6x} + 5e^{-6x} = 4e^{-x} \neq 0.$$

so we conclude that S is a fundamental set of solutions of the equation.

Of course, we can perform the same steps with Mathematica. First, we define caps to be the set of functions S.

```
Clear[x, y, caps]

caps = {Exp[-5x], Exp[-x]};
```

To verify that each function in S is a solution of $y'' + 6y' + 5y = 0$, we define a function f. f[y] computes and returns $y'' + 6y' + 5y$. We then use Map (/@) to apply f to each function in caps to see that each function in caps is a solution of $y'' + 6y' + 5y = 0$, confirming the result we obtained previously.

```
Clear[f]

f[y_]:=D[y, {x, 2}] + 6D[y, x] + 5y//Simplify;

f/@caps
```

$\{0, 0\}$

Next, we use `Wronskian` to find the determinant of the Wronskian matrix $\begin{pmatrix} e^{-5x} & e^{-x} \\ -5e^{-5x} & -e^{-x} \end{pmatrix}$ and display wmat in traditional row-and-column form with `MatrixForm`.

```
wmat = {caps, D[caps, x]}//MatrixForm
```

$$\begin{pmatrix} e^{-5x} & e^{-x} \\ -5e^{-5x} & -e^{-x} \end{pmatrix}$$

`Wronskian` is then used to compute $W(S)$.

```
Wronskian[caps, x]
```

$4e^{-6x}$

∎

We use a fundamental set of solutions to create a *general solution* of an nth-order linear homogeneous differential equation.

Theorem 4 (Principle of Superposition). *If $S = \{f_1(x), f_2(x), \ldots, f_k(x)\}$ is a set of solutions of the nth-order linear homogeneous equation (4.5) and $\{c_1, c_2, \ldots, c_k\}$ is a set of k constants, then*

$$f(x) = c_1 f_1(x) + c_2 f_2(x) + \cdots + c_k f_k(x)$$

is also a solution of equation (4.5).

$f(x) = c_1 f_1(x) + c_2 f_2(x) + \cdots + c_k f_k(x)$ is called a **linear combination of functions** in the set $S = \{f_1(x), f_2(x), \ldots, f_k(x)\}$. A consequence of this fact is that the linear combination of the functions in a fundamental set of solutions of the nth-order linear homogeneous differential equation (4.5) is also a solution of the differential equation, and we call this linear combination a **general solution** of the differential equation.

Definition 14 (General Solution). *If* $S = \{f_1(x), f_2(x), \ldots, f_n(x)\}$ *is a fundamental set of solutions of the nth-order linear homogeneous equation*

$$a_n(x)y^{(n)} + a_{n-1}(x)y^{(n-1)} + \cdots + a_1(x)y' + a_0(x)y = 0,$$

then a **general solution** *of the equation is*

$$f(x) = c_1 f_1(x) + c_2 f_2(x) + \cdots + c_n f_n(x)$$

where $\{c_1, c_2, \ldots, c_n\}$ *is a set of n arbitrary constants.*

In other words, if we have a fundamental set of solutions S, then a general solution of the differential equation is formed by taking the linear combination of the functions in S.

EXAMPLE 4.1.5: Show that $S = \{\cos 2x, \sin 2x\}$ is a fundamental set of solutions of the second-order ordinary linear differential equation with constant coefficients $y'' + 4y = 0$.

SOLUTION: First, we verify that both functions are solutions of $y'' + 4y = 0$. Note that we have defined caps to be the set of functions $S = \{\cos 2x, , \sin 2x\}$. Now, we use Map to apply the function $y'' + 4y$ to the functions in caps: the command Map[D[#,{x,2}]+4#&,caps] computes $y'' + 4y$ for each function y in caps. Thus, we see that given an argument #, the command D[#,{x,2}]+4#& computes the sum of the second derivative (with respect to x) of the argument and four times the argument. We conclude that both functions are solutions of $y'' + 4y$ because the result is a list of two zeros.

```
caps = {Cos[2x], Sin[2x]}

Map[D[#, {x, 2}] + 4#&, caps]

{Cos[2x], Sin[2x]}

{0, 0}
```

Next, we compute the Wronskian

```
Wronskian[caps, x]
```

```
2
```

to show that the functions in S are linearly independent.

By the Principle of Superposition, $y(x) = c_1 \cos 2x + c_2 \sin 2x$, where c_1 and c_2 are arbitrary constants, is also a solution of the equation. We now graph $y(x)$ for various values of c_1 and c_2. After defining y, we use Table to create a list obtained by replacing c[1] in y[x] by $-1, 0$, and 1 and c[2] by $-1, 0$, and 1. We name the resulting list toplot. Note that toplot is a list of lists: toplot consists of three elements each of which is a list consisting of three functions.

```
Clear[y]

y[x_] = c[1]Cos[2x] + c[2]Sin[2x];

toplot = Table[y[x], {c[1], −1, 1}, {c[2], −1, 1}];
```

Last, we use Plot to graph the nine functions in toplot for $0 \le x \le 2\pi$ in Figure 4-7.

```
Plot[toplot, {x, 0, 2Pi}, AxesLabel → {x, y}]
```

Alternatively, we can show the graphs individually in a graphics array as shown in Figure 4-8.

```
toshow = Map[Plot[#, {x, 0, 2Pi}, AxesLabel → {x, y}]&,

Flatten[toplot]]; Length[toshow]

9

Show[GraphicsGrid[Partition[toshow, 3]]]
```

■

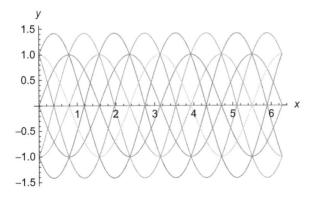

Figure 4-7   Graphs of various *linear combinations* of $\cos 2x$ and $\sin 2x$

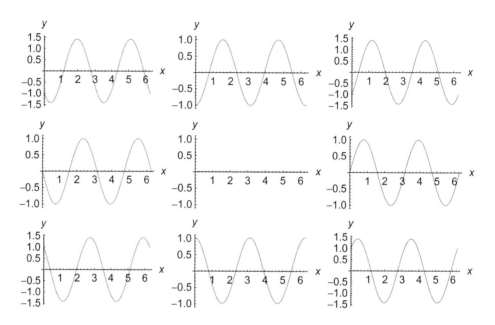

Figure 4-8   Graphs of various *linear combinations* of $\cos 2x$ and $\sin 2x$

The Principle of Superposition is a *very* important property of linear homogeneous equations and is generally not valid for nonlinear equations and *never* valid for nonhomogeneous equations.

EXAMPLE 4.1.6: Is the Principle of Superposition valid for the nonlinear equation $tx'' - 2xx' = 0$?

SOLUTION: We see that DSolve is able to find a general solution of this nonlinear equation.

```
gensol = DSolve[tx"[t] - 2x[t]x'[t] == 0, x[t], t]
```

$$\{\{x[t] \to \frac{1}{2}(-1 + \sqrt{-1 - 8C[1]}\operatorname{Tan}[\frac{1}{2}(\sqrt{-1 - 8C[1]}$$
$$+ \sqrt{-1 - 8C[1]}\operatorname{Log}[t])])\}\}$$

Next, we define sol1 and sol2 to be two solutions to the equation.

```
sol1 = gensol[[1, 1, 2]]/.{C[1] → -1/8, C[2] → 0}
```

$$-\frac{1}{2}$$

```
sol2 = gensol[[1, 1, 2]]/.{C[1] → -1/4, C[2] → 1}
```

$$\frac{1}{2}\left(-1 + \text{Tan}\left[\frac{1}{2}(1 + \text{Log}[t])\right]\right)$$

However, the sum of these two solutions is not a solution to the nonlinear equation because $tf'' - 2ff' \neq 0$; the Principle of Superposition is not valid for this nonlinear equation.

```
y[t_] = sol1 + sol2
```

$$-\frac{1}{2} + \frac{1}{2}\left(-1 + \text{Tan}\left[\frac{1}{2}(1 + \text{Log}[t])\right]\right)$$

```
ty"[t] - 2y[t]y'[t]//Simplify
```

$$\frac{\text{Sec}\left[\frac{1}{2}(1 + \text{Log}[t])\right]^2}{4t}$$

∎

4.1.4 Existence of a Fundamental Set of Solutions

The following two theorems tell us that under reasonable conditions, the nth-order linear homogeneous equation (4.5) has a fundamental set of n solutions.

Theorem 5. *If $a_i(x)$ is continuous on an open interval I for $i = 0, 1, \ldots, n$, and $a_n(x) \neq 0$ for all x in the interval I then the nth-order linear homogeneous equation (4.5),*

$$a_n(x)\frac{d^n y}{dx^n} + a_{n-1}(x)\frac{d^{n-1} y}{dx^{n-1}} + \cdots + a_2(x)\frac{d^2 y}{dx^2} + a_1(x)\frac{dy}{dx} + a_0(x)y = 0,$$

has a fundamental set of n solutions.

Theorem 6. *Any set of $n+1$ solutions of the nth-order linear homogeneous equation (4.5) is linearly dependent.*

We can summarize the results of these theorems by saying that in order to solve the nth-order linear homogeneous differential equation (4.5), we must find a set S of n functions that satisfy the differential equation such that $W(S) \neq 0$.

EXAMPLE 4.1.7: Show that $y = e^{-x}(c_1 \cos 4x + c_2 \sin 4x)$ is a general solution of $y'' + 2y' + 17y = 0$.

SOLUTION: After defining y, we use D to compute the first and second derivatives (with respect to x) of y.

```
Clear[x, y, c1, c2]

y[{c1_, c2_}]:=

  Exp[-x](c1Cos[4x] + c2Sin[4x]);

D[y[{c1, c2}], x]//Simplify
```

$e^{-x}(-(c1 - 4c2)\text{Cos}[4x] - (4c1 + c2)\text{Sin}[4x])$

```
D[y[{c1, c2}], {x, 2}]//Simplify
```

$e^{-x}(-(15c1 + 8c2)\text{Cos}[4x] + (8c1 - 15c2)\text{Sin}[4x])$

We then compute and simplify $y'' + 2y' + 17y$. Because the result is zero and the set of functions S is linearly independent, $y = e^{-x}(c_1 \cos 4x + c_2 \sin 4x)$ is a general solution of the equation.

You should verify that
$S = \{e^{-x}\cos 4x, e^{-x}\sin 4x\}$
is a linearly independent set
of functions.

```
Simplify[D[y[{c1, c2}], {x, 2}]+

  2D[y[{c1, c2}], x] + 17y[{c1, c2}]]
```

0

Next, we define cvals to be the set of ordered pairs consisting of $(0, 1)$, $(1, 0)$, $(2, 1)$, and $(1, -2)$ and use Map (and /@) to compute the value of y for each ordered pair in cvals, naming the resulting set of functions toplot and finally graphing them on the interval $[-1, 2]$ with Plot as shown in Figure 4-9.

```
cvals = {{0, 1}, {1, 0}, {2, 1}, {1, -2}};

toplot = y/@cvals
```

$\{e^{-x}\text{Sin}[4x], e^{-x}\text{Cos}[4x], e^{-x}(2\text{Cos}[4x] + \text{Sin}[4x]),$
$e^{-x}(\text{Cos}[4x] - 2\text{Sin}[4x])\}$

```
toplot = Map[y, cvals]
```

$\{e^{-x}\text{Sin}[4x], e^{-x}\text{Cos}[4x], e^{-x}(2\text{Cos}[4x] + \text{Sin}[4x]),$
$e^{-x}(\text{Cos}[4x] - 2\text{Sin}[4x])\}$

```
Plot[toplot, {x, -1, 2}, AxesLabel → {x, y}]
```

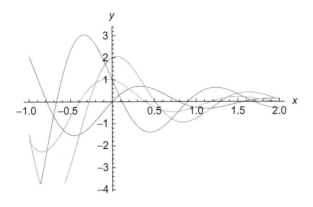

Figure 4-9   Various solutions of $y'' + 2y' + 17y = 0$

∎

4.1.5 Reduction of Order

In the next section, we learn how to find solutions of homogeneous equations with constant coefficients. In doing so, we will find it necessary to determine a second solution from a known solution. We illustrate this procedure, called **reduction of order**, by considering a second-order equation. In certain situations, we can reduce a second-order equation by making an appropriate substitution to convert the second-order equation to a first-order equation (this reduction in order gives the name to the method). Consider the equation

$$y'' + p(x)y' + q(x)y = 0,$$

and suppose that $y_1 = f(x)$ is a solution to this equation. Of course we know from our previous discussion that in order to solve this second-order differential equation, we must have two linearly independent solutions. Hence, we must determine a second linearly independent solution. We accomplish this by attempting to find a solution of the form

$$y_2 = v(x)f(x),$$

where $v(x)$ is *not* a constant function. Differentiating $y_2 = v(x)f(x)$ twice we obtain

$$y_2' = v'f + vf' \quad \text{and} \quad y_2'' = v''f + 2v'f' + vf''.$$

If $v(x)$ were constant, y_1 and y_2 would be linearly dependent.

```
Clear[x, y, f, v]
y[x_] = v[x]f[x];
y'[x]
y"[x]
```

$$v[x]f'[x] + f[x]v'[x]$$

$$2f'[x]v'[x] + v[x]f''[x] + f[x]v''[x]$$

Notice that for convenience, we have omitted the argument of these functions. We now substitute y_2, y'_2, and y''_2 into the equation $y'' + p(x)y' + q(x)y = 0$, which gives us

$f'' + p(x)f' + q(x)f = 0$
because f is a solution to
$y'' + p(x)y' + q(x)y = 0$.

$$
\begin{aligned}
y'' + p(x)y' + q(x)y &= v''f + 2v'f' + vf'' + p(x)\left(v'f + vf'\right) + q(x)vf \\
&= fv'' + \left(2f' + p(x)f\right)v' + v\left(f'' + p(x)f' + q(x)f\right) \\
&= fv'' + \left(2f' + p(x)f\right)v' = 0.
\end{aligned}
$$

```
step1 = Collect[y"[x] + p[x]y'[x] + q[x]y[x],
{v[x], v'[x], v"[x]}]
```

$$\left(f[x]p[x] + 2f'[x]\right)v'[x] + v[x]\left(f[x]q[x] + p[x]f'[x] + f''[x]\right) + f[x]v''[x]$$

```
step2 = step1/. f"[x] + p[x]f'[x] + q[x]f[x] → 0
```

$$\left(f[x]p[x] + 2f'[x]\right)v'[x] + f[x]v''[x]$$

Therefore, we have the equation $fv'' + \left(2f' + p(x)f\right)v' = 0$, which can be written as a first-order equation by letting $w = v'$. Making this substitution gives us the linear first-order equation

$$fw' + \left(2f' + p(x)f\right)w = 0 \qquad \text{or} \qquad f\frac{dw}{dx} + \left(2f' + p(x)f\right)w = 0,$$

which is separable, resulting in the separated equation

$$\frac{1}{w}dw = \left(-2\frac{f'}{f} - p(x)\right)dx.$$

```
step3 = step2/.{v"[x] → w'[x], v'[x] → w[x]}
```

$$w[x]\left(f[x]p[x] + 2f'[x]\right) + f[x]w'[x]$$

We can solve this equation by integrating both sides of the equation to yield

$$\ln |w| = \ln \left(\frac{1}{f^2}\right) - \int p(x)\, dx \qquad \text{so} \qquad w = \frac{1}{f^2} e^{-\int p(x)\, dx}.$$

```
step4 = DSolve[step3 == 0, w[x], x]
```

$$\left\{\left\{w[x] \to e^{\int_1^x \frac{-f[K[1]]p[K[1]] - 2f'[K[1]]}{f[K[1]]} dK[1]} C[1]\right\}\right\}$$

```
step5 = Simplify[step4[[1, 1, 2]]]
```

$$e^{\int_1^x \left(-p[K[1]] - \frac{2f'[K[1]]}{f[K[1]]}\right) dK[1]} C[1]$$

Thus, we have the formula

$$\frac{dv}{dx} = \frac{1}{f^2} e^{-\int p(x)\, dx} \qquad \text{or} \qquad v(x) = \int \frac{1}{[f(x)]^2} e^{-\int p(x)\, dx}.$$

```
step6 = Integrate[step5, x]
```

$$C[1] \int e^{\int_1^x \left(-p[K[1]] - \frac{2f'[K[1]]}{f[K[1]]}\right) dK[1]} dx$$

Therefore, if we have the solution $y_1(x) = f(x)$ of the differential equation $y'' + p(x)y' + q(x)y = 0$, then we can obtain a second linearly independent solution of the form $y_2(x) = v(x)f(x) = v(x)y_1(x)$ where

It is a good exercise to show that y_1 and the solution $y_2 = vy_1$ obtained by reduction of order are linearly independent.

$$v(x) = \int \frac{1}{[y_1(x)]^2} e^{-\int p(x)\, dx} dx \quad \text{and} \quad y_2(x) = y_1(x) \int \frac{1}{[y_1(x)]^2} e^{-\int p(x)\, dx} dx. \quad (4.8)$$

EXAMPLE 4.1.8: Determine a second linearly independent solution to the differential equation $4x^2 y'' + 8xy' + y = 0, x > 0$, given that $y_1 = 1/\sqrt{x}$ is a solution.

SOLUTION: In this case, we must divide by $4x^2$ in order to obtain an equation of the form $y'' + p(x)y' + q(x)y = 0$. This gives us the equation $y'' + 2x^{-1}y' + \frac{1}{4}x^{-2}y = 0$. Therefore, $p(x) = 2x^{-1}$, and $y_1(x) = x^{-1/2}$. Using the formula for v, equation (4.8), we obtain

$$v(x) = \int \frac{1}{[y_1(x)]^2} e^{-\int p(x)\, dx} dx = \int \frac{1}{[x^{-1/2}]^2} e^{-\int 2/x\, dx} dx$$

$$= \int \frac{1}{x^{-1}} e^{-2\ln x} dx = \int \frac{1}{x} dx = \ln x, \quad x > 0.$$

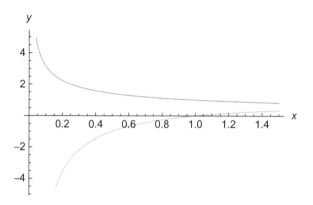

Figure 4-10   Plots of y_1 and y_2

Hence, a second linearly independent solution is $y_2 = x^{-1/2} \ln x$; a general solution is $y = x^{-1/2} (c_1 + c_2 \ln x)$ Of course, we can take advantage of commands like Integrate to carry out the steps encountered here.

```
p[x_] = 2/x;

f[x_] = 1/Sqrt[x];

v[x_] = Integrate[Exp[−Integrate[p[x], x]]/f[x]^2, x]
```

Log[x]

```
y[x_] = v[x]f[x];
```

Some versions of Mathematica might generate error messages, which are not displayed here, when we enter the following Plot command because both $y_1(x)$ and $y_2(x)$ are undefined if $x = 0$. Nevertheless, the resulting graphs are displayed correctly in Figure 4-10.

```
Plot[{f[x], y[x]}, {x, 0, 3/2}, AxesLabel → {x, y}]
```

■

4.2 Solving Homogeneous Equations With Constant Coefficients

We now turn our attention to solving linear homogeneous equations with constant coefficients. Nonhomogeneous equations are considered in the following sections.

4.2.1 Second-Order Equations

Suppose that the coefficient functions of equation $a_2(t)y'' + a_1(t)y' + a_0(t)y = f(t)$ are constants: $a_2(t) = a$, $a_1(t) = b$, and $a_0(t) = c$ and that $f(t)$ is identically the zero function. In this case, the equation $a_2(t)y'' + a_1(t)y' + a_0(t)y = f(t)$ becomes

$$ay'' + by' + cy = 0. \qquad (4.9)$$

Now suppose that $y = e^{kt}$, k constant, is a solution of equation (4.9). Then, $y' = ke^{kt}$ and $y'' = k^2 e^{kt}$. Substitution into equation (4.9) then gives us

$$ay'' + by' + cy = ak^2 e^{kt} + bke^{kt} + ce^{kt}$$
$$= e^{kt}\left(ak^2 + bk + c\right) = 0.$$

Because $e^{kt} \neq 0$, the solutions of equation (4.9) are determined by the solutions of

$$ak^2 + bk + c = 0, \qquad (4.10)$$

called the **characteristic equation** of equation (4.9).

Theorem 7. *Let k_1 and k_2 be the solutions of equation (4.10).*

1. *If $k_1 \neq k_2$ are real and distinct, two linearly independent solutions of equation (4.9) are $y_1 = e^{k_1 t}$ and $y_2 = e^{k_2 t}$; a general solution of equation (4.9) is*

$$y = c_1 e^{k_1 t} + c_2 e^{k_2 t}.$$

2. *If $k_1 = k_2$, two linearly independent solutions of equation (4.9) are $y_1 = e^{k_1 t}$ and $y_2 = te^{k_1 t}$; a general solution of equation (4.9) is*

$$y = c_1 e^{k_1 t} + c_2 te^{k_1 t}.$$

3. *If $k_{1,2} = \alpha \pm \beta i$, $\beta \neq 0$, two linearly independent solutions of equation (4.9) are $y_1 = e^{\alpha t} \cos \beta t$ and $y_2 = e^{\alpha t} \sin \beta t$; a general solution of equation (4.9) is*

$$y = e^{\alpha t} (c_1 \cos \beta t + c_2 \sin \beta t).$$

EXAMPLE 4.2.1: Solve each of the following equations: (a) $6y'' + y' - 2y = 0$; (b) $y'' + 2y' + y = 0$; and (c) $16y'' + 8y' + 145y = 0$.

SOLUTION: (a) The characteristic equation is $6k^2 + k - 2 = (3k + 2)(2k - 1) = 0$ with solutions $k = -2/3$ and $k = 1/2$. We check with either `Factor` or `Solve`.

> `Factor[6k^2 + k − 2]`
>
> `Solve[6k^2 + k − 2 == 0]`
>
> $(-1 + 2k)(2 + 3k)$
>
> $$\left\{\left\{k \to -\frac{2}{3}\right\}, \left\{k \to \frac{1}{2}\right\}\right\}$$

Then, a fundamental set of solutions is $\left\{e^{-2t/3}, e^{t/2}\right\}$ and a general solution is

$$y = c_1 e^{-2t/3} + c_2 e^{t/2}.$$

Of course, we obtain the same result with `DSolve`.

> `DSolve[6y"[t] + y'[t] − 2y[t] == 0, y[t], t]`
>
> $$\left\{\left\{y[t] \to e^{-2t/3} C[1] + e^{t/2} C[2]\right\}\right\}$$

(b) The characteristic equation is $k^2 + 2k + 1 = (k+1)^2 = 0$ with solution $k = -1$, which has multiplicity two, so a fundamental set of solutions is $\left\{e^{-t}, te^{-t}\right\}$ and a general solution is

$$y = c_1 e^{-t} + c_2 te^{-t}.$$

We check the calculation in the same way as in (a).

> `Factor[k^2 + 2k + 1]`
>
> `Solve[k^2 + 2k + 1 == 0]`
>
> `DSolve[y"[t] + 2y'[t] + y[t] == 0, y[t], t]`
>
> $(1 + k)^2$
>
> $\{\{k \to -1\}, \{k \to -1\}\}$
>
> $$\left\{\left\{y[t] \to e^{-t} C[1] + e^{-t} t C[2]\right\}\right\}$$

(c) The characteristic equation is $16k^2 + 8k + 145 = 0$ with solutions $k_{1,2} = -\frac{1}{4} \pm 3i$ so a fundamental set of solutions Is $\left\{e^{-t/4} \cos 3t, e^{-t/4} \sin 3t\right\}$ and a general solution is

$$y = e^{-t/4}(c_1 \cos 3t + c_2 \sin 3t).$$

The calculation is verified in the same way as in (a) and (b). However, to factor a polynomial over the complex numbers, with `Factor`, we include the option `GaussianIntegers->True`.

> `Factor[16k^2 + 8k + 145,`
>
> > `GaussianIntegers → True]`
>
> `Solve[16k^2 + 8k + 145 == 0]`
>
> `DSolve[16y"[t] + 8y'[t] + 145y[t] == 0, y[t], t]`

> $((1 - 12i) + 4k)((1 + 12i) + 4k)$
>
> $$\left\{\left\{k \to -\frac{1}{4} - 3i\right\}, \left\{k \to -\frac{1}{4} + 3i\right\}\right\}$$
>
> $$\left\{\left\{y[t] \to e^{-t/4}C[2]\text{Cos}[3t] + e^{-t/4}C[1]\text{Sin}[3t]\right\}\right\}$$

■

EXAMPLE 4.2.2: Solve

$$64\frac{d^2y}{dt^2} + 16\frac{dy}{dt} + 1025y = 0, \quad y(0) = 1, \quad \frac{dy}{dt}(0) = 3.$$

SOLUTION: A general solution of $64y'' + 16y' + 1025y = 0$ is $y = e^{-t/8}(c_1 \sin 4t + c_2 \cos 4t)$.

> `gensol = DSolve[64y"[t] + 16y'[t] + 1025y[t] == 0,`
>
> > `y[t], t]`
>
> $$\left\{\left\{y[t] \to e^{-t/8}C[2]\text{Cos}[4t] + e^{-t/8}C[1]\text{Sin}[4t]\right\}\right\}$$

Applying $y(0) = 1$ shows us that $c_2 = 1$.

> `e1 = y[t]/.gensol[[1]]/.t → 0`
>
> $C[2]$

Computing y'

```
D[y[t]/.gensol[[1]], t]
```

$$4e^{-t/8}C[1]Cos[4t] - \frac{1}{8}e^{-t/8}C[2]Cos[4t]$$
$$-\frac{1}{8}e^{-t/8}C[1]Sin[4t] - 4e^{-t/8}C[2]Sin[4t]$$

and then $y'(0)$, shows us that $-4c_1 - \frac{1}{8}c_2 = 2$.

```
e2 = D[y[t]/.gensol[[1]], t]/.t → 0
```

$$4C[1] - \frac{C[2]}{8}$$

Solving for c_1 and c_2 with Solve shows us that $c_1 = -25/32$ and $c_1 = 1$.

```
cvals = Solve[{e1 == 1, e2 == 3}]
```

$$\left\{\left\{C[1] \to \frac{25}{32},\ C[2] \to 1\right\}\right\}$$

Thus, $y = e^{-t/8}\left(\frac{17}{32}\sin 4t + \cos 4t\right)$, which we graph with Plot in Figure 4-11.

```
sol = y[t]/.gensol[[1]]/.cvals[[1]]
```

$$e^{-t/8}Cos[4t] + \frac{25}{32}e^{-t/8}Sin[4t]$$

```
Plot[sol, {t, 0, 8Pi}, AxesLabel → {t, y},

    PlotStyle->CMYKColor[0, 0.89, 0.94, 0.28]]
```

We verify the calculation with DSolve.

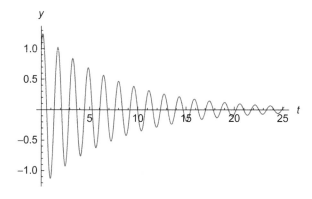

Figure 4-11   The solution to the initial-value problem tends to 0 as $t \to \infty$

```
DSolve[{64y"[t] + 16y'[t] + 1025y[t] == 0, y[0] == 1,
  y'[0] == 3},
  y[t], t]
```

$$\left\{\left\{y[t] \rightarrow \frac{1}{32} e^{-t/8}(32\text{Cos}[4t] + 25\text{Sin}[4t])\right\}\right\}$$

∎

4.2.2 Higher-Order Equations

As with second-order equations, solutions of any nth-order linear homogeneous differential equation with constant coefficients are determined by the solutions of the *characteristic equation*, which is obtained by assuming that $y = e^{kt}$.

Definition 15 (Characteristic Equation). *The equation*

$$a_n k + a_{n-1}k^{n-1} + \cdots + a_2 k^2 + a_1 k + a_0 = 0 \tag{4.11}$$

*is called the **characteristic equation** of the nth-order linear homogeneous differential equation with constant coefficients*

$$a_n y^{(n)} + a_{n-1} y^{(n-1)} + \cdots + a_2 y'' + a_1 y' + a_0 y = 0.$$

In order to explain the process of finding a general solution of any nth-order linear homogeneous differential equations with constant coefficients, we state the following definition.

Definition 16 (Multiplicity). *Suppose that the characteristic equation $a_n k + a_{n-1}k^{n-1} + \cdots + a_2 k^2 + a_1 k + a_0 = 0$ can be written in factored form as*

$$(k - k_1)^{m_1} (k - k_2)^{m_2} \cdots (k - k_r)^{m_r},$$

*where $k_i \neq k_j$ for $i \neq j$ and $m_1 + m_2 + \cdots + m_r = n$. Then the roots of the characteristic equation are $k = k_1, k = k_2, \ldots,$ and $k = k_r$ where the roots have **multiplicity** $m_1, m_2, \ldots,$ and m_r, respectively.*

In the same manner as in the case for a second-order homogeneous equation with real constant coefficients, a general solution of an nth-order linear homogeneous equation with real constant coefficients is determined by the solutions of its characteristic equation. Hence, we state the following rules for finding a general solution of an nth-order linear homogeneous equation for the many situations that may be encountered.

1. If a solution k of equation (4.11) has multiplicity m, m linearly independent solutions corresponding to k are

$$e^{kt}, te^{kt}, \ldots, t^{m-1}e^{kt}.$$

2. If a solution $k = \alpha + \beta i$, $\beta \neq 0$, of equation (4.11) has multiplicity m, $2m$ linearly independent solutions corresponding to $k = \alpha + \beta i$ (and $k = \alpha - \beta i$) are

$$e^{\alpha t} \cos \beta t, e^{\alpha t} \sin \beta t, te^{\alpha t} \cos \beta t, te^{\alpha t} \sin \beta t, \ldots, t^{m-1}e^{\alpha t} \cos \beta t, t^{m-1}e^{\alpha t} \sin \beta t.$$

Notice that the key to the process is identifying each root of the characteristic equation and the associated solution(s).

EXAMPLE 4.2.3: Solve $12y''' - 5y'' - 6y' - y = 0$.

SOLUTION: The characteristic equation is

$$12k^3 - 5k^2 - 6k - 1 = (k - 1)(3k + 1)(4k + 1) = 0$$

Factor[expression] attempts to factor expression.

with solutions $k_1 = -1/3$, $k_2 = -1/4$, and $k_3 = 1$.

> `Factor[12k^3 - 5k^2 - 6k - 1]`

> $(-1 + k)(1 + 3k)(1 + 4k)$

Thus, three linearly independent solutions of the equation are $y_1 = e^{-t/3}$, $y_2 = e^{-t/4}$, and $y_3 = e^t$; a general solution is $y = c_1 e^{-t/3} + c_2 e^{-t/4} + c_3 e^t$. We check with DSolve.

> `DSolve[12y"'[t] - 5y"[t] - 6y'[t] - y[t] == 0,`
>
> `y[t], t]`

> $\left\{ \left\{ y[t] \to e^{-t/4} C[1] + e^{-t/3} C[2] + e^t C[3] \right\} \right\}$

■

EXAMPLE 4.2.4: Solve $y''' + 4y' = 0$, $y(0) = 0$, $y'(0) = 1$, $y''(0) = -1$.

SOLUTION: The characteristic equation is $k^3 + 4k = k(k^2 + 4) = 0$ with solutions $k_1 = 0$ and $k_{2,3} = \pm 2i$ that are found with `Solve` or with `Factor` together with the option `GaussianIntegers->True`.

Enter `?Solve` to obtain basic help regarding the `Solve` function.

```
Solve[k^3 + 4k == 0]
```

$\{\{k \rightarrow 0\}, \{k \rightarrow -2i\}, \{k \rightarrow 2i\}\}$

```
Factor[k^3 + 4k, GaussianIntegers → True]
```

$k(-2i + k)(2i + k)$

Three linearly independent solutions of the equation are $y_1 = 1$, $y_2 = \cos 2t$, and $y_3 = \sin 2t$. A general solution is $y = c_1 + c_2 \sin 2t + c_3 \cos 2t$.

```
gensol = DSolve[y"'[t] + 4y'[t] == 0, y[t], t]
```

$$\left\{\left\{y[t] \rightarrow C[3] - \frac{1}{2}C[2]\cos[2t] + \frac{1}{2}C[1]\sin[2t]\right\}\right\}$$

Application of the initial conditions shows us that $c_1 = 1$, $c_2 = -1/2$, and $c_3 = -1/4$ so the solution to the initial-value problem is $y = -\frac{1}{4} + \frac{1}{2}\sin 2t + \frac{1}{4}\cos 2t$. We verify the computation with `DSolve` and graph the result with `Plot` in Figure 4-12.

```
e1 = y[t]/.gensol[[1]]/.t → 0
```

$$-\frac{C[2]}{2} + C[3]$$

```
e2 = D[y[t]/.gensol[[1]], t]/.t → 0
e3 = D[y[t]/.gensol[[1]], {t, 2}]/.t → 0
```

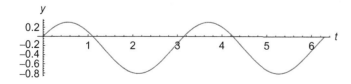

Figure 4-12   Graph of $y = -\frac{1}{4} + \frac{1}{2}\sin 2t + \frac{1}{4}\cos 2t$

$C[1]$

$2C[2]$

```
cvals = Solve[{e1 == 0, e2 == 1, e3 == -1}]
```

$$\left\{\left\{C[1] \to 1,\ C[2] \to -\frac{1}{2},\ C[3] \to -\frac{1}{4}\right\}\right\}$$

```
partsol = DSolve[{y"'[t] + 4y'[t] == 0,
    y[0] == 0, y'[0] == 1, y"[0] == -1}, y[t], t]
```

$$\left\{\left\{y[t] \to \frac{1}{4}(-1 + \text{Cos}[2t] + 2\text{Sin}[2t])\right\}\right\}$$

```
Plot[y[t]/.partsol, {t, 0, 2Pi},
    AxesLabel → {t, y}, PlotStyle->CMYKColor[0, 0.89, 0.94, 0.28],
    AspectRatio → Automatic]
```

∎

EXAMPLE 4.2.5: Solve (a) $4y^{(4)} + 12y''' + 49y'' + 42y' + 10y = 0$ and (b) $y^{(4)} + 4y''' + 24y'' + 40y' + 100y = 0$.

SOLUTION: (a) The characteristic equation of $4y^{(4)} + 12y''' + 49y'' + 42y' + 10y = 0$ is $4k^4 + 12k^3 + 49k^2 + 42k + 10 = 0$. We use `Factor` to try to factor the characteristic polynomial, but see that Mathematica does not completely factor the polynomial,

```
Factor[4k^4 + 12k^3 + 49k^2 + 42k + 10]
```

$(1 + 2k)^2 \left(10 + 2k + k^2\right)$

unless we include the option `GaussianIntegers->True` in the `Factor` command.

```
Factor[4k^4 + 12k^3 + 49k^2 + 42k + 10,
    GaussianIntegers → True]
```

$((1 - 3i) + k)((1 + 3i) + k)(1 + 2k)^2$

From the results, we see that the solutions of the characteristic equation are $k = -1 \pm 3i$ and $k = -1/2$ with multiplicity 2. Of course, we obtain the same results with `Solve`.

```
Solve[4k^4 + 12k^3 + 49k^2 + 42k + 10 == 0]
```

$$\left\{ \{k \to -1 - 3i\}, \{k \to -1 + 3i\}, \left\{k \to -\frac{1}{2}\right\}, \left\{k \to -\frac{1}{2}\right\}\right\}$$

Four linearly independent solutions of the equation are then given by $y_1 = e^{-x}\cos 3x$, $y_2 = e^{-x}\sin 3x$, $y_3 = e^{-x/2}$, and $y_4 = xe^{-x/2}$. This tells us that a general solution is given by

$$y = e^{-x}\left(c_1 \cos 3x + c_2 \sin 3x\right) + e^{-x/2}\left(c_3 + c_4 x\right).$$

We obtain the same result with DSolve. The formula for the general solution is extracted from gensolc with gensolc[[1,1,2]].

```
gensolc = DSolve[4D[y[x], {x, 4}] + 12y'''[x]+

49y''[x] + 42y'[x] + 10y[x] == 0, y[x], x]
```

$$\left\{\left\{ y[x] \to e^{-x/2}C[3] + e^{-x/2}xC[4] + e^{-x}C[2]\text{Cos}[3x] \right.\right.$$
$$\left.\left. + e^{-x}C[1]\text{Sin}[3x]\right\}\right\}$$

In this case, we will graph the general solution for $(c_1, c_2, c_3, c_4) = (1, 0, 1, 0)$, $(0, 1, 0, 1)$, $(1, 1, 0, 1)$, $(1, -1, 1, 2)$, $(0, 2, 1, -2)$, and $(1, -2, 1, 2)$. We accomplish this by applying the pure function

```
gensolc[[1,1,2]] /. {C[1]->#[[1]],
C[2]->#[[2]], C[3]->#[[3]],C[4]->#[[4]]} &
```

to the set of ordered quadruples

```
{{1,0,1,0},{0,1,0,1},{1,1,0,1},{1,-1,1,2},
{0,2,1,-2}, {1,-2,1,2}}
```

with Map. Namely, given an argument #, the function

```
gensolc[[1,1,2]] /. {C[1]->#[[1]],
C[2]->#[[2]], C[3]->#[[3]],C[4]->#[[4]]} &
```

replaces C[1] in gensolc[[1,1,2]] by the first part of the argument, C[2] by the second part, C[3] by the third part, and C[4] by the fourth part.

```
toplot = ((gensolc[[1, 1, 2]])/.

{C[1] → (#1[[1]]), C[2] → (#1[[2]]),

C[3]->(#1[[3]]), C[4]->(#1[[4]])}&)/@

{{1, 0, 1, 0}, {0, 1, 0, 1}, {1, 1, 0, 1},
```

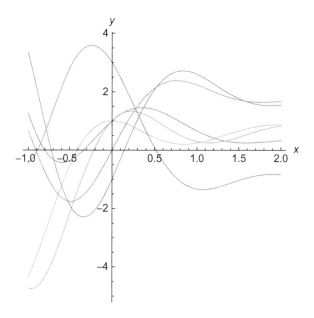

Figure 4-13   Various solutions of $4y^{(4)} + 12y''' + 49y'' + 42y' + 10y = 0$

```
{1, -1, 1, 2}, {0, 2, 1, -2},
{1, -2, 1, 2}};
```

We then graph the set of functions toplot on the interval $[-1, 2]$ with Plot in Figure 4-13.

```
Plot[toplot, {x, -1, 2}, AspectRatio → 1,
    AxesLabel → {x, y}]
```

(b) The characteristic equation of $y^{(4)} + 4y''' + 24y'' + 40y' + 100y = 0$ is $k^4 + 4k^3 + 24k^2 + 40k + 100 = 0$ which we can solve by factoring $k^4 + 4k^3 + 24k^2 + 40k + 100$ using Factor together with the options GaussianIntegers->True

```
Factor[k^4 + 4k^3 + 24k^2 + 40k + 100,
    GaussianIntegers → True]
```

$$((1 - 3i) + k)^2 ((1 + 3i) + k)^2$$

or using Solve.

```
Solve[k^4 + 4k^3 + 24k^2 + 40k + 100 == 0]
```

$$\{\{k \to -1-3i\}, \{k \to -1-3i\}, \{k \to -1+3i\}, \{k \to -1+3i\}\}$$

Thus, we see that the solutions of the characteristic equation are $k = -1 + 3i$ and $k = -1 - 3i$, each with multiplicity 2, so the corresponding solutions are $y_1 = e^{-x} \cos 3x$, $y_2 = e^{-x} \sin 3x$, $y_3 = xe^{-x} \cos 3x$, and $y_4 = xe^{-x} \sin 3x$. This tells us that a general solution is given by

$$y = e^{-x} [(c_1 + c_2 x) \cos 3x + (c_3 + c_4 x) \sin 3x].$$

We obtain the same result using `DSolve`.

```
gensold = DSolve[D[y[x], {x, 4}] + 4D[y[x], {x, 3}]+
    24y''[x] + 40y'[x] + 100y[x] == 0, y[x], x]
```

$$\{\{y[x] \to e^{-x}C[3]\text{Cos}[3x] + e^{-x}xC[4]\text{Cos}[3x] + e^{-x}C[1]\text{Sin}[3x]$$
$$+ e^{-x}xC[2]\text{Sin}[3x]\}\}$$

To graph the solution for various values of the constants, we proceed in the same manner as in (a). First, we define a list of ordered quadruples, `vals`.

```
vals = {{5, 0, 1, 0}, {0, 1, 0, -1},
    {1, 3, 0, 1}, {1, -1, 1, 2}, {0, 2, 1, -2},
    {1, -2, 5, 2}, {0, -3, 0, 2}, {3, 0, 0, 3},
    {1, 1, 1, 1}};
```

We then use `Map` to replace `C[1]` in `gensold[[1,1,2]]`, which represents the formula for the solution, by the first part of each quadruple in `vals`, `C[2]` by the second part, `C[3]` by the third part, and `C[4]` by the fourth part.

```
toplot = Map[gensold[[1, 1, 2]]/.
    {C[1] -> (#1[[1]]), C[2] -> (#1[[2]]),
    C[3]->(#1[[3]]), C[4]->(#1[[4]])}&,
    vals];
```

Next `Table` and `Plot` are used to graph each function in `toplot` on the interval $[-1/2, 3/2]$, naming the resulting list of nine graphics objects `ninegraphs`. Finally, we use `Partition` to partition the set `ninegraphs` into three element subsets, naming the resulting 3×3 array of graphics objects `todisplay`. We then display the array of graphics objects `todisplay` using `Show` together with `GraphicsGrid` in Figure 4-14.

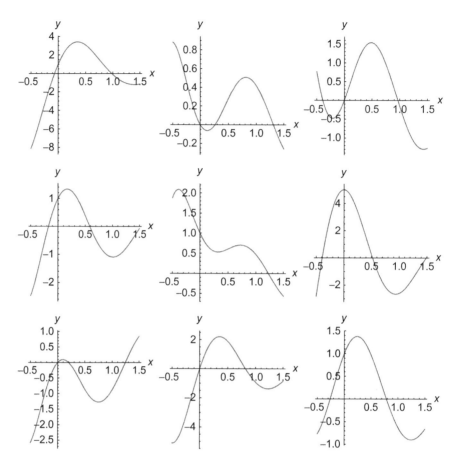

Figure 4-14   Various solutions of $y^{(4)} + 4y''' + 24y'' + 40y' + 100y = 0$

```
todisplay = Partition[ninegraphs, 3];

Show[GraphicsGrid[todisplay]]
```

■

EXAMPLE 4.2.6: Find a differential equation with general solution
$y = c_1 e^{-2t/3} + c_2 t e^{-2t/3} + c_3 t^2 e^{-2t/3} + c_4 \cos t + c_5 \sin t + c_6 t \cos t + c_7 t \sin t + c_8 t^2 \cos t + c_9 t^2 \sin t$.

SOLUTION: A linear homogeneous differential equation with constant coefficients that has this general solution has fundamental set of solutions

$$S = \left\{ e^{-2t/3}, te^{-2t/3}, t^2 e^{-2t/3}, \cos t, \sin t, t \cos t, t \sin t, t^2 \cos t, t^2 \sin t \right\}.$$

Hence, in the characteristic equation $k = -2/3$ has multiplicity 3 while $k = \pm i$ has multiplicity 3. The characteristic equation is

$$\left(k + \frac{2}{3}\right)^3 (k-i)^3 (k+i)^3 = k^9 + 2k^8 + \frac{13}{3}k^7 + \frac{170}{27}k^6 + 7k^5 + \frac{62}{9}k^4$$
$$+ 5k^3 + \frac{26}{9}k^2 + \frac{4}{3}k + \frac{8}{27},$$

where we use Mathematica to compute the multiplication with Expand.

```
Expand[27(k + 2/3)^3(k^2 + 1)^3]
```

$$8 + 36k + 78k^2 + 135k^3 + 186k^4 + 189k^5 + 170k^6 + 117k^7 + 54k^8 + 27k^9$$

```
Expand[(k + 2/3)^3(k^2 + 1)^3]
```

$$\frac{8}{27} + \frac{4k}{3} + \frac{26k^2}{9} + 5k^3 + \frac{62k^4}{9} + 7k^5 + \frac{170k^6}{27} + \frac{13k^7}{3} + 2k^8 + k^9$$

Thus, a differential equation obtained with the indicated general solution is

$$\frac{d^9 y}{dt^9} + 2\frac{d^8 y}{dt^8} + \frac{13}{3}\frac{d^7 y}{dt^7} + \frac{170}{27}\frac{d^6 y}{dt^6} + 7\frac{d^5 y}{dt^5} + \frac{62}{9}\frac{d^4 y}{dt^4} + 5\frac{d^3 y}{dt^3} + \frac{26}{9}\frac{d^2 y}{dt^2}$$
$$+ \frac{4}{3}\frac{dy}{dt} + \frac{8}{27}y = 0.$$

■

4.3 Introduction to Solving Nonhomogeneous Equations

In the previous section, we learned how to solve the nth-order linear homogeneous equation with real constant coefficients. These techniques are also useful in solving nonhomogeneous equations of the form

$$a_n y^{(n)} + a_{n-1} y^{(n-1)} + \cdots + a_1 y' + a_0 y = f(x), \tag{4.12}$$

where the a_i's are constant and $a_n \neq 0$. Before describing how to obtain solutions of some nonhomogeneous equations, we need to describe what is meant by a *general solution of a linear nonhomogeneous equation.*

Definition 17 (Particular Solution).   *A **particular solution**, $y_p(x)$, of the linear differential equation*

$$a_n(x)y^{(n)} + a_{n-1}(x)y^{(n-1)} + \cdots a_2(x)y'' + a_1(x)y' + a_0(x)y = f(x)$$

is a specific function that contains no arbitrary constants and satisfies the differential equation.

EXAMPLE 4.3.1: Verify that $y_p(x) = -\frac{3}{2}\sin x$ is a particular solution of $y'' - 2y' + y = 3\cos x$.

SOLUTION: After defining $y_p(x) = -\frac{3}{2}\sin x$,

```
yp[x_] = -3/2Sin[x];
```

Be sure to substitute y_p into the nonhomogeneous equation.

we compute and simplify $y_p'' - 2y_p' + y_p$

```
yp"[x] - 2yp'[x] + yp[x]
```

```
3Cos[x]
```

and see that the result is identically equal to $3\cos x$.

■

Suppose that y is *any* solution and that y_p is a particular solution of the nonhomogeneous equation

$$a_n(x)y^{(n)} + a_{n-1}(x)y^{(n-1)} + \cdots + a_2(x)y'' + a_1(x)y' + a_0(x)y = f(x)$$

and that y_h is a general solution of the corresponding homogeneous equation

$$a_n(x)y^{(n)} + a_{n-1}(x)y^{(n-1)} + \cdots + a_2(x)y'' + a_1(x)y' + a_0(x)y = 0.$$

Then, $y - y_p$ is a solution of the corresponding homogeneous equation so

$$y - y_p = y_h \qquad \text{or} \qquad y = y_h + y_p.$$

Thus, *any* solution of the nonhomogeneous equation can be written as the sum of particular solution to the nonhomogeneous equation added to the general solution of the corresponding homogeneous equation.

Definition 18 (General Solution of a Nonhomogeneous Equation). *A general solution of the nonhomogeneous equation*

$$a_n(x)y^{(n)} + a_{n-1}(x)y^{(n-1)} + \cdots + a_2(x)y'' + a_1(x)y' + a_0(x)y = f(x)$$

is

$$y = y_h + y_p,$$

where y_h is a general solution of the corresponding homogeneous equation

$$a_n(x)y^{(n)} + a_{n-1}(x)y^{(n-1)} + \cdots + a_2(x)y'' + a_1(x)y' + a_0(x)y = 0$$

and y_p is a particular solution to the nonhomogeneous equation.

EXAMPLE 4.3.2: Find a general solution of $y'' + 6y' + 13y = 2e^{-2x}\sin x$ if

$$y_p = e^{-2x}\left(-\frac{1}{5}\cos x + \frac{2}{5}\sin x\right)$$

is a particular solution to the nonhomogeneous equation and

$$y_h = e^{-3x}(c_1\cos 2x + c_2\sin 2x)$$

is a general solution of the corresponding homogeneous equation.

SOLUTION: We first show that $y_p = e^{-2x}\left(-\frac{1}{5}\cos x + \frac{2}{5}\sin x\right)$ is a particular solution of $y'' + 6y' + 13y = 2e^{-2x}\sin x$. After defining y_p, we calculate $y_p'' + 6y_p' + 13y_p$

```
yp[x_] = Exp[-2x](-1/5Cos[x] + 2/5Sin[x]);
yp"[x] + 6yp'[x] + 13yp[x]//Simplify
```

```
2e^-2xSin[x]
```

and see that the result is $2e^{-2x}\sin x$. We see that $y_h = e^{-3x}(c_1\cos 2x + c_2\sin 2x)$ is a general solution of the corresponding homogeneous equation $y'' + 6y' + 13y = 0$ with DSolve.

```
Clear[y]
DSolve[y"[x] + 6y'[x] + 13y[x] == 0,
  y[x], x]
```

$$\left\{\left\{y[x] \rightarrow e^{-3x}C[2]\text{Cos}[2x] + e^{-3x}C[1]\text{Sin}[2x]\right\}\right\}$$

Thus, a general solution of the equation is $y = y_h + y_p = e^{-3x}(c_1 \cos 2x + c_2 \sin 2x) + e^{-2x}\left(-\frac{1}{5}\cos x + \frac{2}{5}\sin x\right)$. We now graph the general solution for various values of the arbitrary constant. To do so, we define $y_h = e^{-3x}(c_1 \cos 2x + c_2 \sin 2x)$ and $y(x) = y_h(x) + y_p(x)$.

```
yh[x_] = Exp[-3x](c1Cos[2x] + c2Sin[2x]);
```

```
y[x_] = yh[x] + yp[x];
```

Then, we use Table to create a list of functions obtained by replacing c_1 in $y(x)$ by $-1, 0$, and 1 and c_2 by $-1, 0$, and 1. The resulting list of functions toplot is graphed with Plot in Figure 4-15.

```
toplot = Table[y[x], {c1, -1, 1},

   {c2, -1, 1}];
```

```
Plot[toplot, {x, 0, 2}, PlotRange → All,

   AspectRatio → 1, AxesLabel → {x, y}]
```

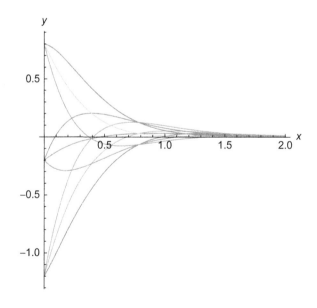

Figure 4-15   Various solutions to a nonhomogeneous equation

∎

Techniques for solving nonhomogeneous equations with constant coefficients are discussed in the next two sections. In addition, you can often use DSolve to find a general solution of a linear nonhomogeneous equation.

EXAMPLE 4.3.3: Solve the initial-value problem

$$\begin{cases} y'' + y = \cos \omega x \\ y(0) = y'(0) = 0 \end{cases}.$$

Graph the solution for various values of ω, including $\omega = 1$.

SOLUTION: We first use DSolve to solve the initial value problem. Note that the result is not valid if $\omega = 1$.

```
Clear[y]

DSolve[{y"[x] + y[x] == Cos[ωx],

   y[0] == 0, y'[0] == 0}, y[x], x]//

   Simplify
```

$$\left\{\left\{y[x] \to \frac{\text{Cos}[x] - \text{Cos}[x\omega]}{-1 + \omega^2}\right\}\right\}$$

In the following commands, observe that we "miss" the value $\omega = 1$ by our choice of the stepvalues in the Table command.

```
graphs = Table[Plot[ (Cos[x] - Cos[xω])/(-1 + ω²), {x, 0, 12Pi},

   Ticks → {{0, 12Pi}, {-1, 1}}, AxesLabel → {x, y},

   PlotStyle->CMYKColor[0, 0.89, 0.94, 0.28],

   AspectRatio → 1],

   {ω, 0, 2.1, 2.1/8}];
```

From the graphs shown in Figure 4-16, we see that the solution to the initial-value problem is bounded and periodic if $\omega \neq 1$.

```
toshow = Partition[graphs, 3];

Show[GraphicsGrid[toshow]]
```

We consider $\omega = 1$ separately.

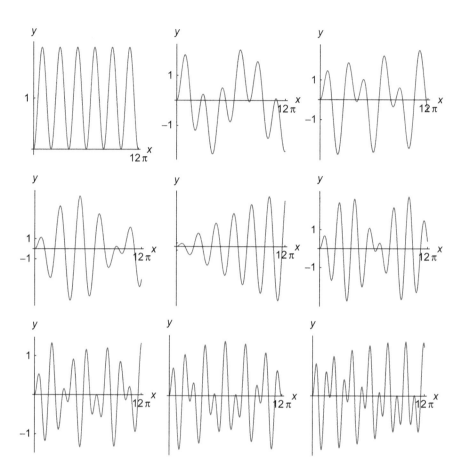

Figure 4-16   Solutions of $y'' + y = \cos \omega x$, $y(0) = y'(0) = 0$ for $\omega \neq 1$

```
sol = DSolve[{y"[x] + y[x] == Cos[x], y[0] == 0,
    y'[0] == 0}, y[x], x]
```

$$\left\{ \left\{ y[x] \to \frac{1}{4} \left(-2\text{Cos}[x] + 2\text{Cos}[x]^3 + 2x\text{Sin}[x] + \text{Sin}[x]\text{Sin}[2x] \right) \right\} \right\}$$

```
FullSimplify[sol]
```

$$\left\{ \left\{ y[x] \to \frac{1}{2} x\text{Sin}[x] \right\} \right\}$$

```
Plot[1/2 xSin[x], {x, 0, 24Pi}, AxesLabel → {x, y},
    PlotStyle->CMYKColor[0, 0.89, 0.94, 0.28],
    AspectRatio → 1]
```

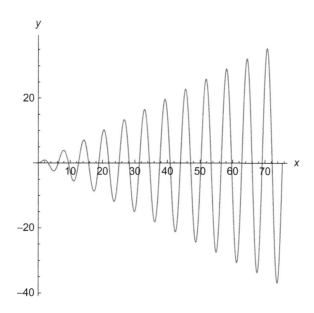

Figure 4-17 Solution of $y'' + y = \cos \omega x$, $y(0) = y'(0) = 0$ for $\omega = 1$

In Figure 4-17, we see that if $\omega = 1$ the solution is unbounded and not periodic.

∎

4.4 Nonhomogeneous Equations With Constant Coefficients: The Method of Undetermined Coefficients

Consider the nth-order linear differential equation with constant coefficients

$$a_n y^{(n)} + a_{n-1} y^{(n-1)} + \cdots + a_2 y'' + a_1 y' + a_0 y = f(x).$$

We know that a general solution of this differential equation is given by $y = y_h + y_p$ where y_p is a particular solution of the nonhomogeneous equation and y_h is a solution of the corresponding homogeneous equation

$$a_n y^{(n)} + a_{n-1} y^{(n-1)} + \cdots + a_2 y'' + a_1 y' + a_0 y = 0.$$

If $f(x)$ is a linear combination of the functions $1, x, x^2, \ldots, e^{kx}, xe^{kx}, x^2e^{kx}, \ldots,$ $e^{\alpha x}\cos\beta x, e^{\alpha x}\sin\beta x, xe^{\alpha x}\cos\beta x, xe^{\alpha x}\sin\beta x, x^2e^{\alpha x}\cos\beta x, x^2e^{\alpha x}\sin\beta x, \ldots$ the *Method of Undetermined Coefficients* provides a method that we can use to determine a particular solution of the nonhomogeneous equation.

Outline of the Method of Undetermined Coefficients

1. Solve the corresponding homogeneous equation for $y_h(x)$.
2. Determine the form of a particular solution $y_p(x)$. (See **Determining the Form** of $y_p(x)$ next.)
3. Determine the unknown coefficients in $y_p(x)$ by substituting $y_p(x)$ into the nonhomogeneous equation and equating the coefficients of like terms.
4. Form a general solution with $y(x) = y_h(x) + y_p(x)$.

Determining the Form of $y_p(x)$ (Step 2):

Suppose that $f(x) = b_1 f_1(x) + b_2 f_2(x) + \cdots + b_j f_j(x)$, where $b_1, b_2, \ldots, b_j$ are constants and each $f_i(x)$, $i = 1, 2, \ldots, j$, is a function of the form x^m, $x^m e^{kx}$, $x^m e^{\alpha x}\cos\beta x$, or $x^m e^{\alpha x}\sin\beta x$.

1. If $f_i(x) = x^m$, the associated set of functions is

$$S = \left\{1, x, x^2, \ldots, x^m\right\}.$$

2. If $f_i(x) = x^m e^{kx}$, the associated set of functions is

$$S = \left\{e^{kx}, xe^{kx}, x^2 e^{kx}, \ldots, x^m e^{kx}\right\}.$$

3. If $f_i(x) = x^m e^{\alpha x}\cos\beta x$ or $f_i(x) = x^m e^{\alpha x}\sin\beta x$, the associated set of functions is

$$S = \{e^{\alpha x}\cos\beta x, xe^{\alpha x}\cos\beta x, x^2 e^{\alpha x}\cos\beta x, \ldots, x^m e^{\alpha x}\cos\beta x,$$
$$e^{\alpha x}\sin\beta x, xe^{\alpha x}\sin\beta x, x^2 e^{\alpha x}\sin\beta x, \ldots, x^m e^{\alpha x}\sin\beta x\}.$$

For each function $f_i(x)$ in $f(x)$, determine the associated set of functions S. If any of the functions in S appears in the general solution to the corresponding homogeneous equation, $y_h(x)$, multiply each function in S by x^r to obtain a new set $x^r S$, where r is the smallest positive integer so that each function in $x^r S$ is not a solution of the corresponding homogeneous equation. A particular solution is obtained by taking the linear combination of all functions in the associated sets where repeated functions should appear only once in the particular solution.

4.4.1 Second-Order Equations

EXAMPLE 4.4.1: Solve the nonhomogeneous equations (a) $y'' + 5y' + 6y = 2e^x$ and (b) $y'' + 5y' + 6y = 3e^{-2x}$.

SOLUTION: (a) The corresponding homogeneous equation $y'' + 5y' + 6y = 0$ has general solution $y_h = c_1 e^{-2x} + c_2 e^{-3x}$.

```
homsol = DSolve[y"[x] + 5y'[x] + 6y[x]

    == 0, y[x], x]
```

$$\{\{y[x] \to e^{-3x} C[1] + e^{-2x} C[2]\}\}$$

Next, we determine the form of $y_p(x)$. We choose $S = \{e^x\}$ because $f(x) = 2e^x$. Notice that e^x is not a solution to the homogeneous equation, so we take $y_p(x)$ to be the linear combination of the functions in S. Therefore,

$$y_p(x) = Ae^x.$$

```
yp[x_] = aExp[x]
```

$$ae^x$$

Substituting this solution into $y'' + 5y' + 6y = 2e^x$, we have

$$Ae^x + 5Ae^x + 6Ae^x = 12Ae^x = 2e^x.$$

```
eqn = yp"[x] + 5yp'[x] + 6yp[x] == 2Exp[x]
```

$$12ae^x == 2e^x$$

Equating the coefficients of e^x then gives us $A = 1/6$.

```
aval = SolveAlways[eqn, x]
```

$$\left\{\left\{a \to \frac{1}{6}\right\}, \{x \to -\infty\}\right\}$$

```
eqn2 = eqn/.x → 0
```

$$12a == 2a$$

```
aval = Solve[eqn2]
```

$$\left\{\left\{a \to \frac{1}{6}\right\}\right\}$$

and a general solution of the nonhomogeneous equation is

$$y = y_h + y_p = c_1 e^{-2x} + c_2 e^{-3x} + \frac{1}{6}e^x.$$

```
yp[x_] = aExp[x]/.aval[[1, 1]]
```

$$\frac{e^x}{6}$$

In this case, we find the same general solution with DSolve.

```
gensol =
    DSolve[y''[x] + 5y'[x] + 6y[x] == 2Exp[x],
        y[x], x, GeneratedParameters → c]
```

$$\left\{\left\{y[x] \to \frac{e^x}{6} + e^{-3x}c[1] + e^{-2x}c[2]\right\}\right\}$$

We then graph the general solution for various values of the arbitrary constants in the same way as in other examples. See Figure 4-18.

```
toplot = Table[gensol[[1, 1, 2]],
    {c[1], -1, 1}, {c[2], -1, 1}];

Plot[toplot, {x, -3, 5},
    PlotRange → {-3, 5}, AspectRatio → 1,
    AxesLabel → {x, y}]
```

(b) In this case, we see that $f(x) = 3e^{-2x}$ so the associated set is $S = \{e^{-2x}\}$. However, because $y = e^{-2x}$ is a solution to the corresponding homogeneous equation, we must multiply each element of this set by x^r so that no element is a solution of the corresponding homogeneous equation. We multiply the element of

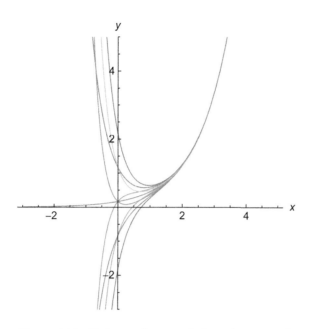

Figure 4-18   Various solutions of $y'' + 5y' + 6y = 2e^x$

S by x to obtain $xS = \{xe^{-2x}\}$ because xe^{-2x} is not a solution of $y'' + 5y' + 6y = 0$. Hence, $y_p(x) = Axe^{-2x}$. Differentiating $y_p(x)$ twice

```
yp[x_] = a x Exp[-2x];
yp'[x]
yp"[x]
```

$ae^{-2x} - 2ae^{-2x}x$

$-4ae^{-2x} + 4ae^{-2x}x$

and substituting into the nonhomogeneous equation yields

$$y'' + 5y' + 6y = -4Ae^{-2x} + 4Axe^{-2x} + 5\left(Ae^{-2x} - 2Axe^{-2x}\right) + 6Axe^{-2x}$$
$$= Ae^{-2x} = 3e^{-2x}$$

```
eqn = yp"[x] + 5yp'[x] + 6yp[x] == 3Exp[-2x]
```

$-4ae^{-2x} + 10ae^{-2x}x + 5\left(ae^{-2x} - 2ae^{-2x}x\right) == 3e^{-2x}$

so $A = 4$ and $y_p(x) = 3xe^{-2x}$.

```
aval = SolveAlways[eqn, x]
```

$\{\{a \rightarrow 3\}\}$

```
eqn2 = eqn/.x → 0
```

$a == 3$

```
aval = Solve[eqn]
```

$\{\{a \rightarrow 3\}\}$

```
yp[x_] = yp[x]/.aval[[1]]
```

$3e^{-2x}x$

A general solution of $y'' + 5y' + 6y = 3e^{-2x}$ is

$$y = y_h + y_p = c_1e^{-2x} + c_2e^{-3x} + 3xe^{-2x}.$$

As in (a), we can use DSolve to obtain equivalent results. For example, entering

```
gensol = DSolve[{y"[x] + 5y'[x] + 6y[x] == 3 Exp[-2x],
    y[0] == a, y'[0] == b}, y[x], x]
```

$$\left\{\left\{y[x] \rightarrow e^{-3x}\left(3 - 2a - b - 3e^x + 3ae^x + be^x + 3e^xx\right)\right\}\right\}$$

solves the equation subject to the initial conditions $y(0) = a$ and $y'(0) = b$ and names the resulting output gensol. Thus entering

```
toplot = Table[gensol[[1, 1, 2]], {a, -1, 1},
    {b, -1, 1}];
Plot[toplot, {x, -2, 3}, PlotRange → {-2, 3},
    AspectRatio → 1, AxesLabel → {x, y}]
```

defines toplot to be the set consisting of nine functions corresponding to the solutions of $y'' + 5y' + 6y = 3e^{-2x}$ that satisfy the initial conditions $y(0) = a$ and $y'(0) = b$ for $a = -1, 0$, and 1 and $b = -1, 0$, and 1; and then graphs the set of functions toplot on the interval $[-2, 3]$ in Figure 4-19.

∎

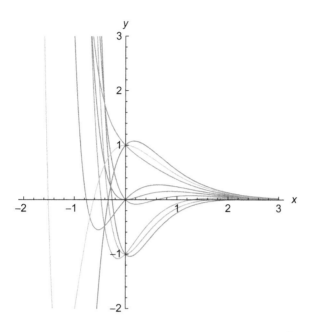

Figure 4-19   Various solutions of $y'' + 5y' + 6y = 3e^{-2x}$

EXAMPLE 4.4.2: Solve

$$4\frac{d^2y}{dt^2} - y = t - 2 - 5\cos t - e^{-t/2}.$$

SOLUTION: The corresponding homogeneous equation is $4y'' - y = 0$ with general solution $y_h = c_1 e^{-t/2} + c_2 e^{t/2}$.

```
DSolve[4y"[t] − y[t] == 0, y[t], t]
```

$$\left\{\left\{y[t] \to e^{t/2} C[1] + e^{-t/2} C[2]\right\}\right\}$$

A fundamental set of solutions for the corresponding homogeneous equation is $S = \left\{e^{-t/2}, e^{t/2}\right\}$. The associated set of functions for $t - 2$ is $F_1 = \{1, t\}$, the associated set of functions for $-5\cos t$ is $F_2 = \{\cos t, \sin t\}$, and the associated set of functions for $-e^{-t/2}$ is $F_3 = \left\{e^{-t/2}\right\}$. Note that $e^{-t/2}$ is an element of S so we multiply F_3 by t resulting in $tF_3 = \left\{te^{-t/2}\right\}$.

No element of F_1 is contained in S and no element of F_2 is contained in S. In other words, the elements of the sets F_1 and F_2 and S are linearly independent.

Then, we search for a particular solution of the form

$$y_p = A + Bt + C\cos t + D\sin t + Ete^{-t/2},$$

where A, B, C, D, and E are constants to be determined.

```
yp[t_] = a + bt + cCos[t] + dSin[t] + etExp[-t/2]
```

$$a + bt + ee^{-t/2}t + c\cos[t] + d\sin[t]$$

Computing y_p' and y_p''

```
dyp = yp'[t]
d2yp = yp"[t]
```

$$b + ee^{-t/2} - \frac{1}{2}ee^{-t/2}t + d\cos[t] - c\sin[t]$$

$$-ee^{-t/2} + \frac{1}{4}ee^{-t/2}t - c\cos[t] - d\sin[t]$$

and substituting into the nonhomogeneous equation results in

$$-A - Bt - 5C\cos t - 5D\sin t - 4Ee^{-t/2} = t - 2 - 5\cos t - e^{-t/2}.$$

```
eqn = 4yp"[t] - yp[t] == t - 2 - 5Cos[t] - Exp[-t/2]//Simplify
```

$$a + 4ee^{-t/2} + t + bt + 5(-1 + c)\cos[t] + 5d\sin[t] == 2 + e^{-t/2}$$

Equating coefficients results in

$$-A = -2 \qquad -B = 1 \qquad -5C = -5 \qquad -5D = 0 \qquad -4E = -1$$

so $A = 2$, $B = -1$, $C = 1$, $D = 0$, and $E = 1/4$.

```
cvals = SolveAlways[eqn, t]
```

```
$Aborted
```

Observe in this case `SolveAlways` is unable to solve the equation for a, b, c, d, and e so that it is true for all values of t so the command is aborted. Instead, we must equate coefficients ourselves and then solve the resulting equations.

```
cvals = Solve[{-a == -2, -b == 1, -5c == -5, -5d == 0,
              -4e == -1}]
```

$$\left\{\left\{a \to 2, b \to -1, c \to 1, d \to 0, e \to \frac{1}{4}\right\}\right\}$$

You can repeat the trial using multiples of $\pi/2$ and obtain the same result. Note that $-A-Bt-5C\cos t-5D\sin t-4Ee^{-t/2} = t-2-5\cos t-e^{-t/2}$ is true for *all* values of t. Evaluating for five different values of t gives us five equations that we then solve for A, B, C, D, and E, resulting in the same solutions as already obtained.

```
e1 = eqn/.t → 0
```

```
e2 = eqn/.t → 1
```

```
e3 = eqn/.t → 2
```

```
e4 = eqn/.t → 3
```

```
e5 = eqn/.t → 4
```

$a + 5(-1 + c) + 4e == 3$

$1 + a + b + \dfrac{4e}{\sqrt{e}} + 5(-1 + c)\text{Cos}[1] + 5d\text{Sin}[1] == 2 + \dfrac{1}{\sqrt{e}}$

$2 + a + 2b + \dfrac{4e}{e} + 5(-1 + c)\text{Cos}[2] + 5d\text{Sin}[2] == 2 + \dfrac{1}{e}$

$3 + a + 3b + \dfrac{4e}{e^{3/2}} + 5(-1 + c)\text{Cos}[3] + 5d\text{Sin}[3] == 2 + \dfrac{1}{e^{3/2}}$

$4 + a + 4b + \dfrac{4e}{e^2} + 5(-1 + c)\text{Cos}[4] + 5d\text{Sin}[4] == 2 + \dfrac{1}{e^2}$

```
cvals = Solve[{e1, e2, e3, e4, e5}, {a, b, c, d, e}]//Simplify
```

$\left\{\left\{a \to 2, b \to -1, c \to 1, d \to 0, e \to \dfrac{1}{4}\right\}\right\}$

y_p is then given by $y_p = 2 - t + \cos t + \frac{1}{4}te^{-t/2}$

```
sol2 = DSolve[4y″[t] − y[t] == t − 2 − 5Cos[t]
```

```
    −Exp[−t/2], y[t], t]//Simplify
```

$\left\{\left\{y[t] \to \dfrac{1}{4}e^{-t/2}\left(1 - 4e^{t/2}(-2 + t) + t + 4e^t C[1] + 4C[2] + 4e^{t/2}\text{Cos}[t]\right)\right\}\right\}$

and a general solution is given by

$$y = y_h + y_p = c_1 e^{-t/2} + c_e^{t/2} + 2 - t + \cos t + \frac{1}{4}te^{-t/2}.$$

■

In order to solve an initial-value problem, first determine a general solution and then use the initial conditions to solve for the unknown constants in the general solution.

EXAMPLE 4.4.3: Solve $y'' + 4y = \cos 2t$, $y(0) = 0$, $y'(0) = 0$.

SOLUTION: A general solution of the corresponding homogeneous equation is $y_h = c_1 \cos 2t + c_2 \sin 2t$.

```
DSolve[y"[t] + 4y[t] == 0, y[t], t]
```

$$\{\{y[t] \to C[1]\text{Cos}[2t] + C[2]\text{Sin}[2t]\}\}$$

For this equation, $F = \{\cos 2t, \sin 2t\}$. Because elements of F are solutions to the corresponding homogeneous equation, we multiply each element of F by t resulting in $tF = \{t \cos 2t, t \sin 2t\}$. Therefore, we assume that a particular solution has the form

This means that elements in F and S are linearly dependent.

This means that elements in tF and S are linearly independent.

$$y_p = At \cos 2t + Bt \sin 2t,$$

where A and B are constants to be determined. Proceeding in the same manner as before, we compute y'_p and y''_p

```
yp[t_] = atCos[2t] + btSin[2t];

yp'[t]

yp"[t]
```

$$a\text{Cos}[2t] + 2bt\text{Cos}[2t] + b\text{Sin}[2t] - 2at\text{Sin}[2t]$$

$$4b\text{Cos}[2t] - 4at\text{Cos}[2t] - 4a\text{Sin}[2t] - 4bt\text{Sin}[2t]$$

and then substitute into the nonhomogeneous equation

```
eqn = yp"[t] + 4yp[t] == Cos[2t]
```

$$4b\text{Cos}[2t] - 4at\text{Cos}[2t] - 4a\text{Sin}[2t] - 4bt\text{Sin}[2t]$$

$$+ 4(at\text{Cos}[2t] + bt\text{Sin}[2t])$$

$$== \text{Cos}[2t]$$

Equating coefficients readily yields $A = 0$ and $B = 1/4$. Alternatively, remember that $-4A \sin 2t + 4B \cos 2t = \cos 2t$ is true for *all* values of t. Evaluating for two values of t and then solving for A and B gives the same result.

```
SolveAlways[eqn, t]
```

$\{\{a \rightarrow 0, b \rightarrow \frac{1}{4}\}, \{a \rightarrow 0, b \rightarrow \frac{1}{4}\}, \{a \rightarrow 0, \text{Cos}[2t] \rightarrow$

$0\}, \{a \rightarrow 0, \text{Sin}[2t] \rightarrow -1\}, \{a \rightarrow 0, \text{Sin}[2t] \rightarrow 1\}, \{b \rightarrow$

$\frac{1}{4}, a \rightarrow 0\}, \{b \rightarrow \frac{1}{4}, \text{Sin}[2t] \rightarrow 0\}, \{b \rightarrow \frac{1}{4}, \text{Sin}[2t] \rightarrow$

$0\}, \{\text{Sin}[2t] \rightarrow 0, b \rightarrow \frac{1}{4}\}, \{\text{Sin}[2t] \rightarrow 0, \text{Cos}[2t] \rightarrow 0\}, \{a \rightarrow$

$-\frac{1}{4}i(-1+4b)\text{Csc}[2t]\sqrt{-1+\text{Sin}[2t]^2}\}, \{a \rightarrow -\frac{1}{4}i(-1+4b)\text{Csc}[2t]$

$\sqrt{-1+\text{Sin}[2t]^2}\}, \{a \rightarrow \frac{1}{4}i(-1+4b)\text{Csc}[2t]\sqrt{-1+\text{Sin}[2t]^2}\}, \{a \rightarrow$

$\frac{1}{4}i(-1+4b)\text{Csc}[2t]\sqrt{-1+\text{Sin}[2t]^2}\}, \{\text{Cos}[2t] \rightarrow \frac{4a\text{Sin}[2t]}{-1+4b}\}, \{a \rightarrow$

$\frac{1}{4}(-1+4b)\text{Cot}[2t]\}, \{b \rightarrow \frac{1}{4}(1+4a\text{Tan}[2t])\}, \{\}, \{\}\}$

```
e1 = eqn/.t → 0

e2 = eqn/.t->Pi/4

cvals = Solve[{e1, e2}]
```

$4b == 1$

$-4a == 0$

$\left\{\left\{a \rightarrow 0, b \rightarrow \frac{1}{4}\right\}\right\}$

It follows that $y_p = \frac{1}{4}t \sin 2t$ and $y = c_1 \cos 2t + c_2 \sin 2t + \frac{1}{4}t \sin 2t$.

```
yp[t]/.cvals[[1]]
```

$\frac{1}{4}t\text{Sin}[2t]$

```
y[t_] = c[1]Cos[2t] + c[2]Sin[2t] + 1/4 tSin[2t];
```

Applying the initial conditions

```
y'[t]//Simplify
```

$\frac{1}{4}(2(t+4c[2])\text{Cos}[2t] + (1 - 8c[1])\text{Sin}[2t])$

```
cvals = Solve[{y[0] == 0, y'[0] == 0}]
```

$\{\{c[1] \rightarrow 0, c[2] \rightarrow 0\}\}$

results in $y = \frac{1}{4}t \sin 2t$, which we graph with Plot in Figure 4-20.

```
y[t]/.cvals[[1]]
```

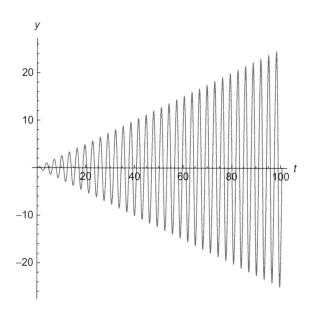

Figure 4-20   The forcing function causes the solution to become unbounded as $t \to \infty$

$\dfrac{1}{4} t \text{Sin}[2t]$

```
Plot[y[t]/.cvals[[1]], {t, 0, 32Pi},

    PlotStyle->CMYKColor[0, 0.89, 0.94, 0.28],

    AxesStyle → Black, AxesLabel → {t, y},

    AspectRatio → 1]
```

We verify the calculation with DSolve.

```
Clear[y]

DSolve[{y″[t] + 4y[t] == Cos[2t], y[0] == 0,

    y′[0] == 0}, y[t], t]//Simplify
```

$$\left\{ \left\{ y[t] \to \frac{1}{4} t \text{Sin}[2t] \right\} \right\}$$

■

Initial-value problems and boundary-value problems can exhibit dramatically different behavior.

EXAMPLE 4.4.4: Show that the boundary-value problem

$$\begin{cases} 4y'' + 4y' + 37y = \cos 3x \\ y(0) = y(\pi) \end{cases}$$

has infinitely many solutions.

SOLUTION: Boundary-value problems can be difficult to solve because the theory that guarantees existence of solutions (and unique solutions) is quite complex. For this example, we first find a general solution of the corresponding homogeneous equation.

```
Clear[x, y]

homsol = DSolve[4y"[x] + 4y'[x] + 37y[x] == 0,

  y[x], x]
```

$$\left\{ \left\{ y[x] \to e^{-x/2} C[2] Cos[3x] + e^{-x/2} C[1] Sin[3x] \right\} \right\}$$

Using the Method of Undetermined Coefficients, we find a particular solution to the nonhomogeneous equation of the form $y_p = A \cos 3x + B \sin 3x$. Substitution into the nonhomogeneous equation yields

```
yp[x_] = aCos[3x] + bSin[3x]
```

$$aCos[3x] + bSin[3x]$$

```
step1 = 4yp"[x] + 4yp'[x] + 37yp[x] == Cos[3x]//

  Simplify
```

$$(-1 + a + 12b)Cos[3x] + (-12a + b)Sin[3x] == 0$$

This equation is true for all values of x. In particular, substituting $x = 0$ and $x = 1$ yields two equations

```
e1 = step1/.x → 0;

e2 = step1/.x → 1;
```

that we then solve for A and B

```
vals = Solve[{e1, e2}, {a, b}]
```

$$\left\{\left\{a \to \frac{1}{145}, b \to \frac{12}{145}\right\}\right\}$$

to see that $A = 1/145$ and $B = 12/145$.

```
yp[x_] = (yp[x]/.vals)[[1]]
```

$$\frac{1}{145}\text{Cos}[3x] + \frac{12}{145}\text{Sin}[3x]$$

```
y[x_] = e^{-x/2}C[2]Cos[3x] + e^{-x/2}C[1]Sin[3x] + yp[x]
```

$$\frac{1}{145}\text{Cos}[3x] + e^{-x/2}C[2]\text{Cos}[3x] + \frac{12}{145}\text{Sin}[3x] + e^{-x/2}C[1]\text{Sin}[3x]$$

Applying the boundary conditions indicates that $\frac{1}{145} + c_2 = -\frac{1}{145} - e^{-\pi/2}c_2$

```
y[0]
```

$$\frac{1}{145} + C[2]$$

```
y[Pi]
```

$$-\frac{1}{145} - e^{-\pi/2}C[2]$$

so $c_2 = \frac{2}{145(1+e^{-\pi/2})}$; c_1 is arbitrary.

```
cval = Solve[y[0] == y[Pi]]
```

$$\left\{\left\{C[2] \to -\frac{2e^{\pi/2}}{145\left(1 + e^{\pi/2}\right)}\right\}\right\}$$

```
N[cval]
```

```
{{C[2] → -0.0114193}}
```

```
y[x_] = (y[x]/.cval)[[1]]
```

$$\frac{1}{145}\text{Cos}[3x] - \frac{2e^{\frac{\pi}{2}-\frac{x}{2}}\text{Cos}[3x]}{145\left(1 + e^{\pi/2}\right)} + \frac{12}{145}\text{Sin}[3x] + e^{-x/2}C[1]\text{Sin}[3x]$$

Note that DSolve is able to solve this boundary-value problem as well.

```
Clear[x, y]

sol = DSolve[{4y''[x] + 4y'[x] + 37y[x] == Cos[3x],

  y[0] == y[Pi]}, y[x], x,

  GeneratedParameters → c]
```

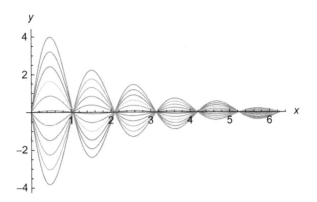

Figure 4-21   The boundary-value problem has infinitely many solutions

$$\{\{y[x] \;\to\; \frac{1}{1740(1+e^{\pi/2})}e^{-x/2}(-24e^{\pi/2}\text{Cos}[3x] \;+\; 12e^{\frac{\pi}{2}+\frac{x}{2}}\text{Cos}$$

$$[3x]\text{Cos}[6x] \;+\; 12e^{x/2}\text{Cos}[3x]\text{Cos}[6x] \;+\; 145e^{\frac{\pi}{2}+\frac{x}{2}}\text{Sin}[3x] \;+$$

$$145e^{x/2}\text{Sin}[3x]+1740c[1]\text{Sin}[3x]+1740e^{\pi/2}c[1]\text{Sin}[3x]+$$

$$e^{\frac{\pi}{2}+\frac{x}{2}}\text{Cos}[6x]\text{Sin}[3x]+e^{x/2}\text{Cos}[6x]\text{Sin}[3x]-e^{\frac{\pi}{2}+\frac{x}{2}}\text{Cos}[3x]\text{Sin}[6x]-$$

$$e^{x/2}\text{Cos}[3x]\text{Sin}[6x] + 12e^{\frac{\pi}{2}+\frac{x}{2}}\text{Sin}[3x]\text{Sin}[6x]$$

$$+ 12e^{x/2}\text{Sin}[3x]\text{Sin}[6x])\}\}$$

Several solutions are then graphed with Plot in Figure 4-21.

```
toplot = Table[sol[[1, 1, 2]], {c[1], -5, 5}];

Plot[toplot, {x, 0, 2Pi}, AxesLabel → {x, y},

  PlotRange → All]
```

■

EXAMPLE 4.4.5: Graph the solution to the initial-value problem

$$\begin{cases} x'' + 4x = \sin \omega t \\ x(0) = 1,\ x'(0) = 0 \end{cases}$$

for various values of ω, including $\omega = 2$.

SOLUTION: First, we find a general solution of the corresponding homogeneous equation.

```
Clear[x, t]
```

```
homsol = DSolve[x"[t] + 4x[t] == 0, x[t], t]
```

$\{\{x[t] \to C[1]\text{Cos}[2t] + C[2]\text{Sin}[2t]\}\}$

If $\omega \neq 2$, we can find a particular solution to the nonhomogeneous equation of the form $x_p = A \cos \omega t + B \sin \omega t$. We substitute this function into the nonhomogeneous equation and simplify the result.

```
xp[t_] = aCos[ωt] + bSin[ωt];
```

```
step1 = xp"[t] + 4xp[t] == Sin[ωt]//

    Simplify
```

$-\left(-4 + \omega^2\right)(a\text{Cos}[t\omega] + b\text{Sin}[t\omega]) == \text{Sin}[t\omega]$

This equation is true for all values of t. In particular, substituting $t = 0$ and $t = \pi/(2\omega)$ yields two equations

```
eq1 = step1/. t → 0
```

$-a\left(-4 + \omega^2\right) == 0$

```
eq2 = step1/. t → 1
```

$-\left(-4 + \omega^2\right)(a\text{Cos}[\omega] + b\text{Sin}[\omega]) == \text{Sin}[\omega]$

that we then solve to determine A and B.

```
coeffs = Solve[{eq1, eq2}, {a, b}]
```

$\left\{\left\{a \to 0, b \to -\dfrac{1}{-4 + \omega^2}\right\}\right\}$

We then form a particular solution, x_p, to the nonhomogeneous equation and a general solution to the nonhomogeneous equation, $x = x_h + x_p$.

```
xp[t_] = xp[t]/.coeffs[[1]];
```

```
x[t_] = homsol[[1, 1, 2]] + xp[t]
```

$C[1]\text{Cos}[2t] + C[2]\text{Sin}[2t] - \dfrac{\text{Sin}[t\omega]}{-4 + \omega^2}$

The solution to the initial-value problem is found by applying the initial conditions

```
cvals = Solve[{x[0] == 1, x'[0] == 0},

  {C[1], C[2]}]
```

$$\left\{\left\{C[1] \to 1, \ C[2] \to \frac{\omega}{2\left(-4 + \omega^2\right)}\right\}\right\}$$

and substituting back into the general solution.

```
sol = x[t]/.cvals[[1]]
```

$$Cos[2\,t] + \frac{\omega Sin[2\,t]}{2\left(-4 + \omega^2\right)} - \frac{Sin[t\omega]}{-4 + \omega^2}$$

If $\omega = 2$, we can find a particular solution to the nonhomogeneous equation of the form $x_p = t\,(A\cos 2t + B\sin 2t)$. We proceed in the same manner as before.

```
Clear[x, xp]

ω = 2;

xp[t_] = t(aCos[ωt] + bSin[ωt]);

step1 = xp"[t] + 4xp[t] == Sin[ωt]//

  Simplify
```

$$4bCos[2\,t] == (1 + 4a)Sin[2\,t]$$

```
eq1 = step1/.t → 0
```

$$4b == 0$$

```
eq2 = step1/.t → 1
```

$$4bCos[2] == (1 + 4a)Sin[2]$$

```
coeffs = Solve[{eq1, eq2}, {a, b}]//FullSimplify
```

$$\left\{\left\{a \to -\frac{1}{4}, \ b \to 0\right\}\right\}$$

```
xp[t_] = xp[t]/.coeffs[[1]]
```

$$-\frac{1}{4}tCos[2\,t]$$

```
x[t_] = homsol[[1, 1, 2]] + xp[t]
```

$$-\frac{1}{4}t\mathrm{Cos}[2t] + C[1]\mathrm{Cos}[2t] + C[2]\mathrm{Sin}[2t]$$

```
cvals = Solve[{x[0] == 1, x'[0] == 0},

    {C[1], C[2]}]
```

$$\left\{\left\{C[1] \rightarrow 1, C[2] \rightarrow \frac{1}{8}\right\}\right\}$$

```
sol = x[t]/.cvals[[1]]
```

$$\mathrm{Cos}[2t] - \frac{1}{4}t\mathrm{Cos}[2t] + \frac{1}{8}\mathrm{Sin}[2t]$$

We see that DSolve is able to solve the initial-value problem as well. Note that the result returned is valid for $\omega \neq 2$.

```
Clear[x, t, ω]

sols = DSolve[{x"[t] + 4x[t] == Sin[ωt],

    x[0] == 1, x'[0] == 0}, x[t], t]//

    Simplify
```

$$\left\{\left\{x[t] \rightarrow \frac{2\left(-4 + \omega^2\right)\mathrm{Cos}[2t] + \omega\mathrm{Sin}[2t] - 2\mathrm{Sin}[t\omega]}{2\left(-4 + \omega^2\right)}\right\}\right\}$$

We use this result to graph the solution for various values of ω in Figure 4-22. Note how we avoid $\omega = 2$ by slightly adjusting the step size.

```
graphs = Table[Plot[sols[[1, 1, 2]], {t, 0, 36Pi},

    PlotStyle->CMYKColor[0, 0.89, 0.94, 0.28],

    AxesLabel → {t, x}, AspectRatio → 1], {ω, 1, 3.1, 2.1/8}];

toshow = Partition[graphs, 3];

Show[GraphicsGrid[toshow]]
```

We also use DSolve to find the solution to the initial-value problem if $\omega = 2$ and then graph the result in Figure 4-23.

```
sol1b =

DSolve[{x"[t] + 4x[t] == Sin[2t],

    x[0] == 1, x'[0] == 0}, x[t], t]//

    Simplify
```

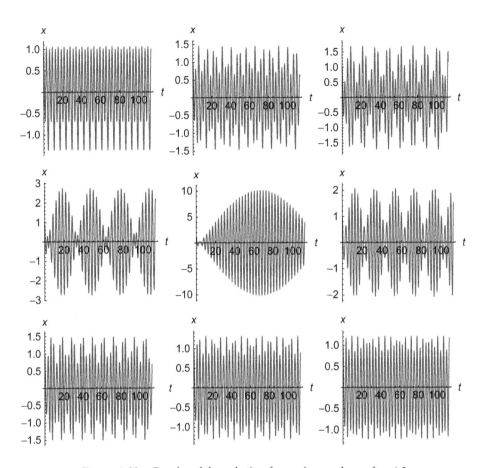

Figure 4-22   Graphs of the solution for various values of $\omega \neq 2$

$$\left\{ \left\{ x[t] \to \frac{1}{8}(-2(-4+t)\text{Cos}[2t] + \text{Sin}[2t]) \right\} \right\}$$

```
Plot[x[t]/.sol1b, {t, 0, 36Pi},
    PlotStyle->CMYKColor[0, 0.89, 0.94, 0.28],
    AxesLabel → {t, x}, AspectRatio → 1]
```

The graphs indicate that if $\omega \neq 2$ the solution to the initial-value problem is bounded and periodic; if $\omega = 2$ the solution is unbounded. We investigate this type of behavior further in Chapter 5.

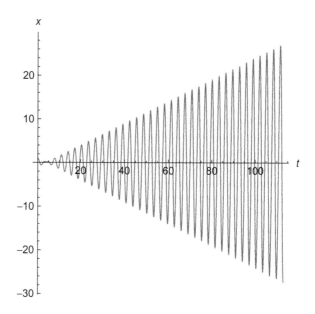

Figure 4-23   If $\omega = 2$, the solution becomes unbounded as $t \to \infty$

■

4.4.2 Higher-Order Equations

Higher-order nonhomogeneous equations are solved in the same way as second-order equations provided that the forcing function involves appropriate terms, although the calculations can be more complicated.

EXAMPLE 4.4.6: Solve

$$\frac{d^3y}{dt^3} + \frac{2}{3}\frac{d^2y}{dt^2} + \frac{145}{9}\frac{dy}{dt} = e^{-t}, \quad y(0) = 1, \quad \frac{dy}{dt}(0) = 2, \quad \frac{d^2y}{dt^2}(0) = -1.$$

SOLUTION: The corresponding homogeneous equation, $y''' + \frac{2}{3}y'' + \frac{145}{9}y' = 0$, has general solution $y_h = c_1 + (c_2 \sin 4t + c_3 \cos 4t)\,e^{-t/3}$ and a fundamental set of solutions for the corresponding homogeneous equation is $S = \left\{1, e^{-t/3}\cos 4t, e^{-t/3}\sin 4t\right\}$.

```
DSolve[y"'[t] + 2/3y"[t] + 145/9y'[t] == 0,

    y[t], t]//Simplify
```

$$\left\{\left\{y[t] \rightarrow C[3] - \frac{3}{145}e^{-t/3}((12C[1] + C[2])\text{Cos}[4t]\right.\right.$$
$$+ (C[1] - 12C[2])\text{Sin}[4t])\}\}$$

For e^{-t}, the associated set of functions is $F = \{e^{-t}\}$. Because no element of F is an element of S, we assume that $y_p = Ae^{-t}$, where A is a constant to be determined. After defining y_p, we compute the necessary derivatives

That is, the functions in F and S are linearly independent.

```
yp[t_] = aExp[-t];

yp'[t]

yp"[t]

yp"'[t]
```

$$-ae^{-t}$$

$$ae^{-t}$$

$$-ae^{-t}$$

and substitute into the nonhomogeneous equation.

```
eqn = yp"'[t] + 2/3yp"[t] + 145/9yp'[t] == Exp[-t]
```

$$-\frac{148}{9}ae^{-t} == e^{-t}$$

Equating coefficients and solving for A gives us $A = -9/148$ so $y_p = -\frac{9}{148}e^{-t}$ and a general solution is $y = y_h + y_p$.

Remark. `SolveAlways[equation, variable]` attempts to solve equation so that it is true for all values of `variable`.

```
SolveAlways[eqn, t]
```

$$\left\{\left\{a \rightarrow -\frac{9}{148}\right\}\right\}$$

We verify the result with `DSolve`.

```
gensol = DSolve[y"'[t] + 2/3y"[t] + 145/9y'[t] == Exp[-t],

    y[t], t]//Simplify
```

$$\left\{\left\{y[t] \rightarrow C[3] - \frac{1}{21460}3e^{-t}\left(435 + 148e^{2t/3}(12C[1] + C[2])\text{Cos}[4t]\right.\right.\right.$$
$$+ 148e^{2t/3}(C[1] - 12C[2])\text{Sin}[4t]\Big)\}\}$$

To apply the initial conditions, we compute $y(0) = 1$, $y'(0) = 2$, and $y''(0) = -1$

```
e1 = gensol[[1, 1, 2]]/.t → 0
```

$$-\frac{3(435 + 148(12C[1] + C[2]))}{21460} + C[3]$$

```
e2 = D[gensol[[1, 1, 2]], t]/.t → 0
```

$$-\frac{3\left(592(C[1] - 12C[2]) + \dfrac{296}{3}(12C[1] + C[2])\right)}{21460} + \frac{3(435 + 148(12C[1] + C[2]))}{21460}$$

```
e3 = D[gensol[[1, 1, 2]], {t, 2}]/.t → 0
```

$$-\frac{3\left(\dfrac{2368}{3}(C[1] - 12C[2]) - \dfrac{20720}{9}(12C[1] + C[2])\right)}{21460} +$$

$$\frac{3\left(592(C[1] - 12C[2]) + \dfrac{296}{3}(12C[1] + C[2])\right)}{10730} - \frac{3(435 + 148(12C[1] + C[2]))}{21460}$$

and solve for c_1, c_2, and c_3.

```
cvals = Solve[{e1 == 1, e2 == 2, e3 == −1}]
```

$$\left\{\left\{C[1] \to -\frac{65}{888},\ C[2] \to \frac{287}{148},\ C[3] \to \frac{157}{145}\right\}\right\}$$

The solution of the initial-value problem is obtained by substituting these values into the general solution.

```
partsol = gensol[[1, 1, 2]]/.cvals[[1]]
```

$$\frac{157}{145} - \frac{3e^{-t}\left(435 + 157e^{2t/3}\mathrm{Cos}[4t] - \dfrac{20729}{6}e^{2t/3}\mathrm{Sin}[4t]\right)}{21460}$$

We check by using DSolve to solve the initial-value problem and graph the result with Plot in Figure 4-24.

```
sol = DSolve[{y'''[t] + 2/3y''[t] + 145/9y'[t] == Exp[−t],

    y[0] == 1, y'[0] == 2, y''[0] == −1},

    y[t], t]//Simplify
```

$$\left\{\left\{y[t] \to \frac{e^{-t}\left(-2610 + 46472e^{t} - 942e^{2t/3}\mathrm{Cos}[4t] + 20729e^{2t/3}\mathrm{Sin}[4t]\right)}{42920}\right\}\right\}$$

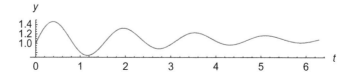

Figure 4-24   The solution of the equation that satisfies $y(0) = 1$, $y'(0) = 2$, and $y''(0) = -1$

```
Plot[y[t]/.sol, {t, 0, 2Pi}, AxesLabel → {t, y},
   AspectRatio → Automatic,
   PlotStyle->CMYKColor[0, 0.89, 0.94, 0.28]]
```

■

EXAMPLE 4.4.7: Solve

$$\frac{d^8y}{dt^8} + \frac{7}{2}\frac{d^7y}{dt^7} + \frac{73}{2}\frac{d^6y}{dt^6} + \frac{229}{2}\frac{d^5y}{dt^5} + \frac{801}{2}\frac{d^4y}{dt^4}$$

$$+ 976\frac{d^3y}{dt^3} + 1168\frac{d^2y}{dt^2} + 640\frac{dy}{dt} + 128y = te^{-t} + \sin 4t + t.$$

SOLUTION: Solving the characteristic equation

```
Solve[k^8 + 7/2k^7 + 73/2k^6 + 229/2k^5
   +801/2k^4 + 976k^3 + 1168k^2
   +640k + 128 == 0]
```

$$\left\{ \{k \to -1\}, \{k \to -1\}, \{k \to -1\}, \left\{k \to -\frac{1}{2}\right\}, \right.$$
$$\left. \{k \to -4i\}, \{k \to 4i\}, \{k \to 4i\}\right\}$$

shows us that the solutions are $k_1 = -1/2$, $k_2 = -1$ with multiplicity 3, and $k_{3,4} = \pm 4i$, each with multiplicity 2. A fundamental set of solutions for the corresponding homogeneous equation is

$$S = \left\{ e^{-t/2}, e^{-t}, te^{-t}, t^2 e^{-t}, \cos 4t, t \cos 4t, \sin 4t, t \sin 4t \right\}.$$

A general solution of the corresponding homogeneous equation is

$$y_h = c_1 e^{-t/2} + \left(c_2 + c_3 t + c_4 t^2\right) e^{-t} + (c_5 + c_7 t) \sin 4t + (c_6 + c_8 t) \cos 4t.$$

```
gensol =
DSolve[D[y[t], {t, 8}] + 7/2D[y[t], {t, 7}] + 73/2D[y[t], {t, 6}]+
   229/2D[y[t], {t, 5}]+
   801/2D[y[t], {t, 4}] + 976y"'[t] + 1168y"[t]+
   +640y'[t] + 128y[t] == 0, y[t], t]
```

$$\{\{y[t] \to e^{-t/2}C[5] + e^{-t}C[6] + e^{-t}tC[7] + e^{-t}t^2C[8] + C[1]Cos[4t]+$$
$$tC[2]Cos[4t] + C[3]Sin[4t] + tC[4]Sin[4t]\}\}$$

The associated set of functions for te^{-t} is $F_1 = \{e^{-t}, te^{-t}\}$. We multiply F_1 by t^r, where r is the smallest nonnegative integer so that no element of $t^r F_1$ is an element of S: $t^3 F_1 = \{t^3 e^{-t}, t^4 e^{-t}\}$. The associated set of functions for $\sin 4t$ is $F_2 = \{\cos 4t, \sin 4t\}$. We multiply F_2 by t^r, where r is the smallest nonnegative integer so that no element of $t^r F_2$ is an element of S: $t^2 F_2 = \{t^2 \cos 4t, t^2 \sin 4t\}$. The associated set of functions for t is $F_3 = \{1, t\}$. No element of F_3 is an element of S.

Thus, we search for a particular solution of the form

$$y_p = A_1 t^3 e^{-t} + A_2 t^4 e^{-t} + A_3 t^2 \cos 4t + A_4 t^2 \sin 4t + A_5 + A_6 t,$$

where the A_i are constants to be determined.

After defining y_p, we compute the necessary derivatives

Remark. We have used `Table` several times for typesetting purposes. You can compute the derivatives using `Table[{n,D[yp[t], {t,n}]},{n,1,8}]`.

yp[t_] = a[1]t^3Exp[−t] + a[2]t^4Exp[−t]+

    a[3]t^2Cos[4 t] + a[4]t^2Sin[4 t] + a[5] + a[6]t

Table[{n, D[yp[t], {t, n}]}, {n, 1, 2}]

$e^{-t} t^3 a[1] + e^{-t} t^4 a[2] + a[5] + t a[6] + t^2 a[3] \text{Cos}[4t] + t^2 a[4] \text{Sin}[4t]$

$\{\{1, 3e^{-t} t^2 a[1] - e^{-t} t^3 a[1] + 4e^{-t} t^3 a[2] - e^{-t} t^4 a[2] + a[6] + 2 t a[3] \text{Cos}[4t] + 4 t^2 a[4] \text{Cos}[4t] - 4 t^2 a[3] \text{Sin}[4t] + 2 t a[4] \text{Sin}[4t]\},$

$\{2, (6e^{-t} t - 6e^{-t} t^2 + e^{-t} t^3) a[1] + (12e^{-t} t^2 - 8e^{-t} t^3 + e^{-t} t^4) a[2] + 2 a[3] \text{Cos}[4t] - 16 t^2 a[3] \text{Cos}[4t] + 16 t a[4] \text{Cos}[4t] - 16 t a[3] \text{Sin}[4t] + 2 a[4] \text{Sin}[4t] - 16 t^2 a[4] \text{Sin}[4t]\}\}$

Table[{n, D[yp[t], {t, n}]}, {n, 3, 4}]

$\{\{3, (6e^{-t} - 18e^{-t} t + 9e^{-t} t^2 - e^{-t} t^3) a[1] + (24e^{-t} t - 36e^{-t} t^2 + 12e^{-t} t^3 - e^{-t} t^4) a[2] - 96 t a[3] \text{Cos}[4t] + 24 a[4] \text{Cos}[4t] - 64 t^2 a[4] \text{Cos}[4t] - 24 a[3] \text{Sin}[4t] + 64 t^2 a[3] \text{Sin}[4t] - 96 t a[4] \text{Sin}[4t]\},$

$\{4, (-24e^{-t} + 36e^{-t} t - 12e^{-t} t^2 + e^{-t} t^3) a[1] + (24e^{-t} - 96e^{-t} t + 72e^{-t} t^2 - 16e^{-t} t^3 + e^{-t} t^4) a[2] - 192 a[3] \text{Cos}[4t] + 256 t^2 a[3] \text{Cos}[4t] - 512 t a[4] \text{Cos}[4t] + 512 t a[3] \text{Sin}[4t] - 192 a[4] \text{Sin}[4t] + 256 t^2 a[4] \text{Sin}[4t]\}\}$

Table[{n, D[yp[t], {t, n}]}, {n, 5, 6}]

$\{\{5, (60e^{-t} - 60e^{-t}t + 15e^{-t}t^2 - e^{-t}t^3)a[1] + (-120e^{-t} + 240e^{-t}t - 120e^{-t}t^2 + 20e^{-t}t^3 - e^{-t}t^4)a[2] + 2560ta[3]Cos[4t] - 1280a[4]Cos[4t] + 1024t^2a[4]Cos[4t] + 1280a[3]Sin[4t] - 1024t^2a[3]Sin[4t] + 2560ta[4]Sin[4t]\}, \{6, (-120e^{-t} + 90e^{-t}t - 18e^{-t}t^2 + e^{-t}t^3)a[1] + (360e^{-t} - 480e^{-t}t + 180e^{-t}t^2 - 24e^{-t}t^3 + e^{-t}t^4)a[2] + 7680a[3]Cos[4t] - 4096t^2a[3]Cos[4t] + 12288ta[4]Cos[4t] - 12288ta[3]Sin[4t] + 7680a[4]Sin[4t] - 4096t^2a[4]Sin[4t]\}\}$

Table[{n, D[yp[t], {t, n}]}, {n, 7, 8}]

$\{\{7, (210e^{-t} - 126e^{-t}t + 21e^{-t}t^2 - e^{-t}t^3)a[1] + (-840e^{-t} + 840e^{-t}t - 252e^{-t}t^2 + 28e^{-t}t^3 - e^{-t}t^4)a[2] - 57344ta[3]Cos[4t] + 43008a[4]Cos[4t] - 16384t^2a[4]Cos[4t] - 43008a[3]Sin[4t] + 16384t^2a[3]Sin[4t] - 57344ta[4]Sin[4t]\}, \{8, (-336e^{-t} + 168e^{-t}t - 24e^{-t}t^2 + e^{-t}t^3)a[1] + (1680e^{-t} - 1344e^{-t}t + 336e^{-t}t^2 - 32e^{-t}t^3 + e^{-t}t^4)a[2] - 229376a[3]Cos[4t] + 65536t^2a[3]Cos[4t] - 262144ta[4]Cos[4t] + 262144ta[3]Sin[4t] - 229376a[4]Sin[4t] + 65536t^2a[4]Sin[4t]\}\}$

and substitute into the nonhomogeneous equation, naming the result eqn. At this point we can either equate coefficients and solve for A_i or use the fact that eqn is true for *all* values of t.

eqn =
    Simplify[D[yp[t], {t, 8}] + 7/2D[yp[t], {t, 7}] + 73/2D[yp[t],
    {t, 6}] + 229/2D[yp[t], {t, 5}]+
    801/2D[yp[t], {t, 4}] + 976yp″′[t] + 1168yp″[t]+
    640yp′[t] + 128yp[t]] == tExp[−t] + Sin[4t] + t

$e^{-t}(-867a[1] + 7752a[2] - 3468ta[2] + 128e^ta[5] + 640e^ta[6] + 128e^tta[6] - 64e^t(369a[3] - 428a[4])Cos[4t] - 64e^t(428a[3] + 369a[4])Sin[4t]) == t + e^{-t}t + Sin[4t]$

We substitute in six values of t

    sysofeqs = Table[eqn/.t → n, {n, 1, 6}]

$\{\frac{1}{e}(-867a[1] + 4284a[2] + 128ea[5] + 768ea[6] - 64e(369a[3] -$

$428a[4])\text{Cos}[4] - 64e(428a[3] + 369a[4])\text{Sin}[4]) == 1 + \frac{1}{e} +$

$\text{Sin}[4], \frac{1}{e^2}(-867a[1] + 816a[2] + 128e^2a[5] + 896e^2a[6] -$

$64e^2(369a[3]-428a[4])\text{Cos}[8]-64e^2(428a[3]+369a[4])\text{Sin}[8]) ==$

$2+\frac{2}{e^2}+\text{Sin}[8], \frac{1}{e^3}(-867a[1]-2652a[2]+128e^3a[5]+1024e^3a[6]-$

$64e^3(369a[3]-428a[4])\text{Cos}[12]-64e^3(428a[3]+369a[4])\text{Sin}[12]) ==$

$3 + \frac{3}{e^3} + \text{Sin}[12], \frac{1}{e^4}(-867a[1] - 6120a[2] + 128e^4a[5] +$

$1152e^4a[6] - 64e^4(369a[3] - 428a[4])\text{Cos}[16] - 64e^4(428a[3] +$

$369a[4])\text{Sin}[16]) == 4 + \frac{4}{e^4} + \text{Sin}[16], \frac{1}{e^5}(-867a[1] - 9588a[2] +$

$128e^5a[5] + 1280e^5a[6] - 64e^5(369a[3] - 428a[4])\text{Cos}[20] -$

$64e^5(428a[3]+369a[4])\text{Sin}[20]) == 5+\frac{5}{e^5}+\text{Sin}[20], \frac{1}{e^6}(-867a[1]-$

$13056a[2]+128e^6a[5]+1408e^6a[6]-64e^6(369a[3]-428a[4])\text{Cos}[24]-$

$64e^6(428a[3] + 369a[4])\text{Sin}[24]) == 6 + \frac{6}{e^6} + \text{Sin}[24]\}$

and then solve for A_i.

```
coeffs = Solve[sysofeqs]//N
```

$\{\{a[1.] \rightarrow -0.00257819, a[2.] \rightarrow -0.000288351, a[3.] \rightarrow$

$-0.0000209413, a[4.] \rightarrow -0.0000180545, a[5.] \rightarrow -0.0390625,$

$a[6.] \rightarrow 0.0078125\}\}$

y_p is obtained by substituting the values for A_i into y_p and a general solution is $y = y_h + y_p$. DSolve is able to find an exact solution.

```
gensol =
```

$\text{DSolve}[D[y[t], \{t, 8\}] + 7/2D[y[t], \{t, 7\}] + 73/2D[y[t], \{t, 6\}]+$

$229/2D[y[t], \{t, 5\}] + 801/2D[y[t], \{t, 4\}] + 976y'''[t] + 1168y''[t]$

$+640y'[t] + 128y[t] == t\text{Exp}[-t] + \text{Sin}[4t] + t, y[t], t]$

$\{\{y[t] \rightarrow e^{-t/2}C[5]+e^{-t}C[6]+e^{-t}tC[7]+e^{-t}t^2C[8]+C[1]\text{Cos}[4t]+$

$tC[2]\text{Cos}[4t]+C[3]\text{Sin}[4t]+tC[4]\text{Sin}[4t]-.(e^{-t}(4935464732590080+$

$1591308177838080e^t+2467732366295040t-317290487255040e^tt+$

$619288250880000t^2+105002679808000t^3+11743720768000t^4-$

$2165610627072e^t\text{Cos}[4t]+1414239094320e^ttC\text{os}[4t]+852881078400$

$e^tt^2\text{Cos}[4t]-451268014080\text{Cos}[4t]^2-401005048080e^t\text{Cos}[4t]^2-$

$446669783040t\text{Cos}[4t]^2-890947302960e^t t\text{Cos}[4t]^2+321047446980e^t$

$\text{Cos}[4t]\text{Cos}[8t]-503767535616e^t\text{Sin}[4t]-2355159306240e^t t\text{Sin}[4t]+$

$735311023200e^t t^2\text{Sin}[4t] + 199758356265e^t\text{Cos}[8t]\text{Sin}[4t] - $

$451268014080\text{Sin}[4t]^2-401005048080e^t\text{Sin}[4t]^2-446669783040t\text{Sin}$

$[4t]^2-890947302960e^t t\text{Sin}[4t]^2-199758356265e^t\text{Cos}[4t]\text{Sin}[8t]+$

$321047446980e^t\text{Sin}[4t]\text{Sin}[8t]))/40727223623424000\}\}$

■

4.5 Nonhomogeneous Equations With Constant Coefficients: Variation of Parameters

4.5.1 Second-Order Equations

Let $S = \{y_1, y_2\}$ be a fundamental set of solutions for equation $y'' + p(t)y'q(t)y = 0$. To solve the nonhomogeneous equation $y'' + p(t)y' + q(t)y = f(t)$, we need to find a particular solution, y_p of equation $y''+p(t)y'+q(t)y = f(t)$. We search for a particular solution of the form

> A **particular solution**, y_p, is a solution that does not contain any arbitrary constants.

$$y_p = u_1(t)y_1(t) + u_2(t)y_2(t), \tag{4.13}$$

where u_1 and u_2 are functions of t. Differentiating equation (4.13) gives us

$$y_p' = u_1'y_1 + u_1y_1' + u_2'y_2 + u_2y_2'.$$

Assuming that

$$y_1u_1' + y_2u_2' = 0 \tag{4.14}$$

results in $y_p' = u_1y_1' + u_2y_2'$. Computing the second derivative then yields

$$y_p'' = u_1'y_1' + u_1y_1'' + u_2'y_2' + u_2y_2''.$$

> Observe that it is pointless to search for solutions of the form $y_p = c_1y_1 + c_2y_2$ where c_1 and c_2 are constants because for every choice of c_1 and c_2, $c_1y_1 + c_2y_2$ is a solution to the corresponding homogeneous equation.

Substituting y_p, y_p', and y_p'' into the equation $y'' + p(t)y' + q(t)y = f(t)$ and using the facts that

$$u_1\left(y_1'' + py_1' + qy_1\right) = 0 \quad \text{and} \quad u_2\left(y_2'' + py_2' + qy_2\right) = 0$$

(because y_1 and y_2 are solutions to the corresponding homogeneous equation, $y'' + p(t)y' + q(t)y = 0$) results in

$$\frac{d^2 y_p}{dt^2} + p(t)\frac{dy_p}{dt} + q(t)y_p = u_1' y_1' + u_1 y_1''$$
$$+ u_2' y_2' + u_2 y_2'' + p(t)\left(u_1 y_1' + u_2 y_2'\right) \qquad (4.15)$$
$$+ q(t)\left(u_1 y_1 + u_2 y_2\right)$$
$$= y_1' u_1' + y_2' u_2' = f(t).$$

Observe that equations (4.14) and (4.15) form a system of two linear equations in the unknowns u_1' and u_2':

$$\begin{aligned} y_1 u_1' + y_2 u_2' &= 0 \\ y_1' u_1' + y_2' u_2' &= f(t). \end{aligned} \qquad (4.16)$$

Applying Cramer's Rule gives us

$$u_1' = \frac{\begin{vmatrix} 0 & y_2 \\ f(t) & y_2' \end{vmatrix}}{\begin{vmatrix} y_1 & y_2 \\ y_1' & y_2' \end{vmatrix}} = -\frac{y_2(t)f(t)}{W(S)} \quad \text{and} \quad u_2' = \frac{\begin{vmatrix} y_1 & 0 \\ y_1' & f(t) \end{vmatrix}}{\begin{vmatrix} y_1 & y_2 \\ y_1' & y_2' \end{vmatrix}} = \frac{y_1(t)f(t)}{W(S)}, \qquad (4.17)$$

where $W(S)$ is the Wronskian, $W(S) = \begin{vmatrix} y_1 & y_2 \\ y_1' & y_2' \end{vmatrix}$. After integrating to obtain u_1 and u_2, we form y_p and then a general solution, $y = y_h + y_p$.

Summary of Variation of Parameters for Second-Order Equations
Given the second-order equation $a_2(t)y'' + a_2(t)y' + a_0(t)y = g(t)$.

1. Divide by $a_2(t)$ to rewrite the equation in standard form, $y'' + p(t)y' + q(t)y = f(t)$.
2. Find a general solution, $y_h = c_1 y_1 + c_2 y_2$, where $S = \{y_1, y_2\}$ is a fundamental set of solutions for the corresponding homogeneous equation, of the corresponding homogeneous equation $y'' + p(t)y' + q(t)y = 0$.
3. Let $W(S) = \begin{vmatrix} y_1 & y_2 \\ y_1' & y_2' \end{vmatrix}$.
4. Let $u_1' = -\dfrac{y_2 f(t)}{W}$ and $u_2' = \dfrac{y_1 f(t)}{W}$.
5. Integrate to obtain u_1 and u_2.
6. A particular solution of $a_2(t)y'' + a_2(t)y' + a_0(t)y = g(t)$ is given by $y_p = u_1 y_1 + u_2 y_2$.
7. A general solution of $a_2(t)y'' + a_2(t)y' + a_0(t)y = g(t)$ is given by $y = y_h + y_p$.

EXAMPLE 4.5.1: Solve $y'' + 9y = \sec 3t$, $y(0) = 0$, $y'(0) = 0$, $0 \le t < \pi/6$.

SOLUTION: The corresponding homogeneous equation is $y'' + 9y = 0$ with general solution $y_h = c_1 \cos 3t + c_2 \sin 3t$. Then, a fundamental set of solutions is $S = \{\cos 3t, \sin 3t\}$ and $W(S) = 3$, as we see using Wronskian.

```
fs = {Cos[3t], Sin[3t]};

wm = {fs, D[fs, t]};

wm//MatrixForm
```

$$\begin{pmatrix} \text{Cos}[3t] & \text{Sin}[3t] \\ -3\text{Sin}[3t] & 3\text{Cos}[3t] \end{pmatrix}$$

```
Wronskian[fs, t]
```

3

We use equation (4.17) to find $u_1 = \frac{1}{9} \ln \cos 3t$ and $u_2 = \frac{1}{3}t$.

```
u1 = Integrate[-Sin[3t]Sec[3t]/3, t]

u2 = Integrate[Cos[3t]Sec[3t]/3, t]
```

$\dfrac{1}{9}\text{Log}[\text{Cos}[3t]]$

$\dfrac{t}{3}$

It follows that a particular solution of the nonhomogeneous equation is

$$y_p = \frac{1}{9} \cos 3t \ln \cos 3t + \frac{1}{3}t \sin 3t$$

and a general solution is

$$y = y_h + y_p = c_1 \cos 3t + c_2 \sin 3t + \frac{1}{9} \cos 3t \ln \cos 3t + \frac{1}{3}t \sin 3t.$$

```
yp = u1Cos[3t] + u2Sin[3t]
```

$$\frac{1}{9}\text{Cos}[3\,t]\text{Log}[\text{Cos}[3\,t]]+\frac{1}{3}t\text{Sin}[3\,t]$$

The negative sign in the
output does not affect the
result because C[1] is
arbitrary.

Identical results are obtained using DSolve.

DSolve[y″[t] + 9y[t] == Sec[3t], y[t], t]

$$\{\{y[t] \;\rightarrow\; C[1]\text{Cos}[3\,t] + C[2]\text{Sin}[3\,t] + \frac{1}{9}(\text{Cos}[3\,t]\text{Log}[\text{Cos}[3\,t]] + 3\,t\text{Sin}[3\,t])\}\}$$

Applying the initial conditions gives us $c_1 = c_2 = 0$ so we conclude that the solution to the initial value problem is

$$y = \frac{1}{9}\cos 3t\,\ln\cos 3t + \frac{1}{3}t\sin 3t.$$

sol = DSolve[{y″[t] + 9y[t] == Sec[3t],

y[0] == 0, y′[0] == 0}, y[t], t]

$$\{\{y[t] \rightarrow \frac{1}{9}(\text{Cos}[3\,t]\text{Log}[\text{Cos}[3\,t]] + 3\,t\text{Sin}[3\,t])\}\}$$

We graph the solution with Plot in Figure 4-25.

Plot[y[t]/.sol, {t, 0, Pi/6}, AxesLabel → {t, y},

PlotStyle->CMYKColor[0, 0.89, 0.94, 0.28]]

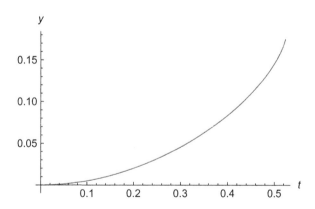

Figure 4-25   The domain of the solution is $-\pi/6 < t < \pi/6$

4.5.2 Higher-Order Nonhomogeneous Equations

In the same way as with second-order equations, we assume that a particular solution of the nth-order linear nonhomogeneous equation

$$y^{(n)} + a_{n-1}(t)y^{(n-1)} + \cdots + a_2(t)y'' + a_1(t)y' + a_0(t)y = f(t)$$

has the form $y_p = u_1(t)y_1 + u_2(t)y_2 + \cdots + u_n(t)y_n$, where $S = \{y_1, y_2, \ldots, y_n\}$ is a fundamental set of solutions to the corresponding homogeneous equation

$$y^{(n)} + a_{n-1}(t)y^{(n-1)} + \cdots + a_2(t)y'' + a_1(t)y' + a_0(t)y = 0.$$

With the assumptions

$$
\begin{aligned}
y_p' &= y_1 u_1' + y_2 u_2' + \cdots + y_n u_n' = 0 \\
y_p'' &= y_1' u_1' + y_2' u_2' + \cdots + y_n' u_n' = 0 \\
&\quad\vdots
\end{aligned}
\tag{4.18}
$$

$$y_p^{(n-1)} = y_1^{(n-2)} u_1' + y_2^{(n-2)} u_2' + \cdots + y_n^{(n-2)} u_n' = 0$$

we obtain the equation

$$y_1^{(n-1)} u_1' + y_2^{(n-1)} u_2' + \cdots + y_n^{(n-1)} u_n' = f(t). \tag{4.19}$$

Equations (4.18) and (4.19) form a system of n linear equations in the unknowns $u_1', u_2', \ldots, u_n'$. Applying Cramer's Rule,

$$u_i' = \frac{W_i(S)}{W(S)}, \tag{4.20}$$

where $W(S)$ is given by equation (4.7),

$$
W(S) = \begin{vmatrix}
y_1 & y_2 & \cdots & y_n \\
y_1' & y_2' & \cdots & y_n' \\
\vdots & \vdots & \cdots & \vdots \\
y_1^{(n-1)} & y_2^{(n-1)} & \cdots & y_n^{(n-1)}
\end{vmatrix},
$$

and $W_i(S)$ is the determinant of the matrix obtained by replacing the ith column of

$$\begin{pmatrix} y_1 & y_2 & \cdots & y_n \\ y_1' & y_2' & \cdots & y_n' \\ \vdots & \vdots & \cdots & \vdots \\ y_1^{(n-1)} & y_2^{(n-1)} & \cdots & y_n^{(n-1)} \end{pmatrix} \quad \text{by} \quad \begin{pmatrix} 0 \\ 0 \\ \vdots \\ f(t) \end{pmatrix}.$$

EXAMPLE 4.5.2: Solve $y^{(3)} + 4y' = \sec 2t$.

SOLUTION: A general solution of the corresponding homogeneous equation is $y_h = c_1 + c_2 \cos 2t + c_3 \sin 2t$; a fundamental set is $S = \{1, \cos 2t, \sin 2t\}$ with Wronskian $W(S) = 8$.

```
yh = DSolve[y'''[t] + 4y'[t] == 0, y[t], t]
```

$$\{\{y[t] \to C[3] - \frac{1}{2}C[2]Cos[2t] + \frac{1}{2}C[1]Sin[2t]\}\}$$

```
s = {1, Cos[2t], Sin[2t]};

Wronskian[s, t]
```

8

Using variation of parameters to find a particular solution of the nonhomogeneous equation, we let $y_1 = 1$, $y_2 = \cos 2t$, and $y_3 = \sin 2t$ and assume that a particular solution has the form $y_p = u_1 y_1 + u_2 y_2 + u_3 y_3$. Using the variation of parameters formula, we obtain

$$u_1' = \frac{1}{8} \begin{vmatrix} 0 & \cos 2t & \sin 2t \\ 0 & -2 \sin 2t & 2 \cos 2t \\ \sec 2t & -4 \cos 2t & -4 \sin 2t \end{vmatrix} = \frac{1}{4} \sec 2t \quad \text{so} \quad u_1 = \frac{1}{8} \ln|\sec 2t + \tan 2t|,$$

$$u_2' = \frac{1}{8} \begin{vmatrix} 1 & 0 & \sin 2t \\ 0 & 0 & 2 \cos 2t \\ 0 & \sec 2t & -4 \sin 2t \end{vmatrix} = -\frac{1}{4} \quad \text{so} \quad u_2 = -\frac{1}{4}t$$

and

$$u_3' = \frac{1}{8} \begin{vmatrix} 1 & \cos 2t & 0 \\ 0 & -2 \sin 2t & 0 \\ 0 & -4 \cos 2t & \sec 2t \end{vmatrix} = -\frac{1}{4} \tan 2t \quad \text{so} \quad u_3 = \frac{1}{8} \ln|\cos 2t|,$$

where we use Det and Integrate to evaluate the determinants and integrals. In the case of u_1, the output given by Mathematica looks different than the result we obtained by hand but using properties of logarithms $(\ln (a/b) = \ln a - \ln b)$ and trigonometric identities $(\cos^2 x + \sin^2 x = 1, \sin 2x = 2 \sin x \cos x, \cos^2 x - \sin^2 x = \cos 2x,$ and the reciprocal identities) shows us that

$$\frac{1}{8} (\ln | \cos t + \sin t| - \ln | \cos t + \sin t|) = \frac{1}{8} \ln \left| \frac{\cos t + \sin t}{\cos t - \sin t} \right|$$

$$= \frac{1}{8} \ln \left| \frac{\cos t + \sin t}{\cos t - \sin t} \cdot \frac{\cos t + \sin t}{\cos t + \sin t} \right|$$

$$= \frac{1}{8} \ln \left| \frac{\cos^2 t + 2 \cos t \sin t + \sin^2 t}{\cos^2 t - \sin^2 t} \right|$$

$$= \frac{1}{8} \ln \left| \frac{1 + \sin 2t}{\cos 2t} \right|$$

$$= \frac{1}{8} \ln \left| \frac{1}{\cos 2t} + \frac{\sin 2t}{\cos 2t} \right| = \frac{1}{8} \ln |\sec 2t + \tan 2t|$$

so the results obtained by hand and with Mathematica are the same.

```
u1p = 1/8Det[{{0, Cos[2t], Sin[2t]},

    {0, -2Sin[2t], 2Cos[2t]},

    {Sec[2t], -4Cos[2t], -4Sin[2t]}}]//Simplify
```

$$\frac{1}{4} Sec[2t]$$

```
u1 = Integrate[u1p, t]
```

$$\frac{1}{4} \left(-\frac{1}{2} Log[Cos[t] - Sin[t]] + \frac{1}{2} Log[Cos[t] + Sin[t]]\right)$$

```
u2p = 1/8Det[{{1, 0, Sin[2t]},

    {0, 0, 2Cos[2t]},

    {0, Sec[2t], -4Sin[2t]}}]//Simplify
```

$$-\frac{1}{4}$$

```
u2 = Integrate[u2p, t]
```

$$-\frac{t}{4}$$

```
u3p = 1/8Det[{{1, Cos[2 t], 0},

     {0, −2Sin[2 t], 0},

     {0, −4Cos[2 t], Sec[2 t]}}]//Simplify
```

$$-\frac{1}{4}\mathrm{Tan}[2\,t]$$

```
u3 = Integrate[u3p, t]
```

$$\frac{1}{8}\mathrm{Log}[\mathrm{Cos}[2\,t]]$$

Thus, a particular solution of the nonhomogeneous equation is

$$y_p = \frac{1}{8}\ln|\sec 2t + \tan 2t| - \frac{1}{4}t\cos 2t + \frac{1}{8}\ln|\cos 2t|\sin 2t$$

and a general solution is $y = y_h + y_p$. We verify the calculations using DSolve, which returns an equivalent solution.

```
gensol = DSolve[y″′[t] + 4y′[t] == Sec[2 t],

     y[t], t]//Simplify
```

$$\{\{y[t] \to \frac{1}{8}(8\,C[3] - 8\,C[2]\mathrm{Cos}[t]^2 - 2\,t\mathrm{Cos}[2\,t] - \mathrm{Log}[\mathrm{Cos}[t] - \mathrm{Sin}[t]] +$$
$$\mathrm{Log}[\mathrm{Cos}[t] + \mathrm{Sin}[t]] + 4\,C[1]\mathrm{Sin}[2\,t] + \mathrm{Log}[\mathrm{Cos}[2\,t]]\mathrm{Sin}[2\,t])\}\}$$

■

4.6 Cauchy-Euler Equations

Generally, solving an arbitrary linear differential equation is a formidable particularly in the case when the coefficients are not constants. However, we are able to solve certain linear equations with variable coefficients using techniques similar to those discussed previously.

Definition 19 (Cauchy-Euler Equation). *A Cauchy-Euler differential equation is an equation of the form*

$$a_n x^n y^{(n)} + a_{n-1} x^{n-1} y^{(n-1)} + \cdots + a_1 xy1 + a_0 xy = f(x), \qquad (4.21)$$

where $a_0, a_1, \ldots, a_n$ are constants.

4.6.1 Second-Order Cauchy-Euler Equations

Consider the second-order homogeneous Cauchy-Euler equation

$$ax^2 y'' + bxy' + cy = 0, \tag{4.22}$$

where $a \neq 0$. Notice that because the coefficient of y'' is zero if $x = 0$, we must restrict our domain to either $x > 0$ or $x < 0$ in order to ensure that the theory of second-order equations stated in Section 4.1 holds.

Suppose that $y = x^m$, $x > 0$, for some constant m. Substitution of $y = x^m$ with derivatives $y' = mx^{m-1}$ and $y'' = m(m-1)x^{m-2}$ into equation (4.22) yields

$$ax^2 y'' + bxy' + cy = am(m-1)x^{m-2} + bmx^{m-1} + cx^m$$
$$= x^m [am(m-1) + bm + c] = 0.$$

Then, $y = x^m$ is a solution of equation (4.22) if m satisfies

$$am(m-1) + bm + c = 0, \tag{4.23}$$

which is called the **characteristic equation** (or **auxiliary equation**) associated with the Cauchy-Euler equation of order two. The solutions of the characteristic equation completely determine the general solution of the homogeneous Cauchy-Euler equation of order two. Let m_1 and m_2 denote the two solutions of the characteristic (or auxiliary) equation (4.23):

$$m_{1,2} = \frac{1}{2a} \left[-(b-a) \pm \sqrt{(b-a)^2 - 4ac} \right].$$

Hence, we can obtain two real roots, one repeated real root, or a complex conjugate pair depending on the values of a, b, and c. We state a general solution that corresponds to the different types of roots.

Theorem 8. *Let m_1 and m_2 be the solutions of equation (4.23).*

1. *If $m_1 \neq m_2$ are real and distinct, two linearly independent solutions of equation (4.22) are $y_1 = x^{m_1}$ and $y_2 = x^{m_2}$; a general solution of (4.22) is*

$$y = c_1 x^{m_1} + c_2 x^{m_2}, \quad x > 0.$$

2. *If $m_1 = m_2$, two linearly independent solutions of equation (4.22) are $y_1 = x^{m_1}$ and $y_2 = x^{m_1} \ln x$; a general solution of (4.22) is*

$$y = c_1 x^{m_1} + c_2 x^{m_1} \ln x, \quad x > 0.$$

3. *If* $m_{1,2} = \alpha \pm \beta i$, $\beta \neq 0$, *two linearly independent solutions of equation (4.22) are* $y_1 = x^{\alpha} \cos{(\beta \ln x)}$ *and* $y_2 = x^{\alpha} \sin{(\beta \ln x)}$; *a general solution of (4.22) is*

$$y = x^{\alpha} \left[c_1 \cos{(\beta \ln x)} + c_2 \sin{(\beta \ln x)} \right], \quad x > 0.$$

EXAMPLE 4.6.1: Solve each of the following equations: (a) $3x^2 y'' - 2xy' + 2y = 0$, $x > 0$; (b) $x^2 y'' - xy' + y = 0$, $x > 0$; (c) $x^2 y'' - 5xy' + 10y = 0$, $x > 0$.

SOLUTION: (a) If $y = x^m$, $y' = mx^{m-1}$, and $y'' = m(m-1)x^{m-2}$, substitution into the differential equation yields

$$
\begin{aligned}
3x^2 y'' - 2xy' + 2y &= 3x^2 \cdot m(m-1)x^{m-2} - 2x \cdot mx^{m-1} + 2x^m \\
&= x^m \left[3m(m-1) - 2m + 2 \right] = 0.
\end{aligned}
$$

Hence, the auxiliary equation is

$$3m(m-1) - 2m + 2 = 3m(m-1) - 2(m-1) = (3m-2)(m-1) = 0$$

with roots $m_1 = 2/3$ and $m_2 = 1$. Therefore, a general solution is $y = c_1 x^{2/3} + c_2 x$. We obtain the same results with DSolve. Entering

```
Clear[x, y]
gensol = DSolve[3x^2y"[x] − 2xy'[x] + 2y[x] == 0,
  y[x], x]
```

$$\{\{y[x] \to x^{2/3} C[1] + xC[2]\}\}$$

finds a general solution of the equation, naming the result gensol, and then entering

```
toplot = Table[gensol[[1, 1, 2]]/.{C[1] → i, C[2] → j},
  {i, −2, 2, 2}, {j, −2, 2, 2}];

p1 = Plot[toplot, {x, 0, 12}, PlotRange → {−6, 6},
  AspectRatio → 1, AxesLabel → {x, y},
  PlotLabel → "(a)"]
```

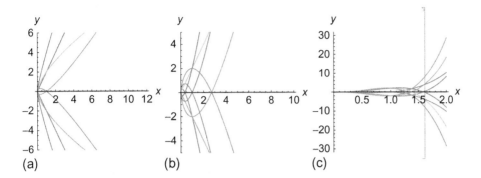

Figure 4-26   (a) Various solutions of $3x^2y'' - 2xy' + 2y = 0$, $x > 0$. (b) Various solutions of $x^2y'' - xy' + y = 0$, $x > 0$. (c) Various solutions of $x^2y'' - 5xy' + 10y = 0$, $x > 0$

defines `toplot` to be the list of functions obtained by replacing `c[1]` in `gensol[[1,1,2]]` by -2, 0, and 2 and `C[2]` in gensol[[1,1,2]] by -2, 0, and 2, and graphs the set of functions `toplot` on the interval $[0, 12]$. See Figure 4-26(a).

(b) In this case, the auxiliary equation is

$$m(m-1) - m + 1 = m(m-1) - (m-1) = (m-1)^2 = 0$$

with root $m = 1$ of multiplicity 2. Hence, a general solution is $y = c_x + c_2x \ln x$. As in the previous example, we see that we obtain the same results with `DSolve`. See Figure 4-26(b).

```
Clear[x, y]

gensol = DSolve[x^2y"[x] - xy'[x] + y[x] == 0,

y[x], x]

{{y[x] → xC[1] + xC[2]Log[x]}}

toplot = Table[gensol[[1, 1, 2]]/.{C[1] → i, C[2] → j},

    {i, -2, 2, 2}, {j, -2, 2, 2}];

p2 = Plot[toplot, {x, 0, 10}, PlotRange → {-5, 5},

    AspectRatio → 1, AxesLabel → {x, y},

    PlotLabel → "(b)"]
```

(c) The auxiliary (characteristic) equation is given by

$$m(m-1) - 5m + 10 = m^2 - 6m + 10 = 0$$

with complex conjugate roots $m_{1,2} = \frac{1}{2}\left(6 \pm \sqrt{36 - 40}\right) = 3 \pm i$. Thus, a general solution is $y = x^3\left[c_1 \cos\left(\ln x\right) + c_2 \sin\left(\ln x\right)\right]$.

Again, we see that we obtain equivalent results with DSolve. First, we find a general solution of the equation, naming the resulting output gensol.

> **Clear[x, y]**
>
> **gensol = DSolve[x^2y″[x] − 5xy′[x] + 10y[x] == 0,**
>
> **y[x], x]**

$$\{\{y[x] \rightarrow x^3\,C[2]\mathrm{Cos}[\mathrm{Log}[x]] + x^3\,C[1]\mathrm{Sin}[\mathrm{Log}[x]]\}\}$$

Now, we define $y(x)$ to be the general solution obtained in gensol. (The same result is obtained with Part by entering y[x_]=gensol[[1,1,2]].)

> **y[x_] = x^3C[2]Cos[Log[x]] + x^3C[1]Sin[Log[x]]**

$$x^3\,C[2]\mathrm{Cos}[\mathrm{Log}[x]] + x^3\,C[1]\mathrm{Sin}[\mathrm{Log}[x]]$$

To find the values of C[1] and C[2] so that the solution satisfies the initial conditions $y(1) = a$ and $y'(1) = b$, we use Solve and name the resulting list cvals.

> **cvals = Solve[{y[1] == a, y′[1] == b}, {C[1], C[2]}]**

$$\{\{C[1] \rightarrow -3a + b, \; C[2] \rightarrow a\}\}$$

The solution to the initial-value problem

$$\begin{cases} x^2 y'' - 5xy' + 10y = 0 \\ y(1) = a, \; y'(1) = b \end{cases}$$

Note that when you enter
the following Plot
command, Mathematica may
display several error
messages because each
solution is undefined if $x = 0$.
Nevertheless, the resulting
graphs are displayed
correctly.

is obtained by replacing C[1] and C[2] in $y(x)$ by the values found in cvals.

> **y[x_] = y[x]/.cvals[[1]]**

$$ax^3\mathrm{Cos}[\mathrm{Log}[x]] + (-3a + b)x^3\mathrm{Sin}[\mathrm{Log}[x]]$$

This solution is then graphed for various initial conditions in Figure 4-26(c).

> **toplot = Table[y[x],**
>
> **{a, −2, 2, 2}, {b, −2, 2, 2}];**

```
p3 = Plot[toplot, {x, 0, 2}, PlotRange → All,

   AspectRatio → 1, AxesLabel → {x, y},

   PlotLabel → "(c)"]

Show[GraphicsRow[{p1, p2, p3}]]
```

∎

4.6.2 Higher-Order Cauchy-Euler Equations

The auxiliary equation of higher order Cauchy-Euler equations is defined in the same way and solutions of higher-order homogeneous Cauchy-Euler equations are determined in the same manner as solutions of higher-order homogeneous differential equations with constant coefficients. In the case of higher-order Cauchy-Euler equations, note that if a real root r of the auxiliary equation is repeated m times, m linearly independent solutions that correspond to r are x^r, $x^r \ln x$, $x^r (\ln x)^2$, $\ldots$, $x^r (\ln x)^{m-1}$; solutions corresponding to repeated complex roots are generated similarly.

EXAMPLE 4.6.2: Solve $2x^3 y''' - 4x^2 y'' - 20xy' = 0$, $x > 0$.

SOLUTION: In this case, if we assume that $y = x^m$ for $x > 0$, we have the derivatives $y' = mx^{m-1}$, $y'' = m(m-1)x^{m-2}$, and $y''' = m(m-1)(m-2)x^{m-3}$. Substitution into the differential equation and simplification then yields $(2m^3 - 10m^2 - 12m) x^m = 0$.

```
Clear[x, y]

eq = 2x^3y'''[x] - 4x^2y''[x] - 20xy'[x] == 0
```

$-20xy'[x] - 4x^2 y''[x] + 2x^3 y^{(3)}[x] == 0$

```
y[x_] = x^m
```

x^m

```
eq
```

$$-20mx^m - 4(-1 + m)mx^m + 2(-2 + m)(-1 + m)mx^m == 0$$

Factor[eq[[1]]]

$$2(-6 + m)m(1 + m)x^m$$

We must solve $2m^3 - 10m^2 - 12m = 2m(m + 1)(m - 6) = 0$ for m because $x^m \neq 0$.

mvals = Solve[eq, m]

$$\{\{m \to -1\}, \{m \to 0\}, \{m \to 6\}\}$$

We see that the solutions are $m_1 = 0$, $m_2 = -1$, and $m_3 = 6$, so a general solution of the equation is $y = c_1 + c_2 x^{-1} + c_3 x^6$. As in the previous examples, we see that we obtain the same results with DSolve.

Clear[x, y]

gensol = DSolve[

    2x^3y″'[x] − 4x^2y″[x] − 20xy'[x] == 0,

    y[x], x]

$$\{\{y[x] \to \frac{1}{6}x^6 C[1] - \frac{C[2]}{x} + C[3]\}\}$$

We graph this solution for various values of the arbitrary constants in the same was as we graph solutions of other equations. See Figure 4-27.

    toplot = Table[gensol[[1, 1, 2]]/.{C[3] → 0, C[1] → i, C[2] → j},

    {i, −1, 1}, {j, −1, 1}];

    Plot[toplot, {x, 0, 2}, PlotRange → {−10, 10},

    AspectRatio → 1, AxesLabel → {x, y}]

■

EXAMPLE 4.6.3: Solve the initial-value problem

$$\begin{cases} x^4 y^{(4)} + 4x^3 y''' + 11x^2 y'' - 9xy' + 9y = 0, \ x > 0 \ . \\ y(1) = 1, \ y'(1) = -9, \ y''(1) = 27, \ y'''(1) = 1 \end{cases}$$

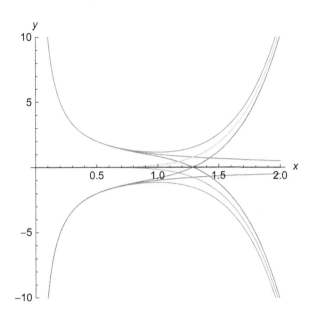

Figure 4-27   Various solutions of $2x^3y''' - 4x^2y'' - 20xy' = 0, x > 0$

SOLUTION: Substitution of $y = x^m$ into the differential equation $x^4y^{(4)} + 4x^3y''' + 11x^2y'' - 9xy' + 9y = 0$ and simplification leads to the equation

$$\left(m^4 - m^3 + 8m^2 - 9m - 9\right)x^m = 0.$$

```
Clear[x, y]
eq = x^4D[y[x], {x, 4}] + 4x^3y'''[x] + 11x^2y''[x]−
   9xy'[x] + 9y[x] == 0;

y[x_] = x^m;

Factor[eq[[1]]]
```

$$(-1 + m)^2(9 + m^2)x^m$$

We solve

$$m^4 - m^3 + 8m^2 - 9m - 9 = \left(m^2 + 9\right)(m - 1)^2 = 0$$

for m because $x^m \neq 0$.

```
Solve[eq, m]
```

$$\{\{m \to -3\,i\},\ \{m \to 3\,i\},\ \{m \to 1\},\ \{m \to 1\}\}$$

Hence, $m_{1,2} = \pm 3i$, and $m_{3,4} = 1$ is a root of multiplicity 2, so a general solution of the differential equation is

$$y = c_1 \cos(3\ln x) + c_2 \sin(3\ln x) + c_3 x + c_4 x \ln x$$

with first, second, and third derivatives computed as follows.

```
y[x_] = c1Cos[3Log[x]] + c2Sin[3Log[x]] + c3x + c4xLog[x];

Simplify[y'[x]]//Together

Simplify[y"[x]]

Simplify[y"'[x]]
```

$$\frac{c3x + c4x + 3c2\text{Cos}[3\text{Log}[x]] + c4x\text{Log}[x] - 3c1\text{Sin}[3\text{Log}[x]]}{x}$$

$$\frac{c4x - 3(3c1 + c2)\text{Cos}[3\text{Log}[x]] + 3(c1 - 3c2)\text{Sin}[3\text{Log}[x]]}{x^2}$$

$$\frac{-c4x + 3(9c1 - 7c2)\text{Cos}[3\text{Log}[x]] + 3(7c1 + 9c2)\text{Sin}[3\text{Log}[x]]}{x^3}$$

Substitution of the initial conditions yields the system of equations,

$$\begin{cases} c_1 + c_3 = 1 \\ 3c_2 + c_3 + c_4 = -9 \\ -9c_1 - 3c_2 + c_4 = 27 \\ 27c_1 - 21c_2 - c_4 = 1 \end{cases},$$

which has the solution $(c_1, c_2, c_3, c_4) = ((-12/5, -89/30, 17/5, -7/2)$.

```
cvals = Solve[{y[1] == 1, y'[1] == -9, y"[1] == 27,

   y"'[1] == 1}]
```

$$\{\{c1 \to -\frac{12}{5},\ c2 \to -\frac{89}{30},\ c3 \to \frac{17}{5},\ c4 \to -\frac{7}{2}\}\}$$

Therefore, the solution to the initial-value problem is

$$y = -\frac{12}{5}\cos(3\ln x) - \frac{89}{30}\sin(3\ln x) + \frac{17}{5}x - \frac{7}{2}x\ln x$$

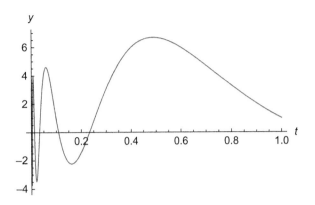

Figure 4-28   Plot of the solution to the initial-value problem

$y[x_] = y[x]/.cvals[[1]]$

$$\frac{17x}{5} - \frac{12}{5}\text{Cos}[3\text{Log}[x]] - \frac{7}{2}x\text{Log}[x] - \frac{89}{30}\text{Sin}[3\text{Log}[x]]$$

We graph this solution with `Plot` in Figure 4-28. From the graph shown in Figure 4-28, we see that it *might appear* to be the case that $\lim_{x\to 0^+} y(x)$ exists.

Mathematica may display several error messages because the solution is undefined if $x = 0$.

```
Plot[y[x], {x, 0, 1}, AxesLabel → {t, y},

    PlotStyle->CMYKColor[0, 0.89, 0.94, 0.28]]
```

However, when we graph the solution on "small" intervals close to the origin as shown in Figure 4-29, we see that $\lim_{x\to 0^+} y(x)$ does not exist.

```
p1 = Plot[y[x], {x, 0.001, .1}, AxesLabel → {t, y},

    PlotStyle->CMYKColor[0, 0.89, 0.94, 0.28]];

p2 = Plot[y[x], {x, 0.00001, .001}, AxesLabel → {t, y},

    PlotStyle->CMYKColor[0, 0.89, 0.94, 0.28]];

p3 = Plot[y[x], {x, 0.0000001, .00001}, AxesLabel → {t, y},

    PlotStyle->CMYKColor[0, 0.89, 0.94, 0.28]];

Show[GraphicsRow[{p1, p2, p3}]]
```

As expected, we see that `DSolve` can be used to solve the initial-value problem directly.

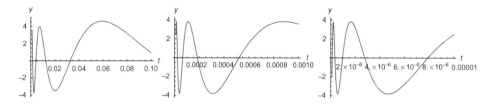

Figure 4-29   Zooming in near $x = 0$ helps convince us that $\lim_{x \to 0^+} y(x)$ does not exist

```
Clear[x, y]
partsol = DSolve[{eq, y[1] == 1, y'[1] == -9, y''[1] == 27,
   y'''[1] == 1}, y[x], x]
```

$$\{\{y[x] \to \frac{1}{30}(102x - 72\text{Cos}[3\text{Log}[x]] - 105x\text{Log}[x] - 89\text{Sin}[3\text{Log}[x]])\}\}$$

∎

4.6.3 Variation of Parameters

Cauchy-Euler equations can be homogenous or nonhomogeneous in which case the method of Variation of Parameters can be used to solve the problem. Before implementing the method to find a particular solution, be sure to write the equation in standard form first.

EXAMPLE 4.6.4: Solve $x^2 y'' - xy' + 5y = x$, $x > 0$.

SOLUTION: We first note that DSolve can be used to find a general solution of the equation directly.

```
Clear[x, y, gensol]
gensol = DSolve[x^2y''[x] - xy'[x] + 5y[x] == x,
   y[x], x]
```

$$\{\{y[x] \to xC[2]\text{Cos}[2\text{Log}[x]] + xC[1]\text{Sin}[2\text{Log}[x]] + \frac{1}{4}(2x\text{Cos}[\text{Log}[x]]^2$$

$$\text{Cos}[2\text{Log}[x]] + x\text{Sin}[2\text{Log}[x]]^2)\}\}$$

Alternatively, we can use Mathematica to help us implement Variation of Parameters. We begin by finding a general solution to the corresponding homogeneous equation $x^2 y'' - xy' + 5y = 0$ with DSolve.

```
homsol = DSolve[x^2y"[x] - xy'[x] + 5y[x] == 0,
    y[x], x]
```

$$\{\{y[x] \to xC[2]Cos[2Log[x]] + xC[1]Sin[2Log[x]]\}\}$$

We see that a general solution of the corresponding homogeneous equation is $y_h = x[c_1 \cos(2 \ln x) + c_2 \sin(2 \ln x)]$. A fundamental set of solutions for the homogeneous equation is $S = \{x \cos(2 \ln x), x \sin(2 \ln x)\}$

```
y1[x_] = xCos[2Log[x]];

y2[x_] = xSin[2Log[x]];

caps = {y1[x], y2[x]};
```

and the Wronskian is $W(S) = 2x$.

```
ws = Wronskian[caps, x]
```

$2x$

To implement Variation of Parameters, we rewrite the equation in standard form

$$y'' - \frac{1}{x}y' + \frac{5}{x^2}y = \frac{1}{x}$$

by dividing by x^2 and identify $f(x) = 1/x$. We then use Integrate to compute

$$u_1 = \int \frac{-y_2(x)f(x)}{2x}dx \quad \text{and} \quad u_2 = \int \frac{y_1(x)f(x)}{2x}dx.$$

```
f[x_] = 1/x;

u1p = -y2[x]f[x]/ws

u2p = y1[x]f[x]/ws
```

$-\dfrac{Sin[2Log[x]]}{2x}$

$\dfrac{Cos[2Log[x]]}{2x}$

```
u1[x_] = Integrate[u1p, x]
```

```
u2[x_] = Integrate[u2p, x]
```

$$\frac{1}{2}\text{Cos}[\text{Log}[x]]^2$$

$$\frac{1}{4}\text{Sin}[2\text{Log}[x]]$$

A particular solution of the nonhomogeneous equation is given by $y_p = y_1 u_1 + y_2 u_2$

```
yp[x_] = u1[x]y1[x] + u2[x]y2[x]//Simplify
```

$$\frac{1}{2}x\text{Cos}[\text{Log}[x]]^2$$

and a general solution is given by $y = y_h + y_p$.

```
y[x_] = homsol[[1, 1, 2]] + yp[x]
```

$$\frac{1}{2}x\text{Cos}[\text{Log}[x]]^2 + xC[2]\text{Cos}[2\text{Log}[x]] + xC[1]\text{Sin}[2\text{Log}[x]]$$

As in previous examples, we graph this general solution for various values of the arbitrary constants. See Figure 4-30.

```
toplot = Table[y[x]/.{C[1] → i, C[2] → j}, {i, -2, 2}, {j, -2, 2}];
```

```
Plot[toplot, {x, 0, 2}, AxesLabel → {x, y}]
```

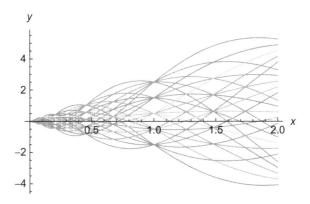

Figure 4-30   Various solutions of a nonhomogeneous Cauchy-Euler equation

■

4.7 Series Solutions

In calculus we learn that Maclaurin and Taylor polynomials can be used to approximate functions. This idea can be extended to approximating the solution of a differential equation. First, we introduce some necessary terminology. In some cases, we can use power series or generalized series to solve some differential equations.

4.7.1 Power Series Solutions About Ordinary Points

Definition 20 (Standard Form, Ordinary, and Singular Points). *Consider the equation $a_2(x)y'' + a_1(x)y' + a_0(x)y = 0$ and let $p(x) = a_1(x)/a_2(x)$ and $q(x) = a_0(x)/a_2(x)$. Then, the equation $a_2(x)y'' + a_1(x)y' + a_0(x)y = 0$ is equivalent to $y'' + p(x)y' + q(x)y = 0$, which is called the **standard form** of the equation. A number x_0 is an **ordinary point** of this differential equation if both $p(x)$ and $q(x)$ are analytic at x_0. If x_0 is not an ordinary point, x_0 is called a **singular point**.*

If x_0 is an ordinary point of the differential equation $y'' + p(x)y' + q(x)y = 0$, we can write $p(x) = \sum_{n=0}^{\infty} b_n (x - x_0)^n$, where $b_n = p^{(n)}(x_0)/n!$, and $q(x) = \sum_{n=0}^{\infty} c_n (x - x_0)^n$, where $c_n = q^{(n)}(x_0)/n!$. Substitution into the equation $y'' + p(x)y' + q(x)y = 0$ results in

$$y'' + y' \sum_{n=0}^{\infty} b_n (x - x_0)^n + y \sum_{n=0}^{\infty} c_n (x - x_0)^n = 0.$$

If we assume that y is analytic at x_0, we can write $y(x) = \sum_{n=0}^{\infty} a_n (x - x_0)^n$. Because a power series can be differentiated term-by-term, we can compute the first and second derivatives of y and substitute back into the equation to calculate the coefficients a_n. Thus, we obtain a power series solution of the equation.

Power Series Solution Method About an Ordinary Point
1. Assume that $y(x) = \sum_{n=0}^{\infty} a_n (x - x_0)^n$.
2. After taking the appropriate derivatives, substitute $y(x) = \sum_{n=0}^{\infty} a_n (x - x_0)^n$ into the differential equation.
3. Find the unknown series coefficients a_n.
4. When applicable, apply any given initial conditions.

Because the differentiation of power series is necessary in this method for solving differential equations, we should make a few observations about this procedure. Consider the Maclaurin series $y = \sum_{n=0}^{\infty} a_n x^n$. Term-by-term differentiation of this series yields $y' = \sum_{n=0}^{\infty} na_n x^{n-1}$. Notice, however, that with the initial index value of $n = 0$, the first term of the series is 0 so we rewrite the series in its equivalent form

$$y' = \sum_{n=1}^{\infty} na_n x^{n-1} = \sum_{n=0}^{\infty} (n+1)a_{n+1} x^n.$$

Similarly,

$$y'' = \sum_{n=1}^{\infty} n(n-1)a_n x^{n-2} = \sum_{n=2}^{\infty} n(n-1)a_n x^{n-2} = \sum_{n=0}^{\infty} (n+1)(n+2)a_{n+2} x^n.$$

We make use of these derivatives throughout the section.

EXAMPLE 4.7.1: (a) Find a general solution of $(4 - x^2)y' + y = 0$.

(b) Solve the initial-value problem $\begin{cases} (4 - x^2)y' + y = 0 \\ y(0) = 1 \end{cases}$.

SOLUTION: (a) Let $y = \sum_{n=0}^{\infty} a_n x^n$. Then term-by-term differentiation yields $y' = dy/dx = \sum_{n=0}^{\infty} na_n x^{n-1}$ and substitution into the differential equation gives us

$$\left(4 - x^2\right)\frac{dy}{dx} + y = \left(4 - x^2\right)\sum_{n=0}^{\infty} na_n x^{n-1} + \sum_{n=0}^{\infty} a_n x^n$$

$$= \sum_{n=1}^{\infty} 4na_n x^{n-1} - \sum_{n=1}^{\infty} na_n x^{n+1} + \sum_{n=0}^{\infty} a_n x^n = 0.$$

Note that the first term in these three series involves x^0, x^2, and x^0, respectively. Thus, if we pull off the first two terms in the first and third series, all three series will begin with an x^2 term. Doing so, we have

$$(4a_1 + a_0) + (8a_2 + a_1)x + \sum_{n=3}^{\infty} 4na_n x^{n-1} - \sum_{n=1}^{\infty} na_n x^{n+1} + \sum_{n=2}^{\infty} a_n x^n = 0.$$

The indices of these three series do not match, so we must change two of the three to match the third. Substitution of $n + 1$ for n in $\sum_{n=3}^{\infty} 4na_n x^{n-1}$ yields

$$\sum_{n+1=3}^{\infty} 4(n+1)a_{n+1}x^{n+1-1} = \sum_{n=2}^{\infty} 4(n+1)a_{n+1}x^n.$$

Similarly, substitution of $n-1$ for n in $\sum_{n=1}^{\infty} na_n x^{n+1}$ yields

$$\sum_{n-1=1}^{\infty} (n-1)a_{n-1}x^{n-1+1} = \sum_{n=2}^{\infty}(n-1)a_{n-1}x^n.$$

Therefore, after combining the three series, we have the equation

$$(4a_1 + a_0)+(8a_2 + a_1)x+\sum_{n=2}^{\infty} [a_n + 4(n+1)a_{n+1} - (n-1)a_{n-1}]x^n = 0.$$

Because the sum of the terms on the left-hand side of the equation is zero, each coefficient must be zero. Equating the coefficients of x^0 and x to zero yields $a_1 = -\frac{1}{4}a_0$ and $a_2 = -\frac{1}{8}a_1 = \frac{1}{32}a_0$. When the series coefficient $a_n + 4(n+1)a_{n+1} - (n-1)a_{n-1}$ is set to zero, we obtain the recurrence relation $a_{n+1} = \dfrac{(n-1)a_{n-1} - a_n}{4(n+1)}$ for the indices in the series, $n \geq 2$. After defining the recursively defined function a,

```
Clear[a, n]

a[n_]:=a[n] = ((n − 2)a[n − 2] − a[n − 1])/(4n);

a[0] = a0;

a[1] = −a0/4;
```

we use the formula to determine the values of a_n for $n = 2$, $3, \ldots, 11$, and give these values in the following table. In this case, note that we define a using the form a [n_] :=a [n] =. . . so that Mathematica "remembers" the values of a [n] computed. Thus, for particular values of n, Mathematica need not recompute a [n-1] and a [n-2] to compute a [n] if these values have previously been computed.

```
TableForm[Table[{n, a[n]}, {n, 0, 11}]]
```

0   a0

1   $-\dfrac{a0}{4}$

2   $\dfrac{a0}{32}$

3	$-\dfrac{3a0}{128}$
4	$\dfrac{11a0}{2048}$
5	$-\dfrac{31a0}{8192}$
6	$\dfrac{69a0}{65536}$
7	$-\dfrac{187a0}{262144}$
8	$\dfrac{1843a0}{8388608}$
9	$-\dfrac{4859a0}{33554432}$
10	$\dfrac{12767a0}{268435456}$
11	$-\dfrac{32965a0}{1073741824}$

Therefore,

$$y = a_0 - \frac{1}{4}a_0 x + \frac{1}{32}a_0 x^2 - \frac{3}{128}a_0 x^3 + \frac{11}{2048}a_0 x^4 - \frac{31}{8192}a_0 x^5 + \cdots$$

(b) When we apply the initial condition $y(0) = 1$, we substitute $x = 0$ into the general solution obtained in (a) and set the result equal to 1. Hence, $a_0 = 1$,

```
a0 = 1;
```

```
TableForm[Table[{n, a[n]}, {n, 0, 11}]]
```

0	1
1	$-\dfrac{1}{4}$
2	$\dfrac{1}{32}$
3	$-\dfrac{3}{128}$
4	$\dfrac{11}{2048}$
5	$-\dfrac{31}{8192}$
6	$\dfrac{69}{65536}$
7	$-\dfrac{187}{262144}$
8	$\dfrac{1843}{8388608}$
9	$-\dfrac{4859}{33554432}$
10	$\dfrac{12767}{268435456}$
11	$-\dfrac{32965}{1073741824}$

so the series solution of the initial-value problem is

$$y = a_0 - \frac{1}{4}x + \frac{1}{32}x^2 - \frac{3}{128}x^3 + \frac{11}{2048}x^4 - \frac{31}{8192}x^5 + \cdots$$

Notice that the equation $(4 - x^2)y' + y = 0$ is separable, so we can compute the solution directly with separation of variables by rewriting the equation as $-\frac{1}{y}dy = \frac{1}{4 - x^2}dx$. Integrating yields $\ln y = \frac{1}{4}(\ln|x - 2| - \ln|x + 2|) = \ln\left|\frac{x - 2}{x + 2}\right|^{1/4} + C$. Applying the initial condition $y(0) = 1$ results in $\ln y = \ln\left|\frac{x - 2}{x + 2}\right|^{1/4}$ so $y = \sqrt[4]{\frac{x - 2}{x + 2}}$.

Nearly identical results are obtained with DSolve.

```
Clear[x, y]

exactsol =
    DSolve[{(4 − x^2)y'[x] + y[x]==0, y[0] == 1}, y[x], x]
```

$$\{\{y[x] \rightarrow \frac{(2 - x)^{1/4}}{(2 + x)^{1/4}}\}\}$$

We can approximate the solution of the problem by taking a finite number of terms of the series solution.

```
yapprox = Sum[a[i]x^i, {i, 0, 10}]
```

$$1 - \frac{x}{4} + \frac{x^2}{32} - \frac{3x^3}{128} + \frac{11x^4}{2048} - \frac{31x^5}{8192} + \frac{69x^6}{65536} - \frac{187x^7}{262144} +$$

$$\frac{1843x^8}{8388608} - \frac{4859x^9}{33554432} + \frac{12767x^{10}}{268435456}$$

The graph of the polynomial approximation of degree 10 is shown in Figure 4-31 along with the solution obtained through separation of variables.

```
p1 = Plot[{exactsol[[1, 1, 2]], yapprox}, {x, −2, 2},
    PlotStyle → {{Black}, {Gray}}, AxesLabel → {x, y}]
```

The graph shows that the accuracy of the approximation decreases near $x = \pm 2$, which are the singular points of the differential equation. (The reason for this is discussed in the theorem following this example.)

Alternatively, we can take advantage of Series to help us form a series solution of the problem.

First, we use Series to compute the first few terms of the power series expansion for the left-hand side of the equation about $x = 0$ and name the result serapprox.

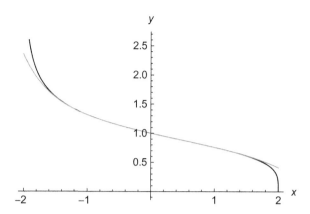

Figure 4-31   Comparison of exact solution to a polynomial approximation of the solution

Clear[x, y]

serapprox = Series[(4 − x^2)y′[x] + y[x], {x, 0, 10}]

$(y[0] + 4y′[0]) + (y′[0] + 4y″[0])x + (−y′[0] + \dfrac{y″[0]}{2} +$

$2y^{(3)}[0])x^2 + \dfrac{1}{6}(−6y″[0] + y^{(3)}[0] + 4y^{(4)}[0])x^3 +$

$(−\dfrac{1}{2}y^{(3)}[0] + \dfrac{1}{24}y^{(4)}[0] + \dfrac{1}{6}y^{(5)}[0])x^4 + (−\dfrac{1}{6}y^{(4)}[0] +$

$\dfrac{1}{120}y^{(5)}[0] + \dfrac{1}{30}y^{(6)}[0])x^5 + (−\dfrac{1}{24}y^{(5)}[0] + \dfrac{1}{720}y^{(6)}[0] +$

$\dfrac{1}{180}y^{(7)}[0])x^6 + (−\dfrac{1}{120}y^{(6)}[0] + \dfrac{y^{(7)}[0]}{5040} + \dfrac{y^{(8)}[0]}{1260})x^7 +$

$(−\dfrac{1}{720}y^{(7)}[0] + \dfrac{y^{(8)}[0]}{40320} + \dfrac{y^{(9)}[0]}{10080})x^8 + (−\dfrac{y^{(8)}[0]}{5040} + \dfrac{y^{(9)}[0]}{362880} +$

$\dfrac{y^{(10)}[0]}{90720})x^9 + (−\dfrac{y^{(9)}[0]}{40320} + \dfrac{y^{(10)}[0]}{3628800} + \dfrac{y^{(11)}[0]}{907200})x^{10} + O[x]^{11}$

Then, we use LogicalExpand to form the system of equations obtained by equating each coefficient in serapprox to zero.

sysofeqs = LogicalExpand[serapprox == 0]

$y[0] + 4y′[0] == 0 \&\& y′[0] + 4y″[0] == 0 \&\& −y′[0] + \dfrac{y″[0]}{2} +$

$2y^{(3)}[0] == 0 \&\& \dfrac{1}{6}(−6y″[0] + y^{(3)}[0] + 4y^{(4)}[0]) == 0 \&\& −$

$\dfrac{1}{2}y^{(3)}[0] + \dfrac{1}{24}y^{(4)}[0] + \dfrac{1}{6}y^{(5)}[0] == 0 \&\& −\dfrac{1}{6}y^{(4)}[0] +$

$\dfrac{1}{120}y^{(5)}[0] + \dfrac{1}{30}y^{(6)}[0] == 0 \&\& −\dfrac{1}{24}y^{(5)}[0] + \dfrac{1}{720}y^{(6)}[0] +$

$$\frac{1}{180}y^{(7)}[0] == 0 \&\& -\frac{1}{120}y^{(6)}[0] + \frac{y^{(7)}[0]}{5040} + \frac{y^{(8)}[0]}{1260} ==$$

$$0 \&\& -\frac{1}{720}y^{(7)}[0] + \frac{y^{(8)}[0]}{40320} + \frac{y^{(9)}[0]}{10080} == 0 \&\& -\frac{y^{(8)}[0]}{5040} +$$

$$\frac{y^{(9)}[0]}{362880} + \frac{y^{(10)}[0]}{90720} == 0 \&\& -\frac{y^{(9)}[0]}{40320} + \frac{y^{(10)}[0]}{3628800} +$$

$$\frac{y^{(11)}[0]}{907200} == 0$$

We want to solve this system of equations for $y'(0)$, $y''(0)$, ..., $y^{(11)}(0)$ so that the results are in terms of $y(0)$. (Note that the symbol $\partial_{\{x,i\}y[x]}$ represents D[y[x],{x,i}] so the same result is obtained by entering

```
vars=Table[D[y[x],{x, i}]/.(x -> 0),{i,1,11}].)
```

```
vars = Table[D[y[x], {x, i}]/.x → 0, {i, 1, 11}]
```

$\{y'[0], y''[0], y^{(3)}[0], y^{(4)}[0], y^{(5)}[0], y^{(6)}[0], y^{(7)}[0], y^{(8)}[0],$

$y^{(9)}[0], y^{(10)}[0], y^{(11)}[0]\}$

We then use `Solve` to solve the system of equations `sysofeqs` for the unknowns specified in `vars`.

```
sols = Solve[sysofeqs, vars]
```

$\{\{y'[0] \to -\frac{y[0]}{4}, y''[0] \to \frac{y[0]}{16}, y^{(3)}[0] \to -\frac{9y[0]}{64}, y^{(4)}[0] \to$

$\frac{33y[0]}{256}, y^{(5)}[0] \to -\frac{465y[0]}{1024}, y^{(6)}[0] \to \frac{3105y[0]}{4096}, y^{(7)}[0] \to$

$-\frac{58905y[0]}{16384}, y^{(8)}[0] \to \frac{580545y[0]}{65536}, y^{(9)}[0] \to -\frac{13775265y[0]}{262144},$

$y^{(10)}[0] \to \frac{180972225y[0]}{1048576}, y^{(11)}[0] \to -\frac{5140067625y[0]}{4194304}\}\}$

The power series solution is formed by substituting these values into the power series for $y(x)$ about $x = 0$ with `ReplaceAll` (/.).

```
sersol = Series[y[x], {x, 0, 11}]/.sols[[1]]
```

$$y[0] - \frac{1}{4}y[0]x + \frac{1}{32}y[0]x^2 - \frac{3}{128}y[0]x^3 + \frac{11y[0]x^4}{2048} -$$

$$\frac{31y[0]x^5}{8192} + \frac{69y[0]x^6}{65536} - \frac{187y[0]x^7}{262144} + \frac{1843y[0]x^8}{8388608} -$$

$$\frac{4859y[0]x^9}{33554432} + \frac{12767y[0]x^{10}}{268435456} - \frac{32965y[0]x^{11}}{1073741824} + O[x]^{12}$$

The solution to the initial-value problem is obtained by replacing each occurrence of $y(0)$ in `sersol` by 1.

```
sol = sersol/.y[0] → 1
```

$$1 - \frac{x}{4} + \frac{x^2}{32} - \frac{3x^3}{128} + \frac{11x^4}{2048} - \frac{31x^5}{8192} + \frac{69x^6}{65536} - \frac{187x^7}{262144} +$$
$$\frac{1843x^8}{8388608} - \frac{4859x^9}{33554432} + \frac{12767x^{10}}{268435456} - \frac{32965x^{11}}{1073741824} + O[x]^{12}$$

Remember that this result cannot be evaluated for particular values of x because of the O-term indicating the omitted higher-order terms of the series. To obtain an approximation of the solution that can be evaluated for particular values of x, use Normal to remove the O-term.

```
polyapprox = Normal[sol]
```

$$1 - \frac{x}{4} + \frac{x^2}{32} - \frac{3x^3}{128} + \frac{11x^4}{2048} - \frac{31x^5}{8192} + \frac{69x^6}{65536} - \frac{187x^7}{262144} +$$
$$\frac{1843x^8}{8388608} - \frac{4859x^9}{33554432} + \frac{12767x^{10}}{268435456} - \frac{32965x^{11}}{1073741824}$$

■

The following theorem explains where the approximation of the solution of the differential equation by the series is valid.

A proof of this theorem can be found in more advanced texts, such as Rabenstein's *Introduction to Ordinary Differential Equations*, [22].

Theorem 9 (Convergence of a Power Series Solution). *Let $x = x_0$ be an ordinary point of the differential equation $a_2(x)y'' + a_1(x)y' + a_0(x)y = 0$ and suppose that R is the distance from $x = x_0$ to the closest singular point of the equation. Then the power series solution $y = \sum_{n=0}^{\infty} a_n (x - x_0)$ converges at least on the interval $(x_0 - R, x_0 + R)$.*

The theorem indicates that the approximation may not be as accurate near singular points of the equation. Hence, we understand why the approximation in Example 4.7.1 breaks down near $x = \pm 2$, the closest singular point to the ordinary point $x = 0$. Of course, $x = 0$ is not an ordinary point for every differential equation. However, because the series $y = \sum_{n=0}^{\infty} a_n (x - x_0)^n$ is easier to work with if $x_0 = 0$, we can always make a transformation so that we can use $y = \sum_{n=0}^{\infty} a_n x^n$ to solve any linear equation. For example, suppose that $x = x_0$ is an ordinary point of a linear equation. Then, if we make the change of variable $t = x - x_0$, then $t = 0$ corresponds to $x = x_0$, so $t = 0$ is an ordinary point of the transformed equation.

EXAMPLE 4.7.2 (Legendre's Equation): **Legendre's equation** is the equation

$$\left(1 - x^2\right) \frac{d^2y}{dx^2} - 2x\frac{dy}{dx} + k(k + 1)y = 0, \tag{4.24}$$

where k is a constant, named after the French mathematician Adrien Marie Legendre (1752–1833). Find a general solution of Legendre's equation.

SOLUTION: In standard form, the equation is

$$\frac{d^2y}{dx^2} - \frac{2x}{1-x^2}\frac{dy}{dx} + \frac{k(k+1)}{1-x^2}y = 0.$$

There is a solution to the equation of the form $y = \sum_{n=0}^{\infty} a_n x^n$ because $x = 0$ is an ordinary point. This solution will converge at least on the interval $(-1, 1)$ because the closest singular points to $x = 0$ are $x = \pm 1$.

Substitution of this function and its derivatives

$$y' = \sum_{n=0}^{\infty}(n+1)a_{n+1}x^n \quad \text{and} \quad y'' = \sum_{n=0}^{\infty}(n+1)(n+2)a_{n+2}x^n$$

into Legendre's equation (4.24) and simplifying the results yields

$$[2a_2 + k(k+1)a_0] + [-2a_1 + k(k+1)a_1 + 6a_3]x + \sum_{n=4}^{\infty}n(n-1)a_n x^{n-2}$$

$$- \sum_{n=2}^{\infty}n(n-1)a_n x^n - \sum_{n=2}^{\infty}2na_n x^n + \sum_{n=2}^{\infty}k(k+1)a_n x^n = 0.$$

After substituting $n + 2$ for each occurrence of n in the first series and simplifying, we have

$$[2a_2 + k(k+1)a_0] + [-2a_1 + k(k+1)a_1 + 6a_3]x + \sum_{n=2}^{\infty}\{(n+2)(n+1)a_{n+2}$$

$$+ [-n(n-1) - 2n + k(k+1)]a_n\}x^n = 0.$$

Equating the coefficients to zero, we find a_2, a_3, and a_{n+2} with Solve.

```
Clear[a, k]
Solve[2 a[2] + k(k + 1)a[0] == 0, a[2]]
```

$$\{\{a[2] \rightarrow -\frac{1}{2}k(1 + k)a[0]\}\}$$

```
Solve[-2 a[1] + k(k + 1)a[1] + 6 a[3] == 0, a[3]]
```

$$\{\{a[3] \rightarrow \frac{1}{6}(2a[1] - ka[1] - k^2 a[1])\}\}$$

```
genform = Solve[(n + 2)(n + 1)a[n + 2] + (-n(n - 1) - 2n + k(k + 1))a[n]
== 0, a[n + 2]]
```

$$\{\{a[2 + n] \rightarrow -\frac{(k + k^2 - n - n^2)a[n]}{(1 + n)(2 + n)}\}\}$$

```
Factor[genform[[1, 1, 2]]]
```

$$\frac{(-k+n)(1+k+n)a[n]}{(1+n)(2+n)}$$

We obtain a formula for a_n by replacing each occurrence of n in a_{n+2} by $n-2$.

```
genform[[1, 1, 2]]/.n → n - 2
```

$$-\frac{(2+k+k^2-(-2+n)^2-n)a[-2+n]}{(-1+n)n}$$

Using this formula, we find several coefficients with `Table`.

```
Clear[a, a1, a0]
```

$$a[n\_]:=a[n] = -\frac{(2+k+k^2-(-2+n)^2-n)a[-2+n]}{(-1+n)n};$$

```
a[1] = a1;
```

```
a[0] = a0;
```

```
Table[{n, a[n]}, {n, 2, 10}]//TableForm
```

2     $-\dfrac{1}{2}a0(k+k^2)$

3     $-\dfrac{1}{6}a1(-2+k+k^2)$

4     $\dfrac{1}{24}a0(-6+k+k^2)(k+k^2)$

5     $\dfrac{1}{120}a1(-12+k+k^2)(-2+k+k^2)$

6     $-\dfrac{1}{720}a0(-20+k+k^2)(-6+k+k^2)(k+k^2)$

7     $-\dfrac{a1(-30+k+k^2)(-12+k+k^2)(-2+k+k^2)}{5040}$

8     $\dfrac{a0(-42+k+k^2)(-20+k+k^2)(-6+k+k^2)(k+k^2)}{40320}$

9     $\dfrac{a1(-56+k+k^2)(-30+k+k^2)(-12+k+k^2)(-2+k+k^2)}{362880}$

10    $-\dfrac{a0(-72+k+k^2)(-42+k+k^2)(-20+k+k^2)(-6+k+k^2)(k+k^2)}{3628800}$

Hence, we have the two linearly independent solutions

$$y_1 = a_0\left(1 - \frac{k(k+1)}{2!}x^2 + \frac{(2-k)(3+k)k(k+1)}{4!}x^4\right.$$
$$\left.\frac{-(4-k)(5+k)(2-k)(3+k)k(k+1)}{6!}x^6 + \cdots\right)$$

and

$$y_2 = a_1\left(x - \frac{(k-1)(k+2)}{3!}x^3 + \frac{(3-k)(4+k)(k-1)(k+2)}{5!}x^5\right.$$
$$\left.-\frac{(5-k)(6+k)(3-k)(4+k)(k-1)(k+2)}{7!}x^7 + \cdots\right)$$

so a general solution of Legendre's equation (4.24) is

$$y = y_1 + y_2$$
$$= a_1 \left(x - \frac{(k-1)(k+2)}{3!} x^3 + \frac{(3-k)(4+k)(k-1)(k+2)}{5!} x^5 \right.$$
$$\left. - \frac{(5-k)(6+k)(3-k)(4+k)(k-1)(k+2)}{7!} x^7 + \cdots \right)$$
$$+ a_1 \left(x - \frac{(k-1)(k+2)}{3!} x^3 + \frac{(3-k)(4+k)(k-1)(k+2)}{5!} x^5 \right.$$
$$\left. - \frac{(5-k)(6+k)(3-k)(4+k)(k-1)(k+2)}{7!} x^7 + \cdots \right).$$

Note that DSolve is able to find a general solution as well—the result is given in terms of the functions LegendreP and LegendreQ, Mathematica's linearly independent solutions of Legendre's equation.

DSolve[(1 − x^2)y″[x] − 2xy′[x]+

k(k + 1)y[x] == 0, y[x], x]

{{y[x] → C[1]LegendreP[k, x] + C[2]LegendreQ[k, x]}}

An interesting observation from the general solution to Legendre's equation is that the series solutions terminate for integer values of k. If k is an even integer, the first series terminates while if k is an odd integer the second series terminates. Therefore, polynomial solutions are found for integer values of k. Because these polynomials are useful and are encountered in numerous applications, we have a special notation for them: $P_n(x)$ is called the **Legendre polynomial of degree** n and represents an nth degree polynomial solution to Legendre's equation. The Mathematica command LegendreP [n, x] returns $P_n(x)$.

We use Table together with LegendreP to list the first few Legendre polynomials as well as the first few LegendreQ functions.

toplot1 = Table[LegendreP[n, x], {n, 0, 5}];

TableForm[toplot]

1

x

$\frac{1}{2}(-1 + 3x^2)$
$\frac{1}{2}(-3x + 5x^3)$
$\frac{1}{8}(3 - 30x^2 + 35x^4)$
$\frac{1}{8}(15x - 70x^3 + 63x^5)$

toplot2 = Table[LegendreQ[n, x], {n, 0, 5}];

TableForm[toplot2]

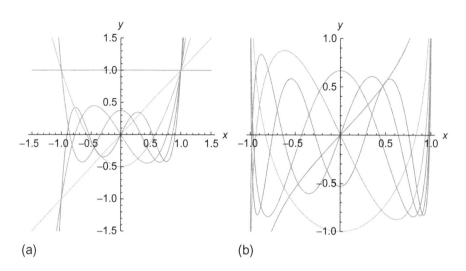

Figure 4-32   (a) Plots of the first few Legendre polynomials, LegendreP. (b) Plots of the second linearly independent solutions of Legendre's equation, LegendreQ

$$-\frac{1}{2}\text{Log}[1-x] + \frac{1}{2}\text{Log}[1+x]$$

$$-1 + x(-\frac{1}{2}\text{Log}[1-x] + \frac{1}{2}\text{Log}[1+x])$$

$$-\frac{3x}{2} + \frac{1}{2}(-1+3x^2)(-\frac{1}{2}\text{Log}[1-x] + \frac{1}{2}\text{Log}[1+x])$$

$$\frac{2}{3} - \frac{5x^2}{2} - \frac{1}{2}x(3-5x^2)(-\frac{1}{2}\text{Log}[1-x] + \frac{1}{2}\text{Log}[1+x])$$

$$\frac{55x}{24} - \frac{35x^3}{8} + \frac{1}{8}(3-30x^2+35x^4)(-\frac{1}{2}\text{Log}[1-x] + \frac{1}{2}\text{Log}[1+x])$$

$$-\frac{8}{15} + \frac{49x^2}{8} - \frac{63x^4}{8} + \frac{1}{8}x(15-70x^2+63x^4)(-\frac{1}{2}\text{Log}[1-x]$$
$$+\frac{1}{2}\text{Log}[1+x])$$

We graph the LegendreP polynomials for $-3/2 \le x \le 3/2$ in Figure 4-32(a) and the LegendreQ functions for $-1 \le x \le 1$ in Figure 4-32(b).

```
p1 = Plot[toplot1, {x, -3/2, 3/2}, PlotRange → {-3/2, 3/2},

    AspectRatio → 1, AxesLabel → {x, y}, PlotLabel → "(a)"]

p2 = Plot[toplot2, {x, -1, 1}, PlotRange → {-1, 1},

    AspectRatio → 1, AxesLabel → {x, y}, PlotLabel → "(b)"]

Show[GraphicsRow[{p1, p2}]]
```

Another interesting observation is about the Legendre polynomials is that they satisfy the relationship $\int_{-1}^{1} P_m(x)P_n(x)\,dx = 0$, $m \ne n$, called an *orthogonality condition*, which we verify with Integrate for $m, n = 0, 1, \ldots, 6$.

```
Table[Integrate[LegendreP[n, x]LegendreP[m, x], {x, -1, 1}],

{n, 0, 6}, {m, 0, 6}]//TableForm
```

2	0	0	0	0	0	0
0	$\frac{2}{3}$	0	0	0	0	0
0	0	$\frac{2}{5}$	0	0	0	0
0	0	0	$\frac{2}{7}$	0	0	0
0	0	0	0	$\frac{2}{9}$	0	0
0	0	0	0	0	$\frac{2}{11}$	0
0	0	0	0	0	0	$\frac{2}{13}$

Note that the entries down the diagonal of this result correspond to the value of $\int_{-1}^{1} [P_n(x)]^2 \, dx$ for $n = 0, 1, \ldots, 6$ and indicate that $\int_{-1}^{1} [P_n(x)]^2 \, dx = 2/(2n + 1)$.

■

4.7.2 Series Solutions About Regular Singular Points

In the previous section, we used a power series expansion about an ordinary point to find (or approximate) the solution of a differential equation. We noted that these series solutions may not converge near the *singular points* of the equation.

In this section, we investigate the problem of obtaining a series expansion about a singular point. We begin with the following classification of singular points.

Definition 21 (Regular and Irregular Singular Points). *Let $x = x_0$ be singular point of $y'' + p(x)y' + q(x)y = 0$. $x = x_0$ is a **regular singular point** of the equation if both $(x - x_0) p(x)$ and $(x - x_0)^2 q(x)$ are analytic at $x = x_0$. If $x = x_0$ is not a regular singular point, then $x = x_0$ is called an **irregular singular point** of the equation.*

Sometimes this definition is difficult to apply. Therefore, we supply the following definition for polynomial coefficients $p(x)$ and $q(x)$ of the equation $y'' + p(x)y' + q(x)y = 0$.

Definition 22 (Singular Points of Equations With Polynomial Coefficients). *Suppose that $p(x)$ and $q(x)$ are polynomials with no common factors. If after reducing $p(x)$ and $q(x)$ to lowest terms, the highest power of $x - x_0$ in the denominator of $p(x)$ is 1* *and the highest power of $x - x_0$ in the denominator of $q(x)$ is 2, then $x = x_0$ is a **regular singular point** of the equation. Otherwise, it is an **irregular singular point**.*

EXAMPLE 4.7.3: Classify the singular points of each of the following equations: (a) $x^2y'' + xy' + (x^2 - \mu^2)y = 0$ (**Bessel's equation**), and (b) $(x^2 - 16)^2 y'' + (x - 4)y' + y = 0$.

SOLUTION: (a) In standard form, Bessel's equation is

$$\frac{d^2y}{dx^2} + \frac{1}{x}\frac{dy}{dx} + \left(1 - \frac{\mu^2}{x^2}\right)y = 0$$

so $x = 0$ is a singular point of this equation because $p(x) = 1/x$ is not analytic at $x = 0$. Because $xp(x) = 1$ and $x^2\left(1 - \frac{\mu^2}{x^2}\right) = x^2 - \mu^2$, $x = 0$ is a regular singular point. We see that DSolve is able to find a general solution of Bessel's equation, although the result is given in terms of the Bessel functions, BesselJ and BesselY.

$$\texttt{DSolve[x\^2y''[x] + xy'[x] + (x\^2 - \mu\^2)y[x] == 0,}$$

$$\texttt{y[x], x]}$$

$$\texttt{\{\{y[x] $\rightarrow$ BesselJ[}\mu\texttt{, x]C[1] + BesselY[}\mu\texttt{, x]C[2]\}\}}$$

(b) In standard form, the equation is

$$\frac{d^2y}{dx^2} + \frac{x-4}{(x^2-16)^2}\frac{dy}{dx} + \frac{1}{(x^2-16)^2}y = 0 \quad \text{or}$$

$$\frac{d^2y}{dx^2} + \frac{1}{(x-4)(x+4)^2}\frac{dy}{dx} + \frac{1}{(x-4)^2(x+4)^2}y = 0.$$

Thus, the singular points are $x = 4$ and $x = -4$. For $x = 4$, we have

$$(x-4)p(x) = (x-4)\frac{1}{(x-4)(x+4)^2} = \frac{1}{(x+4)^2}$$

and

$$(x-4)^2q(x) = (x-4)^2\frac{1}{(x-4)^2(x+4)^2} = \frac{1}{(x+4)^2}.$$

Both of these functions are analytic at $x = 4$, so $x = 4$ is a regular singular point.

For $x = -4$,

$$(x+4)p(x) = (x+4)\frac{1}{(x-4)(x+4)^2} = \frac{1}{(x-4)(x+4)},$$

which is not analytic at $x = -4$. Thus, $x = -4$ is an irregular singular point. DSolve is unable to find a general solution of this equation.

■

4.7.3 Method of Frobenius

Now we illustrate how a series expansion about a regular singular point can be used to solve an equation.

Theorem 10 (Method of Frobenius). *Let $x = x_0$ be a regular singular point of $y'' + p(x)y' + q(x)y = 0$. Then this differential equation has at least one solution of the form*

$$y = \sum_{n=0}^{\infty} a_n (x - x_0)^{n+r},$$

where r is a constant that must be determined. This solution is convergent at least on some interval $|x - x_0| < R, R > 0$.

EXAMPLE 4.7.4: Find a general solution of $xy'' + (1 + x)y' - \dfrac{1}{16x}y = 0$.

SOLUTION: First, we note that in standard form this equation is

$$\frac{d^2y}{dx^2} + \frac{1+x}{x}\frac{dy}{dx} - \frac{1}{16x^2}y = 0.$$

Thus, $x = 0$ is a singular point. Moreover, because $xp(x) = x \cdot \dfrac{1+x}{x} = 1+x$ and $x^2q(x) = x^2 \cdot -\dfrac{1}{16x^2} = -16$ are both analytic at $x = 0$, we classify

$x = 0$ as a regular singular point. By the Method of Frobenius, there is at least one solution of the form $y = \sum_{n=0}^{\infty} a_n x^{n+r}$. Differentiating this function twice, we obtain

$$y' = \sum_{n=0}^{\infty} a_n (n+r) x^{n+r-1} \qquad \text{and} \qquad y'' = \sum_{n=0}^{\infty} a_n (n+r)(n+r-1) x^{n+r-2}.$$

Substituting these series into the differential equation yields

$$x \sum_{n=0}^{\infty} a_n (n+r)(n+r-1) x^{n+r-2} + (1+x) \sum_{n=0}^{\infty} a_n (n+r) x^{n+r-1}$$

$$- \frac{1}{16x} \sum_{n=0}^{\infty} a_n x^{n+r} = 0$$

$$\sum_{n=0}^{\infty} a_n (n+r)(n+r-1) x^{n+r-1} + \sum_{n=0}^{\infty} a_n (n+r) x^{n+r-1} + \sum_{n=0}^{\infty} a_n (n+r) x^{n+r}$$

$$- \sum_{n=0}^{\infty} \frac{1}{16} a_n x^{n+r-1} = 0.$$

Notice that the first term in three of the four series begins with an x^{r-1} term while the first term in $\sum_{n=0}^{\infty} a_n (n+r) x^{n+r}$ begins with an x^r term, so we must pull off the first terms in the other three series so that they match. Hence,

$$\left[r(r-1) + r - \frac{1}{16} \right] a_0 x^{r-1} + \sum_{n=1}^{\infty} a_n (n+r)(n+r-1) x^{n+r-1}$$

$$+ \sum_{n=1}^{\infty} a_n (n+r) x^{n+r-1} + \sum_{n=0}^{\infty} a_n (n+r) x^{n+r} - \sum_{n=1}^{\infty} \frac{1}{16} a_n x^{n+r-1} = 0.$$

Changing the index in the third series by substituting $n - 1$ for each occurrence of n, we have

$$\sum_{n-1=0}^{\infty} a_{n-1} (n-1+r) x^{n-1+r} = \sum_{n=1}^{\infty} a_{n-1} (n+r-1) x^{n+r-1}.$$

After simplification, we have

$$\left[r(r-1)+r-\frac{1}{16}\right]a_0x^{r-1} + \sum_{n=1}^{\infty}\left\{\left[(n+r)(n+r-1)+(n+r)-\frac{1}{16}\right]a_n\right.$$
$$\left. + (n+r-1)a_{n-1}\right\}x^{n+r-1} = 0.$$

We equate the coefficients to zero to find the coefficients and the value of r. Assuming that $a_0 \neq 0$ so that the first term of our series solution is not zero, we have from the first term the equation

$$r(r-1)+r-\frac{1}{16} = 0,$$

called the **indicial equation**, because it yields the value of r. In this case,

```
Solve[r(r − 1) + r − 1/16 == 0]
```

$$\left\{\left\{r \to -\frac{1}{4}\right\}, \left\{r \to \frac{1}{4}\right\}\right\}$$

the roots are $r_1 = 1/4$ and $r_2 = -1/4$. Starting with the *larger* of the two roots, $r_1 = 1/4$, we assume that $y_1 = \sum_{n=0}^{\infty} a_n x^{n+1/4} = x^{1/4}\sum_{n=0}^{\infty} a_n x^n$. Equating the series coefficient to zero, we have

$$\left[\left(n+\frac{1}{4}\right)\left(n+\frac{1}{4}-1\right)+\left(n+\frac{1}{4}\right)-\frac{1}{16}\right]a_n + \left(n+\frac{1}{4}-1\right)a_{n-1} = 0,$$

which we solve for a_n.

```
Clear[a, n]
Solve[((n + 1/4)(n + 1/4 − 1) + (n + 1/4) − 1/16)a[n]+

    (n + 1/4 − 1)a[n − 1] == 0, a[n]]
```

$$\left\{\left\{a[n] \to -\frac{(-3+4n)a[-1+n]}{2n(1+2n)}\right\}\right\}$$

Several of these coefficients are calculated using this formula with `Table`.

```
a[n_]:=a[n] = −(−3 + 4n)a[−1 + n]/2n(1 + 2n);

a[0] = a0;

TableForm[Table[{n, a[n]}, {n, 0, 10}]]
```

0	$a0$
1	$-\dfrac{a0}{6}$
2	$\dfrac{a0}{24}$
3	$-\dfrac{a0}{112}$
4	$\dfrac{13a0}{8064}$
5	$-\dfrac{221a0}{887040}$
6	$\dfrac{17a0}{506880}$
7	$-\dfrac{17a0}{4257792}$
8	$\dfrac{29a0}{68124672}$
9	$-\dfrac{29a0}{706019328}$
10	$\dfrac{1073a0}{296528117760}$

Therefore, one solution to the equation is

$$y_1 = a_0 x^{1/4}\left(1 - \frac{1}{6}x + \frac{1}{24}x^2 - \frac{1}{112}x^3 + \frac{13}{8064}x^4 + \cdots\right).$$

For $r_2 = -1/4$, we assume that $y_2 = \sum_{n=0}^{\infty} a_n x^{n-1/4} = x^{-1/4}\sum_{n=0}^{\infty} a_n x^n$. Then, we have

$$\left[\left(n - \frac{1}{4}\right)\left(n - \frac{1}{4} - 1\right) + \left(n - \frac{1}{4}\right) - \frac{1}{16}\right]b_n + \left(n - \frac{1}{4} - 1\right)b_{n-1} = 0,$$

which we solve for b_n with `Solve`.

```
Clear[b, n]

Solve[((n − 1/4)(n − 1/4 − 1) + (n − 1/4) − 1/16)b[n]+

    (n − 1/4 − 1)b[n − 1] == 0, b[n]]
```

$$\{\{b[n] \to -\frac{(-5 + 4n)b[-1 + n]}{2n(-1 + 2n)}\}\}$$

The value of several coefficients determined with this formula are computed as well.

```
b[n_]:=b[n] = −(−5 + 4n)b[−1 + n] / 2n(−1 + 2n);

b[0] = b0;

TableForm[Table[{n, b[n]}, {n, 0, 10}]]
```

0	$b0$
1	$\dfrac{b0}{2}$
2	$-\dfrac{b0}{8}$
3	$\dfrac{7b0}{240}$
4	$-\dfrac{11b0}{1920}$
5	$\dfrac{11b0}{11520}$
6	$-\dfrac{19b0}{138240}$
7	$\dfrac{437b0}{25159680}$
8	$-\dfrac{437b0}{223641600}$
9	$\dfrac{13547b0}{68434329600}$
10	$-\dfrac{713b0}{39105331200}$

Therefore, a second linearly independent solution of the equation obtained with $r_2 = -1/4$ is

$$y_2 = b_0 x^{-1/4}\left(1 + \frac{1}{2}x - \frac{1}{8}x^2 + \frac{7}{240}x^3 - \frac{11}{1920}x^4 + \cdots\right)$$

and a general solution of the differential equation is $y = c_1 y_1 + c_2 y_2$ where c_1 and c_2 are arbitrary constants. Notice that these two solutions are linearly independent, because they are not scalar multiples of one another.

We see that DSolve is able to find a general solution of the equation as well, although the result is given in terms of the functions HypergeometricU and LaguerreL.

```
gensol1 = DSolve[xy"[x] + (1 + x)y'[x]−
y[x]/(16x) == 0, y[x], x]
```

$$\{\{y[x] \rightarrow e^{-x}x^{1/4}C[1]\text{HypergeometricU}[\frac{5}{4}, \frac{3}{2}, x] + e^{-x}x^{1/4}C[2]$$
$$\text{LaguerreL}[-\frac{5}{4}, \frac{1}{2}, x]\}\}$$

∎

In the previous example, we found the **indicial equation** by direct substitution of the power series solution into the differential equation. In order to derive a general formula for the indicial equation, suppose that $x = 0$ is a regular singular point of the differential equation $y'' + p(x)y' + q(x)y = 0$. Then the functions $xp(x)$

and $x^2 q(x)$ are analytic at $x = 0$, which means that both of these functions have a power series in x with a positive radius of convergence. Hence,

$$xp(x) = p_0 + p_1 x + p_2 x^2 + \cdots \qquad \text{and} \qquad x^2 q(x) = q_0 + q_1 x + q_2 x^2 + \cdots$$

and

$$p(x) = \frac{p_0}{x} + p_1 + p_2 x + p_3 x^2 + \cdots \qquad \text{and} \qquad q(x) = \frac{q_0}{x^2} + \frac{q_1}{x} + q_2 + q_3 x + \cdots$$

Substitution of these series into the differential equation $y'' + p(x)y' + q(x)y = 0$ and multiplying through by the first term in the series for $p(x)$ and $q(x)$, we see that the lowest term in the series involves x^{n+r-2}:

$$\sum_{n=0}^{\infty} a_n(n+r)(n+r-1)x^{n+r-2} + \sum_{n=0}^{\infty} a_n p_0(n+r)x^{n+r-2}$$

$$+ \left(p_1 + p_2 x + p_3 x^2 + \cdots \right) \sum_{n=0}^{\infty} a_n(n+r)x^{n+r-1} + \sum_{n=0}^{\infty} a_n q_0 x^{n+r-2}$$

$$+ \left(\frac{q_1}{x} + q_2 + q_3 x^2 + \cdots \right) \sum_{n=0}^{\infty} a_n x^{n+r} = 0.$$

Then, with $n = 0$, we find that the coefficient of x^{r-2} is

$$-ra_0 + r^2 a_0 + ra_0 p_0 + a_0 q_0 = a_0 \left[r^2 + (p_0 - 1)r + q_0 \right]$$

$$= a_0 \left[r(r-1) + p_0 r + q_0 \right] = 0.$$

Thus, for any equation of the form $y'' + p(x)y' + q(x)y = 0$ with regular singular point $x = 0$, we have the **indicial equation**

$$r(r-1) + p_0 r + q_0 = 0. \tag{4.25}$$

The values of r that satisfy the indicial equation are called the **exponents** or **indicial roots** and are

$$r_{1,2} = \frac{1}{2} \left(1 - p_0 \pm \sqrt{1 - 2p_0 + p_0^2 - 4q_0} \right). \tag{4.26}$$

Note that $r_1 \geq r_2$ and $r_1 - r_2 = \sqrt{1 - 2p_0 + p_0^2 - 4q_0}$.

Several situations can arise when finding the roots of the indicial equation.

1. If $r_1 \neq r_2$ and $\sqrt{1 - 2p_0 + p_0^2 - 4q_0}$ is not an integer, then there are two linearly independent solutions of the equation of the form

$$y_1 = x^{r_1} \sum_{n=0}^{\infty} a_n x^n \quad \text{and} \quad y_2 = x^{r_2} \sum_{n=0}^{\infty} b_n x^n.$$

2. If $r_1 \neq r_2$ and $\sqrt{1 - 2p_0 + p_0^2 - 4q_0}$ is an integer, then there are two linearly independent solutions of the equation of the form

$$y_1 = x^{r_1} \sum_{n=0}^{\infty} a_n x^n \quad \text{and} \quad y_2 = c y_1 \ln x + x^{r_2} \sum_{n=0}^{\infty} b_n x^n.$$

3. If $r_1 - r_2 = \sqrt{1 - 2p_0 + p_0^2 - 4q_0} = 0$, then there are two linearly independent solutions of the problem of the form

$$y_1 = x^{r_1} \sum_{n=0}^{\infty} a_n x^n \quad \text{and} \quad y_2 = y_1 \ln x + x^{r_1} \sum_{n=0}^{\infty} b_n x^n.$$

In any case, if y_1 is a solution of the equation, a second linearly independent solution is given by

$$y_2 = y_1(x) \int \frac{1}{[y_1(x)]^2} e^{-\int p(x)\, dx} dx,$$

which can be obtained through reduction of order.

Note that when solving a differential equation in Case 2, first attempt to find a general solution using $y_2 = x^{r_2} \sum_{n=0}^{\infty} b_n x^n$, where r_2 is the *smaller* of the two roots. However, if the contradiction $a_0 = 0$ is reached, then find solutions of the form $y_1 = x^{r_1} \sum_{n=0}^{\infty} b_n x^n$ and $y_2 = c y_1 \ln x + x^{r_2} \sum_{n=0}^{\infty} b_n x^n$.

The examples here do not illustrate the possibility of complex-valued roots of the indicial equation. When this occurs, the equation is solved using the procedures of Case 1. The solutions that are obtained are complex, so they can be transformed into real solutions by taking the appropriate linear combinations, such as those discussed for complex-valued roots of the characteristic equation of Cauchy-Euler differential equations.

Also, we have not mentioned if a solution can be found with an expansion about an irregular singular point. If $x = x_0$ is an irregular singular point of $y'' + p(x)y' + q(x)y = 0$, there may or may not be a solution of the form $y = x^r \sum_{n=0}^{\infty} a_n x^n$ for some number r.

EXAMPLE 4.7.5 (Bessel's Equation): **Bessel's equation** (of order m), named after the German astronomer Friedrich Wilhelm Bessel, is

$$x^2\frac{d^2y}{dx^2} + x\frac{dy}{dx} + \left(x^2 - \mu^2\right)y = 0, \tag{4.27}$$

where $\mu \geq 0$ is a constant. Solve Bessel's equation.

SOLUTION: To use a series method to solve Bessel's equation, we first write the equation in standard form as

$$\frac{d^2y}{dx^2} + \frac{1}{x}\frac{dy}{dx} + \frac{x^2 - \mu^2}{x^2}y = 0,$$

so $x = 0$ is a regular singular point. Using the Method of Frobenius, we assume that there is a solution of the form $y = \sum_{n=0}^{\infty} a_n x^{n+r}$. We determine the value(s) of r with the indicial equation. Because $xp(x) = x \cdot 1/x = 1$ and $x^2 q(x) = x^2 \cdot \left(x^2 - \mu^2\right)/x^2 = x^2 - \mu^2$, $p_0 = 1$ and $q_0 = -\mu^2$. Hence, the indicial equation is

$$r(r-1) + p_0 r + q_0 = r(r-1) + r - \mu^2 = r^2 - \mu^2 = 0$$

with roots $r_{1,2} = \pm\mu$. Therefore, we assume that $y = \sum_{n=0}^{\infty} a_n x^{n+\mu}$ with derivatives $y' = \sum_{n=0}^{\infty}(n + \mu)a_n x^{n+\mu-1}$ and $y'' = \sum_{n=0}^{\infty}(n + \mu)(n + \mu - 1)a_n x^{n+\mu-2}$. Substitution into Bessel's equation and simplifying the result yields

$$\left[\mu(\mu-1) + \mu - \mu^2\right]a_0 x^{\mu} + \left[(1+\mu)\mu + (1+\mu) - \mu^2\right]a_1 x^{\mu+1}$$

$$+ \sum_{n=2}^{\infty}\left\{\left[(n+\mu)(n+\mu-1) + (n+\mu) - \mu^2\right]a_n + a_{n-2}\right\}x^{n+\mu} = 0.$$

Notice that the coefficient of $a_0 x^{\mu}$ is zero. After simplifying the other coefficients and equating them to zero, we have $(1 + 2\mu)a_1 = 0$ and $\left[(n+\mu)(n+\mu-1) + (n+\mu) - \mu^2\right]a_n + a_{n-2} = 0$, which we solve for a_n.

```
Clear[a]

Solve[((n + μ)(n + μ − 1)+
    (n + μ) − μ^2)a[n] + a[n − 2] == 0, a[n]]
```

$$\{\{a[n] \rightarrow -\frac{a[-2+n]}{n(n+2\mu)}\}\}$$

From the first equation, $a_1 = 0$. Therefore, from $a_n = -\dfrac{a_{n-2}}{n(n+2\mu)}$, $n \geq 2$, so that $a_n = 0$ for all odd n. We use the formula for a_n to calculate several of the coefficients that correspond to even indices.

```
a[n_]:=a[n] = - a[-2 + n] ;
              n(n + 2μ)

a[0] = a0;

Table[{n, a[n]}, {n, 2, 10, 2}]//TableForm
```

2	$-\dfrac{a0}{2(2+2\mu)}$
4	$\dfrac{a0}{8(2+2\mu)(4+2\mu)}$
6	$-\dfrac{a0}{48(2+2\mu)(4+2\mu)(6+2\mu)}$
8	$\dfrac{a0}{384(2+2\mu)(4+2\mu)(6+2\mu)(8+2\mu)}$
10	$-\dfrac{a0}{3840(2+2\mu)(4+2\mu)(6+2\mu)(8+2\mu)(10+2\mu)}$

A general formula for these coefficients is given by

$$a_{2n} = \frac{(-1)^n a_0}{2^{2n}(1+\mu)(2+\mu)\cdots(n+\mu)}, \quad n \geq 2.$$

Our solution can then be written as

$$y_1 = \sum_{n=0}^{\infty} a_{2n}x^{2n+\mu} = \sum_{n=0}^{\infty} \frac{(-1)^n 2^\mu}{(1+\mu)(2+\mu)\cdots(n+\mu)}\left(\frac{x}{2}\right)^{2n+\mu}.$$

If μ is an integer, then by using the gamma function, $\Gamma(x)$, we can write this solution as

$$y_1 = \sum_{n=0}^{\infty} \frac{(-1)^n}{n!\,\Gamma(1+\mu+n)}\left(\frac{x}{2}\right)^{2n+\mu}. \tag{4.28}$$

This function, denoted $J_\mu(x)$, is called the **Bessel function of the first kind** of order μ. The command BesselJ[μ,x] returns $J_\mu(x)$. We use BesselJ to graph $J_\mu(x)$ for $\mu = 0, 1, 2, 3$, and 4 in Figure 4-33(a). Notice that these functions have numerous zeros. We will need to know these values in subsequent sections.

```
toplot1 = Table[BesselJ[μ, x], {μ, 0, 4}];
```

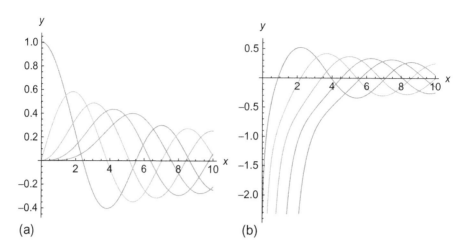

Figure 4-33   (a) The first five Bessel functions of the first kind. (b) The Bessel functions of the second kind tend to $-\infty$ as $x \to 0^+$

```
p1 = Plot[toplot1, {x, 0, 10},

    AxesLabel → {x, y}, AspectRatio → 1,

    PlotLabel → "(a)"]
```

For the other root, $r_2 = -\mu$, of the indicial equation, a similar derivation yields a second linearly independent solution of Bessel's equation,

$$y_1 = \sum_{n=0}^{\infty} \frac{(-1)^n}{n!\,\Gamma(1 - \mu + n)} \left(\frac{x}{2}\right)^{2n - \mu}$$

which is the **Bessel function of the first kind of order** $-\mu$ and is denoted $J_{-\mu}(x)$. Now, we must determine if the functions $J_\mu(x)$ and $J_{-\mu}(x)$ are linearly independent. Notice that if $\mu = 0$, then these two functions are the same. If $\mu > 0$, then $r_1 - r_2 = \mu - (-mu) = 2\mu$. If 2μ is not an integer, then by the Method of Frobenius, the two solutions $J_\mu(x)$ and $J_{-\mu}(x)$ are linearly independent. Also, we can show that if 2μ is an odd integer, $J_\mu(x)$ and $J_{-\mu}(x)$ are linearly independent. In both of these cases, a general solution is given by $y = c_1 J_\mu(x) + c_2 J_{-\mu}(x)$.

If μ is not an integer, we define the **Bessel function of the second kind of order** μ, $Y_\mu(x)$, by the linear combination

$$Y_\mu(x) = \frac{1}{\sin \mu \pi} \left[\cos \mu \pi\, J_\mu(x) - J_{-\mu}(x) \right] \tag{4.29}$$

of the functions $J_\mu(x)$ and $J_{-\mu}(x)$. The command `BesselY[`μ`, x]` returns $Y_\mu(x)$. We can show that $J_\mu(x)$ and $Y_\mu(x)$ are linearly independent, so a general solution of Bessel's equation of order μ can be represented by

$$y = c_1 J_\mu(x) + c_2 Y_\mu(x),$$

which is the form of the general solution returned by using Mathematica's `DSolve` command to solve Bessel's equation.

```
gensol = DSolve[x^2y"[x] + xy'[x]+

(x^2 − μ^2)y[x] == 0, y[x], x]
```

$\{\{y[x] \rightarrow$ BesselJ$[\mu,\ x]C[1]+$ BesselY$[\mu,\ x]C[2]\}\}$

We use `BesselY` to graph the functions $\mu = 0, 1, 2, 3,$ and 4 in Figure 4-33(b). Notice that $\lim_{x \to 0^+} Y_\mu(x) = -\infty$. This property will be important in several applications in later chapters.

```
toplot2 = Table[BesselY[μ, x], {μ, 0, 4}];

p2 = Plot[toplot2, {x, 0, 10},

   AxesLabel → {x, y}, AspectRatio → 1,

   PlotLabel → "(b)"]

Show[GraphicsRow[{p1, p2}]]
```

■

A more general form of Bessel's equation is expressed in the form

$$x^2\frac{d^2y}{dx^2} + x\frac{dy}{dx} + \left(\lambda^2 x^2 - \mu^2\right)y = 0. \tag{4.30}$$

Through a change of variables, we can show that a general solution of this equation defined on the interval $0 < x < \infty$ is

$$y = c_1 J_\mu(\lambda x) + c_2 Y_\mu(\lambda x).$$

```
gensol = DSolve[

   x^2y"[x] + xy'[x] + (λ^2x^2 − μ^2)y[x] == 0,

   y[x], x]
```

$\{\{y[x] \rightarrow$ BesselJ$[\mu,\ x\lambda]C[1]+$ BesselY$[\mu,\ x\lambda]C[2]\}\}$

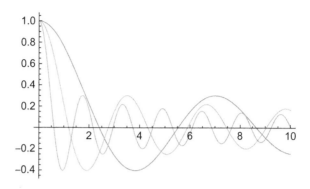

Figure 4-34   Plots of $J_0(x)$, $J_0(2x)$, and $J_0(4x)$

We graph the functions $J_0(x)$, $J_0(2x)$, and $J_0(4x)$ in Figure 4-34. Notice that for larger values of the parameter λ, the graph of the function intersects the x-axis more often.

```
Plot[{BesselJ[0, x], BesselJ[0, 2x],

  BesselJ[0, 4x]}, {x, 0, 10}]
```


EXAMPLE 4.7.6: Find a general solution of each of the following equations: (a) $x^2 y'' + xy' + (x^2 - 16)y = 0$ and (b) $x^2 y'' + xy' + (9x^2 - 4)y = 0$.

SOLUTION: (a) In this case, $\mu = 4$ so a general solution is $y = c_1 J_4(x) + c_2 Y_4(x)$. We graph this solution for various choices of the arbitrary constants in Figure 4-35(a).

```
Clear[x, y]

sol1 = DSolve[x^2y"[x] + xy'[x]

  +(x^2 - 16)y[x] == 0, y[x], x]

{{y[x] → BesselJ[4, x]C[1] + BesselY[4, x]C[2]}}

toplot1 = Table[sol1[[1, 1, 2]]/.

  {C[1] → i, C[2] → j}, {i, -1, 1}, {j, -1, 1}];
```

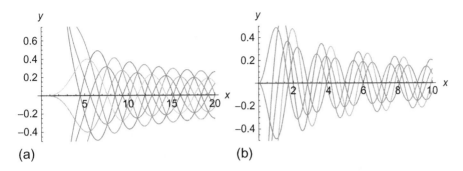

Figure 4-35   (a) Solutions of $x^2y'' + xy' + \left(x^2 - 16\right)y = 0$. (b) Solutions of $x^2y'' + xy' + \left(9x^2 - 4\right)y = 0$

```
p1 = Plot[toplot1, {x, 0, 20},
    PlotRange → {-1/2, 3/4}, AxesLabel → {x, y},
    PlotLabel → "(a)"]
```

(b) Using the parametric Bessel's equation (4.30) with $\lambda = 3$ and $\mu = 2$, we have $y = c_1 J_2(3x) + c_2 J_2(3x)$. We graph this solution for several choices of the arbitrary constants in Figure 4-35(b).

```
Clear[x, y]
sol2 = DSolve[x^2y"[x] + xy'[x]
    +(9x^2 - 4)y[x] == 0, y[x], x]

{{y[x] → BesselJ[2, 3x]C[1] + BesselY[2, 3x]C[2]}}

toplot2 = Table[sol2[[1, 1, 2]]/.
    {C[1] → i, C[2] → j}, {i, -1, 1}, {j, 0, 1}];

p2 = Plot[toplot2, {x, 0, 10},
    PlotRange → {-1/2, 1/2}, AxesLabel → {x, y},
    PlotLabel → "(b)"]

Show[GraphicsRow[{p1, p2}]]
```

■

Application: Zeros of the Bessel Functions of the First Kind

As indicated earlier, zeros of the Bessel functions of the first kind will be used in applications in later chapters. Here, we graph the first nine Bessel function of the first kind on the interval $[0, 40]$ and show all nine graphs together as a GraphicsArray in Figure 4-36.

```
besselarray = Table[Plot[BesselJ[n, x], {x, 0, 40},

    PlotStyle->CMYKColor[0, 0.89, 0.94, 0.28],

    AxesLabel → {x, y}, AspectRatio → 1,

    PlotRange → All], {n, 0, 8}];


toshow = Partition[besselarray, 3];

Show[GraphicsGrid[toshow]]
```

To approximate the zeros, we take advantage of the BesselJZero function. We obtain information about BesselJZero using Mathematica's help facility, as indicated in the following screen shot.

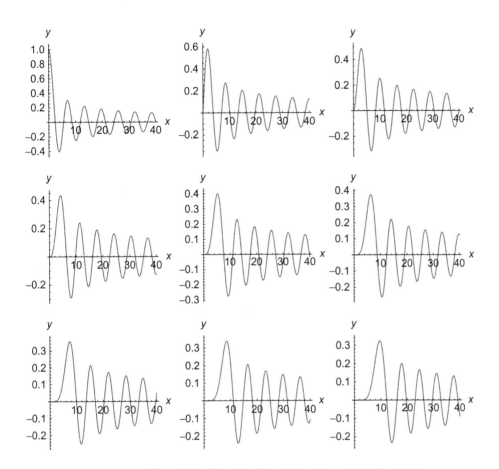

Figure 4-36   Plots of the first nine Bessel functions of the first kind

Thus, entering

```
Table[BesselJZero[μ, x]//N, {μ, 0, 4}, {x, 1, 5}]//TableForm
```

2.40483	5.52008	8.65373	11.7915	14.9309
3.83171	7.01559	10.1735	13.3237	16.4706
5.13562	8.41724	11.6198	14.796	17.9598
6.38016	9.76102	13.0152	16.2235	19.4094
7.58834	11.0647	14.3725	17.616	20.8269

returns a table of the first five zeros of the Bessel functions $J_\mu(x)$ for $\mu = 0, 1, 2, 3,$ and 4. (The first row corresponds to the zeros of $J_0(x)$, the second row to the zeros of $J_1(x)$, and so on.)

For a classic approach to the
subject see Graff's *Wave
Motion in Elastic Solids*, [13].

Application: The Wave Equation on a Circular Plate

The vibrations of a circular plate satisfy the equation

$$D \nabla^4 w(r,\theta,t) + \rho h \frac{\partial^2 w(r,\theta,t)}{\partial t^2} = q(r,\theta,t), \tag{4.31}$$

where $\nabla^4 w = \nabla^2 \nabla^2 w$ and ∇^2 is the **Laplacian in polar coordinates**, which is defined by

$$\nabla^2 = \frac{1}{r}\frac{\partial}{\partial r}\left(r\frac{\partial}{\partial r}\right) + \frac{1}{r^2}\frac{\partial^2}{\partial \theta^2} = \frac{\partial^2}{\partial r^2} + \frac{1}{r}\frac{\partial}{\partial r} + \frac{1}{r^2}\frac{\partial^2}{\partial \theta^2}.$$

Assuming no forcing so that $q(r,\theta,t) = 0$ and $w(r,\theta,t) = W(r,\theta)e^{-i\omega t}$, equation (4.31) can be written as

$$\nabla^4 W(r,\theta) - \beta^4 W(r,\theta) = 0, \quad \beta^4 = \omega^2 \rho h/D. \tag{4.32}$$

For a clamped plate, the boundary conditions are $W(a,\theta) = \partial W(a,\theta)/\partial r = 0$ and after *much work* (see [13]) the **normal modes** are found to be

$$W_{nm}(r,\theta) = \left[J_n(\beta_{nm}r) - \frac{J_n(\beta_{nm}a)}{I_n(\beta_{nm}a)} I_n(\beta_{nm}r) \right] \begin{pmatrix} \sin n\theta \\ \cos n\theta \end{pmatrix}. \tag{4.33}$$

In equation (4.33), $\beta_{nm} = \lambda_{nm}/a$ where λ_{nm} is the mth solution of

$$I_n(x)J_n{}'(x) - J_n(x)I_n{}'(x) = 0, \tag{4.34}$$

where $J_n(x)$ is the Bessel function of the first kind of order n and $I_n(x)$ is the **modified Bessel function of the first kind** of order n, related to $J_n(x)$ by $i^n I_n(x) = J_n(ix)$.

The Mathematica command `BesselI [n,x]` returns $I_n(x)$.

EXAMPLE 4.7.7: Graph the first few normal modes of the clamped circular plate.

SOLUTION: We must determine the value of λ_{nm} for several values of n and m. First, we determine the relationship between J_n and I_n and their derivatives.

```
D[BesselJ[n, x], x]
```

$$\frac{1}{2}(\text{BesselJ}[-1+n, x] - \text{BesselJ}[1+n, x])$$

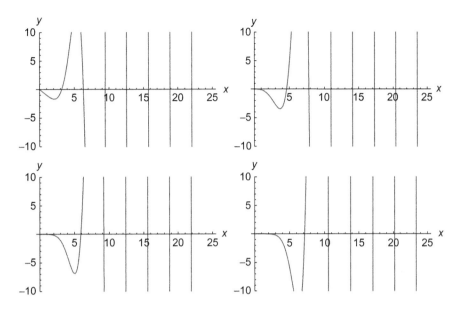

Figure 4-37   Plot of $I_n(x)J_n'(x) - J_n(x)I_n'(x)$ for $n = 0$ and 1 in the first row; $n = 2$ and 3 in the second row

```
D[BesselI[n, x], x]
```

$$\frac{1}{2}(\text{BesselI}[-1+n, x] + \text{BesselI}[1+n, x])$$

Next, using the relationships between the derivatives obtained above we define eqn [n] [x] to be $I_n(x)J_n'(x) - J_n(x)I_n'(x)$. The mth solution of equation (4.34) corresponds to the mth zero of the graph of eqn [n] [x] so we graph eqn [n] [x] for $n = 0, 1, 2,$ and 3 with Plot in Figure 4-37.

```
eqn[n_][x_]:=BesselI[n, x](1/2(BesselJ[-1+n, x]
    -BesselJ[1+n, x])) - BesselJ[n, x](1/2(BesselI[-1+n, x]
    +BesselI[1+n, x]));
```

The result of the Table and Plot command is a list of length four

```
p1 = Table[Plot[eqn[n][x], {x, 0, 25}, PlotRange → {-10, 10},
    AxesLabel → {x, y}, PlotStyle->
    CMYKColor[0, 0.89, 0.94, 0.28]], {n, 0, 3}];
```

so we use Partition to create a 2×2 array of graphics which is displayed using Show and GraphicsGrid.

```
p2 = Partition[p1, 2];
```

```
Show[GraphicsGrid[p2]]
```

To determine λ_{nm} we use `FindRoot`. Recall that to use `FindRoot` to solve an equation an initial approximation of the solution must be given. For example,

```
lambda01 = FindRoot[eqn[0][x] == 0, {x, 3.04}]
```

$\{x \to 3.19622\}$

approximates λ_{01}, the first solution of equation (4.34) if $n = 0$. However, the result of `FindRoot` is a list. The specific value of the solution is the second part of the first part of the list, `lambda01`, extracted from the list with `Part ([[...]])`.

```
lambda01[[1, 2]]
```

```
3.19622
```

We use the graphs in Figure 4-37 to obtain initial approximations of each solution.

Thus,

```
λ0s = Map[FindRoot[eqn[0][x] == 0, {x, #}][[1, 2]]&,
    {3.04, 6.2, 9.36, 12.5, 15.7}]
```

```
{3.19622, 6.30644, 9.4395, 12.5771, 15.7164}
```

approximates the first five solutions of equation (4.34) if $n = 0$ and then returns the specific value of each solution. We use the same steps to approximate the first five solutions of equation (4.34) if $n = 1, 2$, and 3.

```
λ1s = Map[FindRoot[eqn[1][x] == 0, {x, #}][[1, 2]]&,
    {4.59, 7.75, 10.9, 14.1, 17.2}]
```

```
{4.6109, 7.79927, 10.9581, 14.1086, 17.2557}
```

```
λ2s = Map[FindRoot[eqn[2][x] == 0, {x, #}][[1, 2]]&,
    {5.78, 9.19, 12.4, 15.5, 18.7}]
```

```
{5.90568, 9.19688, 12.4022, 15.5795, 18.744}
```

```
λ3s = Map[FindRoot[eqn[3][x] == 0, {x, #}][[1, 2]]&,
    {7.14, 10.5, 13.8, 17., 20.2}]
```

{7.14353, 10.5367, 13.7951, 17.0053, 20.1923}

All four lists are combined together in λs.

```
λs = {λ0s, λ1s, λ2s, λ3s}
```

{{3.19622, 6.30644, 9.4395, 12.5771, 15.7164}, {4.6109, 7.79927,

10.9581, 14.1086, 17.2557}, {5.90568, 9.19688, 12.4022, 15.5795,

18.744}, {7.14353, 10.5367, 13.7951, 17.0053, 20.1923}}

For $n = 0, 1, 2$, and 3 and $m = 1, 2, 3, 4$, and 5, λ_{nm} is the mth part of the $(n + 1)$st part of λs.

Observe that the value of a does not affect the shape of the graphs of the normal modes so we use $a = 1$ and then define β_{nm}.

```
a = 1;
β[n_, m_]:=λs[[n + 1, m]]/a;
```

ws is defined to be the sine part of equation (4.33)

```
ws[n_, m_][r_, θ_]:=(BesselJ[n, β[n, m]r]−

    BesselJ[n, β[n, m]a]/BesselI[n, β[n, m]a]

    BesselJ[n, β[n, m]r]) * Sin[nθ];
```

and wc to be the cosine part.

```
wc[n_, m_][r_, θ_]:=(BesselJ[n, β[n, m]r]−

    BesselJ[n, β[n, m]a]/BesselI[n, β[n, m]a]

    BesselJ[n, β[n, m]r]) * Cos[nθ];
```

We will use ParametricPlot3D to plot ws and wc. For example,

```
ParametricPlot3D[{rCos[θ], rSin[θ], ws[3, 4][r, θ]},

    {r, 0, 1}, {θ, −π, π}, PlotPoints → 60]
```

graphs the sine part of $W_{34}(r, \theta)$ shown in Figure 4-38. We use Table together with ParametricPlot3D followed by Show and GraphicsGrid to graph the sine part of $W_{nm}(r, \theta)$ for $n = 0, 1, 2$, and 3 and $m = 1, 2, 3$, and 4 shown in Figure 4-39.

```
ms = Table[ParametricPlot3D[{rCos[θ], rSin[θ], ws[n, m][r, θ]},

    {r, 0, 1}, {θ, −π, π}, PlotPoints → 30, BoxRatios → {1, 1, 1}],

    {n, 0, 3}, {m, 1, 4}];
```

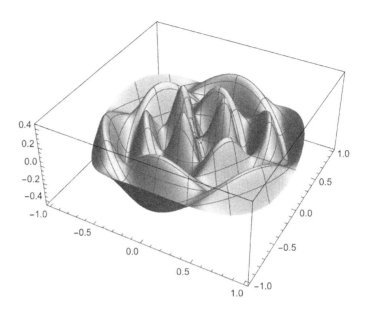

Figure 4-38   The sine part of $W_{34}(r, \theta)$

```
Show[GraphicsGrid[ms]]
```

Identical steps are followed to graph the cosine part shown in Figure 4-40.

```
mc = Table[ParametricPlot3D[{rCos[θ], rSin[θ], wc[n, m][r, θ]},
    {r, 0, 1}, {θ, -π, π}, PlotPoints → 30, BoxRatios → {1, 1, 1}],
    {n, 0, 3}, {m, 1, 4}];

Show[GraphicsGrid[mc]]
```

∎

4.8 Nonlinear Equations

Generally, rigorous results regarding nonlinear equations are very difficult to obtain. In many cases, analysis is best carried out numerically and/or graphically. In other situations, rewriting the equation as a system can be of benefit, which is discussed in Chapter 6.

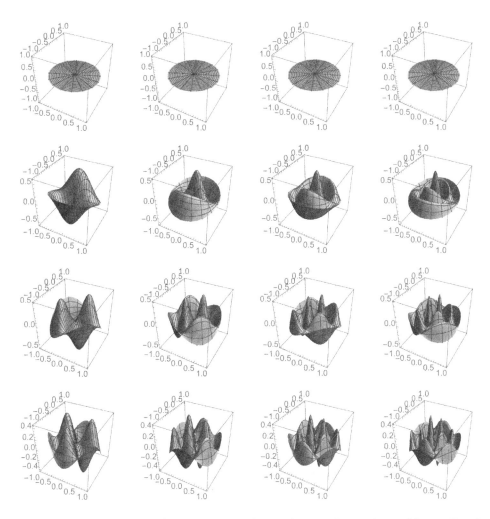

Figure 4-39 The sine part of $W_{nm}(r, \theta)$: $n = 0$ in row 1, $n = 1$, in row 2, $n = 2$ in row 3 and $n = 3$ in row 4 ($m = 1$ to 4 from left to right in each row)

However, if a nonlinear equation can be solved with currently known techniques, Mathematica can often find a solution for you.

EXAMPLE 4.8.1: Solve $4y \left(y'\right)^2 y'' = \left(y'\right)^4 + 3$.

SOLUTION: Mathematica can solve this nonlinear equation with `DSolve`.

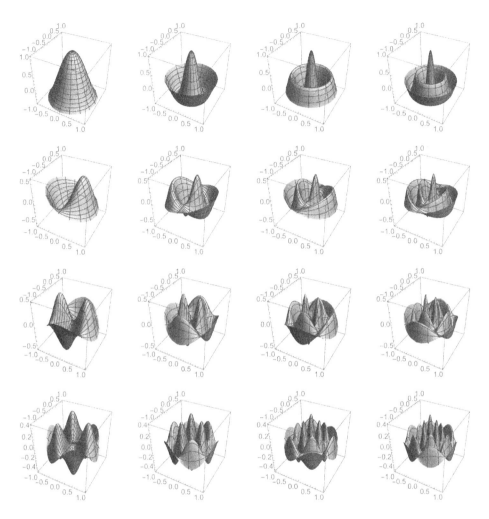

Figure 4-40   The cosine part of $W_{nm}(r,\theta)$: $n = 0$ in row 1, $n = 1$ in row 2, $n = 2$ in row 3, and $n = 3$ in row 4 ($m = 1$ to 4 from left to right in each row)

```
DSolve[4y[x]y'[x]^2y''[x] == y'[x]^4 + 3,

    y[x], x]
```

$\{\{y[x] \rightarrow -\dfrac{3}{8}e^{-4C[1]}(-8 + 6^{1/3}e^{4C[1]}x(-e^{4C[1]}(x + C[2]))^{1/3} +$
$6^{1/3}e^{4C[1]}C[2](-e^{4C[1]}$
$(x+C[2]))^{1/3})\}, \{y[x] \rightarrow \dfrac{1}{8}(24e^{-4C[1]}-3i6^{1/3}(x+C[2])(-ie^{4C[1]}(x+$
$C[2]))^{1/3})\}, \{y[x] \rightarrow \dfrac{1}{8}(24e^{-4C[1]} + 3i6^{1/3}(x + C[2])(ie^{4C[1]}(x +$

$C[2]))^{1/3})\}$, $\{y[x] \to \frac{3}{8}e^{-4C[1]}(8 + 6^{1/3}e^{4C[1]}x(e^{4C[1]}(x + C[2]))^{1/3} + 6^{1/3}e^{4C[1]}C[2](e^{4C[1]}(x + C[2]))^{1/3})\}\}$

Proceeding by hand, let $p = y'$. Then,

$$y'' = p' = \frac{dp}{dx} = \frac{dy}{dx}\frac{dp}{dy} = p\frac{dp}{dy}.$$

With this substitution, we obtain a first-order separable equation.

$$4p^3 y\frac{dp}{dy} = 3 + p^4$$

$$\frac{4p^3}{3 + p^4}dp = \frac{1}{y}dy$$

$$\ln\left(3 + p^4\right) = \ln|y| + c_1$$

$$3 + p^4 = c_1 y$$

$$p = \pm(c_1 y - 3)^{1/4}.$$

Because $p = dy/dx$,

$$\pm\frac{1}{(c_1 y - 3)^{1/4}}dy = dx$$

and integrating and simplifying the result gives us

$$\frac{4}{3c_1}(c_1 y - 3)^{3/4} = x + c_2$$

$$\frac{256}{81c_4}(c_1 y - 3)^3 = (x + c_2)^4.$$

```
Integrate[4p^3/(3 + p^4), p]
```

$\text{Log}[3 + p^4]$

```
Integrate[1/y, y]
```

$\text{Log}[y]$

```
Integrate[(c1y − 3)^(−1/4), y]
```

$$\frac{4(-3 + c1y)^{3/4}}{3c1}$$

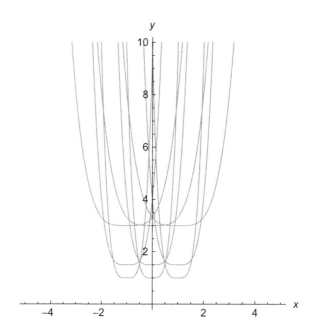

Figure 4-41   Various solutions of $4y\left(y'\right)^2 y'' = \left(y'\right)^4 + 3$ shown together

We plot various solutions by graphing level curves of

$$f(x, y) = \frac{256}{81c_4}\left(c_1 y - 3\right)^3 - \left(x + c_2\right)^4.$$

corresponding to 0 for various values of c_1 and c_2 in Figure 4-41.

```
g1 = Table[−(c2 + x)^4 + 256(c1y − 3)/(81c1^4),

  {c1, 1, 3}, {c2, −1, 1}]//Flatten;

g2 = Map[ContourPlot[#, {x, −5, 5}, {y, 0, 10},

  ContourShading → False, Contours → {0},

  PlotPoints → 240]&, g1];

Show[g2, Frame → False, Axes → Automatic,

  AxesOrigin → {0, 0}, AxesLabel → {x, y}]
```

■

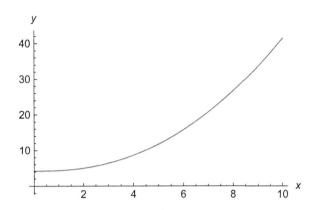

Figure 4-42   Plot of the solution to the initial-value problem

EXAMPLE 4.8.2: Solve $\begin{cases} x^2 y'' + (y')^2 - 2xy' = 0 \\ y(2) = 5,\ y'(2) = 1 \end{cases}$.

SOLUTION: Mathematica can find the solution to the initial-value problem, which we then graph with `Plot` in Figure 4-42.

```
sol = DSolve[{x^2y''[x] + y'[x]^2 - 2xy'[x] == 0,

   y[2] == 5, y'[2] == 1}, y[x], x]
```

$$\left\{ \left\{ y[x] \to \frac{1}{2}\left(2 - 4x + x^2 - 8\text{Log}\left[\frac{4}{e^{3/2}}\right] + 8\text{Log}[2 + x] \right) \right\} \right\}$$

```
Plot[y[x]/.sol, {x, 0, 10},

   PlotStyle->CMYKColor[0, 0.89, 0.94, 0.28],

   AxesLabel → {x, y}]
```

By hand, we proceed as before by letting $p = y'$. Then $p' = y''$ and equation $x^2 y'' + (y')^2 - 2xy' = 0$ becomes

$$x^2 \frac{dp}{dx} + p^2 - 2xp = 0$$
$$x^2\, dp + \left(p^2 - 2xp \right) dx = 0,$$

which is first-order homogeneous of degree 2. Solving for p

DSolve[x^2p'[x] + (p[x]^2 − 2xp[x]) == 0, p[x], x]

$$\left\{\left\{p[x] \to \frac{x^2}{x + C[1]}\right\}\right\}$$

and then integrating the result gives us

$$p = \frac{dy}{dx} = \frac{x^2}{x + c_1}$$

$$dy = \frac{x^2}{x + c_1}dx$$

$$y = \frac{1}{2}x^2 - c_1x + c_1{}^2 \ln|x + c_1| + c_2.$$

y = Integrate$\left[\dfrac{x^2}{x + c1}, x\right]$ + c2

$$c2 - c1x + \frac{x^2}{2} + c1^2 Log[c1 + x]$$

Applying the initial conditions gives us the nonlinear system

$$2 - 2c_1 + c_2 + c_1^2 \ln|2 + c_1| = 5$$

$$2 - c_1 + \frac{c_1{}^2}{2 + c_1} = 1.$$

f1 = y/.x → 2

$$2 - 2c1 + c2 + c1^2 Log[2 + c1]$$

f2 = D[y, x]/.x → 2

$$2 - c1 + \frac{c1^2}{2 + c1}$$

We can see the solution to this system by graphing each equation with ContourPlot as shown in Figure 4-43.

cp1 = ContourPlot[f1, {c1, −3, 3}, {c2, −5, 5},

    Contours → {5}, ContourShading->False,

    PlotPoints → 120, ContourStyle → Black];

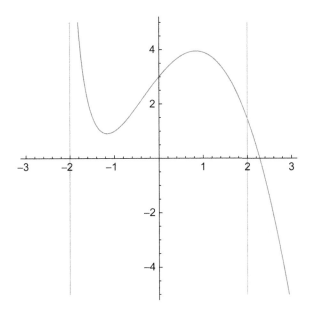

Figure 4-43   The nonlinear system of equations has a unique solution

```
cp2 = ContourPlot[f2, {c1, -3, 3}, {c2, -5, 5},

   Contours → {1}, ContourShading->False,

   PlotPoints → 120, ContourStyle → Gray];

Show[cp1, cp2, Frame → False, Axes → Automatic,

   AxesOrigin → {0, 0}]
```

By hand, solving the second equation for c_1 gives us $c_1 = 2$. Substituting into the first equation and solving for c_2 gives us $c_2 = 7 - 4\ln 4$. We confirm the result with Solve.

```
cvals = Solve[{f1 == 5, f2 == 1}]
```

$\{\{c1 \rightarrow 2,\ c2 \rightarrow 7 - 4\text{Log}[4]\}\}$

```
y/.cvals[[1]]
```

$7 - 2x + \dfrac{x^2}{2} - 4\text{Log}[4] + 4\text{Log}[2 + x]$

■

Of course, in many cases numerical results are most meaningful.

Sources: See texts like
Jordan and Smith's *Nonlinear
Ordinary Differential Equations*,
[17].

EXAMPLE 4.8.3 (Duffing's Equation): Duffing's equation is the second-order nonlinear equation

$$\frac{d^2x}{dt^2} + k\frac{dx}{dt} - x + x^3 = \Gamma \cos \omega t, \tag{4.35}$$

where k, Γ, and ω are positive constants. Depending upon the values of the parameters, solutions to Duffing's equation can exhibit *very* interesting behavior.

SOLUTION: To investigate solutions we define the function $\mathtt{duffingplot}$. Given k, Γ, and ω,

$$\mathtt{duffingplot[k,\Gamma,\omega][\{x0,y0\},\{t,a,b\}]}$$

graphs the solution to the initial-value problem

$$\begin{cases} x'' + kx' - x + x^3 = \Gamma \cos \omega t \\ x(0) = x_0, \ x'(0) = y_0 \end{cases} \tag{4.36}$$

for $a \leq t \leq b$. If $\{\mathtt{t,a,b}\}$ is omitted, the default is $900 \leq t \leq 1000$. Any options included are passed to the $\mathtt{Plot}$ command.

```
Clear[duffingplot]

duffingplot[k_, Γ_, ω_][{x0_, y0_},

  ts_:{t, 900, 1000}, opts___]:=

Module[{numsol},

numsol = NDSolve[{x"[t] + kx'[t] − x[t]+

  x[t]^3 == ΓCos[ωt], x[0] == x0,

  x'[0] == y0}, x[t], ts, MaxSteps → 100000];

Plot[x[t]/.numsol, ts,

  PlotStyle->CMYKColor[0, 0.89, 0.94, 0.28], opts]

]
```

For example, entering

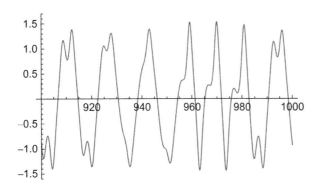

Figure 4-44   Solution to Duffing's equation if $k = 0.3$, $\Gamma = 0.5$, $\omega = 1.2$, and $x_0 = y_0 = 0$

```
duffingplot[0.3, 0.5, 1.2][{0, 0}]
```

plots the solution to the initial-value problem (4.36) shown in Figure 4-44 if $k = 0.3$, $\Gamma = 0.5$, $\omega = 1.2$, and $x_0 = y_0 = 0$.

You can use duffingplot to see how varying the parameters affects the solutions. For example, suppose that $k = 0.3$, $\omega = 1.2$, $x_0 = 0$, and $y_0 = 1$. To see how the solutions vary depending on the value of Γ, we define kvals to be a list of 12 equally spaced numbers between 0 and 0.8 and then use Map to apply duffingplot to the list kvals. In this case, we generate a short-term plot for $0 \leq t \leq 50$. The resulting graphics are not displayed because we include the option DisplayFunction->Identity in the duffingplot command.

```
kvals = Table[k, {k, 0, 0.8, .8/11}];

toshow = Map[duffingplot[0.3, #, 1.2][{0, 1},

  {t, 0, 50}, PlotRange → {-3, 3}]&, kvals];
```

We then use Partition, Show, and GraphicsGrid to display the list of graphics toshow in Figure 4-45.

```
Show[GraphicsGrid[Partition[toshow, 3]]]
```

We enter nearly identical commands to generate the long-term plot shown in Figure 4-46.

```
kvals = Table[k, {k, 0, 0.8, .8/11}];

toshow = Map[duffingplot[0.3, #, 1.2][{0, 1},

  {t, 900, 1000}, PlotRange → {-3, 3}]&, kvals];
```

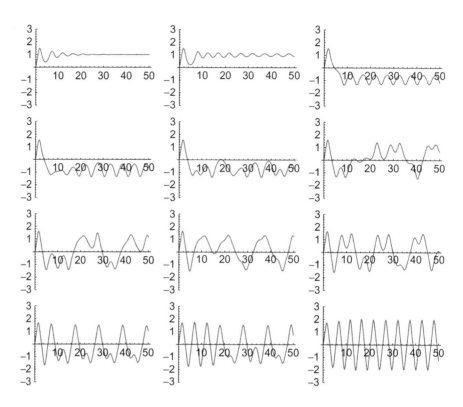

Figure 4-45   Short-term plot: depending upon the value of Γ, some solutions to Duffing's equation exhibit chaotic behavior

```
Show[GraphicsGrid[Partition[toshow, 3]]]
```

Problems such as this that investigate the behavior of solutions depending on various parameters can be explored using the `Manipulate` function.

In the following we use `Manipulate` to investigate the behavior of the solutions to the differential equation depending on different parameter values. Use the slider bars to adjust the values to see how the behavior of the solution changes depending on the parameter values (Figure 4-47).

```
Manipulate[

numsol = NDSolve[{x''[t] + kx'[t] − x[t]+

    x[t]^3 == ΓCos[ωt], x[0] == 0,

        x'[0] == 0}, x[t], {t, tmin, tmax}, MaxSteps → 100000];
```

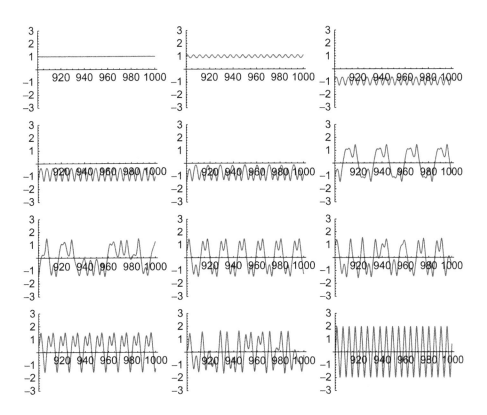

Figure 4-46 Long-term plot: depending upon the value of Γ, some solutions to Duffing's equation exhibit chaotic behavior

```
Plot[x[t]/.numsol,{t, tmin, tmax},
PlotStyle->CMYKColor[0, 0.89, 0.94, 0.28], PlotRange → {-3, 3}],
    {{k, 0.3}, 0, 1}, {{Γ, 0.5}, 0, 2}, {{ω, 1.2}, 0, 5},
    {{tmin, 900}, 0, 1000}, {{tmax, 1000}, 100, 1100}
]
```

The **Fourier transform**, X_k ($k = 1, 2, \ldots, N$) of N equally spaced values of a time series $\texttt{list} = \{x_1, x_2, \ldots, x_N\}$ is

$$X_k = \frac{1}{\sqrt{N}} \sum_{n=1}^{N} x_n e^{2\pi i (n-1)(k-1)/N}. \tag{4.37}$$

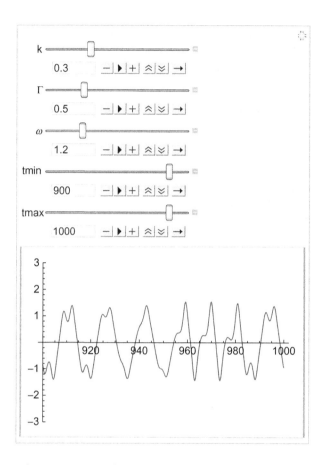

Figure 4-47 `Manipulate` can often help provide insight as to how problems change depending upon parameter values because it allows you to experiment and change those values spontaneously

The Mathematica command `Fourier[list]` computes the Fourier transform of `list`. The **power spectrum**, $P(\omega_k)$ $(k = 1, 2, \ldots, N)$, of the list $\{X_1, X_2, \ldots, X_N\}$ is

$$P(\omega_k) = X_k \bar{X}_k = |x_k|^2. \tag{4.38}$$

The power spectrum helps detect dominant frequencies. See Jordan and Smith [17].

We define the function `duffingpower` to compute the power spectrum of Duffing's equation. Given the appropriate parameter values and initial conditions, duffingpower returns $P(\omega_{2000})$. The 2000 sample points are the value of $x(t_n)$ for $t_n = 0.5n$, $n = 1, \ldots, 2000$.

```
Clear[duffingpower, s2, s3]

   duffingpower[k_, Γ_, ω_][{x0_, y0_},

 omegak_:2000]:=

   Module[{numsol, s2, s3},

      numsol = NDSolve[{x″[t] + kx′[t] − x[t]+

      x[t]^3 == Γ Cos[ωt], x[0] == x0,

        x′[0] == y0}, x[t], {t, 0, 1000}, MaxSteps → 100000];

   s2 = Table[x[t]/.numsol[[1]],

      {t, 0.5, 1000, 0.5}];

   s3 = Map[Abs[#]^2&, Fourier[s2]][[omegak]]

   ]
```

As an illustration, we set $k = 0.3$, $\Gamma = 0.5$, and $x_0 = y_0 = 0$ and then compute the power spectrum for 300 equally spaced values of ω between 0 and 3.

```
t1 = Table[{ω, duffingpower[0.3, 0.5, ω][{0, 0}]},

   {ω, 0, 3., 3./299}];
```

We use ListLogPlot, to plot the list of points t1 so that Mathematica uses a logarithmic scale on the y-axis (the vertical axis). See Figure 4-48.

```
ListLogPlot[t1, Joined → True, PlotRange → All,

   PlotStyle->CMYKColor[0, 0.89, 0.94, 0.28]]
```

For a second-order equation like this, it is often desirable to generate a parametric plot of $x(t)$ versus $x'(t)$. To do so, we set $y = x'$. Then, $y' = x''$ and we see that Duffing's equation (4.35) can be rewritten as the nonlinear system

$$x' = y$$
$$y' + ky - x + x^3 = \Gamma \cos \omega t$$

We define the function duffingparamplot to graph solutions of the initial-value problem

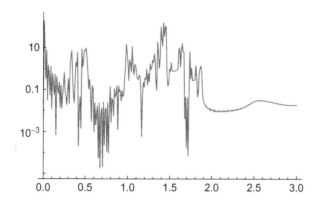

Figure 4-48   Power spectrum of Duffing's equation using $k = 0.3$, $\Gamma = 0.5$, and $x_0 = y_0 = 0$: the horizontal axis corresponds to ω; the vertical axis to the power spectrum $P(\omega_{2000})$

$$\begin{cases} x' = y \\ y' + ky - x + x^3 = \Gamma \cos \omega t \\ x(0) = x_0,\; y(0) = y_0 \end{cases}$$

in the same way as we defined duffingplot.

```
Clear[duffingparamplot, x, y]

duffingparamplot[k_ , Γ_ , ω_][{x0_ , y0_}, ts_:{t, 800, 1000},
  opts___]:=
    Module[{numsol},
      numsol = NDSolve[{y'[t] + ky[t] - x[t]+
      x[t]^3 == ΓCos[ωt], x'[t] == y[t], x[0] == x0,
      y[0] == y0}, {x[t], y[t]}, ts, MaxSteps → 100000];
      ParametricPlot[{x[t], y[t]}/.numsol, ts,
      PlotStyle->CMYKColor[0, 0.89, 0.94, 0.28], AxesLabel
    → {x, y}, opts]]
```

For example, entering

```
duffingparamplot[0.3, 0.5, 0.2][{0, 1}, {t, 800, 1000}]
```

plots $x(t)$ versus $x'(t)$ if $k = 0.3$, $\Gamma = 0.5$, $\omega = 0.2$, $x(0) = 0$, and $y(0) = x'(0) = 1$ as shown in Figure 4-49.

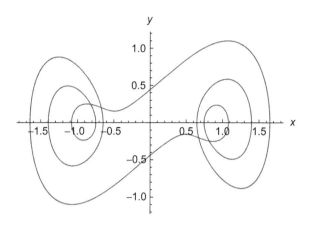

Figure 4-49   A parametric plot of $x(t)$ versus $x'(t)$ for a solution to Duffing's equation

With the following commands, we set $k = 0.3$, $\Gamma = 0.5$, $x(0) = 0$, and $y(0) = x'(0) = 1$. We then plot $x(t)$ versus $x'(t)$ for 12 equally spaced values of ω between 0 and 1.5.

```
kvals = Table[k, {k, 0.1, 1.5, 1.5/11}];
```

The results are shown as an array in Figure 4-50.

```
toshow = Map[duffingparamplot[0.3, 0.5, #][{0, 1},

  {t, 0, 50}, PlotRange → {-3, 3}]&, kvals];

Show[GraphicsGrid[Partition[toshow, 3]]]
```

The long-term plot shown in Figure 4-51 is generated with nearly identical commands.

```
toshow = Map[duffingparamplot[0.3, 0.5, #][{0, 1},

  {t, 800, 900}, PlotRange → {-3, 3}]&, kvals];

Show[GraphicsGrid[Partition[toshow, 3]]]
```

The **Poincaré plots** (or **returns**) are obtained by plotting

$$x = x\,(2n\pi/\omega))$$
$$x' = y = (2n\pi/\omega))\quad.$$

We define the function `duffingpoincareplot` to generate Poincaré plots for Duffing's equation.

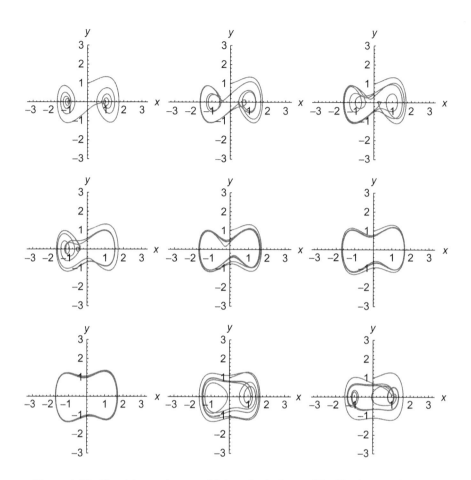

Figure 4-50   Short-term plot: sensitivity of solutions of Duffing's equation to ω

```
Clear[duffiningpoincareplot, x, y, t1]
duffiningpoincareplot[k_, Γ_, ω_][{x0_, y0_},
  ns_:{n, 1, 2000}, opts___]:=
  Module[{numsol, t1},
    numsol = NDSolve[{y'[t] + ky[t] − x[t]+
    x[t]^3 == ΓCos[ωt], x'[t] == y[t], x[0] == x0,
    y[0] == y0}, {x[t], y[t]}, {t, 0, 12000}, MaxSteps → 1000000];
  t1 = Flatten[Table[Evaluate[{x[t], y[t]}/.numsol]/. t → 2nπ/ω, ns], 1];
```

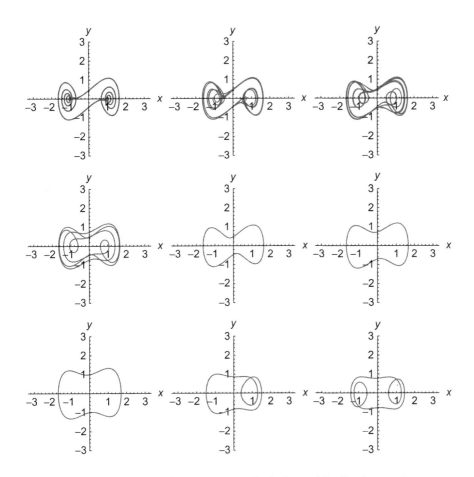

Figure 4-51   Long-term plot: sensitivity of solutions of Duffing's equation to ω

```
ListPlot[t1, PlotStyle->CMYKColor[0, 0.89, 0.94, 0.28],

PlotRange → {{−3/2, 3/2}, {−3/2, 3/2}},

AxesLabel → {x, y}, AspectRatio → 1]

]
```

In Figure 4-52, we use `duffingpoincareplot` to generate a Poincaré plot for Duffing's equation if $k = 0.3$, $\Gamma = 0.4$, $\omega = 1.2$, $x(0) = 0$, and $y(0) = x'(0) = 1$.

```
duffiningpoincareplot[0.3, 0.5, 1.2][{0, 1}]
```

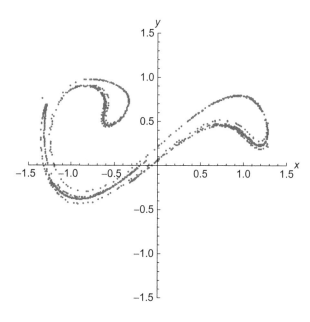

Figure 4-52   A Poincaré plot for Duffing's equation if $k = 0.3$, $\Gamma = 0.4$, $\omega = 1.2$, $x(0) = 0$,
and $y(0) = x'(0) = 1$

Equations such as Duffing's equation are well-suited for Mathematica's `Manipulate` function when you want to experiment with how different parameter values and initial conditions affect the behavior of solutions.

A basic `Manipulate` function for investigating the behavior of Duffing's Poincaré plot follows. See Figure 4-53.

```
Manipulate[
  numsol = NDSolve[{y'[t] + ky[t] - x[t]+
    x[t]^3 == ΓCos[ωt], x'[t] == y[t], x[0] == 0,
    y[0] == 0}, {x[t], y[t]}, {t, 0, 12000}, MaxSteps → 1000000];
  t1 = Flatten[Table[Evaluate[{x[t], y[t]}/.numsol]/. t → 2nπ/
  ω, {n, 1, 2000}], 1];
    ListPlot[t1, PlotStyle->CMYKColor[0, 0.89, 0.94, 0.28],
    PlotRange → {{-3/2, 3/2}, {-3/2, 3/2}},
```

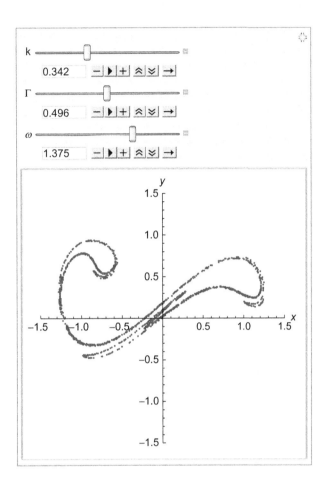

Figure 4-53   Using Manipulate to investigate the behavior of Poincaré plots to Duffing's equation

```
        AxesLabel → {x, y}, AspectRatio → 1],

     {{k, 0.3}, 0, 1}, {{Γ, 0.5}, 0, 1}, {{ω, 1.2}, 0, 2}

  ]

■
```

Applications of Higher-Order Differential Equations

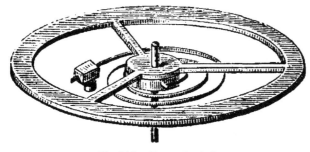

Fig. 820. – Ressort spiral.

The balance wheel at the core of many mechanical clocks and watches depends on *Hooke's Law*. Since the torque generated by the coiled spring is proportional to the angle turned by the wheel, its oscillations have a nearly constant period. Hooke's Law is named after the British physicist Robert Hooke (1635–1703).

In Chapter 4, we discussed several techniques for solving higher-order differential equations. In this chapter, we illustrate how some of these methods can be used to solve initial-value problems that model physical situations.

5.1 Harmonic Motion

5.1.1 Simple Harmonic Motion

Suppose that a mass is attached to an elastic spring that is suspended from a rigid support such as a ceiling. According to Hooke's law, the spring exerts a

Differential Equations with Mathematica. http://dx.doi.org/10.1016/B978-0-12-804776-7.00005-X

restoring force in the upward direction that is proportional to the displacement of the spring.

>   **Hooke's Law:** $F = ks$, where $k > 0$ is the constant of proportionality or spring constant, and s is the displacement of the spring.

A spring has natural length b. When a mass is attached to the spring, it is stretched s units past its natural length to the equilibrium position $x = 0$. When the system is put into motion, the displacement from $x = 0$ at time t is given by $x(t)$.

By Newton's Second Law of Motion, $F = ma = m\,d^2x/dt^2$, where m represents mass and a represents acceleration. If we assume that there are no other forces acting on the mass, then we determine the differential equation that models this situation in the following way:

$$m\frac{d^2x}{dt^2} = \sum \left(\text{forces acting on the system}\right)$$
$$= -k(s + x) + mg$$
$$= -ks - kx + mg.$$

At equilibrium $ks = mg$, so after simplification, we obtain the differential equation

$$m\frac{d^2x}{dt^2} = -kx \quad \text{or} \quad m\frac{d^2x}{dt^2} + kx = 0.$$

The two initial conditions that are used with this problem are the initial displacement $x(0) = \alpha$ and initial velocity $dx/dt(0) = \beta$. Hence, the function $x(t)$ that describes the displacement of the mass with respect to the equilibrium position is found by solving the initial-value problem

$$\begin{cases} m\dfrac{d^2x}{dt^2} + kx = 0 \\ x(0) = \alpha, \ \dfrac{dx}{dt}(0) = \beta \end{cases}. \tag{5.1}$$

The differential equation in initial-value problem (5.1) disregards all retarding forces acting on the motion of the mass.

The solution $x(t)$ to this problem represents the displacement of the mass at time t. Based on the assumptions made in deriving the differential equation (the positive direction is down), positive values of $x(t)$ indicate that the mass is beneath the equilibrium position while negative values of $x(t)$ indicate that the mass is above the equilibrium position.

EXAMPLE 5.1.1: A mass weighing 60 lb. stretches a spring 6 inches. Determine the function $x(t)$ that describes the displacement of the mass if the mass is released from rest 12 inches below the equilibrium position.

SOLUTION: First, the spring constant k must be determined from the given information. By Hooke's law, $F = ks$, so we have $60 = k \cdot 0.5$. Therefore, $k = 120$ lb/ft. Next, the mass, m, must be determined using $F = mg$. In this case, $60 = m \cdot 32$, so $m = 15/8$ slugs. Because $k/m = 64$ and 12 inches equals 1 foot, the initial-value problem that needs to be solved is

$$\begin{cases} x'' + 64x = 0 \\ x(0) = 1,\ x'(0) = 0 \end{cases}.$$

This problem is now solved with DSolve, and the resulting output is named de1.

```
Clear[x, t, de1]
de1 = DSolve[{x''[t] + 64x[t]==0, x[0]==1, x'[0]==0}, x[t], t]
```

$$\{\{x[t] \to \text{Cos}[8t]\}\}$$

We graph the solution with Plot in Figure 5-1.

```
Plot[x[t]/.de1, {t, 0, π/2}, PlotStyle->CMYKColor[0, 0.89,
    0.94, 0.28], .AxesLabel → {t, x}]
```

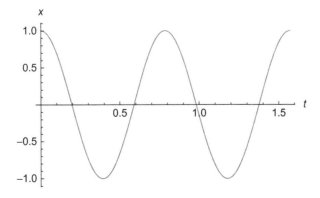

Figure 5-1   Illustrating simple harmonic motion

In order to better understand the relationship between the formula obtained in this example and the motion of the mass on the spring, an alternate approach is taken here. We begin by defining `sol` to be the solution to the initial-value problem: given t, `sol[t]` returns the value of $\cos 8t$.

```
Clear[sol]
```

```
sol[t_] = del[[1, 1, 2]]
```

```
Cos[8t]
```

Then, the function `zigzag` is defined to produce a list of points joined by line segments to represent the graphics of a spring. Given ordered pairs (a, b) and (c, d), a positive integer n, and a "small" number ϵ, `zigzag[{a,b},{c,d},n,eps]` connects the set of points

$$(a, b), \left(a - \epsilon, b + \frac{d-b}{n}\right), \left(a - \epsilon, b + 2\frac{d-b}{n}\right), \ldots,$$

$$\left(a + (-1)^i \epsilon, b + i\frac{d-b}{n}\right), \ldots \left(a + (-1)^{n-1}\epsilon, b + (n-1)\frac{d-b}{n}\right), (c, d)$$

with line segments.

Note that we will always have $a = c$.

```
Clear[spring, zigzag, length, points, pairs]
zigzag[{a_, b_}, {c_, d_}, n_, ε_]:=
    Module[{length, points, pairs}, length = d - b;
        points = Table[b + i length / n, {i, 1, n - 1}];
        pairs = Table[{a + (-1)^i ε, points[[i]]}, {i, 1, n - 1}];
        PrependTo[pairs, {a, b}]; AppendTo[pairs, {c, d}];
        Line[pairs]]
```

The function `spring` produces the graphics of a point (the mass attached to the end of the spring) as well as that of the spring obtained with `zigzag`. The result of entering `spring[t]` when displayed with `Show` looks like a spring with a mass attached.

```
spring[t_]:=
    Show[Graphics[{zigzag[{0, -sol[t]},
        {0, 1}, 20, .05], PointSize[.075],
```

```
      Point[{0, -sol[t]}]}], Axes → Automatic,
      AxesStyle → Black, Ticks → None,
                               3   3
      PlotRange → {{-1, 1}, {- -, -}}, AspectRatio → 1];
                               2   2
```

A list of graphics is produced in somegraphs for values of t from $t = 0$ to $t = \pi/2$ using increments of $\pi/16$.

```
      somegraphs = Table[spring[t], {t, 0, π/2, π/16}];
```

This list of nine graphics objects is then partitioned into groups of three with Partition in toshow for use with GraphicsGrid.

```
      toshow = Partition[somegraphs, 3];
```

We then display the array of graphics objects toshow with Show and GraphicsArray in Figure 5-2. We see that the plots displayed show

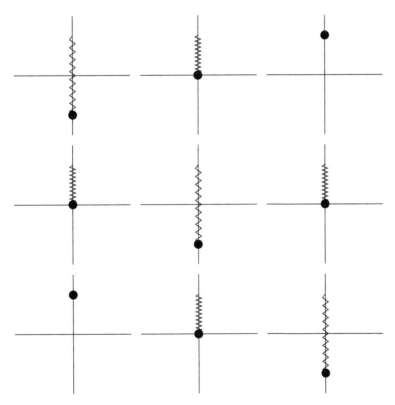

Figure 5-2   Simple harmonic motion: a spring

the displacement of the mass at the values of time from $t = 0$ to $t = \pi/2$ using increments of $\pi/16$.

```
Show[GraphicsGrid[toshow]]
```

In order to achieve an animation so that we can *see* the motion of the spring, we use a `Animate`. Mathematica's Help for `Animate` follows.

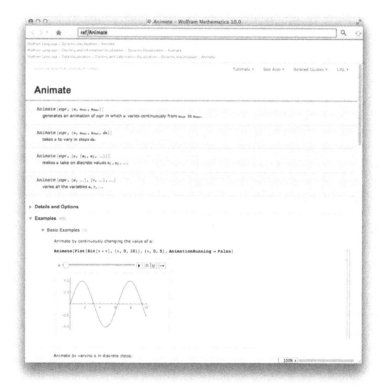

For example, entering

```
Animate[spring[t], {t, 0, π/2, π/32}]
```

displays `spring[t]` for *t*-values from $t = 0$ to $t = \pi/2$ using increments of $\pi/118$. To animate these graphs, click on the control buttons.

When these graphs are animated, as indicated in the following figure, we can see the motion of the spring.

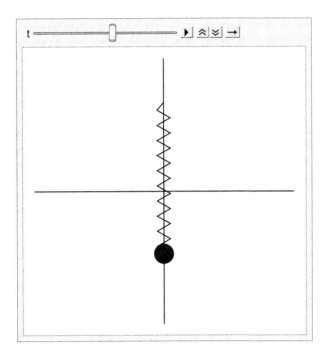

Remember that positive values of $x(t)$ indicate that the mass is beneath the equilibrium position while negative values of $x(t)$ indicate that the mass is above the equilibrium position. To see this, we graph $x(t)$ in Figure 5-3.

```
graph = Plot[sol[t], {t, 0, π/2}, .
    PlotStyle->CMYKColor[0, 0.89, 0.94, 0.28],
    AxesStyle → Black, Ticks → {{1}, {-1, 1}}, PlotRange →
    {-3/2, 3/2}, AspectRatio → 1, AxesLabel → {t, x}];
```

Then, we define p. Given t, p[t] generates a graphics object consisting of the graph of $x(t)$ on the interval $[0, \pi/2]$, which is named graph, and a "small" point placed at $(t, x(t))$.

```
Clear[p]

p[t_]:=Module[{dp},
    dp = Graphics[{PointSize[0.07], Point[{t, sol[t]}]}];
Show[graph, dp]]
```

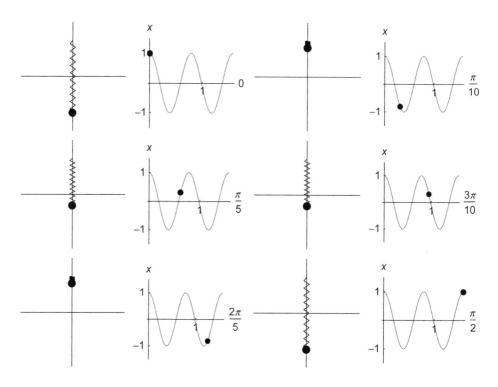

Figure 5-3   Simple harmonic motion illustrated with a spring and a plot

We then use `Table` and `GraphicsGrid` to generate a set of graphics objects consisting of the graphs of `spring[t]` and `p[t]`, shown side-by-side, for t-values from $t = 0$ to $t = \pi/2$ using increments of $\pi/10$.

$$\texttt{moregraphs = Table[\{spring[t], p[t]\}, \{t, 0, } \frac{\pi}{2}, \frac{\pi}{10}\}];$$

The list `moregraphs` is then partitioned into two element subsets and displayed using `Show` and `GraphicsArray` in Figure 5-3.

```
toshow = Flatten[Partition[moregraphs, 2], 2];
```

```
Show[GraphicsGrid[Partition[toshow, 4]]]
```

Notice that the displacement function $x(t) = \cos 8t$ indicates that the spring-mass system never comes to rest once it is set into motion. The solution is periodic, so the mass moves vertically, retracing its motion. Hence, motion of this type is called **simple harmonic motion**.

EXAMPLE 5.1.2: An object with mass $m = 1$ slug is attached to a spring with spring constant $k = 4$. (a) Determine the displacement function of the object if $x(0) = \alpha$ and $x'(0) = 0$. Plot the solution for $\alpha = 1, 4, -2$. How does varying the value of α affect the solution? Does it change the values of t at which the mass passes through the equilibrium position? (b) Determine the displacement function of the object if $x(0) = 0$ and $x'(0) = \beta$. Plot the solution for $\beta = 1, 4, -2$. How does varying the value of β affect the solution? Does it change the values of t at which the mass passes through the equilibrium position?

SOLUTION: For (a), the initial-value problem we need to solve is

$$\begin{cases} x'' + 4x = 0 \\ x(0) = \alpha, \ x'(0) = 0 \end{cases},$$

for $\alpha = 1, 4, -2$. We now determine the solution to each of the three problems with DSolve. For the first problem, entering

```
Clear[x]

de2 = DSolve[{x''[t] + 4x[t]==0, x[0]==1, x'[0]==0}, x[t], t]
```

$\{\{x[t] \to Cos[2t]\}\}$

solves the initial-value problem if $\alpha = 1$ and names the result de2. Note that the formula for the solution is the second part of the first part of the first part of de2 and is extracted from de2 with Part ([[...]]) by entering de2[[1,1,2]]. Alternatively, if you are using Version 10 of Mathematica you can select and copy the formula in the output and paste it to any location. Similarly, entering

```
de3 = DSolve[{x''[t] + 4x[t]==0, x[0]==4, x'[0]==0}, x[t], t]
```

$\{\{x[t] \to 4Cos[2t]\}\}$

```
de4 = DSolve[{x''[t] + 4x[t]==0, x[0]== - 2, x'[0]==0}, x[t], t]
```

$\{\{x[t] \to -2Cos[2t]\}\}$

solves

$$\begin{cases} x'' + 4x = 0 \\ x(0) = 4, \ x'(0) = 0 \end{cases} \quad \text{and} \quad \begin{cases} x'' + 4x = 0 \\ x(0) = -2, \ x'(0) = 0, \end{cases}$$

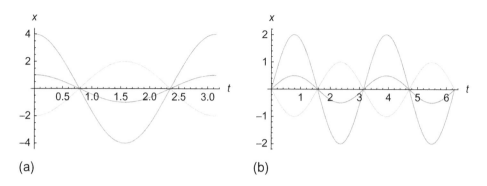

Figure 5-4 (a) Simple harmonic motion: varying the initial displacement. (b) Simple harmonic motion: varying the initial velocity

naming the results de3 and de4, respectively. We graph the solutions on the interval $[0, \pi]$ with Plot in Figure 5-4(a). Note how we use Map to extract the formula for each solution from de2, de3, and de4.

```
toplot = Map[#[[1, 1, 2]]&, {de2, de3, de4}]
```

```
{Cos[2t], 4Cos[2t], −2Cos[2t]}
```

```
p1 = Plot[Evaluate[toplot], {t, 0, π},

    AxesLabel → {t, x}, PlotLabel → "(a)"]
```

We see that the initial position affects only the amplitude of the function (and direction in the case of the negative initial position). The mass passes through the equilibrium position ($x = 0$) at the same time in all three cases.

For (b), we need to solve the initial-value problem

$$\begin{cases} x'' + 4x = 0 \\ x(0) = 0, \ x'(0) = \beta \end{cases},$$

for $\beta = 1, 4, -2$. In this case, we define a procedure d that, given β, returns the solution to the initial-value problem.

Be sure to use (lower case) d instead of (upper case) D to avoid conflict with the built-in function D.

```
d[β_]:=Module[{},

    DSolve[{x''[t] + 4x[t]==0, x[0]==0, x'[0]==β}, x[t], t]]
```

We then use Map to apply d to the list of numbers {1,4,-2} and name the resulting output solutions. (Note that the same result is obtained by using the keyboard shortcut for Map, /@, and entering solutions=d/@{1,4,-2}.)

```
solutions = Map[d, {1, 4, -2}]
```

$$\{\{\{x[t] \to \frac{1}{2}\text{Sin}[2t]\}\}, \{\{x[t] \to 2\text{Sin}[2t]\}\}, \{\{x[t] \to -\text{Sin}[2t]\}\}\}$$

We see that `solutions` consists of three lists. For example, the solution to the initial-value problem when $\beta = -2$ is contained in the third list in `solutions`. We now extract the formula for the `solution` with `Part ([[...]])`.

```
solutions[[3, 1, 1, 2]]
```

```
-Sin[2t]
```

All three solutions are graphed together on $[0, 2\pi]$ with `Plot` in Figure 5-4(b).

```
p2 = Plot[Evaluate[x[t]/.solutions], {t, 0, 2π},

    AxesLabel → {t, x}, PlotLabel → "(b)"]
```

```
Show[GraphicsRow[{p1, p2}]]
```

In Figure 5-4(b), notice that varying the initial velocity affects the amplitude (and direction in the case of the negative initial velocity) of each function. The mass passes through the equilibrium position at the same time in all three cases.

∎

5.1.2 Damped Motion

Equation (5.1) disregards all retarding forces acting on the motion of the mass and a more realistic model which takes these forces into account is needed.

Studies in mechanics reveal that resistive forces due to damping are functions of the velocity of the motion. Hence, for $c > 0$, $F_R = c\, dx/dt$, $F_R = c\, (dx/dt)^3$, or $F_R = c\, \text{sgn}\, (dx/dt)$, where

$$\text{sgn}\left(\frac{dx}{dt}\right) = \begin{cases} 1, & dx/dt > 0 \\ 0, & dx/dt = 0 \\ -1, & dx/dt < 0 \end{cases},$$

are typically used to represent the damping force. Incorporating damping into equation (5.1) and assuming that $F_R = c\, dx/dt$, the displacement function, $x(t)$, is found by solving the initial-value problem

$$\begin{cases} m\dfrac{d^2x}{dt^2} + c\dfrac{dx}{dt} + kx = 0 \\[2mm] x(0) = \alpha, \ \dfrac{dx}{dt}(0) = \beta \end{cases} \qquad (5.2)$$

From our experience with second-order ordinary differential equations with constant coefficients in Chapter 4, the solutions to initial-value problems of this type greatly depend on the values of m, k, and c.

This calculation is identical to those followed in Chapter 4 for second-order linear homogeneous equations with constant coefficients.

Suppose we assume that solutions of the differential equation have the form $x(t) = e^{rt}$. Because $x' = re^{rt}$ and $x'' = r^2e^{rt}$, we have by substitution into the differential equation $mr^2e^{kt} + cre^{rt} + ke^{rt} = 0$, so $e^{rt}\left(mr^2 + cr + k\right) = 0$. The solutions to the characteristic equation are

$$r = \frac{-c \pm \sqrt{c^2 - 4mk}}{2a}.$$

Hence, the solution depends on the value of the quantity $c^2 - 4mk$. In fact, problems of this type are characterized by the value of $c^2 - 4mk$ as follows.

1. $c^2 - 4mk > 0$. This situation is said to be **overdamped** because the damping coefficient c is large in comparison to the spring constant k.
2. $c^2 - 4mk = 0$. This situation is described as **critically damped** because the resulting motion is oscillatory with a slight decrease in the damping coefficient c.
3. $c^2 - 4mk < 0$. This situation is called **underdamped** because the damping coefficient c is small in comparison with the spring constant k.

EXAMPLE 5.1.3: Classify the following differential equations as overdamped, underdamped, or critically damped. Also, solve the corresponding initial-value problem using the given initial conditions and investigate the behavior of the solutions.

(a) $\dfrac{d^2x}{dt^2} + 8\dfrac{dx}{dt} + 16x = 0$ subject to $x(0) = 0$ and $\dfrac{dx}{dt}(0) = 1$; and

(b) $\dfrac{d^2x}{dt^2} + 5\dfrac{dx}{dt} + 4x = 0$ subject to $x(0) = 1$ and $\dfrac{dx}{dt}(0) = 1$.

SOLUTION: For (a), we identify $m = 1$, $c = 8$, and $k = 16$ so that $c^2 - 4mk = 0$, which means that the differential equation $x'' + 8x' + 16x = 0$ is critically damped. After defining del, we solve the equation subject

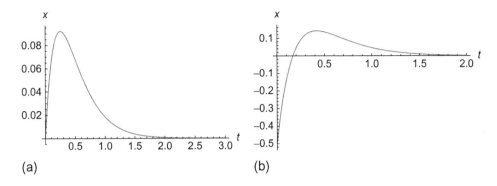

Figure 5-5 Critically damped motion

to the initial conditions and name the resulting output soll. We then graph the solution shown in Figure 5-5(a) by selecting and copying the result given in soll to the subsequent Plot command. If you prefer working with **InputForm**, the formula for the solution to the initial-value problem is extracted from soll with soll[[1,1,2]]. Thus, entering Plot[sol[[1,1,2]],{t,0,4}] displays the same graph as that obtained with the following Plot command. Note that replacing soll[[1,1,2]] with Evaluate[x[t]/.soll] in the Plot command also produces the same result.

```
Clear[x]

eq = x″[t] + 8x′[t] + 16x[t]==0;

de = DSolve[{eq, x[0]==0, x′[0]==1}, x[t], t]
```

$$\{\{x[t] \to e^{-4t}t\}\}$$

```
p1 = Plot[x[t]/.de, {t, 0, 3}, PlotStyle->CMYKColor[0, 0.89,
    0.94, 0.28], AxesLabel → {t, x}, PlotLabel → "(a)"]

de = DSolve[{eq, x[0]== - 0.5, x′[0]==5}, x[t], t]
```

$$\{\{x[t] \to 3.e^{-4t}(-0.166667 + 1.t)\}\}$$

```
p2 = Plot[x[t]/.de, {t, 0, 2}, PlotRange → All,
    PlotStyle->CMYKColor[0, 0.89, 0.94, 0.28],
    AxesLabel → {t, x}, PlotLabel → "(b)"]
```

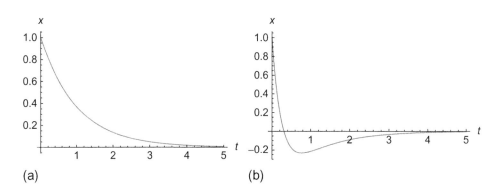

Figure 5-6   Overdamped motion

```
Show[GraphicsRow[{p1, p2}]]
```

For (b), we proceed in the same manner. We identify $m = 1$, $c = 5$, and $k = 4$ so that $c^2 - 4mk = 9$ and the equation $x'' + 5x' + 4x = 0$ is overdamped. We then define de2 to be the equation and the solution to the initial-value problem obtained with DSolve, sol2 and then graph $x(t)$ on the interval $[0, 4]$ in Figure 5-6.

```
eq = x''[t] + 5x'[t] + 4x[t]==0;
```

```
de = DSolve[{eq, x[0]==1, x'[0]== - 1}, x[t], t]
```

$$\{\{x[t] \to e^{-t}\}\}$$

```
p1 = Plot[x[t]/.de, {t, 0, 5}, PlotStyle->CMYKColor[0, 0.89,
   0.94, 0.28], AxesLabel → {t, x}, PlotLabel → "(a)"]
```

```
de = DSolve[{eq, x[0]==1, x'[0]== - 6}, x[t], t]
```

$$\{\{x[t] \to -\frac{1}{3}e^{-4t}(-5 + 2e^{3t})\}\}$$

```
p2 = Plot[x[t]/.de, {t, 0, 5}, PlotRange → All,
   PlotStyle->CMYKColor[0, 0.89, 0.94, 0.28],
   AxesLabel → {t, x}, PlotLabel → "(b)"]
```

```
Show[GraphicsRow[{p1, p2}]]
```

■

EXAMPLE 5.1.4: A 16-pound weight stretches a spring 2 feet. Determine the displacement function if the resistive force due to damping is $F_R = \frac{1}{2} dx/dt$ and the mass is released from the equilibrium position with a downward velocity of 1 ft/s.

SOLUTION: Because $F = 16$ lb, the spring constant is determined with $16 = k \cdot 2$. Hence, $k = 8$ lb/ft. Also, $m = 16/32 = 1/2$ slug. Therefore, the differential equation is $\frac{1}{2}x'' + \frac{1}{2}x' + 8x = 0$ or $x'' + x' + 16x = 0$. The initial position is $x(0) = 0$ and the initial velocity is $x'(0) = 1$. Thus, we must solve the initial-value problem

$$\begin{cases} x'' + x' + 16x = 0 \\ x(0) = 0, \ x'(0) = 1 \end{cases},$$

which is now solved with DSolve.

```
Clear[x, t, deq, sol]
deq = DSolve[{x''[t] + x'[t] + 16x[t]==0,
  x[0]==0, x'[0]==1}, x[t], t]
```

$$\{\{x[t] \to \frac{2e^{-t/2}\mathrm{Sin}[\frac{3\sqrt{7}t}{2}]}{3\sqrt{7}}\}\}$$

```
sol[t_] =
```
$$\frac{2E^{-\frac{t}{2}}\mathrm{Sin}[\frac{3}{2}\sqrt{7}t]}{3\sqrt{7}};$$

Solutions of this type have several interesting properties. First, the trigonometric component of the solution causes the motion to oscillate. Also, the exponential portion forces the solution to approach zero as t approaches infinity. These qualities are illustrated in the plot of the solution shown in Figure 5-7(a).

```
p1 = Plot[sol[t], {t, 0, 2π},
  PlotStyle->CMYKColor[0, 0.89, 0.94, 0.28],
  AxesLabel → {t, x}, PlotLabel → "(a)"]
```

Physically, the displacement of the mass in this case oscillates about the equilibrium position and eventually comes to rest in the equilibrium position. Of course, with our model the displacement function $x(t) \to 0$ as $t \to \infty$, but there is no number T such that $x(t) = 0$ for $t > T$

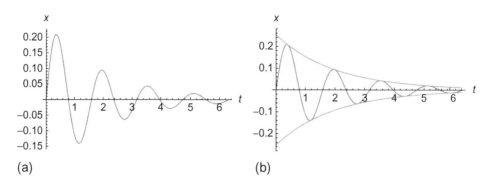

Figure 5-7   (a) Underdamped motion. (b) The solution shown with its envelope functions

as we might expect from the physical situation. Hence, our model only approximates the behavior of the mass. Notice also that the solution is bounded above and below by the exponential term of the solution $e^{-t/2}$ and its reflection through the horizontal axis, $-e^{-t/2}$. This is illustrated with the simultaneous display of these functions in Figure 5-7(b).

```
p2 = Plot[{sol[t], ──────────── , −────────────}, {t, 0, 2π},
                    3√7̄Exp[─]      3√7̄Exp[─]
                          2              2
                        2              2
     PlotStyle → {CMYKColor[0, 0.89, 0.94, 0.28],

     CMYKColor[0.62, 0.57, 0.23, 0],

     CMYKColor[0.62, 0.57, 0.23, 0]},

     AxesLabel → {t, x}, PlotLabel → "(b)"]

  Show[GraphicsRow[{p1, p2}]]
```

Other questions of interest include: (1) When does the mass first pass through its equilibrium point? (2) What is the maximum displacement of the spring?

The time at which the mass passes through $x = 0$ can be determined in several ways. The solution equals zero at the time that $\sin\left(\frac{3}{2}\sqrt{7}t\right)$ first equals zero after $t = 0$ which occurs when $\frac{3}{2}\sqrt{7}t = \pi$. We use Solve to solve this equation for t and then use N to approximate the time. (Note that \% refers to the most recent output.)

```
     Solve[─────── ==π , t]
            3√7̄t
            ────
             2

     N[%]
```

$$\{\{t \to \frac{2\pi}{3\sqrt{7}}\}\}$$

$$\{\{t \to 0.791607\}\}$$

Alternatively, we can approximate the time with `FindRoot`.

```
FindRoot[sol[t]==0, {t, 0.7}]
```

$\{t \to 0.791607\}$

Similarly, the maximum displacement of the spring is found by finding the first value of t for which the derivative of the solution is equal to zero as done here with `FindRoot`.

```
cp1 = FindRoot[sol'[t]==0, {t, 0.4}]
```

$\{t \to 0.364224\}$

The maximum displacement is then given by evaluating the solution for the value of t obtained with `FindRoot`.

```
N[sol[t]/.cp1]
```

```
0.208377
```

Another interesting characteristic of solutions to undamped problems is the time between successive maxima and minima of the solution, called the **quasiperiod**. This quantity is found by first determining the time at which the second maximum occurs with `FindRoot`. Then, the difference between these values of t is taken to obtain the value 1.58321.

```
cp2 = FindRoot[sol'[t]==0, {t, 2}]
```

$\{t \to 1.94744\}$

```
cp2[[1, 2]] - cp1[[1, 2]]
```

```
1.58321
```

To investigate the solution further, an animation can be created with the `zigzag` and `spring` commands, which were defined previously. We redefine `zigzag` and `spring`.

```
Clear[spring, zigzag, length, points, pairs]

zigzag[{a_, b_}, {c_, d_}, n_, ε_]:=Module[{length,
    points, pairs}, length = d - b;
    points = Table[b + ilength/n, {i, 1, n - 1}];
```

```
    pairs = Table[{a + (−1)ⁱϵ, points[[i]]}, {i, 1, n − 1}];

    PrependTo[pairs, {a, b}];

    AppendTo[pairs, {c, d}];

    Line[pairs]]

spring[t_]:=Show[Graphics[{zigzag[{0, −sol[t]},

  {0, .25}, 20, .05],CMYKColor[0.62, 0.57, 0.23, 0],

PointSize[.075], Point[{0, −sol[t]}]}], Axes → Automatic,

AxesStyle → GrayLevel[.5],

Ticks → None, PlotRange → {{−1, 1}, {−.25, .25}},

AspectRatio → 1,DisplayFunction → Identity]
```

Next, we display a graphics array consisting of spring[t] for *t*-values from $t = 0$ to $t = 4$ using increments of $4/15$ in Figure 5-8.

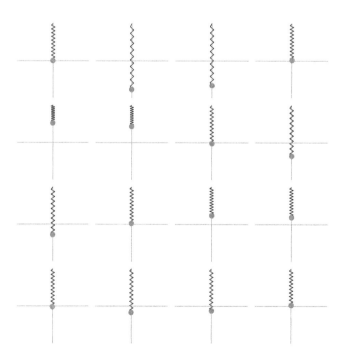

Figure 5-8   The motion of an underdamped spring

```
somegraphs = Table[spring[t], {t, 0, 4, 4/15}];

toshow = Partition[somegraphs, 4];

Show[GraphicsGrid[toshow]]
```

To generate an animation, we use `Animate`.

```
Animate[spring[t], {t, 0, 6, 6/99}]
```

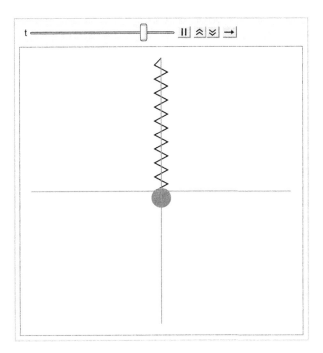

We can also compare the motion of the spring to the graph of the solution as shown in Figure 5-9.

```
Clear[graph, p, toshow]

graph = Plot[sol[t], {t, 0, 2π}, PlotStyle → CMYKColor[0,

0.89, 0.94, 0.28], AxesStyle → Black, Ticks → {{2, 4, 6},

{-0.2, 0.2}}, PlotRange → {-0.25, 0.25}, AspectRatio → 1];

p[t_]:=Module[{dp},

    dp = Graphics[{PointSize[.07], Point[{t, sol[t]}]}];

    Show[graph, dp]]
```

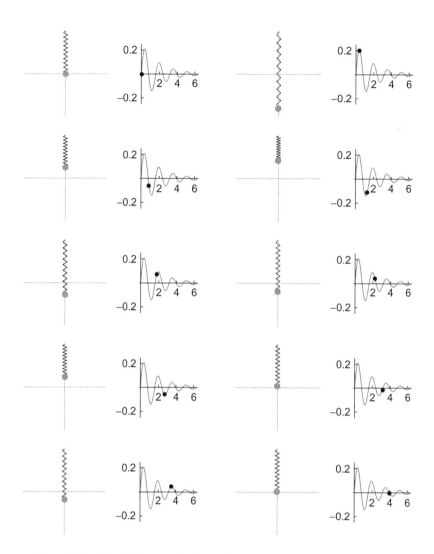

Figure 5-9   Visualizing underdamped motion with a spring and a plot

```
toshow = Partition[Table[Show[GraphicsRow[{spring[t], p[t]}]],
       4
{t, 0, 4, ─}], 2];
       9

Show[GraphicsGrid[toshow], AspectRatio → 1]
```

Alternatively, we can use `Animate` as before to generate graphics and animate the result.

■

EXAMPLE 5.1.5: Given the initial-value problem

$$\begin{cases} x'' + cx' + 6x = 0 \\ x(0) = 0,\ x'(0) = 1 \end{cases},$$ (5.3)

where $c = 2\sqrt{6},\ 4\sqrt{6}$, and $\sqrt{6}$. Determine how the value of c affects the solution of the initial-value problem.

SOLUTION: We begin by defining the function d. Given c, d[c] solves the initial-value problem (5.3).

Be sure to use (lower case) d instead of (upper case) D to avoid conflict with the built-in function D.

```
Clear[x, t, d]

d[c_]:=DSolve[{x''[t] + cx'[t] + 6x[t]==0, x[0]==0, x'[0]==1},
    x[t], t]
```

We then use Map and d to find the solution of the initial-value problem for each value of c, naming the resulting list somesols.

```
somesols = d/@{2√6, 4√6, √6}
```

$$\{\{\{x[t] \to e^{-\sqrt{6}\,t}\,t\}\},$$
$$\{\{x[t] \to -\frac{e^{(-3\sqrt{2}-2\sqrt{6})t} - e^{(3\sqrt{2}-2\sqrt{6})t}}{6\sqrt{2}}\}\},$$
$$\{\{x[t] \to \frac{1}{3}\sqrt{2}\,e^{-\sqrt{\frac{3}{2}}\,t}\,\mathrm{Sin}[\frac{3t}{\sqrt{2}}]\}\}\}$$

Note that each case results in a different classification: $c = 2\sqrt{6}$, critically damped; $c = 4\sqrt{6}$, overdamped; and $c = \sqrt{6}$, underdamped.

All three solutions are graphed together on the interval $[0, 4]$ in Figure 5-10.

```
Plot[Evaluate[x[t]/.somesols], {t, 0, 4},
    AxesLabel → {t, x}]
```

∎

EXAMPLE 5.1.6: Consider the system

$$\begin{cases} x'' + f(t)x' + \frac{5}{4}x = 0,\ t > 0 \\ x(0) = 0,\ x'(0) = 1 \end{cases},$$

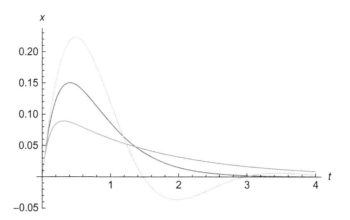

Figure 5-10 Depending on the value of $c > 0$, the motion can be critically damped, overdamped, or underdamped

where

$$f(t) = \begin{cases} 1, \ 0 \le t < \pi \\ 3, \ \pi \le t < 2\pi \end{cases}, \ f(t) = f(t - 2\pi), \ t \ge 2\pi.$$

in which damping oscillates periodically: the rate at which energy is taken away from the system fluctuates periodically. Find the displacement $x(t)$ for $0 \le t \le 4\pi$.

SOLUTION: For $0 < t < \pi$, the solution to the initial-value problem is found by solving

$$\begin{cases} x'' + x' + \frac{5}{4}x = 0 \\ x(0) = 0, \ x'(0) = 1 \end{cases},$$

which is found with DSolve and named y1. Similarly, for $\pi < t < 2\pi$, we solve

$$\begin{cases} x'' + 3x' + \frac{5}{4}x = 0 \\ x(\pi) = a, \ x'(\pi) = b \end{cases},$$

where a=x1[Pi] and b=x1'[Pi] and name the result y2.

```
Clear[sol]
```

$$sol = DSolve[\{x''[t] + x'[t] + \frac{5x[t]}{4} == 0, x[0]==0, x'[0]==1\},$$

```
x[t], t]
```

$$\{\{x[t] \rightarrow e^{-t/2} Sin[t]\}\}$$

```
Clear[x1, x, sol2, a, b]
x1[t_] = sol[[1, 1, 2]];
a = x1[π];
b = x1'[π];
```

$$sol2 = DSolve[\{x''[t] + 3x'[t] + \frac{5x[t]}{4} == 0, x[π]==a, x'[π]==b\},$$

```
x[t], t]
```

$$\{\{x[t] \rightarrow -\frac{1}{2} e^{-5t/2}(-e^{2\pi} + e^{2t})\}\}$$

In a similar way, we find the solution for $2\pi < t < 3\pi$ in y3 and the solution for $3\pi < t < 4\pi$ in y4.

```
Clear[x2, x, sol3, a, b]
x2[t_] = sol2[[1, 1, 2]];
a = x2[2π];
b = x2'[2π];
```

$$sol3 = DSolve[\{x''[t] - x'[t] + \frac{5x[t]}{4} == 0, x[2π]==a, x'[2π]==b\},$$

```
x[t], t]
```

$$\{\{x[t] \rightarrow -\frac{1}{2} e^{-4\pi + \frac{t}{2}}(-Cos[t] + e^{2\pi} Cos[t] + 3Sin[t] - e^{2\pi} Sin[t])\}\}$$

```
Clear[x3, x, sol4, a, b]
x3[t_] = sol3[[1, 1, 2]];
a = x3[3π];
b = x3'[3π];
```

$$sol4 = DSolve[\{x''[t] + 3x'[t] + \frac{5x[t]}{4} == 0, x[3π]==a, x'[3π]==b\},$$

```
x[t], t]
```

$$\{\{x[t] \rightarrow \frac{1}{2} e^{\pi - \frac{5t}{2}}(-e^{4\pi} + e^{2t})\}\}$$

```
x4[t_] = sol4[[1, 1, 2]];
```

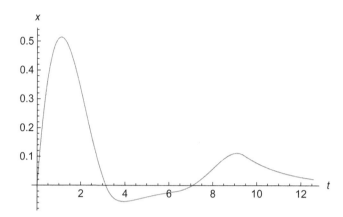

Figure 5-11   Harmonic motion with periodic damping

We see the damped motion of the system by graphing the pieces of the solution individually and then displaying them together with Show in Figure 5-11.

```
plot1 = Plot[x1[t], {t, 0, π},
    PlotStyle->CMYKColor[0, 0.89, 0.94, 0.28]];
plot2 = Plot[x2[t], {t, π, 2π},
    PlotStyle->CMYKColor[0, 0.89, 0.94, 0.28]];
plot3 = Plot[x3[t], {t, 2π, 3π},
    PlotStyle->CMYKColor[0, 0.89, 0.94, 0.28]];
plot4 = Plot[x4[t], {t, 3π, 4π},
    PlotStyle->CMYKColor[0, 0.89, 0.94, 0.28]];
show4 = Show[plot1, plot2, plot3, plot4, AxesLabel → {t, x},
    PlotRange → {{0, 4Pi}, Automatic}
]
```

∎

5.1.3 Forced Motion

In some cases, the motion of the spring is influenced by an external driving force, $f(t)$. Mathematically, this force is included in the differential equation that models the situation as follows:

$$m\frac{d^2x}{dt^2} = -kx - c\frac{dx}{dt} + f(t).$$

The resulting initial-value problem is

$$\begin{cases} mx'' + cx' + kx = f(t) \\ x(0) = \alpha, \ x'(0) = \beta \end{cases} \tag{5.4}$$

Therefore, differential equations modeling forced motion are nonhomogeneous and require the Method of Undetermined Coefficients or Variation of Parameters for solution. We first consider forced motion that is undamped.

EXAMPLE 5.1.7: An object of mass $m = 1$ slug is attached to a spring with spring constant $k = 4$. Assuming there is no damping and that the object is released from rest in the equilibrium position, determine the position function of the object if it is subjected to an external force of (a) $f(t) = 0$, (b) $f(t) = 1$, (c) $f(t) = \cos t$, and (d) $f(t) = \sin t$.

SOLUTION: First, we note that we must solve the initial-value problem

$$\begin{cases} x'' + 4x = f(t) \\ x(0) = 0, \ x'(0) = 0 \end{cases}$$

for each of the forcing functions in (a), (b), (c), and (d). Because we will be solving this initial-value problem for various forcing functions, we begin by defining the function fm. Given a function $f = f(t)$, fm[f] returns the formula for the solution to this initial-value problem.

```
Clear[x, t]

fm[f_]:=DSolve[{x''[t] + 4x[t]==f, x[0]==0,

   x'[0]==0}, x[t], t][[1, 1, 2]]
```

Next, we define fs to be the forcing functions in (a)–(d).

```
fs = {0, 1, Cos[t], Sin[t]};
```

We then use Map to apply fm to fs and name the resulting list of functions somesols.

```
somesols = Map[fm, fs]
```

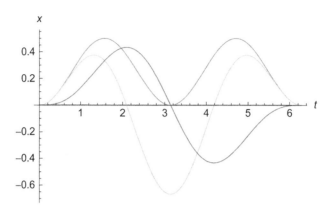

Figure 5-12   Forced motion without damping

$$\{0, \frac{1}{4}(1 - Cos[2t]), \frac{1}{12}(-4Cos[2t] + 4Cos[t]^3Cos[2t] + 3Sin[t]$$

$$Sin[2t] + Sin[2t]Sin[3t]), \frac{1}{12}(-4Cos[2t]Sin[t]^3 - 2Sin[2t] +$$

$$3Cos[t]Sin[2t] - Cos[3t]Sin[2t])\}$$

From the result, we see that for (a) the solution is $x(t) = 0$. Physically, this solution indicates that because there is no forcing function, no initial displacement from the equilibrium position, and no initial velocity, the object does not move from the equilibrium position.

The nontrivial solutions in (b), (c), and (d) are then graphed on the interval $[0, 2\pi]$ with `Plot` in Figure 5-12.

```
Plot[somesols, {t, 0, 2π}, PlotRange → All,

    AxesLabel → {t, x}]
```

Negative values of x indicate that the mass is *above* the equilibrium position; positive values indicate that the mass is *below* the equilibrium position.

From the graph, we see that for (b) the object never moves above the equilibrium position. This makes sense because $0 \leq \cos 2t \leq 1$: $x(t) = \frac{1}{4}(1 - \cos 2t)$ for all t. For (c), we see that the mass passes through the equilibrium position twice (near $t = 2$ and $t = 4$) over the period. For (d), we again see that the resulting motion is periodic, although different from that observed in (c).

■

When studying nonhomogeneous linear equations, we considered equations in which the forcing function was a solution of the corresponding homogeneous equation. In this case, the situation is modeled by the initial-value problem

$$\begin{cases} x'' + \omega^2 x = F_1 \cos \omega t + F_2 \sin \omega t + G(t) \\ x(0) = \alpha, \ x'(0) = \beta \end{cases}, \quad (5.5)$$

where $\omega > 0$, F_1, and F_2 are constants and $G = G(t)$ is a function of t. In this case, we say that ω is the **natural frequency** of the system because the solution of the corresponding homogeneous equation, $x'' + \omega^2 x = 0$, is $x_h = c_1 \cos \omega t + c_2 \sin \omega t$.

Note that one of the constants F_1 or F_2 can equal zero and $G = G(t)$ can be identically the zero function.

EXAMPLE 5.1.8: Investigate the effect that the forcing functions (a) $f(t) = \cos 2t$ and (b) $f(t) = \sin 2t$ have on the solution of the initial-value problem

$$\begin{cases} x'' + 4x = f(t) \\ x(0) = 0, \ x'(0) = 0 \end{cases}.$$

SOLUTION: We take advantage of the function fm defined in Example 5.1.7. In the same manner as in Example 5.1.7, we use Map to apply fm to each of the forcing functions in (a) and (b). (Note that entering

```
moresols=fm/@{Cos[2 t],Sin[2 t]}
```

produces the same result.)

```
moresols = Map[fm, {Cos[2t], Sin[2t]}]
```

$\{\dfrac{1}{16}(-\text{Cos}[2t] + \text{Cos}[2t]\text{Cos}[4t] + 4t\text{Sin}[2t] + \text{Sin}[2t]\text{Sin}[4t]),$
$\dfrac{1}{16}(-4t\text{Cos}[2t] + \text{Sin}[2t] - \text{Cos}[4t]\text{Sin}[2t] + \text{Cos}[2t]\text{Sin}[4t])\}$

From the result, we see that the nonperiodic function $y = t \sin 2t$ appears in the result for (a) while the nonperiodic function $y = t \cos 2t$ appears in the result for (b). In each case, we see that the amplitude increases without bound as t increases, as illustrated in Figure 5-13. This indicates that the spring-mass system will encounter a serious problem in that the mass will eventually hit its support (like a ceiling or beam) or its lower boundary (like the ground or floor).

```
p1 = Plot[moresols, {t, 0, 2π}, AxesLabel → {t, x},

    PlotLabel → "(a)"]
```

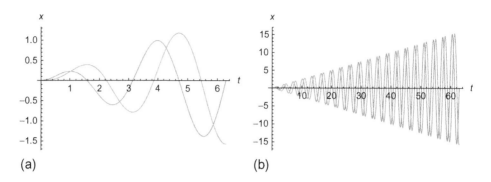

Figure 5-13   Resonance: eventually the object will break

```
p2 = Plot[moresols, {t, 0, 20π}, AxesLabel → {t, x},

   PlotLabel → "(b)"]

Show[GraphicsRow[{p1, p2}]]
```

■

The phenomenon illustrated in Example 5.1.8 is called **resonance** and can be extended to other situations such as vibrations of an aircraft wing, skyscraper, glass, or bridge. Some of the sources of excitation that lead to the vibration of these structures include unbalanced rotating devices, vortex shedding, strong winds, rough surfaces, and moving vehicles. Therefore, the engineer has to overcome many problems when structures and machines are subjected to forced vibrations.

EXAMPLE 5.1.9: How does slightly changing the value of the argument of the forcing function change the solution of the initial-value problem given in Example 5.1.8? Use the functions (a) $f(t) = \cos 1.9t$ and (b) $f(t) = \cos 2.1t$ with the initial-value problem.

SOLUTION: As in Example 5.1.8, we take advantage of the function fm defined in Example 5.1.7. (Note that entering

```
moresols=Map[fm,{Cos[1.9 t],Sin[2.1 t]}]
```

produces the same result as that obtained using /@, the keyboard shortcut for Map.)

```
moresols = Map[fm, {Cos[1.9 t], Sin[2.1 t]}]
```

{2.5(−1.02564Cos[2.t]+1.Cos[0.1t]Cos[2.t]+

0.025641Cos[2.t]Cos[3.9t]+1.Sin[0.1t]Sin[2.t]+

0.025641Sin[2.t]Sin[3.9t]), −2.5(1.Cos[2.t]Sin[0.1t]−

1.02439Sin[2.t]+1.Cos[0.1t]Sin[2.t]+0.0243902Cos[4.1t]

Sin[2.t]−0.0243902Cos[2.t]Sin[4.1t])}

moresols = fm/@{Cos[1.9t], Sin[2.1t]}

{2.5(−1.02564Cos[2.t]+1.Cos[0.1t]Cos[2.t]+0.025641

Cos[2.t]Cos[3.9t]+1.Sin[0.1t]Sin[2.t]+0.025641

Sin[2.t]Sin[3.9t]), −2.5(1.Cos[2.t]Sin[0.1t]−1.02439

Sin[2.t]+1.Cos[0.1t]Sin[2.t]+0.0243902Cos[4.1t]Sin[2.t]

−0.0243902Cos[2.t]Sin[4.1t])}

The result shows that each solution is periodic and bounded. These solutions are then graphed in Figure 5-14 to reveal the behavior of the curves. If the solutions are plotted over only a small interval, however, resonance *seems* to be present.

Compare Figure 5-14 to the graph generated in Example 5.1.8.

p1 = Plot[Evaluate[moresols], {t, 0, 2π},

  AxesLabel → {t, x}, PlotLabel → "(a)"]

However, the functions obtained with fm clearly indicate that there is no resonance. This is further indicated by graphing the solutions over a longer time interval in Figure 5-14(b).

p2 = Plot[moresols[[1]], {t, 0, 40π}, PlotPoints → 200,

  AxesLabel → {t, x}, PlotLabel → "(b)"]

p3 = Plot[moresols[[2]], {t, 0, 40π}, PlotPoints → 200,

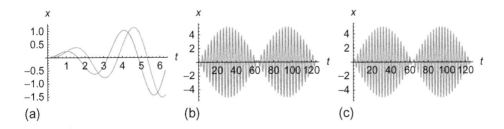

(a) (b) (c)

Figure 5-14 (a) Resonance? (b) and (c) No resonance: the solution is periodic

```
AxesLabel → {t, x}, PlotLabel → "(c)"]

Show[GraphicsRow[{p1, p2, p3}]]
```

■

Let us investigate in detail initial-value problems of the form

$$\begin{cases} x'' + \omega^2 x = F \cos \beta t, \ \omega > 0, \ \omega \neq \beta, \ \beta > 0 \\ x(0) = 0, \ x'(0) = 0 \end{cases} \tag{5.6}$$

A general solution of the corresponding homogeneous equation is $x_h = c_1 \cos \omega t + c_2 \sin \omega t$. Using the Method of Undetermined Coefficients, we assume that there is a particular solution to the nonhomogeneous equation of the form $x_p = A \cos \beta t + B \sin \beta t$.

```
Clear[a, b]

xp[t_] = aCos[β t] + bSin[β t];
```

Next, we calculate the corresponding derivatives of this solution

```
x'p[t]
```

$b\beta \text{Cos}[t\beta] - a\beta \text{Sin}[t\beta]$

```
x''p[t]
```

$-a\beta^2 \text{Cos}[t\beta] - b\beta^2 \text{Sin}[t\beta]$

and substitute into the nonhomogeneous equation $x'' + \omega^2 x = F \cos \beta t$.

```
step1 = Simplify[x''p[t] + ω² xp[t]]==fCos[β t]
```

$-(\beta^2 - \omega^2)(a\text{Cos}[t\beta] + b\text{Sin}[t\beta]) == f\text{Cos}[t\beta]$

This equation is true for all values of t. In particular, substituting $t = 0$ and $t = \pi/(2\beta)$ yields two equations

```
eq1 = step1/.t->0
```

$-a(\beta^2 - \omega^2) == f$

```
eq2 = step1/.t->\frac{π}{2β}
```

$-b(\beta^2 - \omega^2) == 0$

```
vals = Solve[{eq1, eq2}, {a, b}]
```

$$\{\{a \rightarrow -\frac{f}{\beta^2 - \omega^2}, \ b \rightarrow 0\}\}$$

that we then solve for A and B to see that $A = \dfrac{F}{\omega^2 - \beta^2}$ and $B = 0$ and a general solution of the nonhomogeneous equation is

$$x = c_1 \cos \omega t + c_2 \sin \omega t + \frac{F}{\omega^2 - \beta^2} \cos \beta t.$$

Application of the initial conditions yields the solution

$$x = \frac{F}{\omega^2 - \beta^2} (\cos \beta t - \cos \omega t) = \frac{F}{\beta^2 - \omega^2} (\cos \omega t - \cos \beta t).$$

We can use DSolve and Simplify to solve the initial-value problem (5.6) as well.

```
DSolve[{x"[t] + ω²x[t]==fCos[βt], x[0]==0,

x'[0]==0}, x[t], t]//Simplify
```

$$\{\{x[t] \rightarrow \frac{f(-\text{Cos}[t\beta] + \text{Cos}[t\omega])}{(\beta - \omega)(\beta + \omega)}\}\}$$

Using the trigonometric identity $\frac{1}{2}[\cos(\alpha - \beta) - \cos(\alpha + \beta)] = \sin \alpha \sin \beta$, we have

$$x = \frac{2F}{\omega^2 - \beta^2} \sin\left(\frac{\omega + \beta}{2}t\right) \sin\left(\frac{\omega - \beta}{2}t\right).$$

These solutions are of interest because of what they indicate about the motion of the spring under consideration. Notice that the solution can be represented as

$$x = A(t) \sin\left(\frac{\omega + \beta}{2}t\right), \quad \text{where} \quad A(t) = \frac{2F}{\omega^2 - \beta^2} \sin\left(\frac{\omega - \beta}{2}t\right).$$

Therefore, if the quantity $\omega - \beta$ is small, $\omega + \beta$ is relatively large in comparison. Thus, the function $\sin\left(\dfrac{\omega + \beta}{2}t\right)$ oscillates quite frequently because it has period $\pi/(\omega + \beta)$. Meanwhile, the function $\sin\left(\dfrac{\omega - \beta}{2}t\right)$ oscillates relatively slowly because it has period $\pi/|\omega - \beta|$, so the functions $\pm\dfrac{2F}{\omega^2 - \beta^2} \sin\left(\dfrac{\omega - \beta}{2}t\right)$ form an **envelope** for the solution.

EXAMPLE 5.1.10: Solve the initial-value problem

$$\begin{cases} x'' + 4x = f(t) \\ x(0) = 0,\ x'(0) = 0 \end{cases}$$

with (a) $f(t) = \cos 3t$ and (b) $f(t) = \cos 5t$.

SOLUTION: Again, we use fm, defined in Example 5.1.7, to solve the initial-value problem in each case.

```
Clear[x, t]

fm[f_]:=DSolve[{x''[t] + 4x[t]==f, x[0]==0, x'[0]==0}, x[t],

t][[1, 1, 2]]

fs = {Cos[3 t], Cos[5 t]};

somesols = Map[fm, fs]
```

$\{\dfrac{1}{20}(4\text{Cos}[2t] - 5\text{Cos}[t]\text{Cos}[2t] + \text{Cos}[2t]\text{Cos}[5t] + 5\text{Sin}[t]$

$\text{Sin}[2t] \ + \ \text{Sin}[2t]\text{Sin}[5t]), \dfrac{1}{84}(4\text{Cos}[2t] \ - \ 7\text{Cos}[2t]\text{Cos}[3t] \ +$

$3\text{Cos}[2t]\text{Cos}[7t] + 7\text{Sin}[2t]\text{Sin}[3t] + 3\text{Sin}[2t]\text{Sin}[7t])\}$

The solution for (a) is graphed in Figure 5-15 and named p1 for later use.

```
p1 = Plot[somesols[[1]], {t, 0, 6π},

AxesLabel → {t, x}, PlotStyle → CMYKColor[0, 0.77, 0.87, 0]]
```

Using the formula obtained earlier for the functions that "envelope" the solution, we have $x(t) = \pm\dfrac{2}{5}\sin\left(\frac{1}{2}t\right)$. These functions are graphed in p2 and displayed with p1 with Show in Figure 5-16.

```
p2 = Plot[{2/5 Sin[t/2], -2/5 Sin[t/2]}, {t, 0, 6π},

PlotStyle → CMYKColor[0.99, 0, 0.52, 0]];

q1 = Show[p1, p2, AxesLabel → {t, x}, PlotRange → All,

PlotLabel → "(a)"]

q2 = Plot[{somesols[[2]], 2/21 Sin[3t/2], -2/21 Sin[3t/2]}, .

{t, 0, 4π}, PlotStyle → {CMYKColor[0, 0.77, 0.87, 0],
```

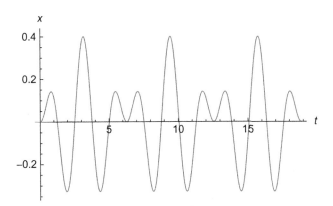

Figure 5-15   The forcing function causes *beats*

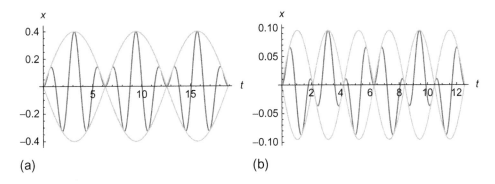

(a)   (b)

Figure 5-16   The envelope functions show the beats more clearly

```
CMYKColor[0.99, 0, 0.52, 0], CMYKColor[0.99, 0, 0.52, 0]},

AxesLabel → {t, x}, PlotLabel → "(b)"]
```

```
Show[GraphicsRow[{q1, q2}]]
```

For (b), the graph of the solution is as follows. See Figure 5-17.

```
dell = Simplify[fm[Exp[-t]Cos[2 t]]]
```

$$\frac{1}{34} e^{-t}(-2(-1 + e^t)\text{Cos}[2 t] + (-8 + 9 e^t)\text{Sin}[2 t])$$

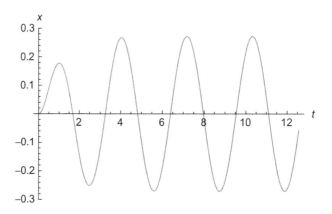

Figure 5-17   Beats

```
Plot[de11, {t, 0, 4π}, AxesLabel → {t, x},
    PlotStyle->CMYKColor[0, 0.89, 0.94, 0.28]
  ]
```

■

Oscillations like those illustrated in the previous example are called **beats** because of the periodic variation of amplitude. This phenomenon is commonly encountered when two musicians (especially bad ones) try to simultaneously tune their instruments or when two tuning forks with almost equivalent frequencies are played at the same time.

See the *Application* at the end of the section for a discussion of how you can listen to beats and resonance with Mathematica.

We now consider spring problems that involve forces due to damping as well as external forces. In particular, consider the following initial-value problem:

$$\begin{cases} mx'' + cx' + kx = \rho \cos \lambda t \\ x(0) = \alpha, \, x'(0) = \beta \end{cases}. \tag{5.7}$$

Problems of this nature have solutions of the form $x(t) = h(t) + s(t)$, where $\lim_{t \to \infty} h(t) = 0$ and $s(t) = c_1 \cos \lambda t + c_2 \sin \lambda t$.

The function $h(t)$ is called the **transient solution** while $s(t)$ is called the **steady-state solution**. Therefore, as t approaches infinity, the solution $x(t)$ approaches the steady-state solution. Note that the steady-state solution corresponds to the particular solution obtained through the Method of Undetermined Coefficients or Variation of Parameters.

EXAMPLE 5.1.11: Solve the initial-value problem

$$\begin{cases} x'' + 4x' + 13x = \cos t \\ x(0) = 0, \ x'(0) = 1 \end{cases},$$

which models the motion of an object of mass $m = 1$ attached to a spring with spring constant $k = 13$ that is subjected to a resistive force of $F_R = 4x'$ and an external force of $f(t) = \cos t$. Identify the transient and steady-state solutions.

SOLUTION: First, DSolve is used to obtain the solution of this non-homogeneous problem.

The Method of Undetermined Coefficients could be used to find this solution as well.

```
deq = Simplify[DSolve[{x''[t] + 4x'[t] + 13x[t]==Cos[t],

    x[0]==0, x'[0]==1}, x[t], t]]
```

$$\left\{\left\{x[t] \rightarrow \frac{1}{40} e^{-2t}(3 e^{2t}\text{Cos}[t] - 3\text{Cos}[3t] + e^{2t}\text{Sin}[t] + 11\text{Sin}[3t])\right\}\right\}$$

The solution is then graphed over the interval $[0, 5\pi]$ in plot1 to illustrate the behavior of this solution. See Figure 5-18(a).

```
p1 = Plot[x[t]/.deq, {t, 0, 5π},

    PlotStyle->CMYKColor[0, 0.89, 0.94, 0.28],

    AxesLabel → {t, x}, PlotLabel → "(a)"]
```

The transient solution is $h = e^{-2t}\left(-\frac{3}{40}\cos 3t + \frac{11}{40}\sin 3t\right)$ and the steady-state solution is $s = \frac{3}{40}\cos t + \frac{1}{40}\sin t$. We graph the steady-state

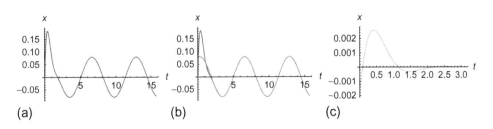

(a)   (b)   (c)

Figure 5-18   (a) Forced motion with damping. (b) Forced motion with damping shown with its steady-state solution. (c) The transient solution quickly tends to 0

solution over the same interval so that it can be compared to plot1 and then we show the two graphs together with Show in Figure 5-18(b).

$$ss[t\_] = \frac{1}{40}(3Cos[t] + Sin[t]);$$

ssplot = Plot[ss[t], {t, 0, 5π},

   PlotStyle->CMYKColor[0.62, 0.57, 0.23, 0]];

p2 = Show[p1, ssplot, AxesLabel → {t, x},

   PlotLabel → "(b)"]

Notice that the two curves appear identical for $t > 2.5$. The reason for this is shown in the plot of the transient solution in Figure 5-18(c), which becomes quite small near $t = 2.5$.

$$p3 = Plot[\frac{1}{40}Exp[-2t](-\frac{3Cos[3t]}{40} + \frac{11Sin[3t]}{40}), \{t, 0, π\},$$

   PlotStyle->CMYKColor[0.64, 0, 0.95, 0.40],

   AxesLabel → {t, x}, PlotLabel → "(c)"]

Show[GraphicsRow[{p1, p2, p3}]]

Notice also that the steady-state solution corresponds to a particular solution to the nonhomogeneous differential equation as verified here with Simplify.

Simplify[ss''[t] + 4ss'[t] + 13ss[t]]

Cos[t]

■

Instead of solving initial value problems that model the motion of damped and undamped systems as functions of time only, we can consider problems that involve an arbitrary parameter. In doing this, we can obtain a new understanding of the phenomena of resonance and beats.

EXAMPLE 5.1.12: Solve (a) $\begin{cases} x'' + 4x' + 13x = \cos \omega t \\ x(0) = 0, \, x'(0) = 0 \end{cases}$;

(b) $\begin{cases} x'' + 4x = \cos \omega t \\ x(0) = 0, \, x'(0) = 0 \end{cases}$. Plot the solution for various values of ω near the natural frequency of the system.

SOLUTION: (a) We solve the initial-value problem and simplify the result with Simplify for arbitrary ω in sol, extract the solution with Part ([[...]]), and define it to be u[t, ω].

```
Clear[sol]

sol = DSolve[{x″[t] + 4x′[t] + 13x[t]==Cos[ωt],

   x[0]==0, x′[0]==0}, x[t], t]//Simplify
```

$$\{\{x[t] \rightarrow \frac{1}{6(169 - 10\omega^2 + \omega^4)} e^{-2t}(6(-13 + \omega^2)\text{Cos}[3t]$$
$$- 6e^{2t}(-13 + \omega^2)\text{Cos}[t\omega] - 4((13 + \omega^2)\text{Sin}[3t]$$
$$- 6e^{2t}\omega\text{Sin}[t\omega]))\}\}$$

```
u[t_, ω_] = sol[[1, 1, 2]];
```

We animate the solution for $0 \leq \omega \leq 6$ using increments of 0.25 with the Animate function in order to animate the resulting plots. We show a frame from the resulting animation.

```
Animate[Plot[u[t, ω], {t, 0, 10},

   PlotPoints → 30, PlotRange → {−.1, .1}], {ω, 0, 6, 0.5}]
```

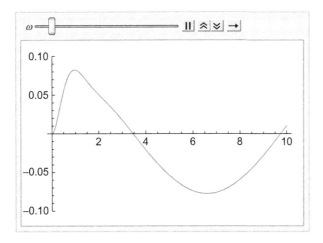

We can also observe how the motion approaches and then moves away from resonance using a GraphicsGrid as shown in Figure 5-19.

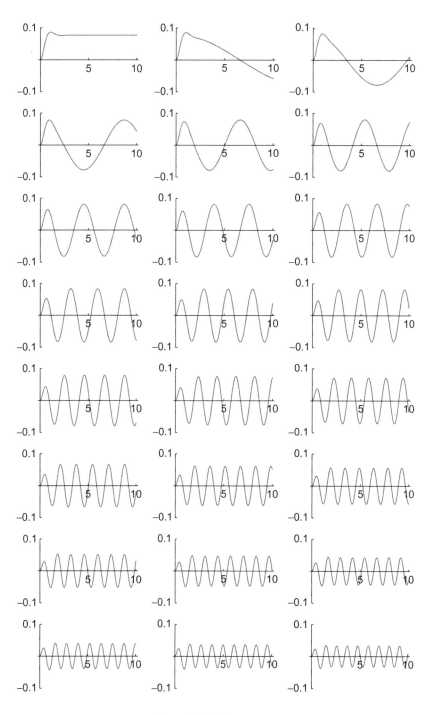

Figure 5-19   Varying ω

```
ωgraph[ω_]:=Plot[u[t, ω], {t, 0, 10}, PlotPoints → 30,

PlotRange → {-0.1, 0.1}, Ticks → {{5, 10}, {-.1, 0.1}}];

graphs = Table[ωgraph[ω], {ω, 0, 6, 0.25}];

toshow = Partition[graphs, 3];

Show[GraphicsGrid[toshow]]
```

On the other hand, we can graph the three-dimensional surface $u[t,\omega]$ to see how the motion depends on the value of ω. See Figure 5-20.

```
Plot3D[u[t, ω], {t, 0, 10}, {ω, 0, 6}, PlotPoints → 30]
```

(b) In a similar way, we solve $\begin{cases} x'' + 4x = \cos \omega t \\ x(0) = 0, \ x'(0) = 0 \end{cases}$ for arbitrary ω in

sol.

```
Clear[u, sol]

sol = DSolve[{x''[t] + 4x[t]==Cos[ωt],

   x[0]==0, x'[0]==0}, x[t], t]//Simplify
```

$$\{\{x[t] \to \frac{\mathrm{Cos}[2t] - \mathrm{Cos}[t\omega]}{-4 + \omega^2}\}\}$$

```
u[t_, ω_] = sol[[1, 1, 2]];
```

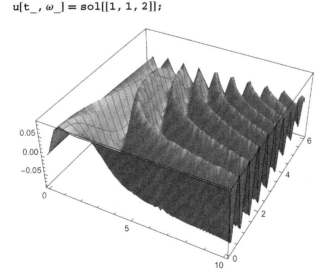

Figure 5-20 Cross-sections of the three-dimensional plot are solutions of the initial-value problem

As in (a), we animate the solution for $0 \leq \omega \leq 3$ using a stepsize of 0.1 with Animate to observe how the solution behaves as ω approaches the natural frequency of the system, 2. Note that Mathematica generates several error messages when it encounters $\omega = 2$ because the solution obtained with DSolve is not defined if $\omega = 2$. Nevertheless, Mathematica accurately displays the graphs of the solutions for $\omega \neq 2$.

```
Animate[Plot[u[t, ω], {t, 0, 10},

    PlotPoints → 30, PlotRange → {−2, 2}], {ω, 0, 4, 0.25}]
```

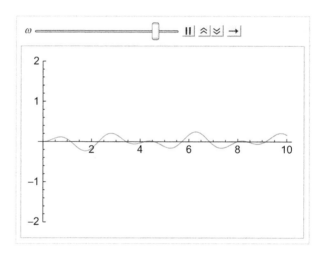

In addition, we can use a GraphicsGrid to observe the behavior of the function as shown in Figure 5-21.

```
Clear[ωgraph]

    ωgraph[ω_]:=Plot[u[t, ω], {t, 0, 10}, PlotPoints → 30,

    PlotRange → {−2, 2},

    Ticks → {{5, 10}, {−2, 2}}, DisplayFunction → Identity]

graphs = Table[ωgraph[ω], {ω, 0, 4, 4/11}];

toshow = Partition[graphs, 3];

Show[GraphicsGrid[toshow]]
```

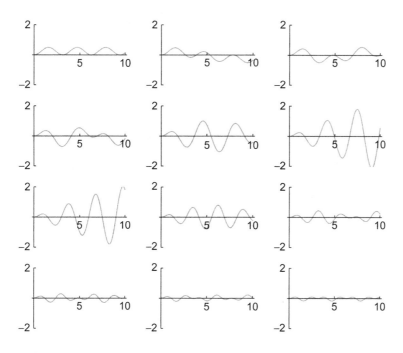

Figure 5-21   The solution is periodic unless $\omega = 2$

We can see this behavior in the three-dimensional graph of u[t,w] in Figure 5-22 as well.

```
Plot3D[u[t, ω], {t, 0, 10}, {ω, 0, 3}, PlotPoints → 30]
```

∎

5.1.4 Soft Springs

In the case of a soft spring, the spring force weakens with compression or extension. For springs of this type, we model the physical system with

$$
\begin{cases}
x'' + cx' + kx - jx^3 = f(t) \\
x(0) = \alpha, \; x'(0) = \beta
\end{cases},
\tag{5.8}
$$

where j is a positive constant.

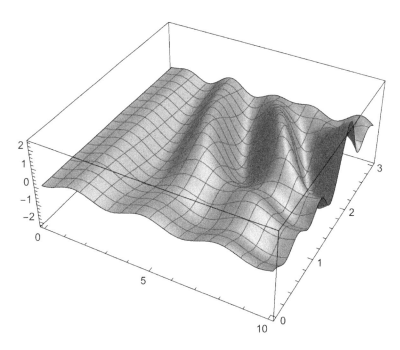

Figure 5-22   Cross-sections of the three-dimensional plot are solutions of the initial-value problem for various values of ω

EXAMPLE 5.1.13: Approximate the solution to

$$\begin{cases} x'' + 0.2x' + 10kx - 0.2x^3 = -9.8 \\ x(0) = \alpha, \, x'(0) = \beta \end{cases},$$

for various values of α and β in the initial conditions.

SOLUTION: After stating this nonlinear differential equation in eq, we define the function $s[\alpha, \beta]$ to approximate the solution to the initial-value problem with NDSolve for specified values of α and β.

```
Clear[eq]

eq = x''[t] + 0.2x'[t] + 10x[t] - 0.2x[t]^3 == -9.8;

s[α_, β_]:=NDSolve[{eq, x[0]==α, x'[0]==β}, x[t],
    {t, 0, 15}];
```

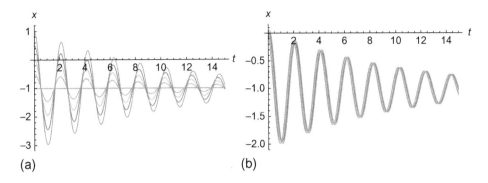

Figure 5-23 (a) The spring does not converge to its equilibrium position. (b) Varying the initial velocity in a soft spring

We then define values for α in `vals` so that we can solve the initial-value problem using $x(0) = \alpha$ for the numbers in `vals` and $x'(0) = 0$ in `sols`. The results are graphed in Figure 5-23(a). We notice that $x(t) \to -1$ as $t \to \infty$.

```
vals = {-1, -0.5, 0, 0.5, 1};

sols = Map[s[#, 0]&, vals];

one = Plot[Evaluate[x[t]/.sols], {t, 0, 15},

    AxesLabel → {t, x}, PlotLabel → "(a)"]
```

Similarly in `sols2`, we use the numbers in `vals` as the initial velocity in the initial-value problem. We graph these approximate solutions in Figure 5-23(b).

```
sols2 = Map[s[0, #1]&, vals];

two = Plot[Evaluate[x[t]/.sols2], {t, 0, 15},

    AxesLabel → {t, x}, PlotLabel → "(b)"]

Show[GraphicsRow[{one, two}]]
```

In each of the two previous sets of initial conditions, we see that $x(t)$ approaches a limit as $t \to \infty$. However, this is not **always** the case. If we consider larger values of α as defined in `vals2`, we find that solutions are unbounded. Because of this, we must use a smaller

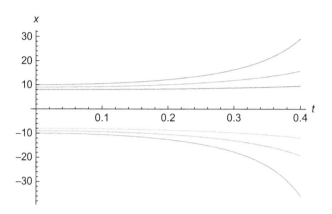

Figure 5-24   The spring becomes weak

interval for t in the NDSolve command in s2[α, β]. Otherwise, we do not obtain meaningful results. The approximate results are graphed in Figure 5-24.

```
s2[α_, β_]:=NDSolve[{eq, x[0]==α, x'[0]==β}, x[t], {t, 0, 0.4}];

vals2 = {-10, -9, -8, 8, 9, 10};

sols3 = s2[#1, 0]&/@vals2;

three = Plot[Evaluate[x[t]/.sols3], {t, 0, .4},

   AxesLabel → {t, x}]
```

■

5.1.5 Hard Springs

In the case of a hard spring, the spring force strengthens with compression or extension. For springs of this type, we model the physical system with

$$\begin{cases} x'' + cx' + kx + jx^3 = f(t) \\ x(0) = \alpha, \ x'(0) = \beta \end{cases}, \tag{5.9}$$

where j is a positive constant.

EXAMPLE 5.1.14: Approximate the solution to

$$\begin{cases} x'' + 0.3x + 0.04x^3 = 0 \\ x(0) = \alpha, \ x'(0) = \beta \end{cases},$$

for various values of α and β in the initial conditions.

SOLUTION: First, we define the undamped nonlinear differential equation in eq. Then, we define s[α, β] to numerically approximate the solution to the initial-value problem for given values of α and β.

```
Clear[eq]
```

```
eq = x″[t] + 0.3x[t] + 0.04x[t]³==0;
```

```
Clear[s]
```

```
s[α_, β_]:=NDSolve[{eq, x[0]==α, x'[0]==β}, x[t], {t, 0, 15}];
```

We approximate the solution using the constants defined in vals4 as the initial displacement, $x(0) = \alpha$. These numerical solutions are then graphed in Figure 5-25(a). Notice that solutions with larger amplitudes have smaller periods as expected with a hard spring.

```
vals4 = {1, 2, 3, 4, 5};
```

```
sols4 = Table[s[vals4[[i]], 0][[1, 1, 2]], {i, 1, 5}];
```

```
five = Plot[sols4, {t, 0, 15},

    AxesLabel → {t, x}, PlotLabel → "(a)"]
```

In a similar manner, we use the values in vals4 as the initial velocity $x'(0) = \beta$. In Figure 5-25(b), we see that when the amplitude is large, the spring strengthens so that the period of the motion is decreased.

```
sols5 = Table[s[0, vals4[[i]]][[1, 1, 2]], {i, 1, 5}];
```

```
six = Plot[Evaluate[sols5], {t, 0, 15},

    AxesLabel → {t, x}, PlotLabel → "(b)"]
```

```
Show[GraphicsRow[{five, six}]]
```

■

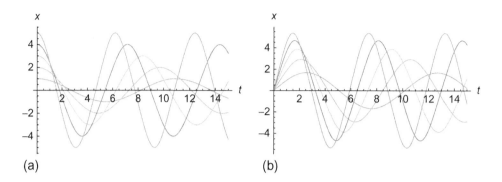

Figure 5-25   (a) Varying the initial displacement of a hard spring. (b) Varying the initial velocity of a hard spring

5.1.6 Aging Springs

In the case of an **aging spring**, the spring constant weakens with time. For springs of this type, we model the physical system with

$$\begin{cases} x'' + cx' + k(t)x = f(t) \\ x(0) = \alpha, \, x'(0) = \beta \end{cases},$$ (5.10)

where $k(t) \to 0$ as $t \to \infty$.

EXAMPLE 5.1.15: Approximate the solution to

$$\begin{cases} x'' + 4e^{-t/4}x = 0 \\ x(0) = \alpha, \, x'(0) = \beta \end{cases},$$

for various values of α and β in the initial conditions.

SOLUTION: First, we state the differential equation in eq and then we define s[α, β] to solve the initial value problem for given values of α and β. Using the numbers in vals4 as the initial displacement and using 0 as the initial velocity, we approximate the solution to five initial value problems in sols6. We graph these numerical solutions in

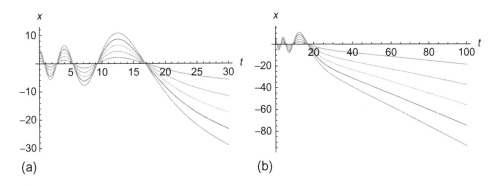

Figure 5-26 (a) Varying the initial displacement of an aging spring. (b) An aging spring eventually stops working

Figure 5-26(a). Notice that the period of the oscillations increases over time due to the diminishing value of the spring constant.

```
Clear[eq]

eq = x''[t] + 4 Exp[-t/4] x[t] == 0;

Clear[s]

s[α_, β_] := NDSolve[{eq, x[0] == α, x'[0] == β}, x[t], {t, 0, 30}];

vals4 = {1, 2, 3, 4, 5};

sols6 = Table[s[vals4[[i]], 0][[1, 1, 2]], {i, 1, 5}];

seven = Plot[sols6, {t, 0, 30},

    AxesLabel → {t, x}, PlotLabel → "(a)"]

Clear[s2]

s2[α_, β_] := NDSolve[{eq, x[0] == α, x'[0] == β}, x[t], {t, 0, 100}];

vals4 = {1, 2, 3, 4, 5};

sols7 = Table[s2[vals4[[i]], 0][[1, 1, 2]], {i, 1, 5}];

eight = Plot[sols7, {t, 0, 100},

    AxesLabel → {t, x}, PlotLabel → "(b)"]

Show[GraphicsRow[{seven, eight}]]
```

Choosing a longer time interval in the NDSolve command as we do in s2[α, β], we see that eventually the motion is not oscillatory. See Figure 5-26(b).

■

Application: Hearing Beats and Resonance
In order to *hear* beats and resonance, we solve the initial-value problem

$$\begin{cases} x'' + \omega^2 x = F \cos \beta t \\ x(0) = \alpha, \ x'(0) = \beta \end{cases}, \qquad (5.11)$$

for each of the following parameter values: (a) $\omega^2 = 6000^2$, $\beta = 5991.62$, $F = 2$; and (a) $\omega^2 = 6000^2$, $\beta = 6000$, $F = 2$.

First, we define the function sol which given the parameters, solves the initial-value problem (5.11).

```
Clear[x, t, f, sol]

sol[ω_, β_, f_]:=
  DSolve[{x"[t]+ω^2x[t] == fCos[βt], x[0] == 0,
    x'[0] == 0}, x[t], t][[1, 1, 2]]
```

Thus, our solution for (a) is obtained by entering

```
a = sol[6000, 5991.62, 2]
```

```
0.0000198886(−1.0007Cos[6000.t]+1.Cos[8.38t]Cos[6000.t]+

0.000698821Cos[6000.t]Cos[11991.6t]+1.Sin[8.38t]Sin[6000.t]+

0.000698821Sin[6000.t]Sin[11991.6t])
```

To *hear* the function we use Play in the same way that we use Plot to *see* functions. The values of a correspond to the amplitude of the sound as a function of time. See Figure 5-27.

```
Play[a, {t, 0, 6}]
```

Similarly, the solution for (b) is obtained by entering

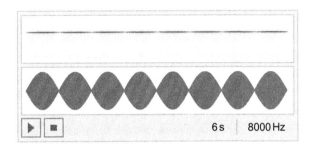

Figure 5-27   Hearing and seeing beats

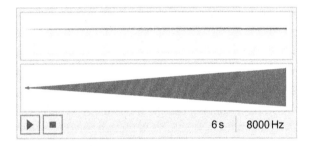

Figure 5-28   Hearing and seeing resonance

$b = \mathtt{sol}[6000, 6000, 2]$

$$\frac{-\mathrm{Cos}[6000t] + \mathrm{Cos}[6000t]\mathrm{Cos}[12000t] + 12000t\mathrm{Sin}[6000t] + \mathrm{Sin}[6000t]\mathrm{Sin}[12000t]}{72000000}$$

We hear resonance with `Play`. See Figure 5-28.

`Play[b, {t, 0, 6}]`

5.2 The Pendulum Problem

Suppose that a mass m is attached to the end of a rod of length L, the weight of which is negligible.

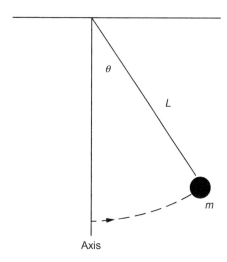

We want to determine the equation that describes the motion of the mass in terms of the displacement $\theta(t)$ which is measured counterclockwise in radians from the axis shown above. This is possible if we are given an initial displacement and an initial velocity of the mass. A force diagram for this situation is shown as follows.

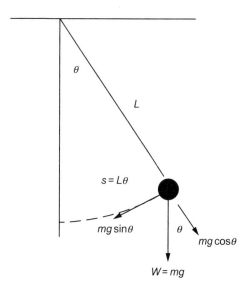

Notice that the forces are determined with trigonometry using the diagram. Here, $\cos\theta = mg/x$ and $\sin\theta = mg/y$, so we obtain the forces $x = mg\cos\theta$ and $y = mg\sin\theta$, indicated as follows.

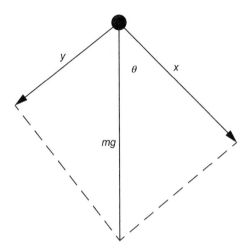

Because the momentum of the mass is given by $m\,ds/dt$, the rate of change of the momentum is

$$\frac{d}{dt}\left(m\frac{ds}{dt}\right) = m\frac{d^2s}{dt^2},$$

where s represents the length of the arc formed by the motion of the mass. Then, because the force $y = mg\sin\theta$ acts in the opposite direction of the motion of the mass, we have the equation

$$m\frac{d^2s}{dt^2} = -mg\sin\theta. \tag{5.12}$$

Using the relationship from geometry between the length of the arc, the length of the rod, and the angle θ, $s = L\theta$, we have the relationship

$$\frac{d^2x}{dt^2} = \frac{d^2}{dt^2}(L\theta) = L\frac{d^2\theta}{dt^2}.$$

Hence, the displacement $\theta(t)$ satisfies $mL\,d^2\theta/dt^2 = -mg\sin\theta$ or

$$mL\frac{d^2\theta}{dt^2} + mg\sin\theta = 0, \tag{5.13}$$

which is a nonlinear equation. However, because we are only concerned with small displacements, we note from the Maclaurin series for $\sin\theta$, $\sin\theta = \theta - \frac{1}{3!}\theta^3 + \frac{1}{5!}\theta^5 - \cdots$, that for small values of θ, $\sin\theta \approx \theta$. Therefore, with this approximation, we obtain the linear equation

$$mL\frac{d^2\theta}{dt^2} + mg\theta = 0 \quad \text{or} \quad \frac{d^2\theta}{dt^2} + \frac{g}{L}\theta = 0, \tag{5.14}$$

which approximates the original equation (5.13). If the initial displacement is given by $\theta(0) = \theta_0$ and the initial velocity is given by $\theta'(0) = v_0$, then we have the initial-value problem

$$\begin{cases} \dfrac{d^2\theta}{dt^2} + \dfrac{g}{L}\theta = 0 \\ \theta(0) = \theta_0, \ \dfrac{d\theta}{dt}(0) = v_0 \end{cases} \tag{5.15}$$

to find the displacement function $\theta(t)$.

Suppose that $\omega^2 = g/L$ so that the differential equation becomes $\theta'' + \omega^2\theta = 0$, which has general solution

$$\theta(t) = c_1 \cos \omega t + c_2 \sin \omega t.$$

Application of the initial conditions $\theta(0) = \theta_0$ and $\theta'(0) = v_0$ shows us that

$$\theta(t) = \theta_0 \cos \omega t + \frac{v_0}{\omega} \sin \omega t \tag{5.16}$$

is the solution of equation (5.15), where $\omega = \sqrt{g/L}$. We can write this function solely in terms of a cosine function that includes a phase shift with

$$\theta(t) = \sqrt{\theta_0^2 + \frac{v_0^2}{\omega^2}} \cos(\omega t - \phi), \tag{5.17}$$

where

$$\phi = \cos^{-1}\left(\frac{\theta_0}{\sqrt{\theta_0^2 + \frac{v_0^2}{\omega^2}}}\right) \quad \text{and} \quad \omega = \sqrt{\frac{g}{L}}.$$

Note that the approximate period of $\theta(t)$ is $T = 2\frac{\pi}{\omega} = 2\pi\sqrt{\frac{L}{g}}$.

EXAMPLE 5.2.1: Determine the displacement of a pendulum of length $L = 32$ feet if $\theta(0) = 0$ and $\theta'(0) = 1/2$ using both the linear and nonlinear models. What is the period? If the pendulum is part of a clock that ticks once for each time the pendulum makes a complete swing, how many ticks does the clock make in one minute?

SOLUTION: The linear initial-value problem that models this situation is

$$\begin{cases} \theta'' + \theta = 0 \\ \theta(0) = 0,\ \theta'(0) = 1/2 \end{cases}$$

because $g/L = 32/32 = 1$. We use DSolve to find a general solution of the equation

    gensol = DSolve[θ''[t] + θ[t] == 0, θ[t], t]

    {{θ[t] → C[1]Cos[t] + C[2]Sin[t]}}

and the solution to the initial-value problem

$$\begin{cases} \theta'' + \theta = 0 \\ \theta(0) = a,\ \theta'(0) = b \end{cases}.$$

    eq = DSolve[{θ''[t] + θ[t] == 0, θ[0] == a, θ'[0] == b},

      θ[t], t]

    {{θ[t] → aCos[t] + bSin[t]}}

In this case, we have that $a = 0$ and $b = 1/2$ so substituting these values into eq[[1,1,2]] results in the solution to the initial-value problem.

    pen = eq[[1, 1, 2]]/.{a → 0, b → 1/2};

The period of this function is

$$T = 2\pi\sqrt{\frac{L}{g}} = 2\pi\sqrt{\frac{32 \text{ ft}}{32\text{ft/s}^2}} = 2\pi \text{ s.}$$

Therefore, the number ticks made by the clock per minute is calculated with the conversion

$$\frac{1 \text{ rev}}{2\pi \text{ s}} \times \frac{1 \text{ tick}}{1 \text{ rev}} \times \frac{60 \text{ s}}{1 \text{ min}} \approx 9.55 \text{ ticks/min.}$$

Hence, the clock makes approximately 9.55 ticks in one minute. To solve the nonlinear equation, we use NDSolve to generate a numerical solution to the initial-value problem valid for $0 \le t \le 20$.

```
numsol = NDSolve[{x"[t] + Sin[x[t]] == 0,

    x[0] == 0, x'[0] == 1/2}, x[t], {t, 0, 20}];
```

We then graph this solution on the interval $[0, 20]$ in Figure 5-29(a).

```
plot1 = Plot[x[t]/.numsol, {t, 0, 20},

    PlotRange → All, AxesLabel → {t, x},

    PlotStyle->CMYKColor[0, 0.89, 0.94, 0.28],

    PlotLabel → "(a)"]
```

The solution pen is also graphed on the interval $[0, 20]$, the resulting graph is named plot2, and then plot1 and plot2 are displayed together with Show in Figure 5-29(b).

```
plot2a = Plot[pen, {t, 0, 20},

    PlotStyle → CMYKColor[0.64, 0, 0.95, 0.40],

    PlotRange → All, AxesLabel → {t, x}];

plot2 = Show[plot1, plot2a, PlotLabel → "(b)"]
```

The graphs in Figure 5-29(b) indicate that the error between the two functions increases as t increases, which is confirmed by graphing the absolute value of the difference of the two functions shown in Figure 5-29(c).

```
plot3 = Plot[Abs[pen − x[t]]/.numsol, {t, 0, 20},

    PlotStyle → CMYKColor[0.62, 0.57, 0.23, 0],
```

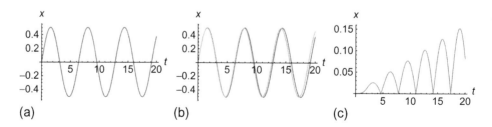

(a)　　　　　　　　　　(b)　　　　　　　　　　(c)

Figure 5-29　(a) Plot of the solution to the nonlinear initial-value problem. (b) Solution of the linear and nonlinear initial-value problems shown together. (c) The linear approximation approximates the nonlinear solution very well until t becomes large

```
            PlotRange → All, AxesLabel → {t, x},

            PlotLabel → "(c)"]

        Show[GraphicsRow[{plot1, plot2, plot3}]]
```

■

Suppose that the pendulum undergoes a damping force that is proportional to the instantaneous velocity. Then, the force due to damping is given as $F_R = b\, d\theta/dt$. Incorporating this force into the sum of the forces acting on the pendulum, we have the nonlinear equation $L\theta'' + b\theta' + g\sin\theta = 0$. Again, using the approximation $\sin\theta \approx \theta$ for small values of t, we obtain the linear equation $L\theta'' + b\theta' + g\theta = 0$ which approximates the situation. Thus, we solve the initial-value problem

$$\begin{cases} L\dfrac{d^2\theta}{dt^2} + b\dfrac{d\theta}{dt} + g\theta = 0 \\[2mm] \theta(0) = \theta_0, \quad \dfrac{d\theta}{dt}(0) = v_0 \end{cases} \tag{5.18}$$

to find the displacement function $\theta(t)$.

EXAMPLE 5.2.2: A pendulum of length $L = 8/5\,\text{ft}$ is subjected to the resistive force $F_R = 32/5\, d\theta/dt$ due to damping. Determine the displacement function if $\theta(0) = 1$ and $\theta'(0) = 2$.

SOLUTION: The initial-value problem that models this situation is

$$\begin{cases} \dfrac{8}{5}\dfrac{d^2\theta}{dt^2} + \dfrac{32}{5}\dfrac{d\theta}{dt} + 32\theta = 0 \\[2mm] \theta(0) = 0, \quad \dfrac{d\theta}{dt}(0) = 2 \end{cases} .$$

Simplifying the differential equation, we obtain $\theta'' + 4\theta' + 20\theta = 0$, and then using DSolve, we find the solution to the initial-value problem,

```
        Clear[θ]

        sol = DSolve[{θ''[t] + 4θ'[t] + 20θ[t]==0,

            θ[0]==1, θ'[0]==2}, θ[t], t]

        {{θ[t] → e⁻²ᵗ(Cos[4t] + Sin[4t])}}
```

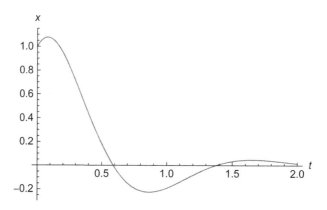

Figure 5-30   A solution to the damped pendulum equation

which is then graphed with `Plot` in Figure 5-30.

$$\theta[t\_] = E^{-2t}(\text{Cos}[4\,t] + \text{Sin}[4\,t]);$$

```
Plot[θ[t], {t, 0, 2},

    PlotRange → All, AxesLabel → {t, x},

    PlotStyle->CMYKColor[0, 0.89, 0.94, 0.28]]
```

Notice that the damping causes the displacement of the pendulum to decrease over time.

To see the pendulum move, we define the procedure pen. Given t, len, and opts, where opts are any options of the Show command, pen[t,len,opts] declares the variable pt1 to be local to the procedure pen, defines pt1 to be the point

$$\left(\text{len} \cos \left(\frac{3}{2}\pi t + \theta(t) \right), \text{len} \sin \left(\frac{3}{2}\pi t + \theta(t) \right) \right),$$

and connects the points pt1 and $(0,0)$ with a line segment. Note that PointSize is used so that pt1 is slightly enlarged in the resulting graphics object. The resulting graphics object *looks* like the pendulum of length $L = $ len at time t.

```
Clear[pen]

pen[t_, len_, opts___]:=Module[{pt1},
```
$$pt1 = \{\text{lenCos}[\frac{3\pi}{2} + \theta[t]], \text{lenSin}[\frac{3\pi}{2} + \theta[t]]\};$$

```
Show[Graphics[{CMYKColor[0,0.89,0.94,0.28],

Line[{{0,0},pt1}],CMYKColor[0.62,0.57,0.23,0],

PointSize[.05],Point[pt1]}],Axes → Automatic,

Ticks → None,AxesStyle → GrayLevel[.5],

PlotRange → {{-2,2},{-2,0}},opts]]
```

For example, entering

$$pen[1,\frac{8}{5}];$$

produces a graphics object corresponding to a pendulum of length $L = 1$ at time $t = 1$. The resulting graphics object is not displayed because a semi-colon is included at the end of the pen command. On the other hand,

$$pen[1,\frac{8}{5}]$$

produces and displays a graphics object corresponding to a pendulum of length $L = 8/5$ at time $t = 1$ as shown in Figure 5-31.

To see the pendulum at various times, we use Table and pen to generate a table consisting of graphics corresponding to a pendulum of length $L = 8/5$ at time t from $t = 0$ to $t = \pi/2$ using increments of $2/15$. The resulting list of sixteen graphics object is then partitioned into four element subsets with Partition and the array of graphics objects toshow is displayed with Show and GraphicsGrid in Figure 5-32.

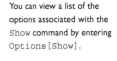

You can view a list of the options associated with the Show command by entering Options [Show].

$$somegraphs = Table[pen[t,\frac{8}{5}],\{t,0,2,\frac{2}{15}\}];$$

```
toshow = Partition[somegraphs,4];

Show[GraphicsGrid[toshow]]
```

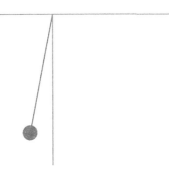

Figure 5-31   A pendulum

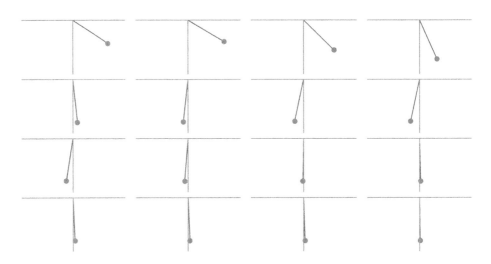

Figure 5-32   A damped pendulum comes to rest

On the other hand, to see the pendulum move, we can use a Animate to generate several graphs and then animate the result. We show one frame of the resulting animation.

$$\texttt{Animate[pen[}t, \frac{8}{5}\texttt{], \{}t, 0, 2, \frac{2}{119}\texttt{\}]}$$

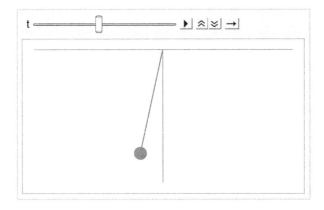

Notice that from our approximate solution, the displacement of the pendulum becomes very close to zero near $t = 2$, which was our observation from the graph of $\theta(t) = e^{-2t}(\cos 4t + 2\sin 4t)$ in Figure 5-30.

∎

Our last example investigates the properties of the nonlinear differential equation.

EXAMPLE 5.2.3: Graph the solution to the initial-value problem

$$\begin{cases} \dfrac{d^2\theta}{dt^2} + 0.5\dfrac{d\theta}{dt} + \theta = 0 \\ \theta(0) = \theta_0, \ \dfrac{d\theta}{dt}(0) = v_0 \end{cases} \tag{5.19}$$

subject to the following initial conditions.

θ_0	v_0	θ_0	v_0	θ_0	v_0	θ_0	v_0
-1	0	-0.5	0	0.5	0	1	0
0	-2	0	-1	0	1	0	2
1	1	1	-1	-1	1	-1	-1
1	2	1	3	-1	4	-1	5
-1	2	-1	3	1	-4	1	-5

SOLUTION: We begin by defining eq to be $\theta'' + 0.5\theta' + \sin\theta = 0$.

```
Clear[eq, t, θ, s]

eq = θ''[t] + 0.5θ'[t] + Sin[θ[t]]==0;
```

To avoid retyping the same commands, we define the procedure s. Given an ordered pair (θ_0, v_0) and any options opts of the Show command, s[{theta0,v0},opts] first declares the variables numsol and numgraph local to the procedure s, uses NDSolve to define numsol to be a numerical solution of the initial-value problem (5.19) valid for $0 \le t \le 15$, generates, but does not display, a graph of the resulting numerical solution on the interval $[0, 15]$, and then displays the result with Show using any options opts passed through the s command.

```
s[{theta0_, v0_}, opts___]:=Module[{numsol},

    numsol = NDSolve[{eq, θ[0]==theta0, θ'[0]==v0},

    θ[t], {t, 0, 15}];

    numgraph = Plot[θ[t]/.numsol, {t, 0, 15},
```

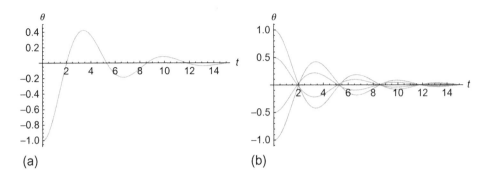

Figure 5-33   (a) Plot of the solution of equation (5.19) that satisfies $\theta_0 = -1$ and $v_0 = 0$. (b) Varying the initial displacement in the pendulum equation

```
            PlotRange → All, opts];

        Show[numgraph, AxesLabel → {t, θ}, opts]]
```

Thus, we see that entering

```
    s[{-1, 0}];
```

does not display the graph of the solution to equation (5.19) if $\theta_0 = -1$ and $v_0 = 0$ but entering

```
    q1 = s[{-1, 0}, PlotRange → All,

        PlotLabel → "(a)"]
```

displays the graph of the solution shown in Figure 5-33(a). Thus, to graph the solutions that satisfy the initial conditions

θ_0	v_0	θ_0	v_0	θ_0	v_0	θ_0	v_0
-1	0	-0.5	0	0.5	0	1	0

we first define t1 to be the initial conditions, use Map to apply s to t1, and then use Show to display the resulting graphs in Figure 5-33(b).

```
    t1 = {{-1, 0}, {-0.5, 0}, {0.5, 0}, {1, 0}};

    toshow1 = Map[s, t1];

    q2 = Show[toshow1, PlotLabel → "(b)"]


Show[GraphicsRow[{q1, q2}]]
```

Similarly, entering

```
t2 = {{0, -2}, {0, -1}, {0, 1}, {0, 2}};

toshow2 = s/@t2;

p1 = Show[toshow2, PlotLabel → "(a)"]
```

defines t2 to be the list of ordered pairs corresponding to the initial conditions

θ_0	v_0	θ_0	v_0	θ_0	v_0	θ_0	v_0
0	-2	0	-1	0	1	0	2

toshow2 to be the resulting list of graphics objects obtained by applying s to each ordered pair in t2, and then displays the list of graphics toshow2 together in Figure 5-34. The solutions that satisfy the remaining initial conditions

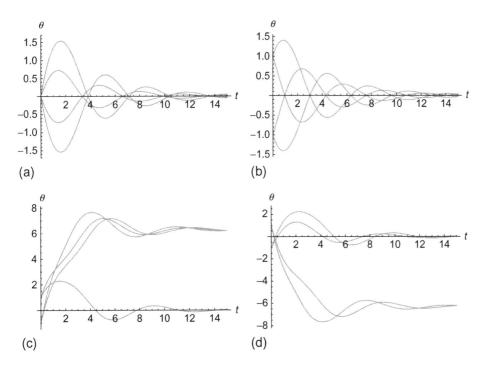

(a)

(b)

(c)

(d)

Figure 5-34　Varying the initial velocities and displacements in the pendulum equation

θ_0	v_0	θ_0	v_0	θ_0	v_0	θ_0	v_0
1	1	1	-1	-1	1	-1	-1
1	2	1	3	-1	4	-1	5
-1	2	-1	3	1	-4	1	-5

are graphed in the same manner in Figure 5-34(b)–(d).

```
t3 = {{1, 1}, {1, −1}, {−1, 1}, {−1, −1}};

toshow3 = s/@t3;

p2 = Show[toshow3, PlotLabel → "(b)"]


t4 = {{1, 2}, {1, 3}, {−1, 4}, {−1, 5}};

toshow4 = s/@t4;

p3 = Show[toshow4, PlotLabel → "(c)"]


t5 = {{−1, 2}, {−1, 3}, {1, −4}, {1, −5}};

toshow5 = s/@t5;

p4 = Show[toshow5, PlotLabel → "(d)"]


Show[GraphicsGrid[{{p1, p2}, {p3, p4}}]]
```

■

5.3 Other Applications

5.3.1 *L-R-C* Circuits

Second-order nonhomogeneous linear ordinary differential equations arise in the study of electrical circuits after the application of *Kirchhoff's law*. Suppose that $I(t)$ is the current in the *L-R-C* series electrical circuit where L, R, and C represent the inductance, resistance, and capacitance of the circuit, respectively.

The voltage drops across the circuit elements shown in the following table have been obtained from experimental data where Q is the charge of the capacitor and $dQ/dt = I$.

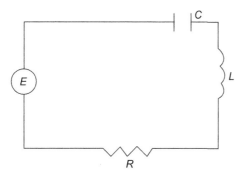

Circuit Element	Voltage Drop
Inductor	$L\dfrac{dI}{dt}$
Resistor	RI
Capacitor	$\dfrac{1}{C}Q$
Voltage Source	$-V(t)$

Our goal is to model this physical situation with an initial-value problem so that we can determine the current and charge in the circuit. For convenience, the terminology used in this section is summarized in the following table.

Electrical Quantities	Units
Inductance (L)	Henrys (H)
Resistance (R)	Ohms (Ω)
Capacitance (C)	Farads (F)
Charge (Q)	Coulombs (C)
Current (I)	Amperes (A)

The physical principle needed to derive the differential equation that models the *L-R-C* series circuit is stated as follows.

> **Kirchhoff's Law:** The sum of the voltage drops across the circuit elements is equivalent to the voltage $E(t)$ impressed on the circuit.

Applying Kirchhoff's law, therefore, yields the differential equation

$$L\frac{dI}{dt} + RI + \frac{1}{C}Q = E(t).$$

Using the fact that $dQ/dt = I$, we also have $d^2Q/dt^2 = dI/dt$. Therefore, the equation becomes

$$L\frac{d^2Q}{dt^2} + R\frac{dQ}{dt} + \frac{1}{C}Q = E(t),$$

which can be solved by the Method of Undetermined Coefficients or the Method of Variation of Parameters. Hence, if the initial charge and current are $Q(0) = Q_0$ and $I(0) = Q'(0) = I_0$, then we must solve the initial-value problem

$$\begin{cases} L\dfrac{d^2Q}{dt^2} + R\dfrac{dQ}{dt} + \dfrac{1}{C}Q = E(t) \\ Q(0) = Q_0,\ I(0) = \dfrac{dQ}{dt}(0) = I_0 \end{cases} \tag{5.20}$$

for the charge $Q(t)$. This solution can then be differentiated to find the current $I(t)$.

EXAMPLE 5.3.1: Consider the *L-R-C* circuit with $L = 1$ henry, $R = 40$ ohms, $C = 4000$ farads, and $E(t) = 24$ volts. Determine the current in this circuit if there is zero initial current and zero initial charge.

SOLUTION: Using the indicated values, the initial-value problem that we must solve is

$$\begin{cases} Q'' + 40Q' + 4000Q = 24 \\ Q(0) = 0,\ I(0) = Q'(0) = 0 \end{cases}.$$

Note that we use lower-case letters to avoid any possible ambiguity with built-in Mathematica functions.

DSolve is used to obtain the solution to the nonhomogeneous problem in cir1.

```
Clear[q]

cir1 = DSolve[{q''[t] + 40q'[t] + 4000q[t]==24,
    q[0]==0, q'[0]==0}, q[t], t]
```

$$\left\{\left\{q[t] \to \frac{1}{500}e^{-20t}(3e^{20t} - 3\text{Cos}[60t] - \text{Sin}[60t])\right\}\right\}$$

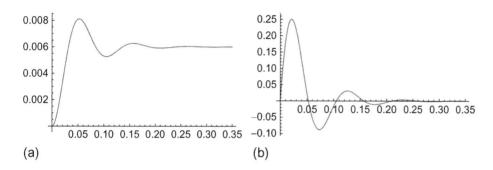

Figure 5-35   (a) Plot of the charge. (b) Plot of the current

These results indicate that in time the charge approaches the constant value of 3/500, which is known as the **steady-state charge**. Also, due to the exponential term, the current approaches zero as t increases. This limit is indicated by the graph of $Q(t)$ in Figure 5-35(a), as well.

```
q[t_] = cir1[[1, 1, 2]];

p1 = Plot[q[t], {t, 0, 0.35}, PlotRange → All,

   PlotStyle->CMYKColor[0, 0.89, 0.94, 0.28],

   PlotLabel → "(a)"]
```

The current, $I(t)$, is obtained by differentiating the charge, $Q(t)$, which is graphed in Figure 5-35(b).

```
q'[t]//Simplify
```

$$\frac{2}{5}e^{-20t}\mathrm{Sin}[60t]$$

```
p2 = Plot[q'[t], {t, 0, 0.35}, PlotRange → All,

   PlotStyle->CMYKColor[0, 0.89, 0.94, 0.28],

   PlotLabel → "(a)"]

Show[GraphicsRow[{p1, p2}]]
```

∎

5.3.2 Deflection of a Beam

An important mechanical model involves the deflection of a long beam that is supported at one or both ends as shown in the following figure.

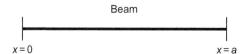

Assuming that in its undeflected form the beam is horizontal, then the deflection of the beam can be expressed as a function of x.

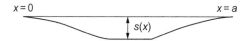

Suppose that the shape of the beam when it is deflected is given by the graph of the function $y(x) = -s(x)$, where x is the distance from one end of the beam and s the measurement of the vertical deflection from the equilibrium position. The boundary value problem that models this situation is derived as follows.

Let $m(x)$ equal the turning moment of the force relative to the point x and $w(x)$ represent the weight distribution of the beam. These two functions are related by the equation

$$\frac{d^2m}{dx^2} = w(x). \tag{5.21}$$

Also, the turning moment is proportional to the curvature of the beam. Hence,

$$m(x) = \frac{EI}{\left[\sqrt{1 + \left(\dfrac{ds}{dx}\right)^2}\,\right]^3} \frac{d^2s}{dx^2}, \tag{5.22}$$

where E and I are constants related to the composition of the beam and the shape and size of a cross-section of the beam, respectively. Notice that this equation is, unfortunately, nonlinear. However, this difficulty is overcome with an approximation. For small values of ds/dx, the denominator of the right-hand side of equation (5.22) can be approximated by the constant 1. With this assumption, equation (5.22) is simplified to

$$m(x) = EI\frac{d^2s}{dx^2}. \tag{5.23}$$

Equation (5.23) is linear and can be differentiated twice to obtain

$$\frac{d^2m}{dx^2} = EI\frac{d^4s}{dx^2}. \tag{5.24}$$

Equation (5.24) can then be used with equation (5.21) relating $m(x)$ and $w(x)$ to obtain the single fourth-order linear nonhomogeneous differential equation

$$EI\frac{d^4s}{dx^4} = w(x). \tag{5.25}$$

Boundary conditions (rather than *initial conditions* for this problem may vary. In most cases, two conditions are given for each end of the beam. Some of these conditions which are specified in pairs. For example, at $x = a$ these include: $s(a) = 0$, $s'(a) = 0$ (fixed end); $s''(a) = 0$, $s'''(a) = 0$ (free end); $s(a) = 0$, $s''(a) = 0$ (simple support); and $s'(a) = 0$, $s'''(a) = 0$ (sliding clamped end).

The following example investigates the effects that a constant weight distribution function $w(x)$ has on the solution to these boundary value problems.

EXAMPLE 5.3.2: Solve the beam equation over the interval $0 \le x \le 1$ if $E = I = 1$, $w(x) = 48$, and the following boundary conditions are used: $s(0) = 0$, $s'(0) = 0$ (fixed end at $x = 0$); and

(a) $s(1) = 0$, $s''(1) = 0$ (simple support at $x = 1$);
(b) $s''(1) = 0$, $s'''(1) = 0$ (free end at $x = 1$);
(c) $s'(1) = 0$, $s'''(1) = 0$ (sliding clamped end at $x = 1$); and
(d) $s(1) = 0$, $s'(1) = 0$ (fixed end at $x = 1$).

SOLUTION: DSolve is used to obtain the solution to this nonhomogeneous problem. In del, the solution that depends on E, I, and w is given

Note that we use (lower case) e to represent E to avoid conflict with the built-in constant E and (lower case) i to represent I to avoid conflict with the built-in constant I.

```
Clear[e, i, w, s]

del = DSolve[{eiD[s[x], {x, 4}]==w, s[0]==0,
    s'[0]==0, s[1]==0, s''[1]==0}, s[x], x]
```

$$\{\{s[x] \to \frac{w(3x^2 - 5x^3 + 2x^4)}{48ei}\}\}$$

We can visualize the shape of the beam by graphing $y = -s(x)$. Thus, we define `toplot1` to be the negative of the solution obtained in `de1`.

```
toplot1 = -de1[[1, 1, 2]]/.

{e → 1, i → 1, w → 48};
```

Similar steps are followed to determine the solution to each of the other three boundary value problems. The corresponding functions to be graphed are named `toplot2`, `toplot3`, and `toplot4`. (Note that $\partial_{\{x,4\}} s[x]$ represents `D[s[x], {x,4}]`, the fourth derivative of $s(x)$.)

```
de2 = DSolve[{ei∂{x,4}s[x]==w, s[0]==0, s'[0]==0,

    s⁽³⁾[1]==0, s''[1]==0}, s[x], x]
```

$$\{\{s[x] \to \frac{w(6x^2 - 4x^3 + x^4)}{24ei}\}\}$$

```
toplot2 = -de2[[1, 1, 2]]/.

{e → 1, i → 1, w → 48};

de3 = DSolve[{ei∂{x,4}s[x]==w, s[0]==0, s'[0]==0,

    s⁽³⁾[1]==0, s'[1]==0}, s[x], x]
```

$$\{\{s[x] \to \frac{4wx^2 - 4wx^3 + wx^4}{24ei}\}\}$$

```
toplot3 = -de3[[1, 1, 2]]/.{e → 1, i → 1, w → 48};

de4 = DSolve[{ei∂{x,4}s[x]==w, s[0]==0,

    s'[0]==0, s[1]==0, s'[1]==0}, s[x], x]
```

$$\{\{s[x] \to \frac{wx^2 - 2wx^3 + wx^4}{24ei}\}\}$$

```
toplot4 = -de4[[1, 1, 2]]/.

{e → 1, i → 1, w → 48};
```

In order to compare the effects that the varying boundary conditions have on the resulting solution, all four functions are graphed together with `Plot` on the interval $[0, 1]$ in Figure 5-36.

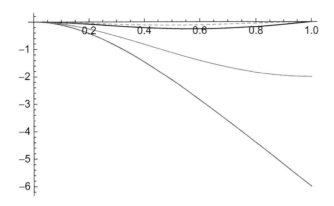

Figure 5-36   Solutions to the beam equation

```
Plot[{toplot1, toplot2, toplot3, toplot4}, {x, 0, 1},
    PlotStyle → {GrayLevel[0],
    GrayLevel[.2], GrayLevel[.4], Dashing[{0.01}]}]]
```

■

5.3.3 Bodé Plots

Consider the differential equation

$$\frac{d^2x}{dt^2} + 2c\frac{dx}{dt} + k^2x = F_0 \sin \omega t, \qquad (5.26)$$

where c and k are positive constants with $c < k$ so that the equation $x'' + 2cx' + k^2x = 0$ is underdamped. To find a particular solution, we can consider the complex exponential form of the forcing function, $F_0 e^{i\omega t}$, which has imaginary part $F_0 \sin \omega t$. Assuming a solution of the form $z_p(t) = Ae^{i\omega t}$, substitution into the differential equation yields $A\left(-\omega^2 + 2ic\omega + k^2\right) = F_0 e^{i\omega t}$. Because $k^2 - \omega^2 + 2ic\omega = 0$ only when $k = \omega$ and $c = 0$, we find that

$$A = \frac{F_0}{k^2 - \omega^2 + 2ic\omega}$$

or

$$A = \frac{F_0}{k^2 - \omega^2 + 2ic\omega} \cdot \frac{k^2 - \omega^2 - 2ic\omega}{k^2 - \omega^2 - 2ic\omega} = \frac{k^2 - \omega^2 - 2ic\omega}{\left(k^2 - \omega^2\right)^2 + 4c^2\omega^2}F_0 = H(i\omega)F_0.$$

Therefore, a particular solution is $z_p(t) = H(i\omega)F_0 e^{i\omega t}$. Now, we can write $H(i\omega)$ in polar form as $H(i\omega) = M(\omega) = e^{i\phi(\omega)}$, where

$$M(\omega) = \frac{1}{\left(k^2 - \omega^2\right)^2 + 4c^2\omega^2} \quad \text{and} \quad \phi(\omega) = \cot^{-1}\left(\frac{\omega^2 - k^2}{2c\omega}\right), \quad -\pi \le \phi \le 0.$$

A particular solution can then be written as

$$z_p(t) = M(\omega)F_0 e^{i\omega t} e^{i\phi(\omega)} = M(\omega)e^{i(\omega t + \phi(\omega))}$$

with imaginary part $M(\omega)F_0 \sin(\omega t + \phi(\omega))$, so we take the particular solution to be $x_p(t) = M(\omega)F_0 \sin(\omega t + \phi(\omega))$. Comparing the forcing function to x_p, we see that the two functions have the same form but with differing amplitudes and phase shifts. The ratio of the amplitude of the particular solution (or steady-state), $M(\omega)F_0$, to that of the forcing function, F_0, is $M(\omega)$ and is called the **gain**. Also, x_p is shifted in time by $|\phi(\omega)|/\omega$ radians to the right, so $\phi(\omega)$ is called the **phase shift**. When we graph the gain and the phase shift against ω (using a $\log_{10}$ scale on the ω-axis) we obtain the **Bodé plots**. Engineers refer to the value of $20 \log_{10} M(\omega)$ as the gain in **decibels**.

EXAMPLE 5.3.3: Solve the initial-value problem

$$\begin{cases} x'' + 2x' + 4x = \sin 2t \\ x(0) = 1/2, \ x'(0) = 1. \end{cases}$$

(a) Graph the solution simultaneously with the forcing function $f(t) = \sin 2t$. Approximate $M(2)$ and $\phi(2)$ using this graph. (b) Graph the corresponding Bodé plots. Compare the values of $M(2)$ and $\phi(2)$ with those obtained in (a).

SOLUTION: (a) First, we define the nonhomogeneous differential equation in eq. Next, we solve the initial-value problem in sol.

```
Clear[eq]

eq = x''[t] + 2x'[t] + 4x[t]==Sin[2 t];

sol = DSolve[{eq, x[0]==1/2, x'[0]==1}, x[t], t]//Simplify
```

$$\{\{x[t] \to \frac{1}{12}e^{-t}(-3e^t Cos[2t] + 9Cos[\sqrt{3}t] + 7\sqrt{3}Sin[\sqrt{3}t])\}\}$$

We extract the formula for the solution with `sol[[1,1,2]]` and graph it simultaneously with $f(t) = \sin 2t$ using a lighter level of gray for the graph of $f(t) = \sin 2t$ in Figure 5-37(a). We use the **Drawing Tools** *Get Coordinates* button

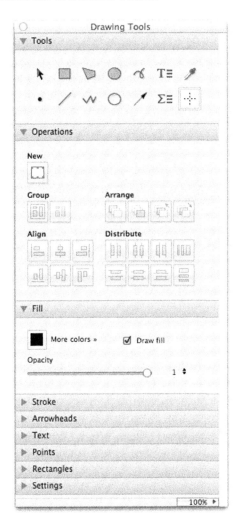

to use the cursor to see that a minimum value of the forcing function occurs near 5.49 and a minimum value of `sol[[1,1,2]]` happens near 6.26. Therefore, the solution is shifted approximately $6.26 - 5.49 = 0.77$ units to the right. Returning to the solution containing $\omega t + \phi(\omega) = 2\left(t + \frac{1}{2}\phi(2)\right)$, we see that $\frac{1}{2}\phi(2) \approx -0.77$, so $\phi(2) \approx -1.54$. Using a similar technique (with the **Command**

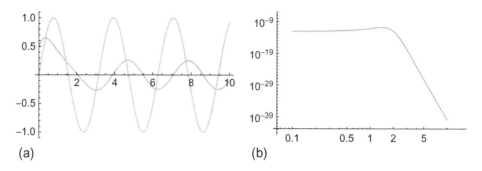

Figure 5-37   (a) Plots of $x(t)$ and $f(t) = \sin 2t$. (b) A log-log plot of $M(\omega)^{20}$

key and cursor), we approximate the amplitude of the steady-state solution (after it dies down) to be 0.255. Therefore from the graph, $M(2) \approx 0.255$.

```
p1 = Plot[{sol[[1, 1, 2]], Sin[2t]}, {t, 0, 10},

PlotLabel → "(a)"]
```

```
6.26 − 5.49
```

```
0.77
```

(b) In the equation $x'' + 2x' + 4x = \sin 2t$, $2c = 2$ and $k^2 = 4$. Therefore, $c = 1$ and $k = 2$. We define the gain function based on these constants in m[w]. Because the graph of $M(\omega)$ is a log-log graph, we load the **Graphics** package to take advantage of the LogLogPlot command. We graph m[w]^20 because engineers are interested in $20 \log_{10} M(\omega) = \log_{10} M(\omega)^{20}$. See Figure 5-37(b). In (a), we obtained $M(2) \approx 0.255$. With the formula for $M(\omega)$, we find that $M(2) = 0.25$.

```
k = 2;
```

```
c = 1;
```

$$m[w\_] := \frac{1}{\sqrt{(k^2 - w^2)^2 + 4c^2 w^2}}$$

```
p2 = LogLogPlot[m[w]^20, {w, 0.1, 10}, .

  PlotLabel → "(b)"]
```

```
Show[GraphicsRow[{p1, p2}]]
```

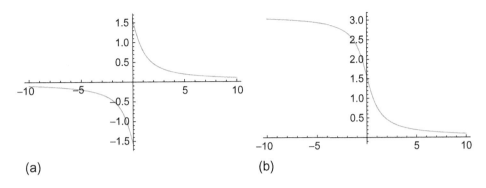

Figure 5-38 (a) Plot of Mathematica's inverse cotangent function. (b) The branch is continuous at $x = 0$

```
N[m[2]]
```

```
0.25
```

The branch of $y = \cot^{-1} x$ used by Mathematica is not continuous at $x = 0$ as seen in Figure 5-38(a).

```
p3 = Plot[ArcCot[x], {x, -10, 10},

    PlotLabel → "(a)"]
```

However, we can construct a function continuous at $x = 0$ as we do in newarccot. See Figure 5-38(b).

```
Clear[newarccot]

newarccot[x_]:=ArcCot[x]/;x ≥ 0

newarccot[x_]:=ArcCot[x] + π/;x < 0

p4 = Plot[newarccot[x], {x, -10, 10},

    PlotLabel → "(b)"]
```

```
Show[GraphicsRow[{p3, p4}]]
```

Using newarccot, we are able to graph $\phi(\omega)$ in Figure 5-39. We define $\phi(\omega)$ so that it returns an angle between $-180°$ and $0°$.

In (a), we found $\phi(2) \approx -1.54$ (radians). Here, we see that $\phi(2) = -90°$. However, $-\pi/2 \approx -1.57$, so the approximations of $M(2)$ and $\phi(2)$ obtained in (a) are quite accurate.

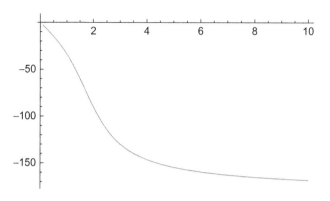

Figure 5-39 Plot of $\phi(\omega)$

```
Clear[φ]
φ[w_]:= (180newarccot[(w^2 - k^2)/(2cw)])/π - 180

Plot[φ[w], {w, 0.1, 10}]

N[φ[2]]
```

$$-90.$$

■

5.3.4 The Catenary

The solution of the second-order nonlinear equation

$$\begin{cases} \dfrac{d^2y}{dx^2} = \dfrac{1}{a}\sqrt{1 + \left(\dfrac{dy}{dx}\right)^2} \\ y(0) = a, \ \dfrac{dy}{dx}(0) = 0 \end{cases}$$

is called a **catenary**.

```
Clear[y, x, c, a, length]
DSolve[{y″[x] == 1/aSqrt[1 + y′[x]^2], y[0] == a,
  y′[0] == 0}, y[x], x]
```

$$\{\{y[x] \to a\text{Cosh}[\tfrac{x}{a}]\}, \{y[x] \to 2a - a\text{Cosh}[\tfrac{x}{a}]\}\}$$

A flexible wire or cable suspended between two poles of the same height takes the shape of the **catenary**,

y = cosh x is defined by
cosh x = ½ (eˣ + e⁻ˣ).

$y = \cosh x$ is defined by
$\cosh x = \frac{1}{2}(e^x + e^{-x})$.

$$y = c + a\cosh\left(\frac{x}{a}\right), \quad a > 0. \tag{5.27}$$

EXAMPLE 5.3.4: A flexible cable with length 150 feet is to be suspended between two poles with height 100 feet. How far apart must the poles be spaced so that the bottom of the cable is 50 feet off the ground?

SOLUTION: Let $2s$ denote the distance the poles must be separated and $f(x, c, a) = c + a\cosh\left(\dfrac{x}{a}\right)$.

```
f[x_, c_, a_] = c + aCosh[x/a]
```

$$c + a\text{Cosh}[\tfrac{x}{a}]$$

At the endpoints, $x = -s$ and $x = s$,

$$f(-s, c, a) = f(s, c, a) = c + a\cosh\left(\frac{s}{a}\right) = 100 \quad \text{or} \quad \cosh^2\left(\frac{s}{a}\right) = \left(\frac{100 - c}{a}\right)^2. \tag{5.28}$$

The minimum of f is attained at $x = 0$ and must be 50:

$$f(0, c, a) = a + c = 50. \tag{5.29}$$

The length of the wire is 150 feet so by the arc length formula

The length, L, of the smooth curve y = f(x) from x = a to x = b is
L = ∫ₐᵇ √(1 + [f'(x)]²) dx.

The **length**, L, of the smooth curve $y = f(x)$ from $x = a$ to $x = b$ is
$L = \int_a^b \sqrt{1 + [f'(x)]^2}\, dx.$

$$\int_{-s}^{s} \sqrt{1 + \left(\frac{df}{dx}\right)^2}\, dx = 2a\sinh\left(\frac{s}{a}\right) = 150 \quad \text{or} \quad \sinh^2\left(\frac{s}{a}\right) = \left(\frac{75}{a}\right)^2. \tag{5.30}$$

```
df = D[f[x, c, a], x]
```

$$\text{Sinh}[\tfrac{x}{a}]$$

```
length = Integrate[Sqrt[1 + df^2],

{x, -s, s}]//PowerExpand
```

$$2\,a\,\text{Sinh}[\frac{s}{a}]$$

```
eq1 = f[s, c, a] == 100
```

```
eq2 = f[0, c, a] == 50
```

```
eq3 = length==150
```

$$c + a\text{Cosh}[\frac{s}{a}] == 100$$

$$a + c == 50$$

$$2\,a\,\text{Sinh}[\frac{s}{a}] == 150$$

Mathematica can solve equations (5.28), (5.29), and (5.30) for s, a, and c as they are written.

```
vals = Solve[{eq1, eq2, eq3}, {s, a, c}]
```

$$\{\{s \to \frac{125}{4}\text{ArcCosh}[\frac{13}{5}],\ a \to \frac{125}{4},\ c \to \frac{75}{4}\}\}$$

```
vals//N
```

$$\{\{s \to 50.2949,\ a \to 31.25,\ c \to 18.75\}\}$$

The system can also be solved by hand if you use the identity $\cosh^2 x - \sinh^2 x = 1$. Subtracting equation (5.30) from equation (5.28) gives us

$$1 = \cosh^2\left(\frac{s}{a}\right) - \sinh^2\left(\frac{s}{a}\right) = \left(\frac{100-c}{a}\right)^2 - \left(\frac{75}{a}\right)^2. \tag{5.31}$$

We use ContourPlot to graph equations (5.29) and (5.31) together in Figure 5-40. The coordinates of the intersection point, (a, c) are the solutions to the system $\{(5.29), (5.31)\}$.

```
p1 = ContourPlot[(4375 - 200c + c^2)/a^2,

    {a, 0.01, 50}, {c, 0, 50}, Contours → {1},

    ContourStyle->CMYKColor[0, 0.89, 0.94, 0.28],

    ContourShading → False];

p2 = ContourPlot[a + c,

    {a, 0.01, 50}, {c, 0, 50}, Contours → {50},
```

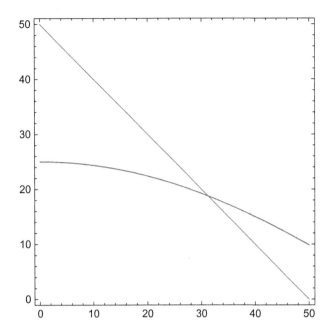

Figure 5-40   Graph of equations (5.29) and (5.31) together

```
ContourStyle->CMYKColor[0.64, 0, 0.95, 0.40],

ContourShading → False];

Show[p1, p2]
```

Solving equations (5.29) and (5.31) for a and c with `Solve` gives us $a = 125/4$ and $c = 75/4$.

```
acvals = Solve[{((100 − c)/a)^2−

    (75/a)^2 == 1, a + c == 50}, {a, c}]
```

$$\{\{a \to \frac{125}{4}, c \to \frac{75}{4}\}\}$$

Substituting these values into equation (5.28) and solving for s gives us $s = \frac{125}{5} \cosh^{-1}(13/5) \approx 50.2949$.

```
eq1b = eq1/.acvals[[1]]
```

$$\frac{75}{4} + \frac{125}{4} \text{Cosh}[\frac{4s}{125}] == 100$$

```
allvals = Solve[eq1b, s]
```

$$\{\{s \to \texttt{ConditionalExpression}[\frac{125}{4}(-\texttt{ArcCosh}[\frac{13}{5}] +$$

$$2 i \pi\, C[1]), C[1] \in \texttt{Integers}]\}, \{s \to \texttt{ConditionalExpression}[\frac{125}{4}$$

$$(\texttt{ArcCosh}[\frac{13}{5}] + 2 i \pi\, C[1]), C[1] \in \texttt{Integers}]\}\}$$

```
allvals/.C[1] → 0//N
```

$$\{\{s \to -50.2949\}, \{s \to 50.2949\}\}$$

Using these values, we visualize the cable and poles in Figure 5-41.

```
p1 = Graphics[{Thickness[0.02],
   Line[{{-50.2949, 0}, {-50.2949, 100}}],
   Line[{{50.2949, 0}, {50.2949, 100}}]}];
p2 = Plot[f[x, 75/4, 125/4], {x, -50.2949,
50.2949}];
Show[p1, p2, Axes → Automatic,
   AxesOrigin → {0, 0}]
```

■

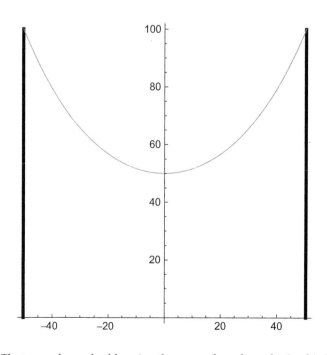

Figure 5-41   The two poles and cable using the s, c, and a values obtained in Example 5.3.4

Using the same notation as Example 5.3.4, if a flexible cable with length 150 feet is suspended between two poles with height 100 feet, the distance between the two poles, $2s$, satisfies $0 < s < 75$. Let h denote the distance from the bottom of the cable to the ground. Then,

$$f(-s, c, a) = f(s, c, a) = 100,$$
$$f(0, c, a) = h, \text{ and} \qquad (5.32)$$
$$\int_{-s}^{s} \sqrt{1 + \left(\frac{df}{dx}\right)^2}\, dx = 2a \sinh\left(\frac{s}{a}\right) = 150.$$

```
eq1 = f[s, c, a] == 100

eq2 = f[0, c, a] == h

eq3 = length == 150
```

$$c + a\text{Cosh}[\frac{s}{a}] == 100$$

$$a + c == h$$

$$2a\text{Sinh}[\frac{s}{a}] == 150$$

We use `Solve` to solve system (5.32) for s, c, and a. Mathematica returns two solutions.

```
posheights = Solve[{eq1, eq2, eq3}, {s, c, a}]
```

$$\left\{\left\{s \to -\frac{(4375 - 200h + h^2)\text{ArcCosh}[\frac{-15625 + 200h - h^2}{4375 - 200h + h^2}]}{2(-100 + h)},\right.\right.$$
$$\left.c \to \frac{-4375 + h^2}{2(-100 + h)}, a \to \frac{4375 - 200h + h^2}{2(-100 + h)}\right\},$$
$$\left\{s \to \frac{(4375 - 200h + h^2)\text{ArcCosh}[\frac{-15625 + 200h - h^2}{4375 - 200h + h^2}]}{2(-100 + h)},\right.$$
$$\left.\left.c \to \frac{-4375 + h^2}{2(-100 + h)}, a \to \frac{4375 - 200h + h^2}{2(-100 + h)}\right\}\right\}$$

We are assuming that a is positive so the meaningful solution is the one for which a is positive. We graph each a, s, and c for each component given in `posheights` in Figure 5-42.

```
p1 = Plot[{a, s, c}/.posheights[[1]],

   {h, 0, 100}, PlotRange → {0, 100}];
```

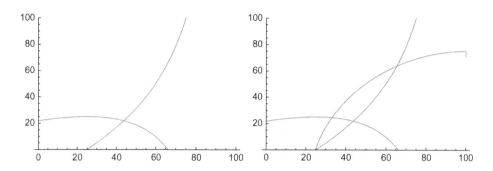

Figure 5-42   Mathematica gives two solutions to system (5.32); the second solution is the meaningful one because $a > 0$

```
p2 = Plot[{a, s, c}/.posheights[[2]],

    {h, 0, 100}, PlotRange → {0, 100}];

Show[GraphicsRow[{p1, p2}]]
```

Using the results of the second solution given by Mathematica, we are able to generate a graphics array illustrating the position of the poles and the cable for various heights, h, in Figure 5-43.

```
Clear[wire]

wire[h_, opts___]:=Module[{a, s, c, p1, p2},
```

$$a = \frac{4375 - 200h + h^2}{2(-100 + h)};$$

$$s = \frac{(4375 - 200h + h^2)\text{ArcCosh}\left[\dfrac{-15625 + 200h - h^2}{4375 - 200h + h^2}\right]}{2(-100 + h)};$$

$$c = \frac{-4375 + h^2}{2(-100 + h)};$$

```
p1 = Graphics[{Thickness[0.02],

    Line[{{-s, 0}, {-s, 100}}],

    Line[{{s, 0}, {s, 100}}]}];

p2 = Plot[c + aCosh[x/a], {x, -s, s}];

Show[p1, p2, Axes → Automatic, AxesOrigin → {0, 0},

    PlotRange → {{-75, 75}, {0, 110}}, opts]

graphs = Table[wire[n], {n, 26, 99, (99 − 26)/15}];

toshow = Partition[graphs, 4];

Show[GraphicsGrid[toshow]]
```

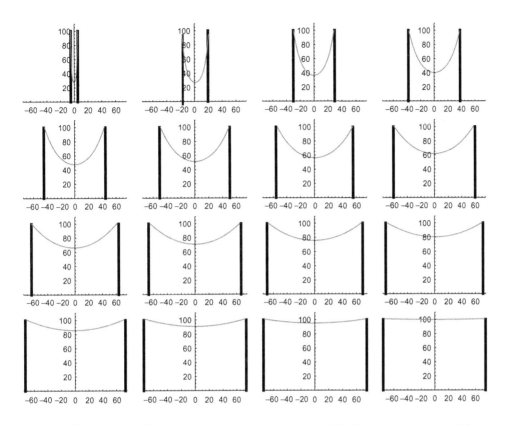

Figure 5-43   An array illustrating how two poles of height 100 feet can be connected by a flexible cable of length 150 feet

EXAMPLE 5.3.5: According to our electric utility, *Excelsior Electric Membership Corp (EMC)*, Metter, Georgia, due to terrain, easements, and so on, the average distance between utility poles ranges from 325 to 340 feet. Each pole is approximately 40 feet long with 6 feet buried so that the length of the pole from the ground to the top of the pole is 34 feet. The *Georgia Department of Transportation* states that the maximum height of a truck using interstates, national, and state routes is 13′ 6″. However, special permits may be granted by the *DOT* for heights up to 18′ 0″. With these restrictions in mind, *EMC* maintains a minimum clearance of 20′ under those lines it installs during cooler months because expansion causes lines to sag during warmer months.

For the obvious reasons, *EMC* prefers that the distance from its lines to the ground is greater than 18′ 6″ at all times.

Find c and a so that $f(x, c, a) = c + a \cosh\left(\dfrac{x}{a}\right)$ models this situation.

SOLUTION: Centering $f(x, c, a) = c + a \cosh\left(\dfrac{x}{a}\right)$ at $x = 0$, we require that

$$f(-170, c, a) = f(170, c, a) = 34, \text{ and}$$
$$f(0, c, a) = 20. \tag{5.33}$$

```
f[x_, c_, a_] = c + aCosh[x/a]
```

$c + a\text{Cosh}[\dfrac{x}{a}]$

```
df = D[f[x, c, a], x]
```

$\text{Sinh}[\dfrac{x}{a}]$

```
length = Integrate[Sqrt[1 + df^2],

{x, -s, s}]//PowerExpand
```

$2\,a\text{Sinh}[\dfrac{s}{a}]$

```
f[170, c, a]
```

$c + a\text{Cosh}[\dfrac{170}{a}]$

```
f[0, c, a]
```

$a + c$

Mathematica cannot solve system (5.33) exactly with `Solve`

```
Solve[{c + aCosh[170/a] == 34,

    a + c == 20}, {a, c}]
```

$\text{Solve}[\{c + a\text{Cosh}[\dfrac{170}{a}] == 34, a + c == 20\}, \{a, c\}]$

but using `ContourPlot` to graph the equations $f(170, c, a) = 34$ and $f(0, c, a) = 20$ together as shown in Figure 5-44 shows us that the system does have a solution.

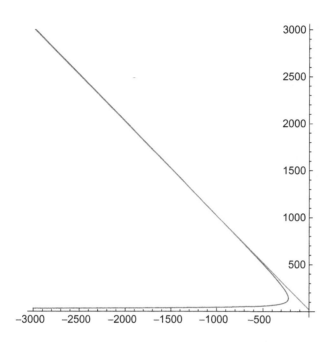

Figure 5-44 Graphs of equations $f(170, c, a) = 40$ and $f(0, c, a) = 20$: the intersection point is difficult to see

```
cp1 = ContourPlot[f[170, c, a],

    {c, -3000, 0}, {a, 1, 3000}, Contours → {34.},

    ContourStyle->CMYKColor[0, 0.89, 0.94, 0.28],

    PlotPoints → 200, AxesOrigin → {0, 0},

    Frame → False, Axes → Automatic,

    ContourShading → False];

cp2 = ContourPlot[a + c,

    {c, -3000, 0}, {a, 1, 3000}, Contours → {20},

    ContourStyle->CMYKColor[0.64, 0, 0.95, 0.40],

    PlotPoints → 200, AxesOrigin → {0, 0},

    Frame → False, Axes → Automatic,

    ContourShading → False];

Show[cp1, cp2]
```

We use `FindRoot` to find the solution.

```
FindRoot[{c + aCosh[170/a] == 34, a + c == 20},
    {a, 1500.}, {c, -1500.}]
```

$$\{a \to 1034.47, c \to -1014.47\}$$

However, using $c = 20 - a$ we graph $f(170, 20 - a, a)$ and $a = 34$ together in Figure 5-45. The solution is much easier to see in Figure 5-45 than in Figure 5-44.

```
Plot[{f[170, 20 - a, a], 34}, {a, 0, 2000},
    PlotRange → {{0, 2000}, {0, 200}},
    AspectRatio → 1]
```

Now, we obtain the same results with `FindRoot` and `Solve` as we did previously with `FindRoot`.

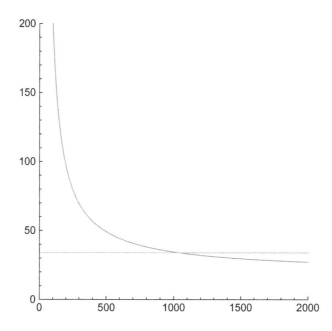

Figure 5-45   Plot of $f(170, 20 - a, a)$ and $a = 34$ together: the solution for a is much easier to see

Figure 5-46 A graphic illustrating how a utility line like an electrical cable may be connected between two poles of equal height

```
aval = FindRoot[f[170, 20 − a, a] == 34.,
{a, 1100.}]

{a → 1034.47}

aval[[1, 2]]

1034.47

cval = 20 − a/.aval[[1]]

−1014.47

length/.{s → 170, aval[[1]]}

341.532

Solve[a + c == 20., c]

{{c → 20. − a}}
```

With the results obtained above, we generate a plot illustrating the hanging wire in Figure 5-46.

```
p1 = Graphics[{Thickness[0.02],
   Line[{{−170, 0}, {−170, 34}}],
   Line[{{170, 0}, {170, 34}}]}];
p2 = Plot[f[x, cval, aval[[1, 2]]], {x, −170, 170}];
Show[p1, p2, Axes → None, AxesOrigin → {0, 0},
   AspectRatio → Automatic]
```

∎

Systems of Ordinary Differential Equations

Because of their importance in the study of systems of linear equations, we now review matrices and the operations associated with them.

6.1 Review of Matrix Algebra and Calculus

6.1.1 Defining Nested Lists, Matrices, and Vectors

In Mathematica, a **matrix** is a list of lists where each list represents a row of the matrix. Therefore, the $m \times n$ matrix

$$\mathbf{A} = \begin{pmatrix} a_{11} & a_{12} & a_{13} & \cdots & a_{1n} \\ a_{21} & a_{22} & a_{23} & \cdots & a_{2n} \\ a_{31} & a_{32} & a_{33} & \cdots & a_{3n} \\ \vdots & \vdots & \vdots & & \vdots \\ a_{m1} & a_{m2} & a_{m3} & \cdots & a_{mn} \end{pmatrix}$$

is entered with

```
A={{a11,a12,...,a1n},{a21,a22,...,a2n},...,{am1,am2,...amn}}.
```

For example, to use Mathematica to define m to be the matrix $\mathbf{A} = \begin{pmatrix} a_{11} & a_{12} \\ a_{21} & a_{22} \end{pmatrix}$ enter the command

```
m={{a11,a12},{a21,a22}}.
```

Differential Equations with Mathematica. http://dx.doi.org/10.1016/B978-0-12-804776-7.00006-1

The command m=Array[a,{2,2}] produces a result equivalent to this. Once a matrix A has been entered, it can be viewed in the traditional row-and-column form using the command MatrixForm[A]. You can quickly construct 2×2 matrices by clicking on the button from the **BasicTypesetting** palette, which is accessed by going to **File** under the Mathematica menu, followed by **Palettes** and then **BasicTypesetting**,

As when using TableForm, the result of using MatrixForm is no longer a list that can be manipulated using Mathematica commands. Use MatrixForm to view a matrix in traditional row-and-column form. Do not attempt to perform matrix operations on a MatrixForm object.

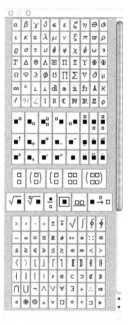

Alternatively, you can construct matrices of any dimension by going to the Mathematica menu under **Input** and selecting **Table/Matrix** followed by **New...**.

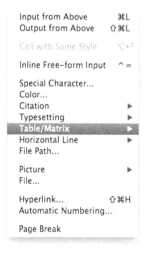

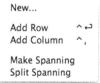

The resulting pop-up window allows you to create tables, matrices, and palettes. To create a matrix, select **Matrix**, enter the number of rows and columns of the matrix, and select any other options. Pressing the **OK** button places the desired matrix at the position of the cursor in the Mathematica notebook.

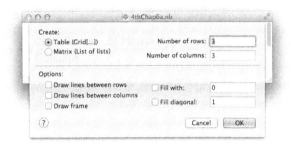

EXAMPLE 6.1.1: Use Mathematica to define the matrices $\begin{pmatrix} a_{11} & a_{12} & a_{13} \\ a_{22} & a_{22} & a_{23} \\ a_{31} & a_{32} & a_{33} \end{pmatrix}$

and $\begin{pmatrix} b_{11} & b_{12} & b_{13} & b_{14} \\ b_{21} & b_{22} & b_{23} & b_{24} \end{pmatrix}$.

SOLUTION: In this case, both `Table[`$a_{i,j}, \{i, 1, 3\}, \{j, 1, 3\}$`]` and `Array[a, {3,3}]` produce equivalent results when we define `matrixa` to be the matrix

$$\begin{pmatrix} a_{11} & a_{12} & a_{13} \\ a_{22} & a_{22} & a_{23} \\ a_{31} & a_{32} & a_{33} \end{pmatrix}.$$

The commands `MatrixForm` or `TableForm` are used to display the results in traditional matrix form.

Note that the results of using `MatrixForm` or `TableForm` *are not* matrices or vectors and cannot be manipulated like them.

```
Clear[a, b, matrixa, matrixb]

matrixa = Table[a[i, j], {i, 1, 3}, {j, 1, 3}]
```

{{a[1, 1], a[1, 2], a[1, 3]}, {a[2, 1], a[2, 2], a[2, 3]}, {a[3, 1], a[3, 2],

a[3, 3]}}

```
MatrixForm[matrixa]
```

$$\begin{pmatrix} a[1, 1] & a[1, 2] & a[1, 3] \\ a[2, 1] & a[2, 2] & a[2, 3] \\ a[3, 1] & a[3, 2] & a[3, 3] \end{pmatrix}$$

matrixa = Array[a, {3, 3}]

{{$a[1, 1]$, $a[1, 2]$, $a[1, 3]$}, {$a[2, 1]$, $a[2, 2]$, $a[2, 3]$}, {$a[3, 1]$, $a[3, 2]$,

$a[3, 3]$}}

MatrixForm[matrixa]

$$\begin{pmatrix} a[1, 1] & a[1, 2] & a[1, 3] \\ a[2, 1] & a[2, 2] & a[2, 3] \\ a[3, 1] & a[3, 2] & a[3, 3] \end{pmatrix}$$

We may also use Mathematica to define nonsquare matrices.

matrixb = Array[b, {2, 4}]

{{$b[1, 1]$, $b[1, 2]$, $b[1, 3]$, $b[1, 4]$}, {$b[2, 1]$, $b[2, 2]$, $b[2, 3]$, $b[2, 4]$}}

MatrixForm[matrixb]

$$\begin{pmatrix} b[1, 1] & b[1, 2] & b[1, 3] & b[1, 4] \\ b[2, 1] & b[2, 2] & b[2, 3] & b[2, 4] \end{pmatrix}$$

Equivalent results would have been obtained by entering
$\text{Table}[b_{i,j}, \{i, 1, 2\}, \{j, 1, 4\}]$.

∎

Note that `IdentityMatrix[n]` returns the $n \times n$ identity matrix $\mathbf{I} =$
$\begin{pmatrix} 1 & 0 & \cdots & 0 \\ 0 & 1 & \cdots & 0 \\ \vdots & \vdots & \vdots & \vdots \\ 0 & 0 & \cdots & 1 \end{pmatrix}$. If you need to generate a matrix with random entries, use
commands like `RandomInteger` or `RandomReal`. For example,

randommatrix1 = RandomInteger[{−5, 5}, {7, 7}];

MatrixForm[randommatrix1]

$$\begin{pmatrix} 3 & -5 & -2 & 0 & 2 & -1 & 1 \\ 5 & -4 & 4 & -3 & 0 & -2 & 0 \\ 5 & 5 & -2 & 5 & 2 & -1 & -5 \\ -1 & -4 & 1 & 4 & -1 & -4 & 3 \\ -5 & -3 & -1 & 0 & 3 & -2 & -5 \\ 3 & 3 & -2 & -3 & 4 & -5 & 2 \\ 2 & -2 & 3 & 4 & 4 & -1 & 4 \end{pmatrix}$$

returns a 7×7 matrix with integer entries between -5 and 5. Similarly,

```
randommatrix2 = RandomReal[{-1, 1}, {5, 5}];
```

```
MatrixForm[randommatrix2]
```

$$\begin{pmatrix} 0.545854 & 0.544879 & -0.0871071 & -0.576673 & 0.38755 \\ 0.872798 & 0.679007 & 0.848618 & 0.0613526 & -0.435581 \\ -0.423375 & -0.664537 & 0.124696 & -0.589354 & 0.531506 \\ -0.564806 & 0.815648 & 0.700856 & -0.299087 & -0.240077 \\ -0.907425 & -0.0737659 & -0.765904 & 0.538299 & 0.226914 \end{pmatrix}$$

returns a 5×5 matrix with entries between -1 and 1.

More generally the commands `Table[f[i,j],{i,imax},{j,jmax}]` and `Array[f,{imax,jmax}]` yield nested lists corresponding to the imax × jmax matrix

$$\begin{pmatrix} f(1,1) & f(1,2) & \cdots & f(1,\text{jmax}) \\ f(2,1) & f(2,2) & \cdots & f(2,\text{jmax}) \\ \vdots & \vdots & \vdots & \vdots \\ f(\text{imax},1) & f(\text{imax},2) & \cdots & f(\text{imax},\text{jmax}) \end{pmatrix}.$$

`Table[f[i,j],{i,imin,imax,istep},{j,jmin,jmax,jstep}]` returns the list of lists

```
{{f[imin,jmin],f[imin,jmin+jstep],...,f[imin,jmax]},
    {f[imin+istep,jmin],...,f[imin+istep,jmax]},
        ...,{f[imax,jmin],...,f[imax,jmax]}}
```

and the command

```
Table[f[i,j,k,...],{i,imin,imax,istep},{j,jmin,jmax,jstep},
  {k,kmin,kmax,kstep},...]
```

calculates a nested list; the list associated with i is outermost. If `istep` is omitted, the stepsize is one.

In Mathematica, a **vector** is a list of numbers and, thus, is entered in the same manner as lists. For example, to use Mathematica to define the row vector `vectorv` to be $(v_1 \quad v_2 \quad v_3)$ enter `vectorv={v1,v2,v3}`. Similarly, to define the column vector `vectorv` to be $\begin{pmatrix} v_1 \\ v_2 \\ v_3 \end{pmatrix}$ enter `vectorv={v1,v2,v3}`

or `vectorv={{v1},{v2},{v3}}`. For a 2×1 vector, you can use the button on the **BasicTypesetting** palette. Generally, with Mathematica you do not need to distinguish between row and column vectors: Mathematica performs computations with vectors and matrices correctly as long as the computations are well-defined.

With Mathematica, you do not need to distinguish between row and column vectors. Provided that computations are well-defined, Mathematica carries them out correctly. Mathematica warns of any ambiguities when they (rarely) occur.

EXAMPLE 6.1.2: Define the vector $\mathbf{w} = \begin{pmatrix} -4 \\ -5 \\ 2 \end{pmatrix}$, `vectorv` to be the vector $(v_1 \quad v_2 \quad v_3 \quad v_4)$ and `zerovec` to be the vector $(0 \quad 0 \quad 0 \quad 0 \quad 0)$.

SOLUTION: To define **w**, we enter

```
w = {-4, -5, 2}
```

{-4, -5, 2}

```
w = {{-4}, {-5}, {2}}
```

{{-4}, {-5}, {2}}

```
MatrixForm[w]
```

$\begin{pmatrix} -4 \\ -5 \\ 2 \end{pmatrix}$

To define `vectorv`, we use `Array`.

```
vectorv = Array[v, 4]
```

$\{v[1], v[2], v[3], v[4]\}$

Equivalent results would have been obtained by entering $\texttt{Table}[v_i, \{i, 1, 4\}]$. To define `zerovec`, we use `Table`.

```
zerovec = Table[0, {5}]
```

$\{0, 0, 0, 0, 0\}$

The same result is obtained by going to **Input** under the Mathematica menu and selecting

$\{0, 0, 0, 0, 0\}$

$\{0, 0, 0, 0, 0\}$

■

6.1.2 Extracting Elements of Matrices

For the 2×2 matrix $\texttt{m} = \{\{a_{1,1}, a_{1,2}\}, \{a_{2,1}, a_{2,2}\}\}$ defined earlier, $\texttt{m[[1]]}$ yields the first element of matrix m which is the list $\{a_{1,1}, a_{1,2}\}$ or the first row of m; $\texttt{m[[2,1]]}$ yields the first element of the second element of matrix m which is $a_{2,1}$. In general, if **m** is an $i \times j$ matrix, $\texttt{m[[i,j]]}$ or $\texttt{Part[m,i,j]}$ returns the unique element in the ith row and jth column of **m**. More specifically, $\texttt{m[[i,j]]}$ yields the jth part of the ith part of **m**; $\texttt{list[[i]]}$ or $\texttt{Part[list,i]}$ yields the ith part of $\texttt{list}$; $\texttt{list[[i,j]]}$ or $\texttt{Part[list,i,j]}$ yields the jth part of the ith part of list, and so on.

EXAMPLE 6.1.3: Define mb to be the matrix $\begin{pmatrix} 10 & -6 & -9 \\ 6 & -5 & -7 \\ -10 & 9 & 12 \end{pmatrix}$. (a)

Extract the third row of mb. (b) Extract the element in the first row and third column of mb. (c) Display mb in traditional matrix form.

SOLUTION: We begin by defining the command mb. $\texttt{mb[[i,j]]}$ yields the (unique) number in the ith row and jth column of mb. Observe how various components of mb (rows and elements) can be extracted and how mb is placed in `MatrixForm`.

mb = {{10, −6, −9}, {6, −5, −7},

{−10, 9, 12}};

MatrixForm[mb]

$$\begin{pmatrix} 10 & -6 & -9 \\ 6 & -5 & -7 \\ -10 & 9 & 12 \end{pmatrix}$$

mb[[3]]

{−10, 9, 12}

mb[[1, 3]]

−9

∎

If m is a matrix, the ith row of m is extracted with m[[i]]. The command Transpose[m] yields the transpose of the matrix m, the matrix obtained by interchanging the rows and columns of m. We extract columns of m by computing Transpose[m] and then using Part to extract rows from the transpose. Namely, if m is a matrix, Transpose[m][[i]] extracts the ith row from the transpose of m which is the same as the ith column of m.

EXAMPLE 6.1.4: Extract the second and third columns from **A** if
$$A = \begin{pmatrix} 0 & -2 & 2 \\ -1 & 1 & -3 \\ 2 & -4 & 1 \end{pmatrix}.$$

SOLUTION: We first define matrixa and then use Transpose to compute the transpose of matrixa, naming the result ta, and then displaying ta in MatrixForm.

matrixa = {{0, −2, 2}, {−1, 1, −3},

{2, −4, 1}};

```
ta = Transpose[matrixa];
```

```
MatrixForm[ta]
```

$$\begin{pmatrix} 0 & -1 & 2 \\ -2 & 1 & -4 \\ 2 & -3 & 1 \end{pmatrix}$$

Next, we extract the second column of `matrixa` using `Transpose` together with `Part` (`[[...]]`). Because we have already defined `ta` to be the transpose of `matrixa`, entering `ta[[2]]` would produce the same result.

```
Transpose[matrixa][[2]]
```

{−2, 1, −4}

To extract the third column, we take advantage of the fact that we have already defined `ta` to be the transpose of `matrixa`. Entering `Transpose[matrixa][[3]]` produces the same result.

```
ta[[3]]
```

{2, −3, 1}

■

6.1.3 Basic Computations With Matrices

Mathematica performs all of the usual operations on matrices. Matrix addition ($\mathbf{A} + \mathbf{B}$), scalar multiplication ($k\mathbf{A}$), matrix multiplication (when defined) ($\mathbf{AB}$), and combinations of these operations are all possible. The **transpose** of $\mathbf{A}$, $\mathbf{A}^t$, is obtained by interchanging the rows and columns of $\mathbf{A}$ and is computed with the command `Transpose[A]`. If $\mathbf{A}$ is a square matrix, the determinant of $\mathbf{A}$ is obtained with `Det[A]`.

If $\mathbf{A}$ and $\mathbf{B}$ are $n \times n$ matrices satisfying $\mathbf{AB} = \mathbf{BA} = \mathbf{I}$, where $\mathbf{I}$ is the $n \times n$ matrix with 1's on the diagonal and 0's elsewhere (the $n \times n$ identity matrix), $\mathbf{B}$ is called the **inverse** of $\mathbf{A}$ and is denoted by $\mathbf{A}^{-1}$. If the inverse of a matrix $\mathbf{A}$ exists, the inverse is found with `Inverse[A]`. Thus, assuming that $\begin{pmatrix} a & b \\ c & d \end{pmatrix}$ has an inverse ($ad - bc \neq 0$), the inverse is

Inverse[{{a, b}, {c, d}}]

$$\{\{\frac{d}{-bc+ad}, -\frac{b}{-bc+ad}\}, \{-\frac{c}{-bc+ad}, \frac{a}{-bc+ad}\}\}$$

EXAMPLE 6.1.5: Let $A = \begin{pmatrix} 3 & -4 & 5 \\ 8 & 0 & -3 \\ 5 & 2 & 1 \end{pmatrix}$ and $B = \begin{pmatrix} 10 & -6 & -9 \\ 6 & -5 & -7 \\ -10 & 9 & 12 \end{pmatrix}$.

Compute (a) $\mathbf{A} + \mathbf{B}$; (b) $\mathbf{B} - 4\mathbf{A}$; (c) the inverse of $\mathbf{AB}$; (d) the transpose of $(\mathbf{A} - 2\mathbf{B})\,\mathbf{B}$; and (e) $\det \mathbf{A} = |\mathbf{A}|$.

SOLUTION: We enter ma (corresponding to **A**) and mb (corresponding to **B**) as nested lists where each element corresponds to a row of the matrix. We suppress the output by ending each command with a semicolon.

$$ma = \{\{3, -4, 5\}, \{8, 0, -3\}, \{5, 2, 1\}\};$$

$$mb = \{\{10, -6, -9\}, \{6, -5, 7\}, \{-10, 9, 12\}\};$$

Entering

ma + mb//MatrixForm

$$\begin{pmatrix} 13 & -10 & -4 \\ 14 & -5 & 4 \\ -5 & 11 & 13 \end{pmatrix}$$

adds matrix ma to mb and expresses the result in traditional matrix form. Entering

mb − 4ma//MatrixForm

$$\begin{pmatrix} -2 & 10 & -29 \\ -26 & -5 & 19 \\ -30 & 1 & 8 \end{pmatrix}$$

subtracts four times matrix ma from mb and expresses the result in traditional matrix form. Entering

```
Inverse[ma.mb]//MatrixForm
```

$$\begin{pmatrix} \dfrac{641}{26220} & -\dfrac{41}{4370} & \dfrac{1567}{26220} \\ \dfrac{1763}{39330} & \dfrac{16}{2185} & \dfrac{2101}{39330} \\ -\dfrac{49}{7866} & -\dfrac{6}{437} & \dfrac{187}{7866} \end{pmatrix}$$

computes the inverse of the matrix product **AB**. Similarly, entering

```
Transpose[(ma − 2mb).mb]//MatrixForm
```

$$\begin{pmatrix} -352 & 190 & 384 \\ 269 & -179 & -277 \\ 485 & -98 & -613 \end{pmatrix}$$

computes the transpose of $(\mathbf{A} - 2\mathbf{B})\,\mathbf{B}$ and entering

```
Det[ma]
```

```
190
```

computes the determinant of ma.

■

EXAMPLE 6.1.6: Compute **AB** and **BA** if $\mathbf{A} = \begin{pmatrix} -1 & -5 & -5 & -4 \\ -3 & 5 & 3 & -2 \\ -4 & 4 & 2 & -3 \end{pmatrix}$

and $\mathbf{B} = \begin{pmatrix} 1 & -2 \\ -4 & 3 \\ 4 & -4 \\ -5 & -3 \end{pmatrix}$.

SOLUTION: Because **A** is a 3×4 matrix and **B** is a 4×2 matrix, **AB** is defined and is a 3×2 matrix. We define matrixa and matrixb with the following commands.

```
matrixa = {{−1, −5, −5, −4}, {−3, 5, 3, −2},

   {−4, 4, 2, −3}};

matrixb = {{1, −2}, {−4, 3}, {4, −4}, {−5, −3}};
```

Matrix products, when defined, are computed by placing a period (.) between the matrices being multiplied. Note that a period is also used to compute the dot product of two vectors, when the dot product is defined.

Remember that you can also define matrices by going to **Input** under the Mathematica menu and selecting **Create Table/Matrix/Palette....** After entering the desired number of rows and columns and pressing the **OK** button, a matrix template is placed at the location of the cursor that you can fill in.

We then compute the product, naming the result ab, and display ab in MatrixForm.

```
ab = matrixa.matrixb;
```

```
MatrixForm[ab]
```

$$\begin{pmatrix} 19 & 19 \\ -1 & 15 \\ 3 & 21 \end{pmatrix}$$

However, the matrix product **BA** is not defined and Mathematica produces error messages (that are not shown here) when we attempt to compute the undefined matrix product.

```
matrixb.matrixa
```

$\{\{1, -2\}, \{-4, 3\}, \{4, -4\}, \{-5, -3\}\}.\{\{-1, -5, -5, -4\},$

$\{-3, 5, 3, -2\}, \{-4, 4, 2, -3\}\}$

∎

6.1.4 Systems of Linear Equations

Mathematica offers numerous options for solving systems of linear equations. Usually, Solve[systemofequations,variables] will quickly solve a linear system. Other times, you may prefer different calculations. For example, the linear system $\mathbf{Ax} = \mathbf{b}$ has a unique solution if $|\mathbf{A}| \neq 0$. In this case $\mathbf{A}$ is invertible and $\mathbf{x} = \mathbf{A}^{-1}\mathbf{b}$. Thus, provided you have defined matrixa and vectorb, x is given by Inverse[matrixa].vectorb. Another option is to form the augmented matrix $(\mathbf{A}|\mathbf{b})$ or $\mathbf{A}$ if $\mathbf{b} = \mathbf{0}$ and then use RowReduce, which row reduces a matrix to reduced row echelon form.

EXAMPLE 6.1.7: Solve each of the following systems. (a) $\begin{cases} 3x-y+3z=0 \\ -6x-y-3z=0; \\ -6x+y-5z=0 \end{cases}$

(b) $\begin{cases} 2x-2y+z=3 \\ x+2y-z=2 \\ x-y+2z=5 \end{cases}$; and (c) $\begin{cases} 2x-y+z=3 \\ x+2y-z=2 \\ 5x+5y-2z=0 \end{cases}$.

SOLUTION: (a) We first use `Solve` to solve the system of equations.

$$\texttt{Solve}[\{3x - y + 3z == 0, -6x - y - 3z == 0, -6x + y - 5z == 0\}]$$

$$\{\{y \to -\frac{3x}{2}, z \to -\frac{3x}{2}\}\}$$

The result indicates that x is a free variable and that y and z depend on x. Because x is free and the authors prefer to avoid fractions, we choose $x = 2t$. Then, $y = -3t$ and $z = -3t$ so $\mathbf{x} = \begin{pmatrix} x \\ y \\ z \end{pmatrix} = \begin{pmatrix} 2t \\ -3t \\ -3t \end{pmatrix} = \begin{pmatrix} 2 \\ -3 \\ -3 \end{pmatrix} t$. This problem has infinitely many solutions. However, notice that all solutions are linearly dependent.

A different way of solving the problem is to notice that the augmented matrix for this linear homogeneous system is $\begin{pmatrix} 3 & -1 & 3 \\ -6 & -1 & -3 \\ -6 & 1 & -5 \end{pmatrix}$. We use `RowReduce` to row reduce the augmented matrix to reduced row echelon form.

$$\texttt{capa} = \{\{2, -1, 3\}, \{-6, -2, -3\}, \{-6, 1, -6\}\};$$

$$\texttt{capb} = \texttt{capa} - (-1)\texttt{IdentityMatrix}[3]$$

$$\{\{3, -1, 3\}, \{-6, -1, -3\}, \{-6, 1, -5\}\}$$

$$\texttt{MatrixForm[capb]}$$

$$\begin{pmatrix} 3 & -1 & 3 \\ -6 & -1 & -3 \\ -6 & 1 & -5 \end{pmatrix}$$

$$\texttt{RowReduce[capb]}$$

$$\{\{1, 0, \frac{2}{3}\}, \{0, 1, -1\}, \{0, 0, 0\}\}$$

Using $\mathbf{x} = \begin{pmatrix} x \\ y \\ z \end{pmatrix}$ notation, the result indicates that $x + \frac{2}{3}z = 0$, $y - z = 0$, and $0 = 0$, which means that there is one free variable. We choose z to be free and then $z = 3t$. With this, we have that $y = 3t$ and $x = -2t$. With

If there are no restrictions on a variable, you can choose any variable to be free. However, be cautious. Sometimes there are conditions that eliminate one variable from being a free variable.

this notation, we obtain that $\mathbf{x} = \begin{pmatrix} x \\ y \\ z \end{pmatrix} = \begin{pmatrix} -2t \\ 3t \\ 3t \end{pmatrix} = \begin{pmatrix} -2 \\ 3 \\ 3 \end{pmatrix} t$, which is

To see that the results are
equivalent, choose $t = -1$.

equivalent to the result obtained previously.

For (b) we proceed in a similar way. With `Solve`,

```
Clear[x, y, z]

Solve[{2x - y + z == 3, x + 2y - z == 2, x - y + 2z == 5}]
```

$\{\{x \to 1, y \to 2, z \to 3\}\}$

we see that the solution to the system of equations is $x = 1$, $y = 2$, and $z = 3$. Alternatively, we write the augmented matrix for the system: $\begin{pmatrix} 2 & -1 & 1 & 3 \\ 1 & 2 & -1 & 2 \\ 1 & -1 & 2 & 5 \end{pmatrix}$ and then use `RowReduce` to row reduce the augmented matrix to reduced row echelon form.

```
a = {{2, -1, 1, 3}, {1, 2, -1, 2}, {1, -1, 2, 5}};

RowReduce[a]
```

$\{\{1, 0, 0, 1\}, \{0, 1, 0, 2\}, \{0, 0, 1, 3\}\}$

Row one indicates that $x = 1$, row two indicates that $y = 2$, and row three indicates that $z = 3$.

For (c), we proceed in the same way as before. First, we use `Solve`. The result indicates that there is not a solution to the system.

```
Clear[x, y, z]

Solve[{2x - y + z == 3, x + 2y - z == 2, 5x + 5y - 2z == 0}, {x, y, z}]
```

$\{\}$

On the other hand, forming the augmented matrix $\begin{pmatrix} 2 & -1 & 1 & 3 \\ 1 & 2 & -1 & 2 \\ 5 & 5 & -2 & 0 \end{pmatrix}$ and using `RowReduce`,

```
step1 = RowReduce[{{2, -1, 1, 3}, {1, 2, -1, 2}, {5, 5, -2, 0}}]
```

$\{\{1, 0, \frac{1}{5}, 0\}, \{0, 1, -\frac{3}{5}, 0\}, \{0, 0, 0, 1\}\}$

```
MatrixForm[step1]
```

$$\begin{pmatrix} 1 & 0 & \dfrac{1}{5} & 0 \\ 0 & 1 & -\dfrac{3}{5} & 0 \\ 0 & 0 & 0 & 1 \end{pmatrix}$$

shows us that $x + \frac{1}{5}y = 0$, $y - \frac{3}{5}z = 0$, and $0 = 1$. The system is inconsistent and does not have a solution.

∎

Usually, you will obtain good results with Solve or with functions like RowReduce or Inverse. However, the different approaches may give you different insight on the problem.

6.1.5 Eigenvalues and Eigenvectors

Let $\mathbf{A}$ be an $n \times n$ matrix. The number λ is an **eigenvalue** of $\mathbf{A}$ if there is a *nonzero* vector, $\mathbf{v}$, called an **eigenvector**, satisfying

$$\mathbf{Av} = \lambda\mathbf{v}. \tag{6.1}$$

We find the eigenvalues of $\mathbf{A}$ by solving the **characteristic polynomial**

$$|\mathbf{A} - \lambda\mathbf{I}| = 0 \tag{6.2}$$

for λ. Once we find the eigenvalues, the corresponding eigenvectors are found by solving

$$(\mathbf{A} - \lambda\mathbf{I})\,\mathbf{v} = \mathbf{0} \tag{6.3}$$

for $\mathbf{v}$.

If $\mathbf{A}$ is a square matrix,

```
Eigenvalues[A]
```

finds the eigenvalues of $\mathbf{A}$,

```
Eigenvectors[A]
```

finds the eigenvectors, and

```
Eigensystem[A]
```

finds the eigenvalues and corresponding eigenvectors.

$$\texttt{CharacteristicPolynomial[A,lambda]}$$

To obtain the λ symbol in Mathematica, press the escape key, type the word "lambda" and then press the escape key to complete the command.
finds the characteristic polynomial of **A** as a function of λ.

EXAMPLE 6.1.8: Find the eigenvalues and corresponding eigenvectors for each of the following matrices. (a) $\mathbf{A} = \begin{pmatrix} -3 & 2 \\ 2 & -3 \end{pmatrix}$; (b) $\mathbf{A} = \begin{pmatrix} 1 & -1 \\ 1 & 3 \end{pmatrix}$; (c) $\mathbf{A} = \begin{pmatrix} 0 & 1 & 1 \\ 1 & 0 & 1 \\ 1 & 1 & 0 \end{pmatrix}$; and (d) $\mathbf{A} = \begin{pmatrix} -1/4 & 2 \\ -8 & -1/4 \end{pmatrix}$.

SOLUTION: (a) We begin by finding the eigenvalues. Solving

$$|\mathbf{A} - \lambda\mathbf{I}| = \begin{vmatrix} -3-\lambda & 2 \\ 2 & -3-\lambda \end{vmatrix} = \lambda^2 + 6\lambda + 5 = 0$$

gives us $\lambda_1 = -5$ and $\lambda_2 = -1$.

   Observe that the same results are obtained using Characteristic Polynomial and Eigenvalues.

```
capa = {{-3, 2}, {2, -3}};

CharacteristicPolynomial[capa, λ]//Factor
```

$$(1+\lambda)(5+\lambda)$$

```
el = Eigenvalues[capa]
```

$$\{-5, -1\}$$

We now find the corresponding eigenvectors. Let $\mathbf{v}_1 = \begin{pmatrix} x_1 \\ y_1 \end{pmatrix}$ be an eigenvector corresponding to λ_1, then

$$(\mathbf{A} - \lambda_1\mathbf{I})\,\mathbf{v}_1 = \mathbf{0}$$

$$\left[\begin{pmatrix} -3 & 2 \\ 2 & -3 \end{pmatrix} - (-5) \begin{pmatrix} 1 & 0 \\ 0 & 1 \end{pmatrix} \right] \begin{pmatrix} x_1 \\ y_1 \end{pmatrix} = \begin{pmatrix} 0 \\ 0 \end{pmatrix}$$

$$\begin{pmatrix} 2 & 2 \\ 2 & 2 \end{pmatrix} \begin{pmatrix} x_1 \\ y_1 \end{pmatrix} = \begin{pmatrix} 0 \\ 0 \end{pmatrix},$$

which row reduces to

$$\begin{pmatrix} 1 & 1 \\ 0 & 0 \end{pmatrix} \begin{pmatrix} x_1 \\ y_1 \end{pmatrix} = \begin{pmatrix} 0 \\ 0 \end{pmatrix}.$$

That is, $x_1 + y_1 = 0$ or $x_1 = -y_1$. Hence, for any value of $y_1 \neq 0$,

$$\mathbf{v}_1 = \begin{pmatrix} x_1 \\ y_1 \end{pmatrix} = \begin{pmatrix} -y_1 \\ y_1 \end{pmatrix} = y_1 \begin{pmatrix} -1 \\ 1 \end{pmatrix}$$

is an eigenvector corresponding to λ_1. Of course, this represents infinitely many vectors. But, they are all linearly dependent. Choosing $y_1 = 1$ yields $\mathbf{v}_1 = \begin{pmatrix} -1 \\ 1 \end{pmatrix}$. Note that you might have chosen $y_1 = -1$ and obtained $\mathbf{v}_1 = \begin{pmatrix} 1 \\ -1 \end{pmatrix}$. However, both of our results are *correct* because these vectors are linearly dependent.

Similarly, letting $\mathbf{v}_2 = \begin{pmatrix} x_2 \\ y_2 \end{pmatrix}$ be an eigenvector corresponding to λ_2 we solve $(\mathbf{A} - \lambda_2\mathbf{I})\,\mathbf{v}_1 = \mathbf{0}$:

$$\begin{pmatrix} -2 & 2 \\ 2 & -2 \end{pmatrix} \begin{pmatrix} x_2 \\ y_2 \end{pmatrix} = \begin{pmatrix} 0 \\ 0 \end{pmatrix} \quad \text{or} \quad \begin{pmatrix} 1 & -1 \\ 0 & 0 \end{pmatrix} \begin{pmatrix} x_2 \\ y_2 \end{pmatrix} = \begin{pmatrix} 0 \\ 0 \end{pmatrix}.$$

Thus, $x_2 - y_2 = 0$ or $x_2 = y_2$. Hence, for any value of $y_2 \neq 0$,

$$\mathbf{v}_2 = \begin{pmatrix} x_2 \\ y_2 \end{pmatrix} = \begin{pmatrix} y_2 \\ y_2 \end{pmatrix} = y_2 \begin{pmatrix} 1 \\ 1 \end{pmatrix}$$

is an eigenvector corresponding to λ_2. Choosing $y_2 = 1$ yields $\mathbf{v}_2 = \begin{pmatrix} 1 \\ 1 \end{pmatrix}$. We confirm these results using `RowReduce`.

```
ev1 = capa − e1[[1]]IdentityMatrix[2]
```

```
{{2, 2}, {2, 2}}
```

```
RowReduce[ev1]
```

{{1, 1}, {0, 0}}

```
ev2 = capa - e1[[2]]IdentityMatrix[2]
```

{{−2, 2}, {2, −2}}

```
RowReduce[ev2]
```

{{1, −1}, {0, 0}}

We obtain the same results using

```
Eigenvectors[capa]
```

{{−1, 1}, {1, 1}}

```
Eigensystem[capa]
```

{{−5, −1}, {{−1, 1}, {1, 1}}}

(b) In this case, we see that $\lambda = 2$ has multiplicity 2. There is only one linearly independent eigenvector, $\mathbf{v} = \begin{pmatrix} -1 \\ 1 \end{pmatrix}$, corresponding to λ.

```
capa = {{1, −1}, {1, 3}};
Factor[CharacteristicPolynomial[capa, λ]]
Eigenvalues[capa]
Eigenvectors[capa]
Eigensystem[capa]
```

$(-2 + \lambda)^2$

{2, 2}

{{−1, 1}, {0, 0}}

{{2, 2}, {{−1, 1}, {0, 0}}}

(c) The eigenvalue $\lambda_1 = 2$ has corresponding eigenvector $\mathbf{v}_1 = \begin{pmatrix} 1 \\ 1 \\ 1 \end{pmatrix}$.

The eigenvalue $\lambda_{2,3} = -1$ has multiplicity 2. In this case, there are two linearly independent eigenvectors corresponding to this

eigenvalue: $\mathbf{v_2} = \begin{pmatrix} -1 \\ 0 \\ 1 \end{pmatrix}$ and $\mathbf{v_3} = \begin{pmatrix} -1 \\ 1 \\ 0 \end{pmatrix}$.

```
capa = {{0, 1, 1}, {1, 0, 1}, {1, 1, 0}};

Factor[CharacteristicPolynomial[capa, λ]]

Eigenvalues[capa]

Eigenvectors[capa]

Eigensystem[capa]
```

$-(-2 + \lambda)(1 + \lambda)^2$

{2, -1, -1}

{{1, 1, 1}, {-1, 0, 1}, {-1, 1, 0}}

{{2, -1, -1}, {{1, 1, 1}, {-1, 0, 1}, {-1, 1, 0}}}

(d) In this case, the eigenvalues $\lambda_{1,2} = -\frac{1}{4} \pm 4i$ are complex conjugates. We see that the eigenvectors $\mathbf{v}_{1,2} = \begin{pmatrix} 0 \\ 2 \end{pmatrix} \pm \begin{pmatrix} 1 \\ 0 \end{pmatrix} i$ are complex conjugates as well.

```
capa = {{-1/4, 2}, {-8, -1/4}};

Factor[CharacteristicPolynomial[capa, λ]]

Eigenvalues[capa]

Eigenvectors[capa]

Eigensystem[capa]
```

$\frac{1}{16}(257 + 8\lambda + 16\lambda^2)$

$\{-\frac{1}{4} + 4i, -\frac{1}{4} - 4i\}$

{{-i, 2}, {i, 2}}

$\{\{-\frac{1}{4} + 4i, -\frac{1}{4} - 4i\}, \{\{-i, 2\}, \{i, 2\}\}\}$

∎

6.1.6 Matrix Calculus

Definition 23 (Derivative and Integral of a Matrix). *The **derivative** of the $m \times n$ matrix*

$$\mathbf{A}(t) = \begin{pmatrix} a_{11}(t) & a_{12}(t) & a_{13}(t) & \cdots & a_{1n}(t) \\ a_{21}(t) & a_{22}(t) & a_{23}(t) & \cdots & a_{2n}(t) \\ a_{31}(t) & a_{32}(t) & a_{33}(t) & \cdots & a_{3n}(t) \\ \vdots & \vdots & \vdots & & \vdots \\ a_{m1}(t) & a_{m2}(t) & a_{m3}(t) & \cdots & a_{mn}(t) \end{pmatrix},$$

where $a_{ij}(t)$ is differentiable for all values of i and j, is

$$\frac{d}{dt}\mathbf{A}(t) = \begin{pmatrix} \frac{d}{dt}a_{11}(t) & \frac{d}{dt}a_{12}(t) & \frac{d}{dt}a_{13}(t) & \cdots & \frac{d}{dt}a_{1n}(t) \\ \frac{d}{dt}a_{21}(t) & \frac{d}{dt}a_{22}(t) & \frac{d}{dt}a_{23}(t) & \cdots & \frac{d}{dt}a_{2n}(t) \\ \frac{d}{dt}a_{31}(t) & \frac{d}{dt}a_{32}(t) & \frac{d}{dt}a_{33}(t) & \cdots & \frac{d}{dt}a_{3n}(t) \\ \vdots & \vdots & \vdots & & \vdots \\ \frac{d}{dt}a_{m1}(t) & \frac{d}{dt}a_{m2}(t) & \frac{d}{dt}a_{m3}(t) & \cdots & \frac{d}{dt}a_{mn}(t) \end{pmatrix}.$$

*The **integral** of $\mathbf{A}(t)$, where $a_{ij}(t)$ is integrable for all values of i and j, is*

$$\int \mathbf{A}(t)\, dt = \begin{pmatrix} \int a_{11}(t)\, dt & \int a_{12}(t)\, dt & \int a_{13}(t)\, dt & \cdots & \int a_{1n}(t)\, dt \\ \int a_{21}(t)\, dt & \int a_{22}(t)\, dt & \int a_{23}(t)\, dt & \cdots & \int a_{2n}(t)\, dt \\ \int a_{31}(t)\, dt & \int a_{32}(t)\, dt & \int a_{33}(t)\, dt & \cdots & \int a_{3n}(t)\, dt \\ \vdots & \vdots & \vdots & & \vdots \\ \int a_{m1}(t)\, dt & \int a_{m2}(t)\, dt & \int a_{m3}(t)\, dt & \cdots & \int a_{mn}(t)\, dt \end{pmatrix}.$$

EXAMPLE 6.1.9: Find $\dfrac{d}{dt}\mathbf{A}(t)$ and $\int \mathbf{A}(t)\, dt$ if $\mathbf{A}(t) = \begin{pmatrix} \cos 3t & \sin 3t & e^{-t} \\ t & t\sin t^2 & e^t \end{pmatrix}$.

SOLUTION: We find $\dfrac{d}{dt}\mathbf{A}(t)$ by differentiating each element of $\mathbf{A}(t)$ with D.

```
a = {{Cos[3 t], Sin[3 t], Exp[-t]}, {t, tSin[t^2], Exp[t]}};
```

```
D[a, t]//MatrixForm
```

$$\begin{pmatrix} -3\sin[3t] & 3\cos[3t] & -e^{-t} \\ 1 & 2t^2\cos[t^2] + \sin[t^2] & e^t \end{pmatrix}$$

Similarly, we find $\int \mathbf{A}(t)\,dt$ by integrating each element of $\mathbf{A}(t)$ with Integrate.

```
Integrate[a, t]//MatrixForm
```

$$\begin{pmatrix} \dfrac{1}{3}\text{Sin}[3\,t] & -\dfrac{1}{3}\text{Cos}[3\,t] & -e^{-t} \\ \dfrac{t^2}{2} & -\dfrac{1}{2}\text{Cos}[t^2] & e^t \end{pmatrix}$$

Note that Mathematica does not include an arbitrary constant of integration with each anti-derivative.

■

6.2 Systems of Equations: Preliminary Definitions and Theory

Up to this point, we have focused our attention on solving differential equations that involve one dependent variable. However, many physical situations are modeled with more than one equation and involve more than one dependent variable. For example, if we want to determine the population of two interacting populations such as foxes and rabbits, we would have two dependent variables which represent the two populations where these populations depend on one independent variable which represents time. Situations like this lead to systems of differential equations which we study in this chapter. For example, we encountered a nonlinear initial-value problem such as the Van-der-Pol initial-value problem,

$$\begin{cases} x'' + \left(x^2 - 1\right)x' + x = 0 \\ x(0) = 1,\ x'(0) = 1 \end{cases}$$

in Chapter 4. If we let $x' = y$, then

$$y' = x'' = -\left[\left(x^2 - 1\right)x' + x\right] = \left(1 - x^2\right)y - x,$$

so the second-order equation $x'' + \left(x^2 - 1\right)x' + x = 0$ is equivalent to the system of first-order differential equations

$$\begin{cases} x' = y \\ y' = \left(1 - x^2\right)y - x \end{cases}$$

and the original initial value problem is equivalent to

$$\begin{cases} x' = y \\ y' = \left(1 - x^2\right) y - x \\ x(0) = 1, \ y(0) = 1 \end{cases}$$

We use NDSolve to generate a numerical solution to this initial-value problem valid for $0 \le t \le 25$.

```
Clear[x, y, z]
numsol = NDSolve[{x'[t] == y[t],
    y'[t] == (1 - x[t]^2)y[t] - x[t],
    x[0] == 1, y[0] == 0}, {x[t], y[t]},
    {t, 0, 25}]
```

$\{\{x[t] \to \text{InterpolatingFunction}[\ ][t], \ y[t] \to \text{InterpolatingFunction}[\ ][t]\}\}$

We can use this result to approximate the solution for various values of t. For example, entering

```
{x[t], y[t]}/.numsol/.t → 1
```

$\{\{1.29848, -0.367035\}\}$

shows us that $x(1) \approx 1.29848$ and $x'(1) = y(1) \approx -0.367035$. We use Plot to graph $x(t)$ and $y(t)$ (the graph of $y(t)$ is in green) in Figure 6-1(a).

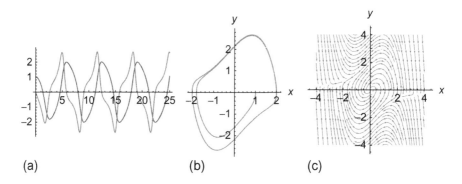

(a) (b) (c)

Figure 6-1   (a) In the limit as $t \to \infty$, the solution is periodic. (b) The solution approaches a *limit cycle*. (c) The direction field associated with the equation

```
p1a = Plot[x[t]/.numsol, {t, 0, 25},

  PlotStyle → CMYKColor[0, 0.89, 0.94, 0.28]];

p1b = Plot[y[t]/.numsol, {t, 0, 25},

  PlotStyle →

  CMYKColor[0.64, 0, 0.95, 0.40]];

p1 = Show[p1a, p1b, PlotLabel → "(a)", PlotRange → All]
```

Because we let $x' = y$, notice that $y(t) > 0$ when $x(t)$ is increasing and $y(t) < 0$ when $x(t)$ is decreasing. The observation that these solutions are periodic is further confirmed by a graph of $x(t)$ (the horizontal axis) versus $y(t)$ (the vertical axis) generated with `ParametricPlot` in Figure 6-1(b). We see that as t increases, the solution approaches a certain fixed path, called a *limit cycle*. To graph the direction field associated with the system, we use `StreamPlot`. See Figure 6-1(c).

```
p2 = ParametricPlot[{x[t], y[t]}/.numsol, {t, 0, 25},

  PlotStyle → CMYKColor[0.62, 0.57, 0.23, 0],

  AxesLabel → {x, y},

  PlotLabel → "(b)"]

p3 = StreamPlot[{y, (1 − x^2)y − x}, {x, −4, 4}, {y, −4, 4},

  StreamStyle->CMYKColor[0.62, 0.57, 0.23, 0],

  Axes → Automatic,

  StreamPoints → Fine, Frame → False, AxesLabel → {x, y},

  PlotLabel → "(c)"]
```

```
Show[GraphicsRow[{p1, p2, p3}]]
```

The Van-der-Pol system is usually written so that it depends on a parameter, μ,

$$
\begin{aligned}
x' &= y \\
y' &= \mu(1 - x^2)y - x.
\end{aligned}
\tag{6.4}
$$

Using the `Manipulate` function can help you investigate how systems such as the Van-der-Pol system behave for different parameter values and different initial conditions. Figure 6-2 shows a frame resulting from the following `Manipulate` command that allows us to investigate the behavior of system (6.4) and the associated direction field for various initial conditions and various values of μ.

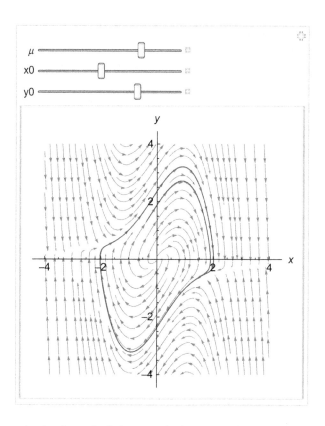

Figure 6-2   Investigating how the behavior of solutions to the Van-der-Pol system change
by varying the parameter μ and initial conditions

```
Clear[x, y, t]

Manipulate[

Clear[x, y, t];

numsol = NDSolve[{x'[t] == y[t],

    y'[t] == μ(1 - x[t]^2)y[t] - x[t],

    x[0] == x0, y[0] == y0}, {x[t], y[t]},

    {t, 0, 25}];

p1 = StreamPlot[{y, μ(1 - x^2)y - x}, {x, -4, 4}, {y, -4, 4},

    StreamStyle->CMYKColor[0.62, 0.57, 0.23, 0],

    StreamPoints → Fine];

p2 = ParametricPlot[{x[t], y[t]}/.numsol, {t, 0, 25},
```

```
PlotStyle->CMYKColor[0, 0.89, 0.94, 0.28],

AxesLabel → {x, y}, PlotRange → {{-4, 4}, {-4, 4}},

AspectRatio → 1];

Show[{p1, p2}, AxesLabel → {x, y},

Axes → Automatic, AxesOrigin → {0, 0},

Frame → False,

PlotRange → {{-4, 4}, {-4, 4}},

AspectRatio → 1], {{μ, 1}, 0, 2},

{{x0, 1}, -3, 3}, {{y0, 0}, -3, 3}]
```

We will find that nonlinear equations are more easily studied when they are written as a system of equations.

6.2.1 Preliminary Theory

Definition 24 (System of Ordinary Differential Equations). *A system of ordinary differential equations is a simultaneous set of equations that involves two or more dependent variables that depend on one independent variable. A **solution** of the system is a set of functions that satisfies each equation on some interval I.*

If the differential equations in the system of differential equations are linear equations, we say that the system is a **system of linear differential equations** or a **linear system**.

EXAMPLE 6.2.1:   Show   that $\begin{cases} x = \frac{1}{5}e^{-t}\left(e^t - \cos 2t - 3\sin 2t\right) \\ y = -e^{-t}\left(\cos 2t - \sin 2t\right) \end{cases}$ is   a

solution to the system $\begin{cases} x' - y = 0 \\ y' + 5x + 2y = 1 \end{cases}$.

SOLUTION: The set of functions is a solution to the system of equations because

```
Clear[x, y, t]

x[t_] = 1/5Exp[-t](Exp[t] - Cos[2t] - 3Sin[2t]);

y[t_] = -Exp[-t](Cos[2t] - Sin[2t]);
```

$x'[t] - y[t]//$Simplify

0

$y'[t] + 5x[t] + 2y[t]//$Simplify

1

We graph this solution in several different ways. First, we graph the solution $\begin{cases} x = x(t) \\ y = y(t) \end{cases}$ parametrically with `ParametricPlot` in Figure 6-3(a). Then, we graph $x(t)$ and $y(t)$ together as functions of t in Figure 6-3(b).

```
p1 = ParametricPlot[{x[t], y[t]}, {t, 0, 3Pi},

    PlotRange → {{-.5, 1}, {-1, .5}}, AspectRatio → 1,

    PlotStyle->CMYKColor[0, 0.89, 0.94, 0.28],

    AxesLabel → {x, y}, PlotLabel → "(a)"]

p2a = Plot[x[t], {t, 0, 3Pi},

    PlotStyle → CMYKColor[0, 0.89, 0.94, 0.28], PlotRange → All];

p2b = Plot[y[t], {t, 0, 3Pi},

    PlotStyle →

    CMYKColor[0.64, 0, 0.95, 0.40], PlotRange → All];

p2 = Show[p2a, p2b, PlotLabel → "(b)", PlotRange → All]
```

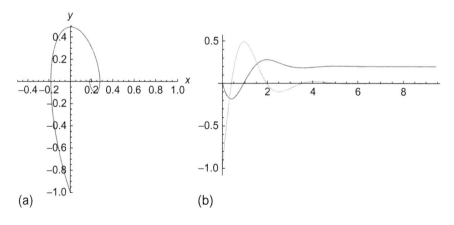

(a)　　　　　　　　　　　　(b)

Figure 6-3　(a) x (on the horizontal axis) versus y (on the vertical axis). (b) x and y as functions of t

Show[GraphicsRow[{p1, p2}]]

Notice that $\lim_{t\to\infty} x(t) = \frac{1}{5}$ and $\lim_{t\to\infty} y(t) = 0$. Therefore, in the parametric plot, the points on the curve approach $(1/5, 0)$ as t increases.

■

We will discuss techniques for solving systems in the following sections. For now, we make the following remarks. First, we saw previously that you can often use NDSolve to generate a numerical solution of a system, which is of particular benefit with nonlinear systems. DSolve can also often be used to find solutions of most linear systems and, in a few special cases, nonlinear systems.

EXAMPLE 6.2.2: Solve $\begin{cases} x' = 2y \\ y' = -\frac{1}{4}x \end{cases}$ and $\begin{cases} x' = 2y \\ y' = -\frac{1}{4}x \\ x(0) = 2, \ y(0) = 1 \end{cases}$.

SOLUTION: DSolve can find a general solution of this linear system.

Clear[x, y, gensol]

gensol = DSolve[{x'[t] == 2y[t],

  y'[t] == -1/4x[t]}, {x[t], y[t]}, t]

$\{\{x[t] \rightarrow C[1]\cos[\frac{t}{\sqrt{2}}] + 2\sqrt{2}C[2]\sin[\frac{t}{\sqrt{2}}], \ y[t] \rightarrow C[2]\cos[\frac{t}{\sqrt{2}}] - \frac{C[1]\sin[\frac{t}{\sqrt{2}}]}{2\sqrt{2}}\}\}$

Similarly, DSolve can solve the initial-value problem. The resulting list is named partsol.

Clear[x, y, partsol]

partsol = DSolve[{x'[t] == 2y[t],

  y'[t] == -1/4x[t], x[0] == 2,

  y[0] == 1}, {x[t], y[t]}, t]

$\{\{x[t] \rightarrow 2(\cos[\frac{t}{\sqrt{2}}] + \sqrt{2}\sin[\frac{t}{\sqrt{2}}]), \ y[t] \rightarrow \frac{1}{2}(2\cos[\frac{t}{\sqrt{2}}] - \sqrt{2}\sin[\frac{t}{\sqrt{2}}])\}\}$

We use `Plot` to graph the x and y components of the solution individually and `ParametricPlot` to graph them parametrically. See Figures 6-4(a) and (b).

```
p1a = Plot[x[t]/.partsol, {t, 0, 4Sqrt[2]Pi},

    PlotStyle → CMYKColor[0, 0.89, 0.94, 0.28]];

p1b = Plot[y[t]/.partsol, {t, 0, 4Sqrt[2]Pi},

    PlotStyle →

    CMYKColor[0.64, 0, 0.95, 0.40]];

p1 = Show[p1a, p1b, PlotLabel → "(a)", PlotRange → All]
```

For an autonomous system like this, we use `StreamPlot` to graph the direction field associated with the system. We use `StreamPlot` to graph the direction field associated with the system and then display the direction field together with the solution to the initial-value problem in Figure 6-4(c).

```
p2 = ParametricPlot[{x[t], y[t]}/.partsol, {t, 0, 4Sqrt[2]Pi},

    PlotStyle → {{Thickness[.01], CMYKColor[0.62, 0.57, 0.23, 0]}},

    AxesLabel → {x, y},

    PlotLabel → "(b)"]

p3a = StreamPlot[{2y, -1/4x}, {x, -6, 6}, {y, -6, 6},

    StreamStyle → Gray, StreamPoints → Fine];

p3 = Show[p2, p3a, Axes → Automatic,

    Frame → False, AxesLabel → {x, y},

    PlotRange → {{-4, 4}, {-4, 4}}, AspectRatio → 1,

    PlotLabel → "(c)"]

Show[GraphicsRow[{p1, p2, p3}]]
```

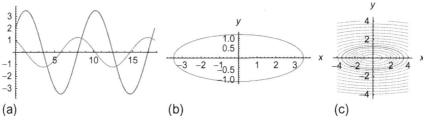

(a)   (b)   (c)

Figure 6-4   (a) Plots of x y. (b) Parametric plot of x versus y. (c) A parametric plot of the solution along with the direction field

In fact, we can show the direction field together with several so-lutions. With the following commands, we use Map to apply a pure function to the list $\{0.5, 1., 1.5, 2., 2.5\}$ that solves the system if $x(0) = 0$ and $y(0) = i$ for $i = 0.5, \ldots, 2.5$.

```
severalsols = Map[DSolve[{x'[t] == 2y[t],
   y'[t] == -1/4x[t], x[0] == 2, y[0] == #}, {x[t], y[t]}, t]&,
   {0.5, 1., 1.5, 2., 2.5}]
```

$$\left\{\left\{\left\{x[t] \to 2.\cos\left[\frac{t}{\sqrt{2}}\right] + 1.41421\sin\left[\frac{t}{\sqrt{2}}\right], y[t] \to 0.5\cos\left[\frac{t}{\sqrt{2}}\right]\right.\right.\right.$$
$$\left.\left.\left. - 0.707107\sin\left[\frac{t}{\sqrt{2}}\right]\right\}\right\},\right.$$

$$\left\{\left\{x[t] \to 2.\cos\left[\frac{t}{\sqrt{2}}\right] + 2.82843\sin\left[\frac{t}{\sqrt{2}}\right], y[t] \to 1.\cos\left[\frac{t}{\sqrt{2}}\right]\right.\right.$$
$$\left.\left. - 0.707107\sin\left[\frac{t}{\sqrt{2}}\right]\right\}\right\},$$

$$\left\{\left\{x[t] \to 2.\cos\left[\frac{t}{\sqrt{2}}\right] + 4.24264\sin\left[\frac{t}{\sqrt{2}}\right], y[t] \to 1.5\cos\left[\frac{t}{\sqrt{2}}\right]\right.\right.$$
$$\left.\left. - 0.707107\sin\left[\frac{t}{\sqrt{2}}\right]\right\}\right\},$$

$$\left\{\left\{x[t] \to 2.\cos\left[\frac{t}{\sqrt{2}}\right] + 5.65685\sin\left[\frac{t}{\sqrt{2}}\right], y[t] \to 2.\cos\left[\frac{t}{\sqrt{2}}\right]\right.\right.$$
$$\left.\left. - 0.707107\sin\left[\frac{t}{\sqrt{2}}\right]\right\}\right\},$$

$$\left\{\left\{x[t] \to 2.\cos\left[\frac{t}{\sqrt{2}}\right] + 7.07107\sin\left[\frac{t}{\sqrt{2}}\right], y[t] \to 2.5\cos\left[\frac{t}{\sqrt{2}}\right]\right.\right.$$
$$\left.\left.\left. - 0.707107\sin\left[\frac{t}{\sqrt{2}}\right]\right\}\right\}\right\}$$

We then use ParametricPlot to graph the solutions obtained in severalsols together and display them with the direction field (p3) in Figure 6-5.

```
p1 = ParametricPlot[{x[t], y[t]}/.severalsols, {t, 0, 4Sqrt[2]Pi},
   PlotStyle → {{Thickness[.01], CMYKColor[0.62, 0.57, 0.23, 0]}}];

Show[p1, p3, PlotRange → {{-6, 6}, {-6, 6}}, AspectRatio → 1,
   AxesLabel → {x, y}]
```

■

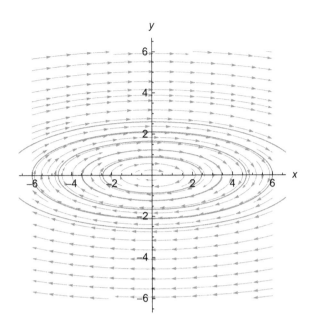

Figure 6-5   Direction field associated with the system together with several solutions of the system

---

EXAMPLE 6.2.3: The **Jacobi elliptic functions** satisfy the nonlinear system

The system is nonlinear because of the products of the dependent variables u, v, and w.

For this system, t is the *independent* variable; $u = u(t)$, $v = v(t)$, and $w = w(t)$ are the *dependent* variables.

$$\begin{cases} du/dt = vw \\ dv/dt = -uw \\ dw/dt = -k^2 uv. \end{cases}$$

Use Mathematica to solve this system.

SOLUTION: Although Mathematica generates several error messages, we see that Mathematica is able to find a general solution of the system, although the result is given in terms of the *Jacobi elliptic function*, JacobiSN.

```
gensol = DSolve[{u'[t] == v[t]w[t],
    v'[t] == -u[t]w[t], w'[t] == -k^2u[t]v[t]},
    {u[t], v[t], w[t]}, t]
```

$\{\{u[t] \rightarrow -\sqrt{2}\sqrt{C[1]}\text{JacobiSN}[\sqrt{2}t\sqrt{C[2]} - \sqrt{2}\sqrt{C[2]}C[3], \dfrac{k^2 C[1]}{C[2]}]$,

$v[t] \rightarrow -\sqrt{2C[1] - 2C[1]\text{JacobiSN}[\sqrt{2}t\sqrt{C[2]} - \sqrt{2}\sqrt{C[2]}C[3], \dfrac{k^2 C[1]}{C[2]}]^2}$,

$$w[t] \to \sqrt{2\,C[2] - 2k^2\,C[1]\mathrm{JacobiSN}[\sqrt{2}\,t\sqrt{C[2]} - \sqrt{2}\sqrt{C[2]}C[3], \frac{k^2\,C[1]}{C[2]}]^2}\},$$

$$\{u[t] \to -\sqrt{2}\sqrt{C[1]}\mathrm{JacobiSN}[\sqrt{2}\,t\sqrt{C[2]} - \sqrt{2}\sqrt{C[2]}C[3], \frac{k^2\,C[1]}{C[2]}],$$

$$v[t] \to \sqrt{2\,C[1] - 2\,C[1]\mathrm{JacobiSN}[\sqrt{2}\,t\sqrt{C[2]} - \sqrt{2}\sqrt{C[2]}C[3], \frac{k^2\,C[1]}{C[2]}]^2},$$

$$w[t] \to -\sqrt{2\,C[2] - 2k^2\,C[1]\mathrm{JacobiSN}[\sqrt{2}\,t\sqrt{C[2]} - \sqrt{2}\sqrt{C[2]}C[3], \frac{k^2\,C[1]}{C[2]}]^2}\},$$

$$\{u[t] \to \sqrt{2}\sqrt{C[1]}\mathrm{JacobiSN}[\sqrt{2}\,t\sqrt{C[2]} - \sqrt{2}\sqrt{C[2]}C[3], \frac{k^2\,C[1]}{C[2]}],$$

$$v[t] \to -\sqrt{2\,C[1] - 2\,C[1]\mathrm{JacobiSN}[\sqrt{2}\,t\sqrt{C[2]} - \sqrt{2}\sqrt{C[2]}C[3], \frac{k^2\,C[1]}{C[2]}]^2},$$

$$w[t] \to -\sqrt{2\,C[2] - 2k^2\,C[1]\mathrm{JacobiSN}[\sqrt{2}\,t\sqrt{C[2]} - \sqrt{2}\sqrt{C[2]}C[3], \frac{k^2\,C[1]}{C[2]}]^2}\},$$

$$\{u[t] \to \sqrt{2}\sqrt{C[1]}\mathrm{JacobiSN}[\sqrt{2}\,t\sqrt{C[2]} - \sqrt{2}\sqrt{C[2]}C[3], \frac{k^2\,C[1]}{C[2]}],$$

$$v[t] \to \sqrt{2\,C[1] - 2\,C[1]\mathrm{JacobiSN}[\sqrt{2}\,t\sqrt{C[2]} - \sqrt{2}\sqrt{C[2]}C[3], \frac{k^2\,C[1]}{C[2]}]^2},$$

$$w[t] \to \sqrt{2\,C[2] - 2k^2\,C[1]\mathrm{JacobiSN}[\sqrt{2}\,t\sqrt{C[2]} - \sqrt{2}\sqrt{C[2]}C[3], \frac{k^2\,C[1]}{C[2]}]^2}\}\}$$

We use the **Help Browser** to obtain information regarding the JacobiSN function as indicated in the following screen shot.

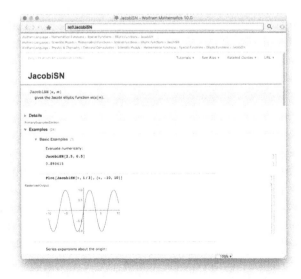

As with other equations, under reasonable conditions, a solution to a system of differential equations always exists.

Theorem 11 (Existence and Uniqueness). *Assume that each of the functions*

$$f_1(t, x_1, x_2, \ldots, x_n), f_2(t, x_1, x_2, \ldots, x_n), \ldots, f_n(t, x_1, x_2, \ldots, x_n)$$

and the partial derivatives $\partial f_1/\partial x_1$, $\partial f_2/\partial x_2$, ..., $\partial f_n/\partial x_n$ *are continuous in a region R containing the point* $(t_0, y_1, y_2, \ldots, y_n)$. *Then, the initial-value problem*

$$\begin{cases} x_1' = f_1(t, x_1, x_2, \ldots, x_n) \\ x_2' = f_2(t, x_1, x_2, \ldots, x_n) \\ \vdots \\ x_n' = f_n(t, x_1, x_2, \ldots, x_n) \\ x_1(t_0) = y_1, \ x_2(t_0) = y_2, \ldots x_n(t_0) = y_n \end{cases} \tag{6.5}$$

has a unique solution

$$\begin{cases} x_1 = \varphi_1(t) \\ x_2 = \varphi_2(t) \\ \vdots \\ x_n = \varphi_n(t) \end{cases} \tag{6.6}$$

on an interval I containing $t = t_0$.

EXAMPLE 6.2.4: Show that the initial-value problem

$$\begin{cases} dx/dt = 2x - xy \\ dy/dt = -3y + xy \\ x(0) = 2, \ y(0) = 3/2 \end{cases}$$

has a unique solution.

SOLUTION: In this case, we identify $f_1(t, x, y) = 2x - xy$ and $f_2(t, x, y) = -3y + xy$ with $\partial f_1/\partial x = 2 - y$ and $\partial f_2/\partial y = -3 + x$. All four of these functions are continuous on a region containing $(0, 2, 3/2)$. Thus, by the Existence and Uniqueness theorem, a unique solution to the initial-value problem exists. In this case, we use NDSolve to approximate the solution to this nonlinear problem valid for $0 \le t \le 10$.

DSolve is not able to find an explicit solution to this nonlinear system.

```
Clear[x, y, t]

numsol = NDSolve[{x'[t] == 2x[t] - x[t]y[t],

  y'[t] == -3y[t] + x[t]y[t], x[0] == 2,

  y[0] == 3/2}, {x[t], y[t]}, {t, 0, 10}]

  {{x[t] → InterpolatingFunction[][t], y[t] →

  InterpolatingFunction[][t]}}
```

Next, we use `Plot` to graph $x(t)$ and $y(t)$ individually in Figure 6-6(a) and `ParametricPlot` to graph the parametric equations $\begin{cases} x = x(t) \\ y = y(t) \end{cases}$ in Figure 6-6(b) for $0 \le t \le 10$.

```
p1a = Plot[x[t]/.numsol, {t, 0, 10},

  PlotStyle → CMYKColor[0, 0.89, 0.94, 0.28]];

p1b = Plot[y[t]/.numsol, {t, 0, 10},

  PlotStyle →

  CMYKColor[0.64, 0, 0.95, 0.40]];

p1 = Show[p1a, p1b, PlotLabel → "(a)", PlotRange → All,

  AxesOrigin → {0, 0}]

p2 = ParametricPlot[{x[t], y[t]}/.numsol, {t, 0, 10},

  PlotStyle → {{Thickness[.01], CMYKColor[0.62, 0.57,
```

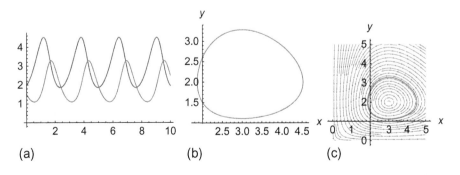

(a)   (b)   (c)

Figure 6-6 (a) $x(t)$ and $y(t)$ as functions of t. (b) $x(t)$ and $y(t)$ are periodic. (c) The direction field indicates that all meaningful solutions are periodic

$$0.23, 0]\}\}, \texttt{AxesLabel} \rightarrow \{x, y\}, \texttt{PlotRange} \rightarrow \{\{0, 5\}, \{0, 5\}\},$$

$$\texttt{PlotLabel} \rightarrow \text{``(b)"}]$$

$$\texttt{p3a} = \texttt{StreamPlot}[\{2x - xy, -3y + xy\}, \{x, 0, 5\}, \{y, 0, 5\},$$

$$\texttt{StreamStyle} \rightarrow \texttt{Gray}, \texttt{StreamPoints} \rightarrow \texttt{Fine}];$$

$$\texttt{p3} = \texttt{Show}[\texttt{p2}, \texttt{p3a}, \texttt{Axes} \rightarrow \texttt{Automatic},$$

$$\texttt{Frame} \rightarrow \texttt{False}, \texttt{AxesLabel} \rightarrow \{x, y\}, \texttt{AxesOrigin} \rightarrow \{0, 0\},$$

$$\texttt{PlotRange} \rightarrow \{\{0, 5\}, \{0, 5\}\}, \texttt{AspectRatio} \rightarrow 1,$$

$$\texttt{PlotLabel} \rightarrow \text{``(c)"}]$$

$$\texttt{Show}[\texttt{GraphicsRow}[\{\texttt{p1}, \texttt{p2}, \texttt{p3}\}]]$$

The graphs illustrate that the solution to the initial-value problem is periodic.

In fact, all *meaningful* (or *interesting*) solutions to the equation are periodic. *Meaningful* (or *interesting*) solutions are ones for which both x and y are greater than 0 and neither is constant. To see this, we use `StreamPlot` to graph the direction field associated with the system. We display the direction field together with the solution to the initial-value problem in Figure 6-6(c).

■

> Later, we will see that a system like this is used to model a basic predator-prey relationship. In such a model, x and y represent population sizes (or ratios) so we are only interested in solutions where both of these quantities are greater than or equal to 0.

Although the Existence and Uniqueness theorem, Theorem 11, guarantees the existence and uniqueness of a solution, the behavior of the solutions of a system can be remarkably complicated, even for systems that appear quite simple.

> See texts like Jordan and Smith's *Nonlinear Ordinary Differential Equations*, [17], for discussions of ways to analyze systems like the Rössler attractor and the Lorenz equations.

EXAMPLE 6.2.5 (Rössler Attractor): The **Rössler attractor** is the system

$$\begin{cases} x' = -y - z \\ y' = x + ay \\ z' = bx - cz + xz \end{cases}. \tag{6.7}$$

Observe that system (6.7) is nonlinear because of the product of the x and z terms in the z' equation.

If $a = 0.4$, $b = 0.3$, $x_0 = 1$, $y_0 = 0.4$, and $z(0) = 0.7$, how does the value of c affect solutions to the initial-value problem

$$\begin{cases} x' = -y - z \\ y' = x + ay \\ z' = bx - cz + xz \\ x(0) = x_0,\ y(0) = y_0,\ z(0) = z_0 \end{cases} \quad ? \qquad (6.8)$$

SOLUTION: By the Existence and Uniqueness theorem, initial-value problem (6.8) will *always* have a unique solution.

We define the function `rosslerplot`:

$$\texttt{rosslerplot[a,b,c][\{x0,y0,z0\},\{t,a,b\}]}$$

1. solves the initial-value problem (6.8) for $a \le t \le b$,
2. generates parametric plots of $x(t)$ versus $y(t)$, $y(t)$ versus $z(t)$, $x(t)$ versus $z(t)$, and $x(t)$ versus $y(t)$ versus $z(t)$, and displays the four graphics as a graphics array, and
3. returns a numerical solution to the initial-value problem (6.8) valid for $a \le t \le b$.

If $\{t,a,b\}$ is omitted from the `rosslerplot` function, the default is $950 \le t \le 1000$. Any options included are passed to the `Show` command.

```
rosslerplot[a_, b_, c_][{x0_, y0_, z0_}, ts_:{t, 950, 1000},
opts___]:=Module[{numsol}, numsol = NDSolve[{x'[t] == −
y[t] − z[t], y'[t] == x[t] + ay[t],   z'[t] == bx[t] − cz[t]+
x[t]z[t], x[0] == x0, y[0] == y0, z[0] == z0}, {x[t], y[t],
z[t]}, ts, MaxSteps → 100000]; p1a = ParametricPlot
[Evaluate[{x[t], y[t]}/.numsol], ts, PlotPoints → 1000,
AspectRatio → 1, AxesLabel → {x, y}]; p1b =
ParametricPlot[Evaluate[{x[t], z[t]}/.numsol], ts,
PlotPoints → 1000, AspectRatio → 1,
AxesLabel → {x, z}, PlotRange → All];
p1c = ParametricPlot[Evaluate[{y[t], z[t]}/.numsol], ts,
PlotPoints → 1000, AspectRatio → 1,
AxesLabel → {y, z}, PlotRange → All];
p1d = ParametricPlot3D[Evaluate[{x[t], y[t], z[t]}/.
```

```
    numsol], ts, PlotPoints → 3000, BoxRatios → {1, 1, 1},
    AxesLabel → {x, y, z}, PlotRange → All];
    Show[GraphicsGrid[{{p1a, p1b}, {p1c, p1d}}], opts]
]
```

For example, entering

```
    rosslerplot[.4, .3, 4.44][{1, .4, .7}, {t, 800, 1000}]
```

generates the plots shown in Figure 6-7, which corresponds to plots for our problem if $c = 4.44$.

For the given values of a, b, x_0. y_0, and z_0, we will vary c by using $c = 1.4, 2.4, 2.6, 3.4$. We then use Map to apply rosslerplot to the list $\{1.4, 2.4, 2.6, 3.4\}$. The resulting list, which corresponds to

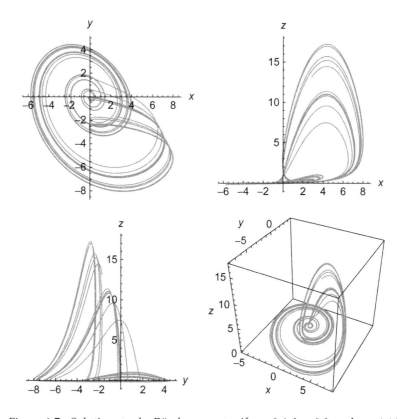

Figure 6-7   Solutions to the Rössler attractor if $a = 0.4$, $b = 0.3$, and $c = 4.44$

the numerical solutions to the initial-value problems is named `r1`; the resulting graphs are shown in Figure 6-7.

```
r1 = Map[rosslerplot[.4, .3, #][{1, .4, .7}, {t, 800, 1000}]&,

{1.4, 2.4, 2.6, 3.4}]; toshow = Partition[r1, 2];

Show[GraphicsGrid[toshow]]
```

In Figure 6-7, we see that the value of c dramatically affects the long-term behavior of the solutions: $c = 1.4$ results in a single limit cycle, $= 2.4$ results in a 2-cycle, $c = 2.6$ results in a 4-cycle, and $c = 3.4$ and $c = 4.44$ (see Figure 6-7) appear to be "chaotic."

We designed the `rossler2` function to return the numerical solutions instead of the graphics in case further manipulation of the numerical solutions is needed. For example, entering

```
rossler2[a_, b_, c_][{x0_, y0_, z0_}, ts_:{t, 950, 1000},

  opts___]:=Module[{numsol}, numsol = NDSolve[{x'[t] ==

  −y[t] − z[t], y'[t] == x[t] + ay[t], z'[t] == bx[t] − cz[t]+

  x[t]z[t], x[0] == x0, y[0] == y0, z[0] == z0}, {x[t], y[t],

  z[t]}, ts, MaxSteps → 100000]

]

r2 = Map[rossler2[.4, .3, #][{1, .4, .7}, {t, 800, 1000}]&,

{1.4, 2.4, 2.6, 3.4}];

r3 = Map[Plot[Evaluate[{x[t], y[t], z[t]}/.#], {t, 950, 1000},

  PlotPoints → 1000]&, r2];

Show[GraphicsGrid[Partition[r3, 2]]]
```

graphs each of the solutions $x(t)$, $y(t)$, and $z(t)$ in `r1`. The resulting array is shown in Figure 6-8.

■

6.2.2 Linear Systems

We now turn our attention to linear systems.

We begin our study of linear systems of ordinary differential equations by introducing several definitions along with some convenient notation.

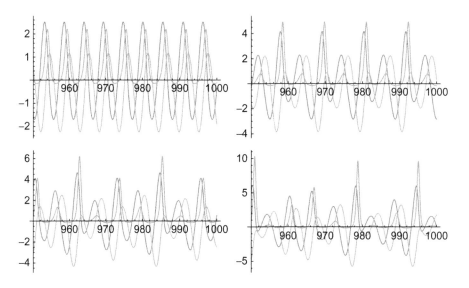

Figure 6-8   Plots of $x(t)$, $y(t)$, and $z(t)$ if $a = 0.4$, $b = 0.3$, and $c = 1.4, 2.4, 2.6, 3.4$

Let

$$\mathbf{X}(t) = \begin{pmatrix} x_1(t) \\ x_2(t) \\ \vdots \\ x_n(t) \end{pmatrix}, \quad \mathbf{A}(t) = \begin{pmatrix} a_{11}(t) & a_{12}(t) & \cdots & a_{1n}(t) \\ a_{21}(t) & a_{22}(t) & \cdots & a_{2n}(t) \\ \vdots & \vdots & \cdots & \vdots \\ a_{n1}(t) & a_{n2}(t) & \cdots & a_{nn}(t) \end{pmatrix}, \quad \text{and} \quad \mathbf{F}(t) = \begin{pmatrix} f_1(t) \\ f_2(t) \\ \vdots \\ f_n(t) \end{pmatrix}.$$

Then, the homogeneous system of first-order linear differential equations

$$\begin{cases} x_1' = a_{11}(t)x_1 + a_{12}(t)x_2 + \cdots + a_{1n}x_n(t) \\ x_2' = a_{21}(t)x_1 + a_{22}(t)x_2 + \cdots + a_{2n}x_n(t) \\ \vdots \\ x_n' = a_{n1}(t)x_1 + a_{n2}(t)x_2 + \cdots + a_{nn}x_n(t) \end{cases} \tag{6.9}$$

is equivalent to

$$\mathbf{X}'(t) = \mathbf{A}(t)\mathbf{X}(t) \tag{6.10}$$

and the nonhomogeneous system

$$\begin{cases} x_1' = a_{11}(t)x_1 + a_{12}(t)x_2 + \cdots + a_{1n}x_n(t) + f_1(t) \\ x_2' = a_{21}(t)x_1 + a_{22}(t)x_2 + \cdots + a_{2n}x_n(t) + f_2(t) \\ \vdots \\ x_n' = a_{n1}(t)x_1 + a_{n2}(t)x_2 + \cdots + a_{nn}x_n(t) + f_n(t) \end{cases} \qquad (6.11)$$

is equivalent to

$$\mathbf{X}'(t) = \mathbf{A}(t)\mathbf{X}(t) + \mathbf{F}(t) \qquad (6.12)$$

For the nonhomogeneous system (6.12), the **corresponding homogeneous system** is system (6.10).

EXAMPLE 6.2.6: (a) Write the homogeneous system $\begin{cases} x' = -5x + 5y \\ y' = -5x + y \end{cases}$ in matrix form. (b) Write the nonhomogeneous system $\begin{cases} x' = x + 2y - \sin t \\ y' = 4x - 3y + t^2 \end{cases}$ in matrix form.

SOLUTION: (a) The homogeneous system $\begin{cases} x' = -5x + 5y \\ y' = -5x + y \end{cases}$ is equiva-

lent to the system $\begin{pmatrix} x \\ y \end{pmatrix}' = \begin{pmatrix} -5 & 5 \\ -5 & 1 \end{pmatrix} \begin{pmatrix} x \\ y \end{pmatrix}$.

With our notation,

$$\begin{pmatrix} x \\ y \end{pmatrix}' = \begin{pmatrix} x' \\ y' \end{pmatrix}$$

(b) The nonhomogeneous system $\begin{cases} x' = x + 2y - \sin t \\ y' = 4x - 3y + t^2 \end{cases}$ is equivalent

to $\begin{pmatrix} x \\ y \end{pmatrix}' = \begin{pmatrix} 1 & 2 \\ 4 & -3 \end{pmatrix} \begin{pmatrix} x \\ y \end{pmatrix} + \begin{pmatrix} -\sin t \\ t^2 \end{pmatrix}$.

∎

The nth-order linear equation

$$y^{(n)}(t) + a_{n-1}(t)y^{(n-1)} + \cdots + a_2(t)y'' + a_1(t)y' + a_0(t)y = f(t), \qquad (6.13)$$

discussed in previous chapters, can be written as a system of first-order equations as well. Let $x_1 = y$, $x_2 = dx_1/dt = y'$, $x_3 = dx_2/dt = y''$, ..., $x_{n-1} =$

The nth-order linear equation is discussed in Chapter 4.

$dx_{n-2}/dt = y^{(n-2)}$, $x_n = dx_{n-1}/dt = y^{(n-1)}$. Then, equation (6.13) is equivalent to the system

$$
\begin{cases}
x_1' = x_2 \\
x_2' = x_3 \\
\vdots \\
x_{n-1}' = x_n \\
x_n' = -a_{n-1}x_n - \cdots - a_2 x_3 - a_1 x_2 - a_0 x_1 + f(t)
\end{cases}
, \qquad (6.14)
$$

which can be written in matrix form as

$$
\begin{pmatrix} x_1 \\ x_2 \\ \vdots \\ x_{n-1} \\ x_n \end{pmatrix}' =
\begin{pmatrix}
0 & 1 & 0 & \cdots & 0 \\
0 & 0 & 1 & \cdots & 0 \\
\vdots & \vdots & \vdots & \cdots & \vdots \\
0 & 0 & 0 & \cdots & 1 \\
-a_0 & -a_1 & -a_2 & \cdots & -a_n
\end{pmatrix}
\begin{pmatrix} x_1 \\ x_2 \\ \vdots \\ x_{n-1} \\ x_n \end{pmatrix}
+
\begin{pmatrix} 0 \\ 0 \\ \vdots \\ 0 \\ f(t) \end{pmatrix}. \qquad (6.15)
$$

EXAMPLE 6.2.7: Write the equation $y'' + 5y' + 6y = \cos t$ as a system of first-order differential equations.

SOLUTION: We let $x_1 = y$ and $x_2 = x_1' = y'$. Then,

$$x_2' = y'' = \cos t - 6y - 5y' = \cos t - 6x_1 - 5x_2$$

so the second-order equation $y'' + 5y' + 6y = \cos t$ is equivalent to the system

$$
\begin{cases}
x_1' = x_2 \\
x_2' = \cos t - 6x_1 - 5x_2
\end{cases},
$$

which can be written in matrix form as

$$
\begin{pmatrix} x_1 \\ x_2 \end{pmatrix}' =
\begin{pmatrix} 0 & 1 \\ -6 & -5 \end{pmatrix}
\begin{pmatrix} x_1 \\ x_2 \end{pmatrix}
+
\begin{pmatrix} 0 \\ \cos t \end{pmatrix}.
$$

∎

At this point, given a system of ordinary differential equations, our goal is to construct either an explicit, numerical, or graphical solution of the system of equations.

We now state the following theorems and terminology which are used in establishing the fundamentals of solving systems of differential equations. In each case, we assume that the matrix $\mathbf{A} = \mathbf{A}(t)$ in the systems $\mathbf{X}'(t) = \mathbf{A}(t)\mathbf{X}(t)$ (equation (6.10)) and $\mathbf{X}'(t) = \mathbf{A}(t)\mathbf{X}(t) + \mathbf{F}(t)$ (equation (6.12)) is an $n \times n$ matrix.

Definition 25 (Solution Vector). *A **solution vector** (or **solution**) of the system* $\mathbf{X}'(t) = \mathbf{A}(t)\mathbf{X}(t) + \mathbf{F}(t)$ *(equation (6.12)) on the interval I is an $n \times 1$ matrix (or vector) of the form*

$$\mathbf{X}(t) = \begin{pmatrix} x_1(t) \\ x_2(t) \\ \vdots \\ x_n(t) \end{pmatrix},$$

where the $x_i(t)$ are differentiable functions that satisfy $\mathbf{X}'(t) = \mathbf{A}(t)\mathbf{X}(t) + \mathbf{F}(t)$ *on I.*

Consider the homogeneous linear system $\mathbf{X}'(t) = \mathbf{A}(t)\mathbf{X}(t)$, where

$$\mathbf{X}(t) = \begin{pmatrix} x_1(t) \\ x_2(t) \\ \vdots \\ x_n(t) \end{pmatrix} \quad \text{and} \quad \mathbf{A}(t) = \begin{pmatrix} a_{11}(t) & a_{12}(t) & \cdots & a_{1n}(t) \\ a_{21}(t) & a_{22}(t) & \cdots & a_{2n}(t) \\ \vdots & \vdots & \cdots & \vdots \\ a_{n1}(t) & a_{n2}(t) & \cdots & a_{nn}(t) \end{pmatrix}$$

for which $a_{ij}(t)$ is continuous for all $1 \le i \le n$ and $1 \le j \le n$.

Let $\{\Phi_i(t)\}_{i=1}^m = \left\{ \begin{pmatrix} \Phi_{1i} \\ \Phi_{2i} \\ \vdots \\ \Phi_{mi} \end{pmatrix} \right\}_{i=1}^m$ be a set of m solutions of $\mathbf{X}'(t) = \mathbf{A}(t)\mathbf{X}(t)$. We

define *linear dependence* and *independence* of the set of vectors $\{\Phi_i(t)\}_{i=1}^m$ in the same way as we define linear dependence and independence of sets of functions. The set $\{\Phi_i(t)\}_{i=1}^m$ is **linearly dependent** on an interval I means that there is a set of constants $\{c_i\}_{i=1}^m$ not all zero such that $\sum_{i=1}^m c_i \Phi_i = \mathbf{0}$; otherwise, the set is **linearly independent**.

Definition 26 (Fundamental Set of Solutions). *A set* $\{\Phi_i(t)\}_{i=1}^n = \left\{ \begin{pmatrix} \Phi_{1i} \\ \Phi_{2i} \\ \vdots \\ \Phi_{ni} \end{pmatrix} \right\}_{i=1}^n$

of n linearly independent solution vectors of $\mathbf{X}'(t) = \mathbf{A}(t)\mathbf{X}(t)$ *on an interval I is called a* **fundamental set of solutions** *of* $\mathbf{X}'(t) = \mathbf{A}(t)\mathbf{X}(t)$ *on I.*

We can determine if a set of solution vectors is linearly independent or linearly dependent by computing the *Wronskian*.

Theorem 12. *The set* $\{\Phi_i(t)\}_{i=1}^n = \left\{ \begin{pmatrix} \Phi_{1i} \\ \Phi_{2i} \\ \vdots \\ \Phi_{ni} \end{pmatrix} \right\}_{i=1}^n$ *of solution vectors to* $\mathbf{X}'(t) = \mathbf{A}(t)\mathbf{X}(t)$

is linearly independent if and only if the **Wronskian**

$$W\left(\{\Phi_i\}_{i=1}^n\right) = \begin{vmatrix} \Phi_1 & \Phi_2 & \cdots & \Phi_n \end{vmatrix} = \begin{vmatrix} \Phi_{11} & \Phi_{12} & \cdots & \Phi_{1n} \\ \Phi_{21} & \Phi_{22} & \cdots & \Phi_{2n} \\ \vdots & \vdots & \cdots & \vdots \\ \Phi_{n1} & \Phi_{n2} & \cdots & \Phi_{nn} \end{vmatrix} \neq 0.$$

EXAMPLE 6.2.8: Which of the following is a fundamental set of solutions for

$$\begin{pmatrix} x \\ y \end{pmatrix}' = \begin{pmatrix} -2 & -8 \\ 1 & 2 \end{pmatrix} \begin{pmatrix} x \\ y \end{pmatrix}?$$

(a) $S_1 = \left\{ \begin{pmatrix} \cos 2t \\ \sin 2t \end{pmatrix}, \begin{pmatrix} \sin 2t \\ \cos 2t \end{pmatrix} \right\}$ (b) $S_2 = \left\{ \begin{pmatrix} -2\sin 2t + 2\cos 2t \\ \sin 2t \end{pmatrix}, \begin{pmatrix} 4\cos 2t \\ \sin 2t - \cos 2t \end{pmatrix} \right\}$

SOLUTION: We first remark that the equation $\begin{pmatrix} x \\ y \end{pmatrix}' = \begin{pmatrix} -2 & -8 \\ 1 & 2 \end{pmatrix} \begin{pmatrix} x \\ y \end{pmatrix}$

is equivalent to the system $\begin{cases} x' = -2x - 8y \\ y' = x + 2y \end{cases}$.

(a) Differentiating we see that

$$\begin{pmatrix} \cos 2t \\ \sin 2t \end{pmatrix}' = \begin{pmatrix} -2\sin 2t \\ 2\cos 2t \end{pmatrix} \neq \begin{pmatrix} -2\cos 2t - 8\sin 2t \\ \cos 2t + 2\sin 2t \end{pmatrix},$$

which shows us that $\begin{pmatrix} \cos 2t \\ \sin 2t \end{pmatrix}$ is not a solution of the system.

```
a = {{-2, -8}, {1, 2}};

v1 = {Cos[2t], Sin[2t]};

D[v1, t]

{-2Sin[2t], 2Cos[2t]}

a.v1

{-2Cos[2t] - 8Sin[2t], Cos[2t] + 2Sin[2t]}
```

Therefore, S_1 is not a fundamental set of solutions.

(b) First we verify that $\begin{pmatrix} -2\sin 2t + 2\cos 2t \\ \sin 2t \end{pmatrix}$ is a solution of the system.

```
v2 = {-2Sin[2t] + 2Cos[2t], Sin[2t]};

D[v2, t]

{-4Cos[2t] - 4Sin[2t], 2Cos[2t]}

a.v2//Simplify

{-4(Cos[2t] + Sin[2t]), 2Cos[2t]}
```

Next, we see that $\begin{pmatrix} 4\cos 2t \\ \sin 2t - \cos 2t \end{pmatrix}$ is a solution of the system.

```
v3 = {4Cos[2t], Sin[2t] - Cos[2t]};

D[v3, t]

{-8Sin[2t], 2Cos[2t] + 2Sin[2t]}

a.v3//Simplify

{-8Sin[2t], 2(Cos[2t] + Sin[2t])}
```

To see that these vectors are linearly independent, we compute the Wronskian.

```
m1 = Transpose[{v2, v3}];
```

```
MatrixForm[m1]
```

$$\begin{pmatrix} 2\text{Cos}[2t] - 2\text{Sin}[2t] & 4\text{Cos}[2t] \\ \text{Sin}[2t] & -\text{Cos}[2t] + \text{Sin}[2t] \end{pmatrix}$$

```
Simplify[Det[m1]]
```

-2

Thus, the set S_2 is a set of two linearly independent solutions of the system and, consequently, a fundamental set of solutions for the equation.

■

The following theorem tells us that a fundamental set of solutions cannot contain more than n vectors because more than n solutions would be linearly dependent.

Theorem 13. *Any $n + 1$ nontrivial solutions of $\mathbf{X}'(t) = \mathbf{A}(t)\mathbf{X}(t)$ are linearly dependent.*

Finally, we state the following theorems, which state that a fundamental set of solutions of $\mathbf{X}'(t) = \mathbf{A}(t)\mathbf{X}(t)$ exists and a general solution can (theoretically) be constructed.

Theorem 14. *There is a set of n nontrivial linearly independent solutions of $\mathbf{X}'(t) = \mathbf{A}(t)\mathbf{X}(t)$.*

Theorem 15 (General Solution). *Let $S = \{\Phi_i\}_{i=1}^{n} = \left\{ \begin{pmatrix} \Phi_{1i} \\ \Phi_{2i} \\ \vdots \\ \Phi_{ni} \end{pmatrix} \right\}_{i=1}^{n}$ be a set of n linearly independent solutions of $\mathbf{X}'(t) = \mathbf{A}(t)\mathbf{X}(t)$. Then every solution of $\mathbf{X}'(t) = \mathbf{A}(t)\mathbf{X}(t)$ is a linear combination of these solutions.*

In this case, S is said to be a **fundamental set of solutions** of $\mathbf{X}'(t) = \mathbf{A}(t)\mathbf{X}(t)$; a **general solution** of $\mathbf{X}'(t) = \mathbf{A}(t)\mathbf{X}(t)$ is

$$\mathbf{X}(t) = c_1\Phi_1(t) + c_2\Phi_2(t) + \cdots + c_n\Phi_n(t).$$

Definition 27 (Fundamental Matrix). *Let* $\{\Phi_i\}_{i=1}^n = \left\{ \begin{pmatrix} \Phi_{1i} \\ \Phi_{2i} \\ \vdots \\ \Phi_{ni} \end{pmatrix} \right\}_{i=1}^n$ *be a fundamental*

set of solutions for $\mathbf{X}'(t) = \mathbf{A}(t)\mathbf{X}(t)$. *Then*

$$\Phi(t) = \begin{pmatrix} \Phi_1 & \Phi_2 & \cdots & \Phi_n \end{pmatrix} = \begin{pmatrix} \Phi_{11} & \Phi_{12} & \cdots & \Phi_{1n} \\ \Phi_{21} & \Phi_{22} & \cdots & \Phi_{2n} \\ \vdots & \vdots & \cdots & \vdots \\ \Phi_{n1} & \Phi_{n2} & \cdots & \Phi_{nn} \end{pmatrix}$$

*is called a **fundamental matrix** of the system* $\mathbf{X}'(t) = \mathbf{A}(t)\mathbf{X}(t)$. *Thus, a general solution*

of the system $\mathbf{X}'(t) = \mathbf{A}(t)\mathbf{X}(t)$ *can be written as* $\mathbf{X}(t) = \Phi(t)\mathbf{C}$, *where* $\mathbf{C} = \begin{pmatrix} c_1 \\ c_2 \\ \vdots \\ c_n \end{pmatrix}$ *is a*

constant vector.

If $\Phi = \begin{pmatrix} \Phi_1 & \Phi_2 & \cdots & \Phi_n \end{pmatrix}$ is a fundamental matrix for $\mathbf{X}' = \mathbf{A}\mathbf{X}$, $\Phi' = \mathbf{A}\Phi$:

$$\begin{aligned} \Phi' &= \begin{pmatrix} \Phi_1' & \Phi_2' & \cdots & \Phi_n' \end{pmatrix} \\ &= \begin{pmatrix} \mathbf{A}\Phi_1 & \mathbf{A}\Phi_2 & \cdots & \mathbf{A}\Phi_n \end{pmatrix} \\ &= \mathbf{A} \begin{pmatrix} \Phi_1 & \Phi_2 & \cdots & \Phi_n \end{pmatrix} \\ &= \mathbf{A}\Phi. \end{aligned}$$

EXAMPLE 6.2.9: Show that $\Phi = \begin{pmatrix} e^{-2t} & -3e^{5t} \\ 2e^{-2t} & e^{5t} \end{pmatrix}$ is a fundamental

matrix for the system $\mathbf{X}'(t) = \begin{pmatrix} 4 & -3 \\ -2 & -1 \end{pmatrix} \mathbf{X}(t)$. Use this fundamental

matrix to find a general solution of $\mathbf{X}'(t) = \begin{pmatrix} 4 & -3 \\ -2 & -1 \end{pmatrix} \mathbf{X}(t)$.

SOLUTION: Because $\begin{pmatrix} e^{-2t} \\ 2e^{-2t} \end{pmatrix}' = \begin{pmatrix} -2e^{-2t} \\ -4e^{-2t} \end{pmatrix} = \begin{pmatrix} 4 & -3 \\ -2 & -1 \end{pmatrix} \begin{pmatrix} e^{-2t} \\ 2e^{-2t} \end{pmatrix}$ and

$\begin{pmatrix} -3e^{5t} \\ e^{5t} \end{pmatrix}' = \begin{pmatrix} -15e^{5t} \\ 5e^{5t} \end{pmatrix} = \begin{pmatrix} 4 & -3 \\ -2 & -1 \end{pmatrix} \begin{pmatrix} -3e^{5t} \\ e^{5t} \end{pmatrix}$, both $\mathbf{X}_1 = \begin{pmatrix} e^{-2t} \\ 2e^{-2t} \end{pmatrix}$ and

$\mathbf{X}_2 = \begin{pmatrix} -3e^{5t} \\ e^{5t} \end{pmatrix}$ are solutions of the system $\mathbf{X}'(t) = \begin{pmatrix} 4 & -3 \\ -2 & -1 \end{pmatrix} \mathbf{X}(t)$.

Alternatively, we show that $\Phi'(t)$ and $\begin{pmatrix} 4 & -3 \\ -2 & -1 \end{pmatrix} \Phi(t)$ are the same.

```
a = {{4, -3}, {-2, -1}};

Φ[t_] = {{Exp[-2 t], -3Exp[5 t]},

{2Exp[-2 t], Exp[5 t]}};

MatrixForm[Φ[t]]
```

$$\begin{pmatrix} e^{-2t} & -3e^{5t} \\ 2e^{-2t} & e^{5t} \end{pmatrix}$$

```
Φ'[t]//MatrixForm
```

$$\begin{pmatrix} -2e^{-2t} & -15e^{5t} \\ -4e^{-2t} & 5e^{5t} \end{pmatrix}$$

```
a.Φ[t]//MatrixForm
```

$$\begin{pmatrix} -2e^{-2t} & -15e^{5t} \\ -4e^{-2t} & 5e^{5t} \end{pmatrix}$$

```
Det[Φ[t]]
```

$7e^{3t}$

A general solution is given by

$$\mathbf{X}(t) = \Phi(t)\mathbf{C} = \begin{pmatrix} e^{-2t} & -3e^{5t} \\ 2e^{-2t} & e^{5t} \end{pmatrix} \begin{pmatrix} c_1 \\ c_2 \end{pmatrix} = \begin{pmatrix} c_1 e^{-2t} - 3c_2 e^{5t} \\ 2c_1 e^{-2t} + c_2 e^{5t} \end{pmatrix}$$

$$= c_1 \begin{pmatrix} e^{-2t} \\ 2e^{-2t} \end{pmatrix} + c_2 \begin{pmatrix} -3e^{5t} \\ e^{5t} \end{pmatrix}.$$

Figure 6-9 shows the direction fields associated with the two systems. For two-dimensional systems, direction fields can often provide insight into the behavior of solutions without actually solving the system.

```
p1 = StreamPlot[{-2x - 8y, x + 2y}, {x, -1, 1}, {y, -1, 1},

StreamPoints → Fine, Frame → False, Axes → Automatic,
```

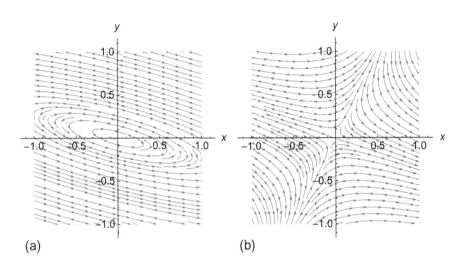

Figure 6-9   Using direction fields to compare the behavior of solutions to linear systems

```
    AxesOrigin → {0, 0}, AxesLabel → {x, y},

    PlotLabel → "(a)"]

p2 = StreamPlot[{4x − 3y, −2x − y}, {x, −1, 1}, {y, −1, 1},

    StreamPoints → Fine, Frame → False, Axes → Automatic,

    AxesOrigin → {0, 0}, AxesLabel → {x, y},

    PlotLabel → "(b)"]

Show[GraphicsRow[{p1, p2}]]
```

■

6.3 Homogeneous Linear Systems With Constant Coefficients

Now that we have covered the necessary terminology, we can turn our attention to solving linear systems with constant coefficients. Let

$$A = \begin{pmatrix} a_{11} & a_{12} & \cdots & a_{1n} \\ a_{21} & a_{22} & \cdots & a_{2n} \\ \vdots & \vdots & \cdots & \vdots \\ a_{n1} & a_{n2} & \cdots & a_{nn} \end{pmatrix}$$

be an $n \times n$ matrix with real-valued entries and let $\{\lambda_k\}$ be the eigenvalues and $\{\mathbf{v}_k\}$ the corresponding eigenvectors of $\mathbf{A}$. Then a general solution of the system $\mathbf{X}' = \mathbf{AX}$ is determined by the eigenvalues and corresponding eigenvectors of $\mathbf{A}$. For the moment, we consider the cases when the eigenvalues of $\mathbf{A}$ are distinct and real or the eigenvalues of $\mathbf{A}$ are distinct and complex. We will consider the case when $\mathbf{A}$ has repeated eigenvalues (eigenvalues of multiplicity greater than one) separately.

6.3.1 Distinct Real Eigenvalues

Let λ be an eigenvalue of $\mathbf{A}$ with corresponding eigenvector $\mathbf{v}$. Then,

$$\lambda \mathbf{v} = \mathbf{Av}$$
$$\lambda \mathbf{v} e^{\lambda t} = \mathbf{Av} e^{\lambda t}$$
$$\frac{d}{dt}\left(\mathbf{v}e^{\lambda t}\right) = \mathbf{A}\left(\mathbf{v}e^{\lambda t}\right),$$

which shows that $\mathbf{v}e^{\lambda t}$ is a solution of $\mathbf{X}'(t) = \mathbf{A}(t)\mathbf{X}(t)$.

If the eigenvalues $\{\lambda_k\}_{k=1}^{n}$ of $\mathbf{A}$ are distinct with corresponding eigenvectors $\{\mathbf{v}_k\}_{k=1}^{n}$,

$$S = \left\{\mathbf{v}_1 e^{\lambda_1 t},\ \mathbf{v}_2 e^{\lambda_2 t},\ \ldots,\ \mathbf{v}_n e^{\lambda_n t}\right\}$$

Eigenvectors corresponding to distinct eigenvalues are linearly independent.

is a fundamental set of solutions for $\mathbf{X}'(t) = \mathbf{A}(t)\mathbf{X}(t)$ because

$$W(S) = \left|\mathbf{v}_1 e^{\lambda_1 t} \quad \mathbf{v}_2 e^{\lambda_2 t} \quad \cdots \quad \mathbf{v}_n e^{\lambda_n t}\right|$$
$$= e^{(\lambda_1 + \lambda_2 + \cdots + \lambda_n)t}\left|\mathbf{v}_1 \quad \mathbf{v}_2 \quad \cdots \quad \mathbf{v}_n\right| \neq 0.$$

Therefore, a general solution of $\mathbf{X}'(t) = \mathbf{A}(t)\mathbf{X}(t)$ is

$$\mathbf{X} = c_1 \mathbf{v}_1 e^{\lambda_1 t} + c_2 \mathbf{v}_2 e^{\lambda_2 t} + \cdots + c_n \mathbf{v}_n e^{\lambda_n t}$$

and a fundamental matrix for $\mathbf{X}'(t) = \mathbf{A}(t)\mathbf{X}(t)$ is

$$\Phi = \left(\mathbf{v}_1 e^{\lambda_1 t} \quad \mathbf{v}_2 e^{\lambda_2 t} \quad \cdots \quad \mathbf{v}_n e^{\lambda_n t} \right).$$

Remark. Use `StreamPlot` to graph direction fields for two dimensional systems. The command

```
StreamPlot[{f[x,y],g[x,y]},{x,a,b},{y,c,d}]
```

generates a basic direction field for the system $\{x' = f(x, y), y' = g(x, y)\}$ for $a \le x \le b$ and $c \le y \le d$.

EXAMPLE 6.3.1: Solve (a) $\begin{cases} x' = 5x - y \\ y' = 3y \end{cases}$ and (b) $\mathbf{X}' = \begin{pmatrix} -1/2 & -1/3 \\ -1/3 & -1/2 \end{pmatrix} \mathbf{X}.$

SOLUTION: (a) In matrix form the system is $\mathbf{X}' = \begin{pmatrix} 5 & -1 \\ 0 & 3 \end{pmatrix} \mathbf{X}$. We find

the eigenvalues of $\mathbf{A} = \begin{pmatrix} 5 & -1 \\ 0 & 3 \end{pmatrix}$ with `Eigensystem`.

```
a = {{5, -1}, {0, 3}};
```

```
Eigensystem[a]
```

```
{{5, 3}, {{1, 0}, {1, 2}}}
```

The results indicate that the eigenvalues are $\lambda_1 = 3$ and $\lambda_2 = 5$ with

corresponding eigenvectors $\mathbf{v}_1 = \begin{pmatrix} 1 \\ 2 \end{pmatrix}$ and $\mathbf{v}_2 = \begin{pmatrix} 1 \\ 0 \end{pmatrix}$, respectively.

Therefore, $S = \left\{ \begin{pmatrix} 1 \\ 2 \end{pmatrix} e^{3t}, \begin{pmatrix} 1 \\ 0 \end{pmatrix} e^{5t} \right\}$ is a fundamental set of solutions of the system, a general solution is

$$\mathbf{X} = c_1 \mathbf{v}_1 e^{\lambda_1 t} + c_2 \mathbf{v}_2 e^{\lambda_2 t} = c_1 \begin{pmatrix} 1 \\ 2 \end{pmatrix} e^{3t} + c_2 \begin{pmatrix} 1 \\ 0 \end{pmatrix} e^{5t},$$

and a fundamental matrix is $\Phi = \begin{pmatrix} e^{3t} & e^{5t} \\ 2e^{3t} & 0 \end{pmatrix}$. We can write the general

solution as $\begin{cases} x = c_1 e^{3t} + c_2 e^{5t} \\ y = 2c_1 e^{3t} \end{cases}$ or as $\mathbf{X} = \begin{pmatrix} e^{3t} & e^{5t} \\ 2e^{3t} & 0 \end{pmatrix} \begin{pmatrix} c_1 \\ c_2 \end{pmatrix}$.

We can use `DSolve` to find a general solution as well.

```
sols = DSolve[{x'[t], y'[t]} == a.{x[t], y[t]}, {x[t], y[t]}, t]
```

$$\{\{x[t] \to e^{5t}C[1] - \frac{1}{2}e^{3t}(-1+e^{2t})C[2], \; y[t] \to e^{3t}C[2]\}\}$$

We can graph the solution parametrically for various values of c_1 and c_2 with `ParametricPlot`. First, we use `Table` to generate a list corresponding to replacing c_1 and c_2 in $\begin{cases} x = c_1 e^{3t} + c_2 e^{5t} \\ y = 2c_1 e^{3t} \end{cases}$ by $-2, -1, 0,$ 1, and 2. The result in `step1`, however, is *not* a list of ordered pairs of functions corresponding to $\begin{cases} x = x(t) \\ y = y(t) \end{cases}$; it is a nested list.

```
step1 = Table[{x[t], y[t]}/.sols/.{C[1] → i, C[2] → j},

    {i, -2, 2}, {j, -2, 2}]
```

$$\{\{\{\{-2e^{5t} + e^{3t}(-1+e^{2t}), \; -2e^{3t}\}\}, \{\{-2e^{5t} + \frac{1}{2}e^{3t}(-1+e^{2t}),$$
$$-e^{3t}\}\}, \{\{-2e^{5t}, 0\}\}, \{\{-2e^{5t} - \frac{1}{2}e^{3t}(-1+e^{2t}), e^{3t}\}\}, \{\{-2e^{5t} -$$
$$e^{3t}(-1+e^{2t}), 2e^{3t}\}\}\}, \{\{\{-e^{5t} + e^{3t}(-1+e^{2t}), \; -2e^{3t}\}\}, \{\{-e^{5t} +$$
$$\frac{1}{2}e^{3t}(-1+e^{2t}), \; -e^{3t}\}\}, \{\{-e^{5t}, 0\}\}, \{\{-e^{5t} - \frac{1}{2}e^{3t}(-1+e^{2t}),$$
$$e^{3t}\}\}, \{\{-e^{5t} - e^{3t}(-1+e^{2t}), 2e^{3t}\}\}\}, \{\{\{e^{3t}(-1+e^{2t}), \; -2e^{3t}\}\},$$
$$\{\{\frac{1}{2}e^{3t}(-1+e^{2t}), \; -e^{3t}\}\}, \{\{0, 0\}\}, \{\{-\frac{1}{2}e^{3t}(-1+e^{2t}), e^{3t}\}\},$$
$$\{\{-e^{3t}(-1+e^{2t}), 2e^{3t}\}\}\}, \{\{\{e^{5t} + e^{3t}(-1+e^{2t}), \; -2e^{3t}\}\}, \{\{e^{5t} +$$
$$\frac{1}{2}e^{3t}(-1+e^{2t}), \; -e^{3t}\}\}, \{\{e^{5t}, 0\}\}, \{\{e^{5t} - \frac{1}{2}e^{3t}(-1+e^{2t}), e^{3t}\}\},$$
$$\{\{e^{5t} - e^{3t}(-1+e^{2t}), 2e^{3t}\}\}\}, \{\{\{2e^{5t} + e^{3t}(-1+e^{2t}), \; -2e^{3t}\}\},$$
$$\{\{2e^{5t} + \frac{1}{2}e^{3t}(-1+e^{2t}), \; -e^{3t}\}\}, \{\{2e^{5t}, 0\}\}, \{\{2e^{5t} - \frac{1}{2}e^{3t}(-1+$$
$$e^{2t}), e^{3t}\}\}, \{\{2e^{5t} - e^{3t}(-1+e^{2t}), 2e^{3t}\}\}\}\}$$

To create a list of ordered pairs of functions that we can graph with `ParametricPlot`, we use `Flatten`.

`Flatten[list,n]`
flattens `list` to level *n*.

```
toplot = Flatten[step1, 1]
```

$$\{\{\{-2e^{5t} + e^{3t}(-1+e^{2t}), \; -2e^{3t}\}\}, \{\{-2e^{5t} + \frac{1}{2}e^{3t}(-1+e^{2t}),$$
$$-e^{3t}\}\}, \{\{-2e^{5t}, 0\}\}, \{\{-2e^{5t} - \frac{1}{2}e^{3t}(-1+e^{2t}), e^{3t}\}\}, \{\{-2e^{5t} -$$
$$e^{3t}(-1+e^{2t}), 2e^{3t}\}\}, \{\{-e^{5t} + e^{3t}(-1+e^{2t}), \; -2e^{3t}\}\}, \{\{-e^{5t} +$$
$$\frac{1}{2}e^{3t}(-1+e^{2t}), \; -e^{3t}\}\}, \{\{-e^{5t}, 0\}\}, \{\{-e^{5t} - \frac{1}{2}e^{3t}(-1+e^{2t}),$$
$$e^{3t}\}\}, \{\{-e^{5t} - e^{3t}(-1+e^{2t}), 2e^{3t}\}\}, \{\{e^{3t}(-1+e^{2t}), \; -2e^{3t}\}\},$$
$$\{\{\frac{1}{2}e^{3t}(-1+e^{2t}), \; -e^{3t}\}\}, \{\{0, 0\}\}, \{\{-\frac{1}{2}e^{3t}(-1+e^{2t}), e^{3t}\}\},$$

$$\{\{-e^{3t}(-1+e^{2t}),\, 2e^{3t}\}\},\, \{\{e^{5t}+e^{3t}(-1+e^{2t}),\, -2e^{3t}\}\},\, \{\{e^{5t}+$$
$$\frac{1}{2}e^{3t}(-1+e^{2t}),\, -e^{3t}\}\},\, \{\{e^{5t},\, 0\}\},\, \{\{e^{5t}-\frac{1}{2}e^{3t}(-1+e^{2t}),\, e^{3t}\}\},$$
$$\{\{e^{5t}-e^{3t}(-1+e^{2t}),\, 2e^{3t}\}\},\, \{\{2e^{5t}+e^{3t}(-1+e^{2t}),\, -2e^{3t}\}\},$$
$$\{\{2e^{5t}+\frac{1}{2}e^{3t}(-1+e^{2t}),\, -e^{3t}\}\},\, \{\{2e^{5t},\, 0\}\},\, \{\{2e^{5t}-\frac{1}{2}e^{3t}(-1+$$
$$e^{2t}),\, e^{3t}\}\},\, \{\{2e^{5t}-e^{3t}(-1+e^{2t}),\, 2e^{3t}\}\}\}$$

Next, we use `ParametricPlot` to graph the list of parametric functions in `toplot` and name the resulting graphics object `pp1`.

```
pp1 = ParametricPlot[toplot, {t, −1, 1}]
```

To show the graphs of the solutions together with the direction field associated with the system, we use `StreamPlot` to graph the direction field associated with the system on the rectangle $[-5,5] \times [-5,5]$, naming the resulting graphics object `pvf`.

```
pvf = StreamPlot[{5x − y, 3y},

{x, −5, 5}, {y, −5, 5}, StreamStyle → Fine]
```

`Show` is then used to display the graphs together in Figure 6-10.

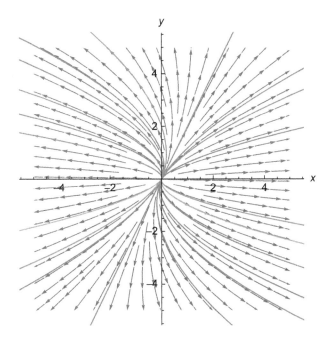

Figure 6-10   All nontrivial solutions move away from the origin as t increases

```
Show[pvf, pp1, PlotRange → {{-5, 5}, {-5, 5}},
    AspectRatio → 1, Frame → False, Axes → Automatic,
    AxesOrigin → {0, 0}, AxesLabel → {x, y}]
```

Notice that each curve corresponds to the parametric plot of the pair $\begin{cases} x = x(t) \\ y = y(t) \end{cases}$. Because both eigenvalues are positive, all solutions move away from the origin as t increases. The arrows on the vectors in the direction field show this behavior.

(b) With `Eigensystem`, we see that the eigenvalues and eigenvectors of
$\mathbf{A} = \begin{pmatrix} -1/2 & -1/3 \\ -1/3 & -1/2 \end{pmatrix}$ are $\lambda_1 = -1/6$ and $\lambda_2 = -5/6$ and $\mathbf{v}_1 = \begin{pmatrix} -1 \\ 1 \end{pmatrix}$ and
$\mathbf{v}_2 = \begin{pmatrix} 1 \\ 1 \end{pmatrix}$, respectively.

```
capa = {{-1/2, -1/3}, {-1/3, -1/2}};

Eigensystem[capa]
```

$$\left\{ \left\{ -\frac{5}{6}, -\frac{1}{6} \right\}, \{\{1, 1\}, \{-1, 1\}\} \right\}$$

Then $\mathbf{X}_1 = \begin{pmatrix} -1 \\ 1 \end{pmatrix} e^{-t/6}$ and $\mathbf{X}_2 = \begin{pmatrix} 1 \\ 1 \end{pmatrix} e^{-5t/6}$ are two linearly independent solutions of the system so a general solution is $\mathbf{X} = \begin{pmatrix} -e^{-t/6} & e^{-5t/6} \\ e^{-t/6} & e^{-5t/6} \end{pmatrix} \begin{pmatrix} c_1 \\ c_2 \end{pmatrix}$; a fundamental matrix is $\Phi = \begin{pmatrix} -e^{-t/6} & e^{-5t/6} \\ e^{-t/6} & e^{-5t/6} \end{pmatrix}$.

We use `DSolve` to find a general solution of the system by entering

```
gensol = DSolve[{x'[t] == -1/2x[t] - 1/3y[t],
    y'[t] == -1/3x[t] - 1/2y[t]}, {x[t], y[t]}, t]
```

$$\left\{ \left\{ x[t] \rightarrow \frac{1}{2}e^{-5t/6}(1 + e^{2t/3})C[1] - \frac{1}{2}e^{-5t/6}(-1 + e^{2t/3})C[2], \ y[t] \rightarrow -\frac{1}{2}e^{-5t/6}(-1 + e^{2t/3})C[1] + \frac{1}{2}e^{-5t/6}(1 + e^{2t/3})C[2] \right\} \right\}$$

Similarly, we use `DSolve` to solve the initial-value problem using the conditions $x(0) = x_0$ and $y(0) = y_0$.

```
partsol = DSolve[{x'[t] == -1/2x[t] - 1/3y[t],
    y'[t] == -1/3x[t] - 1/2y[t], x[0] == x0, y[0] == y0},
    {x[t], y[t]}, t]
```

$$\left\{ \left\{ x[t] \rightarrow \frac{1}{2}e^{-5t/6}(x0 + e^{2t/3}x0 + y0 - e^{2t/3}y0), \ y[t] \rightarrow -\frac{1}{2}e^{-5t/6}(-x0 + e^{2t/3}x0 - y0 - e^{2t/3}y0) \right\} \right\}$$

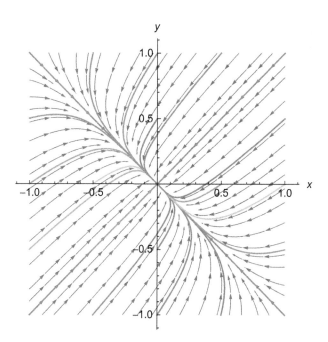

Figure 6-11   Direction field for $\mathbf{X}' = \mathbf{AX}$ along with the graphs of various solutions

We graph the direction field with `StreamPlot`.

```
pvf = StreamPlot[{-1/2x - 1/3y, -1/3x - 1/2y},
    {x, -1, 1}, {y, -1, 1}, StreamStyle → Fine]
```

Several solutions are also graphed with `ParametricPlot` and shown together with the direction field in Figure 6-11.

```
t1 =
    ParametricPlot[
    Evaluate[Flatten[Table[{x[t], y[t]}/.partsol/.
    {x0 → i, y0 → j}, {i, -1, 1, 2/4}, {j, -1, 1, 2/4}],
    1]], {t, 0, 20}]

Show[pvf, t1, PlotRange → {{-1, 1}, {-1, 1}},
    AspectRatio → 1, Frame → False, Axes → Automatic,
    AxesOrigin → {0, 0}, AxesLabel → {x, y}]
```

■

6.3.2 Complex Conjugate Eigenvalues

We use Euler's formula,
$e^{i\theta} = \cos\theta + i\sin\theta$.

If $\mathbf{A}$ has complex conjugate eigenvalues $\lambda_{1,2} = \alpha \pm \beta i$, $\beta \neq 0$, and corresponding eigenvectors $\mathbf{v}_{1,2} = \mathbf{a} \pm \mathbf{b}i$, then one solution of $\mathbf{X}' = \mathbf{A}\mathbf{X}$ is

$$\mathbf{X} = \mathbf{v}_1 e^{\lambda_1 t} = (\mathbf{a} + \mathbf{b}i)e^{(\alpha + \beta i)t} = e^{\alpha t}(\mathbf{a} + \mathbf{b}i)e^{i\beta t} = e^{\alpha t}(\mathbf{a} + \mathbf{b}i)(\cos\beta t + i\sin\beta t)$$
$$= e^{\alpha t}(\mathbf{a}\cos\beta t - \mathbf{b}\sin\beta t) + ie^{\alpha t}(\mathbf{b}\cos\beta t + \mathbf{a}\sin\beta t)$$
$$= \mathbf{X}_1(t) + i\mathbf{X}_2(t).$$

Now, because $\mathbf{X}$ is a solution of the system, $\mathbf{X}' = \mathbf{A}\mathbf{X}$, we have $\mathbf{X}_1' + i\mathbf{X}_2' = \mathbf{A}\mathbf{X}_1 + i\mathbf{A}\mathbf{X}_2$. Equating the real and imaginary parts of this equation yields $\mathbf{X}_1' = \mathbf{A}\mathbf{X}_1$ and $\mathbf{X}_2' = \mathbf{A}\mathbf{X}_2$. Therefore, $\mathbf{X}_1$ and $\mathbf{X}_2$ are solutions of $\mathbf{X}' = \mathbf{A}\mathbf{X}$, so any linear combination of $\mathbf{X}_1$ and $\mathbf{X}_2$ is also a solution. We can show that $\mathbf{X}_1$ and $\mathbf{X}_2$ are linearly independent, so this linear combination forms a portion of a general solution of $\mathbf{X}' = \mathbf{A}\mathbf{X}$.

Theorem 16. *If $\mathbf{A}$ has complex conjugate eigenvalues $\lambda_{1,2} = \alpha \pm \beta i$, $\beta \neq 0$, and corresponding eigenvectors $\mathbf{v}_{1,2} = \mathbf{a} \pm \mathbf{b}i$, then two linearly independent solutions of $\mathbf{X}' = \mathbf{A}\mathbf{X}$ are*

$$\mathbf{X}_1 = e^{\alpha t}(\mathbf{a}\cos\beta t - \mathbf{b}\sin\beta t) \quad \text{and} \quad \mathbf{X}_2 = e^{\alpha t}(\mathbf{b}\cos\beta t + \mathbf{a}\sin\beta t).$$

Notice that in the case of complex conjugate eigenvalues, we are able to obtain two linearly independent solutions from knowing one of the eigenvalues and an eigenvector that corresponds to it.

Observe that our chosen eigenvectors are scalar multiples of the eigenvectors found with Mathematica.

EXAMPLE 6.3.2: Solve (a) $\begin{cases} x' = \frac{1}{2}y \\ y' = -\frac{1}{8}x \end{cases}$ and (b) $\begin{cases} dx/dt = -\frac{1}{4}x + 2y \\ dy/dt = -8x - \frac{1}{4}y \end{cases}$.

SOLUTION: (a) In matrix form the system is equivalent to the system $\mathbf{X}' = \begin{pmatrix} 0 & 1/2 \\ -1/8 & 0 \end{pmatrix} \mathbf{X} = \mathbf{A}\mathbf{X}$, where $\mathbf{A} = \begin{pmatrix} 0 & 1/2 \\ -1/8 & 0 \end{pmatrix}$. As in the previous example we use `Eigensystem` to see that the eigenvalues and eigenvectors of $\mathbf{A} = \begin{pmatrix} 0 & 1/2 \\ -1/8 & 0 \end{pmatrix}$ are $\lambda_{1,2} = 0 \pm \frac{1}{4}i$ and $\mathbf{v}_{1,2} = \begin{pmatrix} 1 \\ 0 \end{pmatrix} \pm \begin{pmatrix} 0 \\ 1/2 \end{pmatrix}i$.

```
capa = {{0, 1/2}, {-1/8, 0}};
```

```
Eigensystem[capa]
```

$$\left\{\left\{\frac{i}{4}, -\frac{i}{4}\right\}, \{\{-2i, 1\}, \{2i, 1\}\}\right\}$$

Two linearly independent solutions of $\mathbf{X}' = \mathbf{AX}$ are then $\mathbf{X}_1 = \begin{pmatrix} 1 \\ 0 \end{pmatrix} \cos\frac{1}{4}t - \begin{pmatrix} 0 \\ 1/2 \end{pmatrix} \sin\frac{1}{4}t = \begin{pmatrix} \cos\frac{1}{4}t \\ -\frac{1}{2}\sin\frac{1}{4}t \end{pmatrix}$ and $\mathbf{X}_2 = \begin{pmatrix} 1 \\ 0 \end{pmatrix} \sin\frac{1}{4}t + \begin{pmatrix} 0 \\ 1/2 \end{pmatrix} \cos\frac{1}{4}t = \begin{pmatrix} \sin\frac{1}{4}t \\ \frac{1}{2}\cos\frac{1}{4}t \end{pmatrix}$ and a general solution is $\mathbf{X} = c_1\mathbf{X}_1 + c_2\mathbf{X}_2 = \begin{pmatrix} \cos\frac{1}{4}t & \sin\frac{1}{4}t \\ -\frac{1}{2}\sin\frac{1}{4}t & \frac{1}{2}\cos\frac{1}{4}t \end{pmatrix}\begin{pmatrix} c_1 \\ c_2 \end{pmatrix}$ or $x = c_1\cos\frac{1}{4}t + c_2\sin\frac{1}{4}t$ and $y = -c_1\frac{1}{2}\sin\frac{1}{4}t + \frac{1}{2}c_2\cos\frac{1}{4}t$.

As before, we use DSolve to find a general solution.

```
gensol = DSolve[{x'[t] == 1/2y[t],

    y'[t] == -1/8x[t]}, {x[t], y[t]}, t]
```

$$\left\{\left\{x[t] \;\rightarrow\; C[1]\mathrm{Cos}[\frac{t}{4}] + 2C[2]\mathrm{Sin}[\frac{t}{4}], \; y[t] \;\rightarrow\; C[2]\mathrm{Cos}[\frac{t}{4}] - \frac{1}{2}C[1]\mathrm{Sin}[\frac{t}{4}]\right\}\right\}$$

Initial value problems for systems are solved in the same way as for other equations. For example, entering

```
partsol = DSolve[{x'[t] == 1/2y[t],

    y'[t] == -1/8x[t], x[0] == 1, y[0] == -1},

    {x[t], y[t]}, t]
```

$$\left\{\left\{x[t] \rightarrow \mathrm{Cos}[\frac{t}{4}] - 2\mathrm{Sin}[\frac{t}{4}], \; y[t] \rightarrow \frac{1}{2}(-2\mathrm{Cos}[\frac{t}{4}] - \mathrm{Sin}[\frac{t}{4}])\right\}\right\}$$

finds the solution to the system that satisfies $x(0) = 1$ and $y(0) = -1$. We graph $x(t)$ and $y(t)$ together as well as parametrically with Plot and ParametricPlot, respectively, in Figure 6-12.

```
p1 = Plot[Evaluate[{x[t], y[t]}/.partsol], {t, 0, 8Pi}]
```

```
p2 = ParametricPlot[{x[t], y[t]}/.partsol, {t, 0, 8Pi}]
```

```
Show[GraphicsRow[{p1, p2}]]
```

We can also use StreamPlot and StreamPlot to graph the direction field and/or various solutions as we do next in Figure 6-13.

```
pvf = StreamPlot[{1/2y, -1/8x},

    {x, -3, 3}, {y, -3, 3}, StreamStyle → Fine]
```

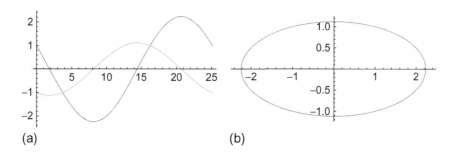

Figure 6-12   (a) Graph of $x(t)$ and $y(t)$. (b) Parametric plot of $x(t)$ versus $y(t)$

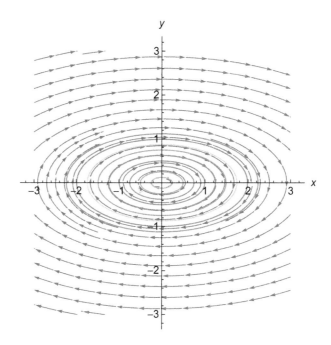

Figure 6-13   Notice that all nontrivial solutions are periodic

```
initsol = DSolve[{x'[t] == 1/2y[t],
    y'[t] == -1/8x[t], x[0] == x0, y[0] == y0},
    {x[t], y[t]}, t]
```

$$\left\{\left\{x[t] \rightarrow x0\cos[\tfrac{t}{4}]+2y0\sin[\tfrac{t}{4}],\ y[t] \rightarrow \tfrac{1}{2}(2y0\cos[\tfrac{t}{4}]-x0\sin[\tfrac{t}{4}])\right\}\right\}$$

```
t1 = ParametricPlot[Evaluate[Table[{x[t], y[t]}/.initsol/.

{x0 → i, y0 → i}, {i, 0, 1, 1/9}]], {t, 0, 8Pi}]

Show[pvf, t1, PlotRange → {{-3, 3}, {-3, 3}},

    AspectRatio → 1, Frame → False, Axes → Automatic,

    AxesOrigin → {0, 0}, AxesLabel → {x, y}]
```

(b) In matrix form, the system is equivalent to the system $\mathbf{X}' = \begin{pmatrix} -\frac{1}{4} & 2 \\ -8 & -\frac{1}{4} \end{pmatrix} \mathbf{X} = \mathbf{AX}$, where $\mathbf{A} = \begin{pmatrix} -\frac{1}{4} & 2 \\ -8 & -\frac{1}{4} \end{pmatrix}$. The eigenvalues and corresponding eigenvectors of $\mathbf{A} = \begin{pmatrix} -\frac{1}{4} & 2 \\ -8 & -\frac{1}{4} \end{pmatrix}$ are found to be

$\lambda_{1,2} = -\frac{1}{4} \pm 4i$ and $\mathbf{v}_{1,2} = \begin{pmatrix} 0 \\ 2 \end{pmatrix} \pm \begin{pmatrix} 1 \\ 0 \end{pmatrix} i$ with Eigensystem.

```
capa = {{-1/4, 2}, {-8, -1/4}};

Eigensystem[capa]
```

$$\left\{\left\{-\frac{1}{4} + 4i, -\frac{1}{4} - 4i\right\}, \{\{-i, 2\}, \{i, 2\}\}\right\}$$

A general solution is then

$$\mathbf{X} = c_1\mathbf{X}_1 + c_2\mathbf{X}_2$$

$$= c_1 e^{-t/4}\left(\begin{pmatrix} 1 \\ 0 \end{pmatrix}\cos 4t - \begin{pmatrix} 0 \\ 2 \end{pmatrix}\sin 4t\right)$$

$$+ c_2 e^{-t/4}\left(\begin{pmatrix} 1 \\ 0 \end{pmatrix}\sin 4t + \begin{pmatrix} 0 \\ 2 \end{pmatrix}\cos 4t\right)$$

$$= e^{-t/4}\left[c_1\begin{pmatrix} \cos 4t \\ -2\sin 4t \end{pmatrix} + c_2\begin{pmatrix} \sin 4t \\ 2\cos 4t \end{pmatrix}\right]$$

$$= e^{-t/4}\begin{pmatrix} \cos 4t & \sin 4t \\ -2\sin 4t & 2\cos 4t \end{pmatrix}\begin{pmatrix} c_1 \\ c_2 \end{pmatrix}$$

or $x = e^{-t/4}(c_1\cos 4t + c_2\sin 4t)$ and $y = e^{-t/4}(2c_2\cos 4t - 2c_1\sin 4t)$. We confirm this result using DSolve.

```
gensol = DSolve[{x'[t] ==

    -1/4x[t] + 2y[t], y'[t] ==

    -8x[t] - 1/4y[t]}, {x[t], y[t]}, t]
```

$$\left\{\left\{x[t] \to e^{-t/4}C[1]\text{Cos}[4t] + \frac{1}{2}e^{-t/4}C[2]\text{Sin}[4t], y[t] \to e^{-t/4}\right.\right.$$
$$\left.\left. C[2]\text{Cos}[4t] - 2e^{-t/4}C[1]\text{Sin}[4t]\right\}\right\}$$

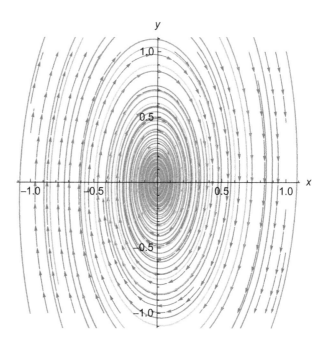

Figure 6-14   Various solutions and direction field associated with the system

We use StreamPlot and ParametricPlot to graph the direction field associated with the system along with various solutions in Figure 6-14.

```
pvf = StreamPlot[{-1/4x + 2y, -8x - 1/4y},
    {x, -1, 1}, {y, -1, 1}, StreamStyle → Fine]

initsol = DSolve[{x'[t] ==
    -1/4x[t] + 2y[t], y'[t] ==
    -8x[t] - 1/4y[t], x[0] == x0, y[0] == y0},
    {x[t], y[t]}, t]
```

$\{\{x[t] \to \frac{1}{2}e^{-t/4}(2x0\text{Cos}[4t]+y0\text{Sin}[4t]), y[t] \to e^{-t/4}(y0\text{Cos}[4t]- 2x0\text{Sin}[4t])\}\}$

```
t1 = ParametricPlot[Evaluate[Table[{x[t], y[t]}/.initsol/.
    {x0 → i, y0 → i}, {i, -1, 1, 2/9}]], {t, 0, 15}]
```

```
Show[pvf, t1, PlotRange → {{-1, 1}, {-1, 1}},

   AspectRatio → 1, Frame → False, Axes → Automatic,

   AxesOrigin → {0, 0}, AxesLabel → {x, y}]
```

Last, we illustrate how to solve an initial-value problem and graph the resulting solutions by finding the solution that satisfies the initial conditions $x(0) = 100$ and $y(0) = 10$ and then graphing the results with Plot and ParametricPlot in Figure 6-15.

```
partsol = DSolve[{x'[t] ==

   -1/4x[t] + 2y[t], y'[t] ==

   -8x[t] - 1/4y[t], x[0] == 100, y[0] == 10},

   {x[t], y[t]}, t]
```

$\{\{x[t] \to 5e^{-t/4}(20\text{Cos}[4t] + \text{Sin}[4t]), y[t] \to 10e^{-t/4}(\text{Cos}[4t] -$
$20\text{Sin}[4t])\}\}$

```
p1 = Plot[Evaluate[{x[t], y[t]}/.partsol],

   {t, 0, 20}, PlotRange → All]
```

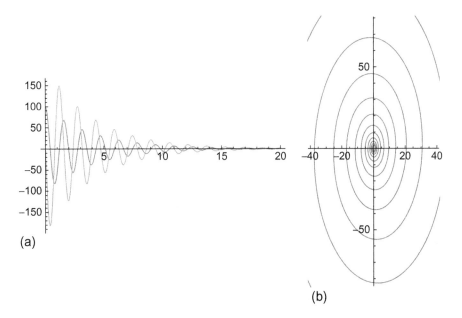

(a)

(b)

Figure 6-15   (a) Graph of $x(t)$ and $y(t)$. (b) Parametric plot of $x(t)$ versus $y(t)$ (For help with Show and GraphicsRow or GraphicsGrid use the **Help Browser**.)

```
p2 = ParametricPlot[{x[t], y[t]}/.partsol, {t, 0, 20},

    AspectRatio → Automatic]

Show[GraphicsRow[{p1, p2}]]
```

Notice the spiraling motion of the vectors in the direction field. This is due to terms in the solution formed by a product of exponential and trigonometric functions.

■

Initial-value problems can be solved through the use of eigenvalues and eigenvectors as well.

EXAMPLE 6.3.3: Solve $\begin{cases} x' = -\frac{1}{2}x - y + 64z \\ y' = -\frac{1}{4}y - 16z \\ z' = y - \frac{1}{4}z \\ x(0) = 1, \ y(0) = -1, \ z(0) = 0 \end{cases}$.

SOLUTION: In matrix form, the system is equivalent to $\mathbf{X}' = \mathbf{AX}$, where $\mathbf{A} = \begin{pmatrix} -1/2 & -1 & 64 \\ 0 & -1/4 & -16 \\ 0 & 1 & -1/4 \end{pmatrix}$. The eigenvalues and corresponding eigenvectors of $\mathbf{A}$ are found with Eigensystem.

```
capa = {{-1/2, -1, 64}, {0, -1/4, -16},

    {0, 1, -1/4}};

Eigensystem[capa]
```

$\{\{-\frac{1}{4}+4i, \ -\frac{1}{4}-4i, \ -\frac{1}{2}\}, \{\{-16i, 4i, 1\}, \{16i, -4i, 1\}, \{1, 0, 0\}\}\}$

These results mean that the eigenvalue $\lambda_1 = -1/2$ has corresponding eigenvector $\mathbf{v}_1 = \begin{pmatrix} 1 \\ 0 \\ 0 \end{pmatrix}$ so one solution of the system is $\mathbf{X}_1 = \mathbf{v}_1 e^{\lambda_1 t} =$

$\begin{pmatrix} 1 \\ 0 \\ 0 \end{pmatrix} e^{-t/2}$. An eigenvector corresponding to $\lambda_2 = -1/4 + 4i$ is $\mathbf{v}_2 =$

$$\begin{pmatrix} -16i \\ 4i \\ 1 \end{pmatrix} = \begin{pmatrix} 0 \\ 0 \\ 1 \end{pmatrix} + \begin{pmatrix} -16 \\ 4 \\ 0 \end{pmatrix} i.$$ Thus, two linearly independent solutions

that correspond to the complex conjugate pair of eigenvalues $\lambda_{2,3} = -1/4 \pm 4i$ are

$$\mathbf{X}_2 = e^{-t/4} \left[\begin{pmatrix} 0 \\ 0 \\ 1 \end{pmatrix} \cos 4t - \begin{pmatrix} -16 \\ 4 \\ 0 \end{pmatrix} \sin 4t \right] = \begin{pmatrix} 16 \sin 4t \\ -4 \sin 4t \\ \cos 4t \end{pmatrix} e^{-t/4}$$

and

$$\mathbf{X}_3 = e^{-t/4} \left[\begin{pmatrix} -16 \\ 4 \\ 0 \end{pmatrix} \cos 4t + \begin{pmatrix} 0 \\ 0 \\ 1 \end{pmatrix} \sin 4t \right] = \begin{pmatrix} -16 \cos 4t \\ 4 \cos 4t \\ \sin 4t \end{pmatrix} e^{-t/4}.$$

Hence, a general solution of the system is

$$\mathbf{X} = c_1 \mathbf{X}_1 + c_2 \mathbf{X}_2 + c_3 \mathbf{X}_3$$

$$= c_1 \begin{pmatrix} 1 \\ 0 \\ 0 \end{pmatrix} e^{-t/2} + c_2 \begin{pmatrix} 16 \sin 4t \\ -4 \sin 4t \\ \cos 4t \end{pmatrix} e^{-t/4} + c_3 \begin{pmatrix} -16 \cos 4t \\ 4 \cos 4t \\ \sin 4t \end{pmatrix} e^{-t/4}$$

$$= \begin{pmatrix} c_1 e^{-t/2} + 16e^{-t/4}(-c_3 \cos 4t + c_2 \sin 4t) \\ 4e^{-t/4}(c_3 \cos 4t - c_2 \sin 4t) \\ e^{-t/4}(c_2 \cos 4t + c_3 \sin 4t) \end{pmatrix};$$

a fundamental matrix is

$$\Phi = \begin{pmatrix} e^{-t/2} & 16e^{-t/4} \sin 4t & -16e^{-t/4} \cos 4t \\ 0 & -4e^{-t/4} \sin 4t & 4e^{-t/4} \cos 4t \\ 0 & e^{-t/4} \cos 4t & e^{-t/4} \sin 4t \end{pmatrix}.$$

```
sol = DSolve[{x'[t], y'[t], z'[t]} == capa.{x[t], y[t], z[t]},
    {x[t], y[t], z[t]}, t]//Simplify
```

$$\left\{ \left\{ x[t] \rightarrow \frac{C[1] + 4C[2]}{\sqrt{e^t}} - 4e^{-t/4}C[2]\text{Cos}[4t] + 16e^{-t/4}C[3]\text{Sin}[4t], \right. \right.$$

$$y[t] \rightarrow e^{-t/4}(C[2]\text{Cos}[4t] - 4C[3]\text{Sin}[4t]), \ z[t] \rightarrow \frac{1}{4}e^{-t/4}(4C[3]$$

$$\left. \left. \text{Cos}[4t] + C[2]\text{Sin}[4t]) \right\} \right\}$$

We solve the initial-value problem by applying the initial condition

$$\mathbf{X}(0) = \begin{pmatrix} 1 \\ -1 \\ 0 \end{pmatrix}$$

```
s1 = x[t]/.sol[[1]]/.t → 0
```

$C[1]$

```
s2 = y[t]/.sol[[1]]/.t → 0
```

$C[2]$

```
s3 = z[t]/.sol[[1]]/.t → 0
```

$C[3]$

```
sysofeqs = {s1 == 1, s2 == -1, s3 == 0}
```

$\{C[1] == 1,\ C[2] == -1,\ C[3] == 0\}$

and solving the resulting system of equations for c_1, c_2, and c_3.

```
cvals = Solve[sysofeqs]
```

$\{\{C[1] \to 1,\ C[2] \to -1,\ C[3] \to 0\}\}$

Substitution of these values into the general solution yields the solution to the initial-value problem.

```
sol2 = sol/.cvals
```

$$\left\{\left\{\left\{x[t] \to -\frac{3}{\sqrt{e^t}} + 4e^{-t/4}\mathrm{Cos}[4t],\ y[t] \to -e^{-t/4}\mathrm{Cos}[4t],\ z[t] \to -\frac{1}{4}e^{-t/4}\mathrm{Sin}[4t]\right\}\right\}\right\}$$

We graph $x(t)$, $y(t)$, and $z(t)$ with Plot in Figure 6-16(a) and a para-

metric plot of $\begin{cases} x = x(t) \\ y = y(t) \\ z = z(t) \end{cases}$ in three dimensions with ParametricPlot3D

in Figure 6-16(b).

```
p1 = Plot[Evaluate[{x[t], y[t], z[t]}/.sol2], {t, 0, 3Pi},

    PlotRange → {-2Pi, Pi}, AspectRatio → 1]

p2 = ParametricPlot3D[{x[t], y[t], z[t]}/.sol2, {t, 0, 3Pi},

    BoxRatios → {1, 1, 1}, PlotPoints → 200]

Show[GraphicsRow[{p1, p2}]]
```

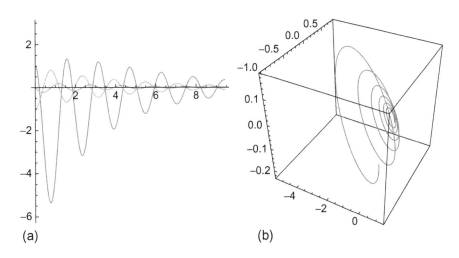

Figure 6-16   (a) $x(t)$, $y(t)$, and $z(t)$. (b) The solution to the initial-value problem tends to $(0,0,0)$ as $t \to \infty$

As in previous examples, we see that DSolve is able to find a general solution of the system as well as solve the initial-value problem, although the results are given in terms of complex exponentials.

```
initsol = DSolve[{{x'[t], y'[t], z'[t]} == capa.{x[t], y[t], z[t]},

  x[0] == x0, y[0] == y0, z[0] == z0},

  {x[t], y[t], z[t]}, t]//Simplify
```

$$\left\{\left\{x[t] \to \frac{x0 + 4y0}{\sqrt{e^t}} - 4e^{-t/4}y0\mathrm{Cos}[4t] + 16e^{-t/4}z0\mathrm{Sin}[4t], \; y[t] \to \right.\right.$$

$$e^{-t/4}(y0\mathrm{Cos}[4t] - 4z0\mathrm{Sin}[4t]), \; z[t] \to \frac{1}{4}e^{-t/4}(4z0\mathrm{Cos}[4t] +$$

$$\left.\left. y0\mathrm{Sin}[4t])\right\}\right\}$$

∎

6.3.3 Alternate Method for Solving Initial-Value Problems

An alternate method can be used to solve initial-value problems.

Let $\Phi(t)$ be a fundamental matrix for the system of equations $\mathbf{X}'(t) = \mathbf{A}(t)\mathbf{X}(t)$. Then, a general solution is $\mathbf{X}'(t) = \Phi(t)\mathbf{C}$, where $\mathbf{C}$ is a constant vector. If the initial condition $\mathbf{X}(0) = \mathbf{X}_0$ is given, then

$$\mathbf{X}(0) = \varPhi(0)\mathbf{C}$$
$$\mathbf{X}_0 = \varPhi(0)\mathbf{C}$$
$$\mathbf{C} = \varPhi^{-1}(0)\mathbf{X}_0.$$

Therefore, the solution to the initial-value problem $\begin{cases} \mathbf{X}'(t) = \mathbf{A}(t)\mathbf{X}(t) \\ \mathbf{X}(0) = \mathbf{X}_0 \end{cases}$ is $\mathbf{X}(t) = \varPhi(t)\varPhi^{-1}(0)\mathbf{X}_0$.

EXAMPLE 6.3.4: Use a fundamental matrix to solve the initial-value problem $\mathbf{X}' = \begin{pmatrix} 1 & 1 \\ 4 & -2 \end{pmatrix}\mathbf{X}$ subject to $\mathbf{X}(0) = \begin{pmatrix} 1 \\ -2 \end{pmatrix}$.

SOLUTION: We first remark that you can use DSolve to solve the initial-value problem directly with the command

```
Clear[x, y]

DSolve[{x'[t] == x[t] + y[t],

    y'[t] == 4x[t] − 2y[t], x[0] == 1,

    y[0] == −2}, {x[t], y[t]}, t]
```

$$\{\{x[t] \to \frac{1}{5}e^{-3t}(3 + 2e^{5t}), \, y[t] \to \frac{2}{5}e^{-3t}(-6 + e^{5t})\}\}$$

The eigenvalues and corresponding eigenvectors of $\mathbf{A} = \begin{pmatrix} 1 & 1 \\ 4 & -2 \end{pmatrix}$ are found with Eigensystem.

```
a = {{1, 1}, {4, −2}};

s1 = Eigensystem[a]
```

$$\{\{-3, 2\}, \{\{-1, 4\}, \{1, 1\}\}\}$$

The results mean that the eigenvalues are $\lambda_1 = -3$ and $\lambda_2 = 2$ with corresponding eigenvectors $\mathbf{v}_1 = \begin{pmatrix} -1 \\ 4 \end{pmatrix}$ and $\mathbf{v}_1 = \begin{pmatrix} 1 \\ 1 \end{pmatrix}$, respectively. A fundamental matrix is then given by $\varPhi(t) = \begin{pmatrix} -e^{3t} & e^{2t} \\ 4e^{3t} & e^{2t} \end{pmatrix}$.

```
phi[t_] = {s1[[2, 1]]Exp[−3t], s1[[2, 2]]Exp[2t]}//Transpose;

MatrixForm[phi[t]]
```

$$\begin{pmatrix} -e^{-3t} & e^{2t} \\ 4e^{-3t} & e^{2t} \end{pmatrix}$$

We calculate $\Phi^{-1}(0)$ with `Inverse`.

Inverse[A] finds the inverse of the square matrix A, if A is invertible.

Inverse[phi[t]]//MatrixForm

$$\begin{pmatrix} -\dfrac{e^{3t}}{5} & \dfrac{e^{3t}}{5} \\ \dfrac{4e^{-2t}}{5} & \dfrac{e^{-2t}}{5} \end{pmatrix}$$

Hence, the solution to the initial-value problem is $\mathbf{X}(t) = \Phi(t)\Phi^{-1}(0)\mathbf{X}_0$.

sol = phi[t].Inverse[phi[0]].{1, −2}//

    Simplify

$$\{\frac{1}{5}e^{-3t}(3 + 2e^{5t}), \frac{2}{5}e^{-3t}(-6 + e^{5t})\}$$

As in the previous examples, we graph $x(t)$ and $y(t)$ together in Figure 6-17(a) and parametrically in Figure 6-17(b).

p1 = Plot[Evaluate[{x[t], y[t]}/.{x[t]->sol[[1]], y[t] → sol[[2]]}],

    {t, −1, 3}, PlotRange → {{−1, 3}, {−2, 2}},

    AspectRatio → 1]

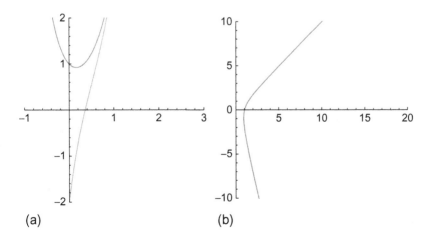

(a)   (b)

Figure 6-17   (a) $x(t)$ and $y(t)$. (b) Parametric plot of $x(t)$ versus $y(t)$

```
p2 = ParametricPlot[Evaluate[{x[t], y[t]}/.{x[t]->sol[[1]],
    y[t] → sol[[2]]}], {t, -1, 3}, PlotRange → {{0, 20},
    {-10, 10}}, AspectRatio → 1]

Show[GraphicsRow[{p1, p2}]]
```

■

6.3.4 Repeated Eigenvalues

We now consider the case of repeated eigenvalues, which is more complicated than the other cases because two situations can arise. An eigenvalue of multiplicity m may have m corresponding linearly independent eigenvectors or it may have fewer than m corresponding linearly independent eigenvectors. In the case of m linearly independent eigenvectors, a general solution is found in the same manner as the case of n distinct eigenvalues.

EXAMPLE 6.3.5: Solve $\mathbf{X}' = \begin{pmatrix} 1 & -3 & 3 \\ 3 & -5 & 3 \\ 6 & -6 & 4 \end{pmatrix} \mathbf{X}$.

SOLUTION: The eigenvalues and corresponding eigenvectors of $\mathbf{A} = \begin{pmatrix} 1 & -3 & 3 \\ 3 & -5 & 3 \\ 6 & -6 & 4 \end{pmatrix}$ are found with `Eigensystem`.

```
Clear[x, y, z, a]

a = {{1, -3, 3}, {3, -5, 3}, {6, -6, 4}};

Eigensystem[a]

{{4, -2, -2}, {{1, 1, 2}, {-1, 0, 1}, {1, 1, 0}}}
```

From the result, we see that the eigenvalue $\lambda_{1,2} = -2$ of multiplicity 2 has two corresponding linearly independent eigenvectors, $\mathbf{v}_1 = \begin{pmatrix} -1 \\ 0 \\ 1 \end{pmatrix}$ and $\mathbf{v}_2 = \begin{pmatrix} 1 \\ 1 \\ 0 \end{pmatrix}$. An eigenvector corresponding to $\lambda_3 = 4$ is

$\mathbf{v}_3 = \begin{pmatrix} 1 \\ 1 \\ 2 \end{pmatrix}$ so a fundamental set of solutions for the system is $S =$

$\left\{ \begin{pmatrix} -1 \\ 0 \\ 1 \end{pmatrix} e^{-2t}, \begin{pmatrix} 1 \\ 1 \\ 0 \end{pmatrix} e^{-2t}, \begin{pmatrix} 1 \\ 1 \\ 2 \end{pmatrix} e^{4t} \right\}$. A general solution is then

$$\mathbf{X} = c_1 \mathbf{v}_1 e^{\lambda_1 t} + c_2 \mathbf{v}_2 e^{\lambda_2 t} + c_3 \mathbf{v}_3 e^{\lambda_3 t}$$

$$= c_1 \begin{pmatrix} -1 \\ 0 \\ 1 \end{pmatrix} e^{-2t} + c_2 \begin{pmatrix} 1 \\ 1 \\ 0 \end{pmatrix} e^{-2t} + c_3 \begin{pmatrix} 1 \\ 1 \\ 2 \end{pmatrix} e^{4t}$$

$$= \begin{pmatrix} (c_1 - c_2) e^{-2t} + c_3 e^{4t} \\ c_1 e^{-2t} + c_3 e^{4t} \\ c_2 e^{-2t} + 2c_3 e^{4t} \end{pmatrix}$$

and a fundamental matrix is

$$\Phi = \begin{pmatrix} e^{-2t} & -e^{-2t} & e^{4t} \\ e^{-2t} & 0 & e^{4t} \\ 0 & e^{-2t} & 2e^{4t} \end{pmatrix}.$$

Of course, DSolve can be used to find a general solution of the system as well, although the form of the solution is slightly different than that obtained above.

```
DSolve[{x'[t], y'[t], z'[t]} == a.{x[t], y[t], z[t]},
   {x[t], y[t], z[t]}, t]//Simplify
```

$\{\{x[t] \to \frac{1}{2} e^{-2t}((1 + e^{6t})C[1] - (-1 + e^{6t})(C[2] - C[3])), y[t] \to$
$\frac{1}{2} e^{-2t}((-1 + e^{6t})C[1] - (-3 + e^{6t})C[2] + (-1 + e^{6t})C[3]), z[t] \to$
$e^{-2t}((-1 + e^{6t})C[1] + C[2] - e^{6t}C[2] + e^{6t}C[3])\}\}$

■

Because an eigenvalue of multiplicity 2 may have only one corresponding eigenvector, let us first restrict our attention to a system where the repeated eigenvalue $\lambda_1 = \lambda_2$ of $\mathbf{A}$ has only one corresponding eigenvector $\mathbf{v}_1$. We obtain one solution, $\mathbf{X}_1 = \mathbf{v}_1 e^{\lambda_1 t}$, to the system $\mathbf{X}' = \mathbf{A}\mathbf{X}$ that corresponds to the eigenvalue λ_1 of $\mathbf{A}$. We now seek a second linearly independent solution corresponding to λ_1 in a manner similar to that considered in the case of repeated characteristic roots of higher-order equations. In this case, however, we suppose that the second linearly independent solution corresponding to λ_1 is of the form

$$X_2 = (v_2 t + w_2) \, e^{\lambda_1 t}.$$

In order to find v_2 and w_2, we substitute X_2 into $X' = AX$. Because $X_2' = \lambda_1 (v_2 t + w_2) \, e^{\lambda_1 t} + v_2 e^{\lambda_1 t}$, we have

$$X_2' = AX_2$$
$$\lambda_1 (v_2 t + w_2) \, e^{\lambda_1 t} + v_2 e^{\lambda_1 t} = A \, (v_2 t + w_2) \, e^{\lambda_1 t}$$
$$\lambda_1 v_2 t + (\lambda_1 w_2 + v_2) = Av_2 t + Aw_2.$$

Equating coefficients yields $\lambda_1 v_2 = Av_2$ and $\lambda_1 w_2 + v_2 = Aw_2$. The equation $\lambda_1 v_2 = Av_2$ indicates that v_2 is an eigenvector of A that corresponds to λ_1, so we choose $v_2 = v_1$. We simplify the equation $\lambda_1 w_2 + v_2 = Aw_2$:

$$\lambda_1 w_2 + v_2 = Aw_2$$
$$v_2 = Aw_2 - \lambda_1 w_2$$
$$v_2 = (A - \lambda_1 I) \, w_2.$$

Because $v_2 = v_1$, w_2 satisfies the equation

$$(A - \lambda_1 I) \, w_2 = v_1.$$

Therefore, a second linearly independent solution corresponding to the eigenvalue λ_2 has the form

$$X_2 = (v_1 t + w_2) \, e^{\lambda_1 t},$$

where w_2 satisfies $(A - \lambda_1 I) \, w_2 = v_1$.

EXAMPLE 6.3.6: Find a general solution of $X' = \begin{pmatrix} -8 & -1 \\ 16 & 0 \end{pmatrix} X$.

SOLUTION: We first note that `DSolve` can find a general solution of the system.

```
gensol = DSolve[{x'[t] == -8x[t] - y[t],
   y'[t] == 16x[t]}, {x[t], y[t]}, t]
{{x[t] → -e^-4t(-1 + 4t)C[1] - e^-4t tC[2], y[t] → 16e^-4t tC[1] +
  e^-4t(1 + 4t)C[2]}}
```

We find the eigenvalues and corresponding eignevectors of $A = \begin{pmatrix} -8 & -1 \\ 16 & 0 \end{pmatrix}$ with `Eigensystem`.

```
a = {{-8, -1}, {16, 0}};

Eigensystem[a]
```

$\{\{-4, -4\}, \{\{-1, 4\}, \{0, 0\}\}\}$

The results mean that $\lambda_{1,2} = -4$ and an eigenvector that corresponds to $\lambda_1 = -4$ is $\mathbf{v}_1 = \begin{pmatrix} -1 \\ 4 \end{pmatrix}$ and one solution to the system is $\mathbf{X}_1 = \begin{pmatrix} -1 \\ 4 \end{pmatrix} e^{-4t}$; there is not a second linearly independent eigenvector corresponding to this repeated eigenvalue.

Therefore, to find $\mathbf{w}_2 = \begin{pmatrix} x_2 \\ y_2 \end{pmatrix}$ in a second linearly independent solution $\mathbf{X}_2 = (\mathbf{v}_1 t + \mathbf{w}_2) e^{\lambda_1 t}$, we solve $(\mathbf{A} - \lambda_1 \mathbf{I}) \mathbf{w}_2 = \mathbf{v}_1$, which in this case is

$$\begin{pmatrix} -4 & -1 \\ 16 & 4 \end{pmatrix} \begin{pmatrix} x_2 \\ y_2 \end{pmatrix} = \begin{pmatrix} -1 \\ 4 \end{pmatrix},$$

with LinearSolve.

LinearSolve[A,b]
solves $\mathbf{A}\mathbf{x} = \mathbf{b}$ for x.

```
LinearSolve[a + 4IdentityMatrix[2], {-1, 4}]
```

$\{\frac{1}{4}, 0\}$

We can use Solve to solve the system as well,

```
Solve[{-4x2 - y2 == -1, 16x2 + 4y2 == 4}, x2]
```

$\{\{x2 \rightarrow \frac{1 - y2}{4}\}\}$

which indicates that $x_2 = \frac{1}{4}(1 - y_2)$. Choosing $y_2 = 0$, $x_2 = 1/4$. With $\mathbf{w}_2 = \begin{pmatrix} 1/4 \\ 0 \end{pmatrix}$, a second linearly independent solution is

$$\mathbf{X}_2 = \left[\begin{pmatrix} -1 \\ 4 \end{pmatrix} t + \begin{pmatrix} 1/4 \\ 0 \end{pmatrix} \right] e^{-4t}$$

and a general solution is

$$\mathbf{X} = c_1 \begin{pmatrix} -1 \\ 4 \end{pmatrix} e^{-4t} + c_2 \left[\begin{pmatrix} -1 \\ 4 \end{pmatrix} t + \begin{pmatrix} 1/4 \\ 0 \end{pmatrix} \right] e^{-4t};$$

a fundamental matrix for the system is

$$\Phi = \begin{pmatrix} -1 & -t + 1/4 \\ 4 & 4t \end{pmatrix} e^{-4t}.$$

We now graph several solutions of the system as well as the direction field associated with the system in Figure 6-18.

```
step1 = Flatten[Table[gensol/.{C[1] → i, C[2] → j}, {i, -8, 8, 2},

  {j, -8, 8, 2}], 1];

pp1 = ParametricPlot[Evaluate[{x[t], y[t]}/.step1], {t, -1, 1}]

pvf = StreamPlot[{-8x - y, 16x},

  {x, -15, 15}, {y, -15, 15}, StreamStyle → Fine]

Show[pvf, pp1, PlotRange → {{-15, 15}, {-15, 15}},

  AspectRatio → 1, Frame → False, Axes → Automatic,

  AxesOrigin → {0, 0}, AxesLabel → {x, y}]
```

In Figure 6-18, notice that the behavior of these solutions differs from those of the other systems solved earlier in the section. This is due to the repeated eigenvalues.

■

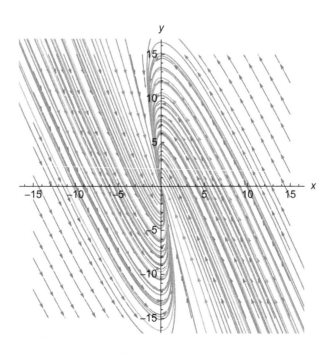

Figure 6-18   All solutions tend to $(0,0)$ as $t \to \infty$

A similar method is carried out in the case that an eigenvalue of $\mathbf{A}$ has multiplicity 3. Suppose that $\lambda_1 = \lambda_2 = \lambda_3$ has only one linearly independent corresponding eigenvector $\mathbf{v}_1$. In this situation, one solution of $\mathbf{X}' = \mathbf{AX}$ is $\mathbf{X}_1 = \mathbf{v}_1 e^{\lambda_1 t}$. We assume that two other linearly independent solutions have the form

$$\mathbf{X}_2 = (\mathbf{v}_2 t + \mathbf{w}_2)\, e^{\lambda_1 t} \quad \text{and} \quad \mathbf{X}_3 = \left(\frac{1}{2}\mathbf{v}_3 t^2 + \mathbf{w}_3 t + \mathbf{u}_3\right) e^{\lambda_1 t}.$$

Substitution of these solutions into the system of differential equations $\mathbf{X}' = \mathbf{AX}$ yields the following system of equations that is solved for the unknown vectors $\mathbf{v}_2, \mathbf{w}_2, \mathbf{v}_3, \mathbf{w}_3$, and $\mathbf{u}_3$:

$$\begin{cases} \lambda_1 \mathbf{v}_2 = \mathbf{Av}_2 \\ (\mathbf{A} - \lambda_1 \mathbf{I})\, \mathbf{w}_2 = \mathbf{v}_2 \\ \lambda_1 \mathbf{v}_3 = \mathbf{Av}_3 \\ (\mathbf{A} - \lambda_1 \mathbf{I})\, \mathbf{w}_3 = \mathbf{v}_3 \\ (\mathbf{A} - \lambda_1 \mathbf{I})\, \mathbf{u}_3 = \mathbf{w}_3 \end{cases}.$$

Similar to the previous case, $\mathbf{v}_1 = \mathbf{v}_2 = \mathbf{v}_3$, $\mathbf{w}_2 = \mathbf{w}_3$, and the vector $\mathbf{u}_3$ is found by solving the system

$$(\mathbf{A} - \lambda_1 \mathbf{I})\, \mathbf{u}_3 = \mathbf{w}_2.$$

Hence, the three solutions have the form

$$\mathbf{X}_1 = \mathbf{v}_1 e^{\lambda_1 t}, \quad \mathbf{X}_2 = (\mathbf{v}_1 t + \mathbf{w}_2)\, e^{\lambda_1 t}, \quad \text{and} \quad \mathbf{X}_3 = \left(\frac{1}{2}\mathbf{v}_1 t^2 + \mathbf{w}_2 t + \mathbf{u}_3\right) e^{\lambda_1 t}.$$

Notice that this method is generalized for instances when the multiplicity of the repeated eigenvalue is greater than 3.

EXAMPLE 6.3.7: Solve $\mathbf{X}' = \begin{pmatrix} 1 & 1 & 1 \\ 2 & 1 & -1 \\ -3 & 2 & 4 \end{pmatrix} \mathbf{X}$.

SOLUTION: After defining $\mathbf{A}$, we can also use DSolve to find a general solution of the system.

```
Clear[a]

a = {{1, 1, 1}, {2, 1, −1}, {−3, 2, 4}};

gensol = DSolve[

    Thread[{x′[t], y′[t], z′[t]} ==

    a.{x[t], y[t], z[t]}],

    {x[t], y[t], z[t]}, t]
```

$\{\{x[t] \rightarrow -e^{2t}(-1 + t)C[1] + e^{2t}tC[2] + e^{2t}tC[3], y[t] \rightarrow -\frac{1}{2}e^{2t}(-4 + t)tC[1] + \frac{1}{2}e^{2t}(2 - 2t + t^2)C[2] + \frac{1}{2}e^{2t}(-2 + t)tC[3], z[t] \rightarrow \frac{1}{2}e^{2t}(-6+t)tC[1] - \frac{1}{2}e^{2t}(-4+t)tC[2] - \frac{1}{2}e^{2t}(-2 - 4t + t^2)C[3]\}\}$

Alternatively, we can use the eigenvalues and corresponding eigenvectors to construct a general solution.

The eigenvalues and corresponding eigenvectors of $\mathbf{A} = \begin{pmatrix} 1 & 1 & 1 \\ 2 & 1 & -1 \\ -3 & 2 & 4 \end{pmatrix}$
are found with `Eigensystem`.

```
Eigensystem[a]
```

$\{\{2, 2, 2\}, \{\{0, -1, 1\}, \{0, 0, 0\}, \{0, 0, 0\}\}\}$

Here, $\lambda_{1,2,3} = 2$ has multiplicity 3 and has one eigenvector, $\mathbf{v}_1 = \begin{pmatrix} 0 \\ -1 \\ 1 \end{pmatrix}$, that corresponds to it; there are not 1 or 2 other linearly independent eigenvectors corresponding to $\lambda = 2$. One solution of the system is $\mathbf{X}_1 = \begin{pmatrix} 0 \\ -1 \\ 1 \end{pmatrix} e^{2t}$. The vector $\mathbf{w}_2 = \begin{pmatrix} x_2 \\ y_2 \\ z_2 \end{pmatrix}$ in the second linearly independent solution of the form $\mathbf{X}_2 = (\mathbf{v}_1 t + \mathbf{w}_2) e^{2t}$ is found by solving the system $(\mathbf{A} - \lambda_1 \mathbf{I}) \mathbf{w}_2 = \mathbf{v}_1$. We can solve this system with `Solve` or `LinearSolve`. Here, we illustrate using `LinearSolve`

```
LinearSolve[a − 2IdentityMatrix[3], {0, −1, 1}]
```

$\{-1, -1, 0\}$

to see that $\mathbf{w}_2 = \begin{pmatrix} -1 \\ -1 \\ 0 \end{pmatrix}$, so $\mathbf{X}_2 = \left(\begin{pmatrix} 0 \\ -1 \\ 1 \end{pmatrix} t + \begin{pmatrix} -1 \\ -1 \\ 0 \end{pmatrix} \right) e^{2t}$. Finally,

we must determine the vector $\mathbf{u}_3 = \begin{pmatrix} x_3 \\ y_3 \\ z_3 \end{pmatrix}$ in the third linearly inde-

pendent solution $\mathbf{X}_3 = \left(\frac{1}{2} \mathbf{v}_1 t^2 + \mathbf{w}_2 t + \mathbf{u}_3 \right) e^{\lambda_1 t}$ by solving the system $(\mathbf{A} - \lambda_1 \mathbf{I}) \mathbf{u}_3 = \mathbf{w}_2$, using Solve in this case.

```
Clear[x1, y1, z1]

Solve[(a − 2IdentityMatrix[3]).{x1, y1, z1}=={−1, −1, 0}]
```

$\{\{x1 \to -2, z1 \to -3 - y1\}\}$

Therefore, $x_3 = -2$ and $y_3 = -3 - z_3$. We select $z_3 = 0$ so $y_3 = -3$.

Hence, $\mathbf{u}_3 = \begin{pmatrix} -2 \\ -3 \\ 0 \end{pmatrix}$ and a third linearly independent solution is

$$\mathbf{X}_3 = \left(\frac{1}{2} \begin{pmatrix} 0 \\ -1 \\ 1 \end{pmatrix} t^2 + \begin{pmatrix} -1 \\ -1 \\ 0 \end{pmatrix} t + \begin{pmatrix} -2 \\ -3 \\ 0 \end{pmatrix} \right) e^{2t}.$$

A general solution is then given by

$$\mathbf{X} = c_1 \mathbf{X}_1 + c_2 \mathbf{X}_2 + c_3 \mathbf{X}_3$$

$$= c_1 \begin{pmatrix} 0 \\ -1 \\ 1 \end{pmatrix} e^{2t} + c_2 \left(\begin{pmatrix} 0 \\ -1 \\ 1 \end{pmatrix} t + \begin{pmatrix} -1 \\ -1 \\ 0 \end{pmatrix} \right) e^{2t} + c_3 \left(\frac{1}{2} \begin{pmatrix} 0 \\ -1 \\ 1 \end{pmatrix} t^2 + \begin{pmatrix} -1 \\ -1 \\ 0 \end{pmatrix} t + \begin{pmatrix} -2 \\ -3 \\ 0 \end{pmatrix} \right) e^{2t}$$

$$= \begin{pmatrix} -c_2 + c_3(-t - 2) \\ -c_1 + c_2(-t - 1) + c_3 \left(-\frac{1}{2} t^2 - t - 3 \right) \\ c_1 + c_2 t + \frac{1}{2} c_3 t^2 \end{pmatrix} e^{2t}.$$

■

6.4 Nonhomogeneous First-Order Systems: Undetermined Coefficients, Variation of Parameters, and the Matrix Exponential

In Chapter 4, we learned how to solve nonhomogeneous differential equations through the use of Undetermined Coefficients and Variation of Parameters. Here we approach the solution of systems of nonhomogeneous equations using those methods.

Let

$$\mathbf{X}(t) = \begin{pmatrix} x_1(t) \\ x_2(t) \\ \vdots \\ x_n(t) \end{pmatrix}, \quad \mathbf{A}(t) = \begin{pmatrix} a_{11}(t) & a_{12}(t) & \cdots & a_{1n}(t) \\ a_{21}(t) & a_{22}(t) & \cdots & a_{2n}(t) \\ \vdots & \vdots & \cdots & \vdots \\ a_{n1}(t) & a_{n2}(t) & \cdots & a_{nn}(t) \end{pmatrix}, \quad \text{and} \quad \mathbf{F}(t) = \begin{pmatrix} f_1(t) \\ f_2(t) \\ \vdots \\ f_n(t) \end{pmatrix}.$$

A general solution of the homogeneous system $\mathbf{X}' = \mathbf{AX}$ is $\mathbf{X} = \Phi(t)\mathbf{C}$, where $\Phi(t) = \begin{pmatrix} \Phi_1 & \Phi_2 & \cdots & \Phi_n \end{pmatrix}$ is a fundamental matrix for the system $\mathbf{X}' = \mathbf{AX}$ and

$$\mathbf{C} = \begin{pmatrix} c_1 \\ c_2 \\ \vdots \\ c_n \end{pmatrix} \text{ is an } n \times 1 \text{ constant vector.}$$

A **particular solution** to a system of ordinary differential equations is a set of functions that satisfy the system but do not contain any arbitrary constants. That is, a particular solution to a system is a set of specific functions, *containing no arbitrary constants*, that satisfy the system.

$\mathbf{X}_h = \Phi\mathbf{C}$ is a general solution of the corresponding homogeneous equation, $\mathbf{X}' = \mathbf{AX}$.

Let $\mathbf{X}$ be *any* solution of $\mathbf{X}' = \mathbf{AX} + \mathbf{F}(t)$, $\mathbf{X}_h = \Phi(t)\mathbf{C}$ a general solution of the corresponding homogeneous system, $\mathbf{X}' = \mathbf{AX}$, and $\mathbf{X}_p$ a particular solution of the nonhomogeneous system.

Then, $\mathbf{X} - \mathbf{X}_p$ is a solution of the corresponding homogeneous system, $\mathbf{X}' = \mathbf{AX}$, so $\mathbf{X} - \mathbf{X}_p = \mathbf{X}_h$ and, consequently, $\mathbf{X} = \mathbf{X}_h + \mathbf{X}_p$.

Thus, to find a general solution of $\mathbf{X}' = \mathbf{AX} + \mathbf{F}(t)$, we note that if $\mathbf{X}_p$ is a particular solution of the equation then all other solutions to the equation can be written in the form

$$\mathbf{X} = \underbrace{\Phi(t)\mathbf{C}}_{\mathbf{X}_h} + \mathbf{X}_p.$$

6.4.1 Undetermined Coefficients

We use the method of undetermined coefficients to find a particular solution of a nonhomogeneous system in much the same way as we approached nonhomogeneous higher-order equations in Chapter 4. The main difference is that the coefficients are *constant vectors* when we work with systems.

EXAMPLE 6.4.1: Solve $\begin{cases} x' = 2x + y + \sin 3t \\ y' = -8x - 2y \\ x(0) = 0, \ y(0) = 1 \end{cases}$.

SOLUTION: In matrix form, the system is equivalent to $\mathbf{X}' = \begin{pmatrix} 2 & 1 \\ -8 & -2 \end{pmatrix} \mathbf{X} + \begin{pmatrix} \sin 3t \\ 0 \end{pmatrix}$. We find a general solution of the corresponding homogeneous system $\mathbf{X}' = \begin{pmatrix} 2 & 1 \\ -8 & -2 \end{pmatrix} \mathbf{X}$ with DSolve.

```
Clear[x, y, xh, xp]
    homsol = DSolve[{x'[t] == 2x[t] + y[t], y'[t] == -8x[t] - 2y[t]},
    {x[t], y[t]}, t]
```

$\{\{x[t] \ \rightarrow \ \frac{1}{2} C[2] \text{Sin}[2t] \ + \ C[1](\text{Cos}[2t] \ + \ \text{Sin}[2t]), \ y[t] \ \rightarrow$
$C[2](\text{Cos}[2t] - \text{Sin}[2t]) - 4 C[1] \text{Sin}[2t]\}\}$

These results indicate that a general solution of the corresponding homogeneous system is

$$\mathbf{X}_h = \begin{pmatrix} \cos 2t + \sin 2t & \frac{1}{2} \sin 2t \\ 4 \sin 2t & \cos 2t - \sin 2t \end{pmatrix} \begin{pmatrix} c_1 \\ c_2 \end{pmatrix}.$$

```
xh[t_] = {{x[t]}, {y[t]}}/.homsol[[1]];
```

Thus, we search for a particular solution of the nonhomogeneous system of the form $\mathbf{X}_p = \mathbf{a} \sin 3t + \mathbf{b} \cos 3t$, where $\mathbf{a} = \begin{pmatrix} a_1 \\ a_2 \end{pmatrix}$ and $\mathbf{b} = \begin{pmatrix} b_1 \\ b_2 \end{pmatrix}$. After defining $\mathbf{A} = \begin{pmatrix} 2 & 1 \\ -8 & -2 \end{pmatrix}$ and $\mathbf{X}_p = \mathbf{a} \sin 3t + \mathbf{b} \cos 3t$, we substitute $\mathbf{X}_p$ into the nonhomogeneous system.

```
capa = {{2, 1}, {-8, -2}};
```

```
MatrixForm[capa]
```

$$\begin{pmatrix} 2 & 1 \\ -8 & -2 \end{pmatrix}$$

```
xp[t_] = {{a1}, {a2}}Sin[3 t] + {{b1}, {b2}}Cos[3 t];
```

```
MatrixForm[xp[t]]
```

$$\begin{pmatrix} b1Cos[3 t] + a1Sin[3 t] \\ b2Cos[3 t] + a2Sin[3 t] \end{pmatrix}$$

```
step1 = xp'[t] == capa.xp[t] + {{Sin[3 t]}, {0}}
```

{{3a1Cos[3 t] − 3b1Sin[3 t]}, {3a2Cos[3 t] − 3b2Sin[3 t]}} ==

{{b2Cos[3 t] + Sin[3 t] + a2Sin[3 t] + 2(b1Cos[3 t] + a1Sin[3 t])},

{−8(b1Cos[3 t] + a1Sin[3 t]) − 2(b2Cos[3 t] + a2Sin[3 t])}}

The result represents a system of equations that is true for all values of t. In particular, substituting $t = 0$ yields

```
eq1 = step1/.t → 0
```

{{3a1}, {3a2}} == {{2b1 + b2}, {−8b1 − 2b2}}

which is equivalent to the system of equations

$$\begin{cases} 3a_1 = 2b_1 + b_2 \\ 3a_2 = -2\left(4b_1 + b_2\right) \end{cases}.$$

Similarly, substituting $t = \pi/2$ results in

```
eq2 = step1/.t → Pi/2
```

{{3b1}, {3b2}} == {{−1 − 2a1 − a2}, {8a1 + 2a2}}

which is equivalent to the system of equations

$$\begin{cases} 3b_1 = -1 - 2a_1 - a_2 \\ 3b_2 - 2\left(-4a_1 - a_2\right) \end{cases}.$$

We now use `Solve` to solve these four equations for a_1, a_2, b_1, and b_2

```
coeffs = Solve[{eq1, eq2}]
```

$$\{\{a1 \to -\frac{2}{5},\ a2 \to \frac{8}{5},\ b1 \to -\frac{3}{5},\ b2 \to 0\}\}$$

and then substitute these values into $\mathbf{X}_p$ to obtain a particular solution to the nonhomogeneous system.

```
xp[t_] = xp[t]/.coeffs[[1]]
```

$$\{\{-\frac{3}{5}\text{Cos}[3t] - \frac{2}{5}\text{Sin}[3t]\},\ \{\frac{8}{5}\text{Sin}[3t]\}\}$$

A general solution to the nonhomogeneous system is then given by $\mathbf{X} = \mathbf{X}_h + \mathbf{X}_p$.

```
xh[t]
```

$$\{\{\frac{1}{2}C[2]\text{Sin}[2t] + C[1](\text{Cos}[2t] + \text{Sin}[2t])\},\ \{C[2](\text{Cos}[2t] - \text{Sin}[2t]) - 4C[1]\text{Sin}[2t]\}\}$$

```
x[t_] = xh[t] + xp[t]
```

$$\{\{-\frac{3}{5}\text{Cos}[3t] + \frac{1}{2}C[2]\text{Sin}[2t] + C[1](\text{Cos}[2t] + \text{Sin}[2t]) - \frac{2}{5}\text{Sin}[3t]\},$$
$$\{C[2](\text{Cos}[2t] - \text{Sin}[2t]) - 4C[1]\text{Sin}[2t] + \frac{8}{5}\text{Sin}[3t]\}\}$$

To solve the initial value problem, we apply the initial condition and solve for the unknown constants.

```
x[0]
```

$$\{\{-\frac{3}{5} + C[1]\},\ \{C[2]\}\}$$

```
cvals = Solve[x[0]=={{0}, {1}}]
```

$$\{\{C[1] \to \frac{3}{5},\ C[2] \to 1\}\}$$

We obtain the solution to the initial value problem by substituting these values back into the general solution.

```
x[t_] = x[t]/.cvals[[1]]//Flatten//Simplify
```

$$\{\frac{1}{10}(6\text{Cos}[2t] - 6\text{Cos}[3t] + 11\text{Sin}[2t] - 4\text{Sin}[3t]),\ \text{Cos}[2t] - \frac{17}{5}\text{Sin}[2t] + \frac{8}{5}\text{Sin}[3t]\}$$

We confirm this result by graphing $x(t)$ and $y(t)$ together in Figure 6-19(a) as well as parametrically in B.

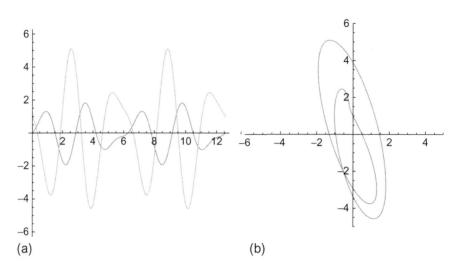

Figure 6-19   (a) $x(t)$ and $y(t)$. (b) Parametric plot of $x(t)$ versus $y(t)$

```
p1 = Plot[Evaluate[x[t]], {t, 0, 4Pi}, PlotRange → {-2Pi, 2Pi},
    AspectRatio → 1, PlotLabel → "(a)"]

p2 = ParametricPlot[x[t], {t, 0, 4Pi},
    PlotRange → {{-6, 5}, {-5, 6}},
    AspectRatio → 1, PlotLabel → "(b)"]
Show[GraphicsRow[{p1, p2}]]
```

Finally, we note that DSolve is able to find a general solution of the nonhomogeneous system

```
Clear[x, y, t]
Clear[x, y, xh, xp]
homsol = DSolve[{x'[t] == 2x[t] + y[t], y'[t] == -8x[t] - 2y[t]},
    {x[t], y[t]}, t]
```

$\{\{x[t]\ \ \rightarrow\ \ \frac{1}{2}C[2]\text{Sin}[2t]\ +\ C[1](\text{Cos}[2t]\ +\ \text{Sin}[2t]),\ y[t]\ \ \rightarrow$
$C[2](\text{Cos}[2t] - \text{Sin}[2t]) - 4C[1]\text{Sin}[2t]\}\}$

as well as solve the initial value problem.

```
Clear[x, y, xh, xp]
homsol = DSolve[{x'[t] == 2x[t] + y[t], y'[t] == -8x[t] - 2y[t],
```

$$x[0] == 0, \, y[0] == 1\},$$

$$\{x[t], \, y[t]\}, \, t]$$

$$\{\{x[t] \to \frac{1}{2} \text{Sin}[2t], \, y[t] \to \text{Cos}[2t] - \text{Sin}[2t]\}\}$$

■

6.4.2 Variation of Parameters

Generally, the method of undetermined coefficients is difficult to implement for nonhomogeneous linear systems as the choice for the particular solution must be very carefully made.

Variation of parameters is implemented in much the same way as for first-order linear equations.

Let Φ be a fundamental matrix for the corresponding homogeneous system. We assume that a particular solution has the form $\mathbf{X}_p = \Phi \mathbf{U}(t) = \Phi \mathbf{U}$. Differentiating $\mathbf{X}_p$ gives us

$$\mathbf{X}_p{}' = \Phi' \mathbf{U} + \Phi \mathbf{U}'.$$

Substituting into equation $\mathbf{X}' = \mathbf{A}\mathbf{X} + \mathbf{F}(t)$ results in

$$\Phi' \mathbf{U} + \Phi \mathbf{U}' = \mathbf{A}\Phi \mathbf{U} + \mathbf{F}$$
$$\Phi \mathbf{U}' = \mathbf{F}(t)$$
$$\mathbf{U}' = \Phi^{-1} \mathbf{F}(t)$$
$$\mathbf{U} = \int \Phi^{-1} \mathbf{F}(t) \, dt,$$

where we have used the fact that $\Phi' \mathbf{U} - \mathbf{A}\Phi \mathbf{U} = \left(\Phi' - \mathbf{A}\Phi\right) \mathbf{U} = \mathbf{0}$. It follows that

$$\mathbf{X}_p = \Phi \int \Phi^{-1} \mathbf{F}(t) \, dt. \tag{6.16}$$

A general solution is then

$$\mathbf{X} = \mathbf{X}_h + \mathbf{X}_p$$
$$= \Phi \mathbf{C} + \Phi \int \Phi^{-1} \mathbf{F}(t) \, dt$$
$$= \Phi \left(\mathbf{C} + \int \Phi^{-1} \mathbf{F}(t) \, dt\right) = \Phi \int \Phi^{-1} \mathbf{F}(t) \, dt,$$

where we have incorporated the constant vector $\mathbf{C}$ into the indefinite integral $\int \Phi^{-1}\mathbf{F}(t)\, dt$.

EXAMPLE 6.4.2: Solve the initial-value problem

$$\mathbf{X}' = \begin{pmatrix} 1 & -1 \\ 10 & -1 \end{pmatrix} \mathbf{X} - \begin{pmatrix} t\cos 3t \\ t\sin t + t\cos 3t \end{pmatrix}, \quad \mathbf{X}(0) = \begin{pmatrix} 1 \\ -1 \end{pmatrix}.$$

*Remark.*In traditional form, the system is equivalent to

$$\begin{cases} x' = x - y - t\cos 3t \\ y' = 10x - y - t\sin t - t\cos 3t \end{cases}, \quad x(0) = 1,\, y(0) = -1.$$

SOLUTION: The corresponding homogeneous system is $\mathbf{X}'_h = \begin{pmatrix} 1 & -1 \\ 10 & -1 \end{pmatrix} \mathbf{X}_h$. The eigenvalues and corresponding eigenvectors of $\mathbf{A} = \begin{pmatrix} 1 & -1 \\ 10 & -1 \end{pmatrix}$ are $\lambda_{1,2} = \pm 3i$ and $\mathbf{v}_{1,2} = \begin{pmatrix} 1 \\ 10 \end{pmatrix} \pm \begin{pmatrix} -3 \\ 0 \end{pmatrix} i$, respectively.

```
capa = {{1, -1}, {10, -1}};

Eigensystem[capa]
```

$$\{\{3i, -3i\}, \{\{1 + 3i, 10\}, \{1 - 3i, 10\}\}\}$$

A fundamental matrix is $\Phi = \begin{pmatrix} \sin 3t & \cos 3t \\ \sin 3t - 3\cos 3t & \cos 3t + 3\sin 3t \end{pmatrix}$ with inverse $\Phi^{-1} = \begin{pmatrix} \frac{1}{3}\cos 3t + \sin 3t & -\frac{1}{3}\cos 3t \\ -\frac{1}{3}\sin 3t + \cos 3t & \frac{1}{3}\sin 3t \end{pmatrix}$.

```
fs = {{Sin[3t], Sin[3t] - 3Cos[3t]},

   {Cos[3t], Cos[3t] + 3Sin[3t]}}//Transpose;

MatrixForm[fs]

fsinv = Inverse[fs]//Simplify;

MatrixForm[fsinv]
```

$$\begin{pmatrix} Sin[3t] & Cos[3t] \\ -3Cos[3t] + Sin[3t] & Cos[3t] + 3Sin[3t] \end{pmatrix}$$

$$\begin{pmatrix} \frac{1}{3}\text{Cos}[3t] + \text{Sin}[3t] & -\frac{1}{3}\text{Cos}[3t] \\ \text{Cos}[3t] - \frac{1}{3}\text{Sin}[3t] & \frac{1}{3}\text{Sin}[3t] \end{pmatrix}$$

We now compute $\Phi^{-1}\mathbf{F}(t)$

```
ft = {-tCos[3 t], -tSin[t] - tCos[3 t]};
```

```
step1 = fsinv.ft;
```

```
MatrixForm[step1]
```

$$\begin{pmatrix} -\frac{1}{3}\text{Cos}[3t](-t\text{Cos}[3t] - t\text{Sin}[t]) - t\text{Cos}[3t](\frac{1}{3}\text{Cos}[3t] + \text{Sin}[3t]) \\ -t\text{Cos}[3t](\text{Cos}[3t] - \frac{1}{3}\text{Sin}[3t]) + \frac{1}{3}(-t\text{Cos}[3t] - t\text{Sin}[t])\text{Sin}[3t] \end{pmatrix}$$

and $\int \Phi^{-1}\mathbf{F}(t)\,dt$.

```
step2 = Integrate[step1, t];
```

```
MatrixForm[step2]
```

$$\begin{pmatrix} \frac{1}{12}t\text{Cos}[2t] + \frac{1}{12}t\text{Cos}[6t] - \frac{1}{24}\text{Sin}[2t] + \frac{1}{6}(-\frac{1}{4}t\text{Cos}[4t] + \frac{1}{16}\text{Sin}[4t]) - \frac{1}{72}\text{Sin}[6t] \\ \frac{1}{288}(-12\text{Cos}[2t] + 3\text{Cos}[4t] - 4(\text{Cos}[6t] + 3t(6t + 2\text{Sin}[2t] - \text{Sin}[4t] + 2\text{Sin}[6t]))) \end{pmatrix}$$

A general solution of the nonhomogeneous system is then $\Phi\left(\int \Phi^{-1}\mathbf{F}(t)\,dt + \mathbf{C}\right)$.

```
fa = Simplify[fs.step2]
```

$$\{\frac{1}{288}(-9\text{Cos}[t] - 4((1 + 18t^2)\text{Cos}[3t] - 9t\text{Sin}[t] + 6t\text{Sin}[3t])),$$

$$\frac{1}{288}(-9(1+4t)\text{Cos}[t]-4(1+18t+18t^2)\text{Cos}[3t]+3(3(-5+4t)\text{Sin}[t] - 4(-1 + 2t + 18t^2)\text{Sin}[3t]))\}$$

Of course, it is easiest to use DSolve to solve the initial value problem directly as we do next.

```
Clear[x, y, t]
```

```
check = DSolve[{x'[t] == x[t] - y[t] - tCos[3 t],
   y'[t] == 10x[t] - y[t] - tSin[t] - tCos[t], x[0] == 1,
   y[0] == -1}, {x[t], y[t]}, t]//Simplify
```

$$\{\{x[t] \rightarrow \frac{1}{864}(27(-1 + 4t)\text{Cos}[t] - 3(-297 + 8t + 72t^2)\text{Cos}[3t] +$$
$$27\text{Sin}[t]+108t\text{Sin}[t]+539\text{Sin}[3t]-72t\text{Sin}[3t]-72t^2\text{Sin}[3t]),$$

$$y[t] \rightarrow \frac{1}{432}(-81\text{Cos}[t] - 3(117 + 40t)\text{Cos}[3t] + 2(27(-1 + 2t)$$
$$\text{Sin}[t] + (821 - 180t^2)\text{Sin}[3t]))\}\}$$

After using ?Evaluate to obtain basic information regarding the Evaluate function, the solutions are graphed with Plot and ParametricPlot in Figure 6-20.

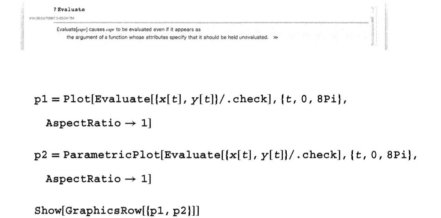

```
p1 = Plot[Evaluate[{x[t], y[t]}/.check], {t, 0, 8Pi},

    AspectRatio → 1]

p2 = ParametricPlot[Evaluate[{x[t], y[t]}/.check], {t, 0, 8Pi},

    AspectRatio → 1]

Show[GraphicsRow[{p1, p2}]]
```

■

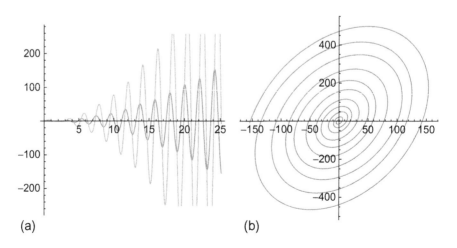

(a)   (b)

Figure 6-20   (a) Graph of $x(t)$ and $y(t)$. (b) Parametric plot of $x(t)$ versus $y(t)$

EXAMPLE 6.4.3: Solve $\begin{cases} x' = -3x + 2y + e^{-t}\sec t \\ y' = -10x + 5y + e^{-t}\csc t, \, 0 < t < \pi. \\ x(\pi/4) = 3, \, y(\pi/4) = -1 \end{cases}$

SOLUTION: To implement the method of Variation of Parameters, we proceed in the same manner as before. First, we find a general solution of the corresponding homogeneous system.

> You do not need to use Thread in the DSolve command if you are using Mathematica Version 10 or later.

```
Clear[x, y, a]

a = {{-3, 2}, {-10, 5}};

homsol = DSolve[Thread[{x'[t], y'[t]} == a.{x[t], y[t]}],

  {x[t], y[t]}, t]//Simplify

{{x[t]  →   e^t(C[1]Cos[2t] + (-2C[1] + C[2])Sin[2t]), y[t]   →

e^t(C[2]Cos[2t] + (-5C[1] + 2C[2])Sin[2t])}}
```

This result means that a general solution of the corresponding homogeneous system is equivalent to

$$\mathbf{X}_h = \begin{pmatrix} e^t(-\cos 2t + 2\sin 2t) & e^t(2\cos 2t + \sin 2t) \\ 5e^t\sin 2t & 5e^t\cos 2t \end{pmatrix}\begin{pmatrix} c_1 \\ c_2 \end{pmatrix},$$

where we have used different constants than those returned by Mathematica, and a fundamental matrix is given by

$$\Phi = \begin{pmatrix} e^t(-\cos 2t + 2\sin 2t) & e^t(2\cos 2t + \sin 2t) \\ 5e^t\sin 2t & 5e^t\cos 2t \end{pmatrix}.$$

```
Φ[t_] = {{Exp[t](-Cos[2t] + 2Sin[2t]), Exp[t](2Cos[2t] + Sin[2t])},

{5Exp[t]Sin[2t], 5Exp[t]Cos[2t]}};

Det[Φ[t]]//Simplify

-5e^{2t}
```

Next, we compute $\mathbf{X}_p(t) = \Phi(t)\int \Phi^{-1}(t)\mathbf{F}(t)\,dt$. The result is *very* lengthy so we suppress the resulting output by including a semi-colon at the end of the command.

```
inverseΦ = Inverse[Φ[t]]//Simplify;

MatrixForm[inverseΦ]
```

$$\begin{pmatrix} -e^{-t}\text{Cos}[2t] & \dfrac{1}{5}e^{-t}(2\text{Cos}[2t]+\text{Sin}[2t]) \\ e^{-t}\text{Sin}[2t] & \dfrac{1}{5}e^{-t}(\text{Cos}[2t]-2\text{Sin}[2t]) \end{pmatrix}$$

```
f[t_] = {{Exp[t]Sec[t]}, {Exp[t]Csc[t]}};
```

```
xp[t_] = Φ[t].Integrate[inverseΦ.f[t], t]//Simplify;
```

However, we view abbreviations of $x(t)$ and $y(t)$ with Short.

```
xp[t][[1]]//Short
```

$\{e^t(\langle\langle 1\rangle\rangle)\}$

```
xp[t][[2]]//Short
```

$\{-e^t(\text{Cos}[2t](\text{Log}[\text{Cos}[\frac{t}{2}]] - \text{Log}[\text{Sin}[\frac{t}{2}]]) - 4\langle\langle 1\rangle\rangle + 2\langle\langle 1\rangle\rangle(4 + (2\text{Log}[\text{Cos}[\frac{t}{2}]] + \langle\langle 5\rangle\rangle))\langle\langle 1\rangle\rangle))\}$

Finally, we form a general solution of the nonhomogeneous system.

```
x[t_] = Φ[t].{{c1}, {c2}} + xp[t]//Simplify;
```

To solve the initial value problem, we substitute $t = \pi/4$ into the general solution

```
x[Pi/4]//FullSimplify
```

$\{\{e^{\pi/4}(-2\sqrt{2}+2c1+c2+\text{ArcSinh}[1])\}, \{e^{\pi/4}(-2\sqrt{2}+5c1+\text{Log}[7+5\sqrt{2}])\}\}$

and solve $\begin{cases} x(\pi/4) = 3 \\ y(\pi/4) = -1 \end{cases}$ for c_1 and c_2.

```
cvals = Solve[x[Pi/4] == {{3}, {-1}}]//FullSimplify
```

$\{\{c1 \rightarrow \dfrac{1}{5}(2\sqrt{2} - e^{-\pi/4} - \text{ArcSinh}[7]), c2 \rightarrow \dfrac{1}{5}(6\sqrt{2} + 17e^{-\pi/4} + \text{Log}[\text{Cot}[\dfrac{\pi}{8}]])\}\}$

This result is rather complicated so we compute approximations with N.

```
numcals = cvals//N
```

$\{\{c1 \rightarrow -0.0543264, c2 \rightarrow 3.42352\}\}$

The solution to the initial value problem is obtained by substituting these numbers back into the general solution.

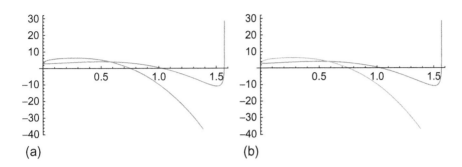

Figure 6-21   (a) $x(t)$ and $y(t)$. (b) When you use `Evaluate`, Mathematica understands that you are plotting a list of functions and shades each differently

`x[t_] = x[t]/.cvals[[1]]//N`

$\{\{2.71828^{t}(-4.\text{Cos}[t] - 1.\text{Cos}[2.t](-6.90137 - 1.\text{Log}[\text{Cos}[0.5t]$

$-1.\text{Sin}[0.5t]] + \text{Log}[\text{Cos}[0.5t] + \text{Sin}[0.5t]]) + (3.31487$

$-1.\text{Log}[\text{Cos}[0.5t]] - 2.\text{Log}[\text{Cos}[0.5t] - 1.\text{Sin}[0.5t]]$

$+\ \text{Log}[\text{Sin}[0.5t]]\ +\ 2.\text{Log}[\text{Cos}[0.5t]\ +\ \text{Sin}[0.5t]])\text{Sin}[2.t])\},$

$\{2.71828^{t}(\text{Cos}[2.t](17.1176 - 1.\text{Log}[\text{Cos}[0.5t]] + \text{Log}[\text{Sin}[0.5t]])$

$+\ 4.\text{Sin}[t] + 2.\text{Cos}[t](-4. + (-0.271632 - 2.\text{Log}[\text{Cos}[0.5t]]$

$-5.\text{Log}[\text{Cos}[0.5t] - 1.\text{Sin}[0.5t]] + 2.\text{Log}[\text{Sin}[0.5t]]$

$+\ 5.\text{Log}[\text{Cos}[0.5t] + \text{Sin}[0.5t]])\text{Sin}[t]))\}\}$

We confirm that the initial conditions are satisfied by graphing $x(t)$ and $y(t)$ on the interval $(0, \pi/2)$ in Figure 6-21.

`p1 = Plot[x[t], {t, 0, Pi/2}, PlotLabel → "(a)"]`

`p2 = Plot[Evaluate[x[t]], {t, 0, Pi/2}, PlotLabel → "(b)"]`

`Show[GraphicsRow[{p1, p2}]]`

■

6.4.3 The Matrix Exponential

Definition 28 (Matrix Exponential). *If At is $n \times n$, the **matrix exponential** is defined by*

$$e^{\mathbf{A}t} = \exp(\mathbf{A}t) = \mathbf{I} + \mathbf{A}t + \frac{1}{2!}\mathbf{A}^2 t^2 + \cdots = \sum_{n=0}^{\infty} \frac{1}{n!}\mathbf{A}^n t^n. \qquad (6.17)$$

Use the command `MatrixExp` to compute the matrix exponential of a matrix. For example, here we use `MatrixExp` to calculate $e^{\mathbf{A}t}$ if $\mathbf{A} = \begin{pmatrix} 1 & 0 \\ -2 & 3 \end{pmatrix}$.

```
s1 = MatrixExp[{{t, 0}, {-2t, 3t}}]
```

$$\left\{\left\{e^t, 0\right\}, \left\{-e^t\left(-1+e^{2t}\right), e^{3t}\right\}\right\}$$

```
MatrixForm[s1]
```

$$\begin{pmatrix} e^t & 0 \\ -e^t\left(-1+e^{2t}\right) & e^{3t} \end{pmatrix}$$

Differentiating the series (6.17) term-by-term shows us that $\dfrac{d}{dt}\left(e^{\mathbf{A}t}\right) = \mathbf{A}e^{\mathbf{A}t}$ so $e^{\mathbf{A}t}$ satisfies the differential equation $\mathbf{X}' = \mathbf{A}\mathbf{X}$. We can use the matrix exponential $e^{\mathbf{A}t}$ to solve the linear first-order system $\mathbf{X}' = \mathbf{A}\mathbf{X} + \mathbf{F}(t)$ in much the same way that we used the integrating factor $e^{\int p(x)\,dx}$ to solve the linear first-order equation $y' + p(x)y = q(x)$. Moreover, $e^{\mathbf{A}t}$ is a fundamental matrix for the homogeneous system; $\left(e^{\mathbf{A}t}\right)^{-1} = e^{-\mathbf{A}t}$; and if $t = 0$, $e^{\mathbf{A}t} = \mathbf{I}$.

To solve the system $\mathbf{X}' = \mathbf{A}\mathbf{X} + \mathbf{F}(t)$, we first rewrite the equation as $\mathbf{X}' - \mathbf{A}\mathbf{X} = \mathbf{F}(t)$. Now, multiply both sides of the equation by $e^{-\mathbf{A}t}$ and integrate:

$$e^{-\mathbf{A}t}\left(\mathbf{X}' - \mathbf{A}\mathbf{X}\right) = e^{-\mathbf{A}t}\mathbf{F}(t)$$

$$\frac{d}{dt}\left(e^{-\mathbf{A}t}\mathbf{X}\right) = e^{-\mathbf{A}t}\mathbf{F}(t)$$

$$e^{-\mathbf{A}t}\mathbf{X} = \int e^{-\mathbf{A}t}\mathbf{F}(t)\,dt + \mathbf{C}$$

$$\mathbf{X} = e^{\mathbf{A}t}\int e^{-\mathbf{A}t}\mathbf{F}(t)\,dt + e^{\mathbf{A}t}\mathbf{C},$$

where C is an arbitrary constant vector.

If, in addition, we are given the initial condition $\mathbf{X}(t_0) = \mathbf{X}_0$, the solution to the initial-value problem $\begin{cases} \mathbf{X}' = \mathbf{A}\mathbf{X} + \mathbf{F}(t) \\ \mathbf{X}(t_0) = \mathbf{X}_0 \end{cases}$ is

$$\mathbf{X} = \int_{t_0}^{t} e^{-\mathbf{A}(t-s)}\mathbf{F}(s)\,ds + e^{\mathbf{A}(t-t_0)}\mathbf{X}_0.$$

EXAMPLE 6.4.4: Solve $X' = \begin{pmatrix} 2 & 5 \\ -4 & -2 \end{pmatrix} X + \begin{pmatrix} \cos 4t \\ \sin 4t \end{pmatrix}$.

SOLUTION: Here, $A = \begin{pmatrix} 2 & 5 \\ -4 & -2 \end{pmatrix}$. We compute e^{At} with `MatrixExp`.

```
Clear[a]

a = {{2, 5}, {-4, -2}};

expa = MatrixExp[at];

MatrixForm[expa]
```

$$\begin{pmatrix} \cos[4t] + \dfrac{1}{2}\sin[4t] & \dfrac{5}{4}\sin[4t] \\ -\sin[4t] & \cos[4t] - \dfrac{1}{2}\sin[4t] \end{pmatrix}$$

A general solution of the system is then given by $X = e^{At} \int e^{-At} F(t)\, dt + e^{At} C$. We compute e^{-At}, $\int e^{-At} F(t)\, dt$, and $e^{At} \int e^{-At} F(t)\, dt$.

```
invexpa = Inverse[expa]//Simplify;

Simplify[invexpa]
```

$$\left\{\left\{\cos[4t] - \frac{1}{2}\sin[4t], -\frac{5}{4}\sin[4t]\right\}, \left\{\sin[4t], \cos[4t] + \frac{1}{2}\sin[4t]\right\}\right\}$$

```
f[t_] = {{Cos[4t]}, {Sin[4t]}};

step1 = invexpa.f[t]//Simplify;

MatrixForm[step1]
```

$$\begin{pmatrix} \dfrac{1}{8}(-1 + 9\cos[8t] - 2\sin[8t]) \\ \dfrac{1}{2}\sin[4t]^2 + \sin[8t] \end{pmatrix}$$

```
step2 = Integrate[step1, t]//Simplify;

MatrixForm[step2]
```

$$\begin{pmatrix} \dfrac{1}{64}(-8t + 2\cos[8t] + 9\sin[8t]) \\ \dfrac{1}{32}(8t - 4\cos[8t] - \sin[8t]) \end{pmatrix}$$

```
step3 = expa.step2//Simplify;

MatrixForm[step3]
```

$$\left(\begin{array}{c} \dfrac{1}{64}((2 - 8t)\text{Cos}[4t] + (9 + 16t)\text{Sin}[4t]) \\ \dfrac{1}{32}((-4 + 8t)\text{Cos}[4t] - \text{Sin}[4t]) \end{array} \right)$$

Then, we form our general solution.

```
gensol = step3 + expa.{{c1}, {c2}}//Simplify
```

$$\{\{(\dfrac{1}{32} + c1 - \dfrac{t}{8})\text{Cos}[4t] + \dfrac{1}{64}(9 + 32c1 + 80c2 + 16t)\text{Sin}[4t]\},$$
$$\{(-\dfrac{1}{8} + c2 + \dfrac{t}{4})\text{Cos}[4t] - \dfrac{1}{32}(1 + 32c1 + 16c2)\text{Sin}[4t]\}\}$$

To graph the solution parametrically for various values of the arbitrary constant, we use Flatten to convert gensol to a list of the form $\{x(t), y(t)\}$.

```
step1 = Flatten[gensol]
```

$$\{(\dfrac{1}{32} + c1 - \dfrac{t}{8})\text{Cos}[4t] + \dfrac{1}{64}(9 + 32c1 + 80c2 + 16t)\text{Sin}[4t],$$
$$(-\dfrac{1}{8} + c2 + \dfrac{t}{4})\text{Cos}[4t] - \dfrac{1}{32}(1 + 32c1 + 16c2)\text{Sin}[4t]\}$$

Next, we use Table together with Flatten to create a set of parametric functions that we will graph with ParametricPlot.

```
toplot = Flatten[Table[step1, {c1, -1, 1},

  {c2, -1, 1}], 1];
```

Now, we define paramgraph. Given a list of the form $\{x(t), y(t)\}$, paramgraph parametrically graphs $\begin{cases} x = x(t) \\ y = y(t) \end{cases}$ for $-3\pi \le t \le 3\pi$. The resulting graphics object is not displayed.

```
paramgraph[list_]:=

  ParametricPlot[list, {t, -3Pi, 3Pi},

  PlotRange → {{-5, 5}, {-5, 5}},

  Ticks → {{-5, 5}, {-5, 5}},

  AspectRatio → 1,

  AxesLabel → {x, y}];
```

We then use Map to apply paramgraph to the list of parametric functions toplot. The resulting list of nine graphics objects is partitioned into three element subsets with Partition. All nine graphs are then shown together using Show and GraphicsArray in Figure 6-22.

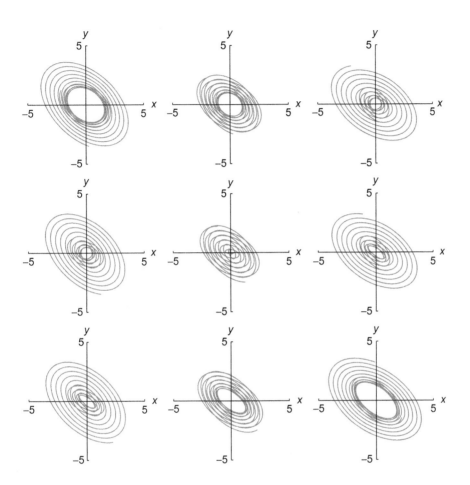

Figure 6-22   Parametric plots of various solutions to the nonhomogeneous system

```
somegraphs = Map[paramgraph, toplot];

toshow = Partition[somegraphs, 3];

Show[GraphicsGrid[toshow]]
```

■

If a system of differential equations contains derivatives of order greater than one, we can often rewrite it as a system of first-order equations.

In Chapter 8, we will also see that Laplace transforms can often be used to solve systems of this type.

EXAMPLE 6.4.5: Solve $\begin{cases} -2\dfrac{d^2x}{dt^2} - 2\dfrac{dy}{dt} = 0 \\ \dfrac{d^2y}{dt^2} + y - \dfrac{dx}{dt} = \cos t \\ x(0) = 2,\ x'(0) = 1,\ y(0) = 1,\ y'(0) = 2 \end{cases}$.

SOLUTION: To rewrite the system as a system of first-order equations, we let $z = dx/dt$ and $w = dy/dt$. Then, $dz/dt = d^2x/dt^2$ and $dw/dt = d^2y/dt^2$. Substituting into the first equation we have $-2\,dz/dt - 2w = 0$ so $dz/dt = -w$. Similarly, substituting into the second equation yields $dw/dt + y - z = \cos t$ so $dw/dt = -y + z + \cos t$. Therefore, the original system is equivalent to the system of first-order equations

$$\begin{cases} dx/dt = z \\ dy/dt = w \\ dz/dt = -w \\ dw/dt = -y + z + \cos t \\ x(0) = 2,\ y(0) = 1,\ z(0) = 1,\ w(0) = 2 \end{cases}.$$

In matrix form, the initial-value problem is equivalent to $\mathbf{X}' = \mathbf{AX} + \mathbf{F}(t)$,

$\mathbf{X}(0) = \begin{pmatrix} 2 \\ 1 \\ 1 \\ 2 \end{pmatrix}$, where

$$\mathbf{X} = \begin{pmatrix} x \\ y \\ z \\ w \end{pmatrix}, \quad \mathbf{A} = \begin{pmatrix} 0 & 0 & 1 & 0 \\ 0 & 0 & 0 & 1 \\ 0 & 0 & 0 & -1 \\ 0 & -1 & 1 & 0 \end{pmatrix}, \quad \text{and} \quad \mathbf{F}(t) = \begin{pmatrix} 0 \\ 0 \\ 0 \\ \cos t \end{pmatrix}.$$

Using the exponential matrix, the solution to the initial-value problem is given by

$$\mathbf{X} = \int_0^t e^{\mathbf{A}(t-s)} \mathbf{F}(s)\, ds + e^{\mathbf{A}t} \begin{pmatrix} 2 \\ 1 \\ 1 \\ 2 \end{pmatrix}.$$

First, we define $\mathbf{A}$ and then use `MatrixExp` together with `ExpToTrig` and `FullSimplify` to compute $e^{\mathbf{A}t}$.

```
a = {{0, 0, 1, 0}, {0, 0, 0, 1},
    {0, 0, 0, -1}, {0, -1, 1, 0}};
expa = MatrixExp[at]//FullSimplify;
MatrixForm[expa]
```

$$
\begin{pmatrix}
1 & \frac{1}{4}(2t - \sqrt{2}\sin[\sqrt{2}t]) & \frac{1}{4}(2t + \sqrt{2}\sin[\sqrt{2}t]) & -\sin[\frac{t}{\sqrt{2}}]^2 \\
0 & \cos[\frac{t}{\sqrt{2}}]^2 & \sin[\frac{t}{\sqrt{2}}]^2 & \frac{\sin[\sqrt{2}t]}{\sqrt{2}} \\
0 & \sin[\frac{t}{\sqrt{2}}]^2 & \cos[\frac{t}{\sqrt{2}}]^2 & -\frac{\sin[\sqrt{2}t]}{\sqrt{2}} \\
0 & -\frac{\sin[\sqrt{2}t]}{\sqrt{2}} & \frac{\sin[\sqrt{2}t]}{\sqrt{2}} & \cos[\sqrt{2}t]
\end{pmatrix}
$$

The matrix $e^{\mathbf{A}(t-s)}$ is obtained by replacing each occurrence of t in $e^{\mathbf{A}t}$ by $t - s$.

```
expats = expa/. t → t - s;
```

Next, we compute $e^{\mathbf{A}(t-s)}\mathbf{F}(s)$ and integrate the result.

```
f[t_] = {{0}, {0}, {0}, {Cos[t]}};
MatrixForm[f[t]]
```

$$
\begin{pmatrix}
0 \\
0 \\
0 \\
\cos[t]
\end{pmatrix}
$$

```
tointegrate = expats. f[s]//Simplify;
MatrixForm[tointegrate]
```

$$
\begin{pmatrix}
-\cos[s]\sin[\frac{-s + t}{\sqrt{2}}]^2 \\
\frac{\cos[s]\sin[\sqrt{2}(-s + t)]}{\sqrt{2}} \\
-\frac{\cos[s]\sin[\sqrt{2}(-s + t)]}{\sqrt{2}} \\
\cos[s]\cos[\sqrt{2}(-s + t)]
\end{pmatrix}
$$

```
step2 = Integrate[tointegrate, {s, 0, t}];
MatrixForm[step2]
```

$$\begin{pmatrix} -\mathrm{Sin}[t] + \dfrac{\mathrm{Sin}[\sqrt{2}\,t]}{\sqrt{2}} \\ \mathrm{Cos}[t] - \mathrm{Cos}[\sqrt{2}\,t] \\ -\mathrm{Cos}[t] + \mathrm{Cos}[\sqrt{2}\,t] \\ -\mathrm{Sin}[t] + \sqrt{2}\,\mathrm{Sin}[\sqrt{2}\,t] \end{pmatrix}$$

Finally, we form the solution to the initial-value problem. Note that the first and second rows correspond to x and y, respectively.

```
x0 = {{2}, {1}, {1}, {2}};
```

```
sol = step2 + expa.x0//Simplify;
```

```
MatrixForm[sol]
```

$$\begin{pmatrix} 2 + t - \mathrm{Sin}[t] - 2\mathrm{Sin}[\dfrac{t}{\sqrt{2}}]^2 + \dfrac{\mathrm{Sin}[\sqrt{2}\,t]}{\sqrt{2}} \\ 1 + \mathrm{Cos}[t] - \mathrm{Cos}[\sqrt{2}\,t] + \sqrt{2}\,\mathrm{Sin}[\sqrt{2}\,t] \\ 1 - \mathrm{Cos}[t] + \mathrm{Cos}[\sqrt{2}\,t] - \sqrt{2}\,\mathrm{Sin}[\sqrt{2}\,t] \\ 2\mathrm{Cos}[\sqrt{2}\,t] - \mathrm{Sin}[t] + \sqrt{2}\,\mathrm{Sin}[\sqrt{2}\,t] \end{pmatrix}$$

```
sol[[2]]
```

$$\{1 + \mathrm{Cos}[t] - \mathrm{Cos}[\sqrt{2}\,t] + \sqrt{2}\,\mathrm{Sin}[\sqrt{2}\,t]\}$$

We confirm that the initial conditions are satisfied by graphing $x(t)$ and $y(t)$ together in Figure 6-23(a) and parametrically in Figure 6-23(b). $z(t)$ and $w(t)$ are graphed together in Figure 6-23(c) and parametrically in Figure 6-23(d).

```
p1 = Plot[Evaluate[{sol[[1, 1]], sol[[2, 1]]}], {t, 0, 12},

    PlotLabel → "(a)"]
```

```
p2 = Plot[Evaluate[{sol[[3, 1]], sol[[4, 1]]}], {t, 0, 12},

    PlotLabel → "(b)"]
```

```
p3 = ParametricPlot[Evaluate[{sol[[1, 1]], sol[[2, 1]]}],

    {t, 0, 12}, PlotLabel → "(c)", AxesLabel → {x, y}]
```

```
p4 = ParametricPlot[Evaluate[{sol[[3, 1]], sol[[4, 1]]}],

    {t, 0, 12}, PlotLabel → "(d)", AxesLabel → {z, w}]
```

```
Show[GraphicsGrid[{{p1, p2}, {p3, p4}}]]
```

■

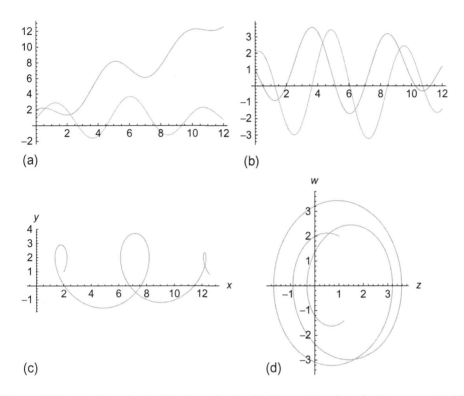

Figure 6-23   (a) $x(t)$ and $y(t)$. (b) $z(t)$ and $w(t)$. (c) Parametric plot of $x(t)$ versus $y(t)$. (d) Parametric plot of $z(t)$ versus $w(t)$

6.5 Numerical Methods

Because it may be difficult or even impossible to construct an explicit solution to some systems of differential equations, we now turn our attention to discussing some numerical methods that are used to construct numerical solutions to systems of differential equations.

6.5.1 Built-In Methods

Numerical approximations of solutions to systems of ordinary differential equations can be obtained with NDSolve. This command is particularly useful when working with nonlinear systems of equations for which DSolve alone is unable to find an explicit or implicit solution.

EXAMPLE 6.5.1: Consider the nonlinear system of equations

$$\begin{cases} x' = \mu x + y - x\left(x^2 + y^2\right) \\ y' = \mu y - x - y\left(x^2 + y^2\right) \end{cases}. \tag{6.18}$$

(a) Graph the direction field associated with the system for $\mu = 2, 1,$ $1/4,$ and $-1/2.$ (b) For each value of μ in (a), approximate the solution that satisfies the initial conditions $x(0) = 0$ and $y(0) = 1/2.$ Use each numerical solution to approximate $x(5)$ and $y(5).$

SOLUTION: We define the function dfield. Given $\mu,$ dfield[μ] graphs the direction field associated with the system (6.18) on the rectangle $[-2, 2] \times [-2, 2].$

```
dfield[μ_]:=StreamPlot[{{μx + y − x(x² + y²), μy − x
    −y(x² + y²)}, {x, −2, 2}, {y, −2, 2},
    Axes → Automatic, AxesOrigin → {0, 0}, StreamPoints
    → Fine, Frame → False, Axes → Automatic, AxesOrigin
    → {0, 0}, AxesLabel → {x, y}];
```

We use dfield to graph the direction field associated with the system for $\mu = 2, 1, 1/4,$ and $-1/2.$

```
step1 = Map[dfield, {2, 1, 1/4, −1/2}];

toshow = Partition[step1, 2];
```

Show together with GraphicsGrid is used to display all four graphs together in Figure 6-24. The direction field indicates that the behavior of the solutions strongly depends on the value of $\mu.$

```
Show[GraphicsGrid[toshow]]
```

Seeing how the solutions to the system behave as a function of μ is well-suited for exploration with the Manipulate function. Next, we use Manipulate to design a function that allows us to explore how the behavior of the system changes as a function of $\mu.$

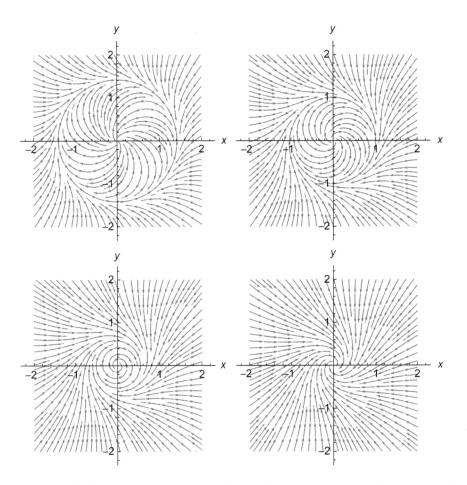

Figure 6-24   The behavior of solutions to the nonlinear system depend strongly on the value of μ

```
Manipulate[
    StreamPlot[{μx + y − x(x² + y²), μy − x − y(x² + y²)}, .
    {x, −2, 2}, {y, −2, 2},
    Axes → Automatic, AxesOrigin → {0, 0}, StreamPoints
    → Fine, Frame → False, Axes → Automatic, AxesOrigin
    → {0, 0}, AxesLabel → {x, y}], {{μ, 0}, −3, 3}]
```

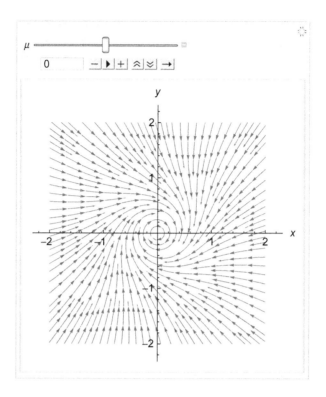

To study the behavior of the system for various values of μ, we define the function sols. Given μ, sols[μ] generates a numerical solution to the initial-value problem. Now, we use sols to generate a numerical approximation to the initial-value problem if $\mu = 2$.

```
sys[μ_]:={x'[t]==μx[t] + y[t] − x[t](x[t]^2 + y[t]^2),

  y'[t] == μy[t] − x[t] − y[t](x[t]^2 + y[t]^2), x[0] == 0,

  y[0] == 1/2}
```

```
sols[μ_]:=NDSolve[sys[μ], {x[t], y[t]}, {t, 0, 10}];
```

```
sola = sols[2]
```

```
{{x[t] → InterpolatingFunction[][t], y[t] →

InterpolatingFunction[][t]}}
```

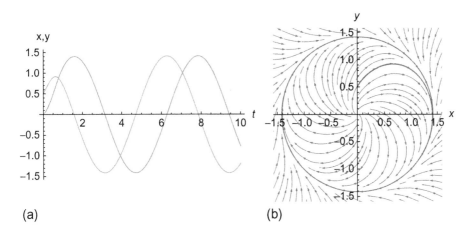

Figure 6-25   (a) If $\mu = 2$, the solution to the initial-value problem approaches a limit cycle. (b) The direction field highlights the location of the limit cycle

We use `ReplaceAll` (`/.`) to see that $x(5) \approx -1.35612$ and $y(5) \approx 0.401167$.

```
sola/.t → 5.
```

```
{{x[5.] → -1.35612, y[5.] → 0.401159}}
```

We use `Plot` to graph $x(t)$ and $y(t)$ for $0 \le t \le 10$ in Figure 6-25(a). Notice that the solution appears to become periodic.

```
p1 = Plot[Evaluate[{x[t], y[t]}/.sola],

  {t, 0, 10}, AxesLabel → {t, "x,y"}]
```

This is further confirmed by graphing the solution parametrically and showing it together with the direction field in Figure 6-25(b).

```
p2a = ParametricPlot[Evaluate[{x[t], y[t]}/.sola],

  {t, 0, 10}, PlotStyle->CMYKColor[0, 0.89, 0.94, 0.28]]

p2b = dfield[2]

p2 = Show[p2a, p2b, AxesLabel → {x, y}]

Show[GraphicsRow[{p1, p2}]]
```

For the remaining values of μ, we use the function `sols` defined previously. We use `sols` to solve each initial-value problem. Note that Mathematica does not display any output because we have included a semi-colon at the end of each command. (You could use `Map[sols[#]&, {1, 1/4, -1/2}]` to obtain an equivalent result.)

```
solb = sols[1];

solc = sols[1/4];

sold = sols[-1/2];
```

As before, we use `ReplaceAll` (`/.`) to approximate the value of each solution if $t = 5$.

```
{solb, solc, sold}/.t → 5
```

$$\{\{\{x[5] \to -0.958858, y[5] \to 0.283643\}\}, \{\{x[5] \to -0.479462,$$

$$y[5] \to 0.141831\}\}, \{\{x[5] \to -0.0321707, y[5]$$

$$\to 0.00951651\}\}\}$$

For each numerical solution, we use `Plot` to graph $x(t)$ and $y(t)$ for $0 \le t \le 10$ in Figure 6-26. Notice that the solutions corresponding to positive values of μ appear to become periodic while the solution corresponding to the negative value of μ appears to tend toward zero.

```
pb = Plot[Evaluate[{x[t], y[t]}/.solb],

  {t, 0, 10}, PlotLabel → "(a)"];

pc = Plot[Evaluate[{x[t], y[t]}/.solc],

  {t, 0, 10}, PlotLabel → "(b)"];

pd = Plot[Evaluate[{x[t], y[t]}/.sold],

  {t, 0, 10}, PlotLabel → "(c)"];


Show[GraphicsRow[{pb, pc, pd}]]
```

These results are further confirmed when we graph each solution parametrically and display the graphs with the direction fields generated in (a) in Figure 6-27.

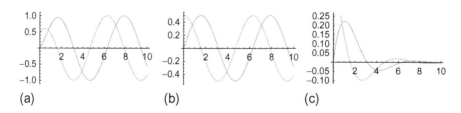

Figure 6-26 If μ is positive, the solutions approach a limit cycle; if μ is negative the solutions tend to 0

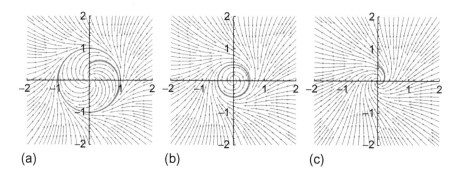

Figure 6-27   If μ is positive, the solutions approach a limit cycle; if μ is negative the solutions tend to 0

```
ppb = ParametricPlot[Evaluate[{x[t], y[t]}/.solb],
    {t, 0, 10}, PlotStyle->CMYKColor[0, 0.89, 0.94, 0.28],
    PlotRange → {{-2, 2}, {-2, 2}},
    PlotLabel → "(a)"];
ppc = ParametricPlot[Evaluate[{x[t], y[t]}/.solc],
    {t, 0, 10}, PlotStyle->CMYKColor[0, 0.89, 0.94, 0.28],
    PlotRange → {{-2, 2}, {-2, 2}},
    PlotLabel → "(b)"];
ppd = ParametricPlot[Evaluate[{x[t], y[t]}/.sold],
    {t, 0, 10}, PlotStyle->CMYKColor[0, 0.89, 0.94, 0.28],
    PlotRange → {{-2, 2}, {-2, 2}},
    PlotLabel → "(c)"];

graphb = Show[ppb, step1[[2]],
    PlotLabel → "(a)"];
graphc = Show[ppc, step1[[3]],
    PlotLabel → "(b)"];
graphd = Show[ppd, step1[[4]],
    PlotLabel → "(c)"];

Show[GraphicsRow[{graphb, graphc, graphd}]]
```

In the cases corresponding to the positive values of μ, we see in the direction field that all solutions appear to tend to a closed curve. Can we find the curve in each case? If $\mu = 2$, we see that the solution that satisfies $x(0) = 0$ and $y(0) = 1.41$ will be periodic. Similarly, if $\mu = 1/4$, we see that the solution that satisfies $x(0) = 0$ and $y(0) = 0.482$ will be periodic. On the other hand, if $\mu = 1$, we need not approximate the solution. From the graph, we see that the solution that satisfies $x(0) = 0$ and $y(0) = 1$ will be periodic. It is relatively easy to verify that the solution that satisfies these initial conditions is $\begin{cases} x = \cos t \\ y = \sin t \end{cases}$.

■

Application: Controlling the Spread of a Disease

Sources: Herbert W. Hethcote, "Three Basic Epidemiological Models," *Applied Mathematical Ecology*, edited by Simon A. Levin, Thomas G. Hallan, and Louis J. Gross, Springer-Verlag (1989), pp. 119–143. Roy M. Anderson and Robert M. May, "Directly Transmitted Infectious Diseases: Control by Vaccination," *Science*, Volume 215 (February 26, 1982), pp. 1053–1060. J. D. Murray, *Mathematical Biology*, Springer-Verlag (1990), pp. 611–618.

If a person becomes immune to a disease after recovering from it and births and deaths in the population are not taken into account, then the percent of persons susceptible to becoming infected with the disease, $S(t)$, the percent of people in the population infected with the disease, $I(t)$, and the percent of the population recovered and immune to the disease, $R(t)$, can be modeled by the system

$$\begin{cases} S' = -\lambda SI \\ I' = \lambda SI - \gamma I \\ R' = \gamma I \\ S(0) = S_0,\ I(0) = I_0,\ R(0) = 0, \end{cases} \tag{6.19}$$

where $' = d/dt$. Because $S(t) + I(t) + R(t) = 1$, once we know $S(t)$ and $I(t)$, we can compute $R(t)$ with $R(t) = 1 - S(t) - I(t)$. This model is called an **SIR model without vital dynamics** because once a person has had the disease the person becomes immune to the disease and because births and deaths are not taken into consideration. This model might be used to model diseases that are **epidemic** to a population: those diseases that persist in a population for short periods of time (less than one year). Such diseases typically include influenza, measles, rubella, and chickenpox.

If $S_0 < \gamma/\lambda$, $I'(0) = \lambda S_0 I_0 - \gamma I_0 < \lambda \frac{\gamma}{\lambda} I_0 - \gamma I_0 = 0$. Thus, the rate of infection immediately begins to decrease; the disease dies out. On the other hand, if $S_0 > \gamma/\lambda$, $I'(0) > \lambda S_0 I_0 - \gamma I_0 < \lambda \frac{\gamma}{\lambda} I_0 - \gamma I_0 = 0$ so the rate of infection first increases; an epidemic results.

Although we cannot find explicit formulas for S, I, and R as functions of t, we can, for example, solve for I in terms of S. The equation $\dfrac{dI}{dS} = -\dfrac{(\lambda S - \gamma)I}{\lambda SI} = -1 + \dfrac{\rho}{S}$, $\rho = \gamma/\lambda$, is separable:

$$\frac{dI}{dS} = -1 + \frac{\rho}{S} \implies dI = \left(-1 + \frac{\rho}{S}\right) dS \implies I = -S + \rho \ln S + C$$

and applying the initial condition results in

$$I_0 = -S_0 + \rho \ln S_0 + C \implies C = I_0 + S_0 - \rho \ln S_0$$

so $I = -S + \rho \ln S + I_0 + S_0 - \rho \ln S_0 \implies I + S - \rho \ln S = I_0 + S_0 - \rho \ln S_0$.

When diseases persist in a population for long periods of time, births and deaths must be taken into consideration. If a person becomes immune to a disease after recovering from it and births and deaths in the population are taken into account, then the percent of persons susceptible to becoming infected with the disease, $S(t)$, and the percent of people in the population infected with the disease, $I(t)$, can be modeled by the system

$$\begin{cases} S' = -\lambda SI + \mu - \mu S \\ I' = \lambda SI - \gamma I - \mu I \\ S(0) = S_0, \, I(0) = I_0 \end{cases} \quad (6.20)$$

This model is called an **SIR model with vital dynamics** because once a person has had the disease the person becomes immune to the disease and because births and deaths are taken into consideration. This model might be used to model diseases that are **endemic** to a population: those diseases that persist in a population for long periods of time (ten or twenty years). Smallpox is an example of a disease that was endemic until it was eliminated in 1977. We use Solve to see that the solutions to the system of equations

$$\begin{cases} -\lambda SI + \mu - \mu S = 0 \\ \lambda SI - \gamma I - \mu I = 0 \end{cases}$$

are $S = 1, I = 0$ and $S = \dfrac{\gamma + \mu}{\lambda}, I = \dfrac{\mu \left[\lambda - (\gamma + \mu)\right]}{\lambda(\gamma + \mu)}$.

```
eq1 = -λsi + μ - μs;

eq2 = λsi - γi - μi;

eqpts = Solve[{eq1==0, eq2==0}, {s, i}]
```

$$\left\{\{s \to 1, \, i \to 0\}, \, \left\{s \to \frac{\gamma + \mu}{\lambda}, \, i \to \frac{-\gamma\mu + \lambda\mu - \mu^2}{\lambda(\gamma + \mu)}\right\}\right\}$$

These two points are called **equilibrium points** because they are constant solutions to the system.

Because $S(t) + I(t) + R(t) = 1$, it follows that $S(t) + I(t) \leq 1$. The following table shows the average infectious period, $1/\gamma$, γ, and typical contact numbers, σ, for several diseases during certain epidemics.

Disease	$1/\gamma$	γ	σ
Measles	6.5	0.153846	14.9667
Chickenpox	10.5	0.0952381	11.3
Mumps	19	0.0526316	8.1
Scarlet fever	17.5	0.0571429	8.5

Let us assume that the average lifetime, $1/\mu$, is 70 so that $\mu = 0.0142857$.

For each of the diseases listed in the previous table, we use the formula $\sigma = \lambda/(\gamma + \mu)$ to calculate the daily contact rate λ.

Disease	λ
Measles	2.51638
Chickenpox	1.23762
Mumps	0.54203
Scarlet fever	0.607143

Diseases like those listed above can be controlled once an effective and inexpensive vaccine has been developed. Since it is virtually impossible to vaccinate everybody against a disease, we would like to know what percentage of a population needs to be vaccinated to eliminate a disease. A population of people has **herd immunity** to a disease means that enough people are immune to the disease so that if it is introduced into the population, it will not spread throughout the population. In order to have herd immunity, an infected person must infect less than one uninfected person during the time the person is infectious. Thus, we must have

$$\sigma S < 1.$$

Since $I + S + R = 1$, when $I = 0$ we have that $S = 1 - R$ and, consequently, herd immunity is achieved when

$$\sigma(1 - R) < 1$$
$$\sigma - \sigma R < 1$$
$$-\sigma R < 1 - \sigma$$
$$R > \frac{\sigma - 1}{\sigma} = 1 - \frac{1}{\sigma}.$$

For each of the diseases listed above, we estimate the minimum percentage of a population that needs to be vaccinated to achieve herd immunity.

Disease	Minimum Value of R to Achieve Herd Immunity
Measles	0.933186
Chickenpox	0.911505
Mumps	0.876544
Scarlet fever	0.882354

Using the values in the previous tables, for each disease we graph the direction field and several solutions $\begin{cases} S = S(t) \\ I = I(t) \end{cases}$ parametrically. For measles, we proceed as follows. After loading the **PlotField** package, we define μ, γ, σ, and λ. For these values, we graph the direction field associated with the system on the rectangle $[0, 1] \times [0, 1]$. Because $S(t) + I(t) \leq 1$, we are only concerned with solutions of the system that are below the line $S + I = 1$.

```
p1 = Plot[1 - x, {x, 0, 1}, PlotStyle → {{Thickness[0.0075],

  CMYKColor[0.64, 0, 0.95, 0.40]}}];

μ = 0.0142857;

γ = 0.153846;

σ = 14.9667;

λ = σ (γ + μ);

eq1 = -λ s i + μ - μ s;

eq2 = λ s i - γ i - μ i;

pvf1 = StreamPlot[{eq1, eq2}, {s, 0, 1}, {i, 0, 1},

StreamPoints → Fine];
```

Next, we define two lists of ordered pairs and use Union to join the two lists. The points in initconds1 are "close" to the S axis while the points in initconds2 are close to the I axis. We will graph the solutions that satisfy these initial conditions.

```
initconds1 = Table[{i/10, 0.01}, {i, 1, 9}];

initconds2 = Table[{1 − i/10, i/10}, {i, 1, 9}];

initconds = Union[initconds1, initconds2];
```

Now we define the function numgraph. Given an ordered pair (S_0, I_0), numgraph generates a numerical solution to the initial-value problem (6.20) and graphs the result.

```
numgraph[{s0_, i0_}]:=Module[{numsol},

    numsol = NDSolve[{s'[t]== − λs[t]i[t] + μ − μs[t],

    i'[t]==λs[t]i[t] − γi[t] − μi[t], s[0]==s0, i[0]==i0}, {s[t], i[t]},

    {t, 0, 20}];

  ParametricPlot[{s[t], i[t]}/.numsol, {t, 0, 20},

    PlotStyle → CMYKColor[0, 0.89, 0.94, 0.28]]

    ]
```

We then use Map to apply numgraph to the list initconds. Show is used to display all three graphics objects together in Figure 6-28. In the result, we see that the (nontrivial) solutions approach the equilibrium point.

```
toshow = Map[numgraph, initconds];

Show[pvf1, toshow, p1, PlotRange->{{0, 1}, {0, 1}},

  AspectRatio->1, Axes->Automatic,

  Frame → False, AxesOrigin → {0, 0},

  AxesLabel → {x, y}]
```

For the remaining three diseases, we change the values of μ, γ, σ, and λ and reenter the code. Here are the results for chickenpox. See Figure 6-29.

```
μ = 0.0142857;

γ = 0.0952381;

σ = 11.3;

λ = σ(γ + μ);

eq1 = −λsi + μ − μs;

eq2 = λsi − γi − μi;

pvf1 = StreamPlot[{eq1, eq2}, {s, 0, 1}, {i, 0, 1}, StreamStyle

→ Fine];
```

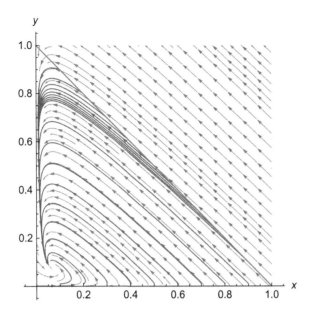

Figure 6-28   Using an *SIR* model to model measles

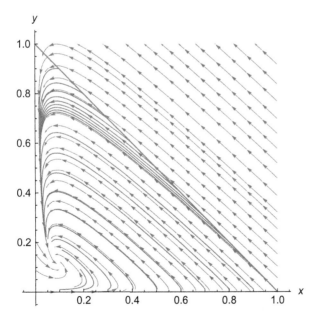

Figure 6-29   Using an *SIR* model to model chickenpox

```
numgraph[{s0_, i0_}]:=Module[{numsol},

  numsol = NDSolve[{s'[t]== − λ s[t]i[t] + μ − μs[t],

   i'[t]==λs[t]i[t] − γ i[t] − μ i[t], s[0]==s0, i[0]==i0}, {s[t], i[t]},

   {t, 0, 20}];

  ParametricPlot[{s[t], i[t]}/.numsol, {t, 0, 20},

    PlotStyle → CMYKColor[0, 0.96, 0.39, 0]]

      ]

toshow = Map[numgraph, initconds];

Show[pvf1, toshow, p1, PlotRange->{{0, 1}, {0, 1}},

  AspectRatio->1, Axes->Automatic,

  Frame → False, AxesOrigin → {0, 0},

  AxesLabel → {x, y}]
```

Similar results are obtained for mumps. See Figure 6-30.

```
μ = 0.0142857;

γ = 0.0526316;

σ = 8.1;
```

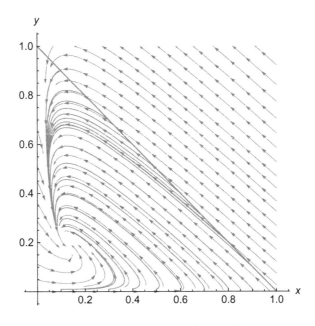

Figure 6-30   Using an *SIR* model to model mumps

```
λ = σ(γ + μ);

eq1 = -λ s i + μ - μ s;

eq2 = λ s i - γ i - μ i; pvf1 = StreamPlot[{eq1, eq2},

{s, 0, 1}, {i, 0, 1}];

numgraph[{s0_, i0_}]:=Module[{numsol},

    numsol = NDSolve[{s'[t]== - λ s[t]i[t] + μ - μ s[t],

        i'[t]==λ s[t]i[t] - γ i[t] - μ i[t],

        s[0]==s0, i[0]==i0}, {s[t], i[t]}, {t, 0, 40}];

    ParametricPlot[{s[t], i[t]}/.numsol, {t, 0, 40},

    PlotStyle → CMYKColor[0.62, 0.57, 0.23, 0]]

    ]

toshow = Map[numgraph, initconds];

Show[pvf1, toshow, p1, PlotRange->{{0, 1}, {0, 1}},

    AspectRatio->1, Axes->Automatic,

    Frame → False, AxesOrigin → {0, 0},

    AxesLabel → {x, y}]
```

Last, we generate graphs for scarlet fever. See Figure 6-31. In all four cases, we see that all solutions approach the equilibrium point, which indicates that although the epidemic runs its course, the disease is never completely removed from the population.

```
μ = 0.0142857;

γ = 0.0571429;

σ = 8.5;

λ = σ(γ + μ);

eq1 = -λ s i + μ - μ s;

eq2 = λ s i - γ i - μ i;

pvf1 = StreamPlot[{eq1, eq2}, {s, 0, 1}, {i, 0, 1},

StreamStyle → Fine];

  numgraph[{s0_, i0_}]:=Module[{numsol},

    numsol = NDSolve[{s'[t]== - λ s[t]i[t] + μ - μ s[t],

        i'[t]==λ s[t]i[t] - γ i[t] - μ i[t],
```

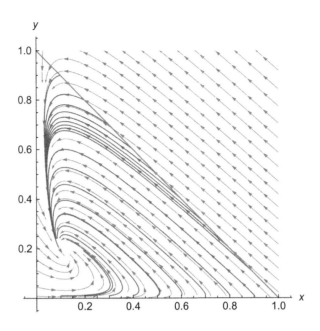

Figure 6-31   Using an *SIR* model to model scarlet fever

```
    s[0]==s0, i[0]==i0}, {s[t], i[t]}, {t, 0, 40}];

  ParametricPlot[{s[t], i[t]}/.numsol, {t, 0, 40},

 PlotStyle->CMYKColor[0, 0.89, 0.94, 0.28]]

    ]

toshow = Map[numgraph, initconds];

Show[pvf1, toshow, p1, PlotRange->{{0, 1}, {0, 1}},

 AspectRatio->1, Axes->Automatic,

 Frame → False, AxesOrigin → {0, 0},

 AxesLabel → {x, y}]
```

NDSolve can be used to generate numerical solutions of systems that involve more than one differential equation as well.

EXAMPLE 6.5.2 (FitzHugh-Nagumo Equation): Under certain assumptions, the **FitzHugh-Nagumo equation** that arises in the study of the impulses in a nerve fiber can be written as the system of ordinary differential equations

$$
\begin{cases}
dV/d\xi = W \\
dW/d\xi = F(V) + R - uW \\
dR/d\xi = \dfrac{\epsilon}{u}(bR - V - a) \\
V(0) = v_0,\ W(0) = W_0,\ R(0) = R_0
\end{cases}
\tag{6.21}
$$

where $F(V) = \frac{1}{3}V^3 - V$. (a) Graph the solution to the FitzHugh-Nagumo equation that satisfies the initial conditions $V(0) = 1$, $W(0) = 0$, and $R(0) = 1$ if $\epsilon = 0.08$, $a = 0.7$, $b = 0$, and $u = 1$. (b) Graph the solution that satisfies the initial conditions $V(0) = 1$, $W(0) = 0.5$, and $R(0) = 0.5$ if $\epsilon = 0.08$, $a = 0.7$, $b = 0.8$, and $u = 0.6$.

SOLUTION: We begin by defining the function fnsol, which given the appropriate parameter values and initial conditions returns a numerical solution of system (6.21). If $\{\xi, a, b\}$ is not included after the initial conditions, the default solution is valid for $0 \le \xi \le 50$; any options included are passed to the NDSolve command. In this case, we use lower-case letters to avoid any ambiguity with built-in Mathematica functions.

```
Clear[fhsol]

fhsol[ε_, a_, b_, u_][{v0_, w0_, r0_}, ξs_:{ξ, 0, 100},

   opts___]:=NDSolve[{v'[ξ] == w[ξ], w'[ξ] ==

   1/3v[ξ]^3 - v[ξ] + r[ξ] - uw[ξ],

   r'[ξ] == ε/u(br[ξ] - v[ξ] - a),

   v[0] == v0, w[0] == w0, r[0] == r0}, {v[ξ], w[ξ],

   r[ξ]}, ξs, opts]
```

For (a), we enter

```
sola = fhsol[0.08, .7, 0, 1][{1, 0, 1}]

{{v[ξ] → InterpolatingFunction[][ξ], w[ξ] →

InterpolatingFunction[][ξ], r[ξ] →

InterpolatingFunction[][ξ]}}
```

We then graph the solution functions parametrically with ParametricPlot3D and then individually with Plot. The option PlotPoints->200 is included in the ParametricPlot3D to

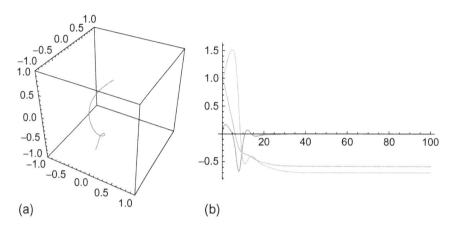

(a)   (b)

Figure 6-32   For $0 \leq \xi \leq 50$: (a) Parametric plot of W versus V versus R. (b) W, V, and R

help assure that the resulting graph is smooth. Using Show and GraphicsRow, both plots are shown in Figure 6-32.

```
pp1 = ParametricPlot3D[Evaluate[{w[ξ], v[ξ], r[ξ]}/.sola],

    {ξ, 0, 100}, PlotRange → {{−1, 1}, {−1, 1}, {−1, 1}},

    BoxRatios → {1, 1, 1}, PlotPoints → 500];

    pa = Plot[Evaluate[{w[ξ], v[ξ], r[ξ]}/.sola], {ξ, 0, 100}];

Show[GraphicsRow[{pp1, pa}]]
```

For (b), we specify that we want the solution to be valid for $0 \leq \xi \leq$ 15 so we enter

```
solb = fhsol[0.08, .7, .8, .6][{1, .5, .5}, {ξ, 0, 15}]

{{v[ξ] → InterpolatingFunction[][ξ],

w[ξ] → InterpolatingFunction[][ξ],

r[ξ] → InterpolatingFunction[][ξ]}}
```

Parametric plots and individual plots are generated in the same way as in (a). See Figure 6-33.

```
pp2 = ParametricPlot3D[Evaluate[{w[ξ], v[ξ], r[ξ]}/.solb],

    {ξ, 0, 15}, PlotRange → {{−1, 1}, {−1, 1}, {−1, 1}},

    BoxRatios → {1, 1, 1}, PlotPoints → 500];

    pb = Plot[Evaluate[{w[ξ], v[ξ], r[ξ]}/.solb], {ξ, 0, 15}];

Show[GraphicsRow[{pp2, pb}]]
```

∎

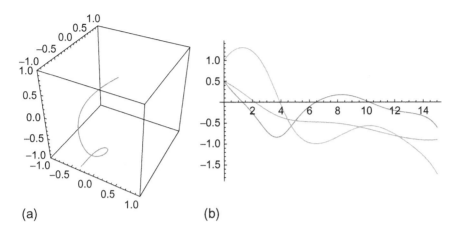

Figure 6-33   For $0 \leq \xi \leq 15$: (a) Parametric plot of W versus V versus R. (b) W, V, and R

In other cases, you may wish to implement your own numerical algorithms to approximate solutions of differential equations. We briefly discuss two familiar methods (Euler's method and the Runge-Kutta method) and illustrate how to implement these algorithms using Mathematica. Details regarding these and other algorithms, including discussions of the error involved in implementing them, can be found in most numerical analysis texts or other references like the Zwillinger's *Handbook of Differential Equations* [29].

6.5.2 Euler's Method

Euler's method for approximation that was discussed for first-order equations may be extended to include systems of first-order equations. The initial-value problem

$$\begin{cases} dx/dt = f(t, x, y) \\ dy/dt = g(t, x, y) \\ x(t_0) = x_0, \; y(t_0) = y_0 \end{cases} \tag{6.22}$$

is approximated at each step by the recursive relationship based on the Taylor expansion of x and y up to order h:

$$\begin{cases} x_{n+1} = x_n + hf(t_n, x_n, x_n) \\ y_{n+1} = y_n + hg(t_n, x_n, x_n) \end{cases}, \tag{6.23}$$

where $t_n = t_0 + nh, n = 0, 1, 2, \ldots$.

EXAMPLE 6.5.3: Use Euler's method with $h = 0.1$ to approximate the solution of the initial-value problem

$$\begin{cases} dx/dt = x - y + 1 \\ dy/dt = x + 3y + e^{-t} \\ x(0) = 0, \ y(0) = 1 \end{cases}.$$

Compare these results to those of the exact solution of the system of equations.

SOLUTION: We use the same notation as in equations (6.22) and (6.23): $f(x, y) = x - y + 1$, $g(x, y) = dy/dt = x + 3y + e^{-t}$, $t_0 = 0$, $x_0 = 0$, and $y_0 = 1$, so we use the formulas

$$\begin{cases} x_{n+1} = x_n + h(x_n - y_n + 1) \\ y_{n+1} = y_n + h\left(x_n + 3y_n + e^{-t_n}\right) \end{cases},$$

where $t_n = 0.1n, n = 0, 1, 2, \ldots$.

For example, if $n = 0$, then

$$\begin{cases} x_1 = x_0 + h(x_0 - y_0 + 1) = 0 \\ y_1 = y_0 + h\left(x_0 + 3y_0 + e^{-t_0}\right) = 1.4 \end{cases}.$$

The exact solution of this initial-value problem is found to be

$$\begin{cases} x(t) = -\frac{3}{4} - \frac{1}{9}e^{-t} + \frac{31}{36}e^{2t} - \frac{11}{6}te^{2t} \\ y(t) = \frac{1}{4} - \frac{2}{9}e^{-t} + \frac{35}{36}e^{2t} + \frac{11}{6}te^{2t} \end{cases}$$

with DSolve.

```
partsol = DSolve[{x'[t]==x[t] − y[t] + 1, y'[t]==x[t] + 3y[t]+

    Exp[−t], x[0]==0, y[0]==1}, {x[t], y[t]}, t]
```

$$\left\{\left\{x[t] \rightarrow -\frac{1}{36}e^{-t}(4 + 27e^t - 31e^{3t} + 66e^{3t}t), \ y[t] \rightarrow \frac{1}{36}e^{-t}(-8 + 9e^t + 35e^{3t} + 66e^{3t}t)\right\}\right\}$$

```
xex[t_] = − 3/4 − Exp[−t]/9 + 31Exp[2 t]/36 − 11/6 tExp[2 t];
```

$$yex[t\_] = \frac{1}{4} - \frac{2Exp[-t]}{9} + \frac{35Exp[2\,t]}{36} + \frac{11}{6}tExp[2\,t];$$

We display the results obtained with this method (in columns three and five) and compare them to the actual function values (in columns four and six).

```
Clear[f, g, t, h, x, y]

f[t_, x_, y_] = x - y + 1;

g[t_, x_, y_] = x + 3y + Exp[-t];

h = 0.1;

t[n_] := = t0 + nh;

t0 = 0;

xe[n_] := = xe[n] = xe[n - 1] + hf[t[n - 1], xe[n - 1], ye[n - 1]];

ye[n_] := = ye[n] = ye[n - 1] + hg[t[n - 1], xe[n - 1], ye[n - 1]];

xe[0] = 0;

ye[0] = 1;

Table[{n, t[n], xe[n], xex[t[n]], ye[n], yex[t[n]]}, {n, 0, 10}]//

TableForm
```

0	0.	0	0.	1	1.
1	0.1	0.	−0.0226978	1.4	1.46032
2	0.2	−0.04	−0.103346	1.91048	2.06545
3	0.3	−0.135048	−0.265432	2.5615	2.85904
4	0.4	−0.304703	−0.540105	3.39053	3.89682
5	0.5	−0.574227	−0.968408	4.44425	5.24975
6	0.6	−0.976074	−1.60412	5.78076	7.00806
7	0.7	−1.55176	−2.51737	7.47226	9.28638
8	0.8	−2.35416	−3.79926	9.60842	12.23
9	0.9	−3.45042	−5.56767	12.3005	16.0232
10	1.	−4.9255	−7.97468	15.6862	20.8987

```
xs = Table[{t[n], xe[n]}, {n, 0, 10}];
```

We also graph the approximation with the actual solution in Figure 6-34.

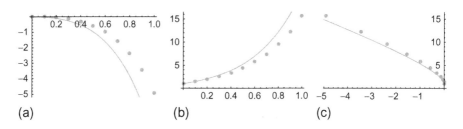

(a)   (b)   (c)

Figure 6-34   Euler's method using $h = 0.1$: (a) Comparison of x_n to $x(t)$. (b) Comparison of y_n to $y(t)$. (c) Comparison of (x_n, y_n) to $(x(t), y(t))$

```
p1 = ListPlot[xs, PlotStyle->PointSize[0.03]];

p2 = Plot[xex[t], {t, 0, 1}];

p3 = Show[p1, p2];

ys = Table[{t[n], ye[n]}, {n, 0, 10}];

p4 = ListPlot[ys, PlotStyle->PointSize[0.03]];

p5 = Plot[yex[t], {t, 0, 1}];

p6 = Show[p4, p5];

both = Table[{xe[n], ye[n]}, {n, 0, 10}];

p7 = ListPlot[both, PlotStyle->PointSize[0.03]];

p8 = ParametricPlot[{xex[t], yex[t]}, {t, 0, 1}];

p9 = Show[p7, p8];

Show[GraphicsRow[{p3, p6, p9}]]
```

Because the accuracy of this approximation diminishes as t increases, we attempt to improve the approximation by decreasing the increment size. We do this next by entering the value $h = 0.05$ and repeating the procedure that was followed above. See Figure 6-35.

```
Clear[f, g, t, h, x, y]

f[t_, x_, y_] = x - y + 1;

g[t_, x_, y_] = x + 3y + Exp[-t];

h = 0.05;
```

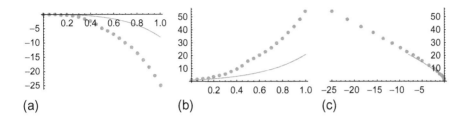

Figure 6-35   Euler's method using $h = 0.05$: (a) Comparison of x_n to $x(t)$. (b) Comparison of y_n to $y(t)$. (c) Comparison of (x_n, y_n) to $(x(t), y(t))$

```
t[n_]:=t0 + nh;

t0 = 0;

xₑ[n_]:=xₑ[n] = xₑ[n − 1] + hf [t[n − 1], xₑ[n − 1], yₑ[n − 1]];

yₑ[n_]:=yₑ[n] = yₑ[n − 1] + hg [t[n − 1], xₑ[n − 1], yₑ[n − 1]];

xₑ[0] = 0;

yₑ[0] = 1;

Table [{n, t[n], xₑ[n], xex[t[n]], yₑ[n], yex[t[n]]},

{n, 0, 20}]//TableForm
```

0	0.	0	0.	1	1.
1	0.05	0.	−0.00532454	1.4	1.21439
2	0.1	−0.04	−0.0226978	1.91048	1.46032
3	0.15	−0.135048	−0.054467	2.5615	1.74231
4	0.2	−0.304703	−0.103346	3.39053	2.06545
5	0.25	−0.574227	−0.172465	4.44425	2.43552
6	0.3	−0.976074	−0.265432	5.78076	2.85904
7	0.35	−1.55176	−0.386392	7.47226	3.34338
8	0.4	−2.35416	−0.540105	9.60842	3.89682
9	0.45	−3.45042	−0.732029	12.3005	4.52876
10	0.5	−4.9255	−0.968408	15.6862	5.24975
11	0.55	−5.90609	−1.25639	17.8232	6.07171

12	0.6	−7.04255	−1.60412	20.2302	7.00806
13	0.65	−8.35619	−2.02091	22.9401	8.07394
14	0.7	−9.871	−2.51737	25.9894	9.28638
15	0.75	−11.614	−3.10558	29.419	10.6645
16	0.8	−13.6157	−3.79926	33.2748	12.23
17	0.85	−15.9102	−4.61405	37.6077	14.0071
18	0.9	−18.5361	−5.56767	42.4747	16.0232
19	0.95	−21.5366	−6.68027	47.9395	18.3088
20	1.	−24.9604	−7.97468	54.0729	20.8987

```
xs = Table[{t[n], xₑ[n]}, {n, 0, 20}];

p1 = ListPlot[xs, PlotStyle->PointSize[0.03]];

p2 = Plot[xex[t], {t, 0, 1}];

p3 = Show[p1, p2];

ys = Table[{t[n], yₑ[n]}, {n, 0, 20}];

p4 = ListPlot[ys, PlotStyle->PointSize[0.03]];

p5 = Plot[yex[t], {t, 0, 1}];

p6 = Show[p4, p5];

both = Table[{xₑ[n], yₑ[n]}, {n, 0, 20}];

p7 = ListPlot[both, PlotStyle->PointSize[0.03]];

p8 = ParametricPlot[{xex[t], yex[t]}, {t, 0, 1}];

p9 = Show[p7, p8];

Show[GraphicsRow[{p3, p6, p9}]]
```

Notice that the approximations are more accurate with the smaller value of h. We also see this in the graphs that compare the approximation with the exact solution.

■

6.5.3 Runge-Kutta Method

Because we would like to be able to improve the approximation without using such a small value for h, we seek to improve the method. As with first-order equations, the Runge-Kutta method can be extended to systems. In this case, the recursive formula at each step is

$$\begin{cases} x_{n+1} = x_n + \frac{1}{6}h\left(k_1 + 2k_2 + 2k_3 + k_4\right) \\ y_{n+1} = y_n + \frac{1}{6}h\left(m_1 + 2m_2 + 2m_3 + m_4\right) \end{cases}, \tag{6.24}$$

where

$$k_1 = f\left(t_n, x_n, y_n\right) \qquad\qquad m_1 = g\left(t_n, x_n, y_n\right) \tag{6.25}$$

$$k_2 = f\left(t_n + \frac{1}{2}h, x_n + \frac{1}{2}hk_1, y_n + \frac{1}{2}hm_1\right) \quad m_2 = g\left(t_n + \frac{1}{2}h, x_n + \frac{1}{2}hk_1, y_n + \frac{1}{2}hm_1\right) \tag{6.26}$$

$$k_3 = f\left(t_n + \frac{1}{2}h, x_n + \frac{1}{2}hk_2, y_n + \frac{1}{2}hm_2\right) \quad m_3 = g\left(t_n + \frac{1}{2}h, x_n + \frac{1}{2}hk_2, y_n + \frac{1}{2}hm_2\right) \tag{6.27}$$

$$k_4 = f\left(t_n + \frac{1}{2}h, x_n + hk_3, y_n + hm_3\right) \qquad m_4 = g\left(t_n + \frac{1}{2}h, x_n + hk_3, y_n + hm_3\right). \tag{6.28}$$

EXAMPLE 6.5.4: Use the Runge-Kutta method to approximate the solution of the initial-value problem from Example 2

$$\begin{cases} dx/dt = x - y + 1 \\ dy/dt = x + 3y + e^{-t} \\ x\left(0\right) = 0, \; y\left(0\right) = 1 \end{cases}$$

using $h = 0.1$. Compare these results to those of the exact solution of the system of equations as well as those obtained with Euler's method.

SOLUTION: We use equations (6.24) and (6.25) with $f(x, y) = x - y + 1$, $g(x, y) = x + 3y + e^{-t}$, $t_0 = 0$, $x_0 = 0$, $y_0 = 1$, and $h = 0.1$.

We show the results obtained with this method and compare them to the exact values.

```
Clear[t0, f, g, x, y, t, k1, k2,

  k3, k4, m1, m2, m3, m4, xr, yr]


f[t_, x_, y_] = x - y + 1;

  g[t_, x_, y_] = x + 3y + Exp[-t];

  t0 = 0;

h = 0.1;

t[n_] := t0 + nh
```

$$x[n\_] := x[n] = x[n-1] + \frac{1}{6} h(k1[n-1] + 2k2[n-1] + 2k3[n-1] +$$

$$k4[n-1]); x[0] = 0;$$

$$y[n\_] := y[n] = y[n-1] + \frac{1}{6} h(m1[n-1] + 2m2[n-1] + 2m3[n-1] +$$

$$m4[n-1]); y[0] = 1;$$

$$k1[n\_] := k1[n] = f[t[n], x[n], y[n]];$$

$$k2[n\_] := k2[n] = f[t[n] + \frac{h}{2}, x[n] + \frac{hk1[n]}{2}, y[n] + \frac{hm1[n]}{2}];$$

$$k3[n\_] := k3[n] = f[t[n] + \frac{h}{2}, x[n] + \frac{hk2[n]}{2}, y[n] + \frac{hm2[n]}{2}];$$

$$k4[n\_] := k4[n] = f[t[n] + h, x[n] + hk3[n], y[n] + hm3[n]];$$

$$m1[n\_] := m1[n] = g[t[n], x[n], y[n]];$$

$$m2[n\_] := m2[n] = g[t[n] + \frac{h}{2}, x[n] + \frac{hk1[n]}{2}, y[n] + \frac{hm1[n]}{2}];$$

$$m3[n\_] := m3[n] = g[t[n] + \frac{h}{2}, x[n] + \frac{hk2[n]}{2}, y[n] + \frac{hm2[n]}{2}];$$

$$m4[n\_] := m4[n] = g[t[n] + h, x[n] + hk3[n], y[n] + hm3[n]];$$

```
Table[{t[n], x[n], xex[t[n]], y[n], yex[t[n]]}, {n, 0, 10}]//

TableForm

 0.    0              0.           1         1.

 0.1   -0.0226878    -0.0226978   1.46031   1.46032

 0.2   -0.10332      -0.103346    2.06541   2.06545
```

0.3	−0.265382	−0.265432	2.85897	2.85904
0.4	−0.540021	−0.540105	3.8967	3.89682
0.5	−0.968273	−0.968408	5.24956	5.24975
0.6	−1.60391	−1.60412	7.00778	7.00806
0.7	−2.51707	−2.51737	9.28596	9.28638
0.8	−3.79882	−3.79926	12.2294	12.23
0.9	−5.56704	−5.56767	16.0223	16.0232
1.	−7.97379	−7.97468	20.8975	20.8987

Notice that the Runge-Kutta method is much more accurate than Euler's method. In fact, the Runge-Kutta method with $h = 0.1$ is more accurate than Euler's method with $h = 0.05$. We also observe the accuracy of the approximation in the graphs that compare the approximation to the exact solution in Figure 6-36.

```
xs = Table[{t[n], x[n]}, {n, 0, 10}];

p1 = ListPlot[xs, PlotStyle->PointSize[0.03]];

p2 = Plot[xex[t], {t, 0, 1}];

p3 = Show[p1, p2];

ys = Table[{t[n], y[n]}, {n, 0, 10}];
```

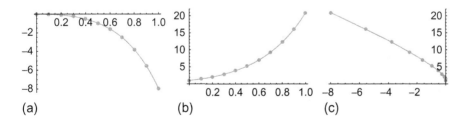

(a)   (b)   (c)

Figure 6-36 Runge-Kutta method using $h = 0.1$: (a) Comparison of x_n to $x(t)$. (b) Comparison of y_n to $y(t)$. (c) Comparison of (x_n, y_n) to $(x(t), y(t))$

```
p4 = ListPlot[ys, PlotStyle->PointSize[0.03]];

p5 = Plot[yex[t], {t, 0, 1}];

p6 = Show[p4, p5];


both = Table[{x[n], y[n]}, {n, 0, 10}];

p7 = ListPlot[both, PlotStyle->PointSize[0.03]];

p8 = ParametricPlot[{xex[t], yex[t]}, {t, 0, 1}];

p9 = Show[p7, p8];


Show[GraphicsRow[{p3, p6, p9}]]
```

■

Remark. Remember that NDSolve can implement these and other numerical methods directly using the Method option.

6.6 Nonlinear Systems, Linearization, and Classification of Equilibrium Points

We now turn our attention to the systems of equations of the form

$$\begin{cases} dx/dt = f(x, y) \\ dy/dt = g(x, y) \end{cases}. \tag{6.29}$$

This system is **autonomous**, because $f(x, y)$ and $g(x, y)$ do not depend explicitly on the independent variable t.

Definition 29 (Equilibrium Point). *A point (x_0, y_0) is an **equilibrium point** of system (6.29) if $f(x_0, y_0) = 0$ and $g(x_0, y_0) = 0$.*

Before discussing nonlinear systems, we first investigate properties of autonomous linear systems of the form

Equilibrium points are also called **rest points**.

$$\begin{cases} dx/dt = ax + by \\ dy/dt = cx + dy \end{cases}, \tag{6.30}$$

where a, b, c, and d are constants with $\begin{vmatrix} a & b \\ c & d \end{vmatrix} = ad - bc \neq 0$, which have only one equilibrium point: $(0, 0)$. We have solved many systems of this type by using the eigenvalues and corresponding eigenvectors of $\mathbf{A} = \begin{pmatrix} a & b \\ c & d \end{pmatrix}$.

6.6.1 Real Distinct Eigenvalues

If λ_1 and λ_2 are real eigenvalues of $\mathbf{A} = \begin{pmatrix} a & b \\ c & d \end{pmatrix}$ where $\lambda_2 < \lambda_1$, with corresponding eigenvectors $\mathbf{v}_1$ and $\mathbf{v}_2$, respectively, a general solution of system (6.30) is

$$\mathbf{X} = \begin{pmatrix} x \\ y \end{pmatrix} = c_1 \mathbf{v}_1 e^{\lambda_1 t} + c_2 \mathbf{v}_2 e^{\lambda_2 t} = e^{\lambda_1 t} \left[c_1 \mathbf{v}_1 + c_2 \mathbf{v}_2 e^{(\lambda_2 - \lambda_1)t} \right]. \tag{6.31}$$

1. Suppose that both eigenvalues are negative. If we assume that $\lambda_2 < \lambda_1 < 0$, then $\lambda_2 - \lambda_1 < 0$. Then $e^{(\lambda_2 - \lambda_1)t}$ and $e^{\lambda_1 t}$ are very small for large values of of t. If $c_1 \neq 0$, then $\lim_{t \to \infty} \mathbf{X} = \mathbf{0}$ in one of the directions determined by $\mathbf{v}_1$ or $-\mathbf{v}_1$. If $c_1 = 0$, then $\mathbf{X} = c_2 \mathbf{v}_2 e^{\lambda_2 t}$. Again, because $\lambda_2 < 0$, $\lim_{t \to \infty} \mathbf{X} = \mathbf{0}$ in the directions determined by $\mathbf{v}_2$ or $-\mathbf{v}_2$. In this case, $(0, 0)$ is a **stable node**.

2. Suppose that both eigenvalues are positive. If $0 < \lambda_2 < \lambda_1$, then $e^{\lambda_1 t}$ and $e^{\lambda_2 t}$ both become unbounded as t increases. If $c_1 \neq 0$, then $\mathbf{X}$ becomes unbounded in either the direction of $\mathbf{v}_1$ or $-\mathbf{v}_1$. If $c_1 = 0$, then $\mathbf{X}$ becomes unbounded in the directions given by $\mathbf{v}_2$ or $-\mathbf{v}_2$. In this case, $(0,0)$ is an **unstable node**.

3. Suppose that the eigenvalues have opposite sign. Then, if $\lambda_2 < 0 < \lambda_1$ and $c_1 \neq 0$, $\mathbf{X}$ becomes unbounded in either the direction of $\mathbf{v}_1$ or $-\mathbf{v}_1$ as it did in (2). However, if $c_1 = 0$, then due to the fact that $\lambda_2 < 0$, $\lim_{t \to \infty} \mathbf{X} = \mathbf{0}$ along the line determined by $\mathbf{v}_2$. If the initial point $\mathbf{X}(0)$ is not on the line determined by $\mathbf{v}_2$, then the line given by $\mathbf{v}_1$ is an asymptote for the solution. We say that $(0,0)$ is a **saddle point** in this case.

EXAMPLE 6.6.1: Classify the equilibrium point $(0,0)$ of the systems:

(a) $\begin{cases} x' = 5x + 3y \\ y' = -4x - 3y \end{cases}$; (b) $\begin{cases} x' = x - 2y \\ y' = 3x - 4y \end{cases}$; (c) $\begin{cases} x' = -x - 2y \\ y' = 3x + 4y \end{cases}$.

SOLUTION: (a) We find the eigenvalues and corresponding eigenvectors of $\mathbf{A} = \begin{pmatrix} 5 & 3 \\ -4 & -3 \end{pmatrix}$ with `Eigensystem`.

```
Clear[a, x, y]

a = ( 5    3
     -4   -3 ) ;

Eigensystem[a]
```

$$\{\{3, -1\}, \{\{-3, 2\}, \{-1, 2\}\}\}$$

Because these eigenvalues have opposite sign, $(0,0)$ is a saddle point. Eigenvectors corresponding to $\lambda_1 = -1$ and $\lambda_2 = 3$ are $\mathbf{v}_1 = \begin{pmatrix} -1 \\ 2 \end{pmatrix}$ and $\mathbf{v}_2 = \begin{pmatrix} -3 \\ 2 \end{pmatrix}$, respectively. Hence the solution becomes unbounded in the directions associated with the positive eigenvalue, $\mathbf{v}_2 = \begin{pmatrix} -3 \\ 2 \end{pmatrix}$ and $-\mathbf{v}_2 = \begin{pmatrix} 3 \\ -2 \end{pmatrix}$. Along the line through $(0,0)$ determined by $\mathbf{v}_1 = \begin{pmatrix} -1 \\ 2 \end{pmatrix}$, the solution approaches $(0,0)$. We see this when we graph various solutions and display the results together with the direction field associated with the system. First, we use `DSolve` to find a general solution of the system.

```
gensol = DSolve[{x'[t]==5x[t] + 3y[t], y'[t]== - 4x[t] - 3y[t]},
{x[t], y[t]}, t]
```

$$\{\{x[t] \rightarrow \frac{1}{2}e^{-t}(-1 + 3e^{4t})C[1] + \frac{3}{4}e^{-t}(-1 + e^{4t})C[2], y[t] \rightarrow$$
$$-e^{-t}(-1 + e^{4t})C[1] - \frac{1}{2}e^{-t}(-3 + e^{4t})C[2]\}\}$$

Then, we use Table and Flatten to create a list of ordered pairs $\{x(t), y(t)\}$, corresponding to the solution for various values of the arbitrary constants. These functions are then graphed with ParametricPlot.

```
toplot = Flatten[
Table[{-E^-tC[1] - 3E^3tC[2], 2E^-tC[1] + 2E^3tC[2]}/.
    {C[1]->i, C[2]->j}, {i, -0.5, 0.5, 0.25}, {j, -0.5,
    0.5, 0.25}], 1];
somegraphs = ParametricPlot[Evaluate[toplot], {t, -3, 3},
    PlotRange → {{-1, 1}, {-1, 1}}, AspectRatio → 1];
p4 = Plot[{-2x, - 2x/3}, {x, -1, 1},
    PlotStyle → {{CMYKColor[0, 0.89, 0.94, 0.28],
    Thickness[.01]}}]
```

We graph the direction field associated with the system with StreamPlot. Last, all graphs are displayed together with Show in Figure 6-37.

```
pvf1 = StreamPlot[{5x + 3y, -4x - 3y}, {x, -1, 1},
    {y, -1, 1}, StreamStyle → Fine];

Show[pvf1, somegraphs, p4, PlotRange → {{-1, 1}, {-1, 1}},
    Axes → Automatic, AxesOrigin → {0, 0}, Frame → False,
    AxesLabel → {x, y}]
```

(b) In this case, the eigenvalues $\lambda_1 = -1$ and $\lambda_2 = -2$ are both negative.

$$a = \begin{pmatrix} 1 & -2 \\ 3 & -4 \end{pmatrix};$$

```
Eigensystem[a]
```

$$\{\{-2, -1\}, \{\{2, 3\}, \{1, 1\}\}\}$$

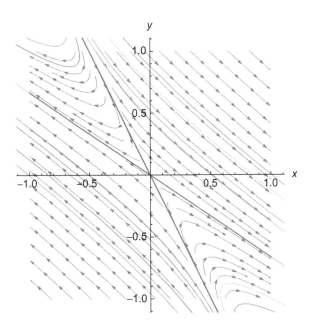

Figure 6-37   The origin is a saddle point

Hence, $(0,0)$ is a stable node. Corresponding eigenvectors are $\mathbf{v}_1 = \begin{pmatrix} 1 \\ 1 \end{pmatrix}$ and $\mathbf{v}_2 = \begin{pmatrix} 2 \\ 3 \end{pmatrix}$. Therefore, the solutions approach $(0,0)$ along the lines through the origin determined by these vectors, $y = x$ and $y = \frac{3}{2}x$. We see this in the graph of the direction field and graphs of several solutions to the system. First, we graph the direction field associated with the system.

```
pvf1 = StreamPlot[{x − 2y, 3x − 4y}, {x, −1, 1},
   {y, −1, 1}, StreamStyle → Fine];
```

Then we use DSolve to solve the initial-value problem $\begin{cases} x' = x - 2y \\ y' = 3x - 4y \\ x(0) = x_0,\ y(0) = y_0 \end{cases}$.

```
gensol = DSolve[{x'[t]==x[t] − 2y[t], y'[t]==3x[t] − 4y[t],
   x[0]==x0, y[0]==y0}, {x[t], y[t]}, t]//Simplify
```

$\{\{x[t] \to e^{-2t}((-2 + 3e^t)x0 - 2(-1 + e^t)y0),\ y[t] \to e^{-2t}(3(-1 + e^t)x0 + (3 - 2e^t)y0)\}\}$

Given an ordered pair $\{x_0, y_0\}$, $\mathtt{sol}[\{x_0, y_0\}]$ returns the solution that satisfies $x(0) = x_0$ and $y(0) = y_0$.

```
sol[{x0_, y0_}] = {E^-t(3x0 - 2y0 + 2E^-t(-x0 + y0)), E^-t(3x0
-2y0 + 3E^-t(-x0 + y0))};
```

We then generate several lists of ordered pairs with `Table`

```
initconds1 = Table[{-1, i}, {i, -1, 1, 2/9}];

initconds2 = Table[{1, i}, {i, -1, 1, 2/9}];

initconds3 = Table[{i, 1}, {i, -1, 1, 2/9}];

initconds4 = Table[{i, -1}, {i, -1, 1, 2/9}];
```

and use `Union` to join them together.

```
initconds = initconds1 ∪ initconds2 ∪ initconds3∪

initconds4;
```

`Map` is used to apply `sol` to the list `initconds`. The resulting list of parametric functions is graphed with `ParametricPlot`.

```
toplot = Map[sol, initconds];

somegraphs = ParametricPlot[Evaluate[toplot], {t, -3, 3},

PlotRange → {{-1, 1}, {-1, 1}}, AspectRatio → 1];

p4 = Plot[{x, 3x/2}, {x, -1, 1},

  PlotStyle → {{CMYKColor[0, 0.89, 0.94, 0.28],

  Thickness[.01]}}];
```

Finally, all the graphics are displayed together with `Show` in Figure 6-38.

```
Show[pvf1, somegraphs, p4, PlotRange → {{-1, 1}, {-1, 1}},

AspectRatio → 1, Axes → Automatic, AxesOrigin → {0, 0},

Frame → False, AxesLabel → {x, y}]
```

(c) Because the eigenvalues $\lambda_1 = 2$ and $\lambda_1 = 1$ are both positive, $(0, 0)$ is an unstable node.

```
a = ( -1  -2
       3   4 );

Eigensystem[a]
```

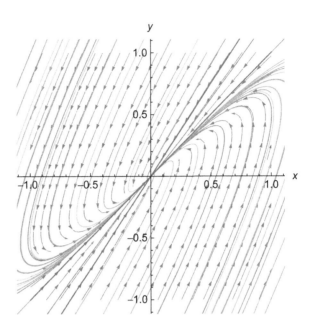

Figure 6-38   The origin is a stable node.

$$\{\{2, 1\}, \{\{-2, 3\}, \{-1, 1\}\}\}$$

Note that the corresponding eigenvectors are $\mathbf{v}_1 = \begin{pmatrix} -2 \\ 3 \end{pmatrix}$ and $\mathbf{v}_2 = \begin{pmatrix} -1 \\ 1 \end{pmatrix}$, respectively. Hence, the solutions become unbounded along the lines passing through the origin determined by these vectors, $y = -\frac{3}{2}x$ and $y = -x$. As before, we see this in the graph of the direction field and various solutions of the system. See Figure 6-39.

```
pvf1 = StreamPlot[{-x - 2y, 3x + 4y}, {x, -1, 1},

{y, -1, 1}, StreamStyle → Fine];

gensol = DSolve[{x'[t]== - x[t] - 2y[t], y'[t]==3x[t] + 4y[t],

x[0]==x0, y[0]==y0}, {x[t], y[t]}, t]

{{x[t] → -e^t(-3x0 + 2e^tx0 - 2y0 + 2e^ty0), y[t] → e^t(-3x0 +

3e^tx0 - 2y0 + 3e^ty0)}}
```

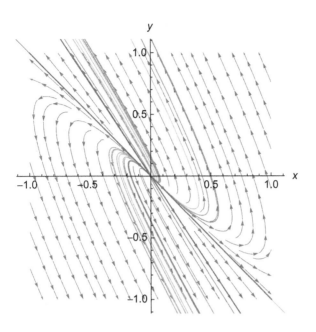

Figure 6-39 The origin is an unstable node

$$\text{sol}[\{x0\_, y0\_\}] = \{-E^t(-3x0 - 2y0 + 2E^t(x0 + y0)),$$

$$E^t(-3x0 - 2y0 + 3E^t(x0 + y0))\};$$

$$\text{initconds} = \text{Table}[\{0.5\,t\text{Cos}[2\pi t], 0.5\,t\text{Sin}[2\pi t]\}, \{t, 0, 1, \frac{1}{24}\}];$$

$$\text{toplot} = \text{Map}[\text{sol}, \text{initconds}];$$

$$\text{somegraphs} = \text{ParametricPlot}[\text{Evaluate}[\text{toplot}], \{t, -3, 3\},$$

$$\quad \text{PlotRange} \to \{\{-1, 1\}, \{-1, 1\}\}, \text{AspectRatio} \to 1];$$

$$\text{p4} = \text{Plot}[\{-x, -\frac{3x}{2}\}, \{x, -1, 1\}, .$$

$$\quad \text{PlotStyle} \to \{\{\text{CMYKColor}[0, 0.89, 0.94, 0.28],$$

$$\quad \text{Thickness}[.01]\}\}];$$

$$\text{Show}[\text{pvf1}, \text{somegraphs}, \text{p4}, \text{PlotRange} \to \{\{-1, 1\}, \{-1, 1\}\},$$

$$\quad \text{AspectRatio} \to 1, \text{Axes} \to \text{Automatic}, \text{AxesOrigin} \to \{0, 0\},$$

$$\quad \text{Frame} \to \text{False}, \text{AxesLabel} \to \{x, y\}]$$

■

6.6.2 Repeated Eigenvalues

We recall from our previous experience with repeated eigenvalues of a 2×2 system that the eigenvalue can have two linearly independent eigenvectors associated with it or only one eigenvector associated with it. Hence, we investigate the behavior of solutions in this case by considering both of these possibilities.

1. Suppose that the eigenvalue $\lambda = \lambda_1 = \lambda_2$ has two corresponding linearly independent eigenvectors $\mathbf{v}_1$ and $\mathbf{v}_2$. Then, a general solution is

$$\mathbf{X} = c_1 \mathbf{v}_1 e^{\lambda t} + c_2 \mathbf{v}_2 e^{\lambda t}.$$

   Hence, if $\lambda > 0$, then $\mathbf{X}$ becomes unbounded along the line through the origin determined by the vector $c_1 \mathbf{v}_1 + c_2 \mathbf{v}_2$ where c_1 and c_2 are arbitrary constants. In this case, we call the equilibrium point a **degenerate unstable node** (or an **unstable star**). On the other hand, if $\lambda < 0$, then $\mathbf{X}$ approaches $(0, 0)$ along these lines, and we call $(0, 0)$ a **degenerate stable node** (or **stable star**). Note that the name "star" was selected due to the shape of the solutions.

2. Suppose that $\lambda = \lambda_1 = \lambda_2$ has only one corresponding eigenvector $\mathbf{v}_1$. Hence, a general solution is

$$\mathbf{X} = c_1 \mathbf{v}_1 e^{\lambda t} + c_2 \left(\mathbf{v}_1 t + \mathbf{w}_2 \right) e^{\lambda t} = \left(c_1 \mathbf{v}_1 + c_2 \mathbf{w}_2 \right) e^{\lambda t} + c_2 \mathbf{v}_1 t e^{\lambda t},$$

   where $(\mathbf{A} - \lambda \mathbf{I}) \mathbf{w}_2 = \mathbf{v}_1$. We can more easily investigate the behavior of this solution if we write this solution as

$$\mathbf{X} = t e^{\lambda t} \left[\frac{1}{t} \left(c_1 \mathbf{v}_1 + c_2 \mathbf{w}_2 \right) + c_2 \mathbf{v}_1 \right].$$

   If $\lambda < 0$, $\lim_{t \to \infty} t e^{\lambda t} = 0$ and $\lim_{t \to \infty} \left[\frac{1}{t} \left(c_1 \mathbf{v}_1 + c_2 \mathbf{w}_2 \right) + c_2 \mathbf{v}_1 \right] = c_2 \mathbf{v}_1$. Hence, the solutions approach $(0, 0)$ along the line determined by $\mathbf{v}_1$, and we call $(0, 0)$ a **degenerate stable node**. If $\lambda > 0$, the solutions become unbounded along this line, and we say that $(0, 0)$ is a **degenerate unstable node**.

EXAMPLE 6.6.2: Classify the equilibrium point $(0, 0)$ in the systems:

(a) $\begin{cases} x' = x + 9y \\ y' = -x - 5y \end{cases}$; (b) $\begin{cases} x' = 2x \\ y' = 2y \end{cases}$.

SOLUTION: (a) Using `Eigensystem`,

$$a = \begin{pmatrix} 1 & 9 \\ -1 & -5 \end{pmatrix};$$

`Eigensystem[a]`

$\{\{-2, -2\}, \{\{-3, 1\}, \{0, 0\}\}\}$

we see that $\lambda_1 = \lambda_2 = -2$ and that there is only one corresponding eigenvector. Therefore, because $\lambda = -2 < 0$, $(0,0)$ is a degenerate stable node. Notice that in the graph of several members of the family of solutions of this system along with the direction field shown in Figure 6-40, which we generate using the same technique as in part (b) of the previous example, the solutions approach $(0,0)$ along the line in the direction of $\mathbf{v}_1 = \begin{pmatrix} -3 \\ 1 \end{pmatrix}$, $y = -\frac{1}{3}x$.

`Clear[x, y]`

`pvf1 = StreamPlot[{x + 9y, -x - 5y}, {x, -1, 1},`

`{y, -1, 1}, StreamStyle → Fine];`

`Simplify[DSolve[{x'[t]==x[t] + 9y[t], y'[t]== - x[t] - 5y[t],`

`x[0]==x0, y[0]==y0}, {x[t], y[t]}, t]]`

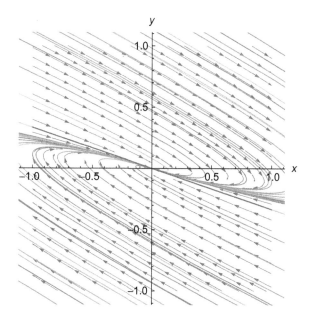

Figure 6-40   The origin is a degenerate stable node

$$\{\{x[t] \rightarrow e^{-2t}(x0 + 3tx0 + 9ty0), \, y[t] \rightarrow e^{-2t}(y0 - t(x0 + 3y0))\}\}$$

```
sol[{x0_, y0_}] = {
```
$$\frac{x0 + 3tx0 + 9ty0}{E^{2t}}, \, \frac{-(tx0) + y0 - 3ty0}{E^{2t}}\};$$

```
initconds1 = Table[{-1, i}, {i, -1, 1, 2/9}];

initconds2 = Table[{1, i}, {i, -1, 1, 2/9}];

initconds3 = Table[{i, 1}, {i, -1, 1, 2/9}];

initconds4 = Table[{i, -1}, {i, -1, 1, 2/9}];

initconds = initconds1 ∪ initconds2 ∪ initconds3∪

initconds4;

toplot = Map[sol, initconds];

somegraphs = ParametricPlot[Evaluate[toplot], {t, -3, 3},

  PlotRange → {{-1, 1}, {-1, 1}}, AspectRatio → 1];
```

```
p4 = Plot[
```
$$\frac{-x}{3}$$
```
, {x, -1, 1}, .

  PlotStyle → {{CMYKColor[0, 0.89, 0.94, 0.28],

  Thickness[.01]}}];

Show[pvf1, somegraphs, p4,

  PlotRange → {{-1, 1}, {-1, 1}}, AspectRatio → 1, Axes →

  Automatic, Frame → False, AxesLabel → {x, y},

  AxesOrigin → {0, 0}]
```

(b) We have $\lambda_1 = \lambda_2 = 2$ and two linearly independent vectors, $\mathbf{v}_1 = \begin{pmatrix} 1 \\ 0 \end{pmatrix}$ and $\mathbf{v}_2 = \begin{pmatrix} 0 \\ 1 \end{pmatrix}$. (Note: The choice of these two vectors does not change the value of the solution, because of the form of the general solution in this case.)

```
a =
```
$$\begin{pmatrix} 2 & 0 \\ 0 & 2 \end{pmatrix};$$
```
Eigensystem[a]

{{2, 2}, {{0, 1}, {1, 0}}}
```

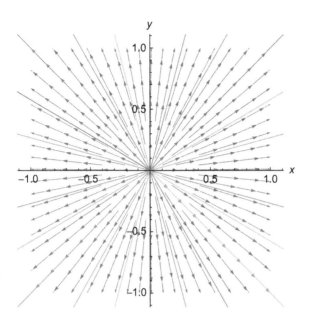

Figure 6-41   The origin is a degenerate unstable node

Because $\lambda = 2 > 0$, we classify (0,0) as a degenerate unstable node (or star). Some of these solutions along with the direction field are graphed in Figure 6-41 in the same manner as in part (c) of the previous example. Notice that they become unbounded in the direction of any vector in the xy-plane because $\mathbf{v}_1 = \begin{pmatrix} 1 \\ 0 \end{pmatrix}$ and $\mathbf{v}_2 = \begin{pmatrix} 0 \\ 1 \end{pmatrix}$.

```
Clear[x, y]

pvf1 = StreamPlot[{2x, 2y}, {x, -1, 1},
   {y, -1, 1}, StreamStyle → Fine];

Simplify[DSolve[{x'[t]==2x[t], y'[t]==2y[t],
   x[0]==x0, y[0]==y0}, {x[t], y[t]}, t]]

{{x[t] → e²ᵗx0, y[t] → e²ᵗy0}}

sol[{x0_, y0_}] = {E²ᵗx0, E²ᵗy0};

initconds = Table[{0.05Cos[2π t], 0.05Sin[2π t]}, {t, 0, 1, 1/24}];
```

```
toplot = Map[sol, initconds];

somegraphs = ParametricPlot[Evaluate[toplot], {t, -3, 3},
    PlotRange → {{-1, 1}, {-1, 1}}, AspectRatio → 1];

Show[pvf1, somegraphs, PlotRange → {{-1, 1}, {-1, 1}},
    AspectRatio → 1, Axes → Automatic,
        Frame → False, AxesLabel → {x, y}, AxesOrigin → {0, 0}]
```

∎

6.6.3 Complex Conjugate Eigenvalues

We have seen that if the eigenvalues of the system (6.30) are $\lambda_{1,2} = \alpha \pm \beta i$, $\beta \neq 0$, with corresponding eigenvectors $\mathbf{v}_{1,2} = \mathbf{a} \pm \mathbf{b}i$, two linearly independent solutions of the system are

$$\mathbf{X}_1 = e^{\alpha t} (\mathbf{a} \cos \beta t - \mathbf{b} \sin \beta t) \quad \text{and} \quad \mathbf{X}_2 = e^{\alpha t} (\mathbf{b} \cos \beta t + \mathbf{a} \sin \beta t).$$

Hence, a general solution is $\mathbf{X} = c_1 \mathbf{X}_1 + c_2 \mathbf{X}_2$, so there are constants A_1, A_2, B_1, and B_2 so that x and y are given by

$$\mathbf{X} = \begin{pmatrix} x \\ y \end{pmatrix} = \begin{pmatrix} A_1 e^{\alpha t} \cos \beta t + A_2 e^{\alpha t} \sin \beta t \\ B_1 e^{\alpha t} \cos \beta t + B_2 e^{\alpha t} \sin \beta t \end{pmatrix}.$$

1. If $\alpha = 0$, the solution is

$$\mathbf{X} = \begin{pmatrix} x \\ y \end{pmatrix} = \begin{pmatrix} A_1 \cos \beta t + A_2 \sin \beta t \\ B_1 \cos \beta t + B_2 \sin \beta t \end{pmatrix}.$$

   Hence, both x and y are periodic and $(0,0)$ is classified as a **center**. Note that the motion around these circles or ellipses is either clockwise or counterclockwise for all solutions.

2. If $\alpha \neq 0$, then $e^{\alpha t}$ is present in the solution. This term causes the solution to spiral around the equilibrium point. If $\alpha > 0$, then the solution spirals away from $(0,0)$, so we classify $(0,0)$ as an **unstable spiral**. Otherwise, if $\alpha < 0$, the solution spirals toward $(0,0)$, so we say that $(0,0)$ is a **stable spiral**.

EXAMPLE 6.6.3: Classify the equilibrium point $(0,0)$ for each of the following systems: (a) $\begin{cases} x' = -y \\ y' = x \end{cases}$; (b) $\begin{cases} x' = \frac{1}{2}x - \frac{153}{32}y \\ y' = 2x - y \end{cases}$.

SOLUTION: (a) The eigenvalues are found to be $\lambda_{1,2} = \pm i$.

```
a = ( 0  -1 );
      1   0
```

```
Eigensystem[a]
```

$\{\{i, -i\}, \{\{i, 1\}, \{-i, 1\}\}\}$

Because these eigenvalues have zero real part (and, hence, are purely imaginary), $(0,0)$ is a center. Several solutions along with the direction field are graphed in Figure 6-42.

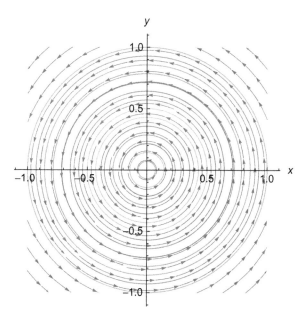

Figure 6-42   The origin is a center

```
Clear[x, y]

pvf1 = StreamPlot[{-y, x}, {x, -1, 1},
  {y, -1, 1}, StreamStyle → Fine];

sol[{x0_, y0_}] = {x0Cos[t] - y0Sin[t], y0Cos[t] + x0Sin[t]};

initconds = Table[{0, i}, {i, 0, 1, 1/14}];

toplot = Map[sol, initconds];

somegraphs = ParametricPlot[Evaluate[toplot], {t, 0, 2π},
  PlotRange → {{-1, 1}, {-1, 1}}, AspectRatio → 1];

Show[pvf1, somegraphs, PlotRange → {{-1, 1}, {-1, 1}},
  AspectRatio → 1, Axes → Automatic,
  Frame → False, AxesLabel → {x, y}, AxesOrigin → {0, 0}]
```

(b) The eigenvalues are found to be $\lambda_{1,2} = -1/4 \pm 3i$.

$$a = \begin{pmatrix} 1/2 & -153/32 \\ 2 & -1 \end{pmatrix};$$

```
Eigensystem[a]
```

$$\left\{\left\{-\frac{1}{4} + 3i, -\frac{1}{4} - 3i\right\}, \left\{\left\{\frac{3}{8} + \frac{3i}{2}, 1\right\}, \left\{\frac{3}{8} - \frac{3i}{2}, 1\right\}\right\}\right\}$$

Thus, $(0, 0)$ is a stable spiral, because $\alpha = -1/4 < 0$. Several solutions along with the direction field are graphed in Figure 6-43 in the same way that we have done before.

```
Clear[x, y]

pvf1 = StreamPlot[{1/2x - 153/32y, 2x - y}, {x, -1, 1},
  {y, -1, 1}, StreamStyle → Fine];

FullSimplify[DSolve[{x'[t]==1/2x[t] - 153/32y[t],
  y'[t]==2x[t] - y[t], x[0]==x0, y[0]==y0},
  {x[t], y[t]}, t]]
```

$$\left\{\left\{x[t] \rightarrow \frac{1}{32} e^{-t/4}(32x0Cos[3t] + (8x0 - 51y0)Sin[3t]), y[t] \rightarrow \frac{1}{12} e^{-t/4}(12y0Cos[3t] + (8x0 - 3y0)Sin[3t])\right\}\right\}$$

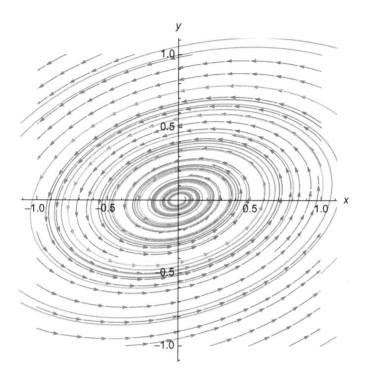

Figure 6-43   The origin is a stable spiral

$$\text{sol}[\{x0\_, y0\_\}] = \{\frac{1}{32} E^{-t/4}(32x0\text{Cos}[3\,t] + (8x0 - 51y0)\text{Sin}[3\,t]),$$

$$\frac{1}{12} E^{-t/4}(12y0\text{Cos}[3\,t] + (8x0 - 3y0)\text{Sin}[3\,t])\};$$

$$\text{initconds} = \text{Table}[\{1, i\}, \{i, -1, 1, 2/4\}];$$

$$\text{toplot} = \text{Map}[\text{sol}, \text{initconds}];$$

$$\text{somegraphs} = \text{ParametricPlot}[\text{Evaluate}[\text{toplot}], \{t, 0, 4\pi\},$$

$$\text{PlotRange} \to \{\{-1, 1\}, \{-1, 1\}\}, \text{AspectRatio} \to 1];$$

$$\text{Show}[\text{pvf1}, \text{somegraphs}, \text{PlotRange} \to \{\{-1, 1\}, \{-1, 1\}\},$$

$$\text{AspectRatio} \to 1, \text{Axes} \to \text{Automatic},$$

$$\text{Frame} \to \text{False}, \text{AxesLabel} \to \{x, y\}, \text{AxesOrigin} \to \{0, 0\}]$$

■

6.6.4 Nonlinear Systems

When working with nonlinear systems, we can often gain a great deal of information concerning the system by making a linear approximation near each equilibrium point of the nonlinear system and solving the linear system. Although the solution to the linearized system only approximates the solution to the nonlinear system, the general behavior of solutions to the nonlinear system near each equilibrium is the same as that of the corresponding linear system in most cases. The first step toward approximating a nonlinear system near each equilibrium point is to find the equilibrium points of the system and the matrix for linearization near each point as defined below.

Recall from multivariate calculus that if $z = F(x, y)$ is a differentiable function, the tangent plane to the surface S given by the graph of $z = F(x, y)$ at the point (x_0, y_0) is

$$z = F_x(x_0, y_0)(x - x_0) + F_y(x_0, y_0)(y - y_0) + F(x_0, y_0).$$

Hence, near each equilibrium point (x_0, y_0) of the nonlinear system

$$\begin{cases} dx/dt = f(x, y) \\ dy/dt = g(x, y) \end{cases},$$

the system can be approximated with

$$\begin{cases} dx/dt = f_x(x_0, y_0)(x - x_0) + f_y(x_0, y_0)(y - y_0) + f(x_0, y_0) \\ dy/dt = g_x(x_0, y_0)(x - x_0) + g_y(x_0, y_0)(y - y_0) + g(x_0, y_0) \end{cases}.$$

Then, because $f(x_0, y_0) = 0$ and $g(x_0, y_0) = 0$, the approximate system is

$$\begin{cases} dx/dt = f_x(x_0, y_0)(x - x_0) + f_y(x_0, y_0)(y - y_0) \\ dy/dt = g_x(x_0, y_0)(x - x_0) + g_y(x_0, y_0)(y - y_0) \end{cases},$$

which can be written in matrix form as

$$\begin{pmatrix} dx/dt \\ dy/dt \end{pmatrix} = \begin{pmatrix} f_x(x_0, y_0) & f_y(x_0, y_0) \\ g_x(x_0, y_0) & g_y(x_0, y_0) \end{pmatrix} \begin{pmatrix} x - x_0 \\ y - y_0 \end{pmatrix}. \tag{6.32}$$

Note that we often call system (6.32) the **linearized system corresponding to the nonlinear system** due to the fact that we have removed the nonlinear terms from the original system. Now that the system is approximated by a system of the form

$$\begin{cases} dx/dt = ax + by \\ dy/dt = cx + dy \end{cases}, \text{ an equilibrium point } (x_0, y_0) \text{ of the system } \begin{cases} dx/dt = f(x, y) \\ dy/dt = g(x, y) \end{cases} \text{ is}$$

classified by the eigenvalues of the matrix

$$\mathbf{J} = \begin{pmatrix} f_x\,(x_0, y_0) & f_y\,(x_0, y_0) \\ g_x\,(x_0, y_0) & g_y\,(x_0, y_0) \end{pmatrix} \tag{6.33}$$

which is called the **Jacobian matrix**. Of course, this linearization must be carried out for each equilibrium point. After determining the matrix for linearization for each equilibrium point, the eigenvalues for the matrix must be found. Then, we classify each equilibrium point according to the following criteria.

*The Jacobian matrix is also called the **variational matrix**.*

Classification of Equilibrium Points

Let (x_0, y_0) be an equilibrium point of the system $\begin{cases} dx/dt = f(x, y) \\ dy/dt = g(x, y) \end{cases}$ and let λ_1 and λ_2 be the eigenvalues of the Jacobian matrix (6.33).

1. Suppose that λ_1 and λ_2 are real. If $\lambda_1 > \lambda_2 > 0$, then (x_0, y_0) is an unstable node; if $\lambda_2 < \lambda_1 < 0$, then (x_0, y_0) is a stable node; and if $\lambda_2 < 0 < \lambda_1$, then (x_0, y_0) is a saddle.
2. Suppose that $\lambda_{1,2} = \alpha \pm \beta i$, $\beta \neq 0$. If $\alpha < 0$, (x_0, y_0) is a stable spiral; if $\alpha > 0$, (x_0, y_0) is an unstable spiral; and if $\alpha = 0$, (x_0, y_0) may be a center, unstable spiral, or stable spiral. Hence, we can draw no conclusion.

We will not discuss the case if the eigenvalues are the same or one eigenvalue is zero.

For analyzing nonlinear systems, we state the following useful theorem.

Theorem 17. *Suppose that (x_0, y_0) is an equilibrium point of the autonomous nonlinear system*

$$\begin{cases} dx/dt = f(x, y) \\ dy/dt = g(x, y) \end{cases}.$$

Then, the relationships in the following table hold for the classification of (x_0, y_0) in the nonlinear system and that in the associated linearized system.

Associated Linearized System	Nonlinear System
Stable Node	Stable Node
Unstable Node	Unstable Node
Stable Spiral	Stable Spiral
Unstable Spiral	Unstable Spiral
Saddle	Saddle
Center	No Conclusion

An **autonomous system** does not explicitly depend on the independent variable, t. That is, if you write the system omitting all arguments, the independent variable (typically t) does not appear.

More generally, for the autonomous system of the form

$$
\begin{aligned}
x_1{}' &= f_1 (x_1, x_2, \ldots, x_n) \\
x_2{}' &= f_2 (x_1, x_2, \ldots, x_n) \\
&\ \ \vdots \\
x_n{}' &= f_n (x_1, x_2, \ldots, x_n),
\end{aligned}
\tag{6.34}
$$

an **equilibrium** (or **rest**) **point**, $E = (x_1{}^*, x_2{}^*, \ldots, x_n{}^*)$, of equation (6.34) is a solution of the system

$$
\begin{aligned}
f_1 (x_1, x_2, \ldots, x_n) &= 0 \\
f_2 (x_1, x_2, \ldots, x_n) &= 0 \\
&\ \ \vdots \\
f_n (x_1, x_2, \ldots, x_n) &= 0.
\end{aligned}
\tag{6.35}
$$

The **Jacobian** of equation (6.34) is

$$
\mathbf{J}(x_1, x_2, \ldots, x_n) =
\begin{pmatrix}
\dfrac{\partial f_1}{\partial x_1} & \dfrac{\partial f_1}{\partial x_2} & \cdots & \dfrac{\partial f_1}{\partial x_n} \\[2mm]
\dfrac{\partial f_2}{\partial x_1} & \dfrac{\partial f_2}{\partial x_2} & \cdots & \dfrac{\partial f_2}{\partial x_n} \\[2mm]
\vdots & \vdots & \cdots & \vdots \\[2mm]
\dfrac{\partial f_n}{\partial x_1} & \dfrac{\partial f_n}{\partial x_2} & \cdots & \dfrac{\partial f_n}{\partial x_n}
\end{pmatrix}.
$$

The rest point, E, is **locally stable** if and only if all the eigenvalues of $\mathbf{J}(E)$ have negative real part. If E is not locally stable, E is **unstable**.

EXAMPLE 6.6.4: Find and classify the equilibrium points of
$$\begin{cases} dx/dt = 1 - y \\ dy/dt = x^2 - y^2 \end{cases}.$$

SOLUTION: We begin by finding the equilibrium points of this non-linear system by solving $\begin{cases} 1 - y = 0 \\ x^2 - y^2 = 0 \end{cases}$.

```
Clear[f, g]

f[x_, y_] = 1 - y;

g[x_, y_] = x² - y²;

Solve[{f[x, y]==0, g[x, y]==0}]
```

$$\{\{x \to -1, y \to 1\}, \{x \to 1, y \to 1\}\}$$

Because $f(x, y) = 1 - y$ and $g(x, y) = x^2 - y^2$, $f_x(x, y) = 0$, $f_y(x, y) = -1$, $g_x(x, y) = 2x$, and $g_y(x, y) = -2y$, so the Jacobian matrix is $\mathbf{J} = \begin{pmatrix} 0 & -1 \\ 2x & -2y \end{pmatrix}$.

```
jac = {{D[f[x, y], x], D[f[x, y], y]},

    {D[g[x, y], x], D[g[x, y], y]}};

MatrixForm[jac]
```

$$\begin{pmatrix} 0 & -1 \\ 2x & -2y \end{pmatrix}$$

Next, we obtain the linearized system about each equilibrium point.

For $(1, 1)$, we obtain $\lambda_{1,2} = -1 \pm i$. Because these eigenvalues are complex-valued with negative real part, we classify $(1, 1)$ as a stable spiral.

```
jac/.{x->1, y->1}//Eigenvalues
```

$$\{-1 + i, -1 - i\}$$

For $(-1, 1)$, we obtain $\lambda_1 = -1 + \sqrt{3} > 0$ and $\lambda_2 = -1 - \sqrt{3} < 0$, so $(-1, 1)$ is a saddle.

```
jac/.{x-> - 1, y->1}//Eigenvalues
```

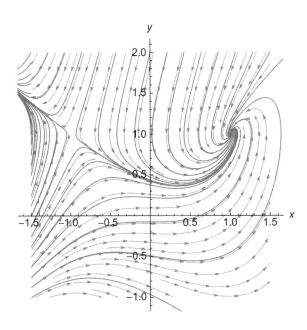

Figure 6-44 Linearization gives us information about the *local* behavior of solutions near equilibrium points but not information about the *global* behavior of solutions

$$\{-1 - \sqrt{3}, \, -1 + \sqrt{3}\}$$

We graph solutions to this nonlinear system obtained using NDSolve together with the direction field associated with the nonlinear system in Figure 6-44. We can see how the solutions move toward and move away from the equilibrium points by observing the arrows on the vectors in the direction field.

```
pvf = StreamPlot[{f[x, y], g[x, y]}, {x, -3/2, 3/2}, {y, -1, 2}];

numgraph[{x0_, y0_}]:=Module[{numsol},

    numsol = NDSolve[{x'[t]==f[x[t], y[t]], y'[t]==g[x[t],

  y[t]], x[0]==x0, y[0]==y0}, {x[t], y[t]}, {t, 0, 15}];

      ParametricPlot[{x[t], y[t]}/.numsol, {t, 0, 15},

    PlotStyle->CMYKColor[0, 0.89, 0.94, 0.28]]

      ]

initconds1 = Table[{-3/2, i}, {i, -1, 2, 3/24}];

initconds2 = Table[{i, 2}, {i, -3/2, 3/2, 3/14}];
```

```
initconds = initconds1 ∪ initconds2;

somegraphs = Map[numgraph, initconds];

Show[pvf, somegraphs, PlotRange->{{-3/2, 3/2}, {-1, 2}},

   AspectRatio → 1, Axes → Automatic, AxesOrigin → {0, 0},

   Frame → False, AxesLabel → {x, y}]
```

∎

EXAMPLE 6.6.5 (Duffing's Equation): Consider the forced **pendulum equation** with damping,

$$x'' + kx' + \omega \sin x = F(t). \tag{6.36}$$

Recall the Maclaurin series for $\sin x$: $\sin x = x - \frac{1}{3!}x^3 + \frac{1}{5!}x^5 - \frac{1}{7!}x^7 + \cdots$.
Using $\sin x \approx x$, equation (6.36) reduces to the linear equation $x'' + kx' + \omega x = F(t)$.

On the other hand, using the approximation $\sin x \approx x - \frac{1}{6}x^3$, we obtain
$x'' + kx' + \omega\left(x - \frac{1}{6}x^3\right) = F(t)$. Adjusting the coefficients of x and x^3 and
assuming that $F(t) = F\cos\omega t$ gives us **Duffing's equation**:

$$x'' + kx' + cx + \epsilon x^3 = F\cos\omega t, \tag{6.37}$$

where k and c are positive constants.

Let $y = x'$. Then, $y' = x'' = F\cos\omega t - kx' - cx - \epsilon x^3 = F\cos\omega t - ky - cx - \epsilon x^3$ and we can write equation (6.37) as the system

$$\begin{aligned} x' &= y \\ y' &= F\cos\omega t - ky - cx - \epsilon x^3 \end{aligned} \tag{6.38}$$

Assuming that $F = 0$ results in the autonomous system

$$\begin{aligned} x' &= y \\ y' &= -cx - \epsilon x^3 - ky. \end{aligned} \tag{6.39}$$

The rest points of system equation (6.39) are found by solving

$$\begin{aligned} y &= 0 \\ -cx - \epsilon x^3 - ky &= 0, \end{aligned}$$

resulting in $E_0 = (0,0)$.

```
Clear[x, y, c, k]
Solve[{y == 0, -cx - cx^3 - ky == 0},
  {x, y}]
```

$\{\{x \to 0, y \to 0\}, \{x \to -i, y \to 0\}, \{x \to i, y \to 0\}\}$

We find the Jacobian of equation (6.39) in `s1`, evaluate the Jacobian at E_0,

```
s1 = {{0, 1}, {-c - 3cx^2, -k}};

s2 = s1/.x → 0
```

$\{\{0, 1\}, \{-c, -k\}\}$

and then compute the eigenvalues with `Eigenvalues`.

```
s3 = Eigenvalues[s2]
```

$\{\frac{1}{2}(-k - \sqrt{-4c + k^2}), \frac{1}{2}(-k + \sqrt{-4c + k^2})\}$

Because k and c are positive, $k^2 - 4c < k^2$ so the real part of each eigenvalue is always negative if $k^2 - 4c \neq 0$. Thus, E_0 is locally stable.

For the autonomous system

$$x' = f(x, y)$$
$$y' = g(x, y),$$

(6.40)

Bendixson's theorem states that if $f_x(x, y) + g_y(x, y)$ is a continuous function that is either always positive or always negative in a particular region R of the plane, then system (6.40) has no limit cycles in R. For equation (6.39) we have

$$\frac{d}{dx}(y) + \frac{d}{dy}\left(-cx - \epsilon x^3 - ky\right) = -k,$$

which is always negative. Hence, equation (6.39) has no limit cycles and it follows that E_0 is globally, asymptotically stable.

```
D[y, x] + D[-cx - cx^3 - ky, y]
```

$-k$

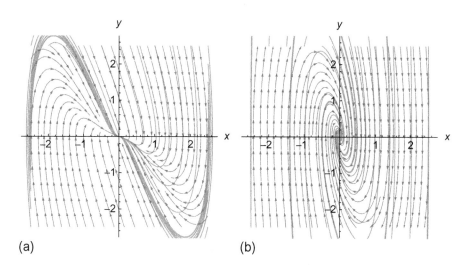

Figure 6-45   (a) The origin is a stable node. (b) The origin is a stable spiral

We use `StreamPlot` and `ParametricPlot` to illustrate two situations that occur. In Figure 6-45(a), we use $c = 1$, $\epsilon = 1/2$, and $k = 3$. In this case, E_0 is a *stable node*. On the other hand, in Figure 6-45(b), we use $c = 10$, $\epsilon = 1/2$, and $k = 3$. In this case, E_0 is a *stable spiral*.

```
pvf1 = StreamPlot[{y, -x - 1/2x^3 - 3y},

  {x, -2.5, 2.5}, {y, -2.5, 2.5},

  StreamStyle → Fine, Frame → False,

  Axes → Automatic, AxesOrigin → {0, 0},

  AxesLabel → {x, y}]

numgraph[init_, c_, opts___] :=

Module[{numsol},

  numsol = NDSolve[{x'[t] == y[t],

  y'[t] == -cx[t] - cx[t]^3 - 3y[t],

  x[0] == init[[1]],

  y[0] == init[[2]]},

  {x[t], y[t]}, {t, 0, 10}];

ParametricPlot[{x[t], y[t]} /. numsol, {t, 0, 10},
```

```
    opts]

  ]

i1 = Table[numgraph[{2.5, i}, 1],

  {i, -2.5, 2.5, .5}];

i2 = Table[numgraph[{-2.5, i}, 1],

  {i, -2.5, 2.5, .5}];

i3 = Table[numgraph[{-i, 2.5}, 1],

  {i, -2.5, 2.5, .5}];

i4 = Table[numgraph[{i, 2.5}, 1],

  {i, -2.5, 2.5, .5}];

c1 = Show[i1, i2, i3, i4, pvf1,

  PlotRange → {{-2.5, 2.5}, {-2.5, 2.5}},

  AspectRatio → 1,

  AxesLabel → {x, y}]

pvf2 = StreamPlot[{y, -10x - 1/2x^3 - 3y},

  {x, -2.5, 2.5}, {y, -2.5, 2.5},

  StreamStyle → Fine, Frame → False,

  Axes → Automatic, AxesOrigin → {0, 0},

  AxesLabel → {x, y}]

i1 = Table[numgraph[{2.5, i}, 10],

  {i, -2.5, 2.5, .5}];

i2 = Table[numgraph[{-2.5, i}, 10],

  {i, -2.5, 2.5, .5}];

i3 = Table[numgraph[{-i, 2.5}, 10],

  {i, -2.5, 2.5, .5}];

i4 = Table[numgraph[{i, 2.5}, 10],

  {i, -2.5, 2.5, .5}];
```

```
c2 = Show[i1, i2, i3, i4, pvf2,

  PlotRange → {{-2.5, 2.5}, {-2.5, 2.5}},

  AspectRatio → 1,

  AxesLabel → {x, y}]


Show[GraphicsRow[{c1, c2}]]
```

Although linearization can help you determine local behavior near rest points, the long-term behavior of solutions to nonlinear systems can be quite complicated, even for deceptively simple looking systems.

EXAMPLE 6.6.6 (Lorenz Equations): The **Lorenz equations** are

See texts like Jordan and Smith's *Nonlinear Ordinary Differential Equations*, [17], for discussions of additional ways to analyze systems like the Rössler attractor and the Lorenz equations.

$$\begin{cases} dx/dt = a(y-x) \\ dy/dt = bx - y - xz \; . \\ dz/dt = xy - cz \end{cases} \quad (6.41)$$

Graph the solutions to the Lorenz equations if $a = 7$, $b = 27.2$, an $c = 3$ if the initial conditions are $x(0) = 3$, $y(0) = 4$, and $z(0) = 2$.

SOLUTION: So that you can experiment with different parameters and initial-conditions, we define the function `lorenzsol`. Given the appropriate parameters and initial conditions,

$$\texttt{lorenzsol[a,b,c] [\{x0,y0,z0\}]}$$

returns a numerical solution of the Lorenz equations (6.41) that satisfies $x(0) = x_0$, $y(0) = y_0$, and $z(0) = z_0$ and is valid for $0 \leq t \leq 1000$. Because the behavior of solutions can be quite intricate, we include the option `MaxSteps->100000` to help Mathematica capture the oscillatory behavior in the long-term solution.

```
lorenzsol[a_, b_, c_][{x0_, y0_, z0_}, ts_:{t, 0, 1000},

  opts___]:=Module[{numsol},

  numsol = NDSolve[{x'[t] == -ax[t] + ay[t], y'[t] == bx[t]-

  y[t] - x[t]z[t], z'[t] == x[t]y[t] - cz[t], x[0] == x0, y[0] == y0,
```

```
z[0] == z0}, {x[t], y[t], z[t]}, ts, MaxSteps → 1000000]

]
```

We then use `lorenzplot` to generate a numerical solution for our parameter values and initial conditions.

```
n1 = lorenzsol[7, 28, 3][{3, 4, 2}]
```

```
{{x[t] → InterpolatingFunction[][t], y[t] →

InterpolatingFunction[][t], z[t] → InterpolatingFunction[][t]}}
```

We generate a short-term plot of the solution in Figure 6-46

```
lorenzplot[a_, b_, c_][{x0_, y0_, z0_}, ts_:{t, 950, 1000},

opts___]:=Module[{numsol},

numsol = NDSolve[{x'[t] == -ax[t] + ay[t], y'[t] == bx[t] - y[t]-

x[t]z[t], z'[t] == x[t]y[t] - cz[t], x[0] == x0, y[0] == y0,

z[0] == z0}, {x[t], y[t], z[t]}, ts, MaxSteps → 1000000];

p1a = ParametricPlot[Evaluate[{x[t], y[t]}/.numsol], ts,

PlotPoints → 1000, AspectRatio → 1, AxesLabel → {x, y}];

p1b = ParametricPlot[Evaluate[{x[t], z[t]}/.numsol], ts,

PlotPoints → 1000, AspectRatio → 1, AxesLabel → {x, z}];

p1c = ParametricPlot[Evaluate[{y[t], z[t]}/.numsol], ts,
```

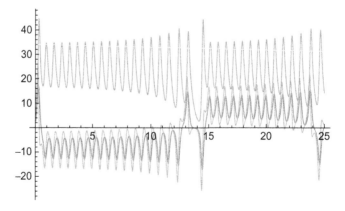

Figure 6-46   Plots of $x(t)$, $y(t)$, and $z(t)$ for $0 \leq t \leq 25$

```
      PlotPoints → 1000, AspectRatio → 1, AxesLabel → {y, z}];

   p1d = ParametricPlot3D[Evaluate[{x[t], y[t], z[t]}/.numsol],

      ts, PlotPoints → 3000, BoxRatios → {1, 1, 1}, AxesLabel

      → {x, y, z}];

   Show[GraphicsGrid[{{p1a, p1b}, {p1c, p1d}}], opts]]

   lorenzplot[7, 28, 3][{3, 4, 2}]

   lorenzplot[7, 27.5, 3][{3, 4, 2}]

   n2 = lorenzsol[7, 27.2, 3][{3, 4, 2}]

   {{x[t] → InterpolatingFunction[][t], y[t] →

   InterpolatingFunction[][t], z[t] →

   InterpolatingFunction[][t]}}

   Plot[Evaluate[{x[t], y[t], z[t]}/.n2], {t, 0, 25}, PlotPoints

      → 1000]
```

and a long-term plot in Figure 6-47.

```
   Plot[Evaluate[{x[t], y[t], z[t]}/.n2], {t, 950, 1000},

      PlotStyle → {GrayLevel[0], GrayLevel[.3],

   Dashing[{0.01}]}, PlotPoints → 1000]
```

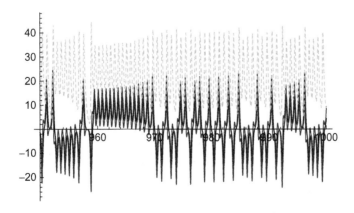

Figure 6-47   Plots of $x(t)$, $y(t)$, and $z(t)$ for $950 \le t \le 1000$

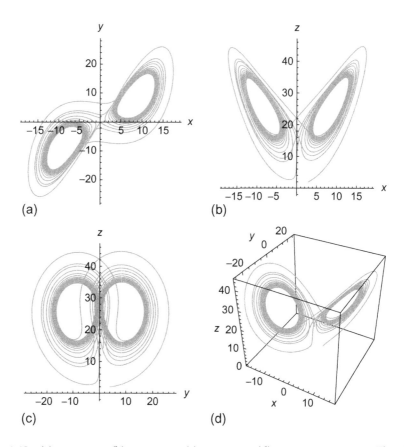

Figure 6-48   (a) x versus y. (b) y versus z. (c) x versus z. (d) x versus y versus z. The chaotic nature of the solutions to the Lorenz equations using these parameter values is seen more clearly in Figure 6-48 than in Figures 6-46 or 6-47

In Figures 6-46 and 6-47 the oscillatory nature of the solutions is very difficult to see. We use `ParametricPlot` and `ParametricPlot3D` to generate parametric plots of the solutions in Figure 6-48.

```
p1a = ParametricPlot[Evaluate[{x[t], y[t]}/.n2], {t, 0, 25},
    PlotPoints → 1000, AspectRatio → 1, AxesLabel → {x, y},
    DisplayFunction → Identity];
p1b = ParametricPlot[Evaluate[{x[t], z[t]}/.n2], {t, 0, 25},
    PlotPoints → 1000, AspectRatio → 1, AxesLabel → {x, z},
    DisplayFunction → Identity];
```

```
p1c = ParametricPlot[Evaluate[{y[t], z[t]}/.n2], {t, 0, 25},
    PlotPoints → 1000, AspectRatio → 1, AxesLabel → {y, z},
    DisplayFunction → Identity];
p1d = ParametricPlot3D[Evaluate[{x[t], y[t], z[t]}/.n2],
    {t, 0, 25}, PlotPoints → 3000, BoxRatios → {1, 1, 1},
    AxesLabel → {x, y, z}, DisplayFunction → Identity];
Show[GraphicsGrid[{{p1a, p1b}, {p1c, p1d}}]]
```

Of course, you could combine all of these commands into a single function. For example, given the appropriate parameter values

```
lorenzplot[a,b,c][{x0,y0,z0}, {t,a,b}]
```

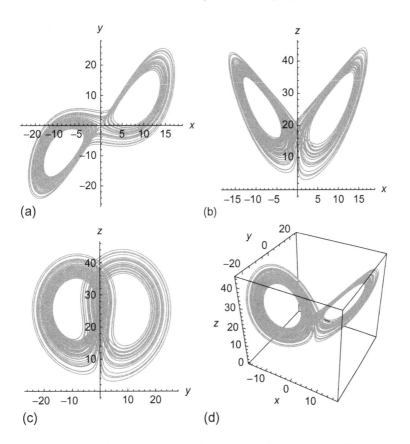

(a)

(b)

(c)

(d)

Figure 6-49   Changing b from 27.2 to 28: (a) x versus y. (b) y versus z. (c) x versus z. (d) x versus y versus z

solves the Lorenz system using initial conditions $x(0) = x_0$, $y(0) = y_0$, and $z(0) = z_0$ for $a \leq t \leq b$, generates parametric plots of x versus y, y versus z, x versus z, and x versus y versus z, and displays the four resulting plots as a graphics array. If $\{t, a, b\}$ is omitted, the default is $950 \leq t \leq 1000$.

For example, entering

```
lorenzplot[7, 27.2, 3][{3, 4, 2}]
```

generates the four plots shown in Figure 6-49, corresponding to changing b from 27.2 to 28. Again, we obtain a chaotic solution.

∎

Applications of Systems of Ordinary Differential Equations

In Chapter 7, we discuss several applications of systems of differential equations. These include standard linear applications that involve mechanical and electrical systems as well as diffusion and population problems. We also discuss several nonlinear applications including predator-prey interactions, food chains in the chemostat, and curvature, which is a topic from differential geometry.

7.1 Mechanical and Electrical Problems With First-Order Linear Systems

7.1.1 *L-R-C* Circuits With Loops

As indicated in Chapter 5, an electrical circuit can be modeled with an ordinary differential equation with constant coefficients. In this section, we illustrate how a circuit involving loops can be described as a system of linear ordinary differential equations with constant coefficients. This derivation is based on the following principles.

> **Kirchhoff's Current Law:** The current entering a point of the circuit equals the current leaving the point.

> **Kirchhoff's Voltage Law:** The sum of the changes in voltage around each loop in the circuit is zero.

Gustav Robert Kirchhoff (1824–87) was an outstanding German mathematician and theoretical physicist. In addition to being an outstanding teacher, he is also remembered for his four volume comprehensive work *Vorlesungen 'ber mathematische Physik.*

Differential Equations with Mathematica. http://dx.doi.org/10.1016/B978-0-12-804776-7.00007-3

As was the case in Chapter 5, we use the following standard symbols for the components of the circuit:

$$I(t) = \text{current where}, I(t) = \frac{dQ}{dt}(t),$$

$$Q(t) = \text{charge},$$

$$R = \text{resistance},$$

$$C = \text{capacitance},$$

$$V = \text{voltage, and}$$

$$L = \text{inductance}.$$

The relationships corresponding to the drops in voltage in the various components of the circuit that were stated in Chapter 5 are also given in the following table.

Circuit Element	Voltage Drop
Inductor	$L\dfrac{dI}{dt}$
Resistor	RI
Capacitor	$\dfrac{1}{C}Q$
Voltage Source	$-V(t)$

7.1.2 *L-R-C* Circuit With One Loop

In determining the drops in voltage around the circuit, we consistently add the voltages in the clockwise direction. The positive direction is directed from the negative symbol toward the positive symbol associated with the voltage source. In summing the voltage drops encountered in the circuit, a drop across a component is added to the sum if the positive direction through the component agrees with the clockwise direction. Otherwise, this drop is subtracted. In the case of the following *L-R-C* circuit with one loop involving each type of component, the current is equal around the circuit by Kirchhoff's Current Law as illustrated in Figure 7-1.

Also, by Kirchhoff's Voltage Law, we have the sum

$$RI + L\frac{dI}{dt} + \frac{1}{C}Q - V(t) = 0.$$

Solving this equation for dI/dt and using the relationship between I and Q, $dQ/dt = I$, we have the following system of differential equations with initial conditions on charge and current, respectively:

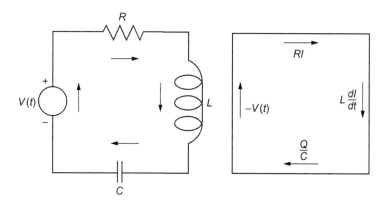

Figure 7-1   A simple *L-R-C* circuit

$$\begin{cases} dQ/dt = I \\ dI/dt = -\dfrac{1}{LC}Q - \dfrac{R}{L}I + \dfrac{V(t)}{L} \\ Q(0) = Q_0,\ I(0) = I_0 \end{cases}.$$

(7.1)

EXAMPLE 7.1.1: Determine the charge and current in an *L-R-C* circuit with $L = 1$, $R = 2$, $C = 4/3$, and $V(t) = e^{-t}$ if $Q(0) = Q_0$ and $I(0) = I_0$.

SOLUTION: We begin by modeling the circuit with the system of differential equations

$$\begin{cases} dQ/dt = I \\ dI/dt = -\tfrac{3}{4}Q - 2I + e^{-t} \end{cases},$$

which can be written in matrix form as

$$\begin{pmatrix} dQ/dt \\ dI/dt \end{pmatrix} = \begin{pmatrix} 0 & 1 \\ -3/4 & -2 \end{pmatrix} \begin{pmatrix} Q \\ I \end{pmatrix} + \begin{pmatrix} 0 \\ e^{-t} \end{pmatrix}.$$

We solve the initial value problem with DSolve, naming the result sol.

```
Clear[q, i]

sol = DSolve[{D[q[t], t]==i[t],
```

$$D[i[t], t] == -3/4q[t] - 2i[t] + \text{Exp}[-t],$$

$$q[0]==q0, i[0]==i0\}, \{q[t], i[t]\}, t]$$

$$\{\{i[t] \to -\frac{1}{4}e^{-3t/2}(12 - 16e^{t/2} + 4e^t - 6i0 + 2e^t i0$$

$$- 3q0 + 3e^t q0), q[t] \to \frac{1}{2}e^{-3t/2}(4 - 8e^{t/2} + 4e^t - 2i0 + 2e^t i0$$

$$- q0 + 3e^t q0)\}\}$$

Next, we select, copy and paste the formulas obtained in sol for Q and I, respectively, and then use Expand to distribute the $e^{-3t/2}$ term through the parentheses.

$$\text{Expand}\left[\frac{1}{2}E^{-3t/2}(-4 - 8E^{t/2} - 2(-4 + i0) + 2E^t(-4 + i0) - q0\right.$$

$$\left. + 3E^t(4 + q0))\right]$$

$$2e^{-3t/2} - 4e^{-t} + 2e^{-t/2} - e^{-3t/2}i0 + e^{-t/2}i0 - \frac{1}{2}e^{-3t/2}q0$$

$$+ \frac{3}{2}e^{-t/2}q0$$

$$\text{Expand}[-\frac{1}{4}E^{-3t/2}(-16E^{t/2} - 6(-4 + i0) + 2E^t(-4 + i0)$$

$$-3(4 + q0) + 3E^t(4 + q0))]$$

$$-3e^{-3t/2} + 4e^{-t} - e^{-t/2} + \frac{3}{2}e^{-3t/2}i0 - \frac{1}{2}e^{-t/2}i0 + \frac{3}{4}e^{-3t/2}q0$$

$$- \frac{3}{4}e^{-t/2}q0$$

The result indicates that $\lim_{t\to\infty} Q(t) = \lim_{t\to\infty} I(t) = 0$ regardless of the values of Q_0 and I_0. This is confirmed by graphing $Q(t)$ (in blue) and $I(t)$ (in orange) together (choosing $Q(0) = I(0) = 1$) in Figure 7-2(a) as well as parametrically in Figure 7-2(b).

$$\text{p1} = \text{Plot}[\text{Evaluate}[\{q[t], i[t]\}/.\text{sol}/.$$

$$\{q0 \to 1, i0 \to 1\}], \{t, 0, 10\}, \text{PlotLabel} \to \text{``(a)''}]$$

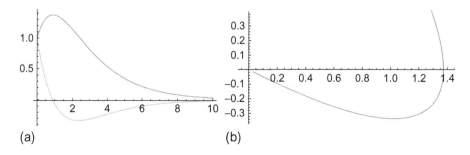

(a)   (b)

Figure 7-2   (a) $Q(t)$ and $I(t)$ for $0 \le t \le 10$. (b) Parametric plot of Q versus I for $0 \le t \le 10$

```
p2 = ParametricPlot[Evaluate[{q[t], i[t]}/.sol/.

    {q0 → 1, i0 → 1}], {t, 0, 10}, PlotLabel → "(b)"]

Show[GraphicsRow[{p1, p2}]]
```

■

7.1.3 *L-R-C* Circuit With Two Loops

The differential equations that model the circuit become more difficult to derive as the number of loops in the circuit increase. For example, consider the circuit in Figure 7-3 that contains two loops.

In this case, the current through the capacitor is equivalent to $I_1 - I_2$. Summing the voltage drops around each loop, we have:

$$\begin{cases} R_1 I_1 + \dfrac{1}{C}Q - V(t) = 0 \\ L\dfrac{dI_2}{dt} + R_2 I_2 - \dfrac{1}{C}Q = 0 \end{cases}. \tag{7.2}$$

Solving the first equation for I_1 we find that $I_1 = \dfrac{1}{R_1}V(t) - \dfrac{1}{R_1 C}Q$ and using the relationship $dQ/dt = I = I_1 - I_2$ we have the following system:

$$\begin{cases} \dfrac{dQ}{dt} = -\dfrac{1}{R_1 C}Q - I_2 + \dfrac{1}{R_1}V(t) \\ \dfrac{dI_2}{dt} = \dfrac{1}{LC}Q - \dfrac{R_2}{L}I_2 \end{cases}. \tag{7.3}$$

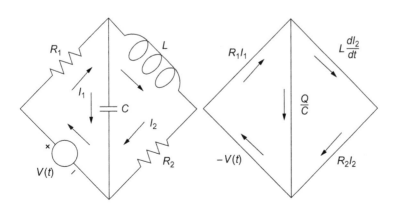

Figure 7-3   A two-loop circuit

EXAMPLE 7.1.2: Find $Q(t)$, $I(t)$, $I_1(t)$, and $I_2(t)$ in the *L-R-C* circuit with two loops given that $R_1 = R_2 = C = 1$ and $V(t) = e^{-t}$ if $Q(0) = 3$ and $I_2(0) = 1$.

SOLUTION: The nonhomogeneous system that models this circuit is

$$\begin{cases} dQ/dt = -Q - I_2 + e^{-t} \\ dI_2/dt = Q - I_2 \end{cases}$$

with initial conditions $Q(0) = 3$ and $I_2(0) = 1$. We solve the initial value problem with DSolve naming the result sol. We define $Q(t)$ and $I_2(t)$ to be the results.

```
Clear[q, i]

sol = DSolve[{D[q[t], t]== - q[t] - i2[t]+

  Exp[-t], D[i2[t], t]==q[t] - i2[t],

  q[0]==3, i2[0]==1}, {q[t], i2[t]}, t]//Simplify
```

$\{\{i2[t] \to e^{-t}(1 + 3\text{Sin}[t]),\ q[t] \to 3e^{-t}\text{Cos}[t]\}\}$

```
q[t_] = 3e^-t Cos[t];

i2[t_] = e^-t(1 + 3Sin[t]);
```

We verify that these functions satisfy the system by substituting back into each equation and simplifying the result with Simplify.

```
D[q[t], t] - (-q[t] - i2[t] + Exp[-t])//Simplify
```

0

```
D[i2[t], t] - (q[t] - i2[t])//Simplify
```

0

We use the relationship $dQ/dt = I$ to find $I(t)$

```
i[t_] = D[q[t], t]
```

$$-3e^{-t}\mathrm{Cos}[t] - 3e^{-t}\mathrm{Sin}[t]$$

and then $I_1(t) = I(t) + I_2(t)$ to find $I_1(t)$.

```
i1[t_] = i[t] + i2[t]
```

$$-3e^{-t}\mathrm{Cos}[t] - 3e^{-t}\mathrm{Sin}[t] + e^{-t}(1 + 3\mathrm{Sin}[t])$$

We graph $Q(t)$, $I(t)$, $I_1(t)$, and $I_2(t)$ with Plot and display the result using Show and GraphicsGrid in Figure 7-4.

```
p1 = Plot[q[t], {t, 0, 5}, PlotRange → All,
   PlotLabel → "(a)"];
p2 = Plot[i[t], {t, 0, 5}, PlotRange → All,
   PlotLabel → "(b)"];
p3 = Plot[i1[t], {t, 0, 5}, PlotRange → All,
   PlotLabel → "(c)"];
p4 = Plot[i2[t], {t, 0, 5}, PlotRange → All,
```

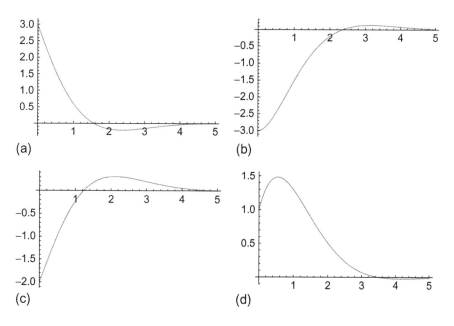

Figure 7-4 $Q(t)$, $I(t)$, $I_1(t)$, and $I_2(t)$ for $0 \leq t \leq 5$

```
      PlotLabel → "(d)"];

   Show[GraphicsGrid[{{p1, p2}, {p3, p4}}]]
```

■

7.1.4 Spring-Mass Systems

The displacement of a mass attached to the end of a spring was modeled with a second-order linear differential equation with constant coefficients in Chapter 5. This situation can then be expressed as a system of first-order ordinary differential equations as well. Recall that if there is no external forcing function, then the second-order differential equation that models this situation is $mx'' + cx' + kx = 0$, where m is the mass attached to the end of the spring, c is the damping coefficient, and k is the spring constant found with Hooke's law. This equation is transformed into a system of equations by letting $x' = y$ so that $y' = x'' = -\dfrac{k}{m}x - \dfrac{c}{m}x'$ and then solving the differential equation for x''. After substitution, we have the system

$$\begin{cases} \dfrac{dx}{dt} = y \\ \dfrac{dy}{dt} = -\dfrac{k}{m}x - \dfrac{c}{m}y \end{cases} \tag{7.4}$$

In previous chapters, the displacement of the spring was illustrated as a function of time. However, problems of this type may also be investigated using the phase plane.

EXAMPLE 7.1.3: Solve the system of differential equations to find the displacement of the mass if $m = 1$, $c = 0$, and $k = 1$.

SOLUTION: In this case, the system is $\begin{cases} dx/dt = y \\ dy/dt = -x \end{cases}$, which in matrix form is $\mathbf{X}' = \begin{pmatrix} 0 & 1 \\ -1 & 0 \end{pmatrix} \mathbf{X}$. The solution that satisfies the initial conditions $x(0) = x_0$ and $y(0) = y_0$ is found with DSolve and named gensol for later use.

```
Clear[x, y]
gensol = DSolve[{D[x[t], t]==y[t],
  D[y[t], t]== - x[t], x[0] == x0,
  y[0] == y0}, {x[t], y[t]}, t]
```

$\{\{x[t] \rightarrow x0\text{Cos}[t] + y0\text{Sin}[t], y[t] \rightarrow y0\text{Cos}[t] - x0\text{Sin}[t]\}\}$

Note that this system is equivalent to the second-order differential equation $x'' + x = 0$, which we solved in Chapters 4 and 5. At that time, we found a general solution to be $x(t) = c_1 \cos t + c_2 \sin t$ which is equivalent to the first component of $\mathbf{X} = \begin{pmatrix} x(t) \\ y(t) \end{pmatrix}$, the result obtained with DSolve. Also notice that $(0,0)$ is the equilibrium point of the system. The eigenvalues of $\mathbf{A} = \begin{pmatrix} 0 & 1 \\ -1 & 0 \end{pmatrix}$ are $\lambda = \pm i$,

```
Eigenvalues [( 0   1 )]
             ( -1  0 )
```

$\{i, -i\}$

so we classify the origin as a center.

We graph several members of the phase plane for this system with ParametricPlot in Figure 7-5.

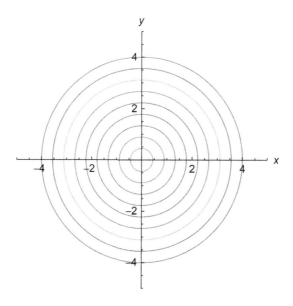

Figure 7-5   The origin is a center

```
toplot = Table[{x[t], y[t]}/.gensol/.{x0 → i, y0 → 0},

    {i, 0, 4, 4/9}];

ParametricPlot[Evaluate[toplot], {t, 0, 2π}, PlotRange →

    {{−5, 5}, {−5, 5}}, AspectRatio → 1, AxesLabel → {x, y}]
```

∎

7.2 Diffusion and Population Problems With First-Order Linear Systems

7.2.1 Diffusion Through a Membrane

Solving problems to determine the diffusion of a substance (such as glucose or salt) in a medium (like a blood cell) also leads to systems of first-order linear ordinary differential equations. For example, suppose that two solutions of a substance are separated by a membrane where the amount of the substance that passes through the membrane is proportional to the difference in the concentrations of the solutions. The constant of proportionality is called the **permeability**, P, of the membrane. Therefore, if we let x and y represent the concentration of each solution, and V_1 and V_2 represent the volume of each solution, respectively, then the system of differential equations is given by

$$\begin{cases} \dfrac{dx}{dt} = \dfrac{P}{V_1}\,(y - x) \\[2mm] \dfrac{dy}{dt} = \dfrac{P}{V_2}\,(x - y) \end{cases}, \tag{7.5}$$

where the initial concentrations of x and y are given.

EXAMPLE 7.2.1: Suppose that two salt concentrations of equal volume V are separated by a membrane of permeability P. Given that $P = V$, determine each concentration at time t if $x(0) = 2$ and $y(0) = 10$.

SOLUTION: In this case, the initial-value problem that models the situation is

$$\begin{cases} dx/dt = y - x \\ dy/dt = x - y \\ x(0) = 2,\ y(0) = 10 \end{cases}.$$

A general solution of the system is found with DSolve and named gensol.

```
Clear[x, y]

gensol = DSolve[{D[x[t], t]==y[t] - x[t],

    D[y[t], t]==x[t] - y[t]}, {x[t], y[t]}, t]
```

$$\{\{x[t] \to \frac{1}{2}e^{-2t}(1 + e^{2t})C[1] + \frac{1}{2}e^{-2t}(-1 + e^{2t})C[2],\ y[t] \to$$
$$\frac{1}{2}e^{-2t}(-1 + e^{2t})C[1] + \frac{1}{2}e^{-2t}(1 + e^{2t})C[2]\}\}$$

The x-component of the solution is the first part of the first part of the second part of gensol and is extracted from gensol with Part

```
xcomponent = gensol[[1, 1, 2]]
```

$$\frac{1}{2}e^{-2t}(1 + e^{2t})C[1] + \frac{1}{2}e^{-2t}(-1 + e^{2t})C[2]$$

Similarly, the y-component of the solution is the first part of the second part of the second part of gensol and is extracted from gensol with Part

```
ycomponent = gensol[[1, 2, 2]]
```

$$\frac{1}{2}e^{-2t}(-1 + e^{2t})C[1] + \frac{1}{2}e^{-2t}(1 + e^{2t})C[2]$$

We then apply the initial conditions and use Solve to determine the values of the arbitrary constants.

```
cvals = Solve[{(xcomponent/. t->0)==2, (ycomponent/.

    t->0)==10}]
```

$$\{\{C[1] \to 2,\ C[2] \to 10\}\}$$

The solution is obtained by substituting these values back into the general solution.

```
sol = gensol/.cvals[[1]]
```

$$\{\{x[t] \to 5e^{-2t}(-1+e^{2t}) + e^{-2t}(1+e^{2t}), \; y[t] \to e^{-2t}(-1+e^{2t})$$
$$+ 5e^{-2t}(1+e^{2t})\}\}$$

Of course, DSolve can be used to solve the initial value problem directly as well.

sol = DSolve[{D[x[t], t]==y[t] − x[t], D[y[t], t]==x[t] − y[t],

 x[0]==2, y[0]==10}, {x[t], y[t]}, t]

$$\{\{x[t] \to 2e^{-2t}(-2+3e^{2t}), \; y[t] \to 2e^{-2t}(2+3e^{2t})\}\}$$

We graph this solution parametrically with ParametricPlot in Figure 7-6(a). We then graph $x(t)$ and $y(t)$ together in Figure 7-6(b). Notice that each concentration approaches 6 which is the average value of the two initial concentrations.

p1a = ParametricPlot[{x[t], y[t]}/.sol, {t, 0, 5},

 PlotRange → {{0, 10}, {0, 10}},

 PlotStyle->CMYKColor[0, 0.89, 0.94, 0.28],

 AspectRatio → 1,

 AxesOrigin → {0, 0}, AxesLabel → {x, y}];

p1b = StreamPlot[{y − x, x − y}, {x, 0, 10}, {y, 0, 10},

 StreamStyle → Fine]

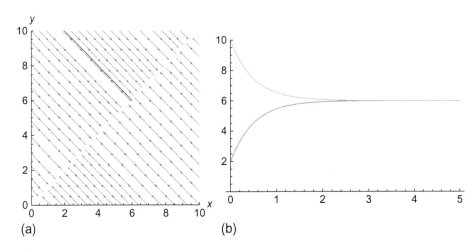

(a)   (b)

Figure 7-6 (a) Parametric plot of x versus y. (b) $x(t)$ and $y(t)$

```
p1 = Show[p1a, p1b, PlotRange → {{0, 10}, {0, 10}},

AspectRatio → 1,

  Frame → False,

  AxesOrigin → {0, 0}, AxesLabel → {x, y}];

p2 = Plot[Evaluate[{x[t], y[t]}/.sol], {t, 0, 5},

  PlotRange → {0, 10}];

Show[GraphicsRow[{p1, p2}]]
```

∎

7.2.2 Diffusion Through a Double-Walled Membrane

Next, consider the situation in which two solutions are separated by a double-walled membrane, where the inner wall has permeability P_1 and the outer wall has permeability P_2 with $0 < P_1 < P_2$. Suppose that the volume of solution within the inner wall is V_1 and that between the two walls is V_2. Let x represent the concentration of the solution within the inner wall and y the concentration between the two walls. Assuming that the concentration of the solution outside the outer wall is constantly C, we have the following system of first-order ordinary differential equations

$$\begin{cases} \dfrac{dx}{dt} = \dfrac{P_1}{V_1}(y - x) \\ \dfrac{dy}{dt} = \dfrac{1}{V_2}[P_2(C - y) + P_1(x - y)] \\ x(0) = x_0, \ y(0) = y_0 \end{cases} \tag{7.6}$$

EXAMPLE 7.2.2: Given that $P_1 = 3$, $P_2 = 8$, $V_1 = 2$, $V_2 = 10$, and $C = 10$, determine x and y if $x(0) = 2$ and $y(0) = 1$.

SOLUTION: In this case, we must solve the initial-value problem

$$\begin{cases} dx/dt = \dfrac{3}{2}(y - x) \\ dy/dt = -\dfrac{11}{10}y + \dfrac{3}{10}x + 8 \\ x(0) = 2, \ y(0) = 1 \end{cases}$$

A general solution of the corresponding homogeneous system is found with DSolve.

```
Clear[x, y]
homsol = DSolve[{x'[t]== 3/2 y[t] - 3/2 x[t],
    y'[t]== - 11/10 y[t] + 3/10 x[t]}, {x[t], y[t]}, t]
```

$$\{\{x[t] \to \frac{1}{14}e^{-2t}(9+5e^{7t/5})C[1]+\frac{15}{14}e^{-2t}(-1+e^{7t/5})C[2], \ y[t] \to$$
$$\frac{3}{14}e^{-2t}(-1+e^{7t/5})C[1]+\frac{1}{14}e^{-2t}(5+9e^{7t/5})C[2]\}\}$$

The result indicates that a fundamental matrix for the corresponding homogeneous system is $\Phi(t) = \begin{pmatrix} -3e^{-2t} & \frac{5}{3}e^{-3t/5} \\ e^{-2t} & e^{-3t/5} \end{pmatrix}$. A different way to see a fundamental matrix for the corresponding homogeneous system is by computing the eigenvalues and eigenvectors for the coefficient matrix of the corresponding homogeneous system.

```
Eigensystem[{{-3/2, 3/2}, {3/10, -11/10}}]
```

$$\{\{-2, -\frac{3}{5}\}, \{\{-3, 1\}, \{\frac{5}{3}, 1\}\}\}$$

```
Φ[t_] = ( -3E^-2t    5/3 E^-3t/5
           E^-2t      E^-3t/5   );
```

Therefore, using the method of variation of parameters, the solution to the initial value problem is given by

$$\mathbf{X}(t) = \Phi(t)\Phi^{-1}(0)\mathbf{X}(0) + \Phi(t)\int_0^t \Phi^{-1}(u)\mathbf{F}(u)\,du.$$

```
sol = Φ[t].(Inverse[Φ[t]]/.t → 0).( 2
                                    1 )

   +Φ[t]. ∫₀ᵗ Inverse[Φ[u]].( 0
                              8 )

du//Simplify; MatrixForm[sol]
```

$$\begin{pmatrix} 10 + \dfrac{9e^{-2t}}{2} - \dfrac{25}{2}e^{-3t/5} \\ 10 - \dfrac{3e^{-2t}}{2} - \dfrac{15}{2}e^{-3t/5} \end{pmatrix}$$

Of course, DSolve can be used to solve the initial value problem directly, as well.

```
sol = DSolve[{x'[t]== 3/2 (y[t] - x[t]),

    y'[t]== - 11y[t]/10 + 3x[t]/10 + 8, x[0]==2, y[0]==1},

    {x[t], y[t]}, t]
```

$$\{\{x[t] \rightarrow \tfrac{1}{2}e^{-2t}(9 - 25e^{7t/5} + 20e^{2t}), \; y[t] \rightarrow \tfrac{1}{2}e^{-2t}(-3 - 15e^{7t/5}$$
$$+ 20e^{2t})\}\}$$

We graph this solution parametrically in addition to graphing the two functions simultaneously in Figure 7-7. Notice that initially $x(t) > y(t)$. However, the two graphs intersect at a value of t near $t \approx 0.2$ so that as the value of the two functions approach 10, which is the concentration of the solution outside the outer wall, as t increases.

```
p1 = ParametricPlot[{x[t], y[t]}/.sol, {t, 0, 7},

    PlotRange → {{0, 10}, {0, 10}},

    AspectRatio → 1, AxesOrigin → {0, 0}, AxesLabel → {x, y}];

p2 = Plot[Evaluate[{x[t], y[t]}/.sol], {t, 0, 7},

    PlotRange → {0, 10}];

Show[GraphicsRow[{p1, p2}]]
```

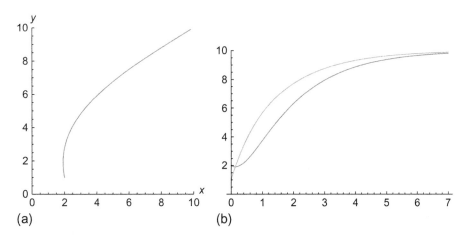

(a)   (b)

Figure 7-7   (a) Parametric plot of x versus y. (b) $x(t)$ and $y(t)$ for $0 \le t \le 7$

Last, we plot the solution parametrically for various initial conditions.

```
sol = DSolve[{x'[t]==(3/2)(y[t] − x[t]),
    y'[t]== − 11y[t]/10 + 3x[t]/10 + 8, x[0]==x0, y[0]==y0},
    {x[t], y[t]}, t]
```

$$\{\{x[t] \rightarrow \frac{1}{14} e^{-2t}(60 - 200e^{7t/5} + 140e^{2t} + 9x0 + 5e^{7t/5}x0$$
$$- 15y0 + 15e^{7t/5}y0), \, y[t] \rightarrow \frac{1}{14} e^{-2t}(-20 - 120e^{7t/5} + 140e^{2t}$$
$$- 3x0 + 3e^{7t/5}x0 + 5y0 + 9e^{7t/5}y0)\}\}$$

Notice how the formulas for $x(t)$ and $y(t)$ are extracted from sol with Part ([[...]]). The formula for $x(t)$ is the second part of the first part of the first part of sol; the formula for $y(t)$ is the second part of the second part of the first part of sol.

```
sol[[1, 1, 2]]
```

```
sol[[1, 2, 2]]
```

$$\frac{1}{14} e^{-2t}(60-200e^{7t/5}+140e^{2t}+9x0+5e^{7t/5}x0-15y0+15e^{7t/5}y0)$$

$$\frac{1}{14} e^{-2t}(-20-120e^{7t/5}+140e^{2t}-3x0+3e^{7t/5}x0+5y0+9e^{7t/5}y0)$$

Then, we use Table and Flatten to construct a list of (pairs of) functions to be plotted with ParametricPlot. Short is used to display an abbreviated portion of toplot.

```
toplot = Flatten[Table[{sol[[1, 1, 2]], sol[[1, 2, 2]]},
    {x0, 0, 10, 2}, {y0, 0, 10, 2}], 1];
```

```
Short[toplot, 2]
```

$$\{\{\frac{1}{14} e^{-2t}(60 - 200e^{7t/5} + 140e^{2t}), \frac{1}{14} e^{-2t}(-20 - 120e^{7t/5}$$
$$+ 140e^{2t})\}, \langle\langle 34 \rangle\rangle, \{10, 10\}\}$$

The list of functions in toplot is then graphed with ParametricPlot for $0 \leq t \leq 7$ in Figure 7-8.

```
ParametricPlot[Evaluate[toplot], {t, 0, 7},
    PlotRange → {{0, 10}, {0, 10}},
    AspectRatio → 1, AxesLabel → {x, y}]
```

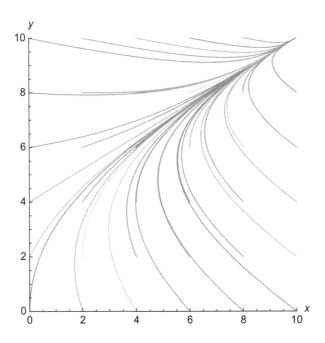

Figure 7-8 Both concentrations approach 10, regardless of the initial conditions

7.2.3 Population Problems

In Chapter 3, population problems were discussed that were based on the principle that the rate at which a population grows (or decays) is proportional to the number present in the population at any time t. Hence, if $x = x(t)$ represents the population at time t, $dx/dt = kx$. for some constant k. This idea can be extended to problems involving more than one population and leads to systems of ordinary differential equations. We illustrate several situations through the following examples. Note that in each problem, we determine the rate at which a population of size P changes with the equation

$$\frac{dP}{dt} = (\text{rate entering}) - (\text{rate leaving}) .$$

We begin by determining the population in two neighboring territories. Suppose that the population x and y of two neighboring territories depends on several factors. The birth rate of x is a_1 while that of y is b_1. The rate at which citizens of x move to y is a_2 while that at which citizens move from y to x is b_2. Finally, the mortality rate of each territory is disregarded. Determine the respective populations of these two territories for any time t.

Using the principles of previous examples, we have that the rate at which population x changes is

$$\frac{dx}{dt} = a_1 x - a_2 x + b_1 y = (a_1 - a_2)x + b_1 y$$

while the rate at which population y changes is

$$\frac{dy}{dt} = b_1 y - b_2 y + a_2 x = (b_1 - b_2)y + a_2 x.$$

Therefore, the system of equations that must be solved is

$$\begin{cases} dx/dt = (a_1 - a_2)x + b_1 y \\ dy/dt = a_2 x + (b_1 - b_2)y \end{cases}, \qquad (7.7)$$

where the initial populations of the two territories $x(0) = x_0$ and $y(0) = y_0$ are given.

EXAMPLE 7.2.3: Determine the populations $x(t)$ and $y(t)$ in each territory if $a_1 = 5$, $a_2 = 4$, $b_1 = 2$, and $b_2 = 3$ given that $x(0) = 60$ and $y(0) = 10$.

SOLUTION: In this example, the initial-value problem that models the situation is

$$\begin{cases} dx/dt = x + y \\ dy/dt = 4x - 2y \\ x(0) = 60,\ y(0) = 10 \end{cases},$$

which we solve with DSolve.

```
Clear[x, y]

sol = DSolve[{x'[t]==x[t] + y[t],

   y'[t]==4x[t] - 2y[t], x[0]==60, y[0]==10},

   {x[t], y[t]}, t]
```

$$\{\{x[t] \to 10e^{-3t}(1 + 5e^{5t}),\ y[t] \to 10e^{-3t}(-4 + 5e^{5t})\}\}$$

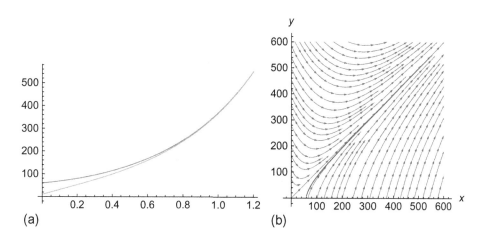

Figure 7-9 (a) As t increases, the two populations are approximately the same. (b) x versus y together with the direction field for the system

We graph these two population functions with Plot in Figure 7-9(a) and parametrically with ParametricPlot in Figure 7-9(b). With the parametric plot, we also display the direction field generated with StreamPlot. Notice that as t increases, the two populations are approximately the same.

```
p1 = Plot[Evaluate[{x[t], y[t]}/.sol], {t, 0, 1.2}]

p2a = ParametricPlot[Evaluate[{x[t], y[t]}/.sol], {t,0,1.2},

   PlotStyle->CMYKColor[0, 0.89, 0.94, 0.28]]

p2b = StreamPlot[{x + y, 4x − 2y}, {x, 0, 600}, {y, 0, 600}]

p2 = Show[p2a, p2b, Frame → False,

   PlotRange → {{0, 600}, {0, 600}},

   Axes → Automatic,AxesOrigin → {0,0},AxesLabel → {x,y}]

Show[GraphicsRow[{p1, p2}]]
```

■

Population problems that involve more than two neighboring populations can be solved with a system of differential equations as well. Suppose that the population of three neighboring territories x, y, and z depends on several factors. The birth rates of x, y, and z are a_1, b_1, and c_1, respectively. The rate at which citizens of x move to y is a_2 while that at which citizens move from x to z is a_3.

Similarly, the rate at which citizens of y move to x is b_2 while that at which citizens move from y to z is b_3. Also, the rate at which citizens of z move to x is c_2 while that at which citizens move from z to y is c_3. Suppose that the mortality rate of each territory is ignored in the model.

The system of equations in this case is similar to that derived in the previous example. The rate at which population x changes is

$$\frac{dx}{dt} = a_1 x - a_2 x - a_3 x + b_2 y + c_2 z = (a_1 - a_2 - a_3) x + b_2 y + c_2 z,$$

while the rate at which population y changes is

$$\frac{dy}{dt} = b_1 y - b_2 y - b_3 y + a_2 x + c_3 z = (b_1 - b_2 - b_3) y + a_2 x + c_3 z,$$

and that of z is

$$\frac{dz}{dt} = c_1 z - c_2 z - c_3 z + a_3 x + b_3 y = (c_1 - c_2 - c_3) z + a_3 x + b_3 y.$$

Hence, we must solve the 3×3 system

$$\begin{cases} dx/dt = (a_1 - a_2 - a_3) x + b_2 y + c_2 z \\ dy/dt = (b_1 - b_2 - b_3) y + a_2 x + c_3 z , \\ dz/dt = (c_1 - c_2 - c_3) z + a_3 x + b_3 y \end{cases} \qquad (7.8)$$

where the initial populations $x(0) = x_0$, $y(0) = y_0$, and $z(0) = z_0$ are given.

EXAMPLE 7.2.4: Determine the population of the three territories if $a_1 = 3$, $a_2 = 0$, $a_3 = 2$, $b_1 = 4$, $b_2 = 2$, $b_3 = 1$, $c_1 = 5$, $c_2 = 3$, and $c_3 = 0$ if $x(0) = 50$, $y(0) = 60$, and $z(0) = 25$.

SOLUTION: We solve the initial-value problem

$$\begin{cases} dx/dt = x + 2y + 3z \\ dy/dt = y \\ dz/dt = 2x + y + 2z \\ x(0) = 50, \ y(0) = 60, \ z(0) = 25 \end{cases}$$

with `DSolve`.

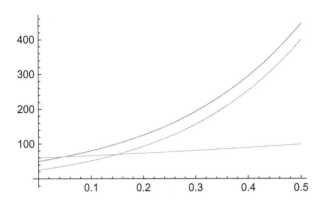

Figure 7-10   Three neighboring territories

```
Clear[x, y]
sol = DSolve[{x'[t]==x[t] + 2y[t] + 3z[t],
  y'[t]==y[t], z'[t]==2x[t] + y[t] + 2z[t],
  x[0]==50, y[0]==60, z[0]==25}, {x[t], y[t], z[t]}, t]
```

$\{\{x[t] \rightarrow e^{-t}(-3 - 10e^{2t} + 63e^{5t}), y[t] \rightarrow 60e^{t}, z[t] \rightarrow$
$e^{-t}(2 - 40e^{2t} + 63e^{5t})\}\}$

The graphs of these three population functions are generated with `Plot` in Figure 7-10. We notice that although y was initially greater than populations x and z, these populations increase at a much higher rate than does y.

```
Plot[Evaluate[{x[t], y[t], z[t]}/.sol], {t, 0, .5}]
```

■

7.3 Applications That Lead to Nonlinear Systems

Several special equations and systems that arise in the study of many areas of applied mathematics can be solved using the techniques of Chapter 6. These include the predator-prey population dynamics problem, the Van der Pol equation that models variable damping in a spring-mass system, and the

Bonhoeffer-Van der Pol (BVP) oscillator. We begin by considering the Lotka-Volterra system, which models the interaction between two populations.

7.3.1 Biological Systems: Predator-Prey Interactions, The Lotka-Volterra System, and Food Chains in the Chemostat

The Lotka-Volterra System

Let $x(t)$ and $y(t)$ represent the number of members at time t of the prey and predator populations, respectively. (Examples of such populations include fox/rabbit and shark/seal.) Suppose that the positive constant a is the birth rate of $x(t)$ so that in the absence of the predator $dx/dt = ax$ and that c is the death rate of y which indicates that $dy/dt = -cy$ in the absence of the prey population. In addition to these factors, the number of interactions between predator and prey affects the number of members in the two populations. Note that an interaction increases the growth of the predator population and decreases the growth of the prey population, because an interaction between the two populations indicates that a predator overtakes a member of the prey population. In order to include these interactions in the model, we assume that the number of interactions is directly proportional to the product of $x(t)$ and $y(t)$. Therefore, the rate at which $x(t)$ changes with respect to time is $dx/dt = ax - bxy$. Similarly, the rate at which $y(t)$ changes with respect to time is $dy/dt = -cy + dxy$. Therefore, we must solve the **Lotka-Volterra system**

$$\begin{cases} dx/dt = ax - bxy \\ dy/dt = -cy + dxy \end{cases} \tag{7.9}$$

subject to the initial populations $x(0) = x_0$ and $y(0) = y_0$.

EXAMPLE 7.3.1: Find and classify the equilibrium points of the Lotka-Volterra system.

SOLUTION: We solve $\begin{cases} ax - bxy = 0 \\ -cy + dxy = 0 \end{cases}$ to see that the equilibrium points are $(0, 0)$ and $(c/d, a/b)$.

```
f[x_, y_] = ax - bxy;

g[x_, y_] = -cy + dxy;

Solve[{f[x, y]==0, g[x, y]==0}, {x, y}]
```

$$\left\{\{x \to 0, y \to 0\}, \left\{x \to \frac{c}{d}, y \to \frac{a}{b}\right\}\right\}$$

To classify these equilibrium points, we first calculate the Jacobian matrix of the nonlinear system.

The Jacobian matrix is also called the **variational matrix**.

$$\texttt{jac} = \begin{pmatrix} D[f[x, y], x] & D[f[x, y], y] \\ D[g[x, y], x] & D[g[x, y], y] \end{pmatrix};$$

```
MatrixForm[jac]
```

$$\begin{pmatrix} a - by & -bx \\ dy & -c + dx \end{pmatrix}$$

```
jac/.{x->0, y->0}//Eigenvalues
```

$$\{a, -c\}$$

Because these eigenvalues are real with opposite sign, we classify $(0, 0)$ as a saddle; $(0, 0)$ is unstable. Similarly, at $(c/d, a/b)$, we have $\mathbf{J}(c/d, a/b) = \begin{pmatrix} 0 & -bc/d \\ ad/b & 0 \end{pmatrix}$ with eigenvalues $\lambda_{1,2} = \pm i\sqrt{ac}$.

```
jac/.{x->c/d, y->a/b}//Eigenvalues
```

$$\{-i\sqrt{a}\sqrt{c}, i\sqrt{a}\sqrt{c}\}$$

Therefore, the point $(c/d, a/b)$ is classified as a center in the linearized system. We show the direction field associated with the system using the values $a = 2$, $b = 1$, $c = 3$, and $d = 1$ in Figure 7-11. The direction field indicates that all solutions oscillate about the center.

```
StreamPlot[{2x - xy, -3y + xy}, {x, 0, 15}, {y, 0, 15},

    Axes->Automatic, AxesOrigin->{0, 0},

    StreamStyle -> Fine, Frame -> False, AxesLabel -> {x, y}]
```

This observation is confirmed by graphing several curves in the phase plane of the system for these values of a, b, c, and d. See Figure 7-12.

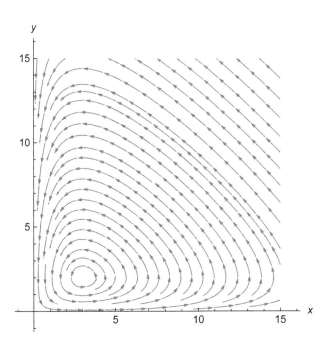

Figure 7-11   Typical direction field associated with the Lotka-Volterra system

```
Clear[x, y, t, s]

graph[s0_]:=Module[{numsol, pp, pxy},

numsol = NDSolve[{x'[t]==2x[t] − x[t]y[t],

  y'[t]== − 3y[t] + x[t]y[t], x[0]==3s0, y[0]==2s0}, {x[t], y[t]},

    {t, 0, 15}];

pp = ParametricPlot[Evaluate[{x[t], y[t]}/.numsol],{t,0,4},

  PlotRange → {{0, 15}, {0, 15}}, Ticks → {{3}, {2}}];

pxy = Plot[Evaluate[{x[t], y[t]}/.numsol],{t, 0, 15},

  PlotRange → {0, 15},

  Ticks->{{5, 10}, {5, 10}}];

  GraphicsRow[{pxy, pp}]]


Clear[x, y, t, s]

graph[s0_]:=Module[{numsol, pp, pxy},

numsol = NDSolve[{x'[t]==2x[t] − x[t]y[t],
```

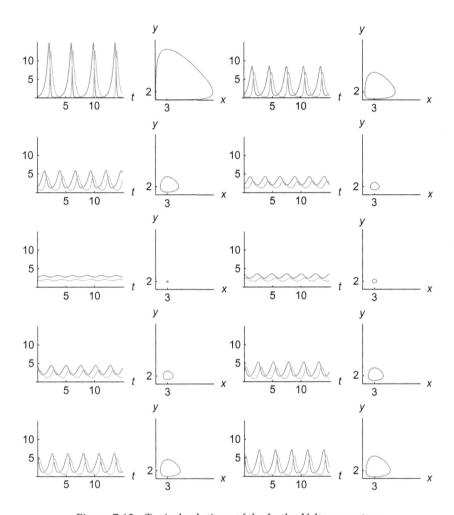

Figure 7-12 Typical solutions of the Lotka-Volterra system

```
    y'[t]== - 3y[t] + x[t]y[t], x[0]==3s0, y[0]==2s0}, {x[t], y[t]},
      {t, 0, 15}];
pp = ParametricPlot[Evaluate[{x[t], y[t]}/.numsol],{t,0,4},
    PlotRange → {{0, 15}, {0, 15}}, Ticks → {{3}, {2}},
    AxesLabel → {x, y}];
pxy = Plot[Evaluate[{x[t], y[t]}/.numsol], {t, 0, 15},
    PlotRange → {0, 15}, Ticks->{{5, 10}, {5, 10}},
    AxesLabel → {t, ""}];
{pxy, pp}]
```

```
graphs = Table[graph[s], {s, 0.1, 2, 1.9/9}];

toshow = Partition[Flatten[graphs], 4];

Show[GraphicsGrid[toshow]]
```

Notice that all of the solutions oscillate about the center. These solutions reveal the relationship between the two populations: prey, $x(t)$, and predator, $y(t)$. As we follow one cycle counterclockwise beginning, for example, near the point $(3, 2)$, we notice that as $x(t)$ increases, then $y(t)$ increases until $y(t)$ becomes overpopulated. Then, because the prey population is too small to supply the predator population, $y(t)$ decreases which leads to an increase in the population of $x(t)$. At this point, because the number of predators becomes too small to control the number in the prey population, $x(t)$ becomes overpopulated and the cycle repeats itself.

∎

An interesting variation of the Lotka-Volterra equations is to assume that a depends strongly on environmental factors and might be given by the differential equation

$$\frac{da}{dt} = -ax + \bar{a} + k\sin(\omega t + \phi), \tag{7.10}$$

where the term $-ax$ represents the loss of nutrients due to species x; $\bar{a}$, k, ω, and ϕ are constants. Observe that incorporating equation (7.10) into system (7.9) results in a nonautonomous system.

EXAMPLE 7.3.2: Suppose that $x(0) = y(0) = a(0) = 0.5$, $b = d = 1$, $c = 0.5$, $\bar{a} = 0.25$, $k = 0.125$ and $\phi = 0$. Plot $x(t)$ and $y(t)$ if $\omega = 0.1, 0.25$, $0.5, 0.75, 1, 1.25, 1.5$, and 2.5.

SOLUTION: Given the appropriate parameter values and initial conditions, `solgraph` solves

$$\begin{cases} dx/dt = ax - bxy \\ dy/dt = -cy + dxy \\ da/dt = -ax + \bar{a} + k\sin(\omega t + \phi) \\ x(0) = x_0, \ y(0) = y_0, \ z(0) = z_0 \end{cases} \tag{7.11}$$

plots $x(t)$ and $y(t)$, parametrically plots $\begin{cases} x = x(t) \\ y = y(t) \end{cases}$, and displays the results side-by-side. Any options included are passed to the Show command. If $\{t, a, b\}$ is omitted from the solgraph command, the default is $\{t, 0, 40\}$.

```
solgraph[b_, d_, c_, abar_, k_, ω_, φ_][{x0_, y0_, a0_},

   ts_:{t, 0, 40}, opts___]:=Module[{numsol, p1, p2},

   numsol = NDSolve[{x'[t] == a[t]x[t] - bx[t]y[t],

   y'[t] == -cy[t] + dx[t]y[t],

   a'[t] == -a[t]x[t] + abar + kSin[ωt + φ],

   x[0] == x0, y[0] == y0, a[0] == a0}, {x[t], y[t], a[t]}, ts];

   p1 = ParametricPlot[{x[t], y[t]}/.numsol, ts,

   PlotRange → {{0, 1}, {0, 1}}, PlotStyle->

      CMYKColor[0, 0.89, 0.94, 0.28],

   AspectRatio → Automatic,

   AxesLabel → {x, y}, Ticks → {{0, 1}, {0, 1}}];

   p2 = Plot[Evaluate[{x[t], y[t]}/.numsol], ts,

      PlotRange → {0, 1},

   DisplayFunction → Identity, AxesLabel → {t, "x,y"},

   Ticks → {Automatic, {0, 1}}];

   Show[GraphicsRow[{p2, p1}], opts]

   ]
```

For example, entering

```
p1 = solgraph[1, 1, .5, .25, .125, .3, 0][{.5, .5, .5}]
```

graphs the solution to the initial-value problem (7.11) for our parameter values and initial conditions if $\omega = 0.3$ shown in Figure 7-13. The command is flexible. For example, entering

```
p2 = solgraph[1, 1, .5, .25, .125, .3, 0][{.5, .5, .5},

   {t, 0, 100}]
```

performs the same calculation but graphs the solutions on the interval $0 \le t \le 100$. We then use Map to apply solgraphgrid to the list of numbers $\{0.01, 0.1, 0.25, 0.5, 0.75, 1, 1.25, 1.5, 2.5\}$.

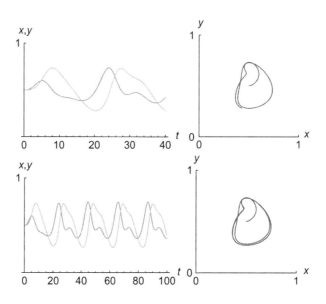

Figure 7-13   Top row: $a = a(t)$, $\omega = 0.3$. Bottom row: the domain is $0 \le t \le 100$

```
solgraphgrid[b_, d_, c_, abar_, k_, ω_, ϕ_][{x0_, y0_, a0_},
  ts_:{t, 0, 40}, opts___]:=Module[{numsol, p1, p2},
  numsol = NDSolve[{x'[t] == a[t]x[t] − bx[t]y[t],
  y'[t] == −cy[t] + dx[t]y[t],
  a'[t] == −a[t]x[t] + abar + kSin[ωt + ϕ],
  x[0] == x0, y[0] == y0, a[0] == a0}, {x[t], y[t], a[t]}, ts];
  p1 = ParametricPlot[{x[t], y[t]}/.numsol, ts,
  PlotRange → {{0, 1}, {0, 1}}, PlotStyle->CMYKColor
    [0, 0.89, 0.94, 0.28],
  AspectRatio → Automatic,
  AxesLabel → {x, y}, Ticks → {{0, 1}, {0, 1}}];
  p2 = Plot[Evaluate[{x[t], y[t]}/.numsol], ts,
    PlotRange → {0, 1},
  DisplayFunction → Identity, AxesLabel → {t, "x,y"},
 Ticks → {Automatic, {0, 1}}];
  {p2, p1}
  ]
```

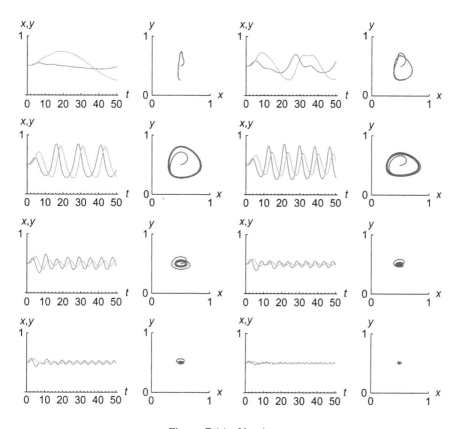

Figure 7-14 Varying ω

Partition is used to partition toshow into two element subsets and the resulting array of graphics is displayed using Show and GraphicsGrid in Figure 7-14.

```
toshowa = Flatten[Map[solgraphgrid[1, 1, .5, .25, .125, #, 0]

[{.5, .5, .5}, {{t, 0, 50}}]&, .1, .25, .5, .75, 1, 1.25,

1.5, 2.5}]];

toshow = Partition[toshow, 4];
```

```
Show[GraphicsGrid[toshow]]
```

From the graphs, we see that larger values of ω appear to stabilize the populations of both species; smaller values of ω appear to cause the size of the populations to oscillate widely.

∎

Simple Food Chain in a Chemostat

See Smith and Waltman's,
*The Theory of the Chemostat:
Dynamics of Microbial
Competition*, [24], for a
detailed discussion of various
chemostat models.
Previously, we discussed
growth in the chemostat in
Example 3.2.6.

The equations that describe a simple food chain in a chemostat are

$$\begin{cases} \dfrac{dS}{dt} = 1 - S - \dfrac{m_1 x S}{a_1 + S} \\[2mm] \dfrac{dx}{dt} = \dfrac{m_1 x S}{a_1 + S} - x - \dfrac{m_2 x y}{a_2 + x} \\[2mm] \dfrac{dy}{dt} = \dfrac{m_2 x y}{a_2 + x} - y \\[2mm] S(0) = S_0,\ x(0) = x_0,\ y(0) = y_0 \end{cases} \tag{7.12}$$

In system (7.12), y (the predator) consumes x (the prey) and x consumes the nutrient S.

Now let $\Sigma = 1 - S - x - y$. Then $\Sigma' = -S' - x' - y' = -(1 + S + x + y) = -\Sigma$ so $\Sigma = \Sigma_0 e^{-t}$ and $\lim_{t \to \infty} \Sigma = 0$. In the limit as $t \to \infty$, $\Sigma = 0 = 1 - S - x - y$ so $S = 1 - x - y$ and system (7.12) becomes

$$\begin{cases} \dfrac{dx}{dt} = \dfrac{m_1 x(1 - x - y)}{1 + a_1 - x - y} - x - \dfrac{m_2 x y}{a_2 + x} \\[2mm] \dfrac{dy}{dt} = \dfrac{m_2 x y}{a_2 + x} - y \\[2mm] x(0) = x_0,\ y(0) = y_0 \end{cases} \tag{7.13}$$

See Chapter 3 of Smith and
Waltman's *The Theory of the
Chemostat: Dynamics of
Microbial Competition*, [24],
for a detailed analysis of
system (7.12).

The analysis of system (7.12) is quite technical and beyond the scope of this text. We illustrate how Mathematica can assist in carrying out a few of the computations needed when analyzing system (7.12).

The rest points of system (7.12) are found by solving

$$\begin{cases} \dfrac{m_1 x(1 - x - y)}{1 + a_1 - x - y} - x - \dfrac{m_2 x y}{a_2 + x} = 0 \\[2mm] \dfrac{m_2 x y}{a_2 + x} - y = 0. \end{cases}$$

```
xeq = x(m1(1 − x − y)/(1 + a1 − x − y) − 1 − m2y/(a2 + x));

yeq = y(m2x/(a2 + x) − 1);

rps = Solve[{xeq == 0, yeq == 0}, {x, y}]//Simplify
```

$$\left\{\left\{x \to \frac{-1 - a1 + m1}{-1 + m1},\ y \to 0\right\},\ \left\{x \to \frac{a2}{-1 + m2},\ y \to \right.\right.$$

$$\frac{1}{2(-1 + m2)}(-1 - 2a2 + a2m1 + a1(-1 + m2) + m2$$

$$\left.\left.+\sqrt{(1 + a2m1 - m2)^2 + a1^2(-1 + m2)^2 + 2a1(-1 + m2)(-1 + a2m1 + m2)})\right\}\right.,$$

$$\{x \to \frac{a2}{-1+m2}, y \to \frac{1}{2(-1+m2)}(-1-2a2+a2m1+a1(-1+m2)+m2$$

$$-\sqrt{(1+a2m1-m2)^2+a1^2(-1+m2)^2+2a1(-1+m2)(-1+a2m1+m2))}\},$$

$$\{x \to 0, y \to 0\}\}$$

From the results, we see that $E_0 = (0,0)$ is a rest point. If the appropriate quantities are positive, another boundary rest point may exist as well as an interior rest point.

In jac, we compute the Jacobian, **J**, of system (7.12).

```
jac = {{D[xeq, x], D[xeq, y]}, {D[yeq, x], D[yeq, y]}}
```

$$\{\{-1+\frac{m1(1-x-y)}{1+a1-x-y}-\frac{m2y}{a2+x}+x(\frac{m1(1-x-y)}{(1+a1-x-y)^2}-\frac{m1}{1+a1-x-y}+$$
$$\frac{m2y}{(a2+x)^2}), x(-\frac{m2}{a2+x}+\frac{m1(1-x-y)}{(1+a1-x-y)^2}-\frac{m1}{1+a1-x-y})\}, \{(-\frac{m2x}{(a2+x)^2}+$$
$$\frac{m2}{a2+x})y, -1+\frac{m2x}{a2+x}\}\}$$

At E_0, **J** (E_0) is

```
j0 = jac/.rps[[4]]//Apart
```

$$\{\{-1+\frac{m1}{1+a1}, 0\}, \{0, -1\}\}$$

with eigenvalues

```
Eigenvalues[j0]
```

$$\left\{\frac{-1-a1}{1+a1}, \frac{-1-a1+m1}{1+a1}\right\}$$

$$\frac{-1-a1+m1}{-1+m1}//\text{Apart}$$

$$1-\frac{a1}{-1+m1}$$

In the context of the problem, it is desirable for E_0 to be unstable. Thus, we require that $m_1 > 1$ and

$$-1+\frac{m_1}{a_1+1} > 0 \quad \text{or, equivalently,} \quad 1-\frac{a_1}{m_1-1} < 1.$$

With this assumption, the boundary point $E_1 = \left(1-\frac{m_1}{a_1+1}, 0\right) = (1-\lambda_1, 0)$ exists. At E_1, **J** (E_1) is given by

We define λ_i to be $\lambda_i = \frac{a_i}{m_i-1}$.

```
j1 = jac/.rps[[1]]//FullSimplify
```

$$\{\{\frac{(1+a1-m1)(-1+m1)}{a1m1}, (1+a1-m1)(\frac{-1+m1}{a1m1}-\frac{m2}{1+a1+a2-(1+a2)m1})\},$$
$$\{0, -1+\frac{(1+a1-m1)m2}{1+a1+a2-(1+a2)m1}\}\}$$

with eigenvalues

```
Eigenvalues[j1]
```

$$\{\frac{(1+a1-m1)(-1+m1)}{a1m1}, \frac{-1-a1-a2+m1+a2m1+m2+a1m2-m1m2}{1+a1+a2-m1-a2m1}\}$$

E_1 may be stable or unstable. It can be shown that E_1 is stable if $\lambda_1 + \lambda_2 > 1$ and a saddle (unstable) if $\lambda_1 + \lambda_2 < 1$. If an interior rest point exists, Mathematica can usually compute the Jacobian as well as the eigenvalues. At E_A, the $\mathbf{J}(E_A)$ is

```
j3 = jac/.rps[[2]]//FullSimplify
```

$$\{\{-((2(a1^2(-1+m2)^2-a1(-1+m2)(a2m1(m1-2m2)+(-2+m1)(-1+m2)$$

$$+\sqrt{(1+a2m1-m2)^2+a1^2(-1+m2)^2+2a1(-1+m2)(-1+a2m1+m2))}$$

$$-(-1+m1)(1+a2m1-m2)(1+a2m1-m2$$

$$+\sqrt{(1+a2m1-m2)^2+a1^2(-1+m2)^2+2a1(-1+m2)(-1+a2m1+m2)})))/$$

$$(m2(1+a1+a2m1-m2-a1m2+$$

$$\sqrt{(1+a2m1-m2)^2+a1^2(-1+m2)^2+2a1(-1+m2)(-1+a2m1+m2)})^2)),$$

$$-((2(a1^2(-1+m2)^2+a1(-1+m2)(-2+2a2m1+2m2)$$

$$-\sqrt{(1+a2m1-m2)^2+a1^2(-1+m2)^2+2a1(-1+m2)(-1+a2m1+m2)}$$

$$+(1+a2m1-m2)(1+a2m1-m2+$$

$$\sqrt{(1+a2m1-m2)^2+a1^2(-1+m2)^2+2a1(-1+m2)(-1+a2m1+m2)})))$$

$$/(1+a1+a2m1-m2-a1m2$$

$$+\sqrt{(1+a2m1-m2)^2+a1^2(-1+m2)^2+2a1(-1+m2)(-1+a2m1+m2)})^2)\},$$

$$\{\frac{1}{2a2m2}(-1+m2)(a2(-2+m1)+(1+a1)(-1+m2)$$

$$+\sqrt{(1+a2m1-m2)^2+a1^2(-1+m2)^2+2a1(-1+m2)(-1+a2m1+m2)}), 0\}\}$$

The command Eigenvalues[j3] returns the eigenvalues of j3; however, the result is very lengthy so it is not shown here for length considerations. Refer to Chapter 3 of Smith and Waltman, [24].

Incorporating a second predator, z, of x into system (7.14) results in

$$\begin{cases} \dfrac{dS}{dt} = 1 - S - \dfrac{m_1 x S}{a_1 + S} \\[2mm] \dfrac{dx}{dt} = \dfrac{m_1 x S}{a_1 + S} - x - \dfrac{m_2 xy}{a_2 + x} - \dfrac{m_3 xz}{a_3 + x} \\[2mm] \dfrac{dy}{dt} = \dfrac{m_2 xy}{a_2 + x} - y \\[2mm] \dfrac{dz}{dt} = \dfrac{m_3 xz}{a_3 + x} - z \\[2mm] S(0) = S_0,\ x(0) = x_0,\ y(0) = y_0,\ z(0) = z_0 \end{cases} \qquad (7.14)$$

In the same way as with system (7.14), we let $\Sigma = 1 - S - x - y - z$. Then, $\Sigma' = -\Sigma$ so $\lim_{t\to\infty} \Sigma = 0$. Substitution of Σ into system (7.14) and taking the limit $t \to \infty$ results in

$$\begin{cases} \dfrac{dx}{dt} = \dfrac{m_1 x\,(1 - x - y - z)}{1 + a_1 - x - y - z} - x - \dfrac{m_2 xy}{a_2 + x} - \dfrac{m_3 xz}{a_3 + x} \\[2mm] \dfrac{dy}{dt} = \dfrac{m_2 xy}{a_2 + x} - y \\[2mm] \dfrac{dz}{dt} = \dfrac{m_3 xz}{a_3 + x} - z \\[2mm] S(0) = S_0,\ x(0) = x_0,\ y(0) = y_0,\ z(0) = z_0 \end{cases} \qquad (7.15)$$

System (7.15) can exhibit *very* interesting behavior.

EXAMPLE 7.3.3: Let $a_1 = .3$, $a_2 = .4$, $m_1 = 8$, $m_2 = 4.5$, and $m_3 = 5.0$. If $x(0) = .1$, $y(0) = .1$, and $z(0) = .3$, how does varying a_3 affect the solutions of system (7.15)?

SOLUTION: We define the function `predplot`:

```
predplot[{a1,a2,a3},{m1,m2,m3}][{x0,y0,z0},
   {t,a,b},opts]
```

solves system (7.15) subject to the initial conditions $x(0) = x_0$, $y(0) = y_0$, and $z(0) = z_0$ for $a \le t \le b$, plots $x(t)$, $y(t)$, and $z(t)$, parametrically plots x versus y versus z, displays the resulting plots side-by-side, and returns a numerical solution to the initial-value problem. Any options included are passed to the `Show` command. If you do not include any options and omit $\{t,a,b\}$, the default is $0 \le t \le 100$.

```
Clear[predplot];
predplot[{a1_, a2_, a3_}, {m1_, m2_, m3_}][{x0_, y0_, z0_},
  ts_:{t, 0, 100}, opts___]:=
Module[{numsol, p1, p2, p3},
  numsol =
  NDSolve[
  {x'[t] ==
  x[t](m1(1 − x[t] − y[t] − z[t])/(a1 + 1 − x[t] − y[t] − z[t])
    −1 − y[t]m2/(a2 + x[t]) − z[t]m3/(a3 + x[t])),
  y'[t] == y[t](m2x[t]/(a2 + x[t]) − 1),
  z'[t] == z[t](m3x[t]/(a3 + x[t]) − 1),
  x[0] == x0, y[0] == y0, z[0] == z0}, {x[t], y[t], z[t]}, ts,
    MaxSteps → 100000];
  p1 = Plot[Evaluate[{x[t], y[t], z[t]}/.numsol], ts,
    PlotRange → {0, 1}, Ticks → {{ts[[2]], ts[[3]]}, {0, 1}},
    AxesLabel → {t, ""}];
  p2 = ParametricPlot3D[Evaluate[{x[t], y[t], z[t]}/.
    numsol], ts, PlotRange → {{0, 1}, {0, 1}, {0, 1}},
    AxesLabel → {x, y, z}, BoxRatios → {1, 1, 1},
    DisplayFunction → Identity, Ticks → {{0, 1}, {0, 1},
    {0,1}}, ViewPoint->{2.210,2.211,1.294}, Boxed → False,
    PlotPoints → 2000];
  {p1, p2}]
```

For example, entering

```
Show[
  GraphicsRow[predplot[{.3, .4, .455}, {8, 4.5, 5.0}][{.1, .1,
    .3}, {t, 50, 60}]]]
```

plots the solutions shown in Figure 7-15 using our parameter values and initial conditions for $50 \le t \le 60$ if $a_3 = 0.455$. We vary a_3 in t1.

```
t1 = Animate[predplot[{.3, .4, a3}, {8, 4.5, 5.0}][{.3, .1, .2},
  {t, 50, 60}], {a3, .35, .55, .2/29}]
```

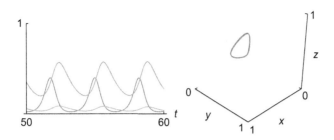

Figure 7-15   If $a_3 = 0.455$, y and z coexist

The resulting graphs result in a striking animation.

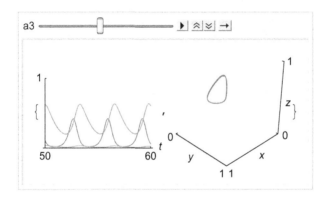

You can also visualize the cycles by displaying all the parametric plots together. In `t1`, each result is an approximate solution that we can use. In the following, we use `ParametricPlot3D` to graph each solution.

```
t1 = Table[predplot[{.3, .4, a3}, {8, 4.5, 5.0}][[{.3, .1, .2},
    {t, 50, 60}], {a3, .35, .55, .2/29}];

Length[t1]
```

30

The results are displayed together with `Show` in Figure 7-16.

```
Show[Table[t1[[i]][[2]], {i, 1, 30}], PlotRange → {{0, 1}, {0, 1},
    {0, 1}}, AxesLabel → {z, y, x}, BoxRatios → {1, 1, 1},
  Ticks → {{0, 1}, {0, 1}, {0, 1}},
  ViewPoint->{1.470, 2.424, 1.847}, Boxed → False]
```

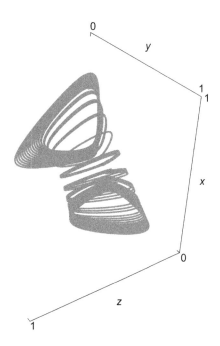

Figure 7-16   In an animation, you can see the limit cycle move from the xz-plane to the xy-plane as a_3 increases

In the plots we see that if a_3 is small, z dominates the predator population, if a_3 is large, y dominates the predator population. For moderate values, y and z coexist.

■

Long Food Chain in a Chemostat

In system (7.12), y predates on x. Incorporating a predator z of y into system (7.12) results in

$$\begin{cases} \dfrac{dS}{dt} = 1 - S - \dfrac{m_1 xS}{a_1 + S} \\[2mm] \dfrac{dx}{dt} = \dfrac{m_1 xS}{a_1 + S} - x - \dfrac{m_2 xy}{a_2 + x} \\[2mm] \dfrac{dy}{dt} = \dfrac{m_2 xy}{a_2 + x} - y - \dfrac{m_3 yz}{a_3 + y} \\[2mm] \dfrac{dz}{dt} = \dfrac{m_3 yz}{a_3 + y} - z \end{cases} . \qquad (7.16)$$

As with system (7.12), in the limit as $t \to \infty$, $S = 1 - x - y - z$ so system (7.16) can be rewritten as

$$\begin{cases} \dfrac{dx}{dt} = x\left[f_1(1 - x - y - z) - 1\right] - yf_2(x) \\[2mm] \dfrac{dy}{dt} = y\left[f_2(x) - 1\right] - zf_3(x) \\[2mm] \dfrac{dz}{dt} = z\left[f_3(y) - 1\right] \end{cases} \qquad (7.17)$$

where

$$f_i(u) = \frac{m_i u}{a_i + u}. \qquad (7.18)$$

Of course, rigorous analysis of system (7.17) is even more complicated than the analysis of system (7.12).

EXAMPLE 7.3.4: Let $a_1 = .08$, $a_2 = .23$, $m_1 = 10$, $m_2 = 4.0$, and $m_3 = 3.5$. If $x(0) = .3$, $y(0) = .1$, and $z(0) = .3$, how does varying a_3 affect the solutions of system (7.17)?

SOLUTION: We define `longchainplot` in the same way as we defined `predplot` in Example 7.3.3.

```
longchainplot[{a1_, a2_, a3_}, {m1_, m2_, m3_}][{x0_, y0_,

   z0_}, ts_:{t, 0, 100}]:=Module[{numsol},

   numsol =

   NDSolve[

   {x'[t] == x[t](m1(1 - x[t] - y[t] - z[t])/(a1 + 1 - x[t] - y[t]

      - z[t]) - 1) - y[t]m2x[t]/(a2 + x[t]),

   y'[t] == y[t](m2x[t]/(a2 + x[t]) - 1) - z[t]m3y[t]/(a3 + y[t]),

   z'[t] == z[t](m3y[t]/(a3 + y[t]) - 1), x[0] == x0, y[0] == y0,

   z[0] == z0}, {x[t], y[t], z[t]}, ts, MaxSteps → 100000];

   p1 = Plot[Evaluate[{x[t], y[t], z[t]}/.numsol], ts,

      PlotRange → {0, 1}, Ticks → {{ts[[2]], ts[[3]]}, {0, 1}},

      AxesLabel → {t, ""}];

   p2 = ParametricPlot3D[Evaluate[{x[t], y[t], z[t]}/.numsol],

      ts, PlotRange → {{0,1},{0,1},{0,1}}, AxesLabel → {x,y,z},

      BoxRatios → {1, 1, 1}, Ticks → {{0, 1}, {0, 1}, {0, 1}},
```

```
ViewPoint->{2.210, 2.211, 1.294}, Boxed → False,

PlotPoints → 2000];

{p1, p2}

]
```

For example, entering

```
Show[GraphicsRow[

longchainplot[{.08, .23, .4}, {10, 4, 3.5}][

{.3, .1, .2}, {t, 50, 60}]]]
```

plots the solution of system (7.17) using our parameter values and initial conditions if $a_3 = 4$ for $50 \leq t \leq 60$ in Figure 7-17. In Figure 7-18, we plot the solutions using the given parameter values

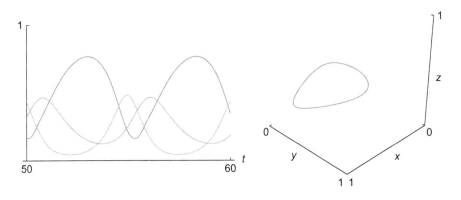

Figure 7-17 $x(t)$, $y(t)$, and $z(t)$ if $a_3 = .4$ for $0 \leq t \leq 60$

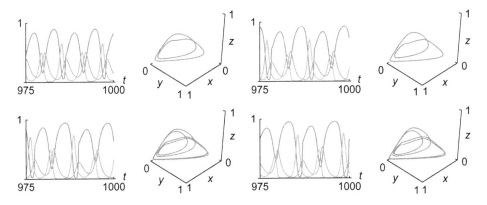

Figure 7-18   Varying a_3

and initial conditions using $a_3 = 0.3, 0.26, 0.24, 0.22$, and 0.2 for $975 \leq t \leq 1000$. In the plots, we see that the solution appears chaotic for $a_3 \approx .2$.

```
r1 = Map[longchainplot[{.08, .23, #}, {10,4,3.5}][{.3,.1,.2},
    {t, 975, 1000}]&, {.3, .26, .24, .22, .2}]//Flatten;

r2 = Partition[r1, 4];

Show[GraphicsGrid[r2]]
```

The apparent chaotic behavior for $a_3 = .2$ is more apparent in Figure 7-19, where we graph the solution for $1100 \leq t \leq 1200$.

```
Show[

  GraphicsRow[longchainplot[{.08, .23, .2}, {10, 4, 3.5}][

    {.3, .1, .2}, {t, 1100, 1200}]]]
```

■

7.3.2 Physical Systems: Variable Damping

In some physical systems, energy is fed into the system when there are small oscillations while energy is taken from the system when there are large oscillations. This indicates that the system undergoes "negative damping" for small oscillations and "positive damping" for large oscillations. A differential equation that models this situation is **Van-der-Pol's equation**.

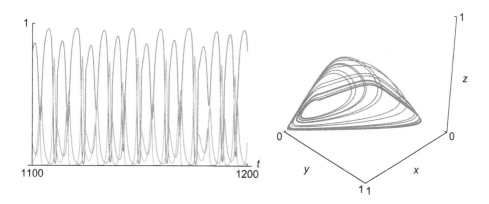

Figure 7-19   The solution appears to be chaotic if $a_3 = .2$

Also see Example 4.1.1.

EXAMPLE 7.3.5: Van-der-Pol's equation In the introduction to Chapter 6, we saw that **Van-der-Pol's equation** $x'' + \mu \left(x^2 - 1\right) x' + x = 0$ is equivalent to the system

$$\begin{cases} x' = y \\ y' = \mu \left(1 - x^2\right) y - x \end{cases}.$$

Classify the equilibrium points, use NDSolve to approximate the solutions to this nonlinear system, and plot the phase plane.

SOLUTION: We find the equilibrium points by solving $\begin{cases} y = 0 \\ \mu \left(1 - x^2\right) y - x = 0 \end{cases}$. From the first equation, we see that $y = 0$. Then, substitution of $y = 0$ into the second equation yields $x = 0$. Therefore, the only equilibrium point is $(0, 0)$. The Jacobian matrix for this system is

$$\mathbf{J}(x, y) = \begin{pmatrix} 0 & 1 \\ -1 - 2\mu xy & -\mu \left(x^2 - 1\right) \end{pmatrix}.$$

The eigenvalues of $\mathbf{J}(0, 0)$ are $\lambda_{1,2} = \frac{1}{2} \left(\mu \pm \sqrt{\mu^2 - 4} \right)$.

```
Clear[f, g]

f[x_, y_] = y;

g[x_, y_] = -x - μ(x^2 - 1)y;

jac = ( D[f[x, y], x]   D[f[x, y], y]
        D[g[x, y], x]   D[g[x, y], y] );

jac/.{x->0, y->0}//Eigenvalues
```

$$\{\frac{1}{2}(\mu - \sqrt{-4 + \mu^2}), \frac{1}{2}(\mu + \sqrt{-4 + \mu^2})\}$$

Notice that if $\mu > 2$, then both eigenvalues are positive and real. Hence, we classify $(0, 0)$ as an **unstable node**. On the other hand, if $0 < \mu < 2$, then the eigenvalues are a complex conjugate pair with a positive real part. Hence, $(0, 0)$ is an **unstable spiral**. (We omit the case $\mu = 2$ because the eigenvalues are repeated.)

We now show several curves in the phase plane that begin at various points for various values of μ. First, we define the function sol, which given μ, x_0, and y_0, generates a numerical solution to the initial-value problem

$$\begin{cases} x' = y \\ y' = \mu \left(1 - x^2\right) y - x . \\ x(0) = x_0 \, y(0) = y_0 \end{cases}$$

and then parametrically graphs the result for $0 \le t \le 20$.

```
Clear[sol]

sol[μ_, {x0_, y0_}, opts___]:=Module[{eqone, eqtwo, solt},

    eqone = x'[t]==y[t];

    eqtwo = y'[t]==μ(1 - x[t]²)y[t] - x[t];

    solt = NDSolve[{eqone, eqtwo, x[0]==x0, y[0]==y0},

        {x[t], y[t]}, {t, 0, 20}];

    ParametricPlot[{x[t], y[t]}/.solt, {t, 0, 20},

    PlotStyle->CMYKColor[0, 0.89, 0.94, 0.28], opts]]
```

We then use `Table` and `Union` to generate a list of ordered pairs `initconds` that will correspond to the initial conditions in the initial-value problem.

```
initconds1 = Table[{0.1Cos[t], 0.1Sin[t]}, {t, 0, 2π, 2π/9}];

initconds2 = Table[{-5, i}, {i, -5, 5, 10/9}];

initconds3 = Table[{5, i}, {i, -5, 5, 10/9}];

initconds4 = Table[{i, 5}, {i, -5, 5, 10/9}];

initconds5 = Table[{i, -5}, {i, -5, 5, 10/9}];


initconds = initconds1 ∪ initconds2 ∪ initconds3∪

    initconds4 ∪ initconds5;
```

Next, we use `Map` to apply `sol` to the list of ordered pairs in `initconds` for $\mu = 1/2$.

```
somegraphs1 = Map[sol[1/2, #, DisplayFunction->Identity]&,

    initconds];

phase1 = Show[somegraphs1, PlotRange → {{-5, 5}, {-5, 5}},

    AspectRatio → 1, Ticks → {{-4, 4}, {-4, 4}},

    PlotLabel → "(a)", AxesLabel → {x, y}]
```

Similarly, we use Map to apply sol to the list of ordered pairs in initconds for $\mu = 1, 3/2$, and 3.

```
somegraphs2 = Map[sol[1, #, DisplayFunction->Identity]&,

    initconds];

phase2 = Show[somegraphs2, PlotRange → {{-5, 5}, {-5, 5}, }

    AspectRatio → 1,

    Ticks → {{-4, 4}, {-4, 4}}, PlotLabel → "(b)",

    AxesLabel → {x, y}]

somegraphs3 = Map[sol[3/2, #, DisplayFunction->Identity]&,

    initconds];

phase3 = Show[somegraphs3, PlotRange → {{-5, 5}, {-5, 5}},

    AspectRatio → 1,

    Ticks → {{-4, 4}, {-4, 4}}, PlotLabel → "(c)",

    AxesLabel → {x, y}]

somegraphs4 = Map[sol[3, #]&, initconds];

phase4 = Show[somegraphs3, PlotRange → {{-5, 5}, {-5, 5}},

    AspectRatio → 1,

    Ticks → {{-4, 4}, {-4, 4}}, PlotLabel → "(d)",

    AxesLabel → {x, y}]
```

We now show all four graphs together in Figure 7-20. In each figure, we see that all of the curves approach a curve called a *limit cycle*. Physically, the fact that the system has a limit cycle indicates that for all oscillations, the motion eventually becomes periodic, which is represented by a closed curve in the phase plane.

```
Show[GraphicsGrid[{{phase1, phase2}, {phase3, phase4}}]]
```

On the other hand, in Figure 7-21 we graph the solutions that satisfies the initial conditions $x(0) = 1$ and $y(0) = 0$ parametrically and individually for various values of μ. Notice that for small values of μ the system more closely approximates that of the harmonic oscillator because the damping coefficient is small. The curves are more circular than those for larger values of μ.

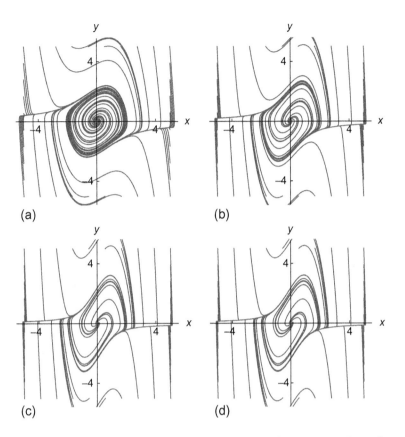

(a)

(b)

(c)

(d)

Figure 7-20   Solutions to the Van-der-Pol equation for various values of μ

```
Clear[x, y, t, s]
graph[μ_]:=Module[{numsol, pp, pxy},
    numsol = NDSolve[{x'[t] == y[t], y'[t] == μ(1 - x[t]²)
        y[t] - x[t], x[0] == 1, y[0] == 0}, {x[t], y[t]}, {t, 0, 20}];
    pp = ParametricPlot[{x[t], y[t]}/.numsol, {t, 0, 20},
    PlotRange → {{-5, 5}, {-5, 5}}, AspectRatio → 1,
    Ticks → {{-4, 4}, {-4, 4}}];
    pxy = Plot[Evaluate[{x[t], y[t]}/.numsol], {t, 0, 20},
    PlotRange → {-5, 5}, AspectRatio → 1,
    Ticks → {{5, 10, 15}, {-4, 4}}];
    {pxy, pp}]
```

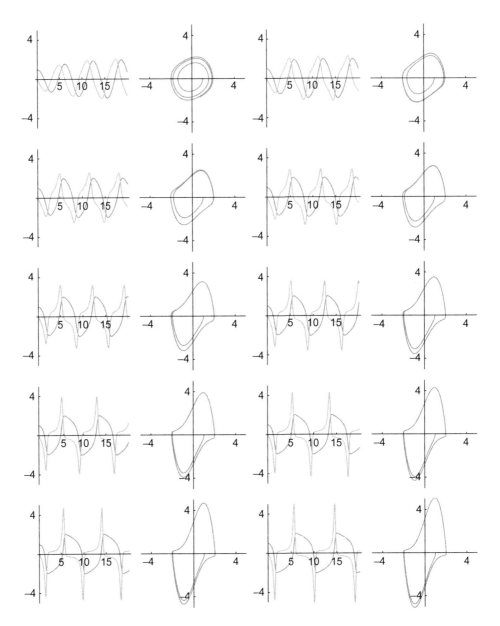

Figure 7-21   The solutions to the Van-der-Pol equation satisfying $x(0) = 1$ and $y(0) = 0$ individually (x in blue and y in orange) for various values of μ

```
graphs = Table[graph[i], {i, 0.25, 3, 2.75/9}]//Flatten;

toshow = Partition[graphs, 4];

Show[GraphicsGrid[toshow]]
```

■

7.3.3 Differential Geometry: Curvature

Let C be a piecewise-smooth curve with parametrization $\mathbf{r}(t) = \langle x(t), y(t) \rangle$, $a \leq t \leq b$. The **unit tangent vector** to C at t is

$$\mathbf{T} = \frac{\mathbf{r}'(t)}{\|\mathbf{r}'(t)\|}. \tag{7.19}$$

The **arc length function**, $s = s(t)$, is defined by

$$s(t) = \int_a^t \|\mathbf{r}'(u)\| \, du = \int_a^t \sqrt{\left(\frac{dx}{du}\right)^2 + \left(\frac{dy}{du}\right)^2} \, du. \tag{7.20}$$

Solving equation (7.20) for t, we have $t = t(s)$ and the **parametrization of C with respect to arc length** is $\mathbf{r}(s) = \langle x(t(s)), y(t(s)) \rangle$. When C is parametrized by arc length, $\|\mathbf{r}'(s)\| = 1$ so the unit tangent vector (7.19) is given by $\mathbf{T}(s) = \mathbf{r}'(s)$. The **curvature** of C, $\kappa(s)$, is

$$\kappa(s) = \left\|\frac{d\mathbf{T}}{ds}\right\|. \tag{7.21}$$

Thus, for the curve C parametrized by arc length, $\kappa(s) = \|\mathbf{r}''(s)\|$.

Conversely, a given curvature function determines a plane curve: the curve C parametrized by arc length with curvature $\kappa(s)$ has parametrization $\mathbf{r}(s) = \langle x(s), y(s) \rangle$ where

$$\begin{cases} dx/ds = \cos\theta \\ dy/ds = \sin\theta \\ d\theta/ds = \kappa \\ x(a) = c, \; y(a) = d, \; \theta(0) = \theta_0 \end{cases} . \tag{7.22}$$

You can often use `NDSolve` to solve system (7.22).

Refer to Gray's outstanding text, *Modern Differential Geometry of Curves and Surfaces*, [14], which incorporates *Mathematica* throughout.

EXAMPLE 7.3.6: Plot the curve C for which $\kappa(s) = e^{-s} + e^{s}$ for $-5 \le s \le 5$ if $x(0) = y(0) = \theta(0) = 0$.

SOLUTION: After defining $\kappa(s) = e^{-s} + e^{s}$,

    $\kappa[s] = \text{Exp}[-s] + \text{Exp}[s]$;

we use NDSolve to solve system (7.22) using the initial conditions $x(0) = y(0) = \theta(0) = 0$ for $-5 \le s \le 5$.

    t1 = NDSolve[{x'[s]==Cos[θ[s]], y'[s] == Sin[θ[s]], θ'[s] == κ[s],

        x[0] == 0, y[0] == 0, θ[0] == 0}, {x[s], y[s], θ[s]}, {s, −5, 5}]

    {{x[s] → InterpolatingFunction[][s], y[s] →

    InterpolatingFunction[][s], θ[s] →

    InterpolatingFunction[][s]}}

We then use ParametricPlot to graph the result in Figure 7-22.

    ParametricPlot[Evaluate[{x[s], y[s]}/.t1], {s, −5, 5},

        PlotStyle->CMYKColor[0, 0.89, 0.94, 0.28],

        AspectRatio → Automatic, PlotRange → All]

■

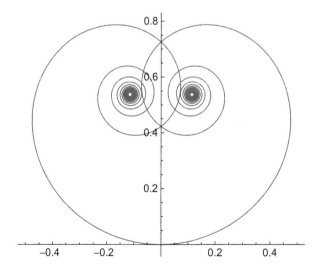

Figure 7-22   For this curve, $\kappa(s) = e^{-s} + e^{s}$

Even relatively simple curvature functions can yield remarkably beautiful curves. To illustrate, we define the function curvek. Given a function $\kappa(s)$, curvek[κ[s], {s, a, b}, opts] solves system (7.22) using the initial conditions $x(0) = y(0) = \theta(0) = 0$ for $a \leq s \leq b$, and parametrically plots the result. Any options are passed to the ParametricPlot command. If you do not include {s, a, b} and do not include any options, the default is $-15 \leq s \leq 15$.

```
Clear[curvek, κ];

curvek[k_, ss_:{s, -15, 15}, opts___]:=
  Module[{numsol}, numsol = NDSolve[{x'[s]==Cos[θ[s]], y'[s] == Sin[θ[s]],
  θ'[s] == k, x[0] == 0, y[0] == 0, θ[0] == 0}, {x[s], y[s], θ[s]}, ss];
  ParametricPlot[Evaluate[{x[s], y[s]}/.numsol], ss,
  PlotStyle->CMYKColor[0, 0.89, 0.94, 0.28], opts,
  AspectRatio → Automatic]
  ]
```

We illustrate the use of curvek using $\kappa(s) = s + \sin s$, $sJ_1(s)$, $sJ_2(s)$, $s\sin(\sin s)$, $s\sin(\sin^2 s^2)$, and $|s\sin(\sin s)|$. All six plots are shown together as an array in Figure 7-23.

```
κ[s_] = s + Sin[s];
p1 = curvek[κ[s], {s, -40, 40}, PlotPoints → 480, AspectRatio → 1];

p2 = curvek[sBesselJ[1, s], {s, -40, 40}, PlotPoints → 120];

p3 = curvek[sBesselJ[2, s], {s, -40, 40}, PlotPoints → 120];

κ[s_] = sSin[Sin[s]];
p4 = curvek[κ[s], {s, -40, 40}, PlotPoints → 480, AspectRatio → 1];

κ[s_] = sSin[Sin[s^2]^2];
p5 = curvek[κ[s], {s, -15, 15}, PlotPoints → 480, AspectRatio → 1,
    PlotRange → {{-3, 3}, {-3, 3}}];

κ[s_] = Abs[sSin[Sin[s]]];
p6 = curvek[κ[s], {s, -40, 40}, PlotPoints → 480, AspectRatio → 1];

Show[GraphicsGrid[{{p1, p2}, {p3, p4}, {p5, p6}}]]
```

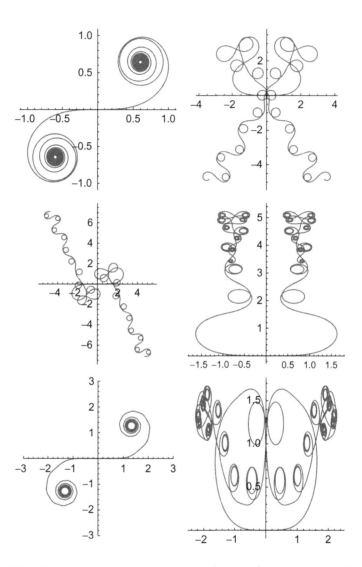

Figure 7-23   You can generate stunning curves by specifying a curvature function

Laplace Transform Methods

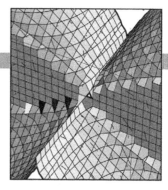

8

In previous chapters we have investigated solving the nth-order linear equation

$$a_n(t)y^{(n)} + a_{n-1}(t)y^{(n-1)} + \cdots + a_2(t)y'' + a_1(t)y' + a_0(t)y = f(t)$$

for y. We have seen that if the coefficients $a_i(t)$ are numbers, we can find a general solution of the equation by first solving the characteristic equation of the corresponding homogeneous equation, forming a general solution of the corresponding homogeneous equation, and then finding a particular solution to the nonhomogeneous equation. If the coefficients $a_i(t)$ are not constants, we have learned that solving above equation may be substantially more difficult that in other cases, such as when the functions $a_i(t)$ are constants. For example, when above equation is a Cauchy-Euler equation, techniques used to solve the case when above equation has constant coefficients can be used to solve the equation. In other situations, we might be able to use a series to find a solution of the equation. Regardless, in all these situations the function $f(t)$ has (typically) been a smooth function. If $f(t)$ is not a smooth function, such as when $f(t)$ is a piecewise-defined or periodic function, solving above equation can be substantially more difficult. In this chapter, we discuss a technique that transforms above equation into an algebraic equation that can sometimes be solved so that a solution to the differential equation can be obtained.

8.1 The Laplace Transform

8.1.1 Definition of the Laplace Transform

Definition 30 (Laplace Transform). *Let $f(t)$ be a function defined on the interval* $[0, \infty)$. *The **Laplace transform** of $f(t)$ is the function (of s)*

$$\mathcal{L}\{f(t)\} = \int_0^\infty e^{-st}f(t)\,dt. \tag{8.21}$$

Pierre-Simon Laplace (1749–1827) is known for his numerous contributions to mathematics, especially theoretical probability, and physics and astronomy, proving the stability of the solar system.

Differential Equations with Mathematica. http://dx.doi.org/10.1016/B978-0-12-804776-7.00008-5

613

The command

$$\texttt{LaplaceTransform[f[t],t,s]}$$

computes the Laplace transform of $f(t)$.

Because the Laplace transform yields a function of s, we often use the notation $\mathcal{L}\{f(t)\} = F(s)$ to denote the Laplace transform of $f(t)$.

EXAMPLE 8.1.1: Compute $\mathcal{L}\{f(t)\}$ if $f(t) = 1$.

SOLUTION: Using the definition, equation (8.21), we have

$$\mathcal{L}\{f(t)\} = \int_0^\infty e^{-st}\, dt = \lim_{M\to\infty} \int_0^M e^{-st}\, dt = \lim_{M\to\infty} \left[-\frac{1}{s} e^{-st} \right]_{t=0}^{t=M}$$

$$= -\frac{1}{s} \lim_{M\to\infty} \left(e^{-sM} - 1 \right) = -\frac{1}{s}(0-1) = \frac{1}{s}, \quad s > 0.$$

Notice that in order for $\lim_{M\to+\infty} e^{-sM} = 0$, we must require that $s > 0$. (Otherwise, the limit does not exist.) We can use Integrate to evaluate this integral as well.

     step1 = Integrate[Exp[−st], {t, 0, capm}]

     $\dfrac{1 - e^{-\text{capm}s}}{s}$

However, Mathematica cannot evaluate $\lim_{M\to\infty} e^{-sM}$ because Mathematica does not assume that $s > 0$ unless we use the Assumptions option to force Mathematica to assume that $s > 0$.

     step2 = Limit[step1, capm → Infinity]

     $\text{Limit}\left[\dfrac{1 - e^{-\text{capm}s}}{s}, \text{capm} \to \infty \right]$

     step2 = Limit[step1, capm → Infinity, Assumptions → s > 0]

     $\dfrac{1}{s}$

Alternatively, we can use Integrate to evaluate the improper integral

     Integrate[Exp[−st], {t, 0, Infinity}]

     $\text{ConditionalExpression}[\dfrac{1}{s}, \text{Re}[s] > 0]$

and understand that Mathematica's result means that the improper integral is $1/s$ for positive real-valued s

```
Integrate[Exp[-st], {t, 0, Infinity}, Assumptions → s > 0]
```

$$\frac{1}{s}$$

or use the command `LaplaceTransform` to compute $\mathcal{L}\{f(t)\}$.

```
LaplaceTransform[1, t, s]
```

$$\frac{1}{s}$$

∎

EXAMPLE 8.1.2: Compute $\mathcal{L}\{f(t)\}$ if $f(t) = e^{at}$.

SOLUTION: As before, we have

$$\mathcal{L}\{f(t)\} = \int_0^\infty e^{-st} f(t)\, dt = \int_0^\infty e^{-st} e^{at}\, dt = \int_0^\infty e^{-(s-a)t}\, dt$$

$$= \lim_{M\to\infty}\left[-\frac{1}{s-a}e^{-(s-a)t}\right]_{t=0}^{t=M} = -\frac{1}{s-a}\lim_{M\to\infty}\left(e^{-(s-a)M} - 1\right)$$

$$= \frac{1}{s-a}, \quad s > 0.$$

Notice that we must require $s > a$ so that $\lim_{M\to\infty} e^{-(s-a)M} = 0$. `LaplaceTransform` can be used to compute the Laplace transform of this function as well.

```
LaplaceTransform[Exp[at], t, s]
```

$$\frac{1}{-a+s}$$

∎

Now that we have proved the formula $\mathcal{L}\{e^{at}\} = \dfrac{1}{s-a}$, it can now be used rather than using the definition.

EXAMPLE 8.1.3: Compute: (a) $\mathcal{L}\{e^{-3t}\}$ and (b) $\mathcal{L}\{e^{5t}\}$.

SOLUTION: We have that (a) $\mathcal{L}\{e^{-3t}\} = \dfrac{1}{s-(-3)} = \dfrac{1}{s+3}, s > -3,$

and (b) $\mathcal{L}\{e^{5t}\} = \dfrac{1}{s-5}, s > 5.$

With Mathematica, we use Map to apply the pure function LaplaceTransform[#,t,s] & to the list of functions $\{e^{-3t}, e^{5t}\}$ to compute both Laplace transforms in a single step.

Map[LaplaceTransform[#, t, s]&, {Exp[−3t], Exp[5t]}]

$$\left\{ \frac{1}{3+s}, \frac{1}{-5+s} \right\}$$

■

In most cases, using the definition of the Laplace transform to calculate the Laplace transform of a function is a difficult and time consuming task.

EXAMPLE 8.1.4: Compute (a) $\mathcal{L}\{t^2\}$; (b) $\mathcal{L}\{\sin at\}$; and (c) $\mathcal{L}\{\cos at\}$.

SOLUTION: To compute $\mathcal{L}\{t^2\}$ by hand requires application of integration by parts three times. Instead, we proceed with Integrate. First we compute $\int_0^M t^3 e^{-st}\, dt$ and then $\int_0^\infty t^2 e^{-st}\, dt = \lim_{M\to\infty} \int_0^M t^2 e^{-st}\, dt$.

Integrate[t^2Exp[−st], {t, 0, capm}]

$$\frac{2 + e^{-\text{capm}s}(-2 - \text{capm}s(2 + \text{capm}s))}{s^3}$$

Observe that Mathematica needs to be reminded that in the context of this discussion s is a positive real number.

Integrate[t^2Exp[−st], {t, 0, Infinity}]

$$\text{ConditionalExpression}\left[\frac{2}{s^3}, \text{Re}[s] > 0\right]$$

Integrate[t^2Exp[−st], {t, 0, Infinity}, Assumptions → s > 0]

$$\frac{2}{s^3}$$

The integrals that result when computing $\mathcal{L}\{\sin at\}$ and $\mathcal{L}\{\cos at\}$ using the definition of the Laplace transform each require the use of Integration by Parts twice. Instead, we use LaplaceTransform to compute each Laplace transform. In the following, Map is used to apply a *pure function* that computes the Laplace transform of a function to the list of functions that consists of the three functions t^2, $\sin at$, and $\cos at$.

Clear[t, s]

Map[LaplaceTransform[#, t, s]&,

{t^2, Sin[at], Cos[at]}]

$$\left\{ \frac{2}{s^3}, \frac{a}{a^2 + s^2}, \frac{s}{a^2 + s^2} \right\}$$

∎

We now discuss the linearity property that enables us to use the transforms that we have found thus far to find the Laplace transform of other functions.

Theorem 18 (Linearity Property). *Let a and b be constants, and suppose that $\mathcal{L}\{f(t)\}$ and $\mathcal{L}\{g(t)\}$ exist. Then,*

$$\mathcal{L}\{af(t) + bg(t)\} = a\mathcal{L}\{f(t)\} + b\mathcal{L}\{g(t)\}.$$

EXAMPLE 8.1.5: Calculate (a) $\mathcal{L}\{6\}$; (b) $\mathcal{L}\{5 - 2e^{-t}\}$.

SOLUTION: Using the results obtained in previous examples, we have for (a)

$$\mathcal{L}\{6\} = 6\mathcal{L}\{1\} = 6 \cdot \frac{1}{s} = \frac{6}{s};$$

and for (b)

$$\mathcal{L}\{5 - 2e^{-t}\} = 5\mathcal{L}\{1\} - 2\mathcal{L}\{e^{-t}\} = 5 \cdot \frac{1}{s} - 2 \cdot \frac{1}{s - (-1)} = \frac{5}{s} - \frac{2}{s+1}.$$

∎

8.1.2 Exponential Order, Jump Discontinuities and Piecewise-Continuous Functions

In our first calculus course, we learn that some improper integrals diverge, which indicates that the Laplace transform may not exist for some functions. Therefore, we present the following definitions and theorems so that we can better understand the types of functions for which the Laplace transform exists.

Definition 31 (Exponential Order). *A function $y = f(t)$ is of **exponential order** b if there are numbers b, M > 0, and T > 0 such that*

$$|f(t)| \le Me^{bt}$$

for t > T.

In the following sections, we will see that the Laplace transform is particularly useful in solving equations involving piecewise or recursively defined functions.

Definition 32 (Jump Discontinuity). *A function $y = f(t)$ has a **jump discontinuity** at $t = c$ on the closed interval $[a, b]$ if the one-sided limits $\lim_{t \to c^+} f(t)$ and $\lim_{t \to c^-} f(t)$ are finite, but unequal, values. The function $y = f(t)$ has a **jump discontinuity** at $t = a$ if $\lim_{t \to a^+} f(t)$ is a finite value different from $f(a)$. The function $y = f(t)$ has a **jump discontinuity** at $t = b$ if $\lim_{t \to b^-} f(t)$ is a finite value different from $f(b)$.*

Definition 33 (Piecewise Continuous). *A function $y = f(t)$ is **piecewise continuous** on the finite interval $[a, b]$ if $y = f(t)$ is continuous at every point in $[a, b]$ except at finitely many points at which $y = f(t)$ has a jump discontinuity.*

  *A function $y = f(t)$ is **piecewise continuous** on $[0, \infty)$ if $y = f(t)$ is piecewise continuous on $[0, N]$ for all N.*

Theorem 19 (Sufficient Condition for Existence of $\mathcal{L}\{f(t)\}$). *Suppose that $y = f(t)$ is a piecewise continuous function on the interval $[0, \infty)$ and that it is of exponential order b for $t > T$. Then, $\mathcal{L}\{f(t)\}$ exists for $s > b$.*

EXAMPLE 8.1.6: Find the Laplace transform of $f(t) = \begin{cases} -1, \ 0 \le t < 4 \\ 1, \ t \ge 4 \end{cases}$.

SOLUTION: Because $y = f(t)$ is a piecewise continuous function on $[0, \infty)$ and of exponential order, $\mathcal{L}\{f(t)\}$ exists. We use the definition and evaluate the integral using the sum of two integrals.

$$\mathcal{L}\{f(t)\} = \int_0^\infty f(t)e^{-st}\, dt = \int_0^4 -1 \cdot e^{-st}\, dt + \int_4^\infty e^{-st}\, dt$$

$$= \left[\frac{1}{s}e^{-st}\right]_{t=0}^{t=4} + \lim_{M \to \infty}\left[-\frac{1}{s}e^{-st}\right]_{t=4}^{t=M}$$

$$= \frac{1}{s}\left(e^{-4s} - 1\right) - \frac{1}{s}\lim_{M \to \infty}\left(e^{-Ms} - e^{-4s}\right) = \frac{1}{s}\left(2e^{-4s} - 1\right).$$

Using Cases (\ ;), we define and graph this piecewise-defined function in Figure 8-1. Because Mathematica uses a "connect the dots" scheme when graphing functions, Mathematica does not automatically detect the discontinuity at $t = 4$. However, when we use the Exclusions option, we instruct Mathematica to avoid $t = 4$.

```
Clear[f]

f[t_]:= - 1/;0 ≤ t ≤ 4

f[t_]:=1/;t > 4
```

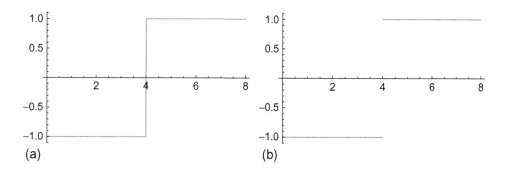

Figure 8-1   Plot of a piecewise-defined function

```
p1 = Plot[f[t], {t, 0, 8},
    PlotLabel → "(a)"]

p2 = Plot[f[t], {t, 0, 8}, Exclusions → {t == 4},
    PlotLabel → "(b)"]

Show[GraphicsRow[{p1, p2}]]
```

However, the LaplaceTransform command is unable to compute the Laplace transform of $y = f(t)$ when f is defined in this manner. To compute the Laplace transform using Mathematica, we take advantage of the UnitStep function, which is defined by

$$\text{UnitStep}[t] = \begin{cases} 0, t < 0 \\ 1, t \geq 0 \end{cases}.$$

Thus, $y = f(t)$ is given by UnitStep[t-4]-UnitStep[4-t]. After defining $y = f(t)$ in this manner, we see that LaplaceTransform is then able to compute $\mathcal{L}\{f(t)\}$.

Note that the commands
UnitStep and
Heaviside have nearly
identical functionality.

```
Clear[f]

f[t_] = UnitStep[t - 4] - UnitStep[4 - t];

LaplaceTransform[f[t], t, s]
```

$$\frac{e^{-4s}}{s} - \frac{1 - e^{-4s}}{s}$$

∎

8.1.3 Properties of the Laplace Transform

The definition of the Laplace transform is not easy to apply to most functions. Therefore, we now discuss several properties of the Laplace transform so that numerous transformations can be made without having to use the definition. Most of the properties discussed here follow directly from our knowledge of integrals.

Theorem 20 (Shifting Property). *If* $\mathcal{L}\{f(t)\} = F(s)$ *exists for* $s > b$*, then*

$$\mathcal{L}\left\{e^{at}f(t)\right\} = F(s - a). \tag{8.22}$$

EXAMPLE 8.1.7: Find the Laplace transform of (a) $f(t) = e^{-2t}\cos t$ and (b) $f(t) = 4te^{3t}$.

SOLUTION: (a) In this case, $f(t) = \cos t$ and $a = -2$. Using $F(s) = \mathcal{L}\{\cos t\} = \dfrac{s}{s^2 + 1}$, we replace each s with $s - a = s + 2$. Therefore,

$$\mathcal{L}\left\{e^{-2t}\cos t\right\} = \frac{s + 2}{(s + 2)^2 + 1} = \frac{s + 2}{s^2 + 4s + 5}.$$

(b) Using the linearity property, we have $\mathcal{L}\left\{4te^{3t}\right\} = 4\mathcal{L}\left\{te^{3t}\right\}$. To apply the Shifting Property we have $f(t) = t$ and $a = 3$, so we replace s in $F(s) = \mathcal{L}\{t\} = s^{-2}$ by $s - a = s - 3$. Therefore,

$$\mathcal{L}\left\{4te^{3t}\right\} = \frac{4}{(s - 3)^2}.$$

Identical results are obtained with `LaplaceTransform`.

`LaplaceTransform[Exp[-2t]Cos[t], t, s]`

$$\frac{2 + s}{1 + (2 + s)^2}$$

`LaplaceTransform[4 t Exp[3 t], t, s]`

$$\frac{4}{(-3 + s)^2}$$

∎

In order to use the Laplace transform to solve differential equations, we will need to be able to compute the Laplace transform of the derivatives of an arbitrary function, provided the Laplace transform of such a function exists.

Theorem 21 (Laplace Transform of the First Derivative). *Suppose that* $y = f(t)$ *is a piecewise continuous function on the interval and that it is of exponential order b for* $t \geq T$. *Then, for* $s > b$

$$\mathcal{L}\{f'(t)\} = s\mathcal{L}\{f(t)\} - f(0). \tag{8.23}$$

```
Clear[f]

LaplaceTransform[f'[t], t, s]

    -f[0] + sLaplaceTransform[f[t], t, s]
```

Using induction, a direct consequence of the theorem is

Theorem 22 (Laplace Transform of the Higher Derivatives). *If* $f^{(i)}(t)$ *is a continuous function on* $[0, \infty)$ *for* $i = 0, 1, \ldots, n-1$ *and* $f^{(n)}(t)$ *is piecewise continuous on* $[0, \infty)$ *and of exponential order b, then for* $s > b$

$$\mathcal{L}\left\{f^{(n)}(t)\right\} = s^n\mathcal{L}\{f(t)\} - s^{n-1}f(0) - \cdots - sf^{(n-2)}(0) - f^{(n-1)}(0). \tag{8.24}$$

We use `LaplaceTransform` to compute the Laplace transform of $f^{(i)}(t)$ for $i = 1, 2, 3, 4, 5$.

```
derivs = Table[D[f[t], {t, n}], {n, 1, 5}]

    {f'[t], f''[t], f^{(3)}[t], f^{(4)}[t], f^{(5)}[t]}


Map[{#, LaplaceTransform[#, t, s]}&, derivs]//TableForm

    f'[t] - f[0] + sLaplaceTransform[f[t], t, s]

    f''[t] - sf[0] + s^2 LaplaceTransform[f[t], t, s] - f'[0]

    f^{(3)}[t] - s^2 f[0] + s^3 LaplaceTransform[f[t], t, s] - sf'[0] - f''[0]

    f^{(4)}[t] - s^3 f[0] + s^4 LaplaceTransform[f[t], t, s] - s^2 f'[0] - sf''[0] - f^{(3)}[0]

    f^{(5)}[t] - s^4 f[0] + s^5 LaplaceTransform[f[t], t, s] - s^3 f'[0] - s^2 f''[0] -
    sf^{(3)}[0] - f^{(4)}[0]
```

We will use this theorem and corollary in solving initial-value problems. However, we can also use them to find the Laplace transform of a function when we know the Laplace transform of the derivative of the function.

EXAMPLE 8.1.8: Find $\mathcal{L}\left\{\sin^2 kt\right\}$.

SOLUTION: We can use the theorem to find the Laplace transform of $f(t) = \sin^2 kt$. Notice that $f'(t) = 2k\sin kt \cos kt = k\sin 2kt$. Then, because $\mathcal{L}\left\{f'(t)\right\} = s\mathcal{L}\left\{f(t)\right\} - f(0)$ and

$$\mathcal{L}\left\{f'(t)\right\} = \mathcal{L}\left\{k\sin 2kt\right\} = k\frac{2k}{s^2 + (2k)^2} = \frac{2k^2}{s^2 + 4k^2},$$

we have $\dfrac{2k^2}{s^2 + 4k^2} = s\mathcal{L}\left\{f(t)\right\} - 0$. Therefore, $\mathcal{L}\left\{f(t)\right\} = \dfrac{2k^2}{s\left(s^2 + 4k^2\right)}$. As in previous examples, we see that the same results are obtained with LaplaceTransform.

> LaplaceTransform[Sin[kt]^2, t, s]

$$\frac{2k^2}{4k^2 s + s^3}$$

■

Theorem 23 (Derivatives of the Laplace Transform). *Suppose that $F(s) = \mathcal{L}\left\{f(t)\right\}$ where $y = f(t)$ is a piecewise continuous function on $[0, \infty)$ and of exponential order b. Then, for $s > b$,*

$$\mathcal{L}\left\{t^n f(t)\right\} = (-1)^n \frac{d^n F}{ds^n}(s). \qquad (8.25)$$

EXAMPLE 8.1.9: Find the Laplace transform of (a) $f(t) = t\cos 2t$ and (b) $f(t) = t^2 e^{-3t}$.

SOLUTION: (a) In this case, $n = 1$ and $F(s) = \mathcal{L}\left\{\cos 2t\right\} = \dfrac{s}{s^2 + 4}$. Then

$$\mathcal{L}\left\{t\cos 2t\right\} = (-1)\frac{d}{ds}\left(\frac{s}{s^2 + 4}\right) = -\frac{(s^2 + 4) - s \cdot 2s}{(s^2 + 4)^2} = \frac{s^2 - 4}{(s^2 + 4)^2}.$$

> LaplaceTransform[tCos[2t], t, s]//Simplify

$$\frac{-4 + s^2}{(4 + s^2)^2}$$

(b) Because $n = 2$ and $F(s) = \mathcal{L}\left\{e^{-3t}\right\} = \dfrac{1}{s+3}$, we have

$$\mathcal{L}\left\{t^2 e^{-3t}\right\} = (-1)^2 \frac{d^2}{ds^2}\left(\frac{1}{s+3}\right) = \frac{2}{(s+3)^2}.$$

```
LaplaceTransform[t^2Exp[-3t], t, s]//Simplify
```

$$\frac{2}{(3+s)^3}$$

∎

EXAMPLE 8.1.10: Find $\mathcal{L}\left\{t^n\right\}$.

SOLUTION: Using the theorem with $\mathcal{L}\left\{t^n\right\} = \mathcal{L}\left\{t^n \cdot 1\right\}$, we have $f(t) = 1$. Then, $F(s) = \mathcal{L}\left\{1\right\} = s^{-1}$. Calculating the derivatives of F, we obtain

$$\frac{dF}{ds}(s) = -\frac{1}{s^2}$$
$$\frac{d^2 F}{ds^2}(s) = \frac{2}{s^3}$$
$$\frac{d^3 F}{ds^3}(s) = -\frac{3 \cdot 2}{s^4}$$
$$\vdots$$
$$\frac{d^n F}{ds^n}(s) = (-1)^n \frac{n! \cdot 2}{s^{n+1}}.$$

Therefore,

$$\mathcal{L}\left\{t^n\right\} = \mathcal{L}\left\{t^n \cdot 1\right\} = (-1)^n (-1)^n \frac{n! \cdot 2}{s^{n+1}} = (-1)^{2n} \frac{n! \cdot 2}{s^{n+1}} = \frac{n! \cdot 2}{s^{n+1}}.$$

Recall that for nonnegative integers n, $\Gamma(n+1) = n!$.

```
LaplaceTransform[t^n, t, s]
```

$$s^{-1-n}\text{Gamma}[1+n]$$

∎

EXAMPLE 8.1.11: Compute the Laplace transform of $f(t)$, $f'(t)$, and $f''(t)$ if $f(t) = (3t - 1)^3$.

SOLUTION: First, $f(t) = (3t - 1)^3 = 27t^3 - 27t^2 + 9t - 1$ and $\mathcal{L}\{t^n\} = \dfrac{n!}{s^{n+1}}$ so

$$\mathcal{L}\{f(t)\} = 27\frac{3!}{s^4} + 27\frac{2!}{s^3} + 9\frac{1}{s^2} - \frac{1}{s} = \frac{1}{s^4}\left(162 - 54s + 9s^2 - s^3\right).$$

```
Clear[f]

f[t_] = (3 t - 1)^3;

lf = LaplaceTransform[f[t], t, s]
```

$$\frac{162}{s^4} - \frac{54}{s^3} + \frac{9}{s^2} - \frac{1}{s}$$

By the previous theorem, $\mathcal{L}\left\{f'(t)\right\} = s\mathcal{L}\{f(t)\} - f(0)$. Hence,

$$\mathcal{L}\left\{f'(t)\right\} = s \cdot \frac{1}{s^4}\left(162 - 54s + 9s^2 - s^3\right) - f(0)$$

$$= \frac{1}{s^3}\left(162 - 54s + 9s^2 - s^3\right) + 1 = \frac{9}{s^3}\left(18 - 6s + s^2\right).$$

```
lfprime = LaplaceTransform[f'[t], t, s]//

  Expand
```

$$\frac{162}{s^3} - \frac{54}{s^2} + \frac{9}{s}$$

```
slf - f[0]//Expand
```

$$\frac{162}{s^3} - \frac{54}{s^2} + \frac{9}{s}$$

Similarly $\mathcal{L}\left\{f''(t)\right\} = s^2\mathcal{L}\{f(t)\} - sf(0) - f'(0)$:

$$\mathcal{L}\left\{f''(t)\right\} = s^2\frac{9}{s^3}\left(18 - 6s + s^2\right) - sf(0) - f'(0) = \frac{54}{s^2}(3 - s).$$

```
lfdoubleprime =

  Expand[LaplaceTransform[f''[t], t, s]]
```

$$\frac{162}{s^2} - \frac{54}{s}$$

```
Expand[s^21f - sf[0] - f'[0]]
```

$$\frac{162}{s^2} - \frac{54}{s}$$

∎

Using the properties of the Laplace transform, we can compute the Laplace transform of a large number of frequently encountered functions. We use Map, LaplaceTransform, and TableForm to compute a table of the Laplace transform of several frequently encountered functions.

```
r1 = {1, Exp[at], Cos[kt], Sin[kt],

Cosh[kt], Sinh[kt], t^n, t^nExp[at],

Exp[at]Cos[kt],

Exp[at]Sin[kt],

Exp[at]Cosh[kt],

Exp[at]Sinh[kt]};
```

```
Map[{#, LaplaceTransform[#, t, s]}&, r1]//TableForm
```

1 $\dfrac{1}{s}$

e^{at} $\dfrac{1}{-a+s}$

$\text{Cos}[kt]$ $\dfrac{s}{k^2+s^2}$

$\text{Sin}[kt]$ $\dfrac{k}{k^2+s^2}$

$\text{Cosh}[kt]$ $\dfrac{s}{-k^2+s^2}$

$\text{Sinh}[kt]$ $\dfrac{k}{-k^2+s^2}$

t^n $s^{-1-n}\text{Gamma}[1+n]$

$e^{at}t^n$ $(-a+s)^{-1-n}\text{Gamma}[1+n]$

$e^{at}\text{Cos}[kt]$ $\dfrac{-a+s}{k^2+(a-s)^2}$

$e^{at}\text{Sin}[kt]$ $\dfrac{k}{k^2+(a-s)^2}$

$e^{at}\text{Cosh}[kt]$ $\dfrac{-a+s}{-k^2+(a-s)^2}$

$e^{at}\text{Sinh}[kt]$ $\dfrac{k}{-k^2+(a-s)^2}$

8.2 The Inverse Laplace Transform

8.2.1 Definition of the Inverse Laplace Transform

In the previous section, we were concerned with finding the Laplace transform of a given function either through the use of the definition of the Laplace transform or with one of the numerous properties of the Laplace transform. At that time, we discussed the sufficient conditions for the existence of the Laplace transform. In this section, we will reverse this process: given a function $F(s)$ we want to find a function $f(t)$ such that $\mathcal{L}\{f(t)\} = F(s)$.

Definition 34 (Inverse Laplace Transform). *The **inverse Laplace transform** of the function $F(s)$ is the unique continuous function $f(t)$ on $[0, \infty)$ that satisfies $\mathcal{L}\{f(t)\} = F(s)$. We denote the inverse Laplace transform of $F(s)$ as*

$$f(t) = \mathcal{L}^{-1}\{F(s)\} \tag{8.26}$$

If the only functions that satisfy this relationship are discontinuous on $[0, \infty)$, we choose a piecewise continuous function on $[0, \infty)$ to be $\mathcal{L}^{-1}\{F(s)\}$.

The table of Laplace transforms listed in the previous section is useful in finding the inverse Laplace transform of a given function. Also, the command

```
InverseLaplaceTransform[F[s],s,t]
```

can often find $\mathcal{L}^{-1}\{F(s)\}$.

EXAMPLE 8.2.1: Find the inverse Laplace transform of (a) $F(s) = \dfrac{1}{s-6}$, (b) $F(s) = \dfrac{2}{s^2+4}$, (c) $F(s) = \dfrac{6}{s^4}$, and (d) $F(s) = \dfrac{6}{(s+2)^4}$.

SOLUTION: (a) Because $\mathcal{L}\{e^{6t}\} = \dfrac{1}{s-6}$, $\mathcal{L}^{-1}\left\{\dfrac{1}{s-6}\right\} = e^{6t}$.

(b) $\mathcal{L}\{\sin 2t\} = \dfrac{2}{s^2+2^2} = \dfrac{2}{s^2+4}$ so $\mathcal{L}^{-1}\left\{\dfrac{2}{s^2+4}\right\} = \sin 2t$. (c) Note that

$\mathcal{L}\{t^3\} = \dfrac{3!}{s^4} = \dfrac{6}{s^4}$ so $\mathcal{L}^{-1}\left\{\dfrac{6}{s^4}\right\} = t^3$. (c) $F(s) = \dfrac{6}{(s+2)^4}$ is obtained from

$F(s) = \dfrac{6}{s^4}$ by substituting $s+2$ for s. Therefore by the shifting property,

$\mathcal{L}\{e^{-2t}t^3\} = \dfrac{6}{(s+2)^4}$, so $\mathcal{L}^{-1}\left\{\dfrac{6}{(s+2)^4}\right\} = e^{-2t}\mathcal{L}^{-1}\left\{\dfrac{6}{s^4}\right\} = e^{-2t}t^3$. In the same way that we use LaplaceTransform to calculate $\mathcal{L}\{f(t)\}$ we use InverseLaplaceTransform to calculate $\mathcal{L}^{-1}\{F(s)\}$.

```
InverseLaplaceTransform[1/(s − 6), s, t]
```

e^{6t}

Here, we use Map to apply the pure function InverseLaplaceTransform [#1,s,t] & to the list of functions $\left\{\dfrac{2}{s^2+4}, \dfrac{6}{s^4}, \dfrac{6}{(s+2)^4}\right\}$.

```
Map[{#1, InverseLaplaceTransform[#1, s, t]}&,
    {2/(s^2 + 4), 6/s^4, 6/(s + 2)^4}]//TableForm
```

$\dfrac{2}{4 + s^2}$	$\mathrm{Sin}[2t]$
$\dfrac{6}{s^4}$	t^3
$\dfrac{6}{(2 + s)^4}$	$e^{-2t}t^3$

■

Theorem 24 (Linearity Property of the Inverse Laplace Transform). *Suppose that $\mathcal{L}^{-1}\{F(s)\}$ and $\mathcal{L}^{-1}\{G(s)\}$ exist and are continuous on $[0,\infty)$. Also, suppose that a and b are constants. Then,*

$$\mathcal{L}^{-1}\{aF(s) + bG(s)\} = a\mathcal{L}^{-1}\{F(s)\} + b\mathcal{L}^{-1}\{G(s)\}. \tag{8.27}$$

EXAMPLE 8.2.2: Find the inverse Laplace transform of (a) $F(s) = \dfrac{1}{s^3}$, (b) $F(s) = -\dfrac{7}{s^2 + 16}$, and (c) $F(s) = \dfrac{5}{s} - \dfrac{2}{s - 10}$.

SOLUTION: (a) $\mathcal{L}^{-1}\left\{\dfrac{1}{s^3}\right\} = \mathcal{L}^{-1}\left\{\dfrac{1}{2}\dfrac{2}{s^3}\right\} = \dfrac{1}{2}\mathcal{L}^{-1}\left\{\dfrac{2}{s^3}\right\} = \dfrac{1}{2}t^2.$

```
InverseLaplaceTransform[1/s^3, s, t]
```

$\dfrac{t^2}{2}$

(b) $\mathcal{L}^{-1}\left\{-\dfrac{7}{s^2+16}\right\}$ $= -7\mathcal{L}^{-1}\left\{\dfrac{1}{s^2+16}\right\}$ $= -7\mathcal{L}^{-1}\left\{\dfrac{1}{4}\dfrac{4}{s^2+4^2}\right\}$

$\qquad\qquad = -\dfrac{7}{4}\mathcal{L}^{-1}\left\{\dfrac{4}{s^2+4^2}\right\} = -\dfrac{7}{4}\sin 4t.$

InverseLaplaceTransform[-7/(s^2 + 16), s, t]

$-\dfrac{7}{4}\mathrm{Sin}[4t]$

(c) $\mathcal{L}^{-1}\left\{\dfrac{5}{s}-\dfrac{2}{s-10}\right\} = 5\mathcal{L}^{-1}\left\{\dfrac{1}{s}\right\} - 2\mathcal{L}^{-1}\left\{\dfrac{1}{s-10}\right\} = 5 - 2e^{10t}.$

InverseLaplaceTransform[5/s - 2/(s - 10), s, t]

$5 - 2e^{10t}$

■

Of course, the functions $F(s)$ that are encountered do not have to be of the forms previously discussed. For example, sometimes we must complete the square in the denominator of $F(s)$ before finding $\mathcal{L}^{-1}\{F(s)\}$.

EXAMPLE 8.2.3: Determine $\mathcal{L}^{-1}\left\{\dfrac{s}{s^2+2s+5}\right\}$.

SOLUTION: Notice that all of the forms of $F(s)$ in the table of Laplace transforms involve a term of the form $s^2 + k^2$ in the denominator. However, through shifting, this term is replaced by $(s - a)^2 + k^2$. We obtain a term of this form in the denominator by completing the square. This yields

$$\frac{s}{s^2+2s+5} = \frac{s}{(s^2+2s+1)+4} = \frac{s}{(s+1)^2+4}.$$

Because the variable appears in the numerator, we must write it in the form $s + 1$ in order to find the inverse Laplace transform. Doing so, we find that

$$\frac{s}{(s+1)^2+4} = \frac{(s+1)-1}{(s+1)^2+4}.$$

Hence,

$$\mathcal{L}^{-1}\left\{\frac{s}{s^2 + 2s + 5}\right\} = \mathcal{L}^{-1}\left\{\frac{(s+1)-1}{(s+1)^2 + 4}\right\}$$

$$= \mathcal{L}^{-1}\left\{\frac{s+1}{(s+1)^2 + 2^2}\right\} - \frac{1}{2}\mathcal{L}^{-1}\left\{\frac{2}{(s+1)^2 + 2^2}\right\}$$

$$= e^{-t}\cos 2t - \frac{1}{2}e^{-t}\sin 2t.$$

As in previous examples, we see that `InverseLaplaceTransform` quickly finds $\mathcal{L}^{-1}\left\{\dfrac{s}{s^2 + 2s + 5}\right\}$.

```
s1 = InverseLaplaceTransform[s/(s^2 + 2s + 5), s, t]
```

$$\frac{1}{4}e^{(-1-2i)t}((2-i) + (2+i)e^{4it})$$

Notice that Mathematica does not automatically simplify the complex exponentials to real valued functions. To do so, we use `ExpToTrig` followed by `Simplify` or `FullSimplify` to convert the exponential functions to trigonometric functions.

```
ExpToTrig[s1]//Simplify
```

$$\frac{1}{2}(2\text{Cos}[2t] - \text{Sin}[2t])(\text{Cosh}[t] - \text{Sinh}[t])$$

```
ExpToTrig[s1]//FullSimplify
```

$$\frac{1}{2}e^{-t}(2\text{Cos}[2t] - \text{Sin}[2t])$$

■

In other cases, partial fractions must be used to obtain terms for which the inverse Laplace transform can be found. Suppose that $F(s) = P(s)/Q(s)$, where $P(s)$ and $Q(s)$ are polynomials of degree m and n, respectively. If $n > m$, the method of partial fractions can be used to expand $F(s)$. Recall from calculus, that there are many possible situations that can be solved through partial fractions. We illustrate three cases in the examples that follow.

We assume that $F(s)$ is reduced to lowest terms.

Linear Factors (Nonrepeated)
In this case, $Q(s)$ can be written as a product of linear factors, so

$$Q(s) = (s - q_1)(s - q_2)\cdots(s - q_n),$$

where $q_1, q_2, \ldots, q_n$ are distinct numbers. Therefore, $F(s)$ can be written as

$$F(s) = \frac{A_1}{s - q_1} + \frac{A_2}{s - q_2} + \cdots + \frac{A_n}{s - q_n},$$

where $A_1, A_2, \ldots, A_n$ are constants that must be determined.

EXAMPLE 8.2.4: Find $\mathcal{L}^{-1}\left\{\dfrac{3s - 4}{s(s - 4)}\right\}$.

SOLUTION: In this case, we have distinct linear factors in the denominator. Hence, we write $F(s)$ as

$$\frac{3s - 4}{s(s - 4)} = \frac{A}{s} + \frac{B}{s - 4}.$$

Multiplying both sides of this equation by the lowest common denominator $s(s - 4)$, we have

$$3s - 4 = A(s - 4) + Bs = (A + B)s - 4A.$$

The set $\{s, 1\}$ is linearly independent.

Equating the coefficients of s as well as the constant terms, we see that the system of equations

$$\begin{cases} A + B = 3 \\ -4A = -4 \end{cases}$$

must be satisfied. Mathematica can solve this system of equations with `Solve` or we can solve the equation $3s - 4 = A(s-4) + Bs = (A+B)s - 4A$ for A and B with `SolveAlways`.

> `SolveAlways[3 s - 4 == (a + b)s - 4a, s]`

> $\{\{a \rightarrow 1, b \rightarrow 2\}\}$

Hence, $A = 1$ and $B = 2$. Therefore,

$$\frac{3s - 4}{s(s - 4)} = \frac{1}{s} + \frac{2}{s - 4},$$

so

$$\mathcal{L}^{-1}\left\{\frac{3s - 4}{s(s - 4)}\right\} = \mathcal{L}^{-1}\left\{\frac{1}{s} + \frac{2}{s - 4}\right\} = 1 + 2e^{4t}$$

or we can use `InverseLaplaceTransform` in the same way as illustrated in the previous examples.

> `InverseLaplaceTransform[(3s − 4)/(s(s − 4)), s, t]`

> $1 + 2e^{4t}$

> `Apart[(3s − 4)/(s(s − 4))]`

> $\dfrac{2}{-4 + s} + \dfrac{1}{s}$

Note that we can compute the partial fraction decomposition of $\dfrac{3s - 4}{s(s - 4)}$ with `Apart`

> `Apart[(5s^2 + 20s + 6)/(s^3 + 2s^2 + s)]`

> $\dfrac{6}{s} + \dfrac{9}{(1 + s)^2} - \dfrac{1}{1 + s}$

`Apart[f[x]]` computes the partial fraction decomposition of the rational function $f(x)$.

or just use `InverseLaplaceTransform` to compute the inverse Laplace transform directly.

> `InverseLaplaceTransform[(5s^2 + 20s + 6)/`

> `(s^3 + 2s^2 + s), s, t]`

> $e^{-t}(-1 + 6e^t + 9t)$

■

Repeated Linear Factors

If $s - q$ is a factor of $Q(s)$ of multiplicity k, the terms in the partial fraction expansion of $F(s)$ that correspond to this factor are

$$\frac{A_1}{s - q} + \frac{A_2}{(s - q)^2} + \cdots + \frac{A_k}{(s - q)^k},$$

where $A_1, A_2, \ldots, A_k$ are constants that must be found.

EXAMPLE 8.2.5: Calculate $\mathcal{L}^{-1} \left\{ \dfrac{5s^2 + 20s + 6}{s^3 + 2s^2 + s} \right\}$.

SOLUTION: After using `Apart`,

`Apart[(2s^3 - 4s - 8)/((s^2 - s)(s^2 + 4))]`

$$-\frac{2}{-1+s} + \frac{2}{s} + \frac{2(2+s)}{4+s^2}$$

we see that

$$\frac{5s^2 + 20s + 6}{s^3 + 2s^2 + s} = \frac{6}{s} - \frac{1}{s+1} + \frac{9}{(s+1)^2}.$$

Therefore,

$$\mathcal{L}^{-1}\left\{\frac{5s^2 + 20s + 6}{s^3 + 2s^2 + s}\right\} = \mathcal{L}^{-1}\left\{\frac{6}{s} - \frac{1}{s+1} + \frac{9}{(s+1)^2}\right\}$$

$$= \mathcal{L}^{-1}\left\{\frac{6}{s} - \frac{1}{s+1} + 9\frac{1}{(s+1)^2}\right\}$$

$$= 6 - e^{-t} + 9te^{-t}.$$

As expected, we obtain the same results using `InverseLaplaceTransform`.

`InverseLaplaceTransform[(2s^3 - 4s - 8)/`

`((s^2 - s)(s^2 + 4)), s, t]`

$2 - 2e^t + 2(\text{Cos}[2t] + \text{Sin}[2t])$

■

Irreducible Quadratic Factors

If $(s - a)^2 + b^2$ is a factor of $Q(s)$ of multiplicity k that cannot be reduced to linear factors, the partial fraction expansion of $F(s)$ corresponding to $(s - a)^2 + b^2$ is

$$\frac{A_1 s + B_1}{(s - a)^2 + b^2} + \frac{A_2 s + B_2}{\left[(s - a)^2 + b^2\right]^2} + \cdots + \frac{A_k s + B_k}{\left[(s - a)^2 + b^2\right]^k}.$$

EXAMPLE 8.2.6: Find $\mathcal{L}^{-1}\left\{\dfrac{2s^3 - 4s - 8}{(s^2 - s)\left(s^2 + 4\right)}\right\}$.

SOLUTION: As in the previous example, we use `Apart`

`Apart[(2s^3 - 4s - 8)/((s^2 - s)(s^2 + 4))]`

$$-\frac{2}{-1+s} + \frac{2}{s} + \frac{2(2+s)}{4+s^2}$$

to obtain the partial fraction decomposition. Thus,

$$\mathcal{L}^{-1}\left\{\frac{2s^3 - 4s - 8}{(s^2 - s)(s^2 + 4)}\right\} = 2\mathcal{L}^{-1}\left\{\frac{1}{s}\right\} - 2\mathcal{L}^{-1}\left\{\frac{1}{s-1}\right\}$$

$$+ 2\mathcal{L}^{-1}\left\{\frac{s}{s^2 + 4}\right\} + 2\mathcal{L}^{-1}\left\{\frac{2}{s^2 + 4}\right\}$$

$$= 2 - 2e^t + 2\cos 2t + 2\sin 2t.$$

```
InverseLaplaceTransform[(2s^3 - 4s - 8)/((s^2 - s)(s^2 + 4)), s, t]
```

$$2 - 2e^t + 2(\text{Cos}[2t] + \text{Sin}[2t])$$

■

8.2.2 Laplace Transform of an Integral

We have seen that the Laplace transform of the derivatives of a given function can be found from the Laplace transform of the function. Similarly, the Laplace transform of the integral of a given function can also be obtained from the Laplace transform of the original function.

Theorem 25 (Laplace Transform of an Integral). *Suppose that $F(s) = \mathcal{L}\{f(t)\}$ where $y = f(t)$ is a piecewise continuous function on $[0, \infty)$ and of exponential order b. Then, for $s > b$,*

$$\mathcal{L}\left\{\int_0^t f(\alpha)\,d\alpha\right\} = \frac{1}{s}\mathcal{L}\{f(t)\}. \tag{8.28}$$

The theorem tells us that

$$\mathcal{L}^{-1}\left\{\frac{1}{s}\mathcal{L}\{f(t)\}\right\} = \int_0^t f(\alpha)\,d\alpha. \tag{8.29}$$

EXAMPLE 8.2.7: Compute $\mathcal{L}^{-1}\left\{\dfrac{1}{s(s+2)}\right\}$.

SOLUTION: In this case, $\dfrac{1}{s(s+2)} = \dfrac{1}{s}\dfrac{1}{s+2}$, so $\mathcal{L}\{f(t)\} = \dfrac{1}{s+2}$.

Therefore, $f(t) = \mathcal{L}^{-1}\left\{\dfrac{1}{s+2}\right\} = e^{-2t}$. Using the theorem, we then have

$$\mathcal{L}^{-1}\left\{\frac{1}{s(s+2)}\right\} = \int_0^t e^{-2\alpha}\,d\alpha = \frac{1}{2}\left(1 - e^{-2t}\right).$$

Note that the same result is obtained with `InverseLaplaceTransform`

> `InverseLaplaceTransform[1/(s(s+2)), s, t]`

$$\frac{1}{2} - \frac{e^{-2t}}{2}$$

or through a partial fraction expansion of $\dfrac{1}{s(s+2)}$: $\dfrac{1}{s(s+2)} = \dfrac{1}{2s} -$

$\dfrac{1}{2(s+2)}$, $\mathcal{L}^{-1}\left\{\dfrac{1}{s(s+2)}\right\} = \mathcal{L}^{-1}\left\{\dfrac{1}{2s} - \dfrac{1}{2(s+2)}\right\} = \dfrac{1}{2} - \dfrac{1}{2}e^{-2t}.$

∎

The following theorem is useful in determining if the inverse Laplace transform of a function $F(s)$ exists.

Theorem 26. *Suppose that* $y = f(t)$ *is a piecewise continuous function on* $[0, \infty)$ *and of exponential order b. Then,*

$$\lim_{s\to\infty} F(s) = \lim_{s\to\infty} \mathcal{L}\{f(t)\} = 0.$$

EXAMPLE 8.2.8: Determine if the inverse Laplace transform of the functions may exist for the functions (a) $F(s) = \dfrac{2s}{s-6}$, and

(b) $F(s) = \dfrac{s^3}{s^2 + 16}$.

SOLUTION: In both cases, we find $\lim_{s\to\infty} F(s)$. If this value is not zero, then $\mathcal{L}^{-1}\{F(s)\}$ cannot be found. (a) $\lim_{s\to\infty} F(s) = \lim_{s\to\infty}\dfrac{2s}{s-6} = 2 \neq 0$, so $\mathcal{L}^{-1}\left\{\dfrac{2s}{s-6}\right\}$ does not exist. (b) $\lim_{s\to\infty} F(s) = \lim_{s\to\infty}\dfrac{s^3}{s^2+16} = \infty \neq 0$. Thus, $\mathcal{L}^{-1}\left\{\dfrac{s^3}{s^2+16}\right\}$ does not exist.

However, in a more abstract setting, the Laplace transform may exist. We will discuss the `DiracDelta` function in Section 8.4.

> `InverseLaplaceTransform[2s/(s - 6), s, t]`

$2(6e^{6t} + \text{DiracDelta}[t])$

```
InverseLaplaceTransform[s^3/(s^2+16), s, t]
```

$-16\text{Cos}[4\,t] + \text{DiracDelta}'[t]$

■

8.3 Solving Initial-Value Problems With the Laplace Transform

Laplace transforms can be used to solve certain initial-value problems. Usually, you will probably find Laplace transform methods most useful when dealing with initial-value problems that involve a discontinuous or piecewise-defined forcing function. Typically, when we use Laplace transforms to solve an initial value problem for a function y, we do the following.

1. Compute the Laplace transform of each term in the differential equation.
2. Solve the resulting equation for $\mathcal{L}\{y(t)\}$.
3. Determine y by computing the inverse Laplace transform of $\mathcal{L}\{y(t)\}$.

The advantage of this method is that through the use of the property

$$\mathcal{L}\left\{f^{(n)}(t)\right\} = s^n \mathcal{L}\{f(t)\} - s^{n-1}f(0) - \cdots - sf^{(n-2)}(0) - f^{(n-1)}(0)$$

we transform a linear differential equation to an algebraic equation.

EXAMPLE 8.3.1: Solve the initial-value problem $y' - 4y = e^{4t}$, $y(0) = 0$.

SOLUTION: We begin by taking the Laplace transform of the both sides of the differential equation and then solving for $\mathcal{L}\{y(t)\} = Y(s)$. Because $\mathcal{L}\{y'(t)\} = sY(s) - y(0) = sY(s)$, we have

$$\mathcal{L}\left\{y' - 4y\right\} = \mathcal{L}\left\{e^{4t}\right\}$$

$$\mathcal{L}\left\{y'\right\} - 4\mathcal{L}\left\{y\right\} = \frac{1}{s-4}$$

$$sY(s) - 4Y(s) = \frac{1}{s-4}$$

$$(s-4)Y(s) = \frac{1}{s-4}$$

$$Y(s) = \frac{1}{(s-4)^2}.$$

We carry out the same steps with Mathematica. After computing the Laplace transform of each side of the equation,

> `step1 = LaplaceTransform[y'[t] − 4y[t] == Exp[4 t], t, s]`

> $-4\text{LaplaceTransform}[y[t],\ t,\ s] + s\text{LaplaceTransform}[y[t],$
> $t,\ s] - y[0] == \dfrac{1}{-4+s}$

we apply the initial condition

> `step2 = step1/.y[0] → 0`

> $-4\text{LaplaceTransform}[y[t],\ t,\ s] + s\text{LaplaceTransform}[y[t],$
> $t,\ s] == \dfrac{1}{-4+s}$

and solve the result equation for $\mathcal{L}\{y(t)\} = Y(s)$.

> `step3 = Solve[step2, LaplaceTransform[y[t], t, s]]`

> $\{\{\text{LaplaceTransform}[y[t],\ t,\ s] \to \dfrac{1}{(-4+s)^2}\}\}$

Hence, by using the shifting property with $\mathcal{L}\{t\} = s^{-2}$, we have

$$y(t) = \mathcal{L}^{-1}\left\{\frac{1}{(s-4)^2}\right\} = te^{4t}.$$

Identical results are obtained using `InverseLaplaceTransform`.

> `sol = InverseLaplaceTransform[step3[[1, 1, 2]], s, t]`

> $e^{4t}t$

We then graph the solution with `Plot` in Figure 8-2.

> `Plot[sol, {t, 0, 1}, AxesLabel → {t, y}]`

We can also use `DSolve` to solve the initial-value problem directly.

> `DSolve[{y'[t] − 4y[t] == Exp[4 t], y[0] == 0},`
>
> `  y[t], t]`

> $\{\{y[t] \to e^{4t}t\}\}$

■

As we can see, Laplace transforms are useful in solving nonhomogeneous equations. Hence, problems in Chapter 4 for which the methods of Undetermined

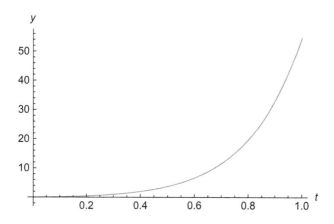

Figure 8-2   In the plot, we see that the initial condition is satisfied

Coefficients or Variation of Parameters were difficult to apply may be more easily solved through the method of Laplace transforms.

EXAMPLE 8.3.2: Use Laplace transforms to solve $y'' + 4y = e^{-t} \cos 2t$ subject to $y(0) = 0$ and $y'(0) = -1$.

SOLUTION: We proceed by computing the Laplace transform of each side of the equation with LaplaceTransform

```
Clear[t, y]

step1 = LaplaceTransform[
  y"[t] + 4y[t] == Exp[-t]Cos[2 t], t, s]
```

$4\text{LaplaceTransform}[y[t], t, s] + s^2\text{LaplaceTransform}[y[t],$

$t, s] - sy[0] - y'[0] == \dfrac{1+s}{4 + (1+s)^2}$

and then applying the initial conditions $y(0) = 0$ and $y'(0) = -1$ with ReplaceAll (/.), naming the result step2.

```
step2 = step1/.{y[0] → 0, y'[0] → -1}
```

$1 + 4\text{LaplaceTransform}[y[t], t, s] + s^2\text{LaplaceTransform}[y[t],$

$t, s] == \dfrac{1+s}{4 + (1+s)^2}$

Next, we solve step2 for the Laplace transform of $y(t)$ and simplify the result, naming the resulting output step3

step3 = Solve[step2, LaplaceTransform[y[t], t, s]]

$$\{\{\text{LaplaceTransform}[y[t], t, s] \to \frac{-4 - s - s^2}{(4 + s^2)(5 + 2s + s^2)}\}\}$$

The formula for the inverse Laplace transform is the first part of the first part of the second part of step3.

step3[[1, 1, 2]]

$$\frac{-4 - s - s^2}{(4 + s^2)(5 + 2s + s^2)}$$

We use InverseLaplaceTransform to compute the inverse Laplace transform of step3, naming the result sol.

sol = Simplify[InverseLaplaceTransform[step3

[[1,1,2]], s, t]]

$$\frac{1}{34}(e^{(-1-2i)t}((1 - 4i) + (1 + 4i)e^{4it}) - 2\text{Cos}[2t] - 8\text{Sin}[2t])$$

Because Mathematica does not automatically simplify the portions of the solution involving complex exponentials, we use FullSimplify to obtain the desired result.

sol = FullSimplify[InverseLaplaceTransform[step3

[[1, 1, 2]], s, t]]

$$\frac{1}{34}e^{-t}(-2(-1 + e^t)\text{Cos}[2t] - 8(1 + e^t)\text{Sin}[2t])$$

Last, we use Plot to graph the solution obtained in sol on the interval $[0, 2\pi]$ in Figure 8-3.

As we have seen in many previous examples, DSolve is able to solve the initial-value problem as well.

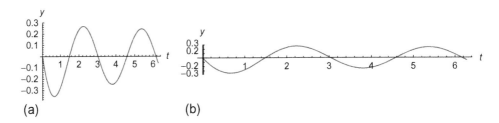

(a)   (b)

Figure 8-3   In the plot of the solution, we see that the initial conditions are satisfied. In (a), Mathematica chooses the aspect ratio. In (b), the plot is scaled $1 - 1$

```
p1 = Plot[sol, {t, 0, 2Pi},
    PlotStyle->CMYKColor[0, 0.89, 0.94, 0.28],
    AxesLabel → {t, y}, PlotLabel → "(a)"]
```

The default aspect ratio is shown in (a). To see the graph plotted to scale, we use the option `AspectRatio->Automatic` in (b).

```
p2 = Plot[sol, {t, 0, 2Pi},
    PlotStyle->CMYKColor[0, 0.89, 0.94, 0.28],
    AxesLabel → {t, y}, PlotLabel → "(b)",
    AspectRatio → Automatic]

Show[GraphicsRow[{p1, p2}]]
```

As in the previous examples, we show that the same results are obtained with `DSolve`.

```
sol = DSolve[{y"[t] + 4y[t] == Exp[-t]Cos[2t],
    y[0] == 0, y'[0] == -1}, y[t], t]//FullSimplify
```

$$\{\{y[t] \to -\frac{1}{17} e^{-t}((-1 + e^t)\text{Cos}[2t] + 4(1 + e^t)\text{Sin}[2t])\}\}$$

∎

Higher-order initial-value problems can be solved with Laplace transforms methods as well.

EXAMPLE 8.3.3: Solve $y''' + y'' - 6y' = \sin 4t$, $y(0) = 2$, $y'(0) = 0$, $y''(0) = -1$.

SOLUTION: We first note that `DSolve` is able to quickly find an explicit solution of the initial-value problem.

```
sol = DSolve[{y^(3)[t] + y"[t] - 6y'[t]==Sin[4t], y[0]==2,
    y'[0]==0, y"[0]== - 1}, y[t], t]

step1 = LaplaceTransform[y^(3)[t] + y"[t] - 6y'[t]==Sin[4t],
    t, s]
```

A graph of the solution is generated with `Plot` in Figure 8-4.

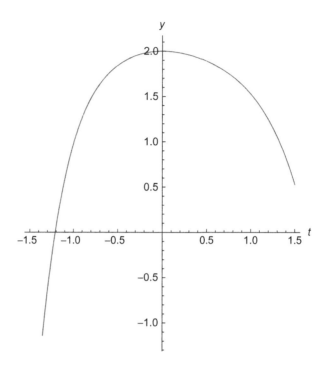

Figure 8-4   Plot of the solution to a third-order initial-value problem

```
Plot[sol, {t, - 3/2, 3/2}, .

    PlotStyle->CMYKColor[0, 0.89, 0.94, 0.28],

    AxesLabel → {t, y},

    AspectRatio → Automatic]
```

Alternatively, we can use Mathematica to implement the steps encountered when solving the equation using the method of Laplace transforms, as in the previous two examples. Taking the Laplace transform of both sides of the equation, we find

```
step1 = LaplaceTransform[y″′[t]

    +y″[t] - 6y′[t] == Sin[4t], t, s]
```

s^2LaplaceTransform$[y[t], t, s] + s^3$LaplaceTransform$[y[t],$

$t, s] - 6(s$LaplaceTransform$[y[t], t, s] - y[0]) - sy[0] - s^2 y[0] -$

$y'[0] - sy'[0] - y''[0] == \dfrac{4}{16 + s^2}$

and then we apply the initial conditions, naming the result step2.

> step2 = step1/.{y[0] → 2, y′[0] → 0,
>
> y″[0] → −1}

$1 - 2s - 2s^2 + s^2$ LaplaceTransform[y[t], t, s] + s^3

LaplaceTransform[y[t], t, s]−6(−2+sLaplaceTransform[y[t],

t, s]) == $\dfrac{4}{16 + s^2}$

Solving for $Y(s)$, we obtain

> step3 = Solve[step2, LaplaceTransform[y[t], t, s]]

$$\{\{\text{LaplaceTransform}[y[t],t,s] \to \frac{-204 + 32s + 19s^2 + 2s^3 + 2s^4}{(16 + s^2)(-6s + s^2 + s^3)}\}\}$$

and computing the inverse Laplace transform of step3 with InverseLaplaceTransform yields the solution to the initial-value problem.

> sol = InverseLaplaceTransform[step3[[1, 1, 2]], s, t]//Simplify

$$\frac{2125 - 56e^{-3t} - 80e^{2t} + 11\text{Cos}[4t] - 2\text{Sin}[4t]}{1000}$$

∎

Some initial-value problems that involve differential equations with nonconstant coefficients can also be solved with the Method of Laplace transforms. However, Laplace transforms do not provide a general method for solving equations with nonconstant coefficients.

EXAMPLE 8.3.4: Solve $\begin{cases} y'' + ty' - 4y = 2 \\ y(0) = y'(0) = 0 \end{cases}$.

SOLUTION: DSolve is able to solve this equation.

> sol = DSolve[{y″[t] + 2 ty′[t] − 4y[t]==2,
>
> y[0]==0, y′[0]==0}, y[t], t]

$\{\{y[t] \to t^2\}\}$

Using the method of Laplace transforms, we take the Laplace transform of both sides of the equation.

```
step1 = LaplaceTransform[y''[t] + 2 t y'[t] - 4 y[t]==2, t, s]
```

-4LaplaceTransform$[y[t], t, s] + s^2$LaplaceTransform$[y[t],$

$t, s] + 2$LaplaceTransform$[ty'[t], t, s] - sy[0] - y'[0] == \dfrac{2}{s}$

Next, we apply the initial conditions.

```
step2 = step1/.{y[0]->0, y'[0]->0}
```

-4LaplaceTransform$[y[t], t, s] + s^2$LaplaceTransform$[y[t],$

$t, s] + 2$

LaplaceTransform$[ty'[t], t, s] == \dfrac{2}{s}$

This is a first-order linear equation that we are able to solve with DSolve. First, in step3, we replace LaplaceTransform[y[t],t,s] with capy[s], which represents $Y(s)$, and LaplaceTransform$^{(0,0,1)}$[y[t], t, s] with capy'[s], which represents $Y'(s)$. Then in step4 we use DSolve to solve for capy[s].

```
step3 =
step2/.{LaplaceTransform[y[t], t, s]->capy[s],
    LaplaceTransform^(0,0,1)[y[t], t, s]->capy'[s]}
```

-4capy$[s] + s^2$capy$[s] + 2$LaplaceTransform$[ty'[t], t, s] == \dfrac{2}{s}$

```
Simplify[step3]
```

$(-4 + s^2)$capy$[s] + 2$LaplaceTransform$[ty'[t], t, s] == \dfrac{2}{s}$

```
step4 = DSolve[step3, capy[s], s]
```

$\{\{$capy$[s] \rightarrow -\dfrac{2(-1 + s$LaplaceTransform$[ty'[t], t, s])}{s(-4 + s^2)}\}\}$

These results indicate that $Y(s) = 2s^{-3} + Ce^{\frac{1}{4}s^2 - 3\ln 3}$. Recall that if $\lim_{s\to\infty} Y(s) \neq 0$, $\mathcal{L}^{-1}\{Y(s)\}$ does not exist. Therefore, we must have that $C = 0$. Hence, $y(t) = \mathcal{L}^{-1}\{Y(s)\} = \mathcal{L}^{-1}\{2s^{-3}\} = t^2$.

```
InverseLaplaceTransform[ 2/s^3, s, t]
```

t^2

■

8.4 Laplace Transforms of Step and Periodic Functions

8.4.1 Piecewise-Defined Functions: The Unit Step Function

An important function in modeling many physical situations is the *unit step function*, $\mathcal{U}$.

Definition 35 (Unit Step Function). *The **unit step function**, $\mathcal{U}(t-a) = \mathcal{U}_a(t)$, where a is a number is defined by*

$$\mathcal{U}(t-a) = \mathcal{U}_a(t) = \begin{cases} 0, & t < a \\ 1, & t \geq a \end{cases}. \tag{8.30}$$

We can use the function UnitStep to define the unit step function:

$$\text{UnitStep}[t] = \begin{cases} 0, & t < 0 \\ 1, & t \geq 0 \end{cases}.$$

so $\mathcal{U}_a(t) =$ UnitStep$[t - a]$. *Note:* The functions UnitStep and HeavisideTheta have nearly identical functionality and can usually be used interchangeably.

EXAMPLE 8.4.1: Graph (a) $2\mathcal{U}(t)$, (b) $\frac{1}{2}\mathcal{U}(t-5)$, and (c) $\mathcal{U}(t-2) - \mathcal{U}(t-8)$.

SOLUTION: (a) Here, $2\mathcal{U}(t) = 2\mathcal{U}(t-0)$, so $2\mathcal{U}(t) = 2$ for $t \geq 2$.

(b) In this case, $\frac{1}{2}\mathcal{U}(t-5) = \begin{cases} 0, & t < 5 \\ 1/2, & t \geq 5 \end{cases}$, so the "jump" occurs at $t = 5$.

(c) $\mathcal{U}(t-2) - \mathcal{U}(t-8) = \begin{cases} 0, & t < 2 \text{ or } t \geq 8 \\ 1, & 2 \leq t < 8 \end{cases}$. These functions are graphed using Plot and UnitStep in Figure 8-5.

$$\text{Plot}[\{\frac{\text{UnitStep}[t-5]}{2}, 2\text{UnitStep}[t],$$

$$\text{UnitStep}[t-2] - \text{UnitStep}[t-8]\}, \{t, 0, 10\},$$

$$\text{AxesLabel} \rightarrow \{t, y\}]$$

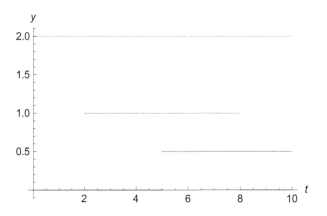

Figure 8-5 Plots of combinations of various step functions

■

The unit step function is useful in defining functions that are piecewise continuous. For example, we can define the function

$$g(t) = \begin{cases} 0, t < a \\ h(t), a \le t < b \\ 0, t \ge b \end{cases}$$

as

$$g(t) = h(t)\left[\mathcal{U}(t - a) - \mathcal{U}(t - b)\right].$$

Similarly, a function like

$$f(t) = \begin{cases} g(t), 0 \le t < a \\ h(t), t \ge a \end{cases}$$

can be written as

$$f(t) = g(t)\left[1 - \mathcal{U}(t - a)\right] + h(t)\mathcal{U}(t - a).$$

The reason for writing piecewise continuous functions in terms of step functions is that we encounter functions of this type in solving initial-value problems. Using our methods in Chapters 4 and 5, we had to solve the problem over each piece of the function. However, the method of Laplace transforms can be used to avoid these complicated calculations.

Theorem 27. *Suppose that* $F(s) = \mathcal{L}\{f(t)\}$ *exists for* $s > b \geq 0$. *If* a *is a positive constant, then*

$$\mathcal{L}\{f(t-a)\mathcal{U}(t-a)\} = e^{-as}F(s). \tag{8.31}$$

EXAMPLE 8.4.2: Find $\mathcal{L}\left\{(t-3)^5\mathcal{U}(t-3)\right\}$.

SOLUTION: In this case, $a = 3$ and $f(t) = t^5$. Thus,

$$\mathcal{L}\left\{(t-3)^5\mathcal{U}(t-3)\right\} = e^{-3s}\mathcal{L}\left\{t^5\right\} = e^{-3s}\frac{5!}{s^6} = \frac{120}{s^6}e^{-3s}.$$

Equivalent results are obtained with Mathematica. Here is the unsimplified the result

```
LaplaceTransform[(t − 3)^5 UnitStep[t − 3], t, s]
```

$$-\frac{243e^{-3s}}{s} + \frac{405e^{-3s}(1+3s)}{s^2} - \frac{270e^{-3s}(2+6s+9s^2)}{s^3}$$
$$+\frac{270e^{-3s}(2+6s+9s^2+9s^3)}{s^4}$$
$$-\frac{45e^{-3s}(8+24s+36s^2+36s^3+27s^4)}{s^5}$$
$$+\frac{3e^{-3s}(40+120s+180s^2+180s^3+135s^4+81s^5)}{s^6}$$

Notice that `FullSimplify` simplifies the result to that you would obtain doing the calculation by hand.

```
LaplaceTransform[(t − 3)^5 UnitStep[t − 3], t, s]//FullSimplify
```

$$\frac{120e^{-3s}}{s^6}$$

∎

In most cases, we must calculate $\mathcal{L}\{g(t)\mathcal{U}(t-a)\}$ instead of $\mathcal{L}\{g(t)\mathcal{U}(t-a)\}$. To solve this problem, we let $g(t) = f(t-a)$, so $f(t) = g(t+a)$. Therefore,

$$\mathcal{L}\{g(t)\mathcal{U}(t-a)\} = e^{-as}\mathcal{L}\{g(t+a)\}. \tag{8.32}$$

EXAMPLE 8.4.3: Calculate $\mathcal{L}\{\sin t\,\mathcal{U}(t-\pi)\}$.

SOLUTION: In this case, $g(t) = \sin t$ and $a = \pi$. Thus,

$$\mathcal{L}\{\sin t\, \mathcal{U}(t - \pi)\} = e^{-\pi s}\mathcal{L}\{\sin(t + \pi)\} = e^{-\pi s}\mathcal{L}\{-\sin t\}$$

$$= -e^{-\pi s}\frac{1}{s^2 + 1} = -\frac{e^{-\pi s}}{s^2 + 1}.$$

The same result is obtained using `LaplaceTransform`.

`LaplaceTransform[Sin[t]UnitStep[t - π], t, s]`

$$-\frac{e^{-\pi s}}{1 + s^2}$$

■

Theorem 28. *Suppose that $F(s) = \mathcal{L}\{f(t)\}$ exists for $s > b \geq 0$. If a is a positive constant and $y = f(t)$ is continuous on $[0, \infty)$, then*

$$\mathcal{L}^{-1}\{e^{-as}F(s)\} = f(t - a)\mathcal{U}(t - a). \tag{8.33}$$

EXAMPLE 8.4.4: Find (a) $\mathcal{L}^{-1}\left\{\dfrac{e^{-4s}}{s^3}\right\}$ and (b) $\mathcal{L}^{-1}\left\{\dfrac{e^{-\pi s/2}}{s^2 + 16}\right\}$.

SOLUTION: (a) If we write the expression $\dfrac{e^{-4s}}{s^3}$ in the form $e^{-as}F(s)$, we see that $a = 4$ and $F(s) = s^{-3}$. Hence, $f(t) = \mathcal{L}^{-1}\{s^{-3}\} = \frac{1}{2}t^2$ and

$$\mathcal{L}^{-1}\left\{\frac{e^{-4s}}{s^3}\right\} = f(t - 4)\mathcal{U}(t - 4) = \frac{1}{2}(t - 4)^2\mathcal{U}(t - 4).$$

(b) In this case, $a = \pi/2$ and $F(s) = \dfrac{1}{s^2 + 1}$. Then, $f(t) = \mathcal{L}^{-1}\left\{\dfrac{1}{s^2 + 1}\right\} = \frac{1}{4}\sin 4t$ and

$$\mathcal{L}^{-1}\left\{\frac{e^{-\pi s/2}}{s^2 + 16}\right\} = f\left(t - \frac{\pi}{2}\right)\mathcal{U}\left(t - \frac{\pi}{2}\right) = \frac{1}{4}\sin\left[4\left(t - \frac{\pi}{2}\right)\right]\mathcal{U}\left(t - \frac{\pi}{2}\right)$$

$$= \frac{1}{4}\sin 4t\,\mathcal{U}\left(t - \frac{\pi}{2}\right).$$

For each of (a) and (b), the same results are obtained using `InverseLaplaceTransform`, although we must use `Simplify` to simplify the result obtained for (b). Observe that several of the results are given in terms of the `HeavisideTheta` function. Observe that `HeavisideTheta`

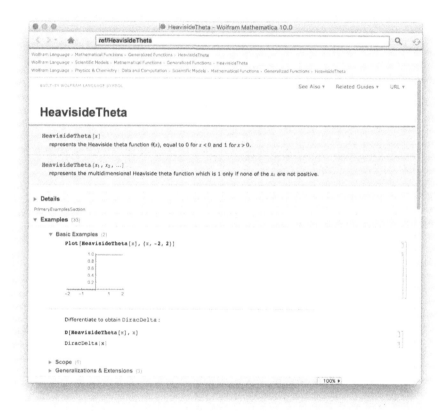

and `UnitStep`
are nearly identical in their utility and scope as mentioned previously.

$$\texttt{InverseLaplaceTransform}\left[\frac{\texttt{Exp}[-4\,s]}{s^3}, s, t\right]$$

$$\frac{1}{2}(-4+t)^2\,\texttt{HeavisideTheta}[-4+t]$$

$$\texttt{step1}=\texttt{InverseLaplaceTransform}\left[\frac{\texttt{Exp}[-\frac{\pi\,s}{2}]}{s^2+16}, s, t\right]$$

$$\frac{1}{4}\text{HeavisideTheta}\left[-\frac{\pi}{2}+t\right]\text{Sin}\left[4(-\frac{\pi}{2}+t)\right]$$

Simplify[step1]

$$\frac{1}{4}\text{HeavisideTheta}\left[-\frac{\pi}{2}+t\right]\text{Sin}[4t]$$

■

8.4.2 Solving Initial-Value Problems With Piecewise-Continuous Forcing Functions

With the unit step function, we can solve initial-value problems that involve piecewise-continuous functions.

EXAMPLE 8.4.5: Solve $y'' + 9y = \begin{cases} 1, \ 0 \leq t < \pi \\ 0, \ t \geq \pi \end{cases}$ subject to $y(0) = y'(0) = 0$.

SOLUTION: In order to solve this initial-value problem, we must compute $\mathcal{L}\{f(t)\}$ where $f(t) = \begin{cases} 1, \ 0 \leq t < \pi \\ 0, \ t \geq \pi \end{cases}$. This is a piecewise continuous function so we write it in terms of the unit step function as

$$f(t) = 1\left[\mathcal{U}(t - 0) - \mathcal{U}(t - \pi)\right] + 0\left[\mathcal{U}(t - \pi)\right] = \mathcal{U}(t) - \mathcal{U}(t - \pi).$$

Then,

$$\mathcal{L}\{f(t)\} = \mathcal{L}\{1 - \mathcal{U}(t - \pi)\} = \frac{1}{s} - \frac{e^{-\pi s}}{s}.$$

Hence,

$$\mathcal{L}\{y''\} + 9\mathcal{L}\{y\} = \mathcal{L}\{f(t)\}$$

$$s^2 Y(s) - sy(0) - y'(0) + 9Y(s) = \frac{1}{s} - \frac{e^{-\pi s}}{s}$$

$$\left(s^2 + 9\right) Y(s) = \frac{1}{s} - \frac{e^{-\pi s}}{s}$$

$$Y(s) = \frac{1}{s\left(s^2 + 9\right)} - \frac{e^{-\pi s}}{s\left(s^2 + 9\right)}.$$

The same steps are performed next with Mathematica. First, we define eq to be the equation $y'' + 9y = \begin{cases} 1, \ 0 \leq t < \pi \\ 0, \ t \geq \pi \end{cases}$.

```
eq = y''[t] + 9y[t]==UnitStep[t] - UnitStep[t - π];
```

Next, we use `LaplaceTransform` to compute the Laplace transform of each side of the equation, naming the resulting equation step1,

```
step1 = LaplaceTransform[eq, t, s]
```

$9\text{LaplaceTransform}[y[t], t, s] + s^2 \text{LaplaceTransform}[y[t],$

$t, s] - sy[0] - y'[0] == \dfrac{1}{s} - \dfrac{e^{-\pi s}}{s}$

apply the initial conditions, naming the result step2,

```
step2 = step1/.{y[0]->0, y'[0]->0}
```

$9\text{LaplaceTransform}[y[t], t, s] + s^2 \text{LaplaceTransform}[y[t],$

$t, s] == \dfrac{1}{s} - \dfrac{e^{-\pi s}}{s}$

and solve step2 for LaplaceTransform[y[t],t,s], naming the result step3.

```
step3 = Solve[step2, LaplaceTransform[y[t], t, s]]
```

$\{\{\text{LaplaceTransform}[y[t], t, s] \rightarrow \dfrac{e^{-\pi s}(-1 + e^{\pi s})}{s(9 + s^2)}\}\}$

Then,

$$y(t) = \mathcal{L}^{-1}\{Y(s)\} = \mathcal{L}^{-1}\left\{\frac{1}{s(s^2 + 9)}\right\} - \mathcal{L}^{-1}\left\{\frac{e^{-\pi s}}{s(s^2 + 9)}\right\}.$$

Consider $\mathcal{L}^{-1}\left\{\dfrac{e^{-\pi s}}{s(s^2 + 9)}\right\}$. In the form of $\mathcal{L}^{-1}\{e^{-as}F(s)\}$, $a = \pi$ and

$F(s) = \dfrac{1}{s(s^2 + 9)}$. Therefore, $f(t) = \mathcal{L}^{-1}\{F(s)\}$ can be found with either a partial fraction expansion or with equation (8.29):

$$f(t) = \mathcal{L}^{-1}\left\{\frac{1}{s(s^2 + 9)}\right\} = \int_0^t \mathcal{L}^{-1}\left\{\frac{1}{s^2 + 9}\right\} d\alpha = \int_0^t \frac{1}{3}\sin 3\alpha \, d\alpha$$

$$= -\frac{1}{3}\left[\frac{1}{3}\cos 3\alpha\right]_0^t = \frac{1}{9} - \frac{1}{9}\cos 3t.$$

Then,

$$\mathcal{L}^{-1}\left\{\frac{e^{-\pi s}}{s(s^2 + 9)}\right\} = \left[\frac{1}{9} - \frac{1}{9}\cos(3(t - \pi))\right]\mathcal{U}(t - \pi)$$

$$= \left[\frac{1}{9} - \frac{1}{9}\cos(3t - 3\pi)\right]\mathcal{U}(t - \pi) = \left[\frac{1}{9} + \frac{1}{9}\cos 3t\right]\mathcal{U}(t - \pi).$$

Combining these results yields the solution

$$y(t) = \mathcal{L}^{-1}\{Y(s)\} = \mathcal{L}^{-1}\left\{\frac{1}{s\left(s^2+9\right)}\right\} - \mathcal{L}^{-1}\left\{\frac{e^{-\pi s}}{s\left(s^2+9\right)}\right\}$$

$$= \frac{1}{9} - \frac{1}{9}\cos 3t - \left[\frac{1}{9} + \frac{1}{9}\cos 3t\right]\mathcal{U}(t-\pi).$$

Equivalent results are obtained with `InverseLaplaceTransform` and `Simplify`.

```
sol = InverseLaplaceTransform[-
```
$$\frac{-1+E^{-\pi s}}{s(9+s^2)}, s, t]$$

$\frac{1}{9}(1 - \text{Cos}[3t] - (1 + \text{Cos}[3t])\text{HeavisideTheta}[-\pi + t])$

We now graph the solution with `Plot` in Figure 8-6.

```
p1 = Plot[sol, {t, 0, 2π}, PlotStyle->
    CMYKColor[0, 0.89, 0.94, 0.28], AxesLabel → {t, y},
    PlotLabel → "(a)"]
```

We obtain the same results using `DSolve`

```
sol = DSolve[{eq, y[0]==0, y'[0]==0}, y[t], t]
```

$\{\{y[t] \rightarrow -\frac{2}{9}\text{Cos}[3t] + \text{UnitStep}[\pi - t](\frac{2}{9}\text{Cos}[3t] + \frac{1}{9}(1 - \text{Cos}[3t])$
$\text{UnitStep}[t])\}\}$

```
p2 = Plot[y[t]/.sol, {t, 0, 3Pi}, PlotStyle->
    CMYKColor[0, 0.89, 0.94, 0.28], AxesLabel → {t, y},
    PlotLabel → "(b)"]
```

Finally, we illustrate the calculation using the method of Laplace transforms.

```
step1 = LaplaceTransform[
    y"[t] + 9y[t]==UnitStep[t] - UnitStep[t - Pi], t, s]
```

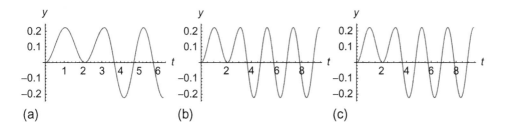

(a)   (b)   (c)

Figure 8-6   Plots of $y(t)$ using Laplace transform methods

$9\mathtt{LaplaceTransform[y[t]},\ t,\ s] + s^2\mathtt{LaplaceTransform[y[t]},$

$t,\ s] - sy[0] - y'[0] == \dfrac{1}{s} - \dfrac{e^{-\pi s}}{s}$

step2 = step1/.{y[0]->0, y'[0]->0}

$9\mathtt{LaplaceTransform[y[t]},\ t,\ s] + s^2\mathtt{LaplaceTransform[y[t]},$

$t,\ s] == \dfrac{1}{s} - \dfrac{e^{-\pi s}}{s}$

step3 = Solve[step2, LaplaceTransform[y[t], t, s]]

$\{\{\mathtt{LaplaceTransform[y[t]},\ t,\ s] \to \dfrac{e^{-\pi s}(-1 + e^{\pi s})}{s(9 + s^2)}\}\}$

sol = InverseLaplaceTransform[step3[[1, 1, 2]], s, t]

$\dfrac{1}{9}(1 - \mathrm{Cos}[3t] - (1 + \mathrm{Cos}[3t])\mathrm{HeavisideTheta}[-\pi + t])$

p3 = Plot[sol, {t, 0, 3Pi}, PlotStyle->CMYKColor[0, 0.89,

0.94, 0.28],

AxesLabel → {t, y}, PlotLabel → "(c)"]

Show[GraphicsRow[{p1, p2, p3}]]

All three plots are shown in Figure 8-6.

∎

8.4.3 Periodic Functions

Another type of function that is encountered in many areas of applied mathematics is the *periodic function*.

Definition 36 (Periodic Function). *A function $y = f(t)$ is **periodic** if there is a positive number T such that $f(t + T) = f(t)$ for all $t \geq 0$. The minimum value of T that satisfies this equation is called the **period** of $y = f(t)$.*

Due to the nature of periodic functions, we can simplify the calculation of the Laplace transform of these functions as indicated in the following theorem.

Theorem 29 (Laplace Transform of Periodic Functions). *Suppose that $y = f(t)$ is a periodic function with period T and that $y = f(t)$ is piecewise continuous on $[0, \infty)$. Then, $\mathcal{L}\{f(t)\}$ exists for $s > 0$ and is given by the definite integral*

$$\mathcal{L}\{f(t)\} = \frac{1}{1 - e^{-sT}} \int_0^T e^{-st}f(t)\,dt. \tag{8.34}$$

EXAMPLE 8.4.6: Find the Laplace transform of the periodic function $f(t) = t, 0 \leq t < 1$, and $f(t+1) = f(t)$.

SOLUTION: The period of $y = f(t)$ is $T = 1$. We use `Plot` to generate a graph of $y = f(t)$ on the interval $[0, 4]$ in Figure 8-7. Observe that $f(t)$ is discontinuous if t is an integer. One way of excluding the testing of integers in a `Plot` command is to use the option `Exclusions->Sin[t Pi]==0` because $\sin t\pi = 0$ if and only if t is an integer.

```
Clear[f]

f[t_]:=f[t − 1]/;t ≥ 1

f[t_]:=t/;0 ≤ t < 1

p1 = Plot[f[t], {t, 0, 4}, PlotLabel → "(a)"]

p2 = Plot[f[t], {t, 0, 4}, Exclusions → Sin[tPi] == 0,

   PlotLabel → "(b)"]

Show[GraphicsRow[{p1, p2}]]
```

We use Integration by Parts,

$$\mathcal{L}\{f(t)\} = \frac{1}{1 - e^{-s}} \int_0^1 te^{-st}\, dt$$

$$= \frac{1}{1 - e^{-s}} \left\{ \left[-\frac{te^{-st}}{s} \right]_{t=0}^{t=1} + \int_0^1 \frac{e^{-st}}{s}\, dt \right\}$$

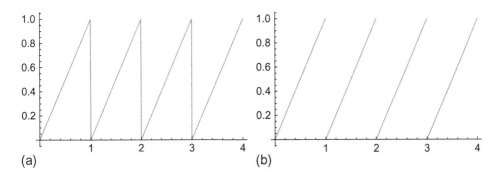

(a)   (b)

Figure 8-7   Plot of $f(t)$ on the interval $[0, 4]$, In (a) Mathematica does not detect the discontinuities. On the other hand, in (b) we use `Exclusions` to instruct Mathematica to detect the discontinuities

$$= \frac{1}{1-e^{-s}} \left\{ -\frac{e^{-s}}{s} - \left[\frac{e^{-st}}{s^2} \right]_{t=0}^{t=1} \right\}$$

$$= \frac{1}{1-e^{-s}} \left(-\frac{e^{-s}}{s} + \frac{1-e^{-s}}{s^2} \right) = \frac{1-(s+1)e^{-s}}{s^2 \left(1-e^{-s}\right)}$$

or Mathematica

$$\texttt{Simplify}[\frac{\int_0^1 \texttt{tExp[-st]dt}}{1-\texttt{Exp[-s]}}]$$

$$\frac{-1+e^s - s}{(-1+e^s)s^2}$$

`term[n_]:=LaplaceTransform[UnitStep[t − n], t, s]`

to compute the Laplace transform. Alternatively, note that

$$f(t) = t\left[\mathcal{U}(t-1)-\mathcal{U}(t)\right]+(t-1)\left[\mathcal{U}(t-2)-\mathcal{U}(t-1)\right]$$
$$+ (t-2)\left[\mathcal{U}(t-3)-\mathcal{U}(t-2)\right]+\cdots$$
$$= t - \mathcal{U}(t-1) - \mathcal{U}(t-2) - \mathcal{U}(t-3) - \mathcal{U}(t-4) - \cdots$$
$$= t - \sum_{n=1}^{\infty}\mathcal{U}(t-n)$$

so

$$\mathcal{L}\left\{f(t)\right\} = \mathcal{L}\left\{t\right\} - \mathcal{L}\left\{ \sum_{n=1}^{\infty}\mathcal{U}(t-n) \right\} = \mathcal{L}\left\{t\right\} - \sum_{n=1}^{\infty}\mathcal{L}\left\{\mathcal{U}(t-n)\right\}.$$

We use `LaplaceTransform` and `Table`

`Table[term[n], {n, 1, 7}]`

$$\{\frac{e^{-s}}{s}, \frac{e^{-2s}}{s}, \frac{e^{-3s}}{s}, \frac{e^{-4s}}{s}, \frac{e^{-5s}}{s}, \frac{e^{-6s}}{s}, \frac{e^{-7s}}{s}\}$$

to see that $\mathcal{L}\left\{\mathcal{U}(t-n)\right\} = \frac{1}{s}e^{-ns}$. Next, we use `Sum` and `Together` to calculate

For the geometric series, $\sum_{n=1}^{\infty} r^n$, if $|r| < 1$, $\sum_{n=1}^{\infty} r^n = \frac{r}{1-r}$.

$$\mathcal{L}\left\{f(t)\right\} = \mathcal{L}\left\{t\right\} - \sum_{n=1}^{\infty}\mathcal{L}\left\{\mathcal{U}(t-n)\right\} = \frac{1}{s^2} - \sum_{n=1}^{\infty}\frac{e^{-ns}}{s}$$

$$= \frac{1}{s}\left(\frac{1}{s} - \sum_{n=1}^{\infty}\left(e^{-s}\right)^n \right) = \frac{1}{s}\left(\frac{1}{s} - \frac{e^{-s}}{1-e^{-s}} \right).$$

$$\texttt{Together[LaplaceTransform[t, t, s]} - \sum\nolimits_{n=1}^{\infty} \frac{1}{\texttt{Exp[ns]s}}\texttt{]}$$

$$\frac{-1 + e^s - s}{(-1 + e^s)s^2}$$

∎

Laplace transforms can now be used to solve initial-value problems with periodic forcing functions more easily.

EXAMPLE 8.4.7: Solve $y'' + y = f(t)$ subject to $y(0) = y'(0) = 0$ if

$$f(t) = \begin{cases} \sin t, & 0 \le t < \pi \\ 0, & \pi \le t < 2\pi \end{cases} \quad \text{and } f(t + 2\pi) = f(t). \text{ (} f(t) \text{ is known as the}$$

half-wave rectification of $\sin t$.)

SOLUTION: To graph $f(t)$, we begin defining $g(t) = \begin{cases} \sin t, & 0 \le t < \pi \\ 0, & \pi \le t < 2\pi \end{cases}$.

$$\texttt{g[t\_] = Sin[t]UnitStep[}\pi\texttt{ - t];}$$

Then,

$$f(t) = g(t)\left[\mathcal{U}(t) - \mathcal{U}(t - 2\pi)\right] + g(t - 2\pi)\left[\mathcal{U}(t - 2\pi) - \mathcal{U}(t - 4\pi)\right] + \cdots$$

$$= \sum_{n=0}^{\infty} g(t - n\pi)\left[\mathcal{U}(t - 2n\pi) - \mathcal{U}(t - 2(n+1)\pi)\right].$$

Thus, the graph of $f(t)$ on the interval $[0, 2k\pi]$, where k represents a positive integer, is obtained by graphing

$$f_k(t) = \sum_{n=0}^{k-1} g(t - n\pi)\left[\mathcal{U}(t - 2n\pi) - \mathcal{U}(t - 2(n+1)\pi)\right]$$

on the interval $[0, 2k\pi]$. For convenience, we define $\texttt{nthterm[n]}$ to be

$$g(t - n\pi)\left[\mathcal{U}(t - 2n\pi) - \mathcal{U}(t - 2(n+1)\pi)\right].$$

Observe that the results are given in terms of the $\texttt{Ceiling}$ and $\texttt{Floor}$ functions. $\texttt{Ceiling[x]}$ returns the smallest integer greater than or equal to x while $\texttt{Floor[x]}$ returns the greatest integer less than or equal to x.

$$\texttt{nthterm[n\_] = g[t - 2n}\pi\texttt{](UnitStep[t - 2n}\pi\texttt{] - UnitStep[t - 2}$$

$$\texttt{(n + 1)}\pi\texttt{]);}$$

$$f[k\_, t\_] = \sum_{n=0}^{k-1} \texttt{nthterm}[n];$$

$$f[2, t]$$

$$-(\{ \begin{array}{ll} 1 & (-t == 0\&\& -2\pi - t \geq 0)\| \\ & (-t \leq 0\&\& -2\pi - t \geq 0\&\& -\pi \\ & \quad +t < 0) \\ -\texttt{Ceiling}\left[\dfrac{-\pi + t}{2\pi}\right] & (\pi - t == 0\&\&\pi + t < 0\&\& -2\pi \\ & \quad -t < 0)\|(\pi - t < 0\&\&\pi + t < 0\&\& \\ & \quad -2\pi - t \leq 0) \\ 1 - \texttt{Ceiling}\left[\dfrac{-\pi + t}{2\pi}\right] + \texttt{Floor}\left[\dfrac{-\pi + t}{2\pi}\right] & \pi - t \leq 0\&\& -\pi - t == 0 \\ 1 - \texttt{Ceiling}\left[\dfrac{-\pi + t}{2\pi}\right] + \texttt{Floor}\left[\dfrac{t}{2\pi}\right] & (\pi - t == 0\&\& -2\pi - t \geq 0)\| \\ & (\pi - t < 0\&\& -2\pi - t > 0) \\ 0 & \texttt{True} \end{array} \right)\texttt{Sin}[t]+$$

$$\{ \begin{array}{ll} 1 & (-t == 0\&\&2\pi - t \geq 0)\| \\ & (-t \leq 0\&\&2\pi - t \geq 0\&\& -\pi + t < 0) \\ 2 & -\pi + t < 0\&\& -t < 0\&\&2\pi - t < 0 \\ 2 - \texttt{Ceiling}\left[\dfrac{-\pi + t}{2\pi}\right] & (\pi - t == 0\&\& -3\pi + t < 0\&\& \\ & 2\pi - t < 0)\|(\pi - t < 0\&\& -3\pi \\ & \quad +t < 0\&\&2\pi - t \leq 0) \\ 1 - \texttt{Ceiling}\left[\dfrac{-\pi + t}{2\pi}\right] + \texttt{Floor}\left[\dfrac{-\pi + t}{2\pi}\right] & \pi - t \leq 0\&\&3\pi - t == 0 \\ 1 - \texttt{Ceiling}\left[\dfrac{-\pi + t}{2\pi}\right] + \texttt{Floor}\left[\dfrac{t}{2\pi}\right] & (\pi - t == 0\&\&2\pi - t \geq 0)\| \\ & (\pi - t < 0\&\&2\pi - t > 0) \\ 0 & \texttt{True} \end{array} \right)\texttt{Sin}[t]$$

We graph $f(t)$ on the interval $[0, 10\pi]$ with `Plot` in Figure 8-8.

```
p1 = Plot[f[5, t], {t, 0, 10Pi},
    PlotStyle → CMYKColor[0.64, 0, 0.95, 0.40],
    PlotLabel → "(a)"]
```

To solve the initial-value problem we must find $\mathcal{L}\{f(t)\}$. Because the period is $T = 2\pi$, we have

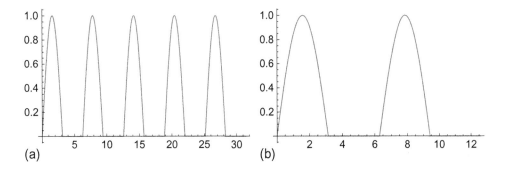

Figure 8-8   Different plots of the half-wave rectification of $\sin t$

$$\mathcal{L}\{f(t)\} = \frac{1}{1 - e^{-2\pi s}} \int_0^{2\pi} e^{-st} f(t)\, dt = \frac{1}{1 - e^{-2\pi s}} \left[\int_0^{\pi} e^{-st} \sin t\, dt \right.$$

$$\left. + \int_{\pi}^{2\pi} e^{-st} \cdot 0\, dt \right] = \frac{1}{1 - e^{-2\pi s}} \int_0^{\pi} e^{-st} \sin t\, dt.$$

We use `Integrate` to evaluate this integral

$$\texttt{step1 = Simplify[} \frac{\int_0^{\pi} \texttt{Exp[}-st\texttt{]Sin[}t\texttt{]}dt}{1 - \texttt{Exp[}-2\pi\, s\texttt{]}} \texttt{]}$$

$$\frac{e^{\pi s}}{(-1 + e^{\pi s})(1 + s^2)}$$

`lapf = step1//ExpandDenominator`

$$\frac{e^{\pi s}}{-1 + e^{\pi s} - s^2 + e^{\pi s} s^2}$$

and see that

$$\mathcal{L}\{f(t)\} = \frac{e^{\pi s}}{(e^{\pi s} - 1)(s^2 + 1)} = \frac{1}{(1 - e^{-\pi s})(s^2 + 1)}.$$

Alternatively, we can use

$$f(t) = \sum_{n=0}^{\infty} g(t - n\pi)\, [\mathcal{U}(t - 2n\pi) - \mathcal{U}(t - 2(n+1)\pi)]$$

to rewrite $f(t)$ as

$$f(t) = \sum_{n=0}^{\infty} (-1)^n \sin t\, \mathcal{U}(t - n\pi).$$

Then,

$$\mathcal{L}\{f(t)\} = \mathcal{L}\left\{\sum_{n=0}^{\infty}(-1)^n \sin t\,\mathcal{U}(t - n\pi)\right\} = \sum_{n=0}^{\infty}\mathcal{L}\{(-1)^n \sin t\,\mathcal{U}(t - n\pi)\}.$$

We use LaplaceTransform and UnitStep to compute $\mathcal{L}\{(-1)^n \sin t\,\mathcal{U}(t - n\pi)\}$, naming the result nthlap,

Clear[h]

h[t_] = Sum[(-1)^nSin[t]UnitStep[t - nPi], {n, 0, 5}]

Sin[t]UnitStep[t] − Sin[t]UnitStep[−5π + t] + Sin[t]

UnitStep[−4π + t] − Sin[t]UnitStep[−3π + t] + Sin[t]

UnitStep[−2π + t] − Sin[t]UnitStep[−π + t]

p2 = Plot[h[t], {t, 0, 4Pi},

  PlotStyle → CMYKColor[0.64, 0, 0.95, 0.40],

  PlotLabel → "(b)"]

Show[GraphicsRow[{p1, p2}]]

Clear[nthlap]

nthlap = LaplaceTransform$\left[(-1)^n\text{Sin}[t]\text{UnitStep}[t - n\pi], t, s\right]$;

TableForm[Table[{n, nthlap}, {n, 0, 8}]]

0	$\dfrac{1}{1 + s^2}$
1	$\dfrac{e^{-\pi s}}{1 + s^2}$
2	$\dfrac{e^{-2\pi s}}{1 + s^2}$
3	$\dfrac{e^{-3\pi s}}{1 + s^2}$
4	$\dfrac{e^{-4\pi s}}{1 + s^2}$
5	$\dfrac{e^{-5\pi s}}{1 + s^2}$
6	$\dfrac{e^{-6\pi s}}{1 + s^2}$
7	$\dfrac{e^{-7\pi s}}{1 + s^2}$
8	$\dfrac{e^{-8\pi s}}{1 + s^2}$

and then use Sum to compute $\sum_{n=0}^{\infty}\mathcal{L}\{(-1)^n \sin t\,\mathcal{U}(t - n\pi)\}$.

$$\frac{\sum_{n=0}^{\infty} \text{Exp}[-n\pi\, s]}{1 + s^2}$$

$$\frac{e^{\pi s}}{(-1 + e^{\pi s})(1 + s^2)}$$

Taking the Laplace transform of both sides of the differential equation, applying the initial conditions, and solving for $Y(s)$ then gives us

$$\mathcal{L}\left\{y''\right\} + \mathcal{L}\left\{y\right\} = \mathcal{L}\left\{f(t)\right\}$$

$$s^2 Y(s) - sy(0) - y'(0) + Y(s) = \frac{1}{\left(1 - e^{-\pi s}\right)\left(s^2 + 1\right)}$$

$$Y(s) = \frac{1}{\left(1 - e^{-\pi s}\right)\left(s^2 + 1\right)^2}.$$

Using `lapf`, we perform the same steps with Mathematica.

step1 = LaplaceTransform[y"[t]+y[t], t, s]==lapf

LaplaceTransform[y[t], t, s] + s²LaplaceTransform[y[t],

t, s] − sy[0] − y'[0] == $\dfrac{e^{\pi s}}{-1 + e^{\pi s} - s^2 + e^{\pi s} s^2}$

step2 = step1/.{y[0]->0, y'[0]->0}

LaplaceTransform[y[t], t, s] + s²LaplaceTransform[y[t],

t, s] == $\dfrac{e^{\pi s}}{-1 + e^{\pi s} - s^2 + e^{\pi s} s^2}$

step3 = Solve[step2, LaplaceTransform[y[t], t, s]]

{{LaplaceTransform[y[t], t, s] → $\dfrac{e^{\pi s}}{(-1 + e^{\pi s})(1 + s^2)^2}$}}

Recall from our work with the geometric series that if $|x| < 1$, then

$$\frac{1}{1 - x} = 1 + x + x^2 + x^3 + \cdots = \sum_{n=0}^{\infty} x^n.$$

Because we do not know the inverse Laplace transform of $\dfrac{1}{\left(1 - e^{-\pi s}\right)\left(s^2 + 1\right)}$, we must use a geometric series expansion of $\dfrac{1}{1 - e^{-\pi s}}$ to obtain terms for which we can calculate the inverse Laplace transform. Using $x = -e^{-\pi s}$, this gives us

$$\frac{1}{1-e^{-\pi s}} = 1 + e^{-\pi s} + e^{-2\pi s} + e^{-3\pi s} + \cdots = \sum_{n=0}^{\infty} e^{-n\pi s},$$

so

$$Y(s) = \left(1 + e^{-\pi s} + e^{-2\pi s} + e^{-3\pi s} + \cdots\right)\frac{1}{\left(s^2+1\right)^2}$$

$$= \frac{1}{\left(s^2+1\right)^2} + \frac{e^{-\pi s}}{\left(s^2+1\right)^2} + \frac{e^{-2\pi s}}{\left(s^2+1\right)^2} + \frac{e^{-3\pi s}}{\left(s^2+1\right)^2} + \cdots$$

$$= \sum_{n=0}^{\infty} \frac{e^{-n\pi s}}{\left(s^2+1\right)^2}.$$

Then,

$$y(t) = \sum_{n=0}^{\infty} \mathcal{L}^{-1}\left\{\frac{e^{-n\pi s}}{\left(s^2+1\right)^2}\right\}$$

Notice that $\mathcal{L}^{-1}\left\{\dfrac{1}{\left(s^2+1\right)^2}\right\}$ is needed to find all of the other terms.
Using InverseLaplaceTransform,

$$\text{Expand[InverseLaplaceTransform[}\frac{1}{\left(s^2+1\right)^2}\text{, s, t]]}$$

$$-\frac{1}{2}t\text{Cos}[t] + \frac{\text{Sin}[t]}{2}$$

we have $\mathcal{L}^{-1}\left\{\dfrac{1}{\left(s^2+1\right)^2}\right\} = \frac{1}{2}(\sin t - t\cos t)$. In fact, we can use
InverseLaplaceTransform together with Table to compute the
inverse Laplace transform of the first few terms of the series.

$$\text{TableForm[Table[\{n, InverseLaplaceTransform[}\frac{\text{Exp}[-n\pi s]}{\left(s^2+1\right)^2},$$

$$\text{s, t]\}, \{n, 0, 5\}]]}$$

0   $\frac{1}{2}(-t\text{Cos}[t] + \text{Sin}[t])$

1   $\frac{1}{2}\text{HeavisideTheta}[-\pi + t]((-\pi + t)\text{Cos}[t] - \text{Sin}[t])$

2   $\frac{1}{2}\text{HeavisideTheta}[-2\pi + t](-(-2\pi + t)\text{Cos}[t] + \text{Sin}[t])$

3   $\frac{1}{2}\text{HeavisideTheta}[-3\pi + t]((-3\pi + t)\text{Cos}[t] - \text{Sin}[t])$

4   $\frac{1}{2}\text{HeavisideTheta}[-4\pi + t](-(-4\pi + t)\text{Cos}[t] + \text{Sin}[t])$

5   $\frac{1}{2}\text{HeavisideTheta}[-5\pi + t]((-5\pi + t)\text{Cos}[t] - \text{Sin}[t])$

Then,

$$
\begin{aligned}
y(t) = \frac{1}{2}\Big\{ & (\sin t - t\cos t) + [\sin(t-\pi) + (t-\pi)\cos(t-\pi)]\mathcal{U}(t-\pi) \\
& + [\sin(t-2\pi) + (t-2\pi)\cos(t-2\pi)]\mathcal{U}(t-2\pi) \\
& + [\sin(t-3\pi) + (t-3\pi)\cos(t-3\pi)]\mathcal{U}(t-3\pi)\Big\} \\
= \frac{1}{2}\sum_{n=0}^{\infty} & [\sin(t-n\pi) + (t-n\pi)\cos(t-n\pi)]\mathcal{U}(t-n\pi).
\end{aligned}
$$

To graph $y(t)$ on the interval $[0, k\pi]$, where k represents a positive integer, we note that

$$[\sin(t-n\pi) + (t-n\pi)\cos(t-n\pi)]\mathcal{U}(t-n\pi) = 0$$

for all values of t in $[0, k\pi]$ if $n \geq k$ so we need to graph

$$\frac{1}{2}\sum_{n=0}^{k-1}[\sin(t-n\pi) + (t-n\pi)\cos(t-n\pi)]\mathcal{U}(t-n\pi)$$

For convenience, we define `nthterm` to represent

$$\frac{1}{2}[\sin(t-n\pi) + (t-n\pi)\cos(t-n\pi)]\mathcal{U}(t-n\pi).$$

```
nthterm[n_] = 1/2 (Sin[t - nπ] - (t - nπ)Cos[t - nπ])UnitStep[t - nπ];
```

Thus, to graph `tograph` on the interval $[0, 5\pi]$, we enter the following commands. See Figure 8-9.

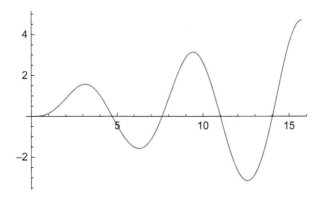

Figure 8-9 Plot of the solution to an initial value problem with a periodic piecewise continuous forcing function

$$tograph = \sum_{n=0}^{4} \text{nthterm}[n];$$

```
Plot[tograph, {t, 0, 5π},

    PlotStyle → CMYKColor[0, 0.89, 0.94, 0.28]]
```

∎

8.4.4 Impulse Functions: The Delta Function

We now consider differential equations of the form $ay'' + by' + cy = f(t)$ where $f(t)$ is large in magnitude over the short interval centered at t_0, $t_0 - \alpha \leq t \leq t_0 + \alpha$, and zero otherwise. Hence, we define the **impulse** delivered by the function $f(t)$ as $I(t) = \int_{t_0 - \alpha}^{t_0 + \alpha} f(t)\,dt$, or because $f(t) = 0$ for t on $(-\infty, t_0 - \alpha) \cup (t_0 + \alpha, \infty)$,

$$I(t) = \int_{-\infty}^{\infty} f(t)\,dt.$$

In order to better understand the *impulse function*, we let $f(t)$ be defined in the following manner:

$$f(t) = \delta_\alpha\,(t - t_0) = \begin{cases} \dfrac{1}{2\alpha}, & t_0 - \alpha \leq t \leq t_0 + \alpha \\ 0, & \text{otherwise} \end{cases}.$$

To graph $\delta_\alpha\,(t - t_0)$ for several values of α and $t_0 = 0$, we define del.

```
del[t_, t0_, α_] := 1/(2α) /; t0 - α ≤ t ≤ t0 + α
del[t_, t0_, α_] := 0 /; t0 - α > t ‖ t > t0 + α
```

For example, entering

```
p1 = Plot[del[t, 0, .25], {t, −1, 1},

    PlotLabel->"(a)"]
```

graphs $\delta_{1/4}(t)$ on the interval $[-1, 1]$. See Figure 8-10(a). Similarly, to graph $\delta_i(t)$ for $i = 0.01, 0.02, 0.03, 0.04$, and 0.05, we first define toplot using Table and then use Plot to graph this set of functions on the interval $[-0.1, 0.1]$. See Figure 8-10(b).

```
toplot = Table[del[t, 0, α], {α, 0.01, 0.05, 0.01}];
p2 = Plot[Evaluate[toplot], {t, −0.1, 0.1},

    PlotLabel->"(b)", PlotRange → All]

Show[GraphicsRow[{p1, p2}]]
```

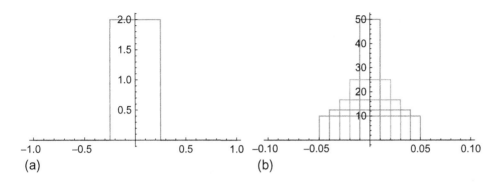

Figure 8-10 (a) Plot of $\delta_{1/4}(t)$ on the interval $[-1, 1]$. (b) Plots of $\delta_i(t)$ for $i = 0.01, 0.02, 0.03, 0.04,$ and 0.05

With this definition, the impulse is given by

$$I(t) = \int_{t_0-\alpha}^{t_0+\alpha} f(t)\, dt = \int_{t_0-\alpha}^{t_0+\alpha} \frac{1}{2\alpha}\, dt = \frac{1}{2\alpha}\left[(t_0 + \alpha) - (t_0 - \alpha)\right] = \frac{1}{2\alpha} \cdot 2\alpha = 1.$$

Notice that the value of this integral does not depend on α as long as $\alpha \neq 0$. We now try to create the *idealized impulse function* by requiring that $\delta_\alpha(t - t_0)$ act on smaller and smaller intervals. From the integral calculation, we have

$$\lim_{\alpha \to 0} I(t) = 1.$$

We also note that

$$\lim_{\alpha \to 0} \delta_\alpha(t - t_0) = 0, \; t \neq t_0.$$

We use these properties to now define the **idealized unit impulse function**.

Definition 37 (Unit Impulse Function). *The idealized unit impulse function (**Dirac delta function**) δ satisfies*

The Dirac delta function is not a real-valued function of a single variable. Objects of this type are called **generalized functions**.

$$\delta(t - t_0) = 0, \; t \neq t_0$$

$$\int_{-\infty}^{\infty} \delta(t - t_0)\, dt = 1. \tag{8.35}$$

The Mathematica function `DiracDelta` represents the Dirac delta function. We now state the following useful theorem involving the unit impulse function.

Theorem 30. *Suppose that* $y = g(t)$ *is a bounded and continuous function. Then,*

$$\int_{-\infty}^{\infty} \delta\left(t - t_0\right) g(t)\, dt = g\left(t_0\right).$$

(8.36)

The Laplace transform of $\delta\left(t - t_0\right)$ is found by using the function $\delta_\alpha\left(t - t_0\right)$ and L'Hopital's rule.

Theorem 31. *For* $t_0 > 0$*,*

$$\mathcal{L}\left\{\delta\left(t - t_0\right)\right\} = e^{-st_0}.$$

(8.37)

EXAMPLE 8.4.8: Find (a) $\mathcal{L}\left\{\delta\left(t - 1\right)\right\}$; (b) $\mathcal{L}\left\{\delta\left(t - \pi\right)\right\}$; and (c) $\mathcal{L}\left\{\delta\left(t\right)\right\}$.

SOLUTION: (a) In this case, $t_0 = 1$, so $\mathcal{L}\left\{\delta\left(t - 1\right)\right\} = e^{-s}$. (b) With $t_0 = \pi$, $\mathcal{L}\left\{\delta\left(t - \pi\right)\right\} = e^{-\pi s}$. (c) Because $t_0 = 0$, $\mathcal{L}\left\{\delta\left(t\right)\right\} = \mathcal{L}\left\{\delta\left(t - 0\right)\right\} = e^{-s \cdot 0} = 1$.

We obtain the same results using `DiracDelta` and `LaplaceTransform` as shown next. We can compute the Laplace transform of each individually.

```
LaplaceTransform[DiracDelta[t − 1], t, s]
```

e^{-s}

Or, we can use `Map` to compute the Laplace transform of all three simultaneously.

```
Map[LaplaceTransform[#, t, s]&, {DiracDelta[t − 1],

    DiracDelta[t − Pi], DiracDelta[t]}]
```

$\{e^{-s}, e^{-\pi s}, 1\}$

∎

EXAMPLE 8.4.9: Solve $y'' + y = \delta(t - \pi) + 1$ subject to $y(0) = y'(0) = 0$.

SOLUTION: As in previous examples, we solve this initial-value problem by taking the Laplace transform of both sides of the differential equation,

```
step1 = LaplaceTransform[y''[t] + y[t]==DiracDelta[t − π]
+1, t, s]
```

$$\text{LaplaceTransform}[y[t], t, s] + s^2 \text{LaplaceTransform}[y[t],$$
$$t, s] - sy[0] - y'[0] == e^{-\pi s} + \frac{1}{s}$$

applying the initial conditions,

```
step2 = step1/.{y[0] → 0, y'[0] → 0}
```

$$\text{LaplaceTransform}[y[t], t, s] + s^2 \text{LaplaceTransform}[y[t],$$
$$t, s] == e^{-\pi s} + \frac{1}{s}$$

and solving for $Y(s)$.

```
step3 = Solve[step2, LaplaceTransform[y[t], t, s]]
```

$$\{\{\text{LaplaceTransform}[y[t], t, s] \rightarrow \frac{e^{-\pi s}(e^{\pi s} + s)}{s(1 + s^2)}\}\}$$

We find $y(t)$ using `InverseLaplaceTransform`.

```
sol = InverseLaplaceTransform[step3[[1, 1, 2]], s, t]
```

$$1 - \text{Cos}[t] - \text{HeavisideTheta}[-\pi + t]\text{Sin}[t]$$

```
p2 = Plot[sol, {t, 0, 3Pi},
    PlotStyle → CMYKColor[0, 0.89, 0.94, 0.28],
    PlotLabel → "(b)"]
```

We can use `DSolve` to find the solution to the initial-value problem as follows. The result is graphed with `Plot` in Figure 8-11(a).

```
sol = DSolve[{y''[t] + y[t]==DiracDelta[t − π] + 1, y[0]==0,
y'[0]==0}, y[t], t]
```

$$\{\{y[t] \rightarrow -\text{Cos}[t] + \text{Cos}[t]^2 - \text{HeavisideTheta}[-\pi + t]\text{Sin}[t] + \text{Sin}[t]^2\}\}$$

```
p1 = Plot[y[t]/.sol, {t, 0, 2π},
    PlotStyle → CMYKColor[0, 0.89, 0.94, 0.28],
    PlotLabel → "(a)"]
```

```
Show[GraphicsRow[{p1, p2}]]
```

■

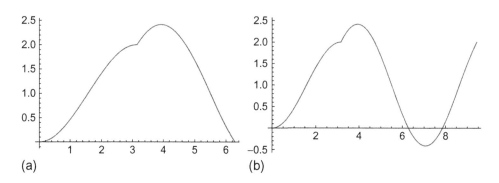

Figure 8-11   At $t = \pi$, an impulse is delivered

The forcing function may involve a combination of functions as illustrated in the following example.

EXAMPLE 8.4.10: Solve $y'' + 2y' + y = 1 + \delta(t - \pi) + \delta(t - 2\pi)$ subject to $y(0) = y'(0) = 0$.

SOLUTION: We first remark that we can use `DSolve` to solve the initial value problem that are then graphed with `Plot` in Figure 8-12.

```
Clear[y, t, sol]

sol = DSolve[{y''[t] + 2y'[t] + y[t]==1 + DiracDelta[t − π]
    +DiracDelta[t − 2π], y[0]==0, y'[0]==0}, y[t], t]
```

$\{\{y[t] \rightarrow e^{-t}(-1 + e^t - t - 2e^{2\pi}\pi\text{HeavisideTheta}[-2\pi + t] +$
$e^{2\pi}t\text{HeavisideTheta}[-2\pi + t] - e^{\pi}\pi\text{HeavisideTheta}[-\pi + t] +$
$e^{\pi}t\text{HeavisideTheta}[-\pi + t])\}\}$

```
p1 = Plot[y[t]/.sol, {t, 0, 4π},

    PlotStyle → CMYKColor[0, 0.89, 0.94, 0.28]]
```

Alternatively, we implement the method of Laplace transforms. After computing the Laplace transform of each side of the equation

```
step1 = LaplaceTransform[y''[t] + 2y'[t] + y[t] ==

    1 + DiracDelta[t − π] + DiracDelta[t − 2π], t, s]
```

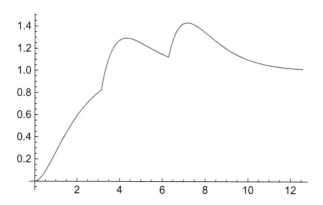

Figure 8-12   Impulses are delivered at $t = \pi$ and $t = 2\pi$

$$\text{LaplaceTransform}[y[t], \, t, \, s] + s^2\text{LaplaceTransform}[y[t],$$

$$t, \, s] + 2(s\text{LaplaceTransform}[y[t], \, t, \, s] - y[0]) - sy[0] - y'[0] ==$$

$$e^{-2\pi s} + e^{-\pi s} + \frac{1}{s}$$

and applying the initial conditions,

```
step2 = step1/.{y[0]->0, y'[0]->0}
```

$$\text{LaplaceTransform}[y[t], \, t, \, s] + 2s\text{LaplaceTransform}[y[t],$$

$$t, \, s] + s^2\text{LaplaceTransform}[y[t], \, t, \, s] == e^{-2\pi s} + e^{-\pi s} + \frac{1}{s}$$

we solve for $Y(s)$

```
step3 = Solve[step2, LaplaceTransform[y[t], t, s]]
```

$$\{\{\text{LaplaceTransform}[y[t], \, t, \, s] \to \frac{e^{-2\pi s}(e^{2\pi s} + s + e^{\pi s}s)}{s(1+s)^2}\}\}$$

and then compute $y(t) = \mathcal{L}^{-1}\{Y(s)\}$.

```
sol = InverseLaplaceTransform[ (1 + E^{-2πs}s + E^{-πs}s)/(s(1+s)^2) , s, t]//
FullSimplify
```

$$e^{-t}(-1 + e^t - t + e^{\pi}(-e^{\pi}(2\pi - t)\text{HeavisideTheta}[-2\pi + t] + (-\pi + t)\text{HeavisideTheta}[-\pi + t]))$$

■

8.5 The Convolution Theorem

8.5.1 The Convolution Theorem

In many cases, we are required to determine the inverse Laplace transform of a product of two functions. Just as in integral calculus when the integral of the product of two functions did not produce the product of the integrals, neither does the inverse Laplace transform of the product yield the product of the inverse Laplace transforms. We state the following theorem.

Theorem 32 (Convolution Theorem). *Suppose that $f(t)$ and $g(t)$ are piecewise continuous on $[0, \infty)$ and both of exponential order b. Further suppose that $\mathcal{L}\{f(t)\} = F(s)$ and $\mathcal{L}\{g(t)\} = G(s)$. Then,*

$$\mathcal{L}^{-1}\{F(s)G(s)\} = \mathcal{L}^{-1}\{\mathcal{L}\{(f * g)(t)\}\} = (f * g)(t) = \int_0^t f(t - v)g(v)\, dv. \quad (8.38)$$

```
Clear[f, g]
LaplaceTransform[∫₀ᵗ f[t − v]g[v]dv, t, s]
```

```
LaplaceTransform[f[t], t, s]LaplaceTransform[g[t], t, s]
```

Note that

$$(f * g)(t) = \int_0^t f(t - v)g(v)\, dv$$

is called the **convolution integral**.

EXAMPLE 8.5.1: Compute $(f * g)(t)$ if $f(t) = e^{-t}$ and $g(t) = \sin t$. Verify the Convolution theorem for these functions.

SOLUTION: We use the definition and integration by parts to obtain

$$(f * g)(t) = \int_0^t f(t - v)g(v)\, dv = \int_0^t e^{-t+v} \sin v\, dv = e^{-t} \int_0^t e^v \sin v\, dv$$

$$= e^{-t}\left[\frac{1}{2}e^v(\sin v - \cos v)\right]_0^v = \frac{1}{2}e^{-t}\left[e^t(\sin t - \cos t) - (\sin 0 - \cos 0)\right]$$

$$= \frac{1}{2}(\sin t - \cos t) + \frac{1}{2}e^{-t}.$$

The same results are obtained with Mathematica. After defining convolution, which computes $(f * g)(t)$,

```
Clear[convolution, f, t, g, v];

convolution[f_, g_]:= ∫₀ᵗ f[t − v]g[v]dv
```

we define $f(t)$ and $g(t)$

```
f[t_] = Exp[−t];

g[t_] = Sin[t];
```

and then use convolution to compute $(f * g)(t)$.

```
convolution[f, g]
```

$$\frac{1}{2}(e^{-t} - \text{Cos}[t] + \text{Sin}[t])$$

Note that $(f * g)(t) = (g * f)(t)$.

```
convolution[g, f]
```

$$\frac{1}{2}(e^{-t} - \text{Cos}[t] + \text{Sin}[t])$$

Now, according to the Convolution theorem, $\mathcal{L}\{f(t)\}\mathcal{L}\{g(t)\} = \mathcal{L}\{(f * g)(t)\}$. In this example, we have

$$F(s) = \mathcal{L}\{f(t)\} = \mathcal{L}\{e^{-t}\} = \frac{1}{s+1} \quad \text{and} \quad G(s) = \mathcal{L}\{g(t)\} = \mathcal{L}\{\sin t\} = \frac{1}{s^2+1}.$$

Hence, $\mathcal{L}^{-1}\{F(s)G(s)\} = \mathcal{L}^{-1}\left\{\frac{1}{s+1} \cdot \frac{1}{s^2+1}\right\}$ should equal $(f * g)(t)$.

We compute $\mathcal{L}^{-1}\left\{\frac{1}{s+1} \cdot \frac{1}{s^2+1}\right\}$ with InverseLaplaceTransform.

```
InverseLaplaceTransform[ 1/((s+1)(s²+1)), s, t]
```

$$\frac{e^{-t}}{2} + \frac{1}{2}(-\text{Cos}[t] + \text{Sin}[t])$$

Hence,

$$\mathcal{L}^{-1}\left\{\frac{1}{s+1} \cdot \frac{1}{s^2+1}\right\} = \frac{1}{2}e^{-t} - \frac{1}{2}\cos t + \frac{1}{2}\sin t,$$

which is the same result as that obtained for $(f * g)(t)$.

∎

EXAMPLE 8.5.2: Use the Convolution theorem to find the Laplace transform of $h(t) = \int_0^t \cos(t - v) \sin v \, dv$.

SOLUTION: Notice that $h(t) = (f * g)(t)$, where $f(t) = \cos t$ and $g(t) = \sin t$. Therefore, by the Convolution theorem, $\mathcal{L}\{(f * g)(t)\} = F(s)G(s)$. Hence,

$$\mathcal{L}\{h(t)\} = \mathcal{L}\{f(t)\} \mathcal{L}\{g(t)\} = \mathcal{L}\{\cos t\} \mathcal{L}\{\sin t\} = \frac{s}{s^2 + 1} \cdot \frac{1}{s^2 + 1} = \frac{s}{\left(s^2 + 1\right)^2}.$$

The same result is obtained with `LaplaceTransform`.

```
LaplaceTransform[∫₀ᵗ Cos[t − v]Sin[v] dv, t, s]//Simplify
```

$$\frac{s}{(1 + s^2)^2}$$

■

8.5.2 Integral and Integrodifferential Equations

The Convolution theorem is useful in solving numerous problems. In particular, this theorem can be employed to solve **integral equations**, which are equations that involve an integral of the unknown function.

EXAMPLE 8.5.3: Use the Convolution theorem to solve the integral equation

$$h(t) = 4t + \int_0^t h(t - v) \sin v \, dv.$$

SOLUTION: We first note that the integral in this equation represents $(h * g)(t)$ where $g(t) = \sin t$. Therefore, if we apply the Laplace transform to both sides of the equation, we obtain

$$\mathcal{L}\{h(t)\} = \mathcal{L}\{4t\} + \mathcal{L}\{h(t)\} \mathcal{L}\{\sin t\}$$

$$H(s) = \frac{4}{s^2} + H(s)\frac{1}{s^2 + 1},$$

where $H(s) = \mathcal{L}\{h(t)\}$. The same result is obtained with `LaplaceTransform`.

```
Clear[h]
```

```
step1 = LaplaceTransform[h[t]==4t + ∫₀ᵗ h[t − v]Sin[v]dv, t, s]
```

$\text{LaplaceTransform}[h[t], t, s] == \dfrac{4}{s^2} + \dfrac{\text{LaplaceTransform}[h[t], t, s]}{1 + s^2}$

Solving for $H(s)$, we have

$$H(s)\left(1 - \frac{1}{s^2 + 1}\right) = \frac{4}{s^2} \quad \text{so} \quad H(s) = \frac{4\left(s^2 + 1\right)}{s^4} = \frac{4}{s^2} + \frac{4}{s^4}.$$

```
step2 = Solve[step1, LaplaceTransform[h[t], t, s]]
```

$\left\{\left\{\text{LaplaceTransform}[h[t], t, s] \rightarrow \dfrac{4(1 + s^2)}{s^4}\right\}\right\}$

Then by computing the inverse Laplace transform,

```
sol = InverseLaplaceTransform[step2[[1, 1, 2]], s, t]
```

$4\left(t + \dfrac{t^3}{6}\right)$

we find that

$$h(t) = \mathcal{L}^{-1}\left\{\frac{4}{s^2} + \frac{4}{s^4}\right\} = 4t + \frac{2}{3}t^3.$$

■

Laplace transforms are helpful in solving problems of other types as well. Next, we illustrate how Laplace transforms can be used to solve an **integrodifferential equation**, an equation that involves a derivative as well as an integral of the dependent variable, the unknown function.

EXAMPLE 8.5.4: Solve $\dfrac{dy}{dt} + y + \int_0^t y(u)\,du = 1$ subject to $y(0) = 0$.

SOLUTION: Because we must take the Laplace transform of both sides of this integrodifferential equation, we first compute

$$\mathcal{L}\left\{\int_0^t y(u)\,du\right\} = \mathcal{L}\{(1 * y)(t)\} = \mathcal{L}\{1\}\mathcal{L}\{y\} = \frac{Y(s)}{s}.$$

Hence,

$$\mathcal{L}\left\{\frac{dy}{dt}\right\} + \mathcal{L}\{y\} + \mathcal{L}\left\{\int_0^t y(u)\,du\right\} = \mathcal{L}\{1\}$$

$$sY(s) - y(0) + Y(s) + \frac{Y(s)}{s} = \frac{1}{s}$$

$$s^2 Y(s) + sY(s) + Y(s) = 1$$

$$Y(s) = \frac{1}{s^2 + s + 1}.$$

The same steps are carried out with Mathematica.

```
step1 = LaplaceTransform[y'[t] + y[t] + ∫₀ᵗ y[u] du==1, t, s]
```

$$\text{LaplaceTransform}[y[t], t, s] + \frac{\text{LaplaceTransform}[y[t], t, s]}{s} +$$
$$s\text{LaplaceTransform}[y[t], t, s] - y[0] == \frac{1}{s}$$

```
step2 = step1/. y[0] → 0
```

$$\text{LaplaceTransform}[y[t], t, s] + \frac{\text{LaplaceTransform}[y[t], t, s]}{s} +$$
$$s\text{LaplaceTransform}[y[t], t, s] == \frac{1}{s}$$

```
step3 = Solve[step2, LaplaceTransform[y[t], t, s]]
```

$$\{\{\text{LaplaceTransform}[y[t], t, s] \to \frac{1}{1 + s + s^2}\}\}$$

Because $Y(s) = \dfrac{1}{s^2 + s + 1} = \dfrac{1}{(s+1/2)^2 + \left(\sqrt{3}/2\right)^2}$, $y(t) =$

$\dfrac{2}{\sqrt{3}} e^{-t/2} \sin \dfrac{\sqrt{3}}{2} t.$

The same solution, which is then graphed on the interval $[0, 3\pi]$ with `Plot` in Figure 8-13, is found with `InverseLaplaceTransform` and named `sol`.

```
sol = InverseLaplaceTransform[step3[[1, 1, 2]], s, t]
```

$$\frac{2e^{-t/2}\text{Sin}[\frac{\sqrt{3}t}{2}]}{\sqrt{3}}$$

```
Plot[sol, {t, 0, 3π}]
```

■

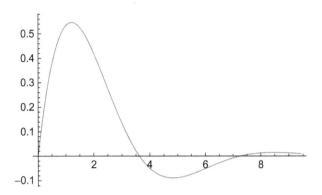

Figure 8-13   Plot of the solution to an integrodifferential equation

8.6 Applications of Laplace Transforms, Part I

8.6.1 Spring-Mass Systems Revisited

Laplace transforms are useful in solving the spring-mass systems that were discussed in earlier sections. Although the method of Laplace transforms can be used to solve all problems discussed in the section on applications of higher-order equations, this method is most useful in alleviating the difficulties associated with problems that involve piecewise-defined forcing functions. Hence, we investigate the use of Laplace transforms to solve the second-order initial value problem that models the motion of a mass attached to the end of a spring. We found in Chapter 5 that without forcing this situation is modeled by the initial-value problem

$$\begin{cases} mx'' + cx' + kx = 0 \\ x(0) = \alpha,\ x'(0) = \beta \end{cases},$$
(8.39)

where m represents the mass, c the damping coefficient, and k the spring constant determined by Hooke's law. We demonstrate how the method of Laplace transforms is used to solve initial-value problems of this type if the forcing function is discontinuous.

EXAMPLE 8.6.1: Suppose that a mass with $m = 1$ is attached to a spring with spring constant $k = 1$. If there is no resistance due to damping determine the displacement of the mass if it is released from its equilibrium position and is subjected to the force

$$f(t) = \begin{cases} \sin t, \, 0 \le t < \pi/2 \\ 0, \, t \ge \pi/2 \end{cases}.$$

SOLUTION: In this case, the constants are $m = k = 1$ and $c = 0$. The initial position is $x(0) = 0$ and the initial velocity is $x'(0) = 0$. Hence, the initial-value problem that models this situation is

$$x'' + x = \begin{cases} \sin t, \, 0 \le t < \pi/2 \\ 0, \, t \ge \pi/2 \end{cases}, \, x(0) = 0, \, x'(0) = 0.$$

Because we will take the Laplace transform of both sides of the differential equation, we write $f(t)$ in terms of the unit step function. This gives us

$$f(t) = [\mathcal{U}(t - 0) - \mathcal{U}(t - \pi/2)] \sin t = [1 - \mathcal{U}(t - \pi/2)] \sin t,$$

which we graph with `Plot` in Figure 8-14(a).

$$\mathbf{p1 = Plot[Sin[t](1 - UnitStep[t - \frac{\pi}{2}]), \{t, 0, \pi\},}$$

$$\mathbf{PlotLabel \to ``(a)''}]$$

Using the Method of Laplace transforms, we compute the Laplace transform of each side of the equation

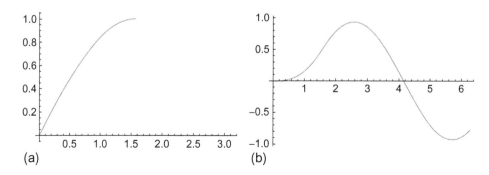

(a)   (b)

Figure 8-14   (a) Plot of the forcing function. (b) For $t \ge \pi/2$, the motion is harmonic

```
step1 = LaplaceTransform[x''[t] + x[t]==Sin[t](1 - UnitStep
```
$[t - \frac{\pi}{2}])$, t, s]

LaplaceTransform[x[t], t, s] + s^2LaplaceTransform[x[t],

t, s] $- sx[0] - x'[0] == \dfrac{1}{1 + s^2} - \dfrac{e^{-\frac{\pi s}{2}} s}{1 + s^2}$

apply the initial conditions

```
step2 = step1/.{x[0] → 0, x'[0] → 0}
```

LaplaceTransform[x[t], t, s] + s^2LaplaceTransform[x[t],

t, s] $== \dfrac{1}{1 + s^2} - \dfrac{e^{-\frac{\pi s}{2}} s}{1 + s^2}$

and solve the resulting equation for $x(t) = \mathcal{L}^{-1}\{X(s)\}$.

```
step3 = Solve[step2, LaplaceTransform[x[t], t, s]]
```

$\{\{\text{LaplaceTransform}[x[t], t, s] \rightarrow \dfrac{e^{-\frac{\pi s}{2}}(e^{\frac{\pi s}{2}} - s)}{(1 + s^2)^2}\}\}$

The solution is obtained with `InverseLaplaceTransform`.

```
sol = InverseLaplaceTransform[
```
$-\dfrac{-1 + E^{-\frac{\pi s}{2}} s}{(1 + s^2)^2}$, s, t]

$\frac{1}{4}(-2t\text{Cos}[t] - (\pi - 2t)\text{Cos}[t]\text{HeavisideTheta}[-\frac{\pi}{2} + t] + 2\text{Sin}[t])$

The same result is obtained with `DSolve`, which we then graph with `Plot` in Figure 8-14(b).

```
Clear[x, t, sol]
sol = DSolve[{x''[t] + x[t]==Sin[t](1 - UnitStep[t - π/2]),
```
     x[0]==0, x'[0]==0}, x[t], t]//Simplify

$\{\{x[t] \rightarrow \frac{1}{4}(-\pi\text{Cos}[t] + 2\text{Sin}[t] + (\pi - 2t)\text{Cos}[t]\text{UnitStep}[\frac{\pi}{2} - t])\}\}$

```
p2 = Plot[x[t]/.sol, {t, 0, 2π}, PlotLabel → "(b)"]

Show[GraphicsRow[{p1, p2}]]
```

Notice that resonance begins on the interval $0 \le t < \pi/2$. Then, for $t \ge \pi/2$, the motion is harmonic. Hence, although the forcing function is zero for $t \geq \pi/2$, the mass continues to follow the path defined by $x(t)$ indefinitely.

∎

EXAMPLE 8.6.2: Suppose that a mass of $m = 1$ is attached to a spring with spring constant $k = 13$. If the mass is subjected to the resistive force due to damping $F_R = 4\,dx/dt$, determine the displacement of the mass if it is released from its equilibrium position and is subjected to the force

$$f(t) = 2t\,[1 - \mathcal{U}(t - 1)] + 2\mathcal{U}(t - 1) + 10\delta(t - 3).$$

SOLUTION: In this case, the initial-value problem is

$$\begin{cases} x'' + 4x' + 13x = 2t\,[1 - \mathcal{U}(t - 1)] + 2\mathcal{U}(t - 1) + 10\delta(t - 3) \\ x(0) = x'(0) = 0 \end{cases}.$$

We first graph $2t\,[1 - \mathcal{U}(t - 1)] + 2\mathcal{U}(t - 1)$ in Figure 8-15(a).

```
p1 = Plot[2 t(1−UnitStep[t−1])+2UnitStep[t−1], {t, 0, 4},

    PlotRange->{0, 4}, AspectRatio->1, PlotLabel → "(a)"]
```

Using the Method of Laplace transforms, we take the Laplace transform of each side of the equation

```
step1 =

LaplaceTransform[x″[t] + 4x′[t] + 13x[t]==2 t(1 − UnitStep

    [t − 1]) + 2UnitStep[t − 1] + 10 textDiracDelta[t − 3], t, s]
```

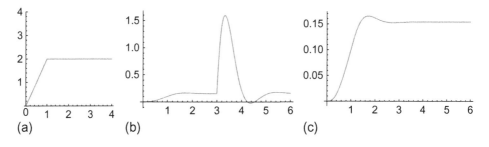

(a)   (b)   (c)

Figure 8-15   (a) Plot of $2t\,[1 - \mathcal{U}(t - 1)] + 2\mathcal{U}(t - 1)$. (b) Note the effect of the impulse delivered at $t = 3$. (c) No impulse is delivered

13LaplaceTransform[x[t], t, s] + s²LaplaceTransform[x[t],

t, s] + 4(sLaplaceTransform[x[t], t, s] − x[0]) − sx[0] − x′[0] ==

$10e^{-3s} + \dfrac{2}{s^2} + \dfrac{2e^{-s}}{s} - \dfrac{2e^{-s}(1+s)}{s^2}$

apply the initial conditions

step2 = step1/.{x[0]->0, x′[0]->0}

13LaplaceTransform[x[t], t, s] + 4sLaplaceTransform[x[t],

t, s] + s²LaplaceTransform[x[t], t, s] == $10e^{-3s} + \dfrac{2}{s^2} + \dfrac{2e^{-s}}{s} -$

$\dfrac{2e^{-s}(1+s)}{s^2}$

and solve for $X(s) = \mathcal{L}\{x(t)\}$.

step3 = Solve[step2, LaplaceTransform[x[t], t, s]]

$\{\{\text{LaplaceTransform}[x[t], t, s] \to \dfrac{2e^{-3s}(-e^{2s} + e^{3s} + 5s^2)}{s^2(13 + 4s + s^2)}\}\}$

The solution to the initial-value problem is obtained with
InverseLaplaceTransform.

sol = InverseLaplaceTransform[$\dfrac{2(1 - E^{-s} + 5E^{-3s}s^2)}{s^2(13 + 4s + s^2)}$, s, t]//

Simplify

$\dfrac{1}{507}(-24 + e^{(-2-3i)t}((12 - 5i) + (12 + 5i)e^{6it}) + 78t -$

$845ie^{(-2-3i)(-3+t)}(-1 + e^{6i(-3+t)})$HeavisideTheta$[-3 + t] -$

$(-102 + (12 - 5i)e^{(-2-3i)(-1+t)} + (12 + 5i)e^{(-2+3i)(-1+t)} +$

$78t)$HeavisideTheta$[-1 + t])$

The same result is obtained with DSolve. The solution is graphed
with Plot in Figure 8-15(b).

sol =

DSolve[

{x″[t] + 4x′[t] + 13x[t]==2t(1 − UnitStep[t − 1]) + 2UnitStep

[t − 1] + 10DiracDelta[t − 3], x[0]==0, x′[0]==0}, x[t], t]//

Simplify

$\{\{x[t] \to \dfrac{2}{507}e^{-2t}(39e^{2t} - 12e^2\text{Cos}[3-3t] + 12\text{Cos}[3t] - 5e^2\text{Sin}[3-$

$3t] - 845e^6$HeavisideTheta$[-3+t]$Sin$[9-3t] - 5\text{Sin}[3t] + (3e^{2t}(-17+$

$13t) + 12e^2\text{Cos}[3 - 3t] + 5e^2\text{Sin}[3 - 3t])$UnitStep$[1 - t])\}\}$

```
p2 = Plot[x[t]/.sol, {t, 0, 6}, PlotRange->All,

     PlotLabel → "(b)"]
```

The graph of the solution shows the effect of the impulse delivered at $t = 3$, which is especially evident when we compare this result to the solution of

$$\begin{cases} x'' + 4x' + 13x = 2t\,[1 - \mathcal{U}(t-1)] + 2\mathcal{U}(t-1) \\ x(0) = x'(0) = 0 \end{cases}.$$

shown in Figure 8-15(c).

```
sol2 =

DSolve[{x''[t] + 4x'[t] + 13x[t]==2t(1 - UnitStep[t - 1])

               +2UnitStep

[t - 1], x[0]==0, x'[0]==0}, x[t], t]//Simplify
```

$$\left\{\left\{x[t] \rightarrow \left\{ \begin{array}{ll} \dfrac{1}{507}e^{-2t}(6e^{2t}(-4+13t)+24\mathrm{Cos}[3t]-10\mathrm{Sin}[3t]) & t \leq 1 \\ \dfrac{2}{507}e^{-2t}(39e^{2t}-12e^2\mathrm{Cos}[3-3t]+12\mathrm{Cos}[3t] \\ \quad -5e^2\mathrm{Sin}[3-3t]-5\mathrm{Sin}[3t]) & \mathrm{True} \end{array} \right. \right\}\right\}$$

```
p3 = Plot[x[t]/.sol2, {t, 0, 6}, PlotRange->All,

     PlotLabel → "(c)"]

Show[GraphicsRow[{p1, p2, p3}]]
```

∎

8.6.2 *L-R-C* Circuits Revisited

Laplace transforms can be used to solve the *L-R-C* circuit problems that were introduced earlier. Recall that the initial-value problem that is used to find the current is

$$\begin{cases} L\dfrac{d^2Q}{dt^2} + R\dfrac{dQ}{dt} + \dfrac{1}{C}Q = E(t) \\ Q(0) = Q_0,\ I(0) = \dfrac{dQ}{dt}(0) = I_0 \end{cases}, \tag{8.40}$$

where L, R, and C represent the inductance, resistance, and capacitance, respectively. Q is the charge of the capacitor and $dQ/dt = I$, where I is the current. $E(t)$ is the voltage supply. In particular, the method of Laplace transforms is most useful when the supplied voltage, $E(t)$, is piecewise defined.

EXAMPLE 8.6.3: Suppose that we consider a circuit with a capacitor C, a resistor R, and a voltage supply

$$E(t) = \begin{cases} 100, \ 0 \le t < 1 \\ 200 - 100t, \ 1 \le t < 2 \\ 0, \ t \ge 2 \end{cases}.$$

If $L = 0$, find $Q(t)$ and $I(t)$ if $Q(0) = 0$, $C = 10^{-2}$ farads, and $R = 100 \ \Omega$.

SOLUTION: Because $L = 0$, we can state the first-order initial-value problem as

$$\begin{cases} 100\dfrac{dQ}{dt} + 100Q = \begin{cases} 100, \ 0 \le t < 1 \\ 200 - 100t, \ 1 \le t < 2 \\ 0, \ t \ge 2 \end{cases}. \\ Q(0) = 0 \end{cases}$$

First, we rewrite $E(t)$ in terms of the unit step functions as

$$E(t) = 100\,[1 - \mathcal{U}(t - 1)] + (200 - 100t)\,[\mathcal{U}(t - 1) - \mathcal{U}(t - 2)]\,.$$

When we use Mathematica to define $E(t)$, we use a lower-case e to avoid ambiguity with E, which represents $e \approx 2.71828$. See Figure 8-16(a).

```
e[t_]=100(1-UnitStep[t-1])+(200-100t)(UnitStep[t-1]-
UnitStep[t-2])
```

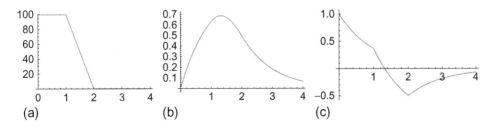

Figure 8-16   (a) Plot of $E(t)$. (b) $Q(t)$. (c) $I(t)$

100(1 − UnitStep[−1 + t]) + (200 − 100t)(−UnitStep[−2 + t] +

UnitStep[−1 + t])

p1 = Plot[e[t], {t, 0, 4}, PlotRange->{0, 100},

  PlotLabel → "(a)"]

Now, we take the Laplace transform of both sides of the differential equation

step1 = LaplaceTransform[100q′[t] + 100q[t]==e[t], t, s]

100LaplaceTransform[q[t], t, s]+100(sLaplaceTransform[q[t],

$t, s] − q[0]) == \dfrac{100}{s} − \dfrac{200e^{-2s}}{s} + \dfrac{100e^{-s}}{s} − \dfrac{100e^{-s}(1 + s)}{s^2}$

$+ \dfrac{100e^{-2s}(1 + 2s)}{s^2}$

apply the initial condition

step2 = step1/. q[0]->0

100LaplaceTransform[q[t], t, s]+100sLaplaceTransform[q[t],

$t, s] == \dfrac{100}{s} − \dfrac{200e^{-2s}}{s} + \dfrac{100e^{-s}}{s} − \dfrac{100e^{-s}(1 + s)}{s^2}$

$+ \dfrac{100e^{-2s}(1 + 2s)}{s^2}$

and solve for $\mathcal{L}\{Q(t)\}$.

step3 = Solve[step2, LaplaceTransform[q[t], t, s]]

$\{\{\text{LaplaceTransform}[q[t], t, s] → \dfrac{e^{-2s}(1 − e^s + e^{2s}s)}{s^2(1 + s)}\}\}$

The solution to the initial-value problem is obtained with InverseLaplaceTransform.

sol = InverseLaplaceTransform[$\dfrac{E^{-2s} − E^{-s} + s}{s^2(1 + s)}$, s, t]

$1 − e^{-t} + (−3 + e^{2-t} + t)$HeavisideTheta[−2 + t] − (−2 + e^{1-t} +

t)HeavisideTheta[−1 + t]

The same result is obtained with DSolve.

sol = DSolve[{100q′[t] + 100q[t]==e[t], q[0]==0}, q[t], t]//Simplify

$$\left\{\left\{q[t] → \left\{\begin{array}{ll} e^{-t}(−1 − e + e^2) & t > 2 \\ 1 − e^{-t} & t \le 1 \\ −e^{-t}(1 + e + e^t(−3 + t)) & \text{True} \end{array}\right.\right\}\right\}$$

We now compute $I = dQ/dt$ and then graph both $Q(t)$ and $I(t)$ on the interval $[0, 4]$ in Figure 8-16(b) and (c).

```
i[t_] = D[q[t]/.sol, t];

pq = Plot[q[t]/.sol, {t, 0, 4}, PlotLabel → "(b)"];

pi = Plot[i[t], {t, 0, 4}, PlotLabel → "(c)"];

Show[GraphicsRow[{p1, pq, pi}]]
```

From the graphs in Figure 8-16, we see that after the voltage source is turned off at $t = 2$, the charge approaches zero.

∎

EXAMPLE 8.6.4: Consider the circuit with no capacitor, $R = 100\,\Omega$, and $L = 100\,H$ if $E(t) = \begin{cases} 100\,V, & 0 \le t < 1 \\ 0, & 1 \le t < 2 \end{cases}$ and $E(t+2) = E(t)$. Find the current $I(t)$ if $I(0) = 0$.

SOLUTION: The differential equation that models this situation is $100Q'' + 100Q' = E(t)$. Now, $Q' = I$, so we can write this equation as $100I' + 100I = E(t)$. Hence, the initial-value problem is

$$\begin{cases} 100\dfrac{dI}{dt} + 100I = E(t) \\ I(0) = 0 \end{cases}.$$

Notice that $E(t)$ is a periodic function, so we first compute

```
Clear[i, step1, step2]

lape = Simplify[∫₀¹ 100Exp[-st]dt / (1 - Exp[-2s])]
```

$$\frac{100e^s}{s + e^s s}$$

and see $\mathcal{L}\{E(t)\} = \dfrac{100}{s\left(1 + e^{-s}\right)}$.

We now compute the Laplace transform of the left side of the equation

```
step1 = LaplaceTransform[100i'[t] + 100i[t], t, s]==lape
```

Note that we use i to represent I instead of I because I represents the imaginary number $i = \sqrt{-1}$.

$$100 \text{LaplaceTransform}[i[t], t, s] + 100(-i[0] + s$$
$$\text{LaplaceTransform}[i[t], t, s]) == \frac{100e^s}{s + e^s s}$$

apply the initial condition

step2 = step1/.i[0]->0

$$100 \text{LaplaceTransform}[i[t], t, s] + 100s$$
$$\text{LaplaceTransform}[i[t], t, s] == \frac{100e^s}{s + e^s s}$$

and solve for $\mathcal{L}\{I(t)\}$.

step3 = Solve[step2, LaplaceTransform[i[t], t, s]]

$$\{\{\text{LaplaceTransform}[i[t], t, s] \rightarrow \frac{e^s}{(1 + e^s)s(1 + s)}\}\}$$

We use
$\frac{1}{1+x} = \sum n = 0^{\infty} (-x)^n = 1 - x + x^2 - x^3 + \cdots.$

As we did before, we write a power series expansion of $\frac{1}{1 + e^{-s}}$:

$$\frac{1}{1 + e^{-s}} = \sum_{n=0}^{\infty} \left(-e^{-s}\right)^n = 1 - e^{-s} + e^{-2s} - e^{-3s} + \cdots.$$

Thus,

$$\mathcal{L}\{I(t)\} = \frac{1}{s(s+1)} \left(1 - e^{-s} + e^{-2s} - e^{-3s} + \cdots\right).$$

Because, $\mathcal{L}^{-1}\left\{\frac{1}{s(s+1)}\right\} = 1 - e^{-t}$,

InverseLaplaceTransform[$\frac{1}{s(s+1)}$, s, t]

$1 - e^{-t}$

we have that

$$I(t) = \left(1 - e^{-t}\right) - \left(1 - e^{-(t-1)}\right)\mathcal{U}(t-1) + \left(1 - e^{-(t-2)}\right)\mathcal{U}(t-2)$$
$$- \left(1 - e^{-(t-3)}\right)\mathcal{U}(t-3) + \cdots.$$

We can write this function as

$$I(t) = \begin{cases} 1 - e^{-t}, \, 0 \leq t < 1 \\ -e^{-t} + e^{-(t-1)}, \, 1 \leq t < 2 \\ 1 - e^{-t} + e^{-(t-1)} - e^{-(t-2)}, \, 2 \leq t < 3 \\ -e^{-t} + e^{-(t-1)} - e^{-(t-2)} + e^{-(t-3)}, \, 3 \leq t < 4 \\ \vdots \end{cases}$$

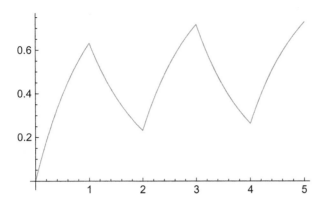

Figure 8-17   Plot of $I(t)$ on the interval $[0,5]$

To graph $I(t)$ on the interval $[0,n]$, we note that $\mathcal{U}(t-n) = 0$ for $t \leq n$ so the graph of $I(t)$ on the interval $[0,n]$ is the same as the graph of

$$\left(1 - e^{-t}\right) - \left(1 - e^{-(t-1)}\right)\mathcal{U}(t-1) + \left(1 - e^{-(t-2)}\right)$$
$$\mathcal{U}(t-2) - \left(1 - e^{-(t-3)}\right)\mathcal{U}(t-3) + \cdots + (-1)^{n-1}\left(1 - e^{-[t-(n-1)]}\right)$$
$$\mathcal{U}(t-(n-1)).$$

```
Clear[i]
i[n_]:=i[n] = i[n − 1] + (−1)ⁿ(1 − Exp[−(t − n)])UnitStep[t − n]
i[0] = 1 − Exp[−t];
```

For example, to graph $I(t)$ on the interval $[0,5]$ we enter

```
i[4]
```

$1 - e^{-t} + (1 - e^{4-t})\mathtt{UnitStep}[-4 + t] - (1 - e^{3-t})\mathtt{UnitStep}[-3 + t] + (1 - e^{2-t})\mathtt{UnitStep}[-2 + t] - (1 - e^{1-t})\mathtt{UnitStep}[-1 + t]$

and then use `Plot`. See Figure 8-17.

```
Plot[i[4], {t, 0, 5}]
```

Notice that $I(t)$ increases over the intervals where $E(t) = 100$ and decreases on those where $E(t) = 0$.

■

We can consider the *L-R-C* circuit in terms of the integrodifferential equation

$$L\frac{dI}{dt} + RI + \frac{1}{C}\int_0^t I(\alpha)\,d\alpha = E(t), \tag{8.41}$$

which is useful when using the method of Laplace transforms to find the current.

EXAMPLE 8.6.5: Find the current $I(t)$ if $L = 1$ henry, $R = 6\,\Omega$, $C = 1/9$ farad, $E(t) = 1$ volt, and $I(0) = 0$.

SOLUTION: In this case, we must solve the initial-value problem

$$\begin{cases} \dfrac{dI}{dt} + 6I + 9\int_0^t I(\alpha)\,d\alpha = 1 \\ I(0) = 0 \end{cases}.$$

First, we compute the Laplace transform of each side of the equation,

```
Clear[i]

step1 = LaplaceTransform[i'[t]+6i[t]+9Integrate[i[alpha],

{alpha, 0, t}]==1, t, s]
```

$-i[0] + 6\,\text{LaplaceTransform}[i[t], t, s]$
$+ \dfrac{9\,\text{LaplaceTransform}[i[t], t, s]}{s} + s\,\text{LaplaceTransform}[i[t],$
$t, s] == \dfrac{1}{s}$

apply the initial condition,

```
step2 = step1/.i[0]->0
```

$6\,\text{LaplaceTransform}[i[t], t, s] + \dfrac{9\,\text{LaplaceTransform}[i[t], t, s]}{s}$
$+ s\,\text{LaplaceTransform}[i[t], t, s] == \dfrac{1}{s}$

and solve for $\mathcal{L}\{I(t)\}$.

```
step3 = Solve[step2, LaplaceTransform[i[t], t, s]]
```

$\{\{\text{LaplaceTransform}[i[t], t, s] \to \dfrac{1}{(3 + s)^2}\}\}$

The solution is obtained with `InverseLaplaceTransform` (Figure 8-18),

```
sol = InverseLaplaceTransform[step3[[1, 1, 2]], s, t]
```

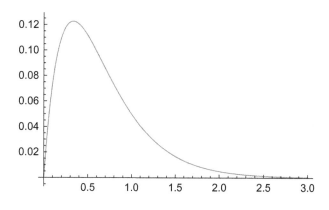

Figure 8-18 $I(t) \to 0$ as $t \to \infty$

$e^{-3t}t$

```
Plot[sol, {t, 0, 3}]
```

■

8.6.3 Population Problems Revisited

Laplace transforms can used to solve the population problems that were discussed as applications of first-order equations and systems. Laplace transforms are especially useful when dealing with piecewise-defined forcing functions, but they are useful in many other cases as well.

EXAMPLE 8.6.6: Let $x(t)$ represent the population of a certain country. The rate at which the population increases and decreases depends on the growth rate of the country as well as the rate at which people are being added to or subtracted from the population due to immigration or emigration. This motivates our study of the population problem

$$\begin{cases} x' + kx = 1000\,(1 + a \sin t) \\ x(0) = x_0 \end{cases}.$$

Solve this problem using Laplace transforms with $k = 3$, $x_0 = 2000$, and $a = 0.2, 0.4, 0.6$, and 0.8. Plot the solution in each case.

SOLUTION: Using the Method of Laplace transforms, we begin by computing the Laplace transform of each side of the equation with `LaplaceTransform`

> `step1 = LaplaceTransform[`
>
> `x'[t] + 3x[t]==1000(1 + aSin[t]), t, s]`
>
> `3LaplaceTransform[x[t], t, s] + sLaplaceTransform[x[t],`
>
> $t, s] - x[0] == 1000(\frac{1}{s} + \frac{a}{1 + s^2})$

apply the initial condition

> `step2 = step1/.x[0]->2000`
>
> $-2000 + 3\text{LaplaceTransform}[x[t], t, s] + s\text{LaplaceTransform}$
>
> $[x[t], t, s] == 1000(\frac{1}{s} + \frac{a}{1 + s^2})$

and then use `Solve` to solve `step2` for $X(s) = \mathcal{L}\{x(t)\}$.

> `step3 = Solve[step2, LaplaceTransform[x[t], t, s]]`
>
> $\{\{\text{LaplaceTransform}[x[t], t, s] \rightarrow \frac{1000(1 + 2s + as + s^2 + 2s^3)}{s(3 + s)(1 + s^2)}\}\}$

To find the solution, we use `InverseLaplaceTransform` and name the result `sol`.

> `sol = InverseLaplaceTransform[step3[[1, 1, 2]], s, t]//`
>
> `Simplify`
>
> $\frac{100}{3}(10 + (50 + 3a)e^{-3t} - 3a(\text{Cos}[t] - 3\text{Sin}[t]))$

We use the result to investigate the population for the values of a using `Plot`. See Figure 8-19.

> `toplot = Table[sol, {a, 0.2, 0.8, .2}];`
>
> `Plot[Evaluate[toplot], {t, 0, 25}, PlotRange->All,`
>
> `AxesOrigin->{0, 0}]`

■

Application: The Tautochrone

Suppose that from rest, a particle slides down a frictionless curve under the force of gravity. What must the shape of the curve be in order for the time of descent to be independent of the starting position of the particle?

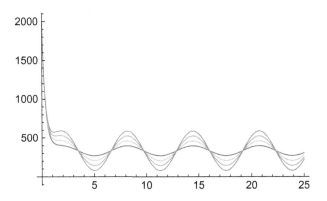

Figure 8-19   Fluctuations in the size of the population are larger for larger values of a

We can determine the shape of the curve using the method of Laplace transforms. Suppose that the particle starts at height y and that its speed is v when it is at a height of z. If m is the mass of the particle and g is the acceleration due to gravity, the speed is found by equating the kinetic and potential energies of the particle with

$$\frac{1}{2}mv^2 = mg(y - z)$$
$$v = \sqrt{2g}\sqrt{y - z}$$

Let σ denote the arc length along the curve from its lowest point to the particle. Then, the time required for the descent is

$$\text{time} = \int_0^{\sigma(y)} \frac{1}{v}\,d\sigma = \int_0^y \frac{1}{v}\frac{d\sigma}{dz} = \int_0^y \frac{1}{v}\phi(z)dz,$$

where $\phi(y) = d\sigma/dy$. The time is constant and $v = \sqrt{2g}\sqrt{y - z}$ so we have

$$\int_0^y \frac{\phi(z)}{\sqrt{y - z}}dz = c_1,$$

where c_1 is a constant. To use a convolution, we multiply by $e^{-sy}\,dy$ and integrate:

$$\int_0^\infty e^{-sy} \int_0^y \frac{\phi(z)}{\sqrt{y - z}}dz\,dy = \int_0^\infty e^{-sy}c_1\,dy$$
$$\mathcal{L}\left\{\phi * y^{-1/2}\right\} = \mathcal{L}\{c_1\}.$$

Using the Convolution theorem, we simplify to obtain

$$\mathcal{L}\{\phi\}\,\mathcal{L}\left\{y^{-1/2}\right\} = \frac{c_1}{s}.$$

```
step1 = LaplaceTransform[∫₀ʸ φ[z]/√(y - z) dz==c1, y, s]
```

$$\frac{\sqrt{\pi}\,\text{LaplaceTransform}[\phi[y],\, y,\, s]}{\sqrt{s}} == \frac{c1}{s}$$

Then, $\mathcal{L}\{\phi\} = \dfrac{c_1}{\sqrt{\pi s}}$.

```
step2 = Solve[step1, LaplaceTransform[φ[y], y, s]]
```

$$\{\{\text{LaplaceTransform}[\phi[y],\, y,\, s] \rightarrow \frac{c1}{\sqrt{\pi}\,\sqrt{s}}\}\}$$

We use InverseLaplaceTransform to compute $\phi = \mathcal{L}^{-1}\left\{\dfrac{c_1}{\sqrt{\pi s}}\right\} = \dfrac{c_1}{\pi}y^{-1/2} = ky^{-1/2}$.

```
step3 = InverseLaplaceTransform[step2[[1, 1, 2]], s, y]
```

$$\frac{c1}{\pi\,\sqrt{y}}$$

Recall that $\phi(y) = d\sigma/dy$ represents arc length. Then, $\phi(y) = d\sigma/dy = \sqrt{1 + (dx/dy)^2}$ and substitution of $\phi = ky^{-1/2}$ into this equation gives us

$$\sqrt{1 + \left(\frac{dx}{dy}\right)^2} = ky^{-1/2} \quad \text{or} \quad 1 + \left(\frac{dx}{dy}\right)^2 = \frac{k^2}{y}.$$

We solve this equation for dx/dy to obtain $\dfrac{dx}{dy} = \sqrt{\dfrac{k^2}{y} - 1}$. With the substitution $y = k^2 \sin^2 \theta$ we obtain

$$dx = \sqrt{\frac{k^2}{k^2 \sin^2 \theta} - 1} \cdot 2k^2 \sin\theta \cos\theta\, d\theta = \sqrt{\frac{k^2\left(1 - \sin^2 \theta\right)}{k^2 \sin^2 \theta}} \cdot 2k^2 \sin\theta \cos\theta\, d\theta$$

$$= \frac{\cos\theta}{\sin\theta} \cdot 2k^2 \sin\theta \cos\theta\, d\theta = 2k^2 \cos\theta\, d\theta$$

and integration results in $x(\theta) = \frac{1}{2}k^2 \left(2\theta + \sin 2\theta\right) + C_1$. To find C_1, we apply the initial condition $x(0) = 0$ to see that $C_1 = 0$ and $x(\theta) = \frac{1}{2}k^2 \left(2\theta + \sin 2\theta\right)$.

```
x[θ_, k_] = ∫ 2k²Cos[θ]² dθ
```

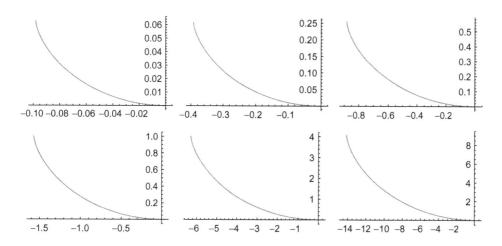

Figure 8-20 Increasing k increases the length of the curve

$$2k^2(\frac{\theta}{2} + \frac{1}{4}\text{Sin}[2\theta])$$

Using the identity $\sin^2\theta = \frac{1}{2}(1 - \cos 2\theta)$ yields $y(\theta) = k^2 \sin^2\theta = \frac{1}{2}k^2(1 - \cos 2\theta)$.

```
y[theta_, k_] = k^2Sin[theta]^2
```

$$k^2\text{Sin}[\text{theta}]^2$$

We use `ParametricPlot` to graph $\begin{cases} x = x(\theta) \\ y = y(\theta) \end{cases}$, $-\pi/2 \le \theta \le 0$ for various values of k in Figure 8-20.

```
somegraphs = Map[ParametricPlot[{x[theta, #], y[theta, #]},

    {theta, -Pi/2, 0}]&, {0.25, 0.5, 0.75, 1, 2, 3}];

toshow = Partition[somegraphs, 3];

Show[GraphicsGrid[toshow]]
```

The graphs illustrate that increasing the value of k increases the length of the curve. The time is independent of the choice of y (that is, the choice of θ). Therefore,

$$\text{time} = \int_0^y \frac{\phi(z)}{\sqrt{y-z}}dz = \int_0^y \frac{ky^{-1/2}}{\sqrt{y-z}}dz = -2k\left[\sqrt{\frac{y-z}{y}}\right]_0^y = -2k \cdot -1 = 2k.$$

8.7 Laplace Transform Methods for Systems

In many cases, Laplace transforms can be used to solve initial-value problems that involve a system of linear differential equations. This method is applied in much the same way that it was in solving initial-value problems involving higher-order differential equations. In the case of systems of differential equations, however, a system of algebraic equations is obtained after taking the Laplace transform of each equation. After solving the algebraic system for the Laplace transform of each of the unknown functions, the inverse Laplace transform is used to find each unknown function in the solution of the system.

EXAMPLE 8.7.1: Solve $\mathbf{X}' = \begin{pmatrix} 0 & 1 \\ 1 & 0 \end{pmatrix}\mathbf{X} + \begin{pmatrix} \sin t \\ 2\cos t \end{pmatrix}$ subject to $\mathbf{X}(0) = \begin{pmatrix} 2 \\ 0 \end{pmatrix}$.

SOLUTION: Let $\mathbf{X}(t) = \begin{pmatrix} x(t) \\ y(t) \end{pmatrix}$. Then, we can rewrite this initial-value problem as

$$\begin{cases} x' = y + \sin t \\ y' = x + 2\cos t \\ x(0) = 2,\ y(0) = 0 \end{cases}.$$

```
Clear[x, y]
sys = {x'[t]==y[t] + Sin[t], y'[t]==x[t] + 2Cos[t]};
```

Taking the Laplace transform of both sides of each equation yields the system

$$\begin{cases} sX(s) - x(0) = Y(s) + \dfrac{1}{s^2 + 1} \\ sY(s) - y(0) = X(s) + \dfrac{2s}{s^2 + 1} \end{cases}$$

```
step1 = LaplaceTransform[sys, t, s]
```

$\{s\text{LaplaceTransform}[x[t], t, s] - x[0] == \dfrac{1}{1 + s^2}$
$+ \text{LaplaceTransform}[y[t], t, s], s\text{LaplaceTransform}[y[t],$
$t, s] - y[0] == \dfrac{2s}{1 + s^2} + \text{LaplaceTransform}[x[t], t, s]\}$

and applying the initial condition results in

$$\begin{cases} sX(s) - Y(s) = \dfrac{1}{s^2 + 1} + 2 \\ -X(s) + sY(s) = \dfrac{2s}{s^2 + 1} \end{cases}.$$

```
step2 = step1/.{x[0]->2, y[0]->1}
```

$\{-2 + s\text{LaplaceTransform}[x[t], t, s] == \dfrac{1}{1 + s^2}$

$+\text{LaplaceTransform}[y[t], t, s], -1+s\text{LaplaceTransform}[y[t],$

$t, s] == \dfrac{2s}{1 + s^2} + \text{LaplaceTransform}[x[t], t, s]\}$

We now use `Solve` to solve this system of algebraic equations for $X(s)$ and $Y(s)$.

```
step3 = Solve[step2, {LaplaceTransform[x[t], t, s],

LaplaceTransform[y[t], t, s]}]
```

$\{\{\text{LaplaceTransform}[x[t], t, s] \to -\dfrac{-1 - 5s - s^2 - 2s^3}{(-1 + s^2)(1 + s^2)},$

$\text{LaplaceTransform}[y[t], t, s] \to -\dfrac{-3 - s - 4s^2 - s^3}{(-1 + s^2)(1 + s^2)}\}\}$

We find $x(t)$ and $y(t)$ with `InverseLaplaceTransform`.

```
x[t_] = InverseLaplaceTransform[-
```
$\dfrac{-1 - 5s - s^2 - 2s^3}{(-1 + s^2)(1 + s^2)}$`, s, t]`

```
//Simplifyy[t_] = InverseLaplaceTransform
```
`[-`$\dfrac{-3 - s - 4s^2 - s^3}{(-1 + s^2)(1 + s^2)}$`, s, t]//Simplify`

$\dfrac{1}{4}(5e^{-t} + 9e^t - 6\text{Cos}[t])$

$\dfrac{1}{4}(-5e^{-t} + 9e^t + 2\text{Sin}[t])$

Last, we graph $x(t)$ and $y(t)$ in Figure 8-21(a) and $\begin{cases} x = x(t) \\ y = y(t) \end{cases}$ in Figure 8-21(b).

```
p1 = Plot[{x[t], y[t]}, {t, -2, 3}, PlotRange->{-1, 4},

    AspectRatio->1, PlotLabel → "(a)"]

p2 = ParametricPlot[{x[t], y[t]}, {t, -2, 3}, PlotRange->{{0,

    10}, {-5, 5}}, AspectRatio->1, PlotLabel → "(b)",

    AxesLabel → {x, y}]
```

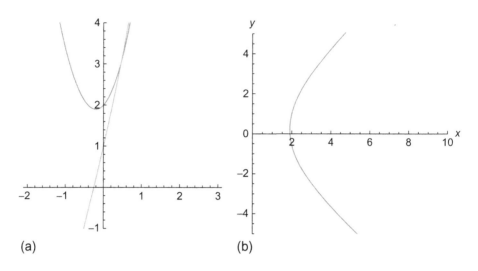

Figure 8-21   (a) $x(t)$ and $y(t)$. (b) Parametric plot of x versus y

```
Show[GraphicsRow[{p1, p2}]]
```

∎

In some cases, systems that involve higher-order differential equations can be solved with Laplace transforms.

EXAMPLE 8.7.2: Solve $\begin{cases} x'' = -2x - 4y - \cos t \\ y'' = -x - 2y + \sin t \\ x(0) = x'(0) = y(0) = y'(0) = 0 \end{cases}$.

SOLUTION: After defining the system of equations in sys, we take the Laplace transform of each equation.

```
Clear[x, y, t]

sys = {x''[t]== - 2x[t] - 4y[t] - Cos[t], y''[t]== - x[t] - 2y[t]
+Sin[t]}

{x''[t] == -Cos[t] - 2x[t] - 4y[t], y''[t] == Sin[t] - x[t] - 2y[t]}

step1 = LaplaceTransform[sys, t, s]
```

$$\{s^2 \text{LaplaceTransform}[x[t], t, s] - sx[0] - x'[0] == -\frac{s}{1+s^2} - 2$$

$\text{LaplaceTransform}[x[t], t, s] - 4\text{LaplaceTransform}[y[t], t, s],$

$$s^2\text{LaplaceTransform}[y[t], t, s] - sy[0] - y'[0] == \frac{1}{1+s^2} -$$

$\text{LaplaceTransform}[x[t], t, s] - 2\text{LaplaceTransform}[y[t], t, s]\}$

We then apply the initial conditions and solve the resulting algebraic system of equations for $X(s)$ and $Y(s)$.

step2 = step1/.{x[0]->0, x'[0]->0, y[0]->0, y'[0]->0}

$$\{s^2\text{LaplaceTransform}[x[t], t, s] == -\frac{s}{1+s^2} - 2$$

$\text{LaplaceTransform}[x[t], t, s] - 4\text{LaplaceTransform}[y[t], t, s],$

$$s^2\text{LaplaceTransform}[y[t], t, s] == \frac{1}{1+s^2} -$$

$\text{LaplaceTransform}[x[t], t, s] - 2\text{LaplaceTransform}[y[t], t, s]\}$

step3 = Solve[step2, {LaplaceTransform[x[t], t, s],

LaplaceTransform[y[t], t, s]}]//Simplify

$$\{\{\text{LaplaceTransform}[x[t], t, s] \to -\frac{4 + 2s + s^3}{4s^2 + 5s^4 + s^6},$$

$$\text{LaplaceTransform}[y[t], t, s] \to \frac{2 + s + s^2}{4s^2 + 5s^4 + s^6}\}\}$$

Finally, we use InverseLaplaceTransform to compute $x(t) = \mathcal{L}^{-1}\{X(s)\}$ and $y(t) = \mathcal{L}^{-1}\{Y(s)\}$.

x[t_] = InverseLaplaceTransform[step3[[1, 1, 2]], s, t]

$$\frac{1}{6}(-3 - 6t + 2\text{Cos}[t] + \text{Cos}[2t] + 8\text{Sin}[t] - \text{Sin}[2t])$$

y[t_] = InverseLaplaceTransform[step3[[1, 2, 2]], s, t]

$$\frac{1}{12}(3 + 6t - 4\text{Cos}[t] + \text{Cos}[2t] - 4\text{Sin}[t] - \text{Sin}[2t])$$

We see that the initial conditions are satisfied by graphing $x(t)$ and $y(t)$ in Figure 8-22(a) and $\begin{cases} x = x(t) \\ y = y(t) \end{cases}$ in Figure 8-22(b).

p1 = Plot[{x[t], y[t]}, {t, -π, 4π}, PlotRange → {-3π, 2π},

AspectRatio → 1, PlotLabel → "(a)"]

p2 = ParametricPlot[{x[t], y[t]}, {t, -π, 4π},

PlotRange → {{-12, 3}, {-3, 12}}, AspectRatio → 1,

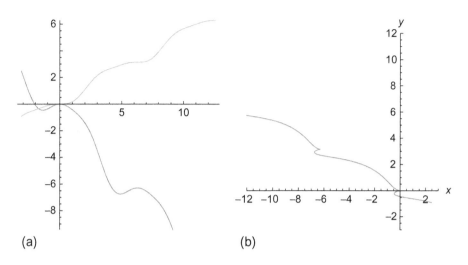

Figure 8-22   (a) $x(t)$ and $y(t)$. (b) Parametric plot of x versus y

```
AxesLabel → {x, y}, PlotLabel → "(b)"]

Show[GraphicsRow[{p1, p2}]]
```

■

Laplace transform methods are especially useful in solving problems that involve piecewise-defined, periodic, or impulse functions.

EXAMPLE 8.7.3:   Solve $\begin{cases} x' = y = 3\delta(t - \pi) \\ y' = -x + 6\delta(t - 2\pi). \\ x(0) = 1,\, y(0) = -1 \end{cases}$

SOLUTION: We proceed in the exact same manner as in the previous examples. After defining the system of equations,

```
Clear[x, y, t]

sys = {x'[t]==y[t] + 3DiracDelta[t − π], y'[t]== − x[t]
+6DiracDelta[t − 2π]}

{x'[t] == 3DiracDelta[−π+t]+y[t], y'[t] == 6DiracDelta[−2π
+ t] − x[t]}
```

we use `LaplaceTransform` to compute the Laplace transform of each equation

`step1 = LaplaceTransform[sys, t, s]`

$\{s$LaplaceTransform$[x[t], t, s] - x[0] == 3e^{-\pi s} +$

LaplaceTransform$[y[t], t, s]$, sLaplaceTransform$[y[t], t, s]$

$- y[0] == 6e^{-2\pi s} -$ LaplaceTransform$[x[t], t, s]\}$

and apply the initial conditions.

`step2 = step1/.{x[0]->1, y[0]-> - 1}`

$\{-1 + s$LaplaceTransform$[x[t], t, s] == 3e^{-\pi s} +$

LaplaceTransform$[y[t], t, s]$, $1+s$LaplaceTransform$[y[t], t, s]$

$== 6e^{-2\pi s} -$ LaplaceTransform$[x[t], t, s]\}$

We then solve the resulting algebraic system of equations for $X(s) = \mathcal{L}\{x(t)\}$ and $Y(s) = \mathcal{L}\{y(t)\}$ and use `InverseLaplaceTransform` to compute $x(t)$ and $y(t)$.

`step3 = Solve[step2, {LaplaceTransform[x[t], t, s],`

`LaplaceTransform[y[t], t, s]}]`

$\{\{$LaplaceTransform$[x[t], t, s] \rightarrow$

$\dfrac{e^{-2\pi s}(6 - e^{2\pi s} + 3e^{\pi s}s + e^{2\pi s}s)}{1 + s^2}$,

LaplaceTransform$[y[t], t, s] \rightarrow$

$-\dfrac{e^{-2\pi s}(3e^{\pi s} + e^{2\pi s} - 6s + e^{2\pi s}s)}{1 + s^2}\}\}$

`x[t_] = InverseLaplaceTransform[step3[[1, 1, 2]], s, t]`

Cos$[t] - 3$Cos$[t]$HeavisideTheta$[-\pi + t] -$ Sin$[t] + 6$

HeavisideTheta$[-2\pi + t]$Sin$[t]$

`y[t_] = InverseLaplaceTransform[step3[[1, 2, 2]], s, t]`

$-$Cos$[t] + 6$Cos$[t]$HeavisideTheta$[-2\pi + t] -$ Sin$[t] + 3$

HeavisideTheta$[-\pi + t]$Sin$[t]$

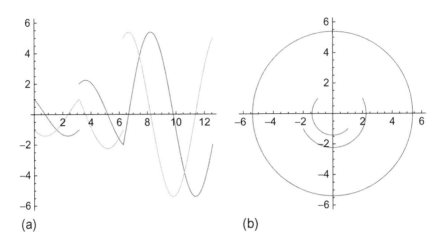

Figure 8-23   (a) $x(t)$ and $y(t)$. (b) Parametric plot of x versus y

We see that the initial conditions are satisfied by graphing $x(t)$ and $y(t)$ in Figure 8-23(a) and $\begin{cases} x = x(t) \\ y = y(t) \end{cases}$ in Figure 8-23(b).

```
p1 = Plot[{x[t], y[t]}, {t, 0, 4π}, PlotRange → {-2π, 2π},

   AspectRatio → 1, PlotLabel → "(a)"]

p2 = ParametricPlot[{x[t], y[t]}, {t, 0, 4π}, PlotRange →

   {{-2π, 2π}, {-2π, 2π}}, AspectRatio → 1, PlotLabel → "(b)"]

Show[GraphicsRow[{p1, p2}]]
```

■

EXAMPLE 8.7.4:   Solve $\begin{cases} x' = -17x + f(t) \\ y' = -\frac{1}{4}x - y - f(t), \quad \text{where} \quad f(t) \quad = \\ x(0) = y(0) = 0 \end{cases}$

$\begin{cases} 1 + t,\, 0 \le t < 1 \\ 3,\, t \ge 1 \end{cases}$.

SOLUTION: We first rewrite $f(t)$ in terms of the unit step function:

$$f(t) = \begin{cases} 1 + t, \ 0 \leq t < 1 \\ 3, \ t \geq 1 \end{cases} = (1 + t)[1 - \mathcal{U}(t - 1)] + 3\mathcal{U}(t - 1).$$

Then, we define and graph $f(t)$ in Figure 8-24(a).

```
Clear[x, y, t, f]

f[t_] = (1 + t)(1 - UnitStep[t - 1]) + 3UnitStep[t - 1]
```

$(1 + t)(1 - \text{UnitStep}[-1 + t]) + 3\text{UnitStep}[-1 + t]$

```
p1 = Plot[f[t], {t, 0, 4}, PlotRange → {0, 4},

AspectRatio → 1, PlotLabel → "(a)"]
```

To solve the initial-value problem, we proceed as in the previous examples. First, we define the system of equations.

```
sys = {x'[t]== - 17y[t] + f[t], y'[t]== x[t]/4 - y[t] - f[t]};
```

Then, we compute the Laplace transform of each equation,

```
step1 = LaplaceTransform[sys, t, s]
```

$\{s\text{LaplaceTransform}[x[t], t, s] - x[0] == \dfrac{1}{s^2} + \dfrac{1}{s} + \dfrac{2e^{-s}}{s} -$

$\dfrac{e^{-s}(1 + s)}{s^2} - 17\text{LaplaceTransform}[y[t], t, s], s$

$\text{LaplaceTransform}[y[t], t, s] - y[0] == -\dfrac{1}{s^2} - \dfrac{1}{s} - \dfrac{2e^{-s}}{s} +$

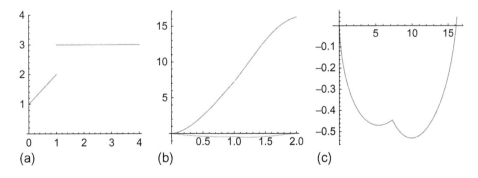

Figure 8-24 (a) A piecewise defined forcing function. (b) $x(t)$ and $y(t)$. (c) Parametric plot of x versus y

$$\frac{e^{-s}(1+s)}{s^2} + \frac{1}{4}\text{LaplaceTransform}[x[t], t, s]$$
$$- \text{LaplaceTransform}[y[t], t, s]\}$$

apply the initial conditions,

step2 = step1/.{x[0]->0, y[0]->0}

$$\{s\text{LaplaceTransform}[x[t], t, s] == \frac{1}{s^2} + \frac{1}{s} + \frac{2e^{-s}}{s} - \frac{e^{-s}(1+s)}{s^2} -$$

$$17\text{LaplaceTransform}[y[t], t, s], s\text{LaplaceTransform}[y[t],$$

$$t, s] == -\frac{1}{s^2} - \frac{1}{s} - \frac{2e^{-s}}{s} + \frac{e^{-s}(1+s)}{s^2} + \frac{1}{4}\text{LaplaceTransform}[x[t],$$

$$t, s] - \text{LaplaceTransform}[y[t], t, s]\}$$

and solve the resulting algebraic system of equations for $X(s) = \mathcal{L}\{x(t)\}$ and $Y(s) = \mathcal{L}\{y(t)\}$.

step3 = Solve[step2, {LaplaceTransform[x[t], t, s],

LaplaceTransform[y[t], t, s]}]

$$\{\{\text{LaplaceTransform}[x[t], t, s] \rightarrow$$

$$\frac{4e^{-s}(18+s)(-1+e^s+s+e^s s)}{s^2(17+4s+4s^2)}, \text{LaplaceTransform}[y[t],$$

$$t, s] \rightarrow -\frac{e^{-s}(-1+4s)(-1+e^s+s+e^s s)}{s^2(17+4s+4s^2)}\}\}$$

The solution is obtained with InverseLaplaceTransform.

x[t_] = InverseLaplaceTransform[step3[[1, 1, 2]], s, t]//

Simplify

$$\frac{1}{1156}e^{-2i-(\frac{1}{2}+2i)t}(e^{2i}((-2008-1437i)-(2008-1437i)e^{4it}$$

$$+16e^{(\frac{1}{2}+2i)t}(251+306t)) + ((-2888+791i)e^{\frac{1}{2}+4i} -$$

$$(2888+791i)e^{\frac{1}{2}+4it} - 16e^{2i+(\frac{1}{2}+2i)t}(-667+306t))$$

HeavisideTheta$[-1+t])$

y[t_] = InverseLaplaceTransform[step3[[1, 2, 2]], s, t]//

Simplify

$$\frac{1}{2312}e^{-2i-(\frac{1}{2}+2i)t}(e^{2i}((220-557i)+(220+557i)e^{4it}+$$

$$8e^{(\frac{1}{2}+2i)t}(-55+17t))-((356+633i)e^{\frac{1}{2}+4i}+(356-633i)e^{\frac{1}{2}+4it}+$$

$$8e^{2i+(\frac{1}{2}+2i)t}(-106+17t))\text{HeavisideTheta}[-1+t])$$

Last, we confirm that the initial conditions are satisfied by graphing $x(t)$ and $y(t)$ in Figure 8-24(b) and $\begin{cases} x = x(t) \\ y = y(t) \end{cases}$ in Figure 8-24(c).

```
p2 = Plot[{x[t], y[t]}, {t, 0, 2}, AspectRatio → 1,

   PlotLabel → "(b)"]

p3 = ParametricPlot[{x[t], y[t]}, {t, 0, 2},

   AspectRatio → 1, PlotLabel → "(c)"]

Show[GraphicsRow[{p1, p2, p3}]]
```

EXAMPLE 8.7.5: Solve $\begin{cases} x' + 2x + 3y = 0 \\ y' - x + 6y = f(t) \\ x(0) = 1, \ y(0) = 0 \end{cases}$, where $f(t) = \begin{cases} 0, \ 0 \le t < 1 \\ 1, \ 1 \le t < 2 \\ 2, \ 2 \le t < 3 \end{cases}$

and $f(t) = f(t - 3), t \ge 3$.

SOLUTION: We begin by defining and graphing $f(t)$ in Figure 8-25.

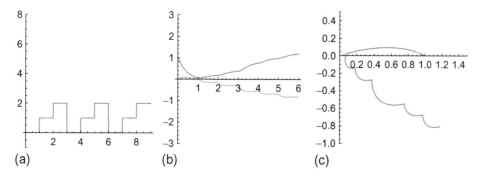

(a)   (b)   (c)

Figure 8-25 (a) A piecewise defined periodic forcing function. (b) $x(t)$ and $y(t)$. (c) Parametric plot of x versus y

```
Clear[x, y, t]

Clear[f]

f[t_]:=0/;0 ≤ t < 1;

f[t_]:=1/;1 ≤ t < 2;

f[t_]:=2/;2 ≤ t < 3

f[t_]:=f[t - 3]/;t ≥ 3;

p1 = Plot[f[t], {t, 0, 9}, PlotRange->{-1, 8},

   AspectRatio->1, PlotLabel → "(a)"]
```

The Laplace transform of the periodic function $f(t)$ is given by equation (8.34). We use Integrate and Simplify to find $\mathcal{L}\{f(t)\}$, naming the result lapf.

$$\text{lapf} = \text{Simplify}[\frac{\int_1^2 \text{Exp}[-st]\,dt + \int_2^3 2\text{Exp}[-st]dt}{1 - \text{Exp}[-3s]}]$$

$$\frac{2 + e^s}{s + e^s s + e^{2s} s}$$

Now we compute the Laplace transform of $x' + 2x + 3y = 0$

```
leq1 = LaplaceTransform[x'[t] + 2x[t] + 3y[t]==0, t, s]/.

{x[0]->1, y[0]->0}

−1+2LaplaceTransform[x[t], t, s]+sLaplaceTransform[x[t],

   t, s] + 3LaplaceTransform[y[t], t, s] == 0
```

and $y' - x + 6y = f(t)$. In each case, we apply the initial conditions as well.

```
leq2 = LaplaceTransform[y'[t] − x[t] + 6y[t], t, s]==lapf/.

{x[0]->1, y[0]->0}

−LaplaceTransform[x[t], t, s]+6LaplaceTransform[y[t], t, s]
```
$$+ s\text{LaplaceTransform}[y[t], t, s] == \frac{2 + e^s}{s + e^s s + e^{2s} s}$$

We then use Solve to solve this system of equations for $X(s)$ and $Y(s)$.

```
Solve[{leq1, leq2}, {LaplaceTransform[x[t], t, s],

LaplaceTransform[y[t], t, s]}]
```

```
{{LaplaceTransform[x[t], t, s] →
```

$$\frac{-6 - 3e^s + 6s + 6e^s s + 6e^{2s}s + s^2 + e^s s^2 + e^{2s}s^2}{(1 + e^s + e^{2s})s(15 + 8s + s^2)},$$

```
LaplaceTransform[y[t], t, s] →
```
$$\frac{4 + 2e^s + 3s + 2e^s s + e^{2s}s}{(1 + e^s + e^{2s})s(15 + 8s + s^2)}\}\}$$

Note that InverseLaplaceTransform cannot be used to compute $x(t) = \mathcal{L}^{-1}\{X(s)\}$ and $y(t) = \mathcal{L}^{-1}\{Y(s)\}$. Instead, we use Apart to rewrite $X(s)$.

$$\text{Apart}\left[\frac{1}{2+s} + \frac{3\left(-1 - \dfrac{(2E^{-2s} + E^{-s})(2+s)}{(1 + E^{-2s} + E^{-s})s}\right)}{(2+s)(15 + 8s + s^2)}\right]$$

$$-\frac{3(2 + e^s)}{(1 + e^s + e^{2s})s(3+s)(5+s)} + \frac{6+s}{(3+s)(5+s)}$$

InverseLaplaceTransform quickly calculates $\mathcal{L}^{-1}\left\{\dfrac{s^2 + 6s - 6}{s(s+3)(s+5)}\right\}$.

```
InverseLaplaceTransform[
```
$$\frac{-6 + 6s + s^2}{s(3+s)(5+s)}, s, t]$$

$$-\frac{2}{5} - \frac{11e^{-5t}}{10} + \frac{5e^{-3t}}{2}$$

To calculate $\mathcal{L}^{-1}\left\{\dfrac{3\left(2 + e^{-s}\right)}{\left(1 + e^{-s} + e^{-2s}\right)s(s+3)(s+5)}\right\}$, we first rewrite the fraction:

$$\frac{3\left(2 + e^{-s}\right)}{\left(1 + e^{-s} + e^{-2s}\right)s(s+3)(s+5)} = \frac{2 + e^{-s}}{1 + e^{-s} + e^{-2s}} \cdot \frac{3}{s(s+3)(s+5)}$$

$$= \frac{2 + e^{-s}}{1 + e^{-s} + e^{-2s}} \cdot \frac{1 - e^{-s}}{1 - e^{-s}} \cdot \frac{3}{s(s+3)(s+5)}$$

$$= \left(\frac{2}{1 - e^{-3s}} + \frac{e^{-s}}{1 - e^{-3s}} + \frac{e^{-2s}}{1 - e^{-3s}}\right) \cdot \frac{3}{s(s+3)(s+5)}$$

and then use the geometric series $\dfrac{1}{1-x} = \sum_{n=0}^{\infty} x^n$:

$$\left(\frac{2}{1 - e^{-3s}} + \frac{e^{-s}}{1 - e^{-3s}} + \frac{e^{-2s}}{1 - e^{-3s}}\right) \cdot \frac{3}{s(s+3)(s+5)} =$$

$$\left(2\sum_{n=0}^{\infty} e^{-3ns} + \sum_{n=0}^{\infty} e^{-(3n+1)s} + \sum_{n=0}^{\infty} e^{-(3n+2)s}\right) \cdot \frac{3}{s(s+3)(s+5)}.$$

Notice that $\mathcal{L}^{-1}\left\{\dfrac{3}{s(s+3)(s+5)}\right\} = \frac{1}{5} + \frac{3}{10}e^{-5t} - \frac{1}{2}e^{-3t}$. We name this function $g(t)$ for later use.

$$g[t\_] = \texttt{InverseLaplaceTransform}[\dfrac{3}{s(3+s)(5+s)}, s, t]$$

$$3(\dfrac{1}{15} + \dfrac{e^{-5t}}{10} - \dfrac{e^{-3t}}{6})$$

Previously, we learned that $\mathcal{L}^{-1}\left\{e^{-as}F(s)\right\} = f(t-a)\mathcal{U}(t-a)$. Thus,

$$\mathcal{L}^{-1}\left\{\left(2\sum_{n=0}^{\infty}e^{-3ns} + \sum_{n=0}^{\infty}e^{-(3n+1)s} + \sum_{n=0}^{\infty}e^{-(3n+2)s}\right) \cdot \dfrac{3}{s(s+3)(s+5)}\right\}$$

$$= 2\sum_{n=0}^{\infty}g(t-3n)\mathcal{U}(t-3n) + \sum_{n=0}^{\infty}g(t-(3n+1))\mathcal{U}(t-(3n+1)) +$$

$$\sum_{n=0}^{\infty}g(t-(3n+2))\mathcal{U}(t-(3n+2))$$

and

$$x(t) = -\dfrac{2}{5} - \dfrac{11}{10}e^{-5t} + \dfrac{5}{2}e^{-3t} + 2\sum_{n=0}^{\infty}g(t-3n)\mathcal{U}(t-3n)$$

$$+ \sum_{n=0}^{\infty}g(t-(3n+1))\mathcal{U}(t-(3n+1)) + \sum_{n=0}^{\infty}g(t-(3n+2))\mathcal{U}(t-(3n+2)).$$

We find $y(t)$ in the same way.

$$\texttt{Apart}\left[-\dfrac{-1 - \dfrac{(2E^{-2s}+E^{-s})(2+s)}{(1+E^{-2s}+E^{-s})s}}{15+8s+s^2}\right]$$

$$\dfrac{1}{(3+s)(5+s)} + \dfrac{(2+e^s)(2+s)}{(1+e^s+e^{2s})s(3+s)(5+s)}$$

We use `InverseLaplaceTransform` to see that

$$\mathcal{L}^{-1}\left\{\dfrac{3s+4}{s(s+3)(s+5)}\right\} = \dfrac{4}{15} - \dfrac{11}{10}e^{-5t} + \dfrac{5}{6}e^{-3t}$$

and

$$\mathcal{L}^{-1}\left\{-\dfrac{s+2}{s(s+3)(s+5)}\right\} = -\dfrac{2}{15} + \dfrac{3}{10}e^{-5t} - \dfrac{1}{6}e^{-3t}.$$

We name the second result $h(t)$ for later use.

$$\text{InverseLaplaceTransform}[\frac{4 + 3s}{s(3 + s)(5 + s)}, \, s, \, t]$$

$$\frac{4}{15} - \frac{11e^{-5t}}{10} + \frac{5e^{-3t}}{6}$$

```
Clear[h]
```

$$h[t\_] = \text{InverseLaplaceTransform}[-\frac{(2 + s)}{s(3 + s)(5 + s)}, \, s, \, t]$$

$$-\frac{2}{15} + \frac{3e^{-5t}}{10} - \frac{e^{-3t}}{6}$$

To calculate $y(t) = \mathcal{L}^{-1}\{Y(s)\}$, we use the results we obtained when calculating $x(t) = \mathcal{L}^{-1}\{X(s)\}$.

$$\mathcal{L}^{-1}\left\{ \frac{2 + e^{-s}}{1 + e^{-s} + e^{-2s}} \cdot \frac{-(s + 2)}{s(s + 3)(s + 5)} \right\} =$$

$$\mathcal{L}^{-1}\left\{ \left(2\sum_{n=0}^{\infty} e^{-3ns} + \sum_{n=0}^{\infty} e^{-(3n+1)s} + \sum_{n=0}^{\infty} e^{-(3n+2)s} \right) \cdot \frac{-(s + 2)}{s(s + 3)(s + 5)} \right\}$$

$$= 2\sum_{n=0}^{\infty} h(t - 3n)\mathcal{U}(t - 3n) + \sum_{n=0}^{\infty} h(t - (3n + 1))\mathcal{U}(t - (3n + 1)) +$$

$$\sum_{n=0}^{\infty} h(t - (3n + 2))\mathcal{U}(t - (3n + 2))$$

and

$$y(t) = \frac{4}{15} - \frac{11}{10}e^{-5t} + \frac{5}{6}e^{-3t} + 2\sum_{n=0}^{\infty} h(t - 3n)\mathcal{U}(t - 3n) +$$

$$\sum_{n=0}^{\infty} h(t - (3n + 1))\mathcal{U}(t - (3n + 1)) + \sum_{n=0}^{\infty} h(t - (3n + 2))\mathcal{U}(t - (3n + 2)).$$

We then graph the solution on the interval $[0, 6]$ in Figure 8-25(b) and parametrically in Figure 8-25(c).

$$\text{xapprox}[t\_] = -\frac{2}{5} - \frac{11}{10}E^{-5t} + \frac{5}{2}E^{-3t} + 2(\sum_{n=0}^{6} g[t - 3n]$$

$$\text{UnitStep}[t - 3n]) + \sum_{n=0}^{6} g[t - (3n + 1)]\text{UnitStep}[t - (3n$$

$$+1)] + \sum_{n=0}^{6} g[t - (3n + 2)]\text{UnitStep}[t - (3n + 2)];$$

$$\text{yapprox}[t\_] = \frac{4}{15} - \frac{11}{10}E^{-5t} + \frac{5}{6}E^{-3t} + 2(\sum_{n=0}^{6} h[t - 3n]$$

$$\text{UnitStep}[t - 3n]) + \sum_{n=0}^{6} h[t - (3n + 1)]\text{UnitStep}[t - (3n$$

$$+1)] + \sum_{n=0}^{6} h[t - (3n + 2)]\text{UnitStep}[t - (3n + 2)];$$

```
p2 = Plot[{xapprox[t], yapprox[t]}, {t, 0, 6},

    PlotRange->{-3, 3}, AspectRatio->1,

    PlotLabel → "(b)"]

p3 = ParametricPlot[{xapprox[t], yapprox[t]}, {t, 0, 6},

    PlotRange->{{0, 1.5}, {-1, 0.5}}, AspectRatio->1,

    PlotLabel → "(c)"]

Show[GraphicsRow[{p1, p2, p3}]]
```

■

8.8 Applications of Laplace Transforms, Part II

8.8.1 Coupled Spring-Mass Systems

The motion of a mass attached to the end of a spring was modeled with a second-order linear differential equation with constant coefficients in Chapter 5. Similarly, if a second spring and mass are attached to the end of the first mass, then the model becomes that of a system of second-order equations. To more precisely state the problem, let masses m_1 and m_2 be attached to the ends of springs S_1 and S_2 having spring constants k_1 and k_2, respectively. Then, spring S_2 is attached to the base of mass m_1.

Suppose that $x(t)$ and $y(t)$ represent the vertical displacement from the equilibrium position of springs S_1 and S_2, respectively. Because spring S_2 undergoes both elongation and compression when the system is in motion (due to the spring S_1 and the mass m_2), then according to Hooke's law, S_2 exerts the force $k_2(y-x)$ on m_2 while S_1 exerts the force $-k_1x$ on m_2. Therefore, the force acting on mass m_1 is the sum $-k_1x + k_2(y - x)$ and that acting on m_2 is $-k_2(y - x)$. Hence, using Newton's second law, $F = ma$, with each mass, we have the system

$$\begin{cases} m_1\dfrac{d^2x}{dt^2} = -k_1x + k_2(y - x) \\ m_2\dfrac{d^2y}{dt^2} = -k_2(y - x). \end{cases} \tag{8.42}$$

The initial position and velocity of the two masses m_1 and m_2 are given by $x(0)$, $x'(0)$, $y(0)$, and $y'(0)$, respectively. If external forces $F_1(t)$ and $F_2(t)$ are applied to the masses, the system (8.42) becomes

$$\begin{cases} m_1 \dfrac{d^2x}{dt^2} = -k_1x + k_2(y - x) + F_1(t) \\ m_2 \dfrac{d^2y}{dt^2} = -k_2(y - x) + F_2(t). \end{cases} \tag{8.43}$$

Therefore, the method of Laplace transforms can be used to solve problems of this type.

EXAMPLE 8.8.1: Consider the spring-mass system with $m_1 = m_2 = 1$, $k_1 = 3$, and $k_2 = 2$. Find the position functions $x(t)$ and $y(t)$ if $x(0) = 0$, $x'(0) = 1$, $y(0) = 1$, and $y'(0) = 0$. (Assume there are no external forces.)

SOLUTION: In order to find $x(t)$ and $y(t)$, we must solve the initial-value problem

$$\begin{cases} \dfrac{d^2x}{dt^2} = -5x + 2y \\ \dfrac{d^2y}{dt^2} = 2x - 2y \\ x(0) = 0,\ x'(0) = 1,\ y(0) = 1,\ y'(0) = 0. \end{cases}$$

We use `LaplaceTransform` to take the Laplace transform of both sides of each equation.

```
Clear[x, y]

eqs = {x''[t]== − 5x[t] + 2y[t], y''[t]==2x[t] − 2y[t]};

step1 = LaplaceTransform[eqs, t, s]
```

$\{s^2\,$LaplaceTransform$[x[t],\ t,\ s] - s\,x[0] - x'[0] == -5$

LaplaceTransform$[x[t],\ t,\ s]$+2LaplaceTransform$[y[t],\ t,\ s]$,

$s^2\,$LaplaceTransform$[y[t],\ t,\ s] - s\,y[0] - y'[0] == 2$

LaplaceTransform$[x[t],\ t,\ s] - 2$LaplaceTransform$[y[t],\ t,\ s]\}$

We then apply the initial conditions.

```
step2 = step1/.{x[0]->0, x'[0]->1, y[0]->1, y'[0]->0}
```

$$\{-1 + s^2\text{LaplaceTransform}[x[t], t, s] == -5$$

$$\text{LaplaceTransform}[x[t], t, s] + 2\text{LaplaceTransform}[y[t], t, s],$$

$$-s + s^2\text{LaplaceTransform}[y[t], t, s] == 2\text{LaplaceTransform}$$

$$[x[t], t, s] - 2\text{LaplaceTransform}[y[t], t, s]\}$$

We solve this system of algebraic equations for $X(s)$ and $Y(s)$ with Solve.

step3 = Solve[step2, {LaplaceTransform[x[t], t, s],

LaplaceTransform[y[t], t, s]}]

$$\{\{\text{LaplaceTransform}[x[t], t, s] \rightarrow -\frac{-2 - 2s - s^2}{6 + 7s^2 + s^4},$$

$$\text{LaplaceTransform}[y[t], t, s] \rightarrow -\frac{-2 - 5s - s^3}{6 + 7s^2 + s^4}\}\}$$

Taking the inverse Laplace transform with InverseLaplaceTransform yields $x(t)$ and $y(t)$.

$$x[t\_] = \text{InverseLaplaceTransform}[-\frac{-2 - 2s - s^2}{6 + 7s^2 + s^4}, s, t]$$

$$\frac{1}{15}(6\text{Cos}[t] - 6\text{Cos}[\sqrt{6}t] + 3\text{Sin}[t] + 2\sqrt{6}\text{Sin}[\sqrt{6}t])$$

$$y[t\_] = \text{InverseLaplaceTransform}[-\frac{-2 - 5s - s^3}{6 + 7s^2 + s^4}, s, t]$$

$$\frac{1}{15}(12\text{Cos}[t] + 3\text{Cos}[\sqrt{6}t] + 6\text{Sin}[t] - \sqrt{6}\text{Sin}[\sqrt{6}t])$$

We graph $x(t)$ and $y(t)$ together in Figure 8-26(a) and then parametrically in Figure 8-26(b). Note that $y(t)$ starts at $(0, 1)$ while $x(t)$ has initial

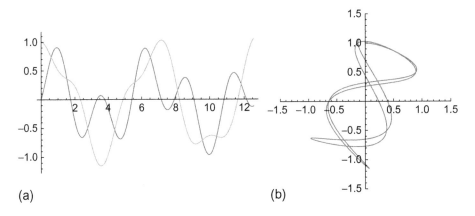

(a)   (b)

Figure 8-26   (a) $x(t)$ and $y(t)$. (b) Parametric plot of x versus y

point $(0, 0)$. Also, the phase plane is different from those discussed in previous sections. One of the reasons for this is that the equations in the system of differential equations are second-order instead of first-order.

```
p1 = Plot[{x[t], y[t]}, {t, 0, 4π}, PlotLabel → "(a)"]
```

```
p2 = ParametricPlot[{x[t], y[t]}, {t, 0, 4π},
```
$$\text{PlotRange} \to \{\{-\frac{3}{2}, \frac{3}{2}\}, \{-\frac{3}{2}, \frac{3}{2}\}\}, .$$
```
   AspectRatio → 1, PlotLabel → "(b)"]
```

```
Show[GraphicsRow[{p1, p2}]]
```

We can illustrate the motion of the spring in nearly the same way as we did in Chapter 5. First, we define the functions `zigzag` and `spring2`.

```
Clear[spring, zigzag, length, points, pairs]

zigzag[{a_, b_}, {c_, d_}, n_, eps_] :=

  Module[{length, points, pairs},

    length = d - b;

    points = Table[b + ilength/n, {i, 1, n - 1}];

    pairs = Table[{a + (-1)^ieps, points[[i]]},

      {i, 1, n - 1}];

    PrependTo[pairs, {a, b}];

    AppendTo[pairs, {c, d}];

    Line[pairs]

    ]

spring2[t_, len1_, len2_] :=

  Show[Graphics[

    {zigzag[{0, -x[t]}, {0, len1}, 20, .025],

      PointSize[.025], Point[{0, len1}],

    zigzag[{0, -y[t] - len2}, {0, -x[t]}, 20, .025],

      PointSize[.075], Point[{0, -x[t]}],

      PointSize[.05], Point[{0, -y[t] - len2}]}],

    Axes->Automatic,

    Ticks->None,
```

```
AspectRatio->1,

PlotRange->{{-1/2, 1/2}, {-2.2, 1.2}}]
```

Next, we define `tvals` to be a list of sixteen evenly spaced numbers between 0 and 4π.

```
tvals = Table[t, {t, 0, 4π, 4π/15}];
```

`Map` is then used to apply `spring2` to the list of numbers in `tvals`.

```
graphs = Map[spring2[#, 1, 1]&, tvals];
```

The resulting list of graphics is partitioned into four element subsets with `Partition` and displayed using `Show` and `GraphicsGrid` in Figure 8-27.

```
toshow = Partition[graphs, 4];

Show[GraphicsGrid[toshow]]
```

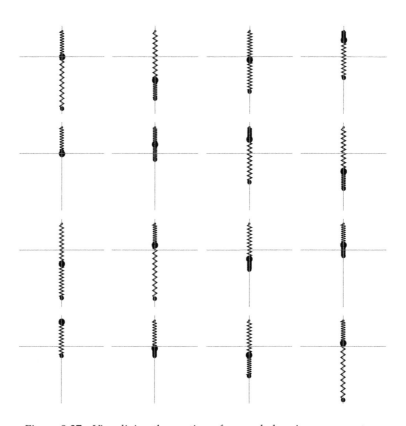

Figure 8-27   Visualizing the motion of a coupled spring-mass system

We can use the `Animate` command to create an animation of the coupled spring-mass system.

```
Animate[spring2[t, 1, 1], {t, 0, 4, 4/29}]
```

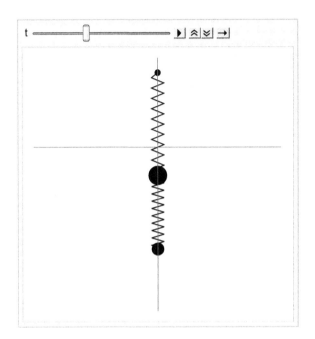

∎

8.8.2 The Double Pendulum

In a method similar to that of the simple pendulum in Chapter 5 and that of the coupled spring-mass system, the motion of a double pendulum as shown in Figure 8-28 is modeled by the following system of equations using the approximation $\sin\theta \approx \theta$ for small displacements

$$\begin{cases} (m_1 + m_2)\,\ell_1{}^2\dfrac{d^2\theta_1}{dt^2} + m_2\ell_1\ell_2\dfrac{d^2\theta_2}{dt^2} + (m_1 + m_2)\,\ell_1 g\theta_1 = 0 \\ m_2\ell_2{}^2\dfrac{d^2\theta_2}{dt^2} + m_2\ell_1\ell_2\dfrac{d^2\theta_1}{dt^2} + m_2\ell_2 g\theta_2 = 0 \end{cases}, \qquad (8.44)$$

where θ_1 represents the displacement of the upper pendulum and θ_2 that of the lower pendulum. Also, m_1 and m_2 represent the mass attached to the upper and lower pendulums, respectively, while the length of each is given by ℓ_1 and ℓ_2.

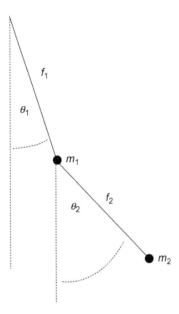

Figure 8-28   A double pendulum

EXAMPLE 8.8.2: Suppose that $m_1 = 3$, $m_2 = 1$, and each pendulum has length 16. If $\theta_1(0) = 1$, $\theta_1'(0) = 0$, $\theta_2(0) = -1$, and $\theta_2'(0) = 0$, solve the double pendulum problem using $g = 32$. Plot the solution.

SOLUTION: In this case, the system to be solved is

$$\begin{cases} 4 \cdot 16^2 \dfrac{d^2\theta_1}{dt^2} + 16^2 \dfrac{d^2\theta_2}{dt^2} + 4 \cdot 16 \cdot 32\theta_1 = 0 \\ 16^2 \dfrac{d^2\theta_2}{dt^2} + 16^2 \dfrac{d^2\theta_1}{dt^2} + 16 \cdot 32\theta_2 = 0 \end{cases},$$

which we simplify to obtain

$$\begin{cases} 4 \dfrac{d^2\theta_1}{dt^2} + \dfrac{d^2\theta_2}{dt^2} + 8\theta_1 = 0 \\ \dfrac{d^2\theta_2}{dt^2} + \dfrac{d^2\theta_1}{dt^2} + 2\theta_2 = 0 \end{cases}.$$

In the following code, we let $x(t)$ and $y(t)$ represent $\theta_1(t)$ and $\theta_2(t)$, respectively. First, we use DSolve to solve the initial-value problem.

```
sol = DSolve[{4x''[t] + y''[t] + 8x[t]==0,
  x''[t] + y''[t] + 2y[t]==0, x[0] == 1,
  x'[0] == 1, y[0] == 0, y'[0] == -1},
  {x[t], y[t]}, t]
```

$$\{\{x[t] \to \frac{1}{8}(4\cos[2t]+4\cos[\frac{2t}{\sqrt{3}}]+3\sin[2t]+\sqrt{3}\sin[\frac{2t}{\sqrt{3}}]),\ y[t] \to$$
$$\frac{1}{4}(-4\cos[2t]+4\cos[\frac{2t}{\sqrt{3}}]-3\sin[2t]+\sqrt{3}\sin[\frac{2t}{\sqrt{3}}])\}\}$$

We define sys to be the system of equations and use LaplaceTransform to compute the Laplace transform of each equation.

```
Clear[x, y]

sys = {4x''[t] + y''[t] + 8x[t]==0, x''[t] + y''[t] + 2y[t]==0};

step1 = LaplaceTransform[sys, t, s]
```

$\{8\text{LaplaceTransform}[x[t],\ t,\ s] + s^2\text{LaplaceTransform}[y[t],$
$t,\ s] - sy[0] + 4(s^2\text{LaplaceTransform}[x[t],\ t,\ s] - sx[0] - x'[0]) -$
$y'[0] == 0,\ s^2\text{LaplaceTransform}[x[t],\ t,\ s] + 2$
$\text{LaplaceTransform}[y[t],\ t,\ s] + s^2\text{LaplaceTransform}[y[t],\ t,\ s]$
$- sx[0] - sy[0] - x'[0] - y'[0] == 0\}$

Next, we apply the initial conditions and solve the resulting system of equations for $\mathcal{L}\{\theta_1(t)\} = X(s)$ and $\mathcal{L}\{\theta_2(t)\} = Y(s)$.

```
step2 = step1/.{x[0]->1, x'[0]->1, y[0]->0, y'[0]-> - 1}
```

$\{1 + 8\text{LaplaceTransform}[x[t],\ t,\ s] + 4(-1 - s + s^2$
$\text{LaplaceTransform}[x[t],\ t,\ s]) + s^2\text{LaplaceTransform}[y[t],\ t,\ s]$
$== 0,\ -s + s^2\text{LaplaceTransform}[x[t],\ t,\ s] + 2$
$\text{LaplaceTransform}[y[t],\ t,\ s] + s^2\text{LaplaceTransform}[y[t],$
$t,\ s] == 0\}$

```
step3 = Solve[step2, {LaplaceTransform[x[t], t, s],
LaplaceTransform[y[t], t, s]}]
```

$$\{\{\text{LaplaceTransform}[x[t],\ t,\ s] \to -\frac{-6 - 8s - 3s^2 - 3s^3}{16 + 16s^2 + 3s^4},$$

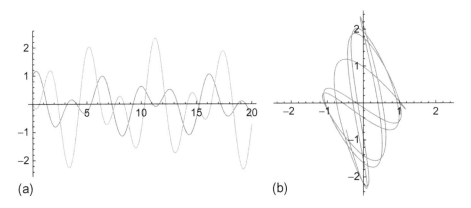

Figure 8-29   (a) $\theta_1(t)$ and $\theta_2(t)$ as functions of t. (b) Parametric plot of $\theta_1(t)$ versus $\theta_2(t)$

$$\text{LaplaceTransform}[y[t], t, s] \rightarrow -\frac{s(-8+3s)}{16+16s^2+3s^4}\}\}$$

InverseLaplaceTransform is then used to find $\theta_1(t)$ and $\theta_2(t)$.

```
x[t_] = InverseLaplaceTransform[-
```
$\dfrac{-6-8s-3s^2-3s^3}{16+16s^2+3s^4}$, s, t]

$$\frac{1}{8}(4\cos[2t]+4\cos[\frac{2t}{\sqrt{3}}]+3\sin[2t]+\sqrt{3}\sin[\frac{2t}{\sqrt{3}}])$$

```
y[t_] = InverseLaplaceTransform[-
```
$\dfrac{-8s+3s^2}{16+16s^2+3s^4}$, s, t]

$$\frac{1}{4}(-4\cos[2t]+4\cos[\frac{2t}{\sqrt{3}}]-3\sin[2t]+\sqrt{3}\sin[\frac{2t}{\sqrt{3}}])$$

These two functions are graphed together in Figure 8-29(a) and parametrically in Figure 8-29(b).

```
p1 = Plot[{x[t], y[t]}, {t, 0, 20},

   PlotLabel → "(a)"]

p2 = ParametricPlot[{x[t], y[t]}, {t, 0, 20},

   PlotRange->{{-5/2, 5/2}, {-5/2, 5/2}}, AspectRatio->1,

   PlotLabel → "(b)"]

Show[GraphicsRow[{p1, p2}]]
```

We can illustrate the motion of the pendulum as follows. First, we define the function pen2.

```
Clear[pen2]

pen2[t_, len1_, len2_]:=

  Module[{pt1, pt2},

    pt1 = {len1Cos[3Pi/2 + x[t]],

      len1Sin[3Pi/2 + x[t]]};

    pt2 = {len1Cos[3Pi/2 + x[t]]+

        len2Cos[3Pi/2 + y[t]],

      len1Sin[3Pi/2 + x[t]]+

        len2Sin[3Pi/2 + y[t]]};

    Show[Graphics[{

      Line[{{0, 0}, pt1}],

      PointSize[.05], Point[pt1],

      Line[{pt1, pt2}],

        PointSize[.05], Point[pt2]}

      ],

    Axes->Automatic,

    Ticks->None,

    PlotRange->{{-32, 32}, {-34, 0}}]]]
```

Next, we define tvals to be a list of sixteen evenly spaced numbers between 0 and 10. Map is then used to apply pen2 to the list of numbers in tvals. The resulting set of graphics is partitioned into four element subsets and displayed using Show and GraphicsGrid in Figure 8-30.

```
tvals = Table[t, {t, 0, 10, 10/15}];

graphs = Map[pen2[#, 16, 16]&, tvals];

toshow = Partition[graphs, 4];
```

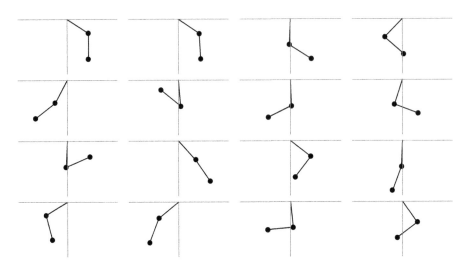

Figure 8-30   The double pendulum for 16 equally spaced values of *t* between 0 and 10

```
Show[GraphicsGrid[toshow]]
```

We can also use `Animate` to see the motion of the double pendulum.

```
Clear[t]
```
```
Animate[pen2[t, 16, 16], {t, 0, 10, 10/29}]
```

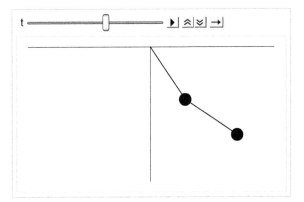

■

Application: Free Vibration of a Three-Story Building

If you have ever gone to the top of a tall building like the Sears Tower, World Trade Center, or Empire State Building on a windy day you may have been acutely aware of the sway of the building. In fact, all buildings sway, or vibrate, naturally. Usually, we are only aware, if ever, of the sway of a building when we are in a very tall building or in a building during an event like an earthquake. In some tall buildings, like the John Hancock Building in Boston, the sway of the building during high winds is reduced by installing a tuned mass damper at the top of the building which oscillates at the same frequency as the building but out of phase. We will investigate the sway of a three-story building and then try to determine how we would investigate the sway of a tall building.

Sources: M. L. James, G. M. Smith, J. C. Wolford, P. W. Whaley, *Vibration of Mechanical and Structural Systems with Microcomputer Applications*, Harper & Row (1989), pp. 282–286. Robert K. Vierck, *Vibration Analysis*, Second Edition, HarperCollins (1979), pp. 266–290.

We make two assumptions to solve this problem. First, we assume that the mass distribution of the building can be represented by the lumped masses at the different levels. Second, we assume that the girders of the structure are infinitely rigid in comparison with the supporting columns. With these assumptions, can we determine the motion of the building by interpreting the columns as springs in parallel.

Assume that the coordinates x_1, x_2, and x_3 as well as the velocities and accelerations are positive to the right. Also assume that $x_3 > x_2 > x_1$.

In applying Newton's second law of motion, recall that we have assumed that acceleration is in the positive direction. Therefore, we sum forces in the same direction as the acceleration positively, and others negatively. With this configuration, Newton's second law on each of the three masses yields the following system of differential equations

$$-k_1 x_1 + k_2 (x_2 - x_1) = m_1 \frac{d^2 x_1}{dt^2} \tag{8.45}$$

$$-k_2 (x_2 - x_1) + k_3 (x_3 - x_2) = m_2 \frac{d^2 x_2}{dt^2}$$

$$-k_3 (x_3 - x_2) = m_3 \frac{d^2 x_3}{dt^2},$$

which we write as

$$m_1 \frac{d^2 x_1}{dt^2} + (k_1 + k_2) x_1 - k_2 x_2 = 0$$

$$m_2 \frac{d^2 x_2}{dt^2} - k_2 x_1 + (k_2 + k_3) x_2 - k_3 x_3 = 0$$

$$m_3 \frac{d^2 x_3}{dt^2} - k_3 x_2 + k_3 x_3 = 0,$$

where m_1, m_2, and m_3 represent the mass of the building on the first, second, and third levels, and k_1, k_2, and k_3, corresponding to the spring constants, represent the total stiffness of the columns supporting a given floor.

If we attempt to find an exact solution with the method of Laplace transforms, we find that each denominator of $\mathcal{L}\{x_1(t)\}$, $\mathcal{L}\{x_2(t)\}$, and $\mathcal{L}\{x_3(t)\}$ is a positive function of s. Therefore, the roots are complex and solutions will involve sines and/or cosines. (Here, we use $x(t)$, $y(t)$, and $z(t)$ in the place of $x_1(t)$, $x_2(t)$, and $x_3(t)$.)

Clear[x, y, z, rule, eq1, eq2]

$\mathbf{eq1} = m_1 x''[t] + (k_1 + k_2)x[t] - k_2 y[t] \mathbf{==0};$

$\mathbf{eq2} = m_2 y''[t] - k_2 x[t] + (k_2 + k_3)y[t] - k_3 z[t] \mathbf{==0};$

$\mathbf{eq3} = m_3 z''[t] - k_3 y[t] + k_3 z[t] \mathbf{==0};$

step1 = LaplaceTransform[{eq1, eq2, eq3}, t, s]

$\{-\text{LaplaceTransform}[y[t], t, s]k_2 + \text{LaplaceTransform}[x[t], t, s]$

$(k_1 + k_2) + m_1 (s^2 \text{LaplaceTransform}[x[t], t, s] - sx[0] - x'[0]) == 0, -$

$\text{LaplaceTransform}[x[t], t, s]k_2 - \text{LaplaceTransform}[z[t], t, s]k_3 +$

$\text{LaplaceTransform}[y[t], t, s](k_2+k_3)+m_2(s^2\text{LaplaceTransform}[y[t],$

$t, s] - sy[0] - y'[0]) == 0, -\text{LaplaceTransform}[y[t], t, s]k_3 +$

$\text{LaplaceTransform}[z[t], t, s]k_3+m_3(s^2\text{LaplaceTransform}[z[t], t, s]-$

$sz[0] - z'[0]) == 0\}$

step2 =
Solve[step1, {LaplaceTransform[x[t], t, s], LaplaceTransform[y[t],
t, s], LaplaceTransform[z[t], t, s]}]//Simplify

$\{\{\text{LaplaceTransform}[x[t], t, s] \rightarrow (m_1(k_3^2 - (k_2 + k_3 + s^2 m_2)(k_3 +$

$s^2 m_3))(sx[0] + x'[0]) - k_2(m_2(k_3 + s^2 m_3)(sy[0] + y'[0]) + k_3 m_3(sz[0] +$

$z'[0])))/(k_2^2(k_3 + s^2 m_3) + (k_1 + k_2 + s^2 m_1)(k_3^2 - (k_2 + k_3 + s^2 m_2)(k_3 +$

$s^2 m_3))), \text{LaplaceTransform}[y[t], t, s] \rightarrow ((k_1 + s^2 m_1)(s^2 m_2 m_3 (sy[0] +$

$y'[0]) + k_3(m_2(sy[0] + y'[0]) + m_3(sz[0] + z'[0]))) + k_2(s^2 m_3(m_1(sx[0] +$

$x'[0]) + m_2(sy[0] + y'[0])) + k_3(m_1(sx[0] + x'[0]) + m_2(sy[0] + y'[0]) +$

$m_3(sz[0] + z'[0])))/(k_1(k_2(k_3 + s^2m_3) + s^2(s^2m_2m_3 + k_3(m_2 + m_3))) +$

$s^2(s^2m_1(s^2m_2m_3 + k_3(m_2 + m_3)) + k_2(s^2(m_1 + m_2)m_3 + k_3(m_1 + m_2 +$

$m_3)))),$ LaplaceTransform$[z[t], t, s] \rightarrow ((k_1 + s^2m_1)(s^2m_2m_3(sz[0] +$

$z'[0]) + k_3(m_2(sy[0] + y'[0]) + m_3(sz[0] + z'[0]))) + k_2((k_1 + s^2(m_1 +$

$m_2))m_3(sz[0]+z'[0]) + k_3(m_1(sx[0]+x'[0])+m_2(sy[0]+y'[0])+m_3(sz[0]+$

$z'[0]))))/(k_1(k_2(k_3+s^2m_3)+s^2(s^2m_2m_3+k_3(m_2+m_3)))+s^2(s^2m_1(s^2m_2m_3+$

$k_3(m_2 + m_3)) + k_2(s^2(m_1 + m_2)m_3 + k_3(m_1 + m_2 + m_3)))))\}\}$

Suppose that $k_1 = 3$, $k_2 = 2$, $k_3 = 1$, $m_1 = 1$, $m_2 = 2$, and $m_3 = 3$ and that the initial conditions are $x(0) = 0$, $x'(0) = 1/4$, $y(0) = 0$, $y'(0) = -1/2$, $z(0) = 0$, and $z'(0) = 1$.

```
step3 =
```

```
step2/.{k₁->3,k₂->2,x[0]->0,k₃->1,m₁->1,m₂->2, m₃->3, x'[0]->1/4, .
```

```
y[0]->0, y'[0]-> - 1/2,
```

```
z[0]->0, z'[0]->1}//Simplify
```

$\{\{$LaplaceTransform$[x[t], t, s] \rightarrow \dfrac{18 - 13s^2 + 6s^4}{4(6 + 45s^2 + 41s^4 + 6s^6)}$,

LaplaceTransform$[y[t], t, s] \rightarrow \dfrac{21 - 23s^2 - 6s^4}{12 + 90s^2 + 82s^4 + 12s^6}$,

LaplaceTransform$[z[t], t, s] \rightarrow \dfrac{57 + 76s^2 + 12s^4}{12 + 90s^2 + 82s^4 + 12s^6}\}\}$

For these values, we use InverseLaplaceTransform to compute $x(t) = \mathcal{L}^{-1}\{X(s)\}$, $y(t) = \mathcal{L}^{-1}\{Y(s)\}$, and $z(t) = \mathcal{L}^{-1}\{Z(s)\}$. First, we compute $x(t)$. The result is very long so we do not display it here.

Instead, we use Short to view a portion of this result. Note that several terms are given in terms of Root.

Root[p[x],k] represents the kth root of the polynomial equation $p(x) = 0$.

```
x[t_] = InverseLaplaceTransform[
```
$\dfrac{18 - 13s^2 + 6s^4}{4(6 + 45s^2 + 41s^4 + 6s^6)}$, s, t];

```
Short[x[t], 3]
```

$$\dfrac{e^{-t\sqrt{\text{Root}[6+45\#1+41\langle\langle 1\rangle\rangle+6\#1^3\&,\,1]-t\langle\langle 1\rangle\rangle-t\sqrt{\langle\langle 1\rangle\rangle}}}(\langle\langle 1\rangle\rangle)}{48\sqrt{\text{Root}[6 + 45\#1 + 41\#1^2 + 6\#1^3\&,\, 1]}\langle\langle 7\rangle\rangle\sqrt{\text{Root}[6 + 45\#1 + 41\#1^2 + 6\#1^3\&,\, 3]}}$$

In this case, we cannot find exact solutions of the equation $6s^6 + 41s^4 + 45s^2 + 6 = 0$. Nevertheless, we can use NRoots to approximate the solutions of this equation.

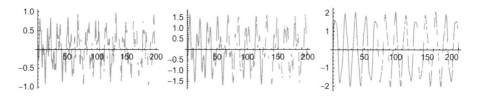

Figure 8-31   The sway of a building is periodic

```
?Root
```

```
NRoots[6 + 45s² + 41s⁴ + 6s⁶ == 0, s]
```

$s == 0. - 0.393222 i \| s == 0. + 0.393222 i \| s == 0. - 1.08402 i \| s ==$

$0. + 1.08402 i \| s == 0. - 2.34598 i \| s == 0. + 2.34598 i$

Now, we use InverseLaplaceTransform to compute $y(t) = \mathcal{L}^{-1}\{Y(s)\}$ and
$z(t) = \mathcal{L}^{-1}\{Z(s)\}$.

$$y[t\_] = \text{InverseLaplaceTransform}[\frac{21 - 23s^2 - 6s^4}{2(6 + 45s^2 + 41s^4 + 6s^6)}, s, t];$$

$$z[t\_] = \text{InverseLaplaceTransform}[\frac{57 + 76s^2 + 12s^4}{2(6 + 45s^2 + 41s^4 + 6s^6)}, s, t];$$

The graphs of $x(t)$, $z(t)$, and $z(t)$ shown in Figure 8-31 indicate that they are
indeed periodic functions.

```
px = Plot[x[t], {t, 0, 200}, PlotPoints → 1000];
```

```
py = Plot[y[t], {t, 0, 200}, PlotPoints → 1000];
```

```
pz = Plot[z[t], {t, 0, 200}, PlotPoints → 1000];
```

```
Show[GraphicsRow[{px, py, pz}]]
```

We can construct an outline of a three-story building and observe its vibration.
The width and height of the floors were selected arbitrarily to be 20 and 1,
respectively. See Figure 8-32.

```
Clear[bldg]
```

```
bldg[t_, opts___] :=
```

```
Show[Graphics[{Line[{{0, 0}, {20, 0}}], PointSize[0.05], Point[{0, 0}],
```

```
Point[{20, 0}], Line[{{0, 0}, {x[t], 1}}], Point[{x[t], 1}],
```

```
Line[{{20, 0}, {20 + x[t], 1}}], Point[{20 + x[t], 1}], Line[{{x[t], 1},
```

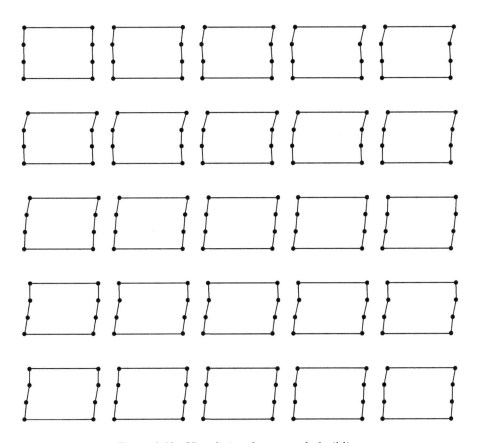

Figure 8-32   Visualizing the sway of a building

```
{y[t], 2}}], Point[{y[t], 2}], Line[{{20 + x[t], 1}, {20 + y[t], 2}}],

Point[{20 + y[t], 2}], Line[{{y[t], 2}, {z[t], 3}}], Point[{z[t], 3}],

Line[{{20 + y[t], 2}, {20 + z[t], 3}}], Point[{20 + z[t], 3}], Line[{{z[t], 3},

{20 + z[t], 3}}]}, opts, Axes → None, Ticks → None, PlotRange → {{-2, 22},

{-1,4}}, AspectRatio → 1]]graphs = Table[bldg[t], {t, 0.1, 6.1, 6.0/24}];

toshow = Partition[graphs, 5];

Show[GraphicsGrid[toshow], AspectRatio → 1]
```

With `Animate`, you can see the sway of the building.

```
Animate[bldg[t], {t, 0.0, 5.0, 5.0/29}]
```

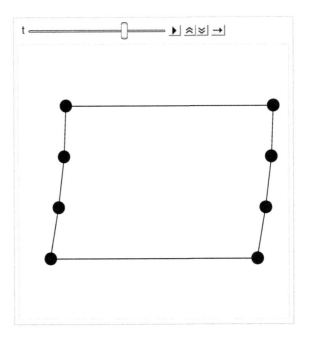

Increasing the number of stories increases the size of the system of differential equations. A five-story building corresponds to a system of five second-order differential equations; a fifty-story building, a system of fifty second-order differential equations, and so on.

Eigenvalue Problems and Fourier Series

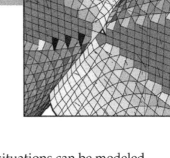

9

In previous chapters, we have seen that many physical situations can be modeled by either ordinary differential equations or systems of ordinary differential equations. However, to understand the motion of a string at a particular location and at a particular time, the temperature in a thin wire at a particular location and a particular time, or the electrostatic potential at a point on a plate, we must solve partial differential equations as each of these quantities depends on (at least) two independent variables.

$$\begin{aligned} \textbf{Wave equation} \quad & c^2 u_{xx} = u_{tt} \\ \textbf{Heat equation} \quad & u_t = c^2 u_{xx} \\ \textbf{Laplace's equation} \quad & u_{xx} + u_{yy} = 0 \end{aligned}$$

In Chapter 10, we introduce a particular method for solving these partial differential equations. In order to carry out this method, however, we introduce the necessary tools in this chapter. We begin with a discussion of boundary-value problems and their solutions.

9.1 Boundary-Value Problems, Eigenvalue Problems, Sturm-Liouville Problems

9.1.1 Boundary-Value Problems

In previous sections, we have solved initial-value problems. However, at this time we will consider boundary-value problems which are solved in much the same way as initial value problems except that the value of the function and its derivatives are given at two values of the independent variable instead of one. The general form of a second-order (two-point) boundary-value problem is

Differential Equations with Mathematica. http://dx.doi.org/10.1016/B978-0-12-804776-7.00009-7

$$\begin{cases} a_2(x)\dfrac{d^2y}{dx^2} + a_1(x)\dfrac{dy}{dx} + a_0(x)y = f(x),\ a < x < b \\[2mm] k_1 y(a) + k_2 \dfrac{dy}{dx}(a) = \alpha,\ h_1 y(b) + h_2 \dfrac{dy}{dx}(b) = \beta \end{cases} \tag{9.1}$$

where $k_1, k_2, \alpha, h_1, h_2$, and β are constants and at least one of k_1, k_2 and at least one of h_1, h_2 is not zero.

Note that if $\alpha = \beta = 0$, then we say the problem has **homogeneous boundary conditions**. We also consider boundary-value problems that include a parameter in the differential equation. We solve these problems, called **eigenvalue problems**, in order to investigate several useful properties associated with their solutions.

EXAMPLE 9.1.1: Solve $\begin{cases} y'' + y = 0,\ 0 < x < \pi \\ y'(0) = 0,\ y'(\pi) = 0 \end{cases}$.

SOLUTION: Because the characteristic equation is $k^2 + 1 = 0$ with roots $k_{1,2} = \pm i$, a general solution of $y'' + y = 0$ is $y = c_1 \cos x + c_2 \sin x$ and it follows that $y' = -c_1 \sin x + c_2 \cos x$. Applying the boundary conditions, we have $y'(0) = c_2 = 0$. Then, $y = c_1 \cos x$. With this solution, we have $y'(\pi) = -c_1 \sin \pi = 0$ for any value of c_1. Therefore, there are infinitely many solutions, $y = c_1 \cos x$, of the boundary-value problem, depending on the choice of c_1. In this case, we are able to use DSolve to solve the boundary-value problem

```
sol = DSolve[{y''[x] + y[x]==0, y'[0]==0, y'[π]==0}, y[x], x]
```

```
{{y[x] → C[1]Cos[x]}}
```

We confirm that the boundary conditions are satisfied for any value of C[1] by graphing several solutions with Plot in Figure 9-1.

```
toplot = Table[y[x]/.sol/.C[1]->i, {i, -5, 5}];
```

```
Plot[Evaluate[toplot], {x, 0, π}]
```

■

From the result in the example, we notice a difference between *initial-value problems* and *boundary-value problems*: an initial-value problem (that meets the hypotheses of the Existence and Uniqueness Theorem) has a unique solution while a boundary-value problem may have more than one solution, one solution, or no solution.

See Chapter 4 and Theorem 2.

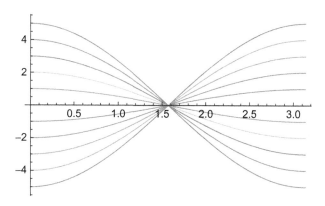

Figure 9-1   The boundary-value problem has infinitely many solutions

EXAMPLE 9.1.2: Solve $\begin{cases} y'' + y = 0, \ 0 < x < \pi \\ y'(0) = 0, \ y'(\pi) = 1 \end{cases}$.

SOLUTION: Using the general solution obtained in the previous example, we have $y = c_1 \cos x + c_2 \sin x$. As before, $y'(0) = c_2 = 0$, so $y = c_1 \cos x$. However, because $y'(\pi) = -c_1 \sin \pi = 0 \neq 1$, the boundary conditions cannot be satisfied with any choice of c_1. Therefore, there is no solution to the boundary-value problem.

■

As indicated in the general form of a boundary-value problem, the boundary conditions in these problems can involve the function and its derivative. However, this modification to the problem does not affect the method of solution.

EXAMPLE 9.1.3: Solve $\begin{cases} y'' - y = 0, \ 0 < x < 1 \\ y'(0) + 3y(0) = 0, \ y'(1) + y(1) = 1 \end{cases}$.

SOLUTION: The characteristic equation is $k^2 - 1 = 0$ with roots $k_{1,2} = \pm 1$. Hence, a general solution is $y = c_1 e^x + c_2 e^{-x}$ with derivative $y' = c_1 e^x - c_2 e^{-x}$. Applying $y'(0) + 3y(0) = 0$ yields $y'(0) + 3y(0) = c_1 - c_2 + 3(c_1 + c_2) = 4c_1 + 2c_2 = 0$. Because $y'(1) + y(1) = 1$,

$$y'(1) + y(1) = c_1 e^1 - c_2 e^{-1} + c_1 e^1 + c_2 e^{-1} = 2c_1 e = 1,$$

so $c_1 = \dfrac{1}{2e}$ and $c_2 = -\dfrac{1}{e}$. Thus, the boundary-value problem has the unique solution $y = \dfrac{1}{2e}e^x - \dfrac{1}{e}e^{-x} = \frac{1}{2}e^{x-1} - e^{-x-1}$, which we confirm with Mathematica. See Figure 9-2.

```
sol = DSolve[{y"[x] − y[x] == 0, y'[0] + 3y[0] == 0,

   y'[1] + y[1] == 1}, y[x], x]

{{y[x] → 1/2 e^(−1−x)(−2 + e^(2x))}}

Plot[y[x]/.sol, {x, 0, 1}]
```

∎

9.1.2 Eigenvalue Problems

We now consider **eigenvalue problems**, boundary-value problems that include a parameter. Values of the parameter for which the boundary-value problem has a nontrivial solution are called **eigenvalues** of the problem. For each eigenvalue, the nontrivial solution that satisfies the problem is called the **corresponding eigenfunction**.

If a value of the parameter leads to the trivial solution, then the value is not considered an eigenvalue of the problem.

In Example 9.1.5 we solve the same differential equation but use the boundary conditions $y'(0) = 0$ and $y'(p) = 0$.

EXAMPLE 9.1.4: Solve the eigenvalue problem $y'' + \lambda y = 0$, $0 < x < p$, subject to $y(0) = 0$ and $y(p) = 0$.

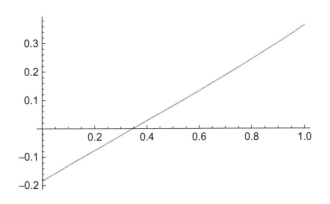

Figure 9-2　The boundary-value problem has a unique solution

SOLUTION: Notice that the differential equation in this problem differs from those solved earlier because it includes the parameter λ. However, we solve it in a similar manner by solving the characteristic equation $k^2 + \lambda = 0$. Of course, the values of k depend on the value of the parameter λ. Hence, we consider the following three cases.

1. ($\lambda = 0$) In this case, the characteristic equation is $k^2 = 0$ with roots $k_{1,2} = 0$, which indicates that a general solution is $y = c_1 x + c_2$. Application of the boundary condition $y(0) = 0$ yields $y(0) = c_1 \cdot 0 + c_2 = 0$, so $c_2 = 0$. For the second condition, $y(p) = c_1 p = 0$, so $c_1 = 0$ and $y = 0$. Because we obtain the trivial solution, $\lambda = 0$ is *not* an eigenvalue.

2. ($\lambda < 0$) To represent λ as a negative value, we let $\lambda = -\mu^2 < 0$. Then, the characteristic equation is $k^2 - \mu^2 = 0$, so $k_{1,2} = \pm\mu$. Therefore, a general solution is, $y = c_1 e^{\mu x} + c_1 e^{-\mu x}$ (or $y = c_1 \cosh \mu x + c_1 \sinh \mu x$). Substitution of the boundary condition $y(0) = 0$ yields $y(0) = c_1 + c_2 = 0$, so $c_2 = -c_1$. Because $y(p) = 0$ indicates that $y = c_1 e^{\mu p} + c_1 e^{-\mu p} = 0$, substitution gives us the equation $y(p) = c_1 e^{\mu p} - c_1 e^{-\mu p} = c_1 \left(e^{\mu p} - e^{-\mu p} \right)$. Notice that $e^{\mu p} - e^{\mu p} = 0$ only if $e^{\mu p} = e^{-\mu p}$ which can only occur if $\mu = 0$ or $p = 0$. If $\mu = 0$, then $\lambda = -\mu^2 = -0^2 = 0$ which contradicts the assumption that $\lambda < 0$. We also assumed that $p > 0$, so $e^{\mu p} - e^{\mu p} > 0$. Hence, $y(p) = c_1 \left(e^{\mu p} - e^{-\mu p} \right)$ implies that $c_1 = 0$, so $c_2 = -c_1 = 0$ as well. Because $\lambda < 0$ leads to the trivial solution $y = 0$, there are no negative eigenvalues.

3. ($\lambda > 0$) To represent λ as a positive value, we let $\lambda = \mu^2 > 0$. Then, we have the characteristic equation $k^2 + \mu^2 = 0$ with complex conjugate roots $k_{1,2} = \pm\mu i$. Thus, a general solution is $y = c_1 \cos \mu x + c_2 \sin \mu x$. Because $y(0) = c_1 \cos \mu \cdot 0 + c_2 \sin \mu \cdot 0 = c_1$, the boundary condition $y(0) = 0$ indicates that $c_1 = 0$. Hence, $y = c_2 \sin \mu x$. Application of $y(p) = 0$ yields $y(p) = c_2 \sin \mu p$, so either $c_2 = 0$ or $\sin \mu p = 0$. Selecting $c_2 = 0$ leads to the trivial solution that we want to avoid, so we determine the values of μ that satisfy $\sin \mu p = 0$. Because $\sin n\pi = 0$ for integer values of n, $\sin \mu p = 0$ if $\mu p = n\pi$, $n = 1, 2, \ldots$ Solving for μ, we have $\mu = n\pi/p$, so the eigenvalues are

$$\lambda = \lambda_n = \mu^2 = \left(\frac{n\pi}{p} \right)^2, \quad n = 1, 2, \ldots.$$

Notice that the subscript n is used to indicate that the parameter depends on the value of n. (Notice also that we omit $n = 0$, because the value $\mu = 0$ was considered in Case 1.) For each eigenvalue, the corresponding eigenfunction is obtained by substitution into $y = c_2 \sin \mu x$. Because c_2 is arbitrary, we choose $c_2 = 1$. Therefore, the eigenvalue $\lambda_n = (n\pi/p)^2$, $n = 1, 2, \ldots$ has corresponding eigenfunction

$$y(x) = y_n(x) = \sin \frac{n\pi x}{p}, \quad n = 1, 2, \ldots.$$

We did not consider negative values of n because $\sin(-n\pi x/p) = -\sin(n\pi x/p)$; the negative sign can be taken into account in the constant; we do not obtain additional eigenvalues or eigenfunctions by using $n = -1, -2, \ldots$.

■

We will find the eigenvalues and eigenfunctions in Example 9.1.4 quite useful in future sections. The following eigenvalue problem will be useful as well.

In Example 9.1.4 we solve the same differential equation but use the boundary conditions $y(0) = 0$ and $y(p) = 0$.

EXAMPLE 9.1.5: Solve $y'' + \lambda y = 0$, $0 < x < p$, subject to $y'(0) = 0$ and $y'(p) = 0$.

SOLUTION: Notice that the only difference between this problem and that in Example 9.1.4 is in the boundary conditions. Again, the characteristic equation is $k^2 + \lambda = 0$, so we must consider the three cases $\lambda = 0$, $\lambda < 0$, and $\lambda > 0$. Note that a general solution in each case is the same as that obtained in Example 9.1.4. However, the final results may differ due to the boundary conditions.

1. ($\lambda = 0$) Because $y = c_1 x + c_2$, $y' = c_1$. Therefore, $y'(0) = c_1 = 0$, so $y = c_2$. Notice that this constant function satisfies $y'(p) = 0$ for all values of c_2. Hence, if we choose $c_2 = 1$, then $\lambda = 0$ is an eigenvalue with corresponding eigenfunction $y = y_0(x) = 1$.

2. ($\lambda < 0$) If $\lambda = -\mu^2 < 0$, then $y = c_1 e^{\mu x} + c_2 e^{-\mu x}$ and $y' = c_1 \mu e^{\mu x} - c_2 \mu e^{-\mu x}$. Applying the first condition results in $y'(0) = c_1 k - c_2 k = 0$, so $c_1 = c_2$. Therefore, $y'(p) = c_1 \mu e^{\mu p} - c_1 \mu e^{-\mu p}$ which is not possible unless $c_1 = 0$, because $\mu \neq 0$ and $p \neq 0$. Thus, $c_1 = c_2 = 0$, so $y = 0$. Because we have the trivial solution, there are no negative eigenvalues.

3. ($\lambda > 0$) By letting $\lambda = \mu^2$, $y = c_1 \cos \mu x + c_2 \sin \mu x$ and $y' = -c_1 \mu \sin \mu x + c_2 \mu \cos \mu x$. Hence, $y'(0) = c_2 \mu = 0$, so $c_2 = 0$. Consequently, $y'(p) = -c_1 \mu \sin \mu p = 0$ which is satisfied if $\mu p = n\pi$, $n = 1, 2, \ldots$ Therefore, the eigenvalues are

$$\lambda = \lambda_n = \left(\frac{n\pi}{p}\right)^2, \quad n = 1, 2, \ldots.$$

Note that we found $c_2 = 0$ in $y = c_1 \cos \mu x + c_2 \sin \mu x$, so choosing $c_1 = 1$, the corresponding eigenfunctions are

$$y = y_n = \cos \frac{n\pi x}{p}, \quad n = 1, 2, \ldots$$

■

EXAMPLE 9.1.6: For the eigenvalue problem $y'' + \lambda y = 0$, $y(0) = 0$, $y(1) + y'(1) = 0$, (a) show that the positive eigenvalues $\lambda = \mu^2$ satisfy the relationship $\mu = -\tan \mu$, and (b) Approximate the first eight positive eigenvalues. Notice that for larger values of μ, the eigenvalues are approximately the vertical asymptotes of $y = \tan \mu$, so $\lambda_n \approx [(2n-1)\pi/2]^2$, $n = 1, 2, \ldots$.

SOLUTION: In order to solve the eigenvalue problem, we consider the three cases.

1. ($\lambda = 0$) The problem $y'' = 0$, $y(0) = 0$, $y(1) + y'(1) = 0$ has the solution $y = 0$, so $\lambda = 0$ is not an eigenvalue.

   ```
   DSolve[{y"[x] == 0, y[0] == 0,

     y[1] + y'[1] == 0}, y[x], x]
   ```

   $\{\{y[x] \to 0\}\}$

2. ($\lambda < 0$) Similarly, $y'' - \mu^2 y = 0$, $y(0) = 0$, $y(1) + y'(1) = 0$ has solution $y = 0$, so there are no negative eigenvalues.

   ```
   DSolve[{y"[x] − μ^2y[x] == 0, y[0] == 0,

     y[1] + y'[1] == 0}, y[x], x]
   ```

   $\{\{y[x] \to 0\}\}$

3. ($\lambda > 0$) If $\lambda = \mu^2 > 0$, we solve $y'' + \mu^2 y = 0$, $y(0) = 0$, $y(1) + y'(1) = 0$. In this case, the result returned by DSolve is incorrect.

   ```
   DSolve[{y"[x] + μ^2y[x] == 0, y[0] == 0,

     y[1] + y'[1] == 0}, y[x], x]
   ```

   $\{\{y[x] \to 0\}\}$

   A general solution of $y'' + \mu^2 y = 0$ is $y = A \cos \mu x + B \sin \mu x$. Applying $y(0) = 0$ indicates that $A = 0$, so $y = B \sin \mu x$. Applying $y(1) + y'(1) = 0$ where $y' = \mu B \cos \mu x$ yields $B \sin \mu + \mu B \cos \mu = 0$. Because we want to avoid requiring that $B = 0$, we note that

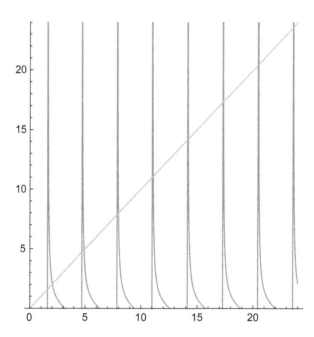

Figure 9-3   The eigenvalues are the x-coordinates of the points of intersection of $y = x$ and $y = -\tan x$

this condition is satisfied if $-\sin \mu = \mu \sin \mu$ or $-\tan \mu = \mu$. To approximate the first eight positive roots of this equation, we graph $y = -tanx$ and $y = x$ simultaneously in Figure 9-3. (We only look for positive roots because $\tan(-\mu) = -\tan \mu$, meaning that no additional eigenvalues are obtained by considering negative values of μ.) The eigenfunctions of this problem are $y = \sin \mu x$ where μ satisfies $-\tan \mu = \mu$.

```
Plot[{-Tan[x], x}, {x, 0, 24}, PlotRange → {0, 24},

    AspectRatio → 1]
```

In Figure 9-3, notice that roots are to the right of the vertical asymptotes of $y = -\tan x$ which are $x = (2n - 1)\pi/2$, n any integer. We use FindRoot to obtain approximations to the roots using initial guesses near the asymptotes. Here, we guess 0.1 unit to the right of $(2n - 1)\pi/2$ for $n = 1, 2, \ldots, 8$.

```
kvals = Table[FindRoot[-Tan[x]==x, {x, (2n - 1)π / 2 + 0.1}],

    {n, 1, 8}]
```

$$\{\{x \rightarrow 2.02876\}, \{x \rightarrow 4.91318\}, \{x \rightarrow 7.97867\},$$

$$\{x \rightarrow 11.0855\},$$

$$\{x \rightarrow 14.2074\}, \{x \rightarrow 17.3364\}, \{x \rightarrow 20.4692\},$$

$$\{x \rightarrow 23.6043\}\}$$

Therefore, the first eight roots are approximately 2.02876, 4.91318, 7.97867, 11.0855, 14.2074, 17.3364, 20.4692, and 23.6043. As x increases, the roots move closer to the value of x at the vertical asymptotes of $y = -\tan x$. We can compare the two approximations by finding a for the first eight vertical asymptotes, $x = a$.

`Table[N[`$\dfrac{(2n-1)\pi}{2}$`], {n, 1, 8}]`

$$\{1.5708, 4.71239, 7.85398, 10.9956, 14.1372, 17.2788,$$

$$20.4204, 23.5619\}$$

The first eight eigenvalues are approximated by squaring the elements of `kvals`. We call this list `evals`.

`evals = Table[kvals[[j, 1, 2]]^2, {j, 1, 8}]`

$$\{4.11586, 24.1393, 63.6591, 122.889, 201.851, 300.55,$$

$$418.987, 557.162\}$$

∎

9.1.3 Sturm-Liouville Problems

Because of the importance of eigenvalue problems, we express these problems in the general form

$$a_2(x)y'' + a_1(x)y' + a_0(x)y + [a_0(x) + \lambda]y = 0, \quad a < x < b, \tag{9.2}$$

where $a_2(x) \neq 0$ on $[a, b]$ and the boundary conditions at the endpoints $x = a$ and $x = b$ can be written as

$$k_1 y(a) + k_2 y'(a) = 0 \quad \text{and} \quad h_1 y(b) + h_2 y'(b) = 0 \tag{9.3}$$

for the constants k_1, k_2 h_1, and h_2 where at least one of h_1, h_2 and at least one of k_1, k_2 is not zero. Equation (9.2) can be rewritten by letting

$$p(x) = e^{\int a_1(x)/a_2(x)\, dx}, \quad q(x) = \frac{a_0(x)}{a_2(x)}p(x), \quad \text{and} \quad s(x) = \frac{p(x)}{a_2(x)}. \tag{9.4}$$

By making this change, equation (9.2) can be rewritten as the equivalent equation

$$\frac{d}{dx}\left(p(x)\frac{dy}{dx}\right) + (q(x) + \lambda s(x))\,y = 0, \tag{9.5}$$

which is called a **Sturm-Liouville equation** and along with appropriate boundary conditions is called a **Sturm-Liouville problem**. This particular form of the equation is known as **self-adjoint form**, which is of interest because of the relationship of the function $s(x)$ and the solutions of the problem.

EXAMPLE 9.1.7: Place the equation $x^2 y'' + 2xy' + \lambda y = 0$, $x > 0$, in self-adjoint form.

SOLUTION: In this case, $a_2(x) = x^2$, $a_1(x) = 2x$, and $a_0(x) = 0$. Hence,

$$p(x) = e^{\int a_1(x)/a_2(x)\,dx} = e^{\int 2x/x^2\,dx} = e^{2\ln x} = x^2, \quad q(x) = \frac{a_0(x)}{a_2(x)}p(x) = 0,$$

and $s(x) = \dfrac{p(x)}{a_2(x)} = \dfrac{x^2}{x^2} = 1$, so the self-adjoint form of the equation is

$\dfrac{d}{dx}\left(x^2\dfrac{dy}{dx}\right) + \lambda y = 0$. We see that our result is correct by differentiating.

■

Solutions of Sturm-Liouville problems have several interesting properties, two of which are included in the following theorem.

Theorem 33 (Linear Independence and Orthogonality of Eigenfunctions). *If $y_m(x)$ and $y_n(x)$ are eigenfunctions of the regular Sturm-Liouville problem*

$$\begin{cases} \dfrac{d}{dx}\left(p(x)\dfrac{dy}{dx}\right) + (q(x) + \lambda s(x))\,y = 0 \\ k_1 y(a) + k_2 y'(a) = 0,\ h_1 y(b) + h_2 y'(b) = 0 \end{cases}, \tag{9.6}$$

*where $m \neq n$, $y_m(x)$ and $y_n(x)$ are **linearly independent** and the **orthogonality** condition $\int_a^b s(x)y_m(x)y_n(x)\,dx = 0$ holds.*

Because we integrate the product of the eigenfunctions with the function $s(x)$ in the orthogonality condition, we call $s(x)$ the **weighting function**.

EXAMPLE 9.1.8: Consider the eigenvalue problem $y'' + \lambda y = 0$, $0 < x < p$, subject to $y(0) = 0$ and $y(p) = 0$ that we solved in Example 9.1.4. Verify that the eigenfunctions $y_1 = \sin(\pi x/p)$ and $y_2 = \sin(2\pi x/p)$ are linearly independent. Also, verify the orthogonality condition.

SOLUTION: We can verify that $y_1 = \sin(\pi x/p)$ and $y_2 = \sin(2\pi x/p)$ are linearly independent by computing the Wronskian.

```
Clear[x, p]

caps = {Sin[Pix/p], Sin[2Pix/p]};

ws = Simplify[Det[{caps, D[caps, x]}]]
```

$$-\frac{2\pi \operatorname{Sin}[\frac{\pi x}{p}]^3}{p}$$

We see that the Wronskian is not the zero function by evaluating it for a particular value of x; we choose $x = p/2$.

```
ws /. x → p/2
```

$$-\frac{2\pi}{p}$$

Because $W\{y_1, y_2\}$ is not zero for all values of x, the two functions are linearly independent. In self-adjoint form, the equation is $y'' + \lambda y = 0$, with $s(x) = 1$. Hence, the orthogonality condition is $\int_0^p y_m(x)y_n(x)\, dx = 0$, $m \neq n$, which we verify for y_1 and y_2.

Of course, these two properties hold for any choices of m and n, $m \neq n$.

```
Integrate[Sin[Pi x/p]Sin[2Pi x/p], {x, 0, p}]
```

0

∎

9.2 Fourier Sine Series and Cosine Series

9.2.1 Fourier Sine Series

Recall the eigenvalue problem $\begin{cases} y'' + \lambda y = 0 \\ y(0) = 0,\ y(p) = 0 \end{cases}$ that was solved in Example 9.1.4. The eigenvalues of this problem are $\lambda = \lambda_n = (n\pi/p)^2$, $n = 1, 2, \ldots$, with corresponding eigenfunctions $\phi_n(x) = \sin(n\pi x/p)$, $n = 1, 2, \ldots$.

We will see that for some functions $y = f(x)$, we can find coefficients c_n so that

$$f(x) = \sum_{n=1}^{\infty} c_n \sin \frac{n\pi x}{p}. \tag{9.7}$$

A series of this form is called a **Fourier sine series**. To make use of these series, we must determine the coefficients c_n. We accomplish this by taking advantage of the orthogonality properties of eigenfunctions stated in Theorem 33.

Because the differential equation $y'' + \lambda y = 0$ is in self-adjoint form, we have that $s(x) = 1$. Therefore, the orthogonality condition is $\int_0^p \sin(n\pi x/p) \sin(m\pi x/p) \, dx$, $m \neq n$. In order to use this condition, multiply both sides of $f(x) = \sum_{n=1}^\infty c_n \sin(n\pi x/p)$ by the eigenfunction $\sin(m\pi x/p)$ and $s(x) = 1$. Then, integrate the result from $x = 0$ to $x = p$ (because the boundary conditions of the corresponding eigenvalue problem are given at these two values of x). This yields

$$\int_0^p f(x) \sin \frac{m\pi x}{p} \, dx = \int_0^p \sum_{n=1}^\infty c_n \sin \frac{n\pi x}{p} \sin \frac{m\pi x}{p} \, dx.$$

Assuming that term-by-term integration is allowed on the right-hand side of the equation, we have

$$\int_0^p f(x) \sin \frac{m\pi x}{p} \, dx = \sum_{n=1}^\infty \int_0^p c_n \sin \frac{n\pi x}{p} \sin \frac{m\pi x}{p} \, dx.$$

Recall that the eigenfunctions $\phi_n(x)$, $n = 1, 2, \ldots$ are orthogonal, so $\int_0^p \sin(n\pi x/p) \sin(m\pi x/p) \, dx = 0$ if $m \neq 0$. On the other hand, if $m = n$,

$$\int_0^p \sin \frac{n\pi x}{p} \sin \frac{m\pi x}{p} \, dx = \int_0^p \sin^2 \frac{n\pi x}{p} \, dx$$

$$= \frac{1}{2} \int_0^p \left(1 - \cos \frac{2n\pi x}{p}\right) dx$$

$$= \frac{1}{2} \left[x - \frac{p}{2n\pi} \sin \frac{2n\pi x}{p}\right]_0^p = \frac{p}{2}.$$

$\int_0^p \mathbf{Sin}[\frac{n\pi\, x}{p}]^2 \, dx$

$\frac{1}{4} p(2 - \frac{\mathbf{Sin}[2n\pi]}{n\pi})$

Therefore, each term in the sum $\sum_{n=1}^\infty c_n \int_0^p \sin(n\pi x/p) \sin(m\pi x/p) \, dx$ equals zero except when $m = n$. Hence, $\int_0^p f(x) \sin(n\pi x/p) \, dx = \frac{1}{2} c_n p$, so the Fourier sine series coefficients are given by

$$c_n = \frac{2}{p} \int_0^p f(x) \sin \frac{n\pi x}{p} \, dx, \tag{9.8}$$

where we assume that $y = f(x)$ is integrable on $[0, p]$.

Sometimes the command `FourierSinSeries` can help you find the Fourier sine series for a function. Keep in mind that the `FourierSinSeries` function is very limited in scope.

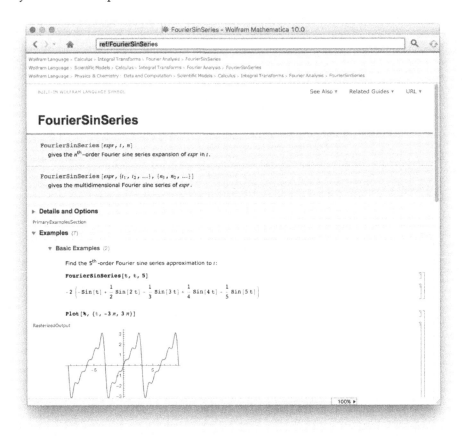

EXAMPLE 9.2.1: Find the Fourier sine series for $f(x) = x$, $0 \le x \le \pi$.

SOLUTION: In this case, $p = \pi$. Using integration by parts we have,

$$c_n = \frac{2}{\pi} \int_0^\pi f(x) \sin \frac{n\pi x}{\pi}\, dx = \frac{2}{\pi} \int_0^\pi x \sin nx\, dx$$

$$= \frac{2}{\pi} \left[-\frac{1}{n} x \cos nx \right]_0^\pi + \frac{2}{\pi} \int_0^\pi \frac{1}{n} \cos nx\, dx = -\frac{2}{n} \cos n\pi + \frac{2}{\pi} \left[\frac{1}{n^2} \sin nx \right]_0^\pi$$

$$= -\frac{2}{n} \cos n\pi + \frac{2}{n^2} (\sin n\pi - \sin 0) = -\frac{2}{n} \cos n\pi.$$

$$\int_0^\pi \frac{2x\text{Sin}[nx]}{\pi}\,dx$$

$$\frac{2(-n\pi\,\text{Cos}[n\pi] + \text{Sin}[n\pi])}{n^2\pi}$$

Observe that n is an integer so $\cos n\pi = (-1)^n$. Hence, $c_n = -\frac{2}{n}(-1)^n = (-1)^{n+1}\frac{2}{n}$, and the Fourier sine series is

$$f(x) = \sum_{n=1}^{\infty} c_n \sin\frac{n\pi x}{\pi} = 2\sum_{n=1}^{\infty}(-1)^{n+1}\frac{1}{n}\sin nx$$

$$= 2\sin x - \sin 2x + \frac{2}{3}\sin 3x - \frac{1}{2}\sin 4x + \cdots.$$

We can use a finite number of terms of the series to obtain a trigonometric polynomial that approximates $f(x) = x$, $0 \le x \le \pi$ as follows. Let $f_k(x) = 2\sum_{n=1}^{k}(-1)^{n+1}\frac{1}{n}\sin nx$. Then, $f_k(x) = f_{k-1}(x) + (-1)^{k+1}\frac{2}{k}\sin kx$. Thus, to calculate the kth partial sum of the Fourier sine series, we need only add $(-1)^{k+1}\frac{2}{k}\sin kx$ to the $(k-1)$st partial sum: we need not recompute all k terms of the kth partial sum if we know the $(k-1)$st partial sum. Using this observation, we define the recursively defined function f to return the kth partial sum of the series. We use the form $f[k\_]:=f[k]=\ldots$ so that Mathematica "remembers" each $f_k(x)$ that is computed. The advantage of doing so is that Mathematica need not recompute $f_k(x)$ to compute $f_{k+1}(x)$.

```
Remove[f]
```

$$f[k\_]:=f[k] = f[k-1] + \frac{2(-1)^{k+1}\text{Sin}[kx]}{k}$$

```
f[1] = 2Sin[x];

t1 = Table[{n, f[n]}, {n, 1, 5}];

TableForm[t1]
```

1    $2\text{Sin}[x]$

2    $2\text{Sin}[x] - \text{Sin}[2x]$

3    $2\text{Sin}[x] - \text{Sin}[2x] + \frac{2}{3}\text{Sin}[3x]$

4    $2\text{Sin}[x] - \text{Sin}[2x] + \frac{2}{3}\text{Sin}[3x] - \frac{1}{2}\text{Sin}[4x]$

5    $2\text{Sin}[x] - \text{Sin}[2x] + \frac{2}{3}\text{Sin}[3x] - \frac{1}{2}\text{Sin}[4x] + \frac{2}{5}\text{Sin}[5x]$

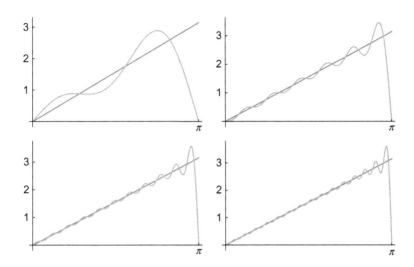

Figure 9-4 $f(x) = x, 0 \leq x \leq \pi$, shown with the 3rd, 12th, 21st, and 30th partial sums of its Fourier sine series

We now graph $f(x)$ on $[0, \pi]$ along with several of the partial sums of the sine series in Figure 9-4. As we increase the number of terms used in approximating $f(x)$, we improve the accuracy. Notice from the graphs that none of the partial sums attain the value of $f(\pi) = \pi$ at $x = \pi$. This is due to the fact that at $x = \pi$, each of the partial sums yields a value of 0. Hence, our approximation can only be reliable on the interval $0 < x < \pi$. In general, however, we are only assured of accuracy at points of continuity of $f(x)$ on the open interval.

```
t1 = Table[{n, f[n]}, {n, 1, 30}];

somegraphs = Table[Plot[{x, f[n]}, {x, 0, π},

   Ticks → {{0, π}, {1, 2, 3}}], {n, 3, 30, 9}];

toshow = Partition[somegraphs, 2];

Show[GraphicsGrid[toshow], AspectRatio → 1]
```

For this problem, the FourierSinSeries function is helpful in verifying our results.

```
FourierSinSeries[x, x, 9]
```

$$-2(-\text{Sin}[x] + \frac{1}{2}\text{Sin}[2x] - \frac{1}{3}\text{Sin}[3x] + \frac{1}{4}\text{Sin}[4x] - \frac{1}{5}\text{Sin}[5x]$$

$$+ \frac{1}{6}\text{Sin}[6x] - \frac{1}{7}\text{Sin}[7x] + \frac{1}{8}\text{Sin}[8x] - \frac{1}{9}\text{Sin}[9x])$$

■

EXAMPLE 9.2.2: Find the Fourier sine series for $f(x) = \begin{cases} 1, & 0 \le x < 1 \\ -1, & 1 \le x \le 2 \end{cases}$.

SOLUTION: Because $f(x)$ is defined on $0 \le x \le 2$, $p = 2$. Hence,

$$c_n = \frac{2}{2}\int_0^2 f(x)\sin\frac{n\pi x}{2}\,dx = \frac{2}{n\pi}\left(-2\cos\frac{n\pi}{2} + \cos n\pi + 1\right).$$

```
cn_ = ∫₀¹ Sin[nπx/2] dx - ∫₁² Sin[nπx/2] dx
```

$$-\frac{2(\text{Cos}[\frac{n\pi}{2}] - \text{Cos}[n\pi])}{n\pi} + \frac{4\text{Sin}[\frac{n\pi}{4}]^2}{n\pi}$$

```
Table[{n, cn}, {n, 1, 15}]//TableForm
```

1	0
2	$\dfrac{4}{\pi}$
3	0
4	0
5	0
6	$\dfrac{4}{3\pi}$
7	0
8	0
9	0

10	$\dfrac{4}{5\pi}$
11	0
12	0
13	0
14	$\dfrac{4}{7\pi}$
15	0

As we can see, most of the coefficients are zero. In fact, only those c_n's where n is an odd multiple of 2 yield a nonzero value. For example,

$$c_6 = c_{2\cdot3} = \frac{2}{6\pi} \cdot 4 = \frac{4}{3\pi}, c_{10} = c_{2\cdot5} = \frac{2}{10\pi} \cdot 4 = \frac{4}{5\pi}, \dots,$$

$$c_{2(2n-1)} = \frac{4}{(2n-1)\pi}, \quad n = 1, 2, \dots$$

so we have the series

$$f(x) = \sum_{n=1}^{\infty} \frac{4}{(2n-1)\pi} \sin \frac{2(2n-1)\pi x}{2} = \sum_{n=1}^{\infty} \frac{4}{(2n-1)\pi} \sin(2n-1)\pi x$$

$$= \frac{4}{\pi} \sin \pi x + \frac{4}{3\pi} \sin 3\pi x + \frac{4}{5\pi} \sin 5\pi x + \cdots$$

As in Example 9.2.1, we graph $f(x)$ with several partial sums of the Fourier sine series in Figure 9-5.

```
Clear[f, g]

g[x_]:=1/;0 ≤ x < 1

g[x_]:= - 1/;1 ≤ x ≤ 2

f[k_]:=f[k] = f[k - 1] + 4Sin[(2k - 1)πx]
                         ─────────────────
                            (2k - 1)π
f[1] = 4Sin[πx];
       ─────────
          π

t1 = Table[{n, f[n]}, {n, 1, 20}];

t2 = Table[{n, f[n]}, {n, 1, 10}]; //TableForm
```

Notice that with a large number of terms the approximation is quite good at values of x where $f(x)$ is continuous.

```
somegraphs = Table[Plot[{g[x], f[n]}, {x, 0, 2}, PlotRange

→ {-1.5, 1.5}, Ticks → {{1, 2}, {-1.5, 1.5}}], {n, 1, 9}];
```

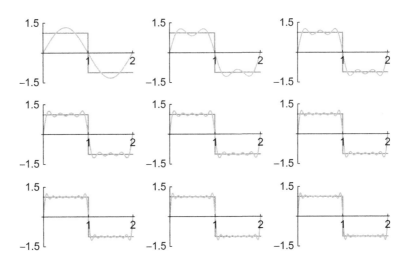

Figure 9-5   At the jump discontinuity at $x = 1$, the Fourier sine series converges to $\frac{1}{2}\left(\lim_{x\to 1^-}f(x) + \lim_{x\to 1^+}f(x)\right) = \frac{1}{2}(1-1) = 0$

```
toshow = Partition[somegraphs, 3];

Show[GraphicsGrid[toshow], AspectRatio → 1]
```

The behavior of the series near points of discontinuity (in that the approximation overshoots the function) is called **Gibbs phenomenon**. The approximation continues to "miss" the function even though more and more terms from the series are used!

In somegraphs, we observe the graph of the error function Abs [g [x] - f [n]] for $n = 1, 2, \ldots, 9$. Notice that the error remains "large" at the points of discontinuity, $x = 0, 1, 2$, even for "large" values of n. See Figure 9-6.

```
somegraphs = Table[Plot[Abs[g[x] − f[n]], {x, 0, 2},

DisplayFunction →

Identity, PlotRange → {0, 1}, Ticks → {{1, 2}, {0, 1}}],

  {n, 1, 9}];

toshow = Partition[somegraphs, 3];

Show[GraphicsGrid[toshow], AspectRatio → 1]
```

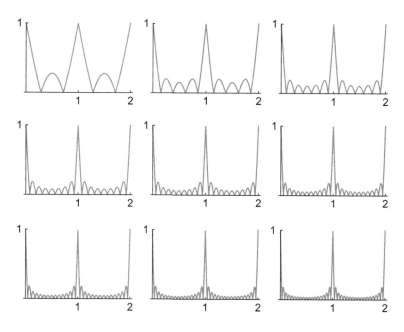

Figure 9-6 The Fourier sine series converges to $f(x)$ on the open intervals where $f(x)$ is continuous

■

9.2.2 Fourier Cosine Series

Another important eigenvalue problem that has useful eigenfunctions is

We solved this eigenvalue problem in Example 9.1.5.

$$\begin{cases} y'' + \lambda y = 0 \\ y'(0) = y'(p) = 0 \end{cases},$$

which has eigenvalues and eigenfunctions given by

$$\lambda_n = \begin{cases} 0, \ n = 0 \\ (n\pi/p)^2, \ n = 1, 2, \ldots \end{cases} \quad \text{and} \quad y_n(x) = \begin{cases} 1, \ n = 0 \\ \cos(n\pi x/p), \ n = 1, 2, \ldots \end{cases}.$$

Therefore, for some functions $f(x)$, we can find a series expansion of the form

$$f(x) = \frac{1}{2}a_0 + \sum_{n=1}^{\infty} a_n \cos \frac{n\pi x}{p}. \tag{9.9}$$

We call this expansion a **Fourier cosine series** where in the first term (associated with $\lambda_0 = 0$), the constant $\frac{1}{2}a_0$ is written in this form for convenience in finding the formula for the coefficients a_n, $n = 1, 2, \ldots$. We find these coefficients in a manner similar to that followed to find the coefficients in the Fourier sine series. Notice that in this case, the orthogonality condition is $\int_0^p \cos(n\pi x/p)\cos(m\pi x/p)\,dx = 0$, $m \neq n$. We use this condition by multiplying both sides of the series expansion by $\cos(m\pi x/p)$ and integrating from $x = 0$ to $x = p$. This yields

$$\int_0^p f(x) \cos \frac{m\pi x}{p}\,dx = \int_0^p \frac{1}{2}a_0 \cos \frac{m\pi x}{p}\,dx + \int_0^p \sum_{n=1}^\infty a_n \cos \frac{n\pi x}{p} \cos \frac{m\pi x}{p}\,dx.$$

Assuming that term-by-term integration is allowed,

$$\int_0^p f(x) \cos \frac{m\pi x}{p}\,dx = \int_0^p \frac{1}{2}a_0 \cos \frac{m\pi x}{p}\,dx + \sum_{n=1}^\infty \int_0^p a_n \cos \frac{n\pi x}{p} \cos \frac{m\pi x}{p}\,dx.$$

If $m = 0$, then this equation reduces to

$$\int_0^p f(x)\,dx = \int_0^p \frac{1}{2}a_0\,dx + \sum_{n=1}^\infty \int_0^p a_n \cos \frac{n\pi x}{p}\,dx.$$

where $\int_0^p \cos(n\pi x/p)\,dx = 0$ and $\int_0^p \frac{1}{2}a_0\,dx = \frac{1}{2}pa_0$. Therefore, $\int_0^p f(x)\,dx = \frac{1}{2}pa_0$, so

$$a_0 = \frac{2}{p} \int_0^p f(x)\,dx. \tag{9.10}$$

If $m > 0$, we note that by the orthogonality property $\int_0^p \cos(n\pi x/p)\cos(m\pi x/p)\,dx = 0$, $m \neq n$. We also note that $\int_0^p \frac{1}{2}a_0 \cos(m\pi x/p)\,dx = 0$ and $\int_0^p \cos^2(n\pi x/p)\,dx = \frac{1}{2}p$. Hence, $\int_0^p f(x)\cos(n\pi x/p)\,dx = 0 + a_n \cdot \frac{1}{2}p$. Solving for a_n, we have

$$a_n = \frac{2}{p} \int_0^p f(x) \cos \frac{n\pi x}{p}\,dx, \quad n = 1, 2, \ldots \tag{9.11}$$

Notice that equation (9.11) is valid for $n = 0$ because $\cos \dfrac{0 \cdot \pi x}{p} = 1$.

Sometimes the command `FourierCosSeries` can help you find the Fourier cosine series for a function. Keep in mind that the `FourierCosSeries` function is very limited in scope.

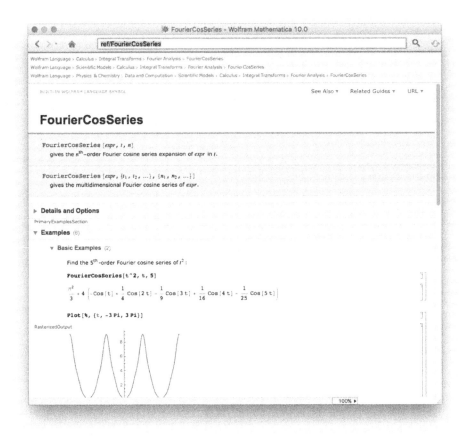

EXAMPLE 9.2.3: Find the Fourier cosine series for $f(x) = x$, $0 \le x \le \pi$.

SOLUTION: In this case, $p = \pi$. Hence,

$$a_0 = \frac{2}{\pi} \int_0^\pi x \, dx = \frac{2}{\pi} \left[\frac{1}{2} x^2 \right]_0^\pi = \pi$$

and using integration by parts we find that

If n is an integer, $\cos n\pi = (-1)^n$ and $\sin n\pi = 0$.

$$a_n = \frac{2}{\pi} \int_0^\pi x \cos \frac{n\pi x}{\pi} \, dx = \frac{2}{\pi} \int_0^\pi x \cos nx \, dx$$

$$= \frac{2}{\pi} \left\{ \left[\frac{1}{n} x \sin nx \right]_0^\pi - \int_0^\pi \frac{1}{n} \sin nx \, dx \right\} = \frac{2}{\pi} \left[\frac{1}{n^2} \cos nx \right]_0^\pi$$

$$= \frac{2}{n^2\pi} (\cos n\pi - 1) = \frac{2}{n^2\pi} \left[(-1)^n - 1 \right].$$

Notice that for even values of n, $(-1)^n - 1 = 0$. Therefore, $a_n = 0$ if n is even. On the other hand, if n is odd, $(-1)^n - 1 = -2$. Hence, $a_1 = -\frac{4}{\pi}$, $a_3 = -\frac{4}{9\pi}$, $a_5 = -\frac{4}{25\pi}$, ...,

$$a_{2n-1} = -\frac{4}{(2n-1)^2\pi},$$

so the Fourier cosine series for $f(x)$ is

$$f(x) = \frac{\pi}{2} - \sum_{n=1}^\infty \frac{2}{n^2\pi} \left[(-1)^n - 1 \right] \cos \frac{n\pi x}{\pi}$$

$$= \frac{\pi}{2} - \frac{4}{\pi} \sum_{n=1}^\infty \frac{1}{(2n-1)^2} \cos(2n-1)x.$$

We plot the function with several terms of the series in Figure 9-7. Compare these results to those obtained when approximating this function with a sine series. Which series yields the better approximation with the fewer number of terms?

```
Clear[f]
f[n_]:=f[n] = f[n-1] - 4Cos[(2n-1)x]/(π(2n-1)^2)
f[0] = π/2;
p1 = Plot[Evaluate[{x, f[1]}], {x, 0, π}];

p2 = Plot[Evaluate[Abs[x - f[1]]], {x, 0, π}];

p3 = Plot[Evaluate[{x, f[2]}], {x, 0, π}];

p4 = Plot[Evaluate[Abs[x - f[2]]], {x, 0, π}];

Show[GraphicsGrid[{{p1, p2}, {p3, p4}}]]
```

FourierCosSeries helps us check the results of our calculation.

```
FourierCosSeries[x, x, 9]
```

$$\frac{\pi}{2} - \frac{4\cos[x]}{\pi} - \frac{4\cos[3x]}{9\pi} - \frac{4\cos[5x]}{25\pi} - \frac{4\cos[7x]}{49\pi} - \frac{4\cos[9x]}{81\pi}$$

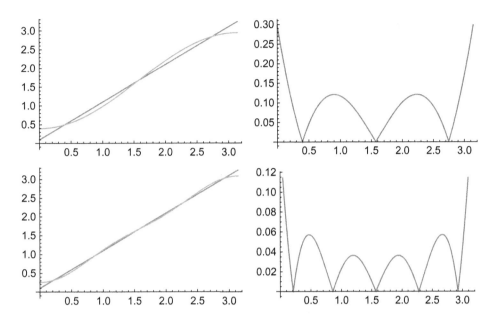

Figure 9-7   Partial sums of the Fourier cosine series shown with $f(x) = x$, $0 \leq x \leq \pi$ on the left and the absolute value of the difference between the two on the right

■

9.3 Fourier Series

9.3.1 Fourier Series

The eigenvalue problem

$$\begin{cases} y'' + \lambda y = 0, \ -p \leq x \leq p \\ y(-p) = y(p), \ y'(-p) = y'(p) \end{cases}$$

has eigenvalues

$$\lambda_n = \begin{cases} 0, \ n = 0 \\ (n\pi/p)^2, \ n = 1, 2 \ldots \end{cases}$$

and eigenfunctions

$$y_n(x) = \begin{cases} 1, \ n = 0 \\ a_n \cos \dfrac{n\pi x}{p} + b_n \sin \dfrac{n\pi x}{p}, \ n = 1, 2 \ldots, \end{cases}$$

so we can consider a series made up of these functions. Hence, we write

$$f(x) = \frac{1}{2}a_0 + \sum_{n=1}^{\infty} \left(a_n \cos \frac{n\pi x}{p} + b_n \sin \frac{n\pi x}{p} \right), \tag{9.12}$$

which is called a **Fourier series**. As was the case with Fourier sine and Fourier cosine series, we must determine the coefficients a_0, a_n ($n = 1, 2, \ldots$), and b_n ($n = 1, 2, \ldots$). Because we use a method similar to that used to find the coefficients in Section 9.2, we state the value of several integrals next.

$$\int_{-p}^{p} \cos \frac{n\pi x}{p} \, dx = 0$$

$$\int_{-p}^{p} \sin \frac{n\pi x}{p} \, dx = 0$$

$$\int_{-p}^{p} \cos \frac{m\pi x}{p} \sin \frac{n\pi x}{p} \, dx = 0 \tag{9.13}$$

$$\int_{-p}^{p} \cos \frac{m\pi x}{p} \cos \frac{n\pi x}{p} \, dx = \begin{cases} 0, & m \neq n \\ p, & m = n \end{cases}$$

$$\int_{-p}^{p} \sin \frac{m\pi x}{p} \sin \frac{n\pi x}{p} \, dx = \begin{cases} 0, & m \neq n \\ p, & m = n. \end{cases}$$

We begin by finding a_0 and a_n ($n = 1, 2, \ldots$). Multiplying both sides of equation (9.12) by $\cos(m\pi x/p)$ and integrating from $x = -p$ to $x = p$ (because of the boundary conditions) yields

$$\int_{-p}^{p} f(x) \cos \frac{m\pi x}{p} \, dx = \int_{-p}^{p} \frac{1}{2} a_0 \cos \frac{m\pi x}{p} \, dx +$$

$$\int_{-p}^{p} \sum_{n=1}^{\infty} \left(a_n \cos \frac{n\pi x}{p} \cos \frac{m\pi x}{p} + b_n \sin \frac{n\pi x}{p} \cos \frac{m\pi x}{p} \right) dx$$

$$= \int_{-p}^{p} \frac{1}{2} a_0 \cos \frac{m\pi x}{p} \, dx +$$

$$\sum_{n=1}^{\infty} \left(\int_{-p}^{p} a_n \cos \frac{n\pi x}{p} \cos \frac{m\pi x}{p} \, dx + \int_{-p}^{p} b_n \sin \frac{n\pi x}{p} \cos \frac{m\pi x}{p} \, dx \right).$$

If $m = 0$, we notice that all of the integrals that we are summing have the value zero. Thus, this equation simplifies to

$$\int_{-p}^{p} f(x)\, dx = \int_{-p}^{p} \frac{1}{2} a_0 \, dx$$

$$\int_{-p}^{p} f(x)\, dx = \frac{1}{2} a_0 \cdot 2p \qquad\qquad (9.14)$$

$$a_0 = \frac{1}{p} \int_{-p}^{p} f(x)\, dx.$$

If $m \neq 0$, only one of the integrals on the right-hand side of the equation yields a value other than zero and this occurs with

$$\int_{-p}^{p} \cos \frac{m\pi x}{p} \cos \frac{n\pi x}{p}\, dx = \begin{cases} 0, & m \neq n \\ p, & m = n \end{cases}$$

if $m = n$. Hence,

$$\int_{-p}^{p} f(x) \cos \frac{n\pi x}{p}\, dx = p \cdot a_n$$

$$a_n = \frac{1}{p} \int_{-p}^{p} f(x) \cos \frac{n\pi x}{p}\, dx, \quad n = 1, 2, \ldots. \qquad (9.15)$$

We find b_n $(n = 1, 2, \ldots)$ by multiplying the series by $\sin(m\pi x/p)$ and integrating from $x = -p$ to $x = p$. This yields

$$\int_{-p}^{p} f(x) \sin \frac{m\pi x}{p}\, dx = \int_{-p}^{p} \frac{1}{2} a_0 \sin \frac{m\pi x}{p}\, dx +$$

$$\int_{-p}^{p} \sum_{n=1}^{\infty} \left(a_n \cos \frac{n\pi x}{p} \sin \frac{m\pi x}{p} + b_n \sin \frac{n\pi x}{p} \sin \frac{m\pi x}{p} \right) dx$$

$$= \int_{-p}^{p} \frac{1}{2} a_0 \sin \frac{m\pi x}{p}\, dx +$$

$$\sum_{n=1}^{\infty} \left(\int_{-p}^{p} a_n \cos \frac{n\pi x}{p} \sin \frac{m\pi x}{p}\, dx + \int_{-p}^{p} b_n \sin \frac{n\pi x}{p} \sin \frac{m\pi x}{p}\, dx \right).$$

Again, we note that only one of the integrals on the right-hand side of the equation is not zero. In this case, we use

$$\int_{-p}^{p} \sin \frac{m\pi x}{p} \sin \frac{n\pi x}{p} \, dx = \begin{cases} 0, & m \neq n \\ p, & m = n \end{cases}$$

to obtain

$$\int_{-p}^{p} f(x) \sin \frac{n\pi x}{p} \, dx = p \cdot b_n$$

$$b_n = \frac{1}{p} \int_{-p}^{p} f(x) \sin \frac{n\pi x}{p} \, dx, \quad n = 1, 2, \ldots. \tag{9.16}$$

Definition 38 (Fourier Series). *Suppose that* $y = f(x)$ *is defined on* $-p \leq x \leq p$. *The Fourier series for* $f(x)$ *is*

$$\frac{1}{2}a_0 + \sum_{n=1}^{\infty} \left(a_n \cos \frac{n\pi x}{p} + b_n \sin \frac{n\pi x}{p} \right), \tag{9.17}$$

where

$$a_0 = \frac{1}{p} \int_{-p}^{p} f(x) \, dx,$$

$$a_n = \frac{1}{p} \int_{-p}^{p} f(x) \cos \frac{n\pi x}{p} \, dx, \quad n = 1, 2, \ldots, \textit{ and} \tag{9.18}$$

$$b_n = \frac{1}{p} \int_{-p}^{p} f(x) \sin \frac{n\pi x}{p} \, dx, \quad n = 1, 2, \ldots.$$

The following theorem tells us that the Fourier series for any function converges to the function except at points of discontinuity.

Theorem 34 (Convergence of Fourier Series). *Suppose that* $f(x)$ *and* $f'(x)$ *are piecewise continuous functions on* $-p \leq x \leq p$. *Then the Fourier series for* $f(x)$ *on* $-p \leq x \leq p$ *converges to* $f(x)$ *at every* x *where* $f(x)$ *is continuous.*

If $f(x)$ *is discontinuous at* $x = a$, *the Fourier series converges to the average*

$$\frac{1}{2} \left(\lim_{x \to a+} f(x) + \lim_{x \to a-} f(x) \right).$$

Sometimes the command `FourierSeries` can help you find the Fourier series for a function. Keep in mind that the `FourierSeries` function is very limited in scope.

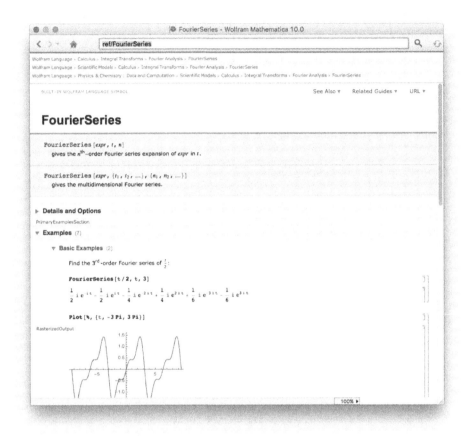

EXAMPLE 9.3.1: Find the Fourier series for $f(x) = \begin{cases} 1, & -2 \leq x < 0 \\ 2, & 0 \leq x < 2 \end{cases}$,

where $f(x+4) = f(x)$.

SOLUTION: In this case, $p = 2$. First we find

$$a_0 = \frac{1}{2}\int_{-2}^{2} f(x)\,dx = \frac{1}{2}\int_{-2}^{0} 1 \cdot dx + \frac{1}{2}\int_{0}^{2} 2 \cdot dx = 3,$$

$$a_0 = \frac{1}{2}\int_{-2}^{0} 1\,dx + \frac{1}{2}\int_{0}^{2} 2\,dx$$

3

$$a_n = \frac{1}{2} \int_{-2}^{2} f(x) \cos \frac{n\pi x}{2} \, dx = \frac{1}{2} \int_{-2}^{0} \cos \frac{n\pi x}{2} \, dx + \frac{1}{2} \int_{0}^{2} 2 \cos \frac{n\pi x}{2} \, dx = 0,$$

$\text{a}_{\text{n\_}} = \frac{1}{2} \int_{-2}^{0} \text{Cos}[\frac{n\pi x}{2}] \, dx + \frac{1}{2} \int_{0}^{2} 2\text{Cos}[\frac{n\pi x}{2}] \, dx$

$\frac{3\text{Sin}[n\pi]}{n\pi}$

and

$$b_n = \frac{1}{2} \int_{-2}^{2} f(x) \sin \frac{n\pi x}{2} \, dx = \frac{1}{2} \int_{-2}^{0} \sin \frac{n\pi x}{2} \, dx + \frac{1}{2} \int_{0}^{2} 2 \sin \frac{n\pi x}{2}$$

$$dx = \frac{1}{n\pi} (1 - \cos n\pi) = \frac{1}{n\pi} \left(1 - (-1)^n\right).$$

$\text{b}_{\text{n\_}} = \frac{1}{2} \int_{-2}^{0} \text{Sin}[\frac{n\pi x}{2}] \, dx + \frac{1}{2} \int_{0}^{2} 2\text{Sin}[\frac{n\pi x}{2}] \, dx //\text{Simplify}$

$\frac{1 - \text{Cos}[n\pi]}{n\pi}$

Therefore, at the values of x for which $f(x)$ is continuous

$$f(x) = \frac{3}{2} + \sum_{n=1}^{\infty} [1 - (-1)^n] \frac{1}{n\pi} \sin \frac{n\pi x}{2}$$

$$= \frac{3}{2} + \frac{2}{\pi} \sin \frac{\pi x}{2} + \frac{2}{3\pi} \sin \frac{3\pi x}{2} + \frac{2}{5\pi} \sin \frac{5\pi x}{2} + \cdots.$$

We now graph $f(x)$ with several partial sums of the Fourier series. First, we define $p_k(x)$ to be the kth partial sum of the Fourier series and then $f(x)$.

$\text{p}_{\text{k\_}}[\text{x\_}] := \frac{3}{2} + \sum_{n=1}^{k} (a_n \text{Cos}[\frac{n\pi x}{2}] + b_n \text{Sin}[\frac{n\pi x}{2}])$

$f[\text{x\_}] := f[x - 4]/; x \geq 2$

$f[\text{x\_}] := 1/; -2 \leq x < 0$

$f[\text{x\_}] := 2/; 0 \leq x < 2$

Given k, the function comp graphs $f(x)$ and $p_k(x)$ on the interval $[-2, 6]$, which corresponds to two periods of $f(x)$, as well as the error, $|f(x) - p_k(x)|$. The resulting two graphics objects are displayed side-by-side.

```
comp[k_]:=Module[{p1, p2},

    p1 = Plot[{f[x], p_k[x]}, {x, -2, 6}, PlotRange → {0, 3}];
```

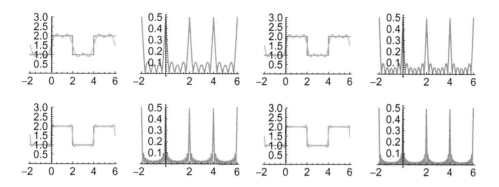

Figure 9-8 The Fourier series converges to $f(x)$ at points of continuity and to the average of the left- and right-hand limits at points of discontinuity

```
p2 = Plot[Abs[f[x] - p_k[x]], {x, -2, 6}, PlotRange → All];

Show[GraphicsRow[{p1, p2}]]]
```

We then use Map to generate these graphs for $k = 3, 5, 11$, and 15. See Figure 9-8.

```
q1 = Map[comp, {3, 5, 11, 15}];

toshow = Partition[q1, 2];

Show[GraphicsGrid[toshow]]
```

The graphs show that if we extend the graph of $f(x)$ over more periods, then the approximation by the Fourier series carries over to those intervals.

■

EXAMPLE 9.3.2: Find the Fourier series for $f(x) = \begin{cases} 0, & -1 \leq x < 0 \\ \sin \pi x, & 0 \leq x < 1 \end{cases}$, where $f(x+2) = f(x)$.

SOLUTION: In this case, $p = 1$, so $a_0 = \int_{-1}^{1} f(x)\,dx = \int_{0}^{1} \sin \pi x\,dx = 2/\pi$ and

```
a_0 = ∫_0^1 Sin[π x] dx
```
$$\frac{2}{\pi}$$

$a_n = \int_{-1}^{1} f(x) \cos n\pi x \, dx = \int_{0}^{1} \sin \pi x \cos n\pi x \, dx$. The value of this integral depends on the value of n. If $n = 1$, we have $a_1 = \int_{0}^{1} \sin \pi x \cos \pi x \, dx = \frac{1}{2} \int_{0}^{1} \sin 2\pi x \, dx = 0$, where we use the identity $\sin \alpha \cos \alpha = \frac{1}{2} \sin 2\alpha$.

$a_1 = \int_0^1 \mathrm{Sin}[\pi x]\mathrm{Cos}[\pi x]dx$

0

If $n \neq 1$, we use the identity $\sin \alpha \cos \beta = \frac{1}{2} [\sin(\alpha - \beta) + \sin(\alpha + \beta)]$ to obtain

$$a_n = \frac{1}{2} \int_0^1 [\sin(1-n)\pi x + \sin(1+n)\pi x] \, dx$$

$$= -\frac{1}{2} \left[\frac{\cos(1-n)\pi x}{(1-n)\pi} + \frac{\cos(1+n)\pi x}{(1+n)\pi} \right]_0^1$$

$$= -\frac{1}{2} \left\{ \left[\frac{\cos(1-n)\pi}{(1-n)\pi} + \frac{\cos(1+n)\pi}{(1+n)\pi} \right] \right.$$

$$\left. - \left[\frac{1}{(1-n)\pi} + \frac{1}{(1+n)\pi} \right] \right\}, \quad n = 2, 3, \ldots$$

$a_n = \int_0^1 \mathrm{Sin}[\pi x]\mathrm{Cos}[n\pi x]dx$
$\dfrac{1 + \mathrm{Cos}[n\pi]}{\pi - n^2\pi}$

Notice that if n is odd, both $1 - n$ and $1 + n$ are even. Hence, $\cos(1 - n)\pi x = \cos(1 + n)\pi x = 1$, so

$$a_n = -\frac{1}{2} \left\{ \left[\frac{1}{(1-n)\pi} + \frac{1}{(1+n)\pi} \right] - \left[\frac{1}{(1-n)\pi} + \frac{1}{(1+n)\pi} \right] \right\} = 0$$

if n is odd. On the other hand, if n is even, $1 - n$ and $1 + n$ are odd. Therefore, $\cos(1 - n)\pi x = \cos(1 + n)\pi x = -1$, so

$$a_n = -\frac{1}{2} \left\{ \left[\frac{-1}{(1-n)\pi} + \frac{-1}{(1+n)\pi} \right] - \left[\frac{1}{(1-n)\pi} + \frac{1}{(1+n)\pi} \right] \right\}$$

$$= \frac{1}{(1-n)\pi} + \frac{1}{(1+n)\pi} = \frac{2}{(1-n)(1+n)\pi} = -\frac{2}{(n-1)(n+1)\pi}$$

if n is even. We confirm this observation by computing several coefficients.

```
Table[{n, aₙ}, {n, 1, 10}]//TableForm
```

1	0
2	$-\dfrac{2}{3\pi}$
3	0
4	$-\dfrac{2}{15\pi}$
5	0
6	$-\dfrac{2}{35\pi}$
7	0
8	$-\dfrac{2}{63\pi}$
9	0
10	$-\dfrac{2}{99\pi}$

Putting this information together, we can write the coefficients as

$$a_{2n} = -\frac{2}{(2n-1)(2n+1)\pi}, \quad n = 1, 2, \ldots$$

Similarly, $b_n = \int_{-1}^{1} f(x) \sin n\pi x \, dx = \int_{0}^{1} \sin \pi x \sin n\pi x \, dx$, so if $n = 1$,

$$b_1 = \int_{0}^{1} \sin^2 \pi x \, dx = \frac{1}{2} \int_{0}^{1} (1 - \cos 2\pi x) dx = \frac{1}{2}\left[x - \frac{1}{2\pi}\sin 2\pi x\right]_0^1 = \frac{1}{2}.$$

```
b₁ = ∫₀¹ Sin[πx]Sin[πx]dx
```

$$\frac{1}{2}$$

If $n \neq 1$, we use $\sin \alpha \sin \beta = \frac{1}{2}[\cos(\alpha - \beta) - \cos(\alpha + \beta)]$. Hence,

$$b_n = \frac{1}{2} \int_{0}^{1} [\cos(1-n)\pi x - \cos(1+n)\pi x] \, dx$$

$$= \frac{1}{2}\left[\frac{\sin(1-n)\pi x}{(1-n)\pi} - \frac{\sin(1+n)\pi x}{(1+n)\pi}\right]_0^1 = 0, \quad n = 2, 3, \ldots.$$

```
bₙ_ = ∫₀¹ Sin[πx]Sin[nπx]dx
```

$$\frac{\text{Sin}[n\pi]}{\pi - n^2\pi}$$

Therefore, we write the Fourier series as

$$f(x) = \frac{1}{\pi} + \frac{1}{2}\sin \pi x - \frac{2}{\pi} \sum_{n=1}^{\infty} \frac{1}{(2n-1)(2n+1)} \cos 2n\pi x.$$

We graph $f(x)$ along with several approximations using this series in the same way as in previous examples. Let

$$p_k(x) = \frac{1}{\pi} + \frac{1}{2}\sin \pi x - \frac{2}{\pi} \sum_{n=1}^{k} \frac{1}{(2n-1)(2n+1)} \cos 2n\pi x.$$

denote the kth partial sum of the Fourier series. Note that

$$p_k(x) = \frac{1}{\pi} + \frac{1}{2}\sin \pi x - \frac{2}{\pi} \sum_{n=1}^{k-1} \frac{1}{(2n-1)(2n+1)}$$

$$\cos 2n\pi x - \frac{2}{\pi}\frac{1}{(2k-1)(2k+1)}$$

$$\cos 2k\pi x = p_{k-1}(x) - \frac{2}{\pi}\frac{1}{(2k-1)(2k+1)} \cos 2k\pi x.$$

Thus, to calculate the kth partial sum of the Fourier series, we need only subtract $\frac{2}{\pi}\frac{1}{(2k-1)(2k+1)} \cos 2k\pi x$ from the $(k-1)$st partial sum: we need not recompute all k terms of the kth partial sum if we know the $(k-1)$st partial sum. Using this observation, we define the recursively defined function p to return the kth partial sum of the series.

    Remove[p]

$$p[k\_] := p[k] = p[k-1] - \frac{2\text{Cos}[2k\pi x]}{\pi((2k-1)(1+2k))}$$

$$p[0] = \frac{1}{\pi} + \frac{\text{Sin}[\pi x]}{2};$$

For illustrative purposes, we compute p[11].

    p[11]

$$\frac{1}{\pi} - \frac{2\text{Cos}[2\pi x]}{3\pi} - \frac{2\text{Cos}[4\pi x]}{15\pi} - \frac{2\text{Cos}[6\pi x]}{35\pi} - \frac{2\text{Cos}[8\pi x]}{63\pi}$$

$$- \frac{2\text{Cos}[10\pi x]}{99\pi} - \frac{2\text{Cos}[12\pi x]}{143\pi} - \frac{2\text{Cos}[14\pi x]}{195\pi} - \frac{2\text{Cos}[16\pi x]}{255\pi}$$

$$- \frac{2\text{Cos}[18\pi x]}{323\pi} - \frac{2\text{Cos}[20\pi x]}{399\pi} - \frac{2\text{Cos}[22\pi x]}{483\pi} + \frac{1}{2}\text{Sin}[\pi x]$$

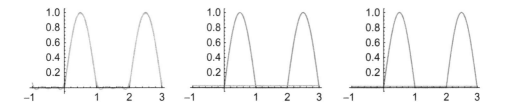

Figure 9-9   The graphs of the 6th and 10th partial sums are virtually indistinguishable from the graph of $f(x)$

We then define $f(x)$.

```
Clear[f]

f[x_]:=0/; -1 ≤ x < 0

f[x_]:=Sin[πx]/; 0 ≤ x < 1

f[x_]:=f[x - 2]/; x ≥ 1
```

We graph $f(x)$ along with the second, sixth, and tenth partial sums of the series in Figure 9-9.

```
Clear[graph]

graph[k_]:=Plot[{f[x], p[k]}, {x, -1, 3}]; |

p1 = Table[Plot[{f[x], p[k]}, {x, -1, 3}], {k, {2, 6, 10}}];

Show[GraphicsRow[p1]]
```

The corresponding errors are graphed in Figure 9-10.

```
Remove[err]

err[k_][x_]:=Plot[Abs[f[x] - p[k]], {x, -1, 3}, PlotRange → All];

somegraphs2 = Map[err[#][x]&, {2, 6, 10}];

Show[GraphicsRow[somegraphs2]]
```

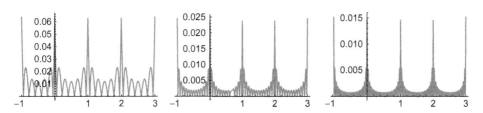

Figure 9-10   The Fourier series converges to $f(x)$ on $(-\infty, \infty)$

∎

9.3.2 Even, Odd, and Periodic Extensions

In the discussion so far in this section, we have assumed that $f(x)$ was defined on the interval $-p < x < p$. However, this is not always the case. Sometimes, we must take a function that is defined on the interval $0 < x < p$ and represent it in terms of trigonometric functions. Three ways of accomplishing this task is to extend $f(x)$ to obtain (a) an **even** function on $-p < x < p$; (b) an **odd** function on $-p < x < p$; (c) a **periodic** function on $-p < x < p$.

We can notice some interesting properties associated with the Fourier series in each of these three cases by noting the properties of even and odd functions. If $f(x)$ is an even function and $g(x)$ is an odd function, then the product $(f \cdot g)(x) = f(x)g(x)$ is an odd function. Similarly, if $f(x)$ is an even function and $g(x)$ is an even function, then $(f \cdot g)(x)$ is an even function, and if $f(x)$ is an odd function and $g(x)$ is an odd function, then $(f \cdot g)(x)$ is an even function. Recall from integral calculus that if $f(x)$ is odd on $-p \le x \le p$, then $\int_{-p}^{p} f(x)\,dx = 0$ while if $g(x)$ is even on $-p \le x \le p$, then $\int_{-p}^{p} g(x)\,dx = 2\int_{0}^{p} g(x)\,dx$. These properties are useful in determining the coefficients in the Fourier series for the even, odd, and periodic extensions of a function, because $\cos(n\pi x/p)$ and $\sin(n\pi x/p)$ are even and odd periodic functions, respectively, on $-p \le x \le p$.

1. The **even extension** $f_{\text{even}}(x)$ of $f(x)$ is an even function. Therefore,

$$a_0 = \frac{1}{p}\int_{-p}^{p} f_{\text{even}}(x)\,dx = \frac{2}{p}\int_{0}^{p} f(x)\,dx$$

$$a_n = \frac{1}{p}\int_{-p}^{p} f_{\text{even}}(x)\cos\frac{n\pi x}{p}\,dx = \frac{2}{p}\int_{0}^{p} f(x)\cos\frac{n\pi x}{p}\,dx, \quad n = 1, 2, \ldots$$

$$b_n = \frac{1}{p}\int_{-p}^{p} f_{\text{even}}(x)\sin\frac{n\pi x}{p}\,dx = 0, \quad n = 1, 2, \ldots$$

$$(9.19)$$

2. The **odd extension** $f_{\text{odd}}(x)$ of $f(x)$ is an odd function, so

$$a_0 = \frac{1}{p}\int_{-p}^{p} f_{\text{odd}}(x)\,dx = 0$$

$$a_n = \frac{1}{p}\int_{-p}^{p} f_{\text{even}}(x)\cos\frac{n\pi x}{p}\,dx = 0, \quad n = 1, 2, \ldots$$

$$b_n = \frac{1}{p}\int_{-p}^{p} f_{\text{even}}(x)\sin\frac{n\pi x}{p}\,dx = \frac{2}{p}\int_{0}^{p} f(x)\sin\frac{n\pi x}{p}\,dx, \quad n = 1, 2, \ldots$$

$$(9.20)$$

3. The **periodic extension** $f_p(x)$ has period p. Because half of the period is $p/2$,

$$a_0 = \frac{2}{p} \int_0^p f(x) \, dx$$

$$a_n = \frac{2}{p} \int_0^p f(x) \cos \frac{2n\pi x}{p} \, dx, \quad n = 1, 2, \dots \qquad (9.21)$$

$$b_n = \frac{2}{p} \int_0^p f(x) \sin \frac{2n\pi x}{p} \, dx, \quad n = 1, 2, \dots$$

EXAMPLE 9.3.3: Let $f(x) = x$ on $(0, 1)$. Find the Fourier series for (a) the even extension of $f(x)$; (b) the odd extension of $f(x)$; and (c) the periodic extension of $f(x)$.

SOLUTION:

(a) Here $p = 1$, so $a_0 = 2 \int_0^1 x \, dx = 1$,

$$a_n = 2 \int_0^1 x \cos n\pi x \, dx = \frac{2}{n^2 \pi^2} (\cos n\pi - 1) = \frac{2}{n^2 \pi^2} \left[(-1)^n - 1 \right],$$

$$n = 1, 2, \dots$$

and $b_n = 0$, $n = 1, 2, \dots$

```
Remove[a, b, f, p]
a₀ = 2 ∫₀¹ x dx
```

```
1
```

```
aₙ_ = 2 ∫₀¹ xCos[nπ x]dx
```

$$\frac{2(-1 + \text{Cos}[n\pi] + n\pi \, \text{Sin}[n\pi])}{n^2 \pi^2}$$

Because $a_n = 0$ if n is even, we can represent the coefficients with odd subscripts as $a_{2n-1} = -\dfrac{4}{(2n-1)^2 \pi^2}$. Therefore, the Fourier cosine series is

$$f_{\text{even}}(x) = \frac{1}{2} - \sum_{n=1}^{\infty} \frac{4}{(2n-1)^2\pi^2} \cos(2n-1)\pi x.$$

We graph the even extension with several terms of the Fourier cosine series by first defining $f(x)$ to be the even extension of $f(x)$ on $(0, 1)$

```
Clear[f]

f[x_]:= - x/; - 1 ≤ x < 0

f[x_]:=x/;0 ≤ x < 1

f[x_]:=f[x - 2]/;x ≥ 1

f[x_]:=f[x + 2]/;x ≤ -1
```

and $p_k(x) = \dfrac{1}{2} - \sum_{n=1}^{k} \dfrac{4}{(2n-1)^2\pi^2} \cos(2n-1)\pi x.$

```
pk_[x_]:=a₀/2 + Σᵏₙ₌₁ a2n-1Cos[(2n - 1)πx]
```

We then graph $f(x)$ together with $p_1(x)$ and $p_5(x)$ in Figure 9-11.

```
p1 = Plot[{f[x], p₁[x]}, {x, -2, 2}];

p2 = Plot[{f[x], p₅[x]}, {x, -2, 2}];

Show[GraphicsRow[{p1, p2}]]
```

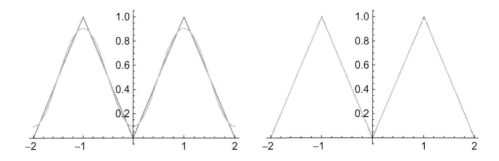

Figure 9-11   Even extension

(b) For the odd extension $f_{\text{odd}}(x)$, we note that $a_0 = 0$, $a_n = 0$, $n = 1, 2,$ $\ldots$, and $b_n = 2 \int_0^1 x \sin \dfrac{n\pi x}{p}, dx = -\dfrac{2}{n\pi} \cos n\pi = (-1)^{n+1} \dfrac{2}{n\pi}$, $n = 1,$ $2, \ldots$.

$$b_{n\_} = 2 \int_0^1 x \text{Sin}[n\pi x] dx$$

$$\frac{2(-n\pi \text{Cos}[n\pi] + \text{Sin}[n\pi])}{n^2 \pi^2}$$

Hence, the Fourier sine series is

$$f_{\text{odd}}(x) = \sum_{n=1}^{\infty} (-1)^{n+1} \frac{2}{n\pi} \sin n\pi x.$$

We graph the odd extension along with several terms of the Fourier sine series in the same manner as in (a). See Figure 9-12.

```
Clear[f]

f[x_]:=x/; -1 ≤ x < 1

f[x_]:=f[x - 2]/; x ≥ 1

f[x_]:=f[x + 2]/; x ≤ -1

pk_[x_]:= ∑(k/n=1) bn Sin[nπ x]

p1 = Plot[{f[x], p5[x]}, {x, -2, 2}];

p2 = Plot[{f[x], p10[x]}, {x, -2, 2}];

Show[GraphicsRow[{p1, p2}]]
```

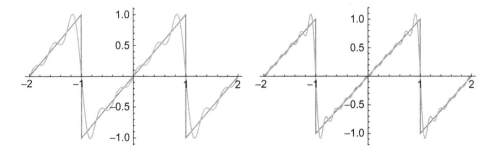

Figure 9-12   Odd extension

(c) The periodic extension has period $2p = 1$, so $p = 1/2$. Thus,

$$a_0 = \frac{1}{1/2} \int_0^1 x \, dx = 2 \int_0^1 x \, dx = 1$$

$$a_n = 2 \int_0^1 x \cos 2n\pi x \, dx = \frac{1}{2} \frac{\cos 2n\pi + 2n\pi \sin 2n\pi}{n^2 \pi^2} - \frac{1}{2n^2 \pi^2} = 0,$$

$n = 1, 2, \ldots$ and

$$b_n = 2 \int_0^1 x \sin 2n\pi x \, dx = -\frac{1}{2} \frac{-\sin 2n\pi + 2n\pi \cos 2n\pi}{n^2 \pi^2}$$

$$= -\frac{2n\pi}{2n^2 \pi^2} = -\frac{1}{n\pi}, \quad n = 1, 2, \ldots$$

```
a₀ = 2 ∫₀¹ x dx
```

```
1
```

```
aₙ_ = 2 ∫₀¹ xCos[2nπx]dx
```

$$\frac{-1 + \cos[2n\pi] + 2n\pi \sin[2n\pi]}{2n^2 \pi^2}$$

```
avals = Table[{n, aₙ}, {n, 1, 20}];
Short[avals]
```

```
{{1, 0}, {2, 0}, {3, 0}, {4, 0}, {5, 0}, {6, 0}, ⟨⟨8⟩⟩, {15, 0},
{16, 0}, {17, 0}, {18, 0}, {19, 0}, {20, 0}}
```

```
bₙ_ = 2 ∫₀¹ xSin[2nπx]dx
```

$$\frac{-2n\pi \cos[2n\pi] + \sin[2n\pi]}{2n^2 \pi^2}$$

```
bvals = Table[{n, bₙ}, {n, 1, 20}];
Short[bvals]
```

$$\left\{ \left\{1, -\frac{1}{\pi}\right\}, \left\{2, -\frac{1}{2\pi}\right\}, \left\{3, -\frac{1}{3\pi}\right\}, \langle\langle 14\rangle\rangle, \left\{18, -\frac{1}{18\pi}\right\}, \left\{19, -\frac{1}{19\pi}\right\},\right.$$
$$\left.\left\{20, -\frac{1}{20\pi}\right\}\right\}$$

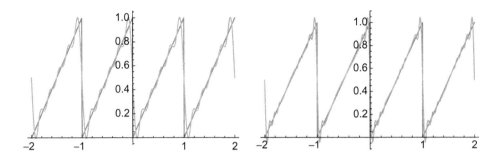

Figure 9-13   Periodic extension

Hence, the Fourier series for the periodic extension is

$$f_p(x) = \frac{1}{2} - \sum_{n=1}^{\infty} \frac{1}{n\pi} \sin 2n\pi x.$$

We graph the periodic extension with several terms of the Fourier series in the same way as in (a) and (b). See Figure 9-13.

```
Clear[f]

f[x_] := x + 1 /; -1 ≤ x < 0

f[x_] := x /; 0 ≤ x < 1

f[x_] := f[x - 2] /; x ≥ 1

f[x_] := f[x + 2] /; x ≤ -1

p[k_][x_] := a₀/2 + Σᵏₙ₌₁ bₙ Sin[2nπx]

p[5][x]
```

$$\frac{1}{2} - \frac{\text{Sin}[2\pi x]}{\pi} - \frac{\text{Sin}[4\pi x]}{2\pi} - \frac{\text{Sin}[6\pi x]}{3\pi}$$

$$- \frac{\text{Sin}[8\pi x]}{4\pi} - \frac{\text{Sin}[10\pi x]}{5\pi}$$

```
p1 = Plot[{f[x], p[5][x]}, {x, -2, 2}];

p2 = Plot[{f[x], p[10][x]}, {x, -2, 2}];

Show[GraphicsRow[{p1, p2}]]
```

■

9.3.3 Differentiation and Integration of Fourier Series

Definition 39 (Piecewise Smooth). *A function $f(x)$, $-p < x < p$ is **piecewise smooth** if $f(x)$ and all of its derivatives are piecewise continuous.*

Theorem 35 (Term-By-Term Differentiation). *Let $f(x)$, $-p < x < p$, be a continuous piecewise smooth function with Fourier series*

$$\frac{1}{2}a_0 + \sum_{n=1}^{\infty}\left(a_n \cos\frac{n\pi x}{p} + b_n \sin\frac{n\pi x}{p}\right).$$

Then, $f'(x)$, $-p < x < p$, has Fourier series

$$\sum_{n=1}^{\infty}\frac{n\pi}{p}\left(-a_n \sin\frac{n\pi x}{p} + b_n \cos\frac{n\pi x}{p}\right).$$

In other words, we differentiate the Fourier series for $f(x)$ term-by-term to obtain the Fourier series for $f'(x)$.

Theorem 36 (Term-By-Term Integration). *Let $f(x)$, $-p < x < p$, be a continuous piecewise smooth function with Fourier series*

$$\frac{1}{2}a_0 + \sum_{n=1}^{\infty}\left(a_n \cos\frac{n\pi x}{p} + b_n \sin\frac{n\pi x}{p}\right).$$

Then, the Fourier series of an antiderivative of $f(x)$ can be found by integrating the Fourier series of $f(x)$ term-by-term.

EXAMPLE 9.3.4: Use the Fourier series for $f(x) = \frac{1}{12}x\left(\pi^2 - x^2\right)$, $-\pi < x < \pi$ to show how term-by-term differentiation and term-by-term integration can used to find the Fourier series of $g(x) = \frac{1}{12}\pi^2 - \frac{1}{4}x^2$, $-\pi < x < \pi$, and $h(x) = \frac{1}{24}\pi^2 x^2\left(1 - \frac{1}{2}x^2\right)$, $-\pi < x < \pi$.

SOLUTION: After defining $f(x)$ and the substitutions in rule to simplify our results, we calculate a_0, a_n, and b_n. (Because $f(x)$ is an odd function, $a_n = 0$, $n \geq 0$.)

```
Clear[f]
f[x_] = 1/12 x(π² - x²);
```

```
rule = {Sin[nπ] → 0, Cos[nπ] → (-1)ⁿ};
```

$$a_0 = \frac{\int_{-\pi}^{\pi} f[x]\,dx}{\pi} \text{ /.rule}$$

0

$$a_{n\_} = \frac{\int_{-\pi}^{\pi} f[x]Cos[nx]\,dx}{\pi} \text{ /.rule}$$

0

$$b_{n\_} = \frac{\int_{-\pi}^{\pi} f[x]Sin[nx]\,dx}{\pi} \text{ /.rule}$$

$$\frac{(-1)^{1+n}}{n^3}$$

We define the nth term of $\sum_{n=1}^{\infty}\left(a_n \cos\frac{n\pi x}{p} + b_n \sin\frac{n\pi x}{p}\right)$ in `fs[n,x]` and the finite sum $\frac{1}{2}a_0 + \sum_{n=1}^{k}\left(a_n \cos\frac{n\pi x}{p} + b_n \sin\frac{n\pi x}{p}\right)$ in `fourier[k]`. We illustrate the use of the function by computing `fourier[4]`.

```
fs[n_] = aₙCos[nx] + bₙSin[nx];
```

```
fourier[k_]:=fourier[k] = fourier[k - 1] + fs[k];
```

$$\texttt{fourier[0]} = \frac{a_0}{2};$$

```
fourier[4]
```

$$Sin[x] - \frac{1}{8}Sin[2x] + \frac{1}{27}Sin[3x] - \frac{1}{64}Sin[4x]$$

We see how quickly the Fourier series converges to $f(x)$ by graphing together with `fourier[1]`, `fourier[2]`, and `fourier[3]`. See Figure 9-14.

```
somegraphs = Table[Plot[{f[x], fourier[k]},
```

```
{x, -π, π}], {k, 1, 3}];
```

```
Show[GraphicsRow[somegraphs]]
```

Notice that $g(x) = \frac{1}{12}\pi^2 - \frac{1}{4}x^2$, $-\pi < x < \pi$, is the derivative of $f(x)$, $-\pi < x < \pi$. Of course, we could compute the Fourier series of $f'(x)$, $-\pi < x < \pi$, directly by applying the integral formulas with $g(x)$ to find the Fourier series coefficients. However, the objective here is to illustrate how term-by-term differentiation of the Fourier series for $f(x)$, $-\pi < x < \pi$, gives us the Fourier series for $f'(x) = g(x)$, $-\pi < x < \pi$. We calculate the derivative of $f(x)$ in `df[x]` in order to make

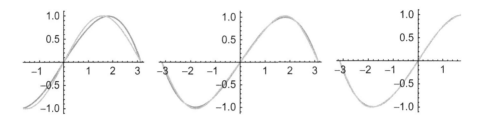

Figure 9-14   The graph of the third partial sum is indistinguishable from the graph of $f(x)$

graphical comparisons. In `dfs[n]`, we determine the derivative of nth term of the Fourier series for $f(x)$, $-\pi < x < \pi$, found above, and in `dfourier[k]`, we calculate the kth partial sum of the Fourier series for $f'(x)$, $-\pi < x < \pi$.

Notice that this series does not include a constant term because the derivative of $\frac{1}{2}a_0$ is zero.

```
df[x_] = D[f[x], x]//Simplify
```

$$\frac{1}{12}(\pi^2 - 3x^2)$$

```
dfs[n_] = D[fs[n], x]
```

$$\frac{(-1)^{1+n}\mathrm{Cos}[nx]}{n^2}$$

```
Remove[dfourier]

dfourier[k_]:=dfourier[k] = dfourier[k − 1] + dfs[k];

dfourier[0] = 0;
```

Next, we graph $f'(x)$, $-\pi < x < \pi$, simultaneously with `dfourier[1]`, `dfourier[2]`, and `dfourier[3]` in Figure 9-15. Again, the convergence of the Fourier series approximations to $f'(x)$, $-\pi < x < \pi$, is quick.

```
Table[dfourier[k], {k, 1, 5}]//TableForm
```

$\mathrm{Cos}[x]$

$\mathrm{Cos}[x] - \frac{1}{4}\mathrm{Cos}[2x]$

$\mathrm{Cos}[x] - \frac{1}{4}\mathrm{Cos}[2x] + \frac{1}{9}\mathrm{Cos}[3x]$

$\mathrm{Cos}[x] - \frac{1}{4}\mathrm{Cos}[2x] + \frac{1}{9}\mathrm{Cos}[3x] - \frac{1}{16}\mathrm{Cos}[4x]$

$\mathrm{Cos}[x] - \frac{1}{4}\mathrm{Cos}[2x] + \frac{1}{9}\mathrm{Cos}[3x] - \frac{1}{16}\mathrm{Cos}[4x] + \frac{1}{25}\mathrm{Cos}[5x]$

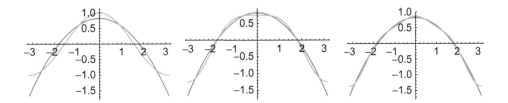

Figure 9-15   Fourier series can be differentiated term-by-term

```
somegraphs = Table[Plot[{df[x], dfourier[k]}, {x, -π, π}], {k, 1, 3}];

Show[GraphicsRow[somegraphs]]
```

Notice that $h(x)$, $-\pi < x < \pi$, is an antiderivative of $f(x)$, $-\pi < x < \pi$. We calculate this antiderivative in `intf[x]`. Of course, this is the antiderivative of $f(x)$ with zero constant of integration because Mathematica does not include an integration constant. When we integrate the terms of the Fourier series of $f(x)$, $-\pi < x < \pi$, a constant term is not included. However, the Fourier series of the even function $h(x) = \frac{1}{24}\pi^2 x^2 \left(1 - \frac{1}{2}x^2\right)$, $-\pi < x < \pi$ should include the constant term $\frac{1}{2}\tilde{a}_0$. We calculate the value of $\tilde{a}_0$ in `inta[0]` with the integral formula $\frac{1}{\pi}\int_{-\pi}^{\pi} h(x)\,dx$.

```
intf[x_] = ∫ f[x] dx
```

$$\frac{1}{12}\left(\frac{\pi^2 x^2}{2} - \frac{x^4}{4}\right)$$

```
inta₀ =
```
$$\text{inta}_0 = \frac{\int_{-\pi}^{\pi} \text{intf}[x]\,dx}{\pi}$$

$$\frac{7\pi^4}{360}$$

In `intfs[n,x]`, we integrate the nth term of the Fourier series of $f(x)$, $-\pi < x < \pi$, found above to determine the coefficients of $\cos nx$ and $\sin nx$ in the Fourier series of $h(x)$, $-\pi < x < \pi$. In `intfourier[k,x]`, we determine the sum of the first k terms of the Fourier series of obtained by adding $\frac{1}{2}\tilde{a}_0$ to the expression obtained through term-by-term integration of the Fourier series of $f(x)$, $-\pi < x < \pi$.

```
intfs[n_] = ∫ fs[n] dx
```

$$\frac{(-1)^{2+n}\text{Cos}[nx]}{n^4}$$

```
Remove[intfourier]
```

$$\texttt{intfourier}[k\_]:=\texttt{intfourier}[k] = \texttt{intfourier}[k-1] + \texttt{intfs}[k];$$

$$\texttt{intfourier}[0] = \frac{\texttt{inta}_0}{2};$$

```
Table[intfourier[k], {k, 1, 5}]//TableForm
```

$$\frac{7\pi^4}{720} - \text{Cos}[x]$$

$$\frac{7\pi^4}{720} - \text{Cos}[x] + \frac{1}{16}\text{Cos}[2x]$$

$$\frac{7\pi^4}{720} - \text{Cos}[x] + \frac{1}{16}\text{Cos}[2x] - \frac{1}{81}\text{Cos}[3x]$$

$$\frac{7\pi^4}{720} - \text{Cos}[x] + \frac{1}{16}\text{Cos}[2x] - \frac{1}{81}\text{Cos}[3x] + \frac{1}{256}\text{Cos}[4x]$$

$$\frac{7\pi^4}{720} - \text{Cos}[x] + \frac{1}{16}\text{Cos}[2x] - \frac{1}{81}\text{Cos}[3x] + \frac{1}{256}\text{Cos}[4x]$$
$$-\frac{1}{625}\text{Cos}[5x]$$

By graphing $h(x)$, $-\pi < x < \pi$, simultaneously with the approximation in `intfourier[k,x]` for $k = 1$, 2, and 3, we see how the graphs of Fourier series approximations obtained through term-by-term integration converge to the graph of $h(x)$, $-\pi < x < \pi$, in `intgraph1`, `intgraph2`, and `intgraph3`. See Figure 9-16.

```
somegraphs = Table[Plot[{intf[x], intfourier[k]}, {x, −π, π}],

{k, 1, 3}]; Show[GraphicsRow[somegraphs]]
```

In this example, `FourierSeries` is useful in helping us check our results. Note that the results are given in terms of complex exponentials.

```
s1 = FourierSeries[1/12x(Pi^2 − x^2), x, 9]
```

$$\frac{1}{2}ie^{-ix} - \frac{1}{2}ie^{ix} - \frac{1}{16}ie^{-2ix} + \frac{1}{16}ie^{2ix} + \frac{1}{54}ie^{-3ix} - \frac{1}{54}ie^{3ix} -$$
$$\frac{1}{128}ie^{-4ix} + \frac{1}{128}ie^{4ix} + \frac{1}{250}ie^{-5ix} - \frac{1}{250}ie^{5ix} - \frac{1}{432}ie^{-6ix} +$$

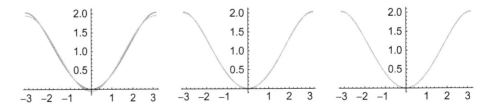

Figure 9-16   Fourier series can be integrated term-by-term

$$\frac{1}{432} i e^{6ix} + \frac{1}{686} i e^{-7ix} - \frac{1}{686} i e^{7ix} - \frac{i e^{-8ix}}{1024} + \frac{i e^{8ix}}{1024} +$$

$$\frac{i e^{-9ix}}{1458} - \frac{i e^{9ix}}{1458}$$

Use `ExpToTrig` to convert the complex exponentials to trigonometric functions if that is the preferred form of your result.

```
s2 = ExpToTrig[s1]
```

$$\mathrm{Sin}[x] - \frac{1}{8} \mathrm{Sin}[2x] + \frac{1}{27} \mathrm{Sin}[3x] - \frac{1}{64} \mathrm{Sin}[4x] + \frac{1}{125} \mathrm{Sin}[5x]$$

$$- \frac{1}{216} \mathrm{Sin}[6x] + \frac{1}{343} \mathrm{Sin}[7x] - \frac{1}{512} \mathrm{Sin}[8x] + \frac{1}{729} \mathrm{Sin}[9x]$$

∎

9.3.4 Parseval's Equality

Let $f(x)$, $-p < x < p$, be a continuous piecewise smooth function with Fourier series

$$\frac{1}{2} a_0 + \sum_{n=1}^{\infty} \left(a_n \cos \frac{n\pi x}{p} + b_n \sin \frac{n\pi x}{p} \right).$$

Parseval's Equality states that

$$\frac{1}{p} \int_{-p}^{p} [f(x)]^2 \, dx = \frac{1}{2} a_0^2 + \sum_{n=1}^{\infty} \left(a_n^2 + b_n^2 \right) = 2A_0^2 + \sum_{n=1}^{\infty} \left(a_n^2 + b_n^2 \right), \tag{9.22}$$

where $A_0 = \frac{1}{2} a_0$ is the constant term in the Fourier series.

EXAMPLE 9.3.5: Verify Parseval's Equality for $f(x) = \frac{1}{12} x \left(\pi^2 - x^2 \right)$, $-\pi < x < \pi$.

SOLUTION: Notice that the function $f(x) = \frac{1}{12} x \left(\pi^2 - x^2 \right)$, $-\pi < x < \pi$, is odd as we see from its graph in Figure 9-17.

```
Clear[f];
f[x_] = 1/12 x(π² - x²);
Plot[f[x], {x, -π, π}]
```

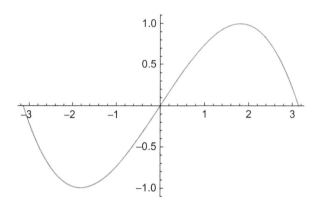

Figure 9-17 $f(x)$ is an odd function

Therefore, the only nonzero coefficients in the Fourier series of $f(x)$ are found in b_n. Notice that we simplify the results by using the substitutions defined in `rule`.

```
rule = {Sin[nπ] → 0, Cos[nπ] → (-1)ⁿ};
```

$$b_{n\_} = \frac{2 \int_0^{\pi} f[x]Sin[nx]dx}{\pi} \text{/.rule}$$

$$\frac{(-1)^{1+n}}{n^3}$$

Next, we evaluate $\frac{1}{\pi} \int_{-\pi}^{\pi} [f(x)]^2 \, dx$.

$$\frac{\int_{-\pi}^{\pi} f[x]^2 \, dx}{\pi}$$

$$\frac{\pi^6}{945}$$

$$N[\frac{\pi^6}{945}]$$

```
1.01734
```

We compare this result with the value of $\sum_{n=1}^{\infty} b_n^2$ by calculating $\sum_{n=1}^{k} b_n^2$ for $k = 1, 2, \ldots, 20$. Notice that this sequence of partial sums converges quickly to 1.01734, an approximation of $\frac{1}{945}\pi^6$.

```
Table[N[∑_{n=1}^{j} b_n²], {j, 1, 20}]
```

```
{1., 1.01563, 1.017, 1.01724, 1.0173, 1.01733, 1.01733,

1.01734, 1.01734, 1.01734, 1.01734, 1.01734, 1.01734,

1.01734, 1.01734, 1.01734, 1.01734, 1.01734, 1.01734,

1.01734}
```

Thus, for the convergent p-series

$$\sum_{n=1}^{\infty} b_n^2 = \sum_{n=1}^{\infty} \left[-\frac{(-1)^n}{n^3} \right]^2 = \sum_{n=1}^{\infty} \frac{1}{n^6} = \frac{1}{945}\pi^6.$$

A p-**series** is a series of the form $\sum_{k=1}^{\infty} k^{-p}$. The p-series converges if $p > 1$ and diverges if $0 < p \leq 1$.

■

9.4 Generalized Fourier Series

In addition to the trigonometric eigenfunctions that were used to form the Fourier series in Sections 9.2 and 9.3, the eigenfunctions of other eigenvalue problems can be used to form what we call **generalized Fourier series**. We will find that these series will assist in solving problems in applied mathematics that involve physical phenomena that cannot be modeled with trigonometric functions.

Recall Bessel's equation of order zero

$$x^2 \frac{d^2 y}{dx^2} + x \frac{dy}{dx} + \lambda^2 x^2 y = 0. \tag{9.23}$$

If we require that the solutions of this differential equation satisfy the boundary conditions $|y(0)| < \infty$ (meaning that the solution is bounded at $x = 0$) and $y(p) = 0$, we can find the eigenvalues of the boundary-value problem

$$\begin{cases} x^2 \dfrac{d^2 y}{dx^2} + x \dfrac{dy}{dx} + \lambda^2 x^2 y = 0 \\ |y(0)| < \infty, \ y(p) = 0 \end{cases}. \tag{9.24}$$

A general solution of Bessel's equation of order zero is $y = c_1 J_0(\lambda x) + c_2 Y_0(\lambda x)$. Because $|y(0)| < \infty$, we must choose $c_2 = 0$ because $\lim_{x\to 0+} Y_0(\lambda x) = -\infty$. Hence, $y(p) = c_1 J_0(\lambda p) = 0$. Just as we did with the eigenvalue problems solved earlier in Section 9.1, we want to avoid choosing $c_1 = 0$, so we must select λ so that $J_0(\lambda p) = 0$.

Let α_n represent the nth zero of the Bessel function of order zero, $J_0(x)$, where $n = 1, 2, \ldots$, which we approximate with `BesselJZeros`. The command `BesselJZeros[m,n]` returns a list of the first n zeros of $J_m(x)$; `BesselJZeros[m,{p,q}]` returns a list of the pth through qth zeros of $J_m(x)$; `BesselJZeros[0,n]` returns a list of the first n zeros of $J_0(x)$.

The function α_n returns the nth zero of $J_0(x)$.

```
αₙ_:=BesselJZero[0,{n, n}][[1]]

Table[{n, BesselJ[1, αₙ], BesselJ[
   2, αₙ]}//N, {n, 1, 5}]//TableForm
```

```
1.   0.519147    0.431755

2.   -0.340265   -0.123283

3.   0.271452    0.0627365

4.   -0.23246    -0.0394283

5.   0.206546    0.0276669
```

Therefore, in trying to find the eigenvalues, we must solve $J_0(\lambda p) = 0$. From our definition of α_n, this equation is satisfied if $\lambda p = \alpha_n$, $n = 1, 2, \ldots$. Hence, the eigenvalues are $\lambda = \lambda_n = \alpha_n/p$, $n = 1, 2, \ldots$, and the corresponding eigenfunctions are

$$y(x) = y_n(x) = J_0(\lambda_n x) = J_0(\alpha_n x/p), \quad n = 1, 2, \ldots$$

As with the trigonometric eigenfunctions that we found in Sections 9.2 and 9.3, $J_0(\alpha_n x/p)$ can be used to build an eigenfunction series expansion of the form

$$f(x) = \sum_{n=1}^{\infty} c_n J_0\left(\frac{\alpha_n x}{p}\right), \tag{9.25}$$

which is called a **Bessel-Fourier series**. We use the orthogonality properties of $J_0(\alpha_n x/p)$ to find the coefficients c_n.

See Theorem 33.   We determine the orthogonality condition by placing Bessel's equation of order zero in the self-adjoint form

$$\frac{d}{dx}\left(x\frac{dy}{dx}\right) + \lambda^2 xy = 0.$$

Because the weighting function is $s(x) = x$, the orthogonality condition is

$$\int_0^p x J_0\left(\frac{\alpha_n x}{p}\right) J_0\left(\frac{\alpha_m x}{p}\right) dx = 0, \quad n \neq m.$$

Multiplying equation (9.25) by $xJ_0(\alpha_m x/p)$ and integrating from $x = 0$ to $x = p$ yields

$$\int_0^p xf(x)J_0\left(\frac{\alpha_m x}{p}\right) dx = \int_0^p \sum_{n=1}^{\infty} c_n xJ_0\left(\frac{\alpha_n x}{p}\right) J_0\left(\frac{\alpha_m x}{p}\right) dx$$

$$= \sum_{n=1}^{\infty} c_n \int_0^p xJ_0\left(\frac{\alpha_n x}{p}\right) J_0\left(\frac{\alpha_m x}{p}\right) dx.$$

However, by the orthogonality condition, each of the integrals on the right-hand side of the equation equals zero except for $m = n$. Therefore,

$$c_n = \frac{\int_0^p xf(x)J_0\left(\frac{\alpha_n x}{p}\right) dx}{\int_0^p x\left[J_0\left(\frac{\alpha_n x}{p}\right)\right]^2 dx}, \quad n = 1, n = 2 \ldots$$

The value of the integral in the denominator can be found through the use of several of the identities associated with the Bessel functions. Because $\lambda_n = \alpha_n/p$, $n = 1, 2, \ldots$, the function $J_0(\alpha_n x/p) = J_0(\lambda_n x)$ satisfies Bessel's equation of order zero:

$$\frac{d}{dx}\left(x\frac{d}{dx}J_0(\lambda_n x)\right) + \lambda_n^2 xJ_0(\lambda_n x) = 0.$$

Multiplying by the factor $2x\dfrac{d}{dx}J_0(\lambda_n x)$, we can write this equation as

$$\frac{d}{dx}\left(x\frac{d}{dx}J_0(\lambda_n x)\right)^2 + \lambda_n^2 x^2 [J_0(\lambda_n x)]^2 = 0.$$

Integrating each side of this equation from $x = 0$ to $x = p$ gives us

$$2\lambda_n^2 \int_0^p x[J_0(\lambda_n x)]^2 \, dx = \lambda_n^2 p^2 \left[J_0'(\lambda_n p)\right]^2 + \lambda_n^2 p^2 [J_0(\lambda_n p)]^2 .$$

With the substitution $\lambda_n p = \alpha_n$ the equation becomes

$$2\lambda_n^2 \int_0^p x[J_0(\lambda_n x)]^2 \, dx = \lambda_n^2 p^2 \left[J_0'(\alpha_n)\right]^2 + \lambda_n^2 p^2 [J_0(\alpha_n)]^2 .$$

Now, $J_0(\alpha_n) = 0$, because α_n is the nth zero of $J_0(x)$. Also, with $n = 0$, the identity $\dfrac{d}{dx}\left(x^{-n}J_n(x)\right) = -x^{-n}J_{n+1}(x)$ indicates that $J_0'(\alpha_n) = -J_1(\alpha_n)$. Therefore,

$$2\lambda_n^2 \int_0^p x\left[J_0(\lambda_n x)\right]^2 \, dx = \lambda_n^2 p^2 \left[-J_1(\alpha_n)\right]^2 + \lambda_n^2 p^2 \cdot 0$$

$$\int_0^p x\left[J_0(\lambda_n x)\right]^2 \, dx = \frac{1}{2}p^2 \left[J_1(\alpha_n)\right]^2.$$

Using this expression in the denominator of c_n, the series coefficients are found with

$$c_n = \frac{2}{p^2 \left[J_1(\alpha_n)\right]^2} \int_0^p x f(x) J_0\left(\frac{\alpha_n x}{p}\right) \, dx, \quad n = 1, 2, \ldots. \tag{9.26}$$

EXAMPLE 9.4.1: Find the Bessel-Fourier series for $f(x) = 1 - x^2$ on $0 < x < 1$.

SOLUTION: In this case, $p = 1$, so

$$c_n = \frac{2}{\left[J_1(\alpha_n)\right]^2} \int_0^1 x\left(1 - x^2\right) J_0\left(\frac{\alpha_n x}{p}\right) \, dx$$

$$= \frac{2}{\left[J_1(\alpha_n)\right]^2} \left\{ \int_0^1 x J_0\left(\frac{\alpha_n x}{p}\right) \, dx - \int_0^1 x^3 J_0\left(\frac{\alpha_n x}{p}\right) \, dx \right\}.$$

Using the formula, $\dfrac{d}{dx}\left(x^n J_n(x)\right) = -x^n J_{n-1}(x)$ with $n = 1$ yields

$$\int_0^1 x J_0\left(\frac{\alpha_n x}{p}\right) \, dx = \left[\frac{1}{\alpha_n} x J_1\left(\alpha_n x\right)\right]_0^1 = \frac{1}{\alpha_n} J_1\left(\alpha_n\right).$$

Integration by parts formula:
$\int u \, dv = uv - \int v \, du.$

Note that the factor $1/\alpha_n$ is due to the chain rule for differentiating the argument of $J_1\left(\alpha_n x\right)$. We use integration by parts with $u = x^2$ and $dv = x J_0\left(\alpha_n x\right)$ to evaluate $\int_0^1 x^3 J_0\left(\frac{\alpha_n x}{p}\right) \, dx$. As in the first integral we obtain

$v = -\dfrac{1}{\alpha_n} x J_1\left(\alpha_n x\right)$. Then, because $du = 2x\,dx$, we have

$$\int_0^1 x^3 J_0\left(\frac{\alpha_n x}{p}\right) dx = \left[\frac{1}{\alpha_n} x^3 J_1\left(\alpha_n x\right)\right]_0^1 - \frac{2}{\alpha_n}\int_0^1 x^2 J_1\left(\alpha_n x\right)\,dx$$

$$= \frac{1}{\alpha_n} J_1\left(\alpha_n\right) - \frac{2}{\alpha_n}\left[\frac{1}{\alpha_n} x^2 J_2\left(\alpha_n x\right)\right]_0^1 = \frac{1}{\alpha_n} J_1\left(\alpha_n\right) - \frac{2}{\alpha_n^2} J_2\left(\alpha_n\right).$$

Thus, the coefficients are

$$c_n = \frac{2}{[J_1(\alpha_n)]^2}\int_0^1 x\left(1 - x^2\right) J_0\left(\frac{\alpha_n x}{p}\right) dx$$

$$= \frac{2}{[J_1(\alpha_n)]^2}\left\{\int_0^1 x J_0\left(\frac{\alpha_n x}{p}\right) dx - \int_0^1 x^3 J_0\left(\frac{\alpha_n x}{p}\right) dx\right\}$$

$$= \frac{2}{[J_1(\alpha_n)]^2}\left[\frac{1}{\alpha_n} J_1\left(\alpha_n\right) - \left(\frac{1}{\alpha_n} J_1\left(\alpha_n\right) - \frac{2}{\alpha_n^2} J_2\left(\alpha_n\right)\right)\right]$$

$$= \frac{4 J_2\left(\alpha_n\right)}{\alpha_n^2\left[J_1\left(\alpha_n\right)\right]^2}, \quad n = 1, 2, \ldots$$

so that the Bessel-Fourier series is

$$f(x) = \sum_{n=1}^{\infty} \frac{4 J_2\left(\alpha_n\right)}{\alpha_n^2\left[J_1\left(\alpha_n\right)\right]^2} J_0\left(\alpha_n x\right).$$

We now graph $f(x)$ along with several terms of the series. To do so, we need to compute the values of $J_1\left(\alpha_n\right)$ and $J_2\left(\alpha_n\right)$ for various values of n. We list the values of $J_1\left(\alpha_n\right)$ and $J_2\left(\alpha_n\right)$ for various values of n.

```
Table[{n, BesselJ[1, αn], BesselJ[
    2, αn]}//N, {n, 1, 5}]//TableForm
```

```
1.   0.519147    0.431755

2.  -0.340265   -0.123283

3.   0.271452    0.0627365

4.  -0.23246    -0.0394283

5.   0.206546    0.0276669
```

Next, we define $f(x) = 1 - x^2$ and $p_k(x) = \sum_{n=1}^{k} \dfrac{4J_2(\alpha_n)}{\alpha_n^2 [J_1(\alpha_n)]^2} J_0(\alpha_n x)$, the kth partial sum of the Bessel-Fourier series. Note that

$$p_k(x) = \sum_{n=1}^{k-1} \frac{4J_2(\alpha_n)}{\alpha_n^2 [J_1(\alpha_n)]^2} J_0(\alpha_n x) + \frac{4J_2(\alpha_k)}{\alpha_k^2 [J_1(\alpha_k)]^2} J_0(\alpha_k x)$$

$$= p_{k-1}(x) + \frac{4J_2(\alpha_k)}{\alpha_k^2 [J_1(\alpha_k)]^2} J_0(\alpha_k x)$$

Thus, to calculate the kth partial sum of the Fourier series, we need only add $\dfrac{4J_2(\alpha_k)}{\alpha_k^2 [J_1(\alpha_k)]^2} J_0(\alpha_k x)$ to the $(k-1)$st partial sum: we need not recompute all k terms of the kth partial sum if we know the $(k-1)$st partial sum. Using this observation, we define the recursively defined function p to return the kth partial sum of the series.

```
f[x_] = 1 - x²;

p[k_]:=p[k] = p[k - 1] + 4BesselJ[2, αₖ]BesselJ[0, αₖx] //N;
                         ───────────────────────────────
                             αₖ²BesselJ[1, αₖ]²

p[0] = 0;
```

The graphs of $f(x)$ and the first three or four partial sums are practically indistinguishable. See Figure 9-18.

```
p2 = Plot[Evaluate[{f[x], p[2]//N}], {x, 0, 1}];

p3 = Plot[Evaluate[{f[x], p[3]//N}], {x, 0, 1}];

p4 = Plot[Evaluate[{f[x], p[4]//N}], {x, 0, 1}];

Show[GraphicsRow[{p2, p3, p4}]]
```

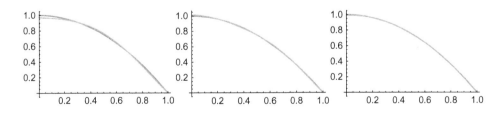

Figure 9-18   The Bessel-Fourier series quickly converges to $f(x)$

As was the case with Fourier series, we can make a statement about the convergence of the Bessel-Fourier series.

Theorem 37 (Convergence of Bessel-Fourier Series). *Suppose that $f(x)$ and $f'(x)$ are piecewise continuous functions on $0 < x < p$. Then the Bessel-Fourier series for $f(x)$ on $0 < x < p$ converges to $f(x)$ at every x where $f(x)$ is continuous. If $f(x)$ is discontinuous at $x = a$, the Bessel-Fourier series converges to the average*

$$\frac{1}{2}\left(\lim_{x \to a^+} f(x) + \lim_{x \to a^-} f(x)\right) = \frac{1}{2}\left(f(a^+) + f(a^-)\right).$$

Series involving the eigenfunctions of other eigenvalue problems can be formed as well.

EXAMPLE 9.4.2: The eigenvalue problem $\begin{cases} y'' + 2y' - (\lambda - 1)y = 0 \\ y(0) = y(2) = 0 \end{cases}$ has
eigenvalues $\lambda_n = -(n\pi/2)^2$ and eigenfunctions $y_n(x) = e^{-x}\sin(n\pi x/2)$.
Use these eigenfunctions to approximate $f(x) = e^{-x}$ for $0 < x < 2$.

SOLUTION: In order to approximate $f(x)$, we need the orthogonality condition for these eigenfunctions. We obtain this condition by placing the differential equation in self-adjoint form using the formulas given in equation (9.5). In the general equation, $a_2(x) = 1$, $a_1(x) = 2$, and $a_0(x) = 0$. Therefore, $p(x) = e^{\int 2\,dx} = e^{2x}$ and $s(x) = p(x)/a_2(x) = e^{2x}$, so in self-adjoint form the equation is

$$\frac{d}{dx}\left(e^{2x}\frac{dy}{dx}\right) - (\lambda - 1)e^{2x}y.$$

This means that the orthogonality condition, $\int_a^b s(x)y_n(x)y_m(x)\,dx = 0$ $(m \neq n)$, is

$$\int_0^2 e^{2x}e^{-x}\sin\frac{m\pi x}{2}e^{-x}\sin\frac{n\pi x}{2}\,dx = \int_0^2 \sin\frac{m\pi x}{2}\sin\frac{n\pi x}{2}\,dx = 0, \; m \neq n.$$

We use this condition to determine the coefficients in the eigenfunction expansion

$$f(x) = \sum_{n=1}^{\infty} c_n y_n(x) = \sum_{n=1}^{\infty} c_n e^{-x}\sin\frac{n\pi x}{2}.$$

Multiplying both sides of this equation by $y_m(x) = e^{-x}\sin(m\pi x/2)$ and $s(x) = e^{2x}$ and then integrating from $x = 0$ to $x = 2$ yields

$$\int_0^2 f(x)e^{2x}e^{-x}\sin\frac{m\pi x}{2}\,dx = \int_0^2 \sum_{n=1}^{\infty} c_n e^{-x}\sin\frac{n\pi x}{2}e^{2x}e^{-x}\sin\frac{m\pi x}{2}\,dx$$

$$\int_0^2 f(x)e^x\sin\frac{m\pi x}{2}\,dx = \sum_{n=1}^{\infty}\int_0^2 c_n\sin\frac{n\pi x}{2}\sin\frac{m\pi x}{2}\,dx.$$

Each integral in the sum on the right-hand side of the equation is zero except if $m = n$. In this case, $\int_0^2 \sin^2(n\pi x/2)\,dx = 1$. Therefore,

$$c_n = \int_0^2 f(x)e^x\sin\frac{n\pi x}{2}\,dx.$$

For $f(x) = e^{-x}$,

$$c_n = \int_0^2 e^{-x}e^x\sin\frac{n\pi x}{2}\,dx = \int_0^2 \sin\frac{n\pi x}{2}\,dx = -\frac{2}{n\pi}(\cos n\pi - 1).$$

$$c_{n\_} = \int_0^2 \mathrm{Sin}[\frac{n\pi x}{2}]\,dx$$

$$\frac{2 - 2\mathrm{Cos}[n\pi]}{n\pi}$$

Because $\cos n\pi = (-1)^n$, we can write the eigenfunction expansion of $f(x)$ as

$$f(x) = \sum_{n=1}^{\infty}-\frac{2}{n\pi}\left((-1)^n - 1\right)e^{-x}\sin\frac{n\pi x}{2}$$

$$= e^{-x}\left(\frac{4}{\pi}\sin\frac{\pi x}{2} + \frac{4}{3\pi}\sin\frac{3\pi x}{2} + \frac{4}{5\pi}\sin\frac{5\pi x}{2} + \cdots\right).$$

We graph $f(x)$ together with

$$p_k(x) = \sum_{n=1}^{k}-\frac{2}{n\pi}\left((-1)^n - 1\right)e^{-x}\sin\frac{n\pi x}{2}$$

for $k = 6, 10$, and 14 in Figure 9-19.

```
f[x_] = Exp[-x];
```

$$p_{k\_}[x\_] := \sum_{n=1}^{k} c_n\mathrm{Exp}[-x]\mathrm{Sin}[\frac{n\pi x}{2}];$$

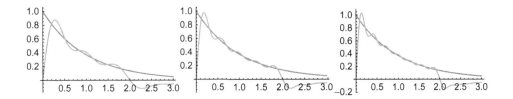

Figure 9-19   Approximating $f(x) = e^{-x}$ with a generalized Fourier series

```
p6 = Plot[{f[x], p6[x]}, {x, 0, 3}];

p10 = Plot[{f[x], p10[x]}, {x, 0, 3}];

p14 = Plot[{f[x], p14[x]}, {x, 0, 3}];

Show[GraphicsRow[{p6, p10, p14}]]
```

∎

EXAMPLE 9.4.3: Use the eigenvalues and eigenfunctions of the eigenvalue problem

$$\begin{cases} y'' + \lambda y = 0 \\ y(0) = 0, \ y(1) + y'(1) = 0 \end{cases}$$

to obtain a generalized Fourier series for $f(x) = x(1 - x)$, $0 < x < 1$.

The eigenvalues and corresponding eigenfunctions for this eigenvalue problem are found in Example 9.1.6.

SOLUTION: In Example 9.1.6, the eigenvalues of this problem, $\lambda = k^2$, were shown to satisfy the relationship $k = -\tan k$. In the example, we approximated the first eight roots of this equation to be 2.02876, 4.91318, 7.97867, 11.0855, 14.2074, 17.3364, 20.4692, and 23.6043 entered in kvals.

```
kvals = Table[FindRoot[-Tan[x]==x, {x, (2n - 1)π/2 + 0.1}],
{n, 1, 8}]

{{x → 2.02876}, {x → 4.91318}, {x → 7.97867},
{x → 11.0855}, {x → 14.2074}, {x → 17.3364},
{x → 20.4692}, {x → 23.6043}}
```

Let k_n represent the nth positive root of $k = -\tan k$. Therefore, the eigenfunctions of $\begin{cases} y'' + \lambda y = 0 \\ y(0) = 0, \ y(1) + y'(1) = 0 \end{cases}$ are $y_n(x) = \sin k_n x$. Because of the orthogonality of the eigenfunctions, we have the orthogonality condition $\int_0^1 \sin k_n x \sin k_m x \, dx = 0$, $m \neq n$. If $m = n$, we have $\int_0^1 \sin^2 k_n x \, dx = \frac{1}{2} - \frac{1}{2k_n} \sin 2k_n$.

$\int_0^1 \texttt{Sin[kx]}^2 \ \texttt{dx}$

$\dfrac{1}{2} - \dfrac{\texttt{Sin[2k]}}{4k}$

Therefore,

$$\int_0^1 \sin^2 k_n x \, dx = \frac{2k_n - \sin 2k_n}{4k_n} = \frac{2k_n - 2\sin k_n \cos k_n}{4k_n}.$$

With the condition $k_n = -\tan k_n$ or $\sin k_n = -k_n \cos k_n$ from the eigenvalue problem, we have

$$\int_0^1 \sin^2 k_n x \, dx = \frac{k_n - \sin k_n \cos k_n}{2k_n} = \frac{k_n - (-k_n \cos k_n)\cos k_n}{2k_n}$$

$$= \frac{1}{2}\left(1 + \cos^2 k_n\right).$$

To determine the coefficients c_n in the generalized Fourier series $f(x) = \sum_{n=1}^{\infty} c_n \sin k_n x$ using the eigenfunctions $y_n(x) = \sin k_n x$ of the eigenvalue problem, we multiply both sides of $f(x) = \sum_{n=1}^{\infty} c_n \sin k_n x$ by $\sin k_m x$ and integrate from $x = 0$ to $x = 1$. This yields

$$\int_0^1 f(x) \sin k_m x \, dx = \int_0^1 \sum_{n=1}^{\infty} c_n \sin k_n x \sin k_m x \, dx.$$

Assuming uniform convergence of the series, we have

$$\int_0^1 f(x) \sin k_m x \, dx = \sum_{n=1}^{\infty} c_n \int_0^1 \sin k_n x \sin k_m x \, dx.$$

All terms on the right are zero except if $m = n$. In this case, we have

$$\int_0^1 f(x) \sin k_n x \, dx = c_n \int_0^1 \sin^2 k_n x \, dx = \frac{1}{2}c_n\left(1 + \cos^2 k_n\right)$$

so that

$$c_n = \frac{2}{1 + \cos^2 k_n} \int_0^1 f(x) \sin k_n x \, dx.$$

We approximate the value of c_n for $n = 1, 2, \ldots, 8$ in cvals using the values of k in kvals.

```
f[x_]:=x(1 − x)cvals = Table
 2NIntegrate[f[x]Sin[(kvals[[j, 1, 2]])x], {x, 0, 1}]
[─────────────────────────────────────────────, {j, 1, 8}]
          1 + Cos[kvals[[j, 1, 2]]]²

{0.213285, 0.104049, −0.0219788, 0.0187303, −0.00834994,

0.00734255, −0.00426841, 0.00387074}
```

We define the sum of the first j terms of $\sum_{n=1}^{\infty} c_n \sin k_n x$ for $f(x) = x(1 − x)$, $0 < x < 1$, with fapprox[x,n] and then create a table of fapprox[x,n] for $n = 1$ to $n = 8$ in funcs.

```
fapprox[x_, j_]:= ∑_{n=1}^{j}(cvals[[n]])Sin[(kvals[[n, 1, 2]])x];

funcs = Table[fapprox[x, j], {j, 1, 8}]

{0.213285Sin[2.02876x], 0.213285Sin[2.02876x]

+ 0.104049Sin[4.91318x], 0.213285Sin[2.02876x] + 0.104049

Sin[4.91318x] − 0.0219788Sin[7.97867x], 0.213285

Sin[2.02876x] + 0.104049Sin[4.91318x] − 0.0219788

Sin[7.97867x] + 0.0187303Sin[11.0855x], 0.213285

Sin[2.02876x] + 0.104049Sin[4.91318x] − 0.0219788

Sin[7.97867x] + 0.0187303Sin[11.0855x] − 0.00834994

Sin[14.2074x], 0.213285Sin[2.02876x] + 0.104049

Sin[4.91318x] − 0.0219788Sin[7.97867x] + 0.0187303

Sin[11.0855x] − 0.00834994Sin[14.2074x] + 0.00734255

Sin[17.3364x], 0.213285Sin[2.02876x] + 0.104049

Sin[4.91318x] − 0.0219788Sin[7.97867x] + 0.0187303

Sin[11.0855x] − 0.00834994Sin[14.2074x] + 0.00734255

Sin[17.3364x] − 0.00426841Sin[20.4692x], 0.213285

Sin[2.02876x] + 0.104049Sin[4.91318x] − 0.0219788
```

$$\text{Sin}[7.97867x] + 0.0187303\text{Sin}[11.0855x] - 0.00834994$$

$$\text{Sin}[14.2074x] + 0.00734255\text{Sin}[17.3364x] - 0.00426841$$

$$\text{Sin}[20.4692x] + 0.00387074\text{Sin}[23.6043x]\}$$

We graph $f(x) = x(1 - x)$, $0 < x < 1$, simultaneously with the first term of the generalized Fourier series, funcs[[1]], to observe the accuracy in Figure 9-20.

funcs[[1]]

$0.213285\text{Sin}[2.02876x]$

Plot[{fapprox[x, 1], f[x]}, {x, 0, 1}]

Next, we plot the approximation using the sum of the first 2, 4, 6, and 8 terms of the generalized Fourier series and display the results in Figure 9-21. We see that the approximation improves as the number of terms increases.

graphs = Table[Plot[{fapprox[x, j], f[x]}, {x, 0, 1},

    PlotRange → {0, 0.3}], {j, 2, 8, 2}];

Show[GraphicsGrid[Partition[graphs, 2]]]

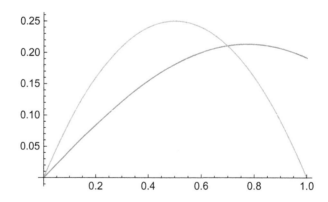

Figure 9-20   Using the first partial sum to approximate $f(x)$ does not result in a very good approximation

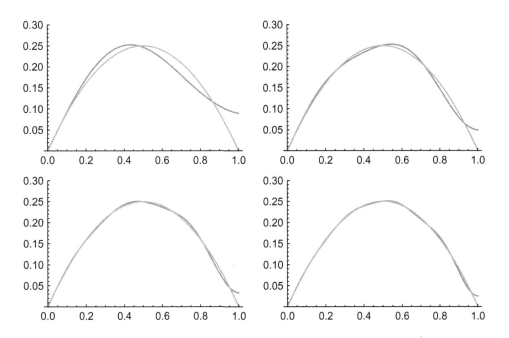

Figure 9-21   The approximation improves when the number of terms in the partial sum is increased

■

Partial Differential Equations

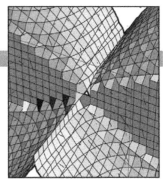

Chapter 10 introduces the separation of variables technique. The emphasis of the chapter is solving the one-dimensional heat equation, the one-dimensional wave equation, D'Alembert's solution to the one-dimensional wave equation, Laplace's equation, Laplace's equation in a circular region, and the wave equation in a circular region.

10.1 Introduction to Partial Differential Equations and Separation of Variables

10.1.1 Introduction

We begin our study of partial differential equations with an introduction of some of the terminology associated with the topic. A **linear second-order partial differential equation (PDE)** in the two independent variables x and y has the form

$$A(x,y)\frac{\partial^2 u}{\partial x^2} + B(x,y)\frac{\partial^2 u}{\partial y \partial x} + C(x,y)\frac{\partial^2 u}{\partial y^2} + D(x,y)\frac{\partial u}{\partial x} + E(x,y)\frac{\partial u}{\partial y} + F(x,y)u = G(x,y),$$

(10.1)

where the solution is $u(x,y)$. If $G(x,y) = 0$ for all x and y, we say that the equation is **homogeneous**. Otherwise, the equation is **nonhomogeneous**.

EXAMPLE 10.1.1: Classify the following partial differential equations: (a) $u_{xx} + u_{yy} = u$; (b) $uu_x = x$.

Differential Equations with Mathematica. http://dx.doi.org/10.1016/B978-0-12-804776-7.00010-3

SOLUTION: (a) This equation satisfies the form of the linear second-order partial differential equation (10.1) with $A = C = 1, F = -1$, and $B = D = E = 0$. Because $G(x, y) = 0$, the equation is homogeneous. (b) This equation is nonlinear, because the coefficient of u_x is a function of u. It is also nonhomogeneous because $G(x, y) = x$.

∎

Definition 40 (Solution of a Partial Differential Equation). *A solution of a partial differential equation in some region R of the space of the independent variables is a function that possesses all of the partial derivatives that are present in the PDE in some region containing R and satisfies the PDE everywhere in R.*

EXAMPLE 10.1.2: Show that $u(x, y) = y^2 - x^2$ and $u(x, y) = e^y \sin x$ are solutions to Laplace's equation $u_{xx} + u_{yy} = 0$.

SOLUTION: For $u(x, y) = y^2 - x^2$, $u_x(x, y) = -2x$, $u_y(x, y) = 2y$, $u_{xx}(x, y) = -2$, and $u_{yy}(x, y) = 2$, so we have that $u_{xx} + u_{yy} = (-2) + 2 = 0$, which we verify with Mathematica.

```
Clear[u]

u[x_, y_] = y^2 - x^2;

D[u[x, y], {x, 2}] + D[u[x, y], {y, 2}]
```

```
0
```

Similarly, for $u(x, y) = e^y \sin x$, we have $u_x = e^y \cos x$, $u_y = e^y \cos x$, $u_{xx} = -e^y \sin x$, and $u_{yy} = e^y \sin x$. Therefore, $u_{xx} + u_{yy} = (-e^y \sin x) + e^y \sin x = 0$, so the equation is satisfied for both functions.

```
Clear[u]

u[x_, y_] = Exp[y]Sin[x];

D[u[x, y], {x, 2}] + D[u[x, y], {y, 2}]
```

```
0
```

We notice that the solutions to Laplace's equation differ in form. This is unlike solutions to homogeneous linear ordinary differential equations. There, we found that solutions were similar in form. (Recall, all solutions could be generated from a general solution.)

∎

Some of the techniques used in constructing solutions of homogeneous linear ordinary differential equations can be extended to the study of partial differential equations as we see with the following theorem.

Theorem 38 (Principle if Superposition). *If $u_1, u_2, \ldots, u_m$ are solutions to a linear homogeneous partial differential equation in a region R, then*

$$c_1 u_1 + c_2 u_2 + \cdots + c_m u_m = \sum_{k=1}^{m} c_k u_k,$$

where $c_1, c_2, \ldots, c_m$ are constants is also a solution in R.

The Principle of Superposition will be used in solving partial differential equations throughout the rest of the chapter. In fact, we will find that equations can have an infinite set of solutions so that we construct another solution in the form of an infinite series.

10.1.2 Separation of Variables

A method that can be used to solve linear partial differential equations is called separation of variables (or the **product method**). Generally, the goal of the method of separation of variables is to transform the partial differential equation into a system of ordinary differential equations each of which depends on only one of the functions in the product form of the solution. Suppose that the function $u(x, y)$ is a solution of a partial differential equation in the independent variables x and y. In separating variables, we assume that $u = u(x, y)$ can be written as the product of a function of x and a function of y. Hence,

$$u(x, y) = X(x)Y(y),$$

and we substitute this product into the partial differential equation to determine $X(x)$ and $Y(y)$. Of course, in order to substitute into the differential equation, we must be able to differentiate this product. However, this is accomplished by following the differentiation rules of multivariate calculus:

$$u_x = X'Y, \quad u_{xx} = X''Y, \quad u_{xy} = X'Y', \quad u_y = XY', \quad and \quad u_{yy} = XY'',$$

where X' represents dX/dx and Y' represents dY/dy. After these substitutions are made and if the equation is separable, we can obtain an ordinary differential equation for X and an ordinary differential equation for Y. These two equations are then solved to find $X(x)$ and $Y(y)$.

EXAMPLE 10.1.3: Use separation of variables to find a solution of $xu_x = u_y$.

SOLUTION: If $u(x, y) = X(x)Y(y)$, then $u_x = X'Y$ and $u_y = XY'$. The equation then becomes

$$xX'Y = XY',$$

which can be written as the separated equation

$$\frac{xX'}{X} = \frac{Y'}{Y}.$$

Notice that the left-hand side of the equation is a function of x while the right-hand side is a function of y. Hence, the only way that this situation can be true is for xX'/X and Y'/Y to both be constant. Therefore,

$$\frac{xX'}{X} = \frac{Y'}{Y} = k,$$

so we obtain the ordinary differential equations $xX' - kX = 0$ and $Y' - ky = 0$. We find X first.

$$xX' - kX = 0$$
$$x\frac{dX}{dx} = kX$$
$$\frac{1}{X}dX = \frac{k}{x}dx$$
$$\ln|X| = k\ln|x| + c_1$$
$$X(x) = e^{c_1}x^k = C_1x^k.$$

Similarly, we find

$$Y' - kY = 0$$
$$\frac{dY}{dy} = kY$$
$$\frac{1}{Y}dY = k\,dy$$
$$\ln|Y| = ky + c_2$$
$$Y(y) = e^{c_2}e^{ky} = C_2e^{ky}.$$

Therefore, a solution is $u(x, y) = X(x)Y(y) = \left(C_1 x^k\right)\left(C_2 e^{ky}\right) = C_3 x^k e^{ky}$ where k and C_3 are arbitrary constants. DSolve can be used to find a solution of this partial differential equation as well.

```
Clear[x, y, u]

DSolve[xD[u[x, y], x] == D[u[x, y], y],

    u[x, y], {x, y}]
```

$\{\{u[x, y] \rightarrow C[1][y + \text{Log}[x]]\}\}$

In this result, the symbol C[1] represents an arbitrary differentiable function. That is, if f is a differentiable function of a single variable, $u(x, y) = f(y + \ln x)$ is a solution to $xu_x = u_y$, which we verify by substituting this result into the partial differential equation.

```
xD[C[1][y + Log[x]], x] == D[C[1][y + Log[x]], y]
```

True

■

10.2 The One-Dimensional Heat Equation

One of the more important partial differential equations is the heat equation,

$$\frac{\partial u}{\partial t} = c^2 \frac{\partial^2 u}{\partial x^2}. \tag{10.2}$$

In one spatial dimension, the solution of the heat equation represents the temperature (at any position x and any time t) in a thin rod or wire of length p. Because the rate at which heat flows through the rod depends on the material that makes up the rod, the constant c^2 which is related to the thermal diffusivity of the material is included in the heat equation. Several different situations can be considered when determining the temperature in the rod. The ends of the wire can be held at a constant temperature, the ends may be insulated, or there can be a combination of these situations.

10.2.1 The Heat Equation With Homogeneous Boundary Conditions

The first problem that we investigate is the situation in which the temperature at the ends of the rod are constantly kept at zero and the initial temperature distribution in the rod is represented as the given function $f(x)$. Hence, the fixed end zero temperature is given in the boundary conditions

$$u(0, t) = u(p, t) = 0$$

while the initial temperature distribution is given by

$$u(x, 0) = f(x).$$

Because the temperature is zero at the endpoints, we say that the problems has **homogeneous boundary conditions**, which are important in finding a solution with separation of variables. We call problems of this type **initial-boundary value problems** (IBVP), because they include initial as well as boundary conditions. Thus, the problem is summarized as

$$\begin{cases} \dfrac{\partial u}{\partial t} = c^2 \dfrac{\partial^2 u}{\partial x^2} \\ u(0, t) = 0,\ u(p, t) = 0,\ t > 0 \\ u(x, 0) = f(x),\ 0 < x < p \end{cases} \tag{10.3}$$

We solve this problem through separation of variables by assuming that

$$u(x, t) = X(x)T(t).$$

Substitution into the heat equation (10.2) yields

$$\frac{T'}{c^2 T} = \frac{X''}{X} = -\lambda$$

where $-\lambda$ is the separation constant. (Note that we selected this constant in order to obtain an eigenvalue problem that was solved in Example 9.1.4). Separating the variables, we have the two equations

$$T' + c^2 \lambda T = 0 \quad \text{and} \quad X'' + \lambda X = 0.$$

Now that we have successfully separated the variables, we turn our attention to the homogeneous boundary conditions. In terms of the functions $X(x)$ and $T(t)$, these boundary conditions become

$$u(0, t) = X(0)T(t) = 0 \quad \text{and} \quad u(p, t) = X(p)T(t) = 0.$$

In each case, we must avoid setting $T(t) = 0$ for all t, because if this were the case, our solution would be the trivial solution $u(x, t) = X(x)T(t) = 0$. Therefore, we have the boundary conditions

$$X(0) = 0 \quad \text{and} \quad X(p) = 0,$$

so we solve the eigenvalue problem

See Example 9.1.4.

$$\begin{cases} X'' + \lambda X = 0 \\ X(0) = 0, \ X(p) = 0 \end{cases}.$$

The eigenvalues of this problem are $\lambda_n = (n\pi/p)^2$ with corresponding eigen-functions $X_n(x) = \sin(n\pi x/p)$. Similarly, a general solution of $T' + c^2\lambda_n T = 0$ is $T_n(t) = Ae^{-c^2\lambda_n t}$, where A is an arbitrary constant and $\lambda_n = (n\pi/p)^2$, $n = 1, 2, \ldots$.

```
DSolve[capt'[t] + c²λₙcapt[t]==0, capt[t], t]
```

$$\{\{capt[t] \to e^{-c^2 t \lambda_n} C[1]\}\}$$

Because $X(x)$ and $T(t)$ both depend on n, the solution $u(x, t) = X(x)T(t)$ does as well. Hence,

$$u_n(x, t) = X_n(t)T_n(t) = c_n \sin \frac{n\pi x}{p} e^{-c^2\lambda_n t},$$

where we have replaced the constant A by one that depends on n. In order to find the value of c_n, we apply the initial condition $u(x, 0) = f(x)$. Notice that

$$u_n(x, 0) = c_n \sin \frac{n\pi x}{p} e^{-c^2\lambda_n \cdot 0} = c_n \sin \frac{n\pi x}{p}$$

is satisfied only by functions of the form $\sin(\pi x/p)$, $\sin(2\pi x/p)$, $\ldots$ (which, in general, is not the case). Therefore, we use the principle of superposition to state that

$$u(x, t) = \sum_{n=1}^{\infty} u_n(x, t) = \sum_{n=1}^{\infty} c_n \sin \frac{n\pi x}{p} e^{-c^2\lambda_n t}$$

is also a solution of the problem, because this solution satisfies the heat equation as well as the boundary conditions. Then, when we apply the initial condition $u(x, 0) = f(x)$, we find that

$$u(x,0) = \sum_{n=1}^{\infty} c_n \sin \frac{n\pi x}{p} e^{-c^2 \lambda_n \cdot 0} = \sum_{n=1}^{\infty} c_n \sin \frac{n\pi x}{p} = f(x).$$

Therefore, c_n represents the Fourier sine series coefficients for $f(x)$, which are given by

$$c_n = \frac{2}{p} \int_0^p f(x) \sin \frac{n\pi x}{p} \, dx, \quad n = 1, 2, \ldots.$$

EXAMPLE 10.2.1: Solve $\begin{cases} u_t = u_{xx}, \, 0 < x < 1, \, t > 0 \\ u(0,t) = 0, \, u(1,t) = 0, \, t > 0 \\ u(x,0) = 50, \, 0 < x < 1 \end{cases}$

SOLUTION: In this case, $c = 1$, $p = 1$, and $f(x) = 50$. Hence,

$$u(x,t) = \sum_{n=1}^{\infty} c_n \sin n\pi x \, e^{-\lambda_n t},$$

If n is an integer, $\cos n\pi = (-1)^n$.

where

$$c_n = \frac{2}{1} \int_0^1 50 \sin n\pi x \, dx = -\frac{100}{n\pi} (\cos n\pi - 1) = -\frac{100}{n\pi} \left((-1)^n - 1 \right),$$

and $\lambda_n = (n\pi)^2$.

$c_{n\_} = 100 \int_0^1 \text{Sin}[n\pi x] dx$

$\dfrac{100(1 - \text{Cos}[n\pi])}{n\pi}$

$\lambda_{n\_} = (n\pi)^2;$

Therefore, because $c_n = 0$ if n is even, we write $u(x,t)$ as

$$u(x,t) = \sum_{n=1}^{\infty} \frac{200}{(2n-1)\pi} \sin(2n-1)\pi x \, e^{-(2n-1)^2 \pi^2 t}.$$

We graph an approximation of $u(x,t)$ at various times by graphing

$$u_k(x,t) = \sum_{n=1}^{k} \frac{200}{(2n-1)\pi} \sin(2n-1)\pi x\, e^{-(2n-1)^2\pi^2 t}$$

if $k = 10$ in Figure 10-1(a).

```
uapprox[x_, t_] = ∑_{n=1}^{50} cₙSin[nπx]Exp[−λₙt];

toplot = Table[uapprox[x, t], {t, 0, 1, .05}];

p1 = Plot[Evaluate[toplot], {x, 0, 1},

   AxesLabel → {x, u}, PlotRange → All,

   PlotLabel → "(a)"]
```

An alternative approach to visualizing the solution is to generate a density plot of $u(x,t)$ with DensityPlot as shown in Figure 10-1(b).

DensityPlot works in the same way as ContourPlot.

```
p2 = DensityPlot[uapprox[x, t], {x, 0, 1}, {t, 0, 0.5},

   PlotPoints->30, Frame → False, Axes → Automatic,

   AxesLabel → {x, t}, PlotLabel → "(b)"]

Show[GraphicsRow[{p1, p2}]]
```

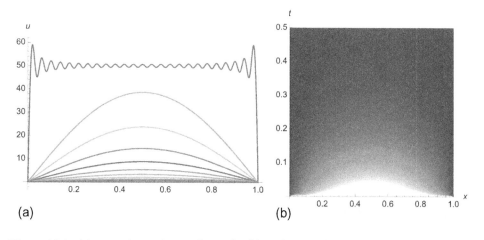

(a) (b)

Figure 10-1   (a) $u(x,t)$ for various values of t. (b) A density plot of $u(x,t)$: t corresponds to the vertical axis and x the horizontal axis

In the density plot, darker shades correspond to smaller values of $u(x, t)$ so we see that as t increases, the temperature throughout the rod approaches zero.

■

10.2.2 Nonhomogeneous Boundary Conditions

The ability to apply the method of separation of variables depends on the presence of homogeneous boundary conditions as we just saw in the previous problem. However, with the heat equation, the temperature at the endpoints may not be held constantly at zero. Instead, consider the case when the temperature at the left-hand endpoint is $T_0 \neq 0$ and at the right-hand endpoint it is $T_1 \neq 0$. Mathematically, we state these **nonhomogeneous boundary conditions** as

$$u(0, t) = T_0 \quad \text{and} \quad u(p, t) = T_1,$$

so we are faced with solving the problem

$$\begin{cases} \dfrac{\partial u}{\partial t} = c^2 \dfrac{\partial^2 u}{\partial x^2} \\ u(0, t) = T_0,\, u(p, t) = T_1,\, t > 0 \\ u(x, 0) = f(x),\, 0 < x < p \end{cases} \tag{10.4}$$

In this case, we must modify the problem in order to introduce homogeneous boundary conditions to the problem. We do this by using the physical observance that as $t \to \infty$, the temperature in the wire does not depend on t. Hence,

$$\lim_{t \to \infty} u(x, t) = S(x), \tag{10.5}$$

where we call $S(x)$ in equation (10.5) the **steady-state temperature**. Therefore, we let

$$u(x, t) = v(x, t) + S(x), \tag{10.6}$$

where $v(x, t)$ is called the **transient temperature**. We use these two functions to obtain two problems that we can solve. In order to substitute $u(x, t)$ into the heat equation, $u_t = c^2 u_{xx}$, we calculate the derivatives

$$u_t(x, t) = v_t(x, t) + 0 \quad \text{and} \quad u_{xx}(x, t) = v_{xx}(x, t) + S''(x).$$

Substitution into the heat equation (10.2) yields

$$\frac{\partial u}{\partial t} = c^2 \frac{\partial^2 u}{\partial x^2}$$

$$\frac{\partial v}{\partial t} = c^2 \frac{\partial^2 v}{\partial x^2} + c^2 S'',$$

so we have the two equations $v_t = c^2 v_{xx}$ and $S'' = 0$. We next consider the boundary conditions. Because

$$u(0, t) = v(0, t) + S(0) = T_0 \quad \text{and} \quad u(p, t) = v(p, t) + S(p) = T_1,$$

we can choose the boundary conditions for S to be the nonhomogeneous conditions

$$S(0) = T_0 \quad \text{and} \quad S(p) = T_1$$

and the boundary conditions for $v(x, t)$ to be the homogeneous boundary conditions

$$v(0, t) = 0 \quad \text{and} \quad v(p, t) = 0.$$

Of course, we have failed to include the initial temperature. Applying this condition, we have $u(x, 0) = v(x, 0) + S(x) = f(x)$, so the initial condition for v is

$$v(x, 0) = f(x) - S(x).$$

Therefore, we have two problems, one for v with homogeneous boundary conditions and one for S that has nonhomogeneous boundary conditions:

$$\begin{cases} S'' = 0, \ 0 < x < p \\ S(0) = T_0, \ S(p) = T_1 \end{cases} \quad \text{and} \quad \begin{cases} v_t = c^2 v_{xx}, \ 0 < x < p, \ t > 0 \\ v(0, t) = 0, \ v(p, t) = 0, \ t > 0 \\ v(x, 0) = f(x) - S(x), \ 0 < x < p \end{cases}.$$

Because S is needed in the determination of v, we begin by finding the steady-state temperature and obtain $S(x) = T_0 + \dfrac{T_1 - T_0}{p} x.$

```
Clear[s, t0]

DSolve[{s"[x] == 0, s[0] == t0,

  s[p] == t1}, s[x], x]
```

$$\left\{\left\{s[x] \to \frac{pt0 - t0x + t1x}{p}\right\}\right\}$$

We are now able to find $v(x, t)$ by solving the heat equation with homogeneous boundary conditions for v. Because we solved this problem at the beginning of this section, we do not need to go through the separation of variables procedure. Instead, we use the formula that we derived there using the initial temperature $f(x) - S(x)$. Therefore,

$$v(x, t) = \sum_{n=1}^{\infty} c_n \sin \frac{n\pi x}{p} e^{-c^2 \lambda_n t}, \tag{10.7}$$

where $v(x, 0) = \sum_{n=1}^{\infty} c_n \sin \frac{n\pi x}{p} = f(x) - S(x)$. This means that c_n represents the Fourier sine series coefficients for the function $f(x) - S(x)$ given by

$$c_n = \frac{2}{p} \int_0^p (f(x) - S(x)) \sin \frac{n\pi x}{p} dx, \quad n = 1, 2, \ldots. \tag{10.8}$$

EXAMPLE 10.2.2: Solve $\begin{cases} u_t = u_{xx}, \ 0 < x < 1, \ t > 0 \\ u(0, t) = 10, \ u(1, t) = 60, \ t > .0 \\ u(x, 0) = 10, \ 0 < x < 1 \end{cases}$

SOLUTION: In this case, $c = 1$, $p = 1$, $T_0 = 10$, $T_1 = 60$, and $f(x) = 10$. Therefore, the steady-state solution is

$$S(x) = T_0 + \frac{T_1 - T_0}{p} x = 10 + \frac{60 - 10}{1} x = 10 + 50x.$$

Then, the initial transient temperature is

$$v(x, 0) = 10 - (10 + 50x) = -50x$$

so that the series coefficients in the solution (10.7) are given by equation (10.8):

$$c_n = \frac{2}{1} \int_0^1 -50x \sin n\pi x \, dx = -100 \int_0^1 x \sin n\pi x \, dx$$

$$= \frac{100}{n\pi} \cos n\pi = \frac{100}{n\pi} (-1)^n, \ \ldots$$

$\text{cn}\_ = -100 \int_0^1 x\text{Sin}[n\pi x]dx$

$$-\frac{100(-n\pi \cos[n\pi] + \sin[n\pi])}{n^2 \pi^2}$$

$$\lambda_{n\_} = (n\pi)^2;$$

so the transient temperature is

$$v(x,t) = \sum_{n=1}^{\infty} c_n \sin \frac{n\pi x}{p} e^{-c^2 \lambda_n t} = \sum_{n=1}^{\infty} (-1)^n \frac{100}{n\pi} \sin n\pi x\, e^{-n^2 \pi^2 t}$$

and

$$u(x,t) = v(x,t) + S(x) = 100 + 50x + \sum_{n=1}^{\infty} (-1)^n \frac{100}{n\pi} \sin n\pi x\, e^{-n^2 \pi^2 t}.$$

We graph an approximation of $u(x,t)$ for several values of t by graphing

$$100 + 50x + \sum_{n=1}^{30} (-1)^n \frac{100}{n\pi} \sin n\pi x\, e^{-n^2 \pi^2 t}.$$

```
uapprox[x_, t_] = 10 + 50x + ∑_{n=1}^{50} cₙSin[nπx]Exp[−λₙt];

toplot = Table[uapprox[x, t], {t, 0, 0.5, 0.5/20}];

p1 = Plot[Evaluate[toplot], {x, 0, 1},

   AxesLabel → {x, u}, PlotLabel → "(a)"]
```

In Figure 10-2(a), notice that as $t \to \infty$, $u(x,t) \to S(x)$.
We generate a density plot of this function in Figure 10-2(b).

```
p2 = DensityPlot[uapprox[x, t], {x, 0, 1}, {t, 0, 0.5},

   PlotPoints->30, Frame → False, Axes → Automatic,

   AxesLabel → {x, t}, PlotLabel → "(b)"]

Show[GraphicsRow[{p1, p2}]]
```

Notice that the temperature throughout the bar approaches the steady-state temperature as t increases.

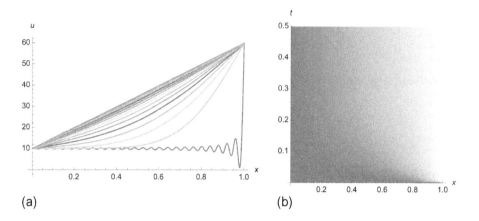

Figure 10-2 (a) An approximation of $u(x, t)$ for 21 equally spaced values of t between 0 and 0.5. (b) Density plot of an approximation of $u(x, t)$

■

10.2.3 Insulated Boundary

Another important situation concerning the flow of heat in a wire involves insulated ends. In this case, heat is not allowed to escape from the ends of the wire. Mathematically, we express these boundary conditions as

$$\frac{\partial u}{\partial x}(0, t) = 0 \quad and \quad \frac{\partial u}{\partial x}(p, t) = 0,$$

because the rate at which the heat changes along the x-axis at the endpoints $x = 0$ and $x = p$ is zero. Therefore, if we want to determine the temperature in a wire of length p with insulated ends, we solve the initial-boundary value problem

$$\begin{cases} \dfrac{\partial u}{\partial t} = c^2 \dfrac{\partial^2 u}{\partial x^2} \\ \dfrac{\partial u}{\partial x}(0, t) = 0, \ \dfrac{\partial u}{\partial x}(p, t) = 0, \ t > 0 \\ u(x, 0) = f(x), \ 0 < x < p \end{cases} \tag{10.9}$$

Notice that the boundary conditions are homogeneous, so we can use separation of variables to find $u(x, t) = X(x)T(t)$. By following the steps taken in the solution of the problem with homogeneous boundary conditions, we obtain the ordinary differential equations

$$T' + c^2 \lambda T = 0 \quad and \quad X'' + \lambda X = 0$$

However, when we consider the boundary conditions

$$u_x(0, t) = X'(0)T(0) = 0 \quad \text{and} \quad u_x(p, t) = X'(p)T(p) = 0,$$

we wish to avoid letting $T(t) = 0$ for all t (which leads to the trivial solution), we have the homogeneous boundary conditions

$$X'(0) = 0 \quad \text{and} \quad X'(p) = 0,$$

Therefore, we solve the eigenvalue problem

We solve this eigenvalue problem in Example 9.1.5.

$$\begin{cases} X'' + \lambda X = 0, \, 0 < x < p \\ X'(0) = 0, \, X'(p) = 0 \end{cases}$$

to find $X(x)$. The eigenvalues and corresponding eigenfunctions of this problem are

We solve this eigenvalue problem in Example 9.1.5.

$$\lambda_n = \begin{cases} 0, \, n = 0 \\ (n\pi/p)^2, \, n = 1, 2, \ldots \end{cases} \quad \text{and} \quad X_n(x) = \begin{cases} 1, \, n = 0 \\ \cos(n\pi x/p), \, n = 1, 2, \ldots. \end{cases}$$

Next, we solve the equation $T' + c^2\lambda_n T = 0$. First, for $\lambda_0 = 0$, we have the equation $T' = 0$ which has the solution $T(t) = A_0$, where A_0 is a constant. Therefore, for $\lambda_0 = 0$, the solution is the product

$$u_0(x, t) = X_0(x)T_0(t) = A_0.$$

For $\lambda_n = (n\pi/p)^2$, $T' + c^2\lambda_n T = 0$ has general solution $T_n(t) = a_n e^{-c^2\lambda_n t}$. For these eigenvalues, we have the solution

$$u_n(x, t) = X_n(x)T_n(t) = a_n \cos\frac{n\pi x}{p} e^{-c^2\lambda_n t}$$

Therefore, by the Principle of Superposition, the solution is

$$u(x, t) = A_0 + \sum_{n=1}^{\infty} a_n \cos\frac{n\pi x}{p} e^{-c^2\lambda_n t}.$$

Application of the initial temperature yields

$$u(x, 0) = A_0 + \sum_{n=1}^{\infty} a_n \cos\frac{n\pi x}{p} = f(x),$$

which is the Fourier cosine series for $f(x)$ where the coefficient A_0 is equivalent to $\frac{1}{2}a_0$ in the original Fourier series given in Section 9.2. Therefore,

$$A_0 = \frac{1}{2}a_0 = \frac{1}{2}\frac{2}{p}\int_0^p f(x)\,dx = \frac{1}{p}\int_0^p f(x)\,dx$$

and   (10.10)

$$a_n = \frac{2}{p}\int_0^p f(x)\cos\frac{n\pi x}{p}\,dx, \quad n = 1, 2, \ldots.$$

EXAMPLE 10.2.3: Solve $\begin{cases} u_t = u_{xx},\ 0 < x < \pi,\ t > 0 \\ u_x(0,t) = 0,\ u_x(p,t) = 0,\ t > .0 \\ u(x,0) = x,\ 0 < x < p \end{cases}$

See Example 9.2.1.

SOLUTION: In this case, $p = \pi$ and $c = 1$. The Fourier cosine series coefficients for $f(x) = x$ are given by

$$A_0 = \frac{1}{2}a_0 = \frac{1}{\pi}\int_0^\pi x\,dx = \frac{\pi}{2}$$

and

$$a_n = \frac{2}{\pi}\int_0^\pi x\cos\frac{n\pi x}{\pi}\,dx = \frac{2}{\pi n^2}\left[(-1)^n - 1\right], n = 1, 2, \ldots.$$

Therefore, the solution is

$$u(x,t) = \frac{\pi}{2} - \sum_{n=1}^\infty \frac{4}{(2n-1)^2\pi}\cos((2n-1)x)\,e^{-(2n-1)^2 t},$$

where we have used the fact that $a_n = 0$ if n is even. We graph an approximation of $u(x,t)$ by graphing

$$\frac{\pi}{2} - \sum_{n=1}^{40} \frac{4}{(2n-1)^2\pi}\cos((2n-1)x)\,e^{-(2n-1)^2 t}$$

in Figure 10-3(a) and then a density plot of this function in Figure 10-3(b).

$$\mathtt{a_{n\_}} = \frac{4}{(2n-1)^2\pi};$$

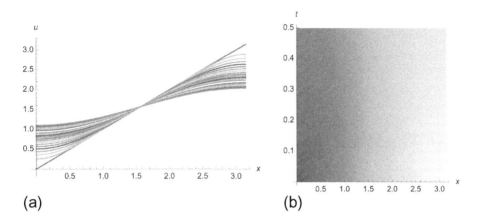

Figure 10-3 (a) An approximation of $u(x, t)$ for 21 equally spaced values of t between 0 and 1. (b) A density plot

```
uapprox[x_, t_] = π/2 - Σ_{n=1}^{40} a_n Cos[(2n-1)x]Exp[-(2n-1)²t];

toplot = Table[uapprox[x, t], {t, 0, 1, 1/20}];

p1 = Plot[Evaluate[toplot], {x, 0, π}, AxesLabel → {x, u},

    PlotLabel → "(a)"]

p2 = DensityPlot[uapprox[x, t], {x, 0, π}, {t, 0, 0.5},

    PlotPoints->30, Frame → False, Axes → Automatic,

    AxesLabel → {x, t}, PlotLabel → "(b)"]

Show[GraphicsRow[{p1, p2}]]
```

Notice that the temperature eventually becomes $A_0 = \pi/2$ throughout the wire. Temperatures to the left of $x = \pi/2$ increase while those to the right decrease.

∎

10.3 The One-Dimensional Wave Equation

The one-dimensional *wave equation* is important in solving an interesting problem.

10.3.1 The Wave Equation

Suppose that we pluck a string (like a guitar or violin string) of length p and constant mass density that is fixed at each end. A question that we might ask is: "What is the position of the string at a particular instance of time?". We answer this question by modeling the physical situation with a partial differential equation, namely the *wave equation* in one spatial variable. We will not go through this derivation as we did with the heat equation, but we point out that it is based on determining the forces that act on a small segment of the string and applying Newton's Second Law of Motion. The partial differential equation that is found is

$$c^2 \frac{\partial^2 u}{\partial x^2} = \frac{\partial^2 u}{\partial t^2}, \tag{10.11}$$

which is called the (one-dimensional) **wave equation**. In this equation $c^2 = T/\rho$, where T is the tension of the string and ρ is the constant mass of the string per unit length. The solution $u = u(x, t)$ represents the displacement of the string from the x-axis at time t. In order to determine $u = u(x, t)$ we must describe the boundary and initial conditions that model the physical situation. At the ends of the string, the displacement from the x-axis is fixed at zero, so we use the homogeneous boundary conditions

$$u(0, t) = 0 \quad \text{and} \quad u(p, t) = 0$$

for $t > 0$. The motion of the string also depends on the displacement and the velocity at each point of the string at $t = 0$. If the initial displacement is given by $f(x)$ and the initial velocity by $g(x)$, we have the initial conditions

$$u(x, 0) = f(x) \quad \text{and} \quad \frac{\partial u}{\partial t}(x, 0) = g(x)$$

for $0 < x < p$. Therefore, we determine the displacement of the string with the initial-boundary value problem

$$\begin{cases} c^2 \dfrac{\partial^2 u}{\partial x^2} = \dfrac{\partial^2 u}{\partial t^2}, \ 0 < x < p, \ t > 0 \\ u(0,t) = 0, \ u(p,t) = 0, \ t > 0 \, . \\ u(x,0) = f(x), \ \dfrac{\partial u}{\partial t}(x,0) = g(x), \ 0 < x < p \end{cases} \qquad (10.12)$$

This problem is solved through separation of variables by assuming that $u(x,t) = X(x)T(t)$. Substitution into the wave equation yields

$$c^2 X'' T = X T''$$

$$\frac{X''}{X} = \frac{T''}{c^2 T} = -\lambda$$

> Notice that wave equation requires two initial conditions where the heat equation only needed one. This is due to the fact that there is a second derivative with respect to t while there is only one derivative with respect to t in the heat equation.

so we obtain the two second-order ordinary differential equations

$$X'' + \lambda X = 0 \quad \text{and} \quad T'' + c^2 \lambda T = 0.$$

At this point, we solve the equation that involves the homogeneous boundary conditions. As was the case with the heat equation, the boundary conditions in terms of $u(x,t) = X(x)T(t)$ are

$$u(0,t) = X(0)T(t) = 0 \quad \text{and} \quad u(p,t) = X(p)T(t) = 0,$$

so we have

$$X(0) = 0 \quad \text{and} \quad X(p) = 0.$$

Therefore, we determine $X(x)$ by solving the eigenvalue problem

$$\begin{cases} X'' + \lambda X = 0, \ 0 < x < p \\ X(0) = 0, \ X(p) = 0 \end{cases},$$

which we encountered when solving the heat equation and solved in Section 10.2. The eigenvalues of this problem are

See Example 9.1.4.

$$\lambda_n = \left(\frac{n\pi}{p}\right)^2, \quad n = 1, 2, \dots$$

with corresponding eigenfunctions

$$X_n(x) = \sin \frac{n\pi x}{p}, \quad n = 1, 2, \dots.$$

Next, we solve the equation $T'' + c^2 \lambda_n T = 0$. A general solution is

$$T_n(t) = a_n \cos \left(c\sqrt{\lambda_n} t \right) + b_n \sin \left(c\sqrt{\lambda_n} t \right) = a_n \cos \frac{cn\pi t}{p} + b_n \sin \frac{cn\pi t}{p},$$

where the coefficients a_n and b_n must be determined. Putting this information together, we obtain

$$u_n(x, t) = \left(a_n \cos \frac{cn\pi t}{p} + b_n \sin \frac{cn\pi t}{p} \right) \sin \frac{n\pi x}{p},$$

so by the Principle of Superposition, we have

$$u(x, t) = \sum_{n=1}^{\infty} \left(a_n \cos \frac{cn\pi t}{p} + b_n \sin \frac{cn\pi t}{p} \right) \sin \frac{n\pi x}{p}.$$

Applying the initial position yields

$$u(x, 0) = \sum_{n=1}^{\infty} a_n \sin \frac{n\pi x}{p} = f(x)$$

so a_n is the Fourier sine series coefficient for $f(x)$, which is given by

$$a_n = \frac{2}{p} \int_0^p f(x) \sin \frac{n\pi x}{p} \, dx, \ n = 1, 2, \ldots \tag{10.13}$$

To determine b_n, we must use the initial velocity. Therefore, we compute

$$\frac{\partial u}{\partial t}(x, t) = \sum_{n=1}^{\infty} \left(-a_n \frac{ncn\pi}{p} \sin \frac{cn\pi t}{p} + b_n \frac{cn\pi}{p} \cos \frac{cn\pi t}{p} \right) \sin \frac{n\pi x}{p}.$$

Then,

$$\frac{\partial u}{\partial t}(x, 0) = \sum_{n=1}^{\infty} b_n \frac{cn\pi}{p} \sin \frac{n\pi x}{p} = g(x),$$

so $b_n \dfrac{cn\pi}{p}$ represents the Fourier sine series coefficient for $g(x)$, which means that

$$b_n = \frac{p}{cn\pi} \frac{2}{p} \int_0^p g(x) \sin \frac{n\pi x}{p} \, dx = \frac{2}{cn\pi} \int_0^p g(x) \sin \frac{n\pi x}{p} \, dx, \quad n = 1, 2, \ldots. \tag{10.14}$$

EXAMPLE 10.3.1: Solve $\begin{cases} u_{xx} = u_{tt}, \ 0 < x < 1, \ t > 0 \\ u(0,t) = 0, \ u(1,t) = 0, \ t > 0 \\ u(x,0) = x(1-x), \ u_t(x,0) = 0, \ 0 < x < 1 \end{cases}$.

SOLUTION: For this problem, $c = p = 1$, $f(x) = x(1-x)$, and $g(x) = 0$. With this information and equation (10.13) we compute

$$a_m = \frac{2}{1} \int_0^1 x(1-x) \sin n\pi x \, dx = -\frac{4}{n^3\pi^3} \cos n\pi + \frac{4}{n^3\pi^3}$$

$$= \frac{4}{n^3\pi^3} \left[1 - (-1)^n \right], \quad n = 1, 2, \ldots.$$

$a_{n\_} = 2 \int_0^1 x(1-x) \text{Sin}[n\pi x] dx$

$\dfrac{2(2 - 2\text{Cos}[n\pi] - n\pi \text{Sin}[n\pi])}{n^3\pi^3}$

With $g(x) = 0$, we use equation (10.14) to see that the coefficients $b_n = 0$ for all n. Using the fact that $a_n = 0$ for even values of n, the solution is

$$u(x,t) = \sum_{n=1}^{\infty} \frac{8}{(2n-1)^3\pi^3} \cos((2n-1)\pi t) \sin((2n-1)\pi x).$$

We illustrate the motion of the string by graphing

$$u_k(x,t) = \sum_{n=1}^{k} \frac{8}{(2n-1)^3\pi^3} \cos((2n-1)\pi t) \sin((2n-1)\pi x)$$

using $k = 10$ for 16 equally spaced values of t between 0 and 1 in Figure 10-4.

```
u[x_, t_] = Σ_{n=1}^{10} (8Cos[(2n-1)π t]Sin[(2n-1)π x])/((2n-1)^3 π^3);

somegraphs = Table[Plot[u[x, t], {x, 0, 1},

   PlotRange → {-0.3, 0.3}, Ticks → {{0, 1},

   {-0.3, 0.3}}], {t, 0, 1, 1/15}];

toshow = Partition[somegraphs, 4];
```

```
Show[GraphicsGrid[toshow]]
```

To see the motion of the string, use `Animate`.

```
Animate[Plot[u[x, t], {x, 0, 1},
    PlotRange → {-0.3, 0.3}], {{t, 0}, 0, 2}]
```

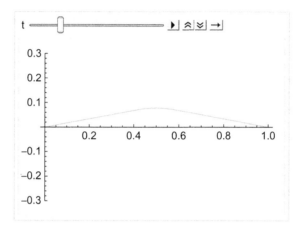

■

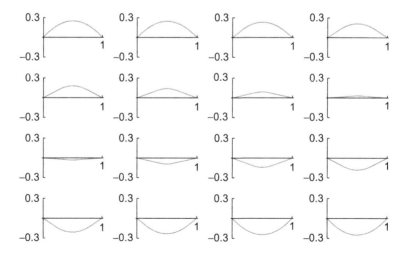

Figure 10-4   Visualizing the motion of a string

EXAMPLE 10.3.2: Solve $\begin{cases} u_{xx} = u_{tt}, \, 0 < x < 1, \, t > 0 \\ u(0,t) = 0, \, u(1,t) = 0, \, t > 0 \, . \\ u(x,0) = \sin \pi x, \, u_t(x,0) = 3x + 1, \, 0 < x < 1 \end{cases}$

SOLUTION: The appropriate parameters and initial conditions are defined first.

```
Remove[a, b];

f[x_] = Sin[πx];

g[x_] = 3x + 1;
```

Next, the functions to determine the coefficients a_n and b_n in the series approximation of the solution $u = u(x,t)$ are defined.

```
aₙ_ = 2 ∫₀¹ f[x]Sin[nπx]dx
```

$$\frac{2\mathrm{Sin}[n\pi]}{\pi - n^2\pi}$$

```
bₙ_ = 2 ∫₀¹ g[x]Sin[nπx]dx
       ─────────────────────
               nπ
```

$$\frac{2(n\pi - 4n\pi\,\mathrm{Cos}[n\pi] + 3\mathrm{Sin}[n\pi])}{n^3\pi^3}$$

Because n represents an integer, these results indicate that $a_n = 0$ for all $n \geq 2$. We use Table to calculate the first 10 values of b_n.

```
Table[{n, bₙ, bₙ//N}, {n, 1, 10}]//TableForm
```

1	$\dfrac{10}{\pi^2}$	1.01321
2	$-\dfrac{3}{2\pi^2}$	−0.151982
3	$\dfrac{10}{9\pi^2}$	0.112579
4	$-\dfrac{3}{8\pi^2}$	−0.0379954
5	$\dfrac{2}{5\pi^2}$	0.0405285
6	$-\dfrac{1}{6\pi^2}$	−0.0168869
7	$\dfrac{10}{49\pi^2}$	0.0206778
8	$-\dfrac{3}{32\pi^2}$	−0.00949886
9	$\dfrac{10}{81\pi^2}$	0.0125088
10	$-\dfrac{3}{50\pi^2}$	−0.00607927

The function u defined next computes the *n*th term in the series expansion. Hence, uapprox determines the approximation of order *k* by summing the first *k* terms of the expansion, as illustrated with approx[10].

Notice that we define uapprox[n] so that Mathematica "remembers" the terms uapprox that are computed. That is, Mathematica need to recompute uapprox[n-1] to compute uapprox[n] provided that uapprox[n-1] has already been computed.

```
Clear[u, uapprox]
```

```
u[n_] = b_n Sin[nπ t]Sin[nπ x];
```

```
uapprox[k_]:=uapprox[k] = uapprox[k - 1] + u[k];
```

```
uapprox[1] = u[1];
```

```
ua[x_, t_] = uapprox[10]
```

$$\frac{10\mathrm{Sin}[\pi t]\mathrm{Sin}[\pi x]}{\pi^2} - \frac{3\mathrm{Sin}[2\pi t]\mathrm{Sin}[2\pi x]}{2\pi^2} + \frac{10\mathrm{Sin}[3\pi t]\mathrm{Sin}[3\pi x]}{9\pi^2} -$$

$$\frac{3\mathrm{Sin}[4\pi t]\mathrm{Sin}[4\pi x]}{8\pi^2} + \frac{2\mathrm{Sin}[5\pi t]\mathrm{Sin}[5\pi x]}{5\pi^2} - \frac{\mathrm{Sin}[6\pi t]\mathrm{Sin}[6\pi x]}{6\pi^2} +$$

$$\frac{10\mathrm{Sin}[7\pi t]\mathrm{Sin}[7\pi x]}{49\pi^2} - \frac{3\mathrm{Sin}[8\pi t]\mathrm{Sin}[8\pi x]}{32\pi^2} + \frac{10\mathrm{Sin}[9\pi t]\mathrm{Sin}[9\pi x]}{81\pi^2} -$$

$$\frac{3\mathrm{Sin}[10\pi t]\mathrm{Sin}[10\pi x]}{50\pi^2}$$

To illustrate the motion of the string, we graph uapprox[10], the tenth partial sum of the series, on the interval [0, 1] for 16 equally spaced values of *t* between 0 and 2 in Figure 10-5.

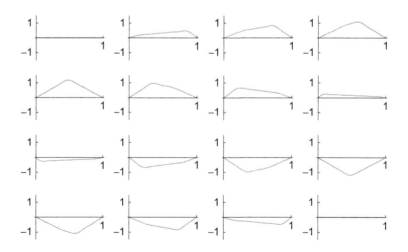

Figure 10-5   Visualizing the motion of a string

```
somegraphs = Table[Plot[Evaluate[uapprox[10]], {x, 0, 1},
    PlotRange → {-3/2, 3/2}, Ticks → {{0, 1}, {-1, 1}}], {t, 0, 2, 2/15}];
toshow = Partition[somegraphs, 4];
Show[GraphicsGrid[toshow]]
```

If instead we wished to see the motion of the string, we use
`Animate`. We show a frame from the resulting animation.

```
Animate[Plot[Evaluate[ua[x, t]], {x, 0, 1},
    PlotRange → {-3/2, 3/2}], {{t, 0}, 0, 2}]
```

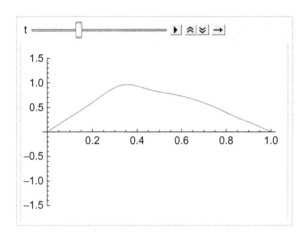

■

10.3.2 D'Alembert's Solution

An interesting version of the wave equation is to consider a string of infinite
length. Therefore, the boundary conditions are no longer of importance. Instead,
we simply work with the wave equation with the initial position and velocity
functions. In order to solve the problem

$$\begin{cases} c^2 \dfrac{\partial^2 u}{\partial x^2} = \dfrac{\partial^2 u}{\partial t^2}, & -\infty < x < \infty, t > 0 \\ u(x,0) = f(x), & \dfrac{\partial u}{\partial t}(x,0) = g(x) \end{cases} \qquad (10.15)$$

we use the change of variables $r = x + ct$ and $s = x - ct$. Using the Chain Rule, we compute the derivatives u_{xx} and u_{tt} in terms of the variables r and s:

$$u_x = u_r r_x + u_s s_x = u_r + u_s,$$
$$u_{xx} = (u_r + u_s)_r\, r_x + (u_r + u_s)_s\, s_x = u_{rr} + 2u_{rs} + u_{ss},$$
$$u_t = u_r r_t + u_s s_t = cu_r - cu_s = c\,(u_r - u_s),$$

and

$$u_{tt} = c\left[(u_r - u_s)_r\, r_t + (u_r - u_s)_s\, s_t\right] = c^2\,(u_{rr} - 2u_{rs} + u_{ss}).$$

Substitution into the wave equation yields

$$c^2 u_{xx} = u_{tt}$$
$$c^2\,(u_{rr} + 2u_{rs} + u_{ss}) = c^2\,(u_{rr} - 2u_{rs} + u_{ss})$$
$$4c^2 u_{rs} = 0$$
$$u_{rs} = 0.$$

The partial differential equation $u_{rs} = 0$ can be solved by first integrating with respect s to obtain

$$u_r = f(r),$$

where $f(r)$ is an arbitrary function of r. Then, integrating with respect to s, we have

$$u(r, s) = F(r) + G(s),$$

where F is an anti-derivative of f and G is an arbitrary function of s. Returning to our original variables then gives us

$$u(x, t) = F(x + ct) + G(x - ct).$$

We see that this is the solution that DSolve returns as well. (Note that C[1] and C[2] represent the arbitrary functions F and G.)

    Clear[u, c]

    DSolve[c^2D[u[x, t], {x, 2}]==D[u[x, t], {t, 2}], u[x, t], {x, t}]

$$\left\{\left\{u[x,\, t] \rightarrow C[1]\Big[t - \frac{x}{\sqrt{c^2}}\Big] + C[2]\Big[t + \frac{x}{\sqrt{c^2}}\Big]\right\}\right\}$$

The functions F and G are determined by the initial conditions which indicate that

$$u(x, 0) = F(x) + G(x) = f(x)$$

and

$$u_t(x, 0) = cF'(x) + cG'(x) = g(x).$$

We can rewrite the second equation by integrating to obtain

$$F'(x) - G'(x) = \frac{1}{c} g(x)$$

$$F(x) - G(x) = \frac{1}{c} \int_0^x g(v) \, dv.$$

Therefore, we solve the system

$$F(x) + G(x) = f(x)$$

$$F(x) - G(x) = \frac{1}{c} \int_0^x g(v) \, dv$$

for $F(x)$ and $G(x)$. Adding these equations yields

$$F(x) = \frac{1}{2} \left(f(x) + \frac{1}{c} \int_0^x g(v) \, dv \right)$$

and subtracting gives us

$$G(x) = \frac{1}{2} \left(f(x) - \frac{1}{c} \int_0^x g(v) \, dv \right).$$

Therefore,

$$F(x + ct) = \frac{1}{2} \left(f(x + ct) + \frac{1}{c} \int_0^{x+ct} g(v) \, dv \right)$$

and

$$G(x - ct) = \frac{1}{2} \left(f(x - ct) - \frac{1}{c} \int_0^{x-ct} g(v) \, dv \right),$$

so the solution is

$$u(x, t) = \frac{1}{2} \left(f(x + ct) + f(x - ct) \right) + \frac{1}{2c} \int_{x-ct}^{x+ct} g(v) \, dv. \qquad (10.16)$$

EXAMPLE 10.3.3: Solve $\begin{cases} u_{xx} = u_{tt}, \ -\infty < x < \infty, \ t > 0 \\ u(x, 0) = 2 \left(1 + x^2 \right)^{-1}, \ u_t(x, 0) = 0 \end{cases}.$

SOLUTION: Using equation (10.16) with $c = 1$, $f(x) = 2\left(1 + x^2\right)^{-1}$, and $g(x) = 0$, we have the solution

$$u(x, t) = \frac{1}{2}\left(f(x + ct) + f(x - ct)\right) = \frac{1}{2}\left[\frac{2}{1 + (x + t)^2} + \frac{2}{1 + (x - t)^2}\right]$$

$$= \frac{1}{1 + (x + t)^2} + \frac{1}{1 + (x - t)^2}.$$

We plot the solution for $t = 0$ to $t = 15$ to illustrate the motion of the string of infinite length in Figure 10-6.

```
Clear[u, x, t]

u[x_, t_] = 1/(1 + (x + t)^2) + 1/(1 + (x - t)^2);

somegraphs = Table[Plot[u[x, t], {x, -20, 20}, PlotRange → {0, 2},

    Ticks → {{-5, 5}, {0, 5}}], {t, 0, 15}];

toshow = Partition[somegraphs, 4];

Show[GraphicsGrid[toshow]]
```

Alternatively, you can use `Animate` to see the motion of the string.

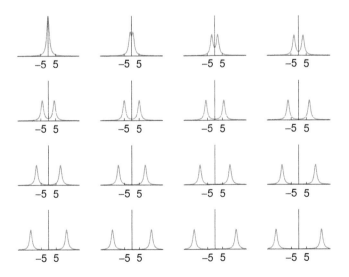

Figure 10-6   A traveling wave solution

```
Animate[Plot[Evaluate[u[x, t]], {x, -20, 20},

  PlotRange → {0, 2}], {{t, 0}, 0, 15}]
```

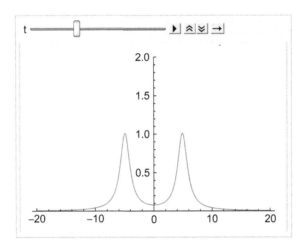

D'Alembert's solution is sometimes referred to as the **traveling wave solution** due to the behavior of its graph. The waves appear to move in opposite directions along the x-axis as t increases as we can see in the graphs.

■

10.4 Problems in Two Dimensions: Laplace's Equation

10.4.1 Laplace's Equation

Laplace's equation, often called the **potential equation**, is given by

$$\frac{\partial^2 u}{\partial x^2} + \frac{\partial^2 u}{\partial y^2} = 0, \tag{10.17}$$

in rectangular coordinates and is one of the most useful partial differential equations in that it arises in many fields of study. These include fluid flows as well as electrostatic and gravitational potential. Because the potential $u = u(x, y)$ does not depend on time, no initial condition is required, so we are faced with solving a pure boundary value problem when working with Laplace's equation. The boundary conditions can be stated in different forms. If the value of the solution is given around the boundary of the region, then the boundary value problem is called a **Dirichlet problem** whereas if the normal derivative of the solution is given around the boundary, the problem is known as a **Neumann problem**. We now investigate the solutions to Laplace's equation in a rectangular region by, first, stating the general form of the Dirichlet problem:

$$\begin{cases} \dfrac{\partial^2 u}{\partial x^2} + \dfrac{\partial^2 u}{\partial y^2} = 0, \ 0 < x < a, \ 0 < y < b \\ u(x, 0) = f_1(x), \ u(x, b) = f_2(x), \ 0 < x < a \\ u(0, y) = g_1(y), \ u(a, y) = g_2(y), \ 0 < y < b \end{cases} \tag{10.18}$$

This boundary-value problem is solved through separation of variables. We begin by considering the problem

$$\begin{cases} \dfrac{\partial^2 u}{\partial x^2} + \dfrac{\partial^2 u}{\partial y^2} = 0, \ 0 < x < a, \ 0 < y < b \\ u(x, 0) = 0, \ u(x, b) = f(x), \ 0 < x < a \\ u(0, y) = 0, \ u(a, y) = 0, \ 0 < y < b \end{cases} \tag{10.19}$$

In this case, we assume that

$$u(x, y) = X(x)Y(y)$$

so substitution into Laplace's equation (10.17) yields

$$X''Y + XY'' = 0$$

$$\frac{X''}{X} = -\frac{Y''}{Y} = -\lambda,$$

where $-\lambda$ is the separation constant. Therefore, we have the ordinary differential equations $X'' + \lambda X = 0$ and $Y'' - \lambda Y = 0$. Notice that the boundary conditions along the lines $x = 0$ and $x = a$ are homogeneous. In fact, because $u(0, y) = X(0)Y(y) = 0$ and $u(a, y) = X(a)Y(y) = 0$, we have $X(0) = 0$ and $X(a) = 0$. Therefore, we first solve the eigenvalue problem

$$\begin{cases} X'' + \lambda X = 0, \ 0 < x < a \\ X(0) = 0, \ X(a) = 0 \end{cases},$$

which was solved with $a = p$ in Section 9.1. There, we found the eigenvalues and See Example 9.1.4. corresponding eigenfunctions to be $\lambda_n = (n\pi/a)^2$ and $X_n(x) = \sin(n\pi x/a)$, $n = 1$, $2, \ldots$. We then solve the equation $Y'' - \lambda Y = 0$. From our experience with second-order equations, we know that $Y_n(y) = a_n e^{\lambda_n y} + b_n e^{-\lambda_n y}$, which can be written in terms of the hyperbolic trigonometric functions as

$$Y_n(y) = A_n \cosh \lambda_n y + B_n \sinh \lambda_n y = A_n \cosh \frac{n\pi y}{a} + B_n \sinh \frac{n\pi y}{a}.$$

Then, using the homogeneous boundary condition $u(x, 0) = X(x)Y(0) = 0$, which indicates that $Y(0) = 0$, we have

$$Y_n(0) = A_n \cosh 0 + B_n \sinh 0 = A_n = 0,$$

so $A_n = 0$ for all n. Therefore, $Y_n(y) = B_n \sinh \lambda_n y$, and a solution of equation (10.19) is

$$u_n(x, y) = B_n \sinh \frac{n\pi y}{a} \sin \frac{n\pi x}{a},$$

so by the Principle of Superposition,

$$u(x, y) = \sum_{n=1}^{\infty} B_n \sinh \frac{n\pi y}{a} \sin \frac{n\pi x}{a}$$

is also a solution, where the coefficients are determined with the boundary condition $u(x, b) = f(x)$. Substitution into the solution yields

$$u(x, b) = \sum_{n=1}^{\infty} B_n \sinh \frac{n\pi b}{a} \sin \frac{n\pi x}{a} = f(x),$$

where $B_n \sinh(n\pi b/a)$ represents the Fourier sine series coefficients given by

$$B_n \sinh \frac{n\pi b}{a} = \frac{2}{a} \int_0^a f(x) \sin \frac{n\pi x}{a} \, dx$$

$$B_n = \frac{2}{a \sinh \dfrac{n\pi b}{a}} \int_0^a f(x) \sin \frac{n\pi x}{a} \, dx. \qquad (10.20)$$

EXAMPLE 10.4.1: Solve
$$\begin{cases} u_{xx} + u_{yy} = 0, \ 0 < x < 1, \ 0 < y < 2 \\ u(x,0) = 0, \ u(x,2) = x(1-x), \ 0 < x < 1. \\ u(0,y) = 0, \ u(1,y) = 0, \ 0 < y < 2 \end{cases}$$

SOLUTION: In this case, $a = 1$, $b = 2$, and $f(x) = x(1-x)$. Therefore,

$$B_n = \frac{2}{\sinh 2n\pi} \int_0^1 x(1-x) \sin n\pi x \, dx = \frac{2}{\sinh 2n\pi} \left(-\frac{2}{n^3\pi^3} \cos n\pi + \frac{2}{n^3\pi^3} \right)$$

$$= \frac{4}{n^3\pi^3 \sinh 2n\pi} \left[1 - (-1)^n \right], \quad n = 1, 2, \ldots.$$

$$\frac{2\int_0^1 x(1-x)\text{Sin}[n\pi x]dx}{\text{Sinh}[2n\pi]}$$

$$\frac{2\text{Csch}[2n\pi](2 - 2\text{Cos}[n\pi] - n\pi\text{Sin}[n\pi])}{n^3\pi^3}$$

so the solution is

$$u(x,y) = \sum_{n=1}^{\infty} B_n \sinh n\pi y \sin n\pi x = \sum_{n=1}^{\infty} \frac{8}{(2n-1)^3\pi^3 \sinh 2(2n-1)\pi}$$

$$\sinh(2n-1)\pi y \sin(2n-1)\pi x.$$

We plot $u(x,y)$ using the first 15 terms of the series solution in Figure 10-7.

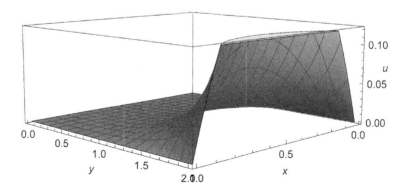

Figure 10-7   Approximating a solution of Laplace's equation

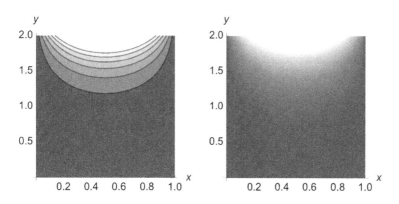

Figure 10-8   On the left, a contour plot and on the right a density plot

```
Clear[u]
```

$$u[x\_, y\_] = \sum_{n=1}^{15} \frac{8\operatorname{Sinh}[(2n-1)\pi y]\operatorname{Sin}[(2n-1)\pi x]}{(2n-1)^3\pi^3\operatorname{Sinh}[2(2n-1)\pi]};$$

```
Plot3D[u[x, y], {x, 0, 1}, {y, 0, 2}, ViewPoint → {2.365,
    2.365, 0.514}, AxesLabel → {x, y, u}, PlotPoints → 40]
```

Alternatively, we can generate a contour or density plot of $u(x, y)$ as shown in Figure 10-8.

```
cplot = ContourPlot[u[x, y], {x, 0, 1}, {y, 0, 2},
    Frame → False, Axes → Automatic, AxesLabel → {x, y},
    PlotPoints->30];
dplot = DensityPlot[u[x, y], {x, 0, 1}, {y, 0, 2},
    Frame → False, Axes → Automatic, AxesLabel → {x, y},
    PlotPoints->30];
Show[GraphicsRow[{cplot, dplot}]]
```

We notice that the value of $u(x, y)$ decreases to zero away from the boundary $y = 2$.

■

Any version of Laplace's equation on a rectangular region can be solved through separation of variables as long as we have a pair of homogeneous boundary conditions in the same variable.

EXAMPLE 10.4.2: Solve
$$\begin{cases} u_{xx} + u_{yy} = 0,\ 0 < x < \pi,\ 0 < y < 1 \\ u(x,0) = 0,\ u(x,1) = 0,\ 0 < x < \pi \\ u(0,y) = \sin 2\pi y,\ u(\pi,y) = 4,\ 0 < y < 1 \end{cases}.$$

SOLUTION: As we did in the previous problem, we assume that $u(x,y) = X(x)Y(y)$. Notice that this problem differs from the previous problem in that the homogeneous boundary conditions are in terms of the variable y. Hence, when we separate variables, we use a different constant of separation. This yields

$$X''Y + XY'' = 0$$
$$\frac{X''}{X} = -\frac{Y''}{Y} = \lambda,$$

so we have the ordinary differential equations $X'' - \lambda X = 0$ and $Y'' + \lambda Y = 0$. Therefore, with the homogeneous boundary conditions $u(x,0) = X(x)Y(0) = 0$ and $u(x,1) = X(x)Y(1) = 0$, we have $Y(0) = 0$ and $Y(1) = 0$. The eigenvalue problem

$$\begin{cases} Y'' + \lambda Y = 0,\ 0 < y < 1 \\ Y(0) = 0,\ Y(1) = 0 \end{cases}$$

has eigenvalues $\lambda_n = (n\pi/1)^2 = n^2\pi^2$, $n = 1, 2, \ldots$ and eigenfunctions $Y_n(y) = \sin n\pi y$, $n = 1, 2, \ldots$. We then solve the equation $X'' - \lambda_n X = 0$ obtaining $X_n(x) = a_n e^{n\pi x} + b_n e^{-n\pi x}$, which can be written in terms of the hyperbolic trigonometric functions as

$$X_n(x) = A_n \cosh n\pi x + B_n \sinh n\pi x.$$

Now, because the boundary conditions on the boundaries $x = 0$ and $x = \pi$ are nonhomogeneous, we use the Principle of Superposition to obtain the solution

$$u(x,t) = \sum_{n=1}^{\infty} (A_n \cosh n\pi x + B_n \sinh n\pi x) \sin n\pi y.$$

Therefore,

$$u(0, y) = \sum_{n=1}^{\infty} A_n \sin n\pi y = \sin 2\pi y,$$

so $A_2 = 1$ and $A_n = 0$ for $n \neq 2$. Similarly,

$$u(\pi, y) = A_2 \cosh 2\pi^2 + \sum_{n=1}^{\infty} B_n \sinh n\pi^2 \sin n\pi y = 4,$$

which indicates that

$$\sum_{n=1}^{\infty} B_n \sinh n\pi^2 \sin n\pi y = 4 - \cosh 2\pi^2.$$

Then, $B_n \sinh n\pi^2$ are the Fourier sine series coefficients for the constant function $4 - \cosh 2\pi^2$ which are given by

$$B_n \sinh n\pi^2 = \frac{2}{1} \int_0^1 \left(4 - \cosh 2\pi^2\right) \sin n\pi y \, dy = -2 \left(4 - \cosh 2\pi^2\right)$$

$$\left[\frac{1}{n\pi} \cos n\pi y\right]_0^1 = -\frac{2\left(4 - \cosh 2\pi^2\right)}{n\pi} \left[(-1)^n - 1\right], \quad n = 1, 2, \ldots$$

From this formula, we see that $B_n = 0$ if n is even. Therefore, we express these coefficients as

$$B_{2n-1} = \frac{4\left(4 - \cosh 2\pi^2\right)}{(2n - 1)\pi}, \quad n = 1, 2, \ldots$$

so that the solution is

$$u(x, y) = \cosh 2\pi x + \sum_{n=1}^{\infty} \frac{4\left(4 - \cosh 2\pi^2\right)}{(2n - 1)\pi} \sin(2n - 1)\pi y.$$

As in the first example, we generate contour and density plots of an approximation of u. See Figure 10-9.

```
Clear[u]

u[x_, y_] = Cosh[2πx] + ∑₍ₙ₌₁₎³⁰ (4(4 - Cosh[2π²])Sin[(2n - 1)πy])/((2n - 1)π);

p1 = ContourPlot[u[x, y], {x, 0, π}, {y, 0, 1},

    Frame → False, Axes → Automatic, AxesLabel → {x, y},
```

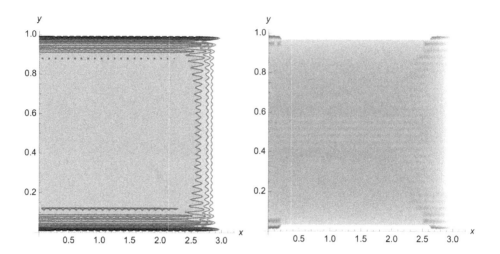

Figure 10-9   On the left, a contour plot and on the right a density plot

```
    PlotPoints → 30];

p2 = DensityPlot[u[x, y], {x, 0, π}, {y, 0, 1},

    Frame → False, Axes → Automatic, AxesLabel → {x, y},

    PlotPoints → 30];

Show[GraphicsRow[{p1, p2}]]
```

■

10.5 Two-Dimensional Problems in a Circular Region

In some situations, the region on which we solve a boundary-value problem or an initial boundary-value problem is not rectangular in shape. For example, we usually do not have rectangular shaped drumheads and the heating elements on top of the stove are not square. Instead, these objects are typically circular in shape, so we find the use of polar coordinates convenient. In this section, we discuss problems of this type by presenting two important problems solved on circular regions, Laplace's equation which is related to the steady-state temperature and the wave equation which is used to find the displacement of a drumhead.

10.5.1 Laplace's Equation in a Circular Region

In calculus, we found that polar coordinates are useful in solving many problems. The same can be said for solving boundary value problems in a circular region. With the change of variables

$$\begin{cases} x = r\cos\theta \\ y = r\sin\theta \end{cases}$$

we transform Laplace's equation in rectangular coordinates, $u_{xx} + u_{yy} = 0$, to polar coordinates

$$\frac{\partial^2 u}{\partial r^2} + \frac{1}{r}\frac{\partial u}{\partial r} + \frac{1}{r^2}\frac{\partial^u}{\partial\theta^2} = 0,\ 0 < r < \rho,\ -\pi < \theta < \pi, \tag{10.21}$$

where ρ is the radius of the drumhead. Recall that for the solution of Laplace's equation in a rectangular region, we had to specify a boundary condition on each of the four boundaries of the rectangle. However, in the case of a circle, there are not four sides, so we must alter the boundary conditions. Because in polar coordinates the points (r, π) and $(r, -\pi)$ are equivalent for the same value of r, we want our solution and its derivative with respect to θ to match at these points (so that the solution is smooth). Therefore, two of the boundary conditions are

$$u(r, -\pi) = u(r, \pi) \quad \text{and} \quad \frac{\partial u}{\partial\theta}(r, -\pi) = \frac{\partial u}{\partial\theta}(r, \pi) \tag{10.22}$$

for $0 < r < \rho$. Also, we want our solution to be bounded at $r = 0$, so another boundary condition is $|u(0, \theta)| < \infty$ for $-\pi < \theta < \pi$. Finally, we specify the value of the solution around the boundary of the circle $r = \rho$ to be $u(\rho, \theta) = f(\theta)$ for $-\pi < \theta < \pi$. Therefore, we solve the following boundary-value problem to solve Laplace's equation (the Dirichlet problem) in a circular region of radius ρ:

$$\begin{cases} \dfrac{\partial^2 u}{\partial r^2} + \dfrac{1}{r}\dfrac{\partial u}{\partial r} + \dfrac{1}{r^2}\dfrac{\partial^u}{\partial\theta^2} = 0,\ 0 < r < \rho,\ -\pi < \theta < \pi \\ u(r, -\pi) = u(r, \pi),\ \dfrac{\partial u}{\partial\theta}(r, -\pi) = \dfrac{\partial u}{\partial\theta}(r, \pi),\ 0 < r < \rho \\ |u(0, \theta)| < \infty,\ u(\rho, \theta) = f(\theta),\ -\pi < \theta < \pi \end{cases} \tag{10.23}$$

Using separation of variables, we assume that $u(r, \theta) = R(r)H(\theta)$. Substitution into Laplace's equation yields

$$R''H + \frac{1}{r}R'H + RH'' = 0$$

$$R''H + \frac{1}{r}R'H = -RH''$$

$$\frac{rR'' + R'}{rR} = -\frac{H''}{H} = \lambda.$$

Therefore, we have the ordinary differential equations

$$H'' + \lambda H = 0 \quad \text{and} \quad r^2 R'' + rR' - \lambda R = 0.$$

Notice that the boundary conditions given by equation (10.22) imply that

$$R(r)H(-\pi) = R(r)H(\pi) \quad \text{and} \quad R(r)H'(-\pi) = R(r)H'(\pi),$$

so that

$$H(-\pi) = H(\pi) \quad \text{and} \quad H'(-\pi) = H'(\pi).$$

This means that we begin by solving the eigenvalue problem

$$\begin{cases} H'' + \lambda H = 0, -\pi < \theta < \pi \\ H(-\pi) = H(\pi), H'(-\pi) = H'(\pi) \end{cases}$$

The eigenvalues and corresponding eigenfunctions of this problem are

$$\lambda_n = \begin{cases} 0, n = 0 \\ n^2, n = 1, 2, \ldots \end{cases} \quad \text{and} \quad H_n(\theta) = \begin{cases} 1, n = 0 \\ a_n \cos n\theta + b_n \sin n\theta, n = 1, 2, \ldots \end{cases}$$

Because $r^2 R'' + rR' - \lambda_n^2 R = 0$ is a Cauchy-Euler equation, we assume that $R(r) = r^m$:

$$m(m - 1)r^2 r^{m-2} + mr r^{m-1} - \lambda_n r^m = 0$$

$$r^m [m(m - 1) + m - \lambda_n] = 0.$$

Therefore,

$$m^2 - \lambda_n^2 = 0 \quad \text{so} \quad m = \pm\lambda_n$$

If $\lambda_0 = 0$, then $R_0(r) = A_0 + B_0 \ln r$. However, because we require that the solution be bounded near $r = 0$ and $\lim_{r \to 0^+} \ln r = -\infty$, we must choose $B_0 = 0$. Therefore,

$R_0(r) = A_0$. On the other hand, if $\lambda_n = n^2$, $n = 1, 2, \ldots$, then $R_n(r) = A_n r^n + B_n r^{-n}$. Similarly, because $\lim_{r \to 0^+} r^{-n} = \infty$, we must let $B_n = 0$, so $R_n(r) = A_n r^n$. By the Principle of Superposition, we have the solution

$$u(r, \theta) = A_0 + \sum_{n=1}^{\infty} r^n \left(A_n \cos n\theta + B_n \sin n\theta\right).$$

We find the coefficients by applying the boundary condition $u(\rho, \theta) = f(\theta)$. This yields

$$u(\rho, \theta) = A_0 + \sum_{n=1}^{\infty} \rho^n \left(A_n \cos n\theta + B_n \sin n\theta\right) = f(\theta).$$

so A_0, A_n, and B_n are related to the Fourier series coefficients in the following way:

$$
\begin{aligned}
A_0 &= \frac{1}{2\pi} \int_{-\pi}^{\pi} f(\theta)\, d\theta, \\
A_n &= \frac{1}{\pi \rho^n} \int_{-\pi}^{\pi} f(\theta) \cos n\theta \, d\theta, \quad n = 1, 2, \ldots, \text{and} \\
B_n &= \frac{1}{\pi \rho^n} \int_{-\pi}^{\pi} f(\theta) \sin n\theta \, d\theta, \quad n = 1, 2, \ldots.
\end{aligned}
\tag{10.24}
$$

EXAMPLE 10.5.1: Solve
$$
\begin{cases}
\dfrac{\partial^2 u}{\partial r^2} + \dfrac{1}{r}\dfrac{\partial u}{\partial r} + \dfrac{1}{r^2}\dfrac{\partial^u}{\partial \theta^2} = 0,\ 0 < r < 2,\ -\pi < \theta < \pi \\[2mm]
u(r, -\pi) = u(r, \pi),\ \dfrac{\partial u}{\partial \theta}(r, -\pi) = \dfrac{\partial u}{\partial \theta}(r, \pi),\ 0 < r < 2 \\[2mm]
|u(0, \theta)| < \infty,\ u(2, \theta) = |\theta|,\ -\pi < \theta < \pi
\end{cases}
$$

SOLUTION: Notice that $f(\theta) = |\theta|$ is an even function on $-\pi < \theta < \pi$. Therefore, $B_n = 0$ for $n = 1, 2, \ldots$, A_0 is given by

$$A_0 = \frac{1}{2\pi} \int_{-\pi}^{\pi} |\theta|\, d\theta = \frac{1}{\pi} \int_0^{\pi} \theta \, d\theta = \frac{\pi}{2},$$

$$\frac{\int_0^{\pi} \theta \, d\theta}{\pi}$$

$$\frac{\pi}{2}$$

and A_n is given by

$$A_n = \frac{1}{2^n\pi}\int_{\pi}^{\pi}|\theta|\cos n\theta\,d\theta = \frac{1}{2^{n-1}\pi}\int_0^{\pi}\theta\cos n\theta\,d\theta$$

$$= \frac{1}{2^n n^2\pi}(\cos n\pi - 1) = \frac{1}{2^n n^2\pi}\left[(-1)^n - 1\right], \quad n = 1, 2, \ldots$$

$$\mathtt{a_{n\_}} = \frac{\int_0^{\pi}\theta\mathtt{Cos[n\theta]d\theta}}{\pi\,2^{n-1}}\,\mathtt{//Simplify}$$

$$\frac{2^{1-n}(-1 + \mathtt{Cos[n\pi]} + n\pi\,\mathtt{Sin[n\pi]})}{n^2\pi}$$

Notice that $A_{2n} = 0$, $n = 1, 2, \ldots$, while $A_{2n-1} = \dfrac{-2}{2^{2n-1}(2n-1)^2\pi}$, $n = 1, 2, \ldots$, so the solution is

$$u(r,\theta) = \frac{\pi}{2} - \sum_{n=1}^{\infty}\frac{1}{2^{2n-2}(2n-1)^2\pi}r^{2n-1}\cos(2n-1)\theta.$$

In the same manner as in previous examples, we graph an approximation of this solution. See Figure 10-10.

```
Clear[u, r, theta]
```

$$\mathtt{u[r\_, \theta\_]} = \frac{\pi}{2} - \sum_{n=1}^{20}\frac{r^{2n-1}\mathtt{Cos[(2n-1)\theta]}}{\pi\,2^{2n-2}(2n-1)^2};$$

```
ParametricPlot3D[{rCos[θ], rSin[θ], u[r, θ]},

  {r, 0, 2}, {θ, -π, π}, Boxed → False, PlotPoints → 35]
```

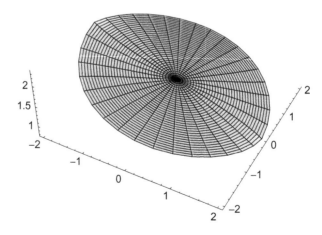

Figure 10-10   Plot of a solution to Laplace's equation in polar coordinates

■

10.5.2 The Wave Equation in a Circular Region

One of the more interesting problems involving two spatial dimensions (x and y) is the wave equation,

$$c^2 \left(\frac{\partial^2 u}{\partial x^2} + \frac{\partial^2 u}{\partial y^2} \right) = \frac{\partial^2 u}{\partial t^2}. \tag{10.25}$$

This is due to the fact that the solution to this problem represents something with which we are all familiar, the displacement of a drumhead. Because drumheads are circular in shape, we investigate the solution of the wave equation in a circular region. Therefore, we transform the wave equation into polar coordinates. Previously, we saw that converting Laplace's equation from rectangular coordinates (x, y) to polar coordinates (r, θ) results in the equation

$$\frac{\partial^2 u}{\partial x^2} + \frac{\partial^2 u}{\partial y^2} = \frac{\partial^2 u}{\partial r^2} + \frac{1}{r} \frac{\partial u}{\partial r} + \frac{1}{r^2} \frac{\partial^u}{\partial \theta^2}$$

so it follows that the wave equation in polar coordinates becomes

$$c^2 \left(\frac{\partial^2 u}{\partial r^2} + \frac{1}{r} \frac{\partial u}{\partial r} + \frac{1}{r^2} \frac{\partial^u}{\partial \theta^2} \right) = \frac{\partial^2 u}{\partial t^2}. \tag{10.26}$$

If we assume that the displacement of the drumhead from the xy-plane at time t is the same at equal distances from the origin, we say that the solution $u = u(r, \theta)$ is **radially symmetric**. Therefore, $\partial^2 u / \partial \theta^2 = 0$, so the wave equation can be expressed in terms of r and t as

If the solution is radially symmetric, the value of u does not depend on the angle θ.

$$c^2 \left(\frac{\partial^2 u}{\partial r^2} + \frac{1}{r} \frac{\partial u}{\partial r} \right) = \frac{\partial^2 u}{\partial t^2}. \tag{10.27}$$

Of course, to find $u = u(r, t)$ we need the appropriate boundary and initial conditions. Because the circular boundary of the drumhead $r = \rho$ must be fixed so that it does not move we say that $u(\rho, t) = 0$ for $t > 0$. Then, as we had in Laplace's equation on a circular region, we require that the solution u be bounded near the origin, so we have the condition $|u(0, t)| < \infty$ for $t > 0$. The initial position and initial velocity functions are given as functions of r as

$$u(r, 0) = f(r) \quad \text{and} \quad \frac{\partial u}{\partial t}(r, 0) = g(r)$$

for $0 < r < \rho$. Therefore, the initial-boundary value problem to find the displacement $u = u(r, t)$ of a circular drumhead (of radius ρ) is given by

$$\begin{cases} c^2 \left(\dfrac{\partial^2 u}{\partial r^2} + \dfrac{1}{r} \dfrac{\partial u}{\partial r} \right) = \dfrac{\partial^2 u}{\partial t^2}, \ 0 < r < \rho, t > 0 \\[2mm] u(\rho, t) = 0, \ |u(0, t)| < \infty, \ t > 0 \\[2mm] u(r, 0) = f(r), \ \dfrac{\partial u}{\partial t}(r, 0) = g(r), \ 0 < r < \rho \end{cases} \qquad (10.28)$$

As with other problems, we are able to use separation of variables to find $u = u(r, t)$ by assuming that $u(r, t) = R(r)T(t)$. Substitution into the wave equation yields

$$c^2 \left(R''T + \frac{1}{r} R'T \right) = RT''$$

$$\frac{rR'' + R'}{rR} = \frac{T''}{c^2 T} = -k^2,$$

where $-k^2$ is the separation constant. Separating the variables, we have the ordinary differential equations

$$r^2 R'' + rR' + k^2 r^2 R = 0 \quad \text{and} \quad T'' + c^2 k^2 T = 0.$$

We recognize the equation $r^2 R'' + rR' + k^2 r^2 R = 0$ as Bessel's equation of order zero that has solution

$$R(r) = c_1 J_0(kr) + c_2 Y_0(kr),$$

where J_0 and Y_0 are the Bessel functions of order zero of the first and second kind, respectively. In terms of R, we express the boundary condition $|u(0, t)| < \infty$ as $|R(0)| < \infty$. Therefore, because $\lim_{r \to 0^+} Y_0(kr) = -\infty$, we must choose $c_2 = 0$. Applying the other boundary condition, $R(\rho) = 0$, we have

$$R(\rho) = c_1 J_0(k\rho) = 0,$$

so to avoid the trivial solution with $c_1 = 0$, we have $k\rho = \alpha_n$, where α_n is the nth zero of $J_0(x)$. Because k depends on n, we write

$$k_n = \frac{\alpha_n}{\rho}.$$

The solution of $T'' + c^2 k_n^2 T = 0$ is

$$T_n(t) = A_n \cos c k_n t + B_n \sin c k_n t,$$

so with the Principle of Superposition, we form the solution

$$u(r, t) = \sum_{n=1}^{\infty} (A_n \cos ck_n t + B_n \sin ck_n t) J_0 (k_n r),$$

where the coefficients A_n and B_n are found through application of the initial position and velocity functions. With

$$u(r, 0) = \sum_{n=1}^{\infty} A_n J_0 (k_n r) = f(r)$$

and the orthogonality conditions of the Bessel functions, we find that

$$A_n = \frac{\int_0^\rho rf(r) J_0 (k_n r) \, dr}{\int_0^\rho r [J_0 (k_n r)]^2 \, dr} = \frac{2}{[J_1 (\alpha_n)]^2} \int_0^\rho rf(r) J_0 (k_n r) \, dr, \quad n = 1, 2, \ldots$$

Similarly, because

$$\frac{\partial u}{\partial t}(r, 0) = \sum_{n=1}^{\infty} (-ck_n A_n \sin ck_n t + ck_n B_n \cos ck_n t) J_0 (k_n r)$$

we have

$$u_t(r, 0) = \sum_{n=1}^{\infty} ck_n B_n J_0 (k_n r) = g(r).$$

Therefore,

$$B_n = \frac{\int_0^\rho rg(r) J_0 (k_n r) \, dr}{ck_n \int_0^\rho r [J_0 (k_n r)]^2 \, dr} = \frac{2}{ck_n [J_1 (\alpha_n)]^2} \int_0^\rho rg(r) J_0 (k_n r) \, dr, \quad n = 1, 2, \ldots$$

As a practical matter, in nearly all cases, these formulas are difficult to evaluate.

EXAMPLE 10.5.2: Solve $\begin{cases} \dfrac{\partial^2 u}{\partial r^2} + \dfrac{1}{r} \dfrac{\partial u}{\partial r} = \dfrac{\partial^2 u}{\partial t^2}, \, 0 < r < 1, t > 0 \\ u(1, t) = 0, \, |u(0, t)| < \infty, \, t > 0 \cdot \\ u(r, 0) = r(r - 1), \, \dfrac{\partial u}{\partial t}(r, 0) = \sin \pi r, \, 0 < r < 1 \end{cases}$

SOLUTION: In this case, $\rho = 1$, $f(r) = r(r-1)$, and $g(r) = \sin \pi r$. To calculate the coefficients, we will need to have approximations of the zeros of the Bessel functions. Using `BesselJZero` we define α_n to be the nth zero of $y = J_0(x)$.

$$\alpha_{\texttt{n\_}} := \alpha_n = \texttt{BesselJZero[0, \{n, n\}][[1]]//N}$$

Next, we define the constants ρ and c and the functions $f(r) = r(r-1)$, $g(r) = \sin \pi r$, and $k_n = \alpha_n / \rho$.

```
c = 1;

ρ = 1;

f[r_] = r(r - 1);

g[r_] = Sin[π r];
```

$$k_{\texttt{n\_}} := k_n = \frac{\alpha_n}{\rho};$$

The formulas for the coefficients A_n and B_n are then defined so that an approximate solution may be determined. (We use lower case letters to avoid any possible ambiguity with built-in Mathematica functions.) Note that we use `NIntegrate` to approximate the coefficients and avoid the difficulties in integration associated with the presence of the Bessel function of order zero.

$$a_{\texttt{n\_}} := a_n = \frac{\texttt{2NIntegrate[rf[r]BesselJ[0, } k_n \texttt{r], \{r, 0, } \rho \texttt{\}]}}{\texttt{BesselJ[1, } \alpha_n \texttt{]}^2};$$

$$b_{\texttt{n\_}} := b_n = \frac{\texttt{2NIntegrate[rg[r]BesselJ[0, } k_n \texttt{r], \{r, 0, } \rho \texttt{\}]}}{c k_n \texttt{BesselJ[1, } \alpha_n \texttt{]}^2}$$

We now compute the first ten values of A_n and B_n. Because a and b are defined using the form $\texttt{a}_{\texttt{n\_}} := \texttt{a}_{\texttt{n}} = \dots$ and $\texttt{b}_{\texttt{n\_}} := \texttt{b}_{\texttt{n}} = \dots$, Mathematica remembers these values for later use.

```
Table[{n, aₙ, bₙ}, {n, 1, 10}]//TableForm
```

1	−0.323503	0.52118
2	0.208466	−0.145776
3	0.00763767	−0.0134216
4	0.0383536	−0.00832269
5	0.00534454	−0.00250503
6	0.0150378	−0.00208315
7	0.00334937	−0.000882012

8	0.00786698	−0.000814719
9	0.00225748	−0.000410202
10	0.00479521	−0.000399219

The nth term of the series solution is defined in u. Then, an approximate solution is obtained in uapprox by summing the first ten terms of u.

$$u[n\_, r\_, t\_] := (a_n \text{Cos}[ck_n t] + b_n \text{Sin}[ck_n t]) \text{BesselJ}[0, k_n r];$$

$$\text{uapprox}[r\_, t\_] = \sum_{n=1}^{10} u[n, r, t];$$

We graph uapprox for several values of t in Figure 10-11.

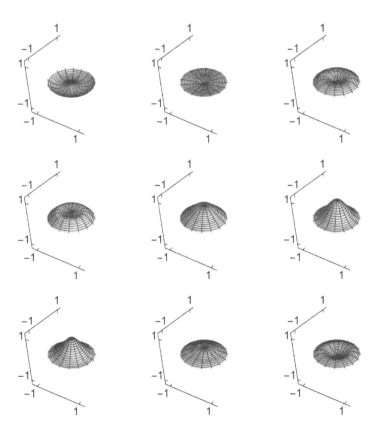

Figure 10-11   The drumhead for 9 equally spaced values of t between 0 and 1.5

```
somegraphs = Table[ParametricPlot3D[{rCos[θ], rSin[θ],

   uapprox[r, t]}, {r, 0, 1}, {θ, -π, π}, Boxed → False,

   PlotRange → {-1.25, 1.25}, BoxRatios → {1, 1, 1},

   Ticks → {{-1, 1}, {-1, 1}, {-1, 1}}], {t, 0, 1.5, 1.5/8}];

toshow = Partition[somegraphs, 3];

Show[GraphicsGrid[toshow]]
```

In order to actually watch the drumhead move, we use Animate.
We show one frame from the animation that results from the following.

```
Animate[ParametricPlot3D[{rCos[θ], rSin[θ], uapprox[r, t]},

   {r, 0, 1}, {θ, -π, π}, Boxed → False, PlotRange → {-1.25, 1.25},

   BoxRatios → {1, 1, 1}, Ticks → {{-1, 1}, {-1, 1}, {-1, 1}}],

   {{t, 0}, 0, 2}]
```

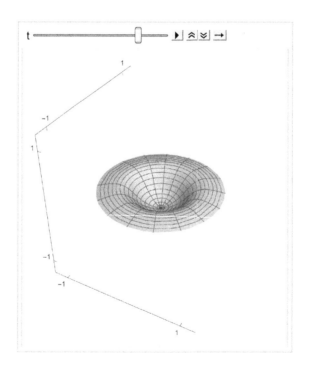

∎

The problem that depends on the angle θ is more complicated to solve. Due to the presence of $\partial^2 u/\partial\theta^2$ we must include two more boundary conditions in order to solve the initial boundary-value problem. So that the solution is a smooth function, we require the "artificial" boundary conditions

$$u(r,\pi,t) = u(r,-\pi,t) \quad \text{and} \quad \frac{\partial u}{\partial\theta}u(r,\pi,t) = \frac{\partial u}{\partial\theta}u(r,-\pi,t)$$

for $0 < r < \rho$ and $t > 0$. Therefore, we solve the problem

$$\begin{cases} c^2\left(\dfrac{\partial^2 u}{\partial r^2} + \dfrac{1}{r}\dfrac{\partial u}{\partial r} + \dfrac{1}{r^2}\dfrac{\partial^2 u}{\partial\theta^2}\right) = \dfrac{\partial^2 u}{\partial t^2}, \ 0 < r < \rho, \ -\pi < \theta < \pi, t > 0 \\[2mm] u(\rho,\theta,t) = 0, \ |u(0,\theta,t)| < \infty, \ -\pi \le \theta \le \pi, t > 0 \\[2mm] u(r,\pi,t) = u(r,-\pi,t), \ \dfrac{\partial u}{\partial\theta}(r,\pi,t) = \dfrac{\partial u}{\partial\theta}(r,-\pi,t), \ 0 < r < \rho, t > 0 \\[2mm] u(r,\theta,0) = f(r,\theta), \ \dfrac{\partial u}{\partial t}(r,\pi,0) = g(r,\theta), \ 0 < r < \rho, \ -\pi < \theta < \pi \end{cases} \quad (10.29)$$

Using separation of variables and assuming that $u(r,\theta,t) = R(t)H(\theta)T(t)$, we obtain that a general solution is given by

$$u(r,\theta,t) = \sum_n a_{0n}J_0\left(\lambda_{0n}r\right)\cos\left(\lambda_{0n}ct\right) + \sum_{m,n} a_{mn}J_m\left(\lambda_{mn}r\right)\cos\left(m\theta\right)\cos\left(\lambda_{mn}ct\right)$$

$$+ \sum_{m,n} b_{mn}J_m\left(\lambda_{mn}r\right)\sin\left(m\theta\right)\cos\left(\lambda_{mn}ct\right) + \sum_n A_{0n}J_0\left(\lambda_{0n}r\right)\sin\left(\lambda_{0n}ct\right)$$

$$+ \sum_{m,n} A_{mn}J_m\left(\lambda_{mn}r\right)\cos\left(m\theta\right)\sin\left(\lambda_{mn}ct\right)$$

$$+ \sum_{m,n} B_{mn}J_m\left(\lambda_{mn}r\right)\sin\left(m\theta\right)\sin\left(\lambda_{mn}ct\right),$$

where J_m represents the mth Bessel function of the first kind, α_{mn} denotes the nth zero of the Bessel function $y = J_m(x)$, and $\lambda_{mn} = \alpha_{mn}/\rho$. The coefficients are given by the following formulas.

$$a_{0n} = \frac{\int_0^{2\pi}\int_0^{\rho} f(r,\theta)J_0\left(\lambda_{0n}r\right)r\,dr\,d\theta}{2\pi\int_0^{\rho}\left[J_0\left(\lambda_{0n}r\right)\right]^2 r\,dr}$$

$$a_{mn} = \frac{\int_0^{2\pi}\int_0^{\rho} f(r,\theta)J_m\left(\lambda_{mn}r\right)\cos\left(m\theta\right)r\,dr\,d\theta}{\pi\int_0^{\rho}\left[J_m\left(\lambda_{mn}r\right)\right]^2 r\,dr}$$

$$b_{mn} = \frac{\int_0^{2\pi}\int_0^{\rho} f(r,\theta)J_m\left(\lambda_{mn}r\right)\sin\left(m\theta\right)r\,dr\,d\theta}{\pi\int_0^{\rho}\left[J_m\left(\lambda_{mn}r\right)\right]^2 r\,dr}$$

$$A_{0n} = \frac{\int_0^{2\pi}\int_0^{\rho} g(r,\theta)J_0\left(\lambda_{0n}r\right)r\,dr\,d\theta}{2\pi\lambda_{0n}c\pi\int_0^{\rho}\left[J_0\left(\lambda_{0n}r\right)\right]^2 r\,dr}$$

$$A_{mn} = \frac{\int_0^{2\pi}\int_0^{\rho} g(r,\theta)J_m\left(\lambda_{mn}r\right)\cos\left(m\theta\right)r\,dr\,d\theta}{\pi\lambda_{mn}c\int_0^{\rho}\left[J_m\left(\lambda_{mn}r\right)\right]^2 r\,dr}$$

$$B_{mn} = \frac{\int_0^{2\pi}\int_0^{\rho} g(r,\theta)J_m\left(\lambda_{mn}r\right)\sin\left(m\theta\right)r\,dr\,d\theta}{\pi\lambda_{mn}c\int_0^{\rho}\left[J_m\left(\lambda_{mn}r\right)\right]^2 r\,dr}.$$

EXAMPLE 10.5.3: Solve

$$\begin{cases} 10^2\left(\dfrac{\partial^2 u}{\partial r^2}+\dfrac{1}{r}\dfrac{\partial u}{\partial r}+\dfrac{1}{r^2}\dfrac{\partial^2 u}{\partial \theta^2}\right)=\dfrac{\partial^2 u}{\partial t^2},\ 0<r<1,\ -\pi<\theta<\pi,\ t>0 \\[2mm] u(1,\theta,t)=0,\ |u(0,\theta,t)|<\infty,\ -\pi\le\theta\le\pi,\ t>0 \\[2mm] u(r,\pi,t)=u(r,-\pi,t),\ \dfrac{\partial u}{\partial\theta}(r,\pi,t)=\dfrac{\partial u}{\partial\theta}(r,-\pi,t),\ 0<r<1,\ t>0 \\[2mm] u(r,\theta,0)=\cos(\pi r/2)\sin\theta,\ \dfrac{\partial u}{\partial t}(r,\pi,0)=(r-1)\cos(\pi\theta/2), \\[2mm] 0<r<1,\ -\pi<\theta<\pi \end{cases}$$

SOLUTION: To calculate the coefficients, we will need to have approximations of the zeros of the Bessel functions. Using `BesselJZero`, we define α_{mn} to be the nth zero of $y=J_m(x)$. We illustrate the use of α_{mn} by using it to compute the first five zeros of $y=J_0(x)$.

$\alpha_{\mathrm{m\_,n\_}}\mathtt{:=}\alpha_{m,n}=\mathtt{BesselJZero[m,\{n,n\}][[1]]//N}$

`Table[`$\alpha_{0,n}$`, {n, 1, 5}]`

`{2.40483, 5.52008, 8.65373, 11.7915, 14.9309}`

The appropriate parameter values as well as the initial condition functions are defined as follows. Notice that the functions describing the initial displacement and velocity are defined as the product of functions. This enables the subsequent calculations to be carried out using `NIntegrate`.

```
Clear[a, f, f1, f2, g1, g2, A, c, g, capa, capb, b]
c = 10;
ρ = 1;
```
$\mathtt{f1[r\_]}=\mathtt{Cos}[\dfrac{\pi\,r}{2}]\mathtt{;}$

$\mathtt{f2[\theta\_]}=\mathtt{Sin}[\theta]\mathtt{;}$

$\mathtt{f[r\_,\theta\_]:=f[r,\theta]}=\mathtt{f1[r]f2[\theta]}\mathtt{;}$

$\mathtt{g1[r\_]}=r-1\mathtt{;}$

$\mathtt{g2[\theta\_]}=\mathtt{Cos}[\dfrac{\pi\,\theta}{2}]\mathtt{;}$

$\mathtt{g[r\_,\theta\_]:=g[r,\theta]}=\mathtt{g1[r]g2[\theta]}\mathtt{;}$

The coefficients a_{0n} are determined with the function a.

```
Clear[a]
```

```
a[n_]:=

  a[n] = N[.(NIntegrate[f1[r]BesselJ[0, α₀,ₙr]r, {r, 0, ρ}]

  NIntegrate[f2[t], {t, 0, 2π}])/.(2πNIntegrate[rBesselJ

  [0, α₀,ₙr]², {r, 0, ρ}])];
```

Hence, as represents a table of the first five values of a_{0n}. Chop is used to round off very small numbers to zero.

```
as = Table[a[n]//Chop, {n, 1, 5}]
```

```
{0, 0, 0, 0, 0}
```

Because the denominator of each integral formula used to find a_{mn} and b_{mn} is the same, the function bjmn which computes this value is defined next. A table of nine values of this coefficient is then determined.

```
bjmn[m_, n_]:=bjmn[m, n] = N[NIntegrate[rBesselJ[m, αₘ,ₙr]²,

{r, 0, ρ}]]Table[Chop[bjmn[m, n]], {m, 1, 3}, {n, 1, 3}]
```

```
{{0.0811076, 0.0450347, 0.0311763}, {0.0576874, 0.0368243,

0.0270149}, {0.0444835, 0.0311044, 0.0238229}}
```

We also note that in evaluating the numerators of a_{mn} and b_{mn} we must compute $\int_0^\rho rf_1(r)J_m(\alpha_{mn}r)\,dr$. This integral is defined in fbjmn and the corresponding values are found for $n = 1, 2, 3$ and $m = 1, 2, 3$.

```
Clear[fbjmn]

fbjmn[m_, n_]:=fbjmn[m, n] = N[NIntegrate[f1[r]BesselJ

[m, αₘ,ₙr]r, {r, 0, ρ}]]Table[Chop[fbjmn[m, n]], {m, 1, 3}, {n, 1, 3}]
```

```
{{0.103574, 0.020514, 0.0103984}, {0.0790948, 0.0275564,

0.0150381}, {0.0628926, 0.0290764, 0.0171999}}
```

The formula to compute a_{mn} is then defined and uses the information calculated in fbjmn and bjmn. As in the previous calculation, the coefficient values for $n = 1, 2, 3$ and $m = 1, 2, 3$ are determined.

```
a[m_, n_]:=a[m, n]
    fbjmn[m, n]NIntegrate[f2[t]Cos[mt], {t, 0, 2π}]
  = N[―――――――――――――――――――――――――――――――――――――];
                    πbjmn[m, n]
Table[Chop[a[m, n]], {m, 1, 3}, {n, 1, 3}]
```

```
{{0, 0, 0}, {0, 0, 0}, {0, 0, 0}}
```

A similar formula is then defined for the computation of b_{mn}.

```
b[m_, n_]:=b[m, n]
= N[(fbjmn[m, n]NIntegrate[f2[t]Sin[mt], {t, 0, 2π}])/(πbjmn[m, n])];
Table[Chop[b[m, n]], {m, 1, 3}, {n, 1, 3}]
```

$\{\{1.277, 0.455514, 0.333537\}, \{0, 0, 0\}, \{0, 0, 0\}\}$

Note that defining the coefficients in this manner a [m_, n_] := a [m, n] = ... and b [m_, n_] :=b [m, n] = ... so that Mathematica "remembers" previously computed values reduces computation time. The values of A_{0n} are found similar to those of a_{0n}. After defining the function capa to calculate these coefficients, a table of values is then found.

```
capa[n_]:=

capa[n] =

    N[(NIntegrate[g1[r]BesselJ[0, α_{0,n}r]r, {r, 0, ρ}]

    NIntegrate[g2[t], {t, 0, 2π}])/.(2π cα_{0,n}NIntegrate[rBesselJ

    [0, α_{0,n}r]^2, {r, 0, ρ}])]; Table[Chop[capa[n]], {n, 1, 6}]
```

$\{0.00142231, 0.0000542518, 0.0000267596,$

$6.419764234815093^*{}^-6, 4.9584284641187775^*{}^-6,$

$1.8858472721004265^*{}^-6\}$

The value of the integral of the component of g, g1, which depends on r and the appropriate Bessel functions, is defined as gbjmn.

```
gbjmn[m_, n_]:=gbjmn[m, n] = NIntegrate[g1[r]*

    BesselJ[m, α_{m,n}r]r, {r, 0, ρ}]//N

Table[gbjmn[m, n]//Chop, {m, 1, 3}, {n, 1, 3}]
```

$\{\{-0.0743906, -0.019491, -0.00989293\}, \{-0.0554379, -0.022$

$7976, -0.013039\}, \{-0.0433614, -0.0226777, -0.0141684\}\}$

Then, A_{mn} is found by taking the product of integrals, gbjmn depending on r and one depending on θ. A table of coefficient values is generated in this case as well.

```
capa[m_, n_]:=capa[m, n]
= N[(gbjmn[m, n]NIntegrate[g2[t]Cos[mt], {t, 0, 2π}])/(πα_{m,n}cbjmn[m, n])];
Table[Chop[capa[m, n]], {m, 1, 3}, {n, 1, 3}]
```

```
{{0.0035096, 0.000904517, 0.000457326}, {−0.00262692,

−0.00103252, −0.000583116}, {−0.000503187, −0.000246002,

− 0.000150499}}
```

Similarly, the B_{mn} are determined.

```
capb[m_, n_]:=capb[m, n]
```
$$= N[\frac{\text{gbjmn}[m, n]\text{NIntegrate}[\text{g2}[t]\text{Sin}[mt], \{t, 0, 2\pi\}]}{\pi\alpha_{m,n}\text{cbjmn}[m, n]}];$$
```
Table[Chop[capb[m, n]], {m, 1, 3}, {n, 1, 3}]
```

```
{{0.00987945, 0.00254619, 0.00128736}, {−0.0147894,

− 0.00581305, −0.00328291}, {−0.00424938, −0.00207747,

− 0.00127095}}
```

Now that the necessary coefficients have been found, we construct an approximate solution to the wave equation by using our results. In the following, `term1` represents those terms of the expansion involving a_{0n}, `term2` those terms involving a_{mn}, `term3` those involving b_{mn}, `term4` those involving A_{0n}, `term5` those involving A_{mn}, and `term6` those involving B_{mn}. Therefore, our approximate solution is given as the sum of these terms as computed in u.

```
Clear[term1, term2, term3, term4, term5, term6]

term1[r_, t_, n_]:=a[n]BesselJ[0, α0,nr]Cos[α0,nct];

term2[r_, t_, θ_, m_, n_]:=a[m, n]BesselJ[m, αm,nr]Cos[mθ]

  Cos[αm,nct];

term3[r_, t_, θ_, m_, n_]:=b[m, n]BesselJ[m, αm,nr]Sin[mθ]

  Cos[αm,nct];

term4[r_, t_, n_]:=capa[n]BesselJ[0, α0,nr]Sin[α0,nct];

term5[r_, t_, θ_, m_, n_]:=capa[m, n]BesselJ[m, αm,nr]Cos[mθ]

  Sin[αm,nct];

term6[r_, t_, θ_, m_, n_]:=capb[m, n]BesselJ[m, αm,nr]Sin[mθ]

  Sin[αm,nct];

Clear[u]
```
$$u[r\_, t\_, th\_] = \sum_{n=1}^{5} \text{term1}[r, t, n] + \sum_{m=1}^{3}\sum_{n=1}^{3}$$
$$\text{term2}[r, t, th, m, n] + \sum_{m=1}^{3}\sum_{n=1}^{3} \text{term3}[r, t, th, m, n]+$$

$$\sum_{n=1}^{5} \text{term4}[r, t, n] + \sum_{m=1}^{3} \sum_{n=1}^{3} \text{term5}[r, t, \text{th}, m, n]$$

$$+ \sum_{m=1}^{3} \sum_{n=1}^{3} \text{term6}[r, t, \text{th}, m, n];$$

The solution is *compiled* in uc. The command Compile is used to compile functions. Compile returns a CompiledFunction which represents the compiled code. Generally, compiled functions take less time to perform computations than uncompiled functions although compiled functions can only be evaluated for numerical arguments.

```
uc = Compile[{r, t, th}, u[r, t, th]]
```

Next, we define the function tplot which uses ParametricPlot3D to produce the graph of the solution for a particular value of *t*. Note that the *x* and *y* coordinates are given in terms of polar coordinates.

```
Clear[tplot]
tplot[t_]:=ParametricPlot3D[{rCos[θ], rSin[θ], uc[r, t, θ]},
    {r, 0, 1}, {θ, -π, π}, PlotPoints → {20, 20}, BoxRatios
    → {1, 1, 1}, Axes → False, Boxed → False]
```

A table of nine plots for nine equally spaced values of *t* from $t = 0$ to $t = 1$ using increments of $1/8$ is then generated. This table of graphs is displayed as a graphics array in Figure 10-12.

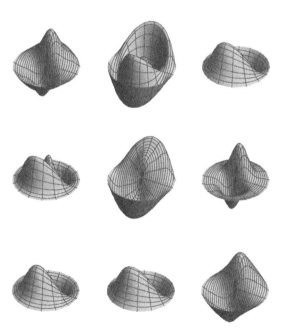

Figure 10-12   The drumhead for nine equally spaced values of *t* from $t = 0$ to $t = 1$

```
somegraphs = Table[tplot[t], {t, 0, 1, 1/8}];

toshow = Partition[somegraphs, 3];

Show[GraphicsGrid[toshow]]
```

Of course, we can animate the result as in the previous example using `Animate`. Be aware, however, that generating many three-dimensional graphics and then animating the results uses a great deal of memory and can take considerable time, even on a relatively powerful computer.

```
Animate[tplot[t], {{t, 0}, 0, 2}]
```

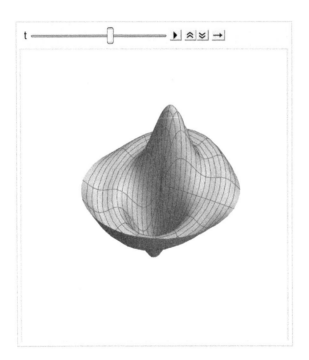

■

Appendix
Getting Started

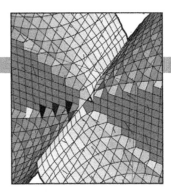

Introduction to Mathematica

Mathematica, first released in 1988 by Wolfram Research, Inc.,

```
http://www.wolfram.com/,
```

is a system for doing mathematics on a computer. Mathematica combines symbolic manipulation, numerical mathematics, outstanding graphics, and a sophisticated programming language. Because of its versatility, Mathematica has established itself as the computer algebra system of choice for many computer users. However, due to its special nature and sophistication, beginning users need to be aware of the special syntax required to make Mathematica perform in the way intended. You will find that calculations and sequences of calculations most frequently used by beginning users are discussed in detail along with many typical examples. In addition, the comprehensive index not only lists a variety of topics but also cross-references commands with frequently used options.

For information, including purchasing information, about Mathematica contact:

Corporate Headquarters:
Wolfram Research, Inc.
100 Trade Center Drive
Champaign, IL 61820
USA

telephone: 217-398-0700
fax: 217-398-0747
email: `info@wolfram.com`
web: `http://www.wolfram.com`

Europe:
Wolfram Research Europe Ltd.
10 Blenheim Office Park
Lower Road, Long Hanborough
Oxfordshire OX8 8LN
UNITED KINGDOM
telephone: +44-(0) 1993-883400
fax: +44-(0) 1993-883800
email: `info-europe@wolfram.com`

Asia:
Wolfram Research Asia Ltd.
Izumi Building 8F
3-2-15 Misaki-cho
Chiyoda-ku, Tokyo 101
JAPAN
telephone: +81-(0)3-5276-0506
fax: +81-(0)3-5276-0509
email: `info-asia@wolfram.com`

A Note Regarding Different Versions of Mathematica

With the release of Version 10 of Mathematica, many new functions and features have been added to Mathematica. We encourage users of earlier versions of Mathematica to update to Version 10 as soon as they can. All examples in *Differential Equations with Mathematica*, Fourth Edition, were completed with Version 10. In most cases, the same results will be obtained if you are using Version 8.0 or later, although the appearance of your results will almost certainly differ from that presented here. Occasionally, however, particular features of Version 10 are used and in those cases, of course, these features are not available in earlier versions. If you are using an earlier or later version of Mathematica, your results may not appear in a form identical to those found in this book: some commands found in Version 10 are not available in earlier versions of Mathematica; in later versions some commands will certainly be changed, new commands added, and obsolete commands removed. For details regarding these changes, please refer to Mathematica's extensive help facility that is discussed next. You can determine the version of Mathematica you are using during a given Mathematica session by entering either the command $Version or the command $VersionNumber. In this text, we assume that Mathematica has been correctly installed on the computer you are using. If you need to install Mathematica on your computer, please refer to the documentation that came with the Mathematica software package.

On-line help for upgrading older versions of Mathematica and installing new versions of Mathematica is available at the Wolfram Research, Inc. website:

$$\text{http://www.wolfram.com/.}$$

mma

Getting Started With Mathematica

We begin by introducing the essentials of Mathematica. The examples presented are taken from algebra, trigonometry, and calculus topics that you are familiar with to assist you in becoming acquainted with the Mathematica computer algebra system.

We assume that Mathematica has been correctly installed on the computer you are using. If you need to install Mathematica on your computer, please refer to the documentation that came with the Mathematica software package.

Start Mathematica on your computer system. Using Windows or Macintosh mouse or keyboard commands, activate the Mathematica program by selecting the Mathematica icon or an existing Mathematica document (or notebook), and then clicking or double-clicking on the icon.

Mathematica

If you start Mathematica by selecting the Mathematica icon, Mathematica's welcome window opens that allows you to perform a variety of actions.

If you select **New Document**, a blank untitled notebook is opened, as illustrated in the following screen shot.

When you start typing, the thin black horizontal line near the top of the window is replaced by what you type.

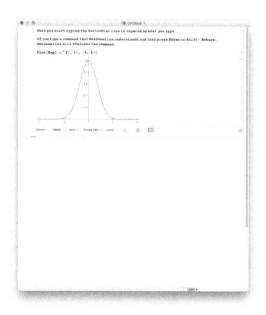

With some operating systems, **Enter** evaluates commands and **Return** yields a new line
The **Basic Math Input** palette:

Once Mathematica has been started, computations can be carried out immediately. Mathematica commands are typed and the black horizontal line is replaced by the command, which is then evaluated by pressing **Enter**. Note that pressing **Enter** or **Return** evaluates commands and pressing **Shift-Return** yields a new line. Output is displayed below input. We illustrate some of the typical steps involved in working with Mathematica in the calculations that follow. In each case, we type the command and press **Enter**. Mathematica evaluates the command, displays the result, and inserts a new horizontal line after the result. For example, typing N[, then pressing the π key on the **Basic Math Input** palette, followed by typing , 50] and pressing the enter key

$N[\pi, 50]$

3.1415926535897932384626433832795028841971693993751

returns a 50-digit approximation of π. Note that both π and Pi represent the mathematical constant π so entering N[Pi, 50] returns the same result.

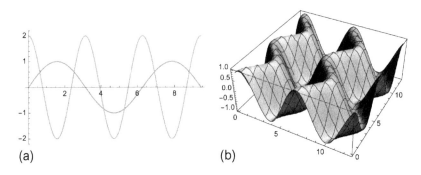

Figure A-1   (a) A two-dimensional plot. (b) A three-dimensional plot

The next calculation can then be typed and entered in the same manner as the first. For example, entering

```
p1 = Plot[{Sin[x], 2Cos[2x]}, {x, 0, 3Pi}]
```

graphs the functions $y = \sin x$ and $y = 2 \cos 2x$ and on the interval $[0, 3\pi]$ shown in Figure A-1(a). Similarly, entering

```
p2 = Plot3D[Sin[x + Cos[y]], {x, 0, 4Pi}, {y, 0, 4Pi}]
```

graphs the function $z = \sin(x + \cos y)$ for $0 \le x \le 4\pi$ and $0 \le y \le 4\pi$ shown in Figure A-1(b). The plots are shown side-by-side as in the figure with

```
Show[GraphicsRow[{p1, p2}]]
```

Notice that both of the following commands

```
Solve[x³ − 2x + 1 == 0]
```

$$\left\{ \{x \to 1\}, \left\{ x \to \frac{1}{2}\left(-1 - \sqrt{5}\right) \right\}, \left\{ x \to \frac{1}{2}\left(-1 + \sqrt{5}\right) \right\} \right\}$$

Notice that every Mathematica command begins with capital letters and the argument is enclosed by square brackets [. . .].

```
Solve[x^3 − 2 * x + 1 == 0]
```

$$\left\{ \{x \to 1\}, \left\{ x \to \frac{1}{2}\left(-1 - \sqrt{5}\right) \right\}, \left\{ x \to \frac{1}{2}\left(-1 + \sqrt{5}\right) \right\} \right\}$$

To type x^3 in Mathematica,

press the ▪ on the **Basic Math Input** palette, type x in the base position, and then click (or tab to) the exponent position and type 3.

solve the equation $x^3 - 3x + 1 = 0$ for x.

In the first case, the input and output are in **Standard Form** while in the second case the input and output are in **Traditional Form**. Move the cursor to the Mathematica menu,

 Mathematica   File   Edit   Insert   Format   Cell   Graphics   Evaluation   Palettes   Window   Help select **Cell**, and then **Convert To**, as illustrated in the following screen shot.

You can change how input and output appear by using **Convert To** or by changing the default settings. Moreover, you can determine the form of input/output by looking at the cell bracket that contains the input/output. For example, even though all three of the following commands look different, all three evaluate $\int_0^{2\pi} x^3 \sin x\, dx$.

Throughout *Differential Equations with Mathematica*, Fourth Edition, we display input and output using **Input Form** or **Standard Form**, unless otherwise stated.

To enter code in **Standard Form**, we often take advantage of the **Basic Typesetting** palette, which is accessed by going to **Palettes** under the Mathematica menu and then selecting **Other**

followed by **Basic Typesetting**.

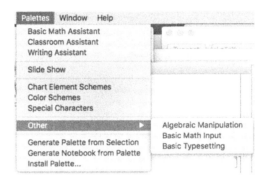

Use the buttons to create templates and enter special characters.

Mathematica sessions are terminated by entering Quit[] or by selecting **Quit** from the **File** menu, or by using a keyboard shortcut, like **command-Q**, as with other applications. They can be saved by referring to **Save** from the **File** menu.

Mathematica allows you to save notebooks (as well as combinations of cells) in a variety of formats, in addition to the standard Mathematica format.

Remark. Input and text regions in notebooks can be edited. Editing input can create a notebook in which the mathematical output does not make sense in the sequence it appears. It is also possible to simply go into a notebook and alter input without doing any recalculation. This also creates misleading notebooks. Hence, common sense and caution should be used when editing the input regions of notebooks. Recalculating all commands in the notebook will clarify any confusion.

Five Basic Rules of Mathematica Syntax

In order for the Mathematica user to take full advantage of this powerful software, an understanding of its syntax is imperative. Although all of the rules of Mathematica syntax are far too numerous to list here, knowledge of the following five rules equips the beginner with the necessary tools to start using the Mathematica program with little trouble.

1. The arguments of *all* functions (both built-in ones and ones that you define) are given in brackets [...]. Parentheses (...) are used for grouping operations; vectors, matrices, and lists are given in braces { ... }; and double square brackets [[...]] are used for indexing lists and tables.
2. Every word of a built-in Mathematica function begins with a capital letter.
3. Multiplication is represented by a * or space between characters. Enter 2*x*y or 2x y to evaluate $2xy$ *not* 2xy.
4. Powers are denoted by a ^. Enter (8*x^3)^(1/3) to evaluate $(8x^3)^{1/3} = 8^{1/3}(x^3)^{1/3} = 2x$ instead of 8x^1/3, which returns 8x/3.

5. Mathematica follows the order of operations *exactly*. Thus, entering `(1+x)^1/x` returns $\dfrac{(1+x)^1}{x}$ while `(1+x)^(1/x)` returns $(1+x)^{1/x}$. Similarly, entering `x^3x` returns $x^3 \cdot x = x^4$ while entering `x^(3x)` returns x^{3x}.

Remark. If you get no response or an incorrect response, you may have entered or executed the command incorrectly. In some cases, the amount of memory allocated to Mathematica can cause a crash. Like people, Mathematica is not perfect and errors can occur.

Getting Help From Mathematica

Becoming competent with Mathematica can take a serious investment of time. Hopefully, messages that result from syntax errors are viewed lightheartedly. Ideally, instead of becoming frustrated, beginning Mathematica users will find it challenging and fun to locate the source of errors. Frequently, Mathematica's error messages indicate where the error(s) has (have) occurred. In this process, it is natural that you will become more proficient with Mathematica. In addition to Mathematica's extensive help facilities, which are described next, a tremendous amount of information is available for all Mathematica users at the Wolfram Research website.

```
http://www.wolfram.com/
```

Note that many of these resources can be accessed from the Welcome screen when starting Mathematica.

One way to obtain information about commands and functions, including user-defined functions, is the command `?`. `?object` gives a basic description and syntax information of the Mathematica object `object`. `??object` yields detailed information regarding syntax and options for the object `object`.

EXAMPLE A.1: Use `?` and `??` to obtain information about the command `Plot`.

Solution: `?Plot` uses basic information about the `Plot` function.

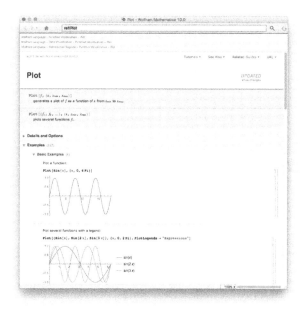

Observe that clicking on the << button accesses Mathematica's extensive help for the Plot function.

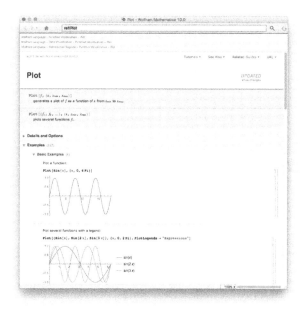

■

Options[object] returns a list of the available options associated with object along with their current settings. This is quite useful when working with a Mathematica command such as ParametricPlot which has many options. Notice that the default value (the value automatically assumed by Mathematica) for each option is given in the output.

EXAMPLE A.2: Use Options to obtain a list of the options and their current settings for the command ParametricPlot.

Solution: The command `Options[ParametricPlot]` lists all the options and their current settings for the command `ParametricPlot`.

■

As indicated above, `??object` or, equivalently, `Information[object]` yields the information on the Mathematica object `object` returned by both `?object` and `Options[object]` in addition to a list of attributes of object. Note that object may either be a user-defined object or a built-in Mathematica object.

EXAMPLE A.3: Use `??` to obtain information about the commands `Solve` and `Map`. Use Information to obtain information about the command `PolynomialLCM`.

Solution: We use `??`to obtain information about the commands `Solve` and `Map` including a list of options and their current settings.

Similarly, we use Information to obtain information about the command PolynomialLCM including a list of options and their current settings.

The command Names["form"] lists all objects that match the pattern defined in form. For example, Names["Plot"] returns Plot, Names["*Plot"] returns all objects that end with the string Plot, and Names["Plot*"] lists all objects that begin with the string Plot, and Names["*Plot*"] lists all objects that contain the string Plot. Names["form",SpellingCorrection->True] finds those symbols that match the pattern defined in form after a spelling correction.

EXAMPLE A.4: Create a list of all built-in functions beginning with the string Plot.

Solution: We use Names to find all object that match the pattern Plot.

```
Names["Plot"]
```

{Plot}

Next, we use Names to create a list of all built-in functions beginning with the string Plot.

```
Names["Plot*"]
```

{Plot, Plot3D, Plot3Matrix, PlotDivision, PlotJoined,

   PlotLabel, PlotLayout, PlotLegends, PlotMarkers,

   PlotPoints, PlotRange, PlotRangeClipping,

   PlotRangeClipPlanesStyle, PlotRangePadding,

   PlotRegion, PlotStyle, PlotTheme}

As indicated above, the ? function can be used in many ways. Entering ?letters* gives all Mathematica objects that begin with the string letters; ?*letters* gives all Mathematica objects that contain the string letters; and ?*letters gives all Mathematica commands that end in the string letters.

EXAMPLE A.5: What are the Mathematica functions that (a) end in the string Cos; (b) contain the string Sin; and (c) begin with the string Polynomial?

Solution: Entering

returns all functions ending with the string Cos, entering

returns all functions containing the string Sin, and entering

returns all functions that begin with the string Polynomial.

■

Mathematica Help

Additional help features are accessed from the Mathematica menu under **Help**. For basic information about Mathematica, go to **Help** and select **Help Browser...**

If you are a beginning Mathematica user, you may choose to select **Welcome Screen...**

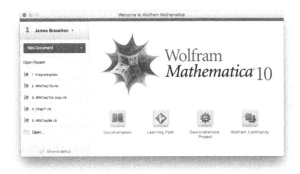

and then select **Wolfram Language and System Documentation Center**.

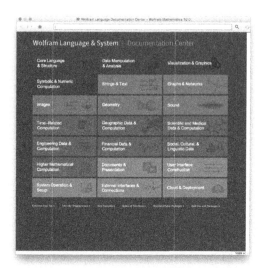

The most comprehensive way to learn Mathematica is probably from the **Mathematica Resources** page.

To obtain information about a particular Mathematica object or function, open the **Help Browser**, type the name of the object, function, or topic and press the **Go** button. Alternatively, you can type the name of a function that you wish to obtain

help about, select it, go to **Help** and then select **Find in Help...** as we do here with
the `DSolve` function.

A typical help window not only contains a detailed description of the command
and its options as well as hyperlinked cross-references to related commands.

The Mathematica Menu

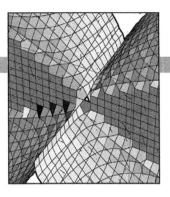

Mathematica   File   Edit   Insert   Format   Cell   Graphics   Evaluation   Palettes   Window   Help

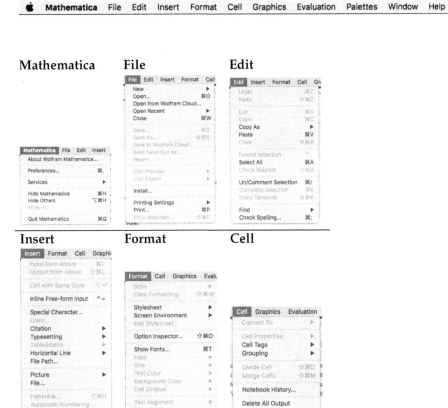

Mathematica

Mathematica	File	Edit	Insert
About Wolfram Mathematica...			
Preferences...	⌘,		
Services	▶		
Hide Mathematica	⌘H		
Hide Others	⌥⌘H		
Show All			
Quit Mathematica	⌘Q		

File

File	Edit	Insert	Format	Cell
New				▶
Open...				⌘O
Open from Wolfram Cloud...				
Open Recent				▶
Close				⌘W
Save...				⌘S
Save As...				⇧⌘S
Save to Wolfram Cloud...				
Save Selection As...				
Revert...				
CDF Preview				▶
CDF Export				▶
Install...				
Printing Settings				▶
Print...				⌘P
Print Selection...				⇧⌘P

Edit

Edit	Insert	Format	Cell	Gr
Undo				⌘Z
Redo				⇧⌘Z
Cut				⌘X
Copy				⌘C
Copy As				▶
Paste				⌘V
Clear				⇧⌘X
Extend Selection				^
Select All				⌘A
Check Balance				⇧⌘B
Un/Comment Selection				⌘/
Complete Selection				⌘K
Make Template				⇧⌘K
Find				▶
Check Spelling...				⌘;

Insert

Insert	Format	Cell	Graphic
Input from Above			⌘L
Output from Above			⇧⌘L
Cell with Same Style			⌥⏎
Inline Free-form Input			^=
Special Character...			
Color...			
Citation			▶
Typesetting			▶
Table/Matrix			▶
Horizontal Line			▶
File Path...			
Picture			▶
File...			
Hyperlink...			⇧⌘H
Automatic Numbering...			
Page Break			

Format

Format	Cell	Graphics	Evalu
Style			▶
Clear Formatting			⇧⌘⌫
Stylesheet			▶
Screen Environment			▶
Edit Stylesheet...			
Option Inspector...			⇧⌘O
Show Fonts...			⌘T
Face			▶
Size			▶
Text Color			▶
Background Color			▶
Cell Dingbat			▶
Text Alignment			▶
Text Justification			▶
Word Wrapping			▶

Cell

Cell	Graphics	Evaluation
Convert To		▶
Cell Properties		▶
Cell Tags		▶
Grouping		▶
Divide Cell		⇧⌘D
Merge Cells		⇧⌘M
Notebook History...		
Delete All Output		
Show Expression		⇧⌘E

Graphics　　Evaluation　　Palettes

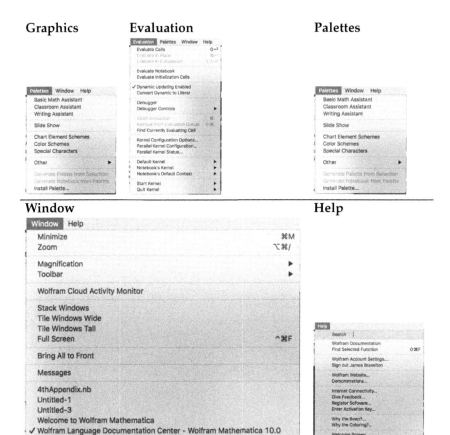

Window　　　　　　　　　　　　　　　　　　　Help

Bibliography

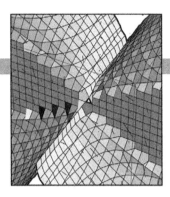

[1] Abell, Martha and Braselton, James, *Mathematica By Example*, Fourth Edition, Academic Press, 2008.

[2] Abell, Martha and Braselton, James, *Introductory Differential Equations*, Fourth Edition, Academic Press, 2014.

[3] Apostol, Tom, *Mathematical Analysis*, Second Edition, Pearson, 1974.

[4] Barnsley, Michael, *Fractals Everywhere: New Edition*, S Dover, 2012.

[5] Boyce, William E. and DiPrima, Richard C., *Elementary Differential Equations and Boundary-Value Problems*, Tenth Edition, Wiley, 2012.

[6] Coddington, Earl and Levinson, Norman, *Theory of Ordinary Differential Equations*, Krieger, 1984.

[7] Corduneanu, C., *Principles of Differential and Integral Equations*, Chelsea Publishing, 1977.

[8] Devaney, Robert L. and Keen, Linda (eds.), *Chaos and Fractals: The Mathematics Behind the Computer Graphics*, Proceedings of Symposia in Applied Mathematics, Volume 39, American Mathematical Society, 1989.

[9] Edwards, C. Henry and Penney, David E., *Calculus*, Sixth Edition, Pearson, 2002.

[10] Edwards, C. Henry and Penney, David E., *Differential Equations and Boundary Value Problems: Computing and Modeling*, Fourth Edition, Pearson, 2007.

[11] Gaylord, Richard J., Kamin, Samuel N., and Wellin, Paul R., *An Introduction to Programming with Mathematica*, Third Edition, Cambridge University Press, 2005.

[12] Giordano, Frank R., Weir, Maurice D., and Fox, William P., *A First Course in Mathematical Modeling*, Fourth Edition, Brooks/Cole, 2008.

[13] Graff, Karl F., *Wave Motion in Elastic Solids,* Oxford University Press/Dover, 1975/1991.

[14] Gray, Alfred, *Modern Differential Geometry of Curves and Surfaces with Mathematica,* Second Edition, CRC Press, 1997.

[15] Gray, John W., *Mastering Mathematica: Programming Methods and Applications,* Second Edition, Academic Press, 1997.

[16] Herriott, Scott R., *College Algebra through Functions and Models,* Preliminary Edition, Brooks Cole, 2002.

[17] Jordan, D.W. and Smith, Peter, *Nonlinear Ordinary Differential Equations: An Introduction to Dynamical Systems,* Third Edition, Oxford Applied and Engineering Mathematics, Oxford University Press, 1999.

[18] Kyreszig, Erwin, *Advanced Engineering Mathematics,* Eighth Edition, John Wiley & Sons, 1999.

[19] Larson, Roland E. and Hostetler, Robert P., *Calculus,* Tenth Edition, Brooks Cole, 2013.

[20] Maeder, Roman E., *The Mathematica Programmer II,* Academic Press, 1996.

[21] Maeder, Roman E., *Programming in Mathematica,* Third Edition, Addison-Wesley, 1997.

[22] Rabenstein, Albert L., *Elementary Differential Equations with Linear Algebra,* Fourth Edition, Cengage Learning, 1992.

[23] Robinson, Clark, *Dynamical Systems: Stability, Symbolic Dynamics, and Chaos,* Second Edition, CRC Press, 1999.

[24] Smith, Hal L. and Waltman, P., *The Theory of the Chemostat: Dynamics of Microbial Competition,* Cambridge University Press, 1995.

[25] Stewart, James, *Calculus,* Seventh Edition, Brooks Cole, 2012.

[26] Weisstein, Eric W., *CRC Concise Encyclopedia of Mathematics,* CRC Press, 1999.

[27] Wolfram, Stephen, *A New Kind of Science,* Wolfram Media, 2002.

[28] Wolfram, Stephen, *The Mathematica Book,* Fourth Edition, Wolfram Media, 2004.

[29] Zwillinger, Daniel, *Handbook of Differential Equations,* Second Edition, Academic Press, 1992.

Index

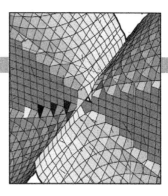

Note: Page numbers followed by *f* indicate figures and *b* indicate boxes.

A

Aging springs, 374–376
  application
    hearing beats, 376–377,
      377*f*
    resonance, 376–377, 377*f*
  initial displacement of, 374,
    375*f*
Animate, 334, 346–347, 365,
  386, 719–720, 808,
  825–826
Antibiotic production, 100–102
Apart, 59, 631, 702
Applications
  antibiotic production,
    100–102
  drug concentration, 88–89*b*
  first-order equations
    free-falling bodies, 177–186
    Newton's law of cooling,
      171–177
    orthogonal trajectories,
      133–145
  harvesting, 163–166
  kidney dialysis, 62–66
  modeling the spread of a
    disease, 108–114
  models of pursuit, 72–76
  oblique trajectories, 142–145
  radioactive decay, 146–147*b*
  series solutions

bessel function, 302–308,
  303*f*
  wave equation, 304–308
Array, 419
AspectRatio, 32, 134
  scaling graphics, 134
Assumptions, 614
Attractor, 448–449*b*
  Rössler, 448–449*b*
Autonomous system, 552
AxesLabel, 35–36
AxesStyle, 32

B

BasicTypesetting palette,
  415–416
Beam, deflection of, 394–397
Beats, 362
Bendixson's theorem, 556
Bessel-Fourier series, 768
Bessel function, 297–298
BesselI, 304
BesselJ, 288, 297–298
BesselJZeros, 302, 767, 824, 828
Bessel's equation, 3, 288*b*, 296*b*,
  767
BesselY, 288, 298–299
Bodé plots, 397–402
Boundary-value problems, 23,
  721–724
Building, vibration, 715–720

C

Catenary, 402–403
Cauchy-Euler equations
  higher-order equations,
    267–272
  parameters variation,
    272–274
  second-order equations,
    263–267
Ceiling function, 655–656
Center, 546
Chain rule, 805–806
Characteristic equation, 211,
  215, 265
Characteristic polynomial,
  429–430
Chemostat
  long food chain, 600–603
  simple food chain, 594
Chemostat growth, 159–162*b*
Chop, 829
Circuit, 390–393, 566, 679*b*
Circular plate, wave equation,
  304–308
Circular region, 816–833
  Laplace's equation, 817–820
  wave equation, 821–833
CMYKColor, 8
Competing species, 36*b*
Compiled function, 832

Complex conjugate
eigenvalues, 468–477,
546–549
Composing functions, 168
Computations matrices,
423–426
Constant coefficients. *See also*
Second-order equations
parameters variation,
nonhomogeneous
equations
higher-order equations,
259–262
second-order equations,
255–258
solving linear homogeneous
equations
higher-order equations,
215–223
second-order equations,
211–215
undetermined method,
nonhomogeneous
equations
second-order equations,
231–248
ContourPlot, 10–11, 68–69, 138,
141–142, 812, 813*f*
AxesOrigin, 81–82
AxesStyle, 81–82
Contours, 68–69, 138
ContourShading, 81–82, 136
ContourStyle, 138–139
Frame, 81–82
PlotPoints, 81–82, 136
Controlling the spread of a
disease, 514–525
Convolution integral, 668
Convolution theorem, 668–672
Cooling, newton's law of, 98*b*,
171–177
Corresponding homogeneous
equation, 82, 193
Cosine Series, 731–742
Coupled spring-mass systems,
704–709
Cramer's rule, 255–262
Curvature, 609–611
Cycle
limit, 41, 437

D

D, 196–197
D'Alembert's solution, 805–809
Damped motion

critically damped, 340
FindRoot, 345
overdamped, 340
sol1, 340
underdamped, 340
Decay
radioactive, 146–147*b*
Decibel, 397–398
Deflection of a beam, 394–397
Degenerate
stable node, 542
unstable node, 542
Delta function, 662–667
Density plot, 788–789, 812, 813*f*
Dependent
linearly, 195
Derivative
of a matrix, 434
Det, 195, 423
Determinant, 423–424
Dialysis, 62–66
Difference equations
logistic, 166–171
Differential equations, 2
boundary conditions, 14
boundary-value problems
auxiliary condition, 20
general solution, 19–20
second-order differential
equation, 23, 24*f*
corresponding homogeneous
equation, 82
exact, 76
forced motion, 352–353
homogeneous, 66
implicit solution, 9
independent variable, 9
initial-value problem, 330
linear, 5, 82–102, 193–199
first-order, 82
homogeneous, 193
L-R-C series, 391
numerical solutions, 7
order, 4
partial, 4
partial wave equation, 12*b*
plot, 8, 9*f*
separable, 52
solution
implicit, 9
independent variable, 9
replacement rule, 16
variation of parameters,
488–507

Differential geometry,
curvature, 609–611
Differentiation, implicit, 10
Diffusion, 574–585
population problems
(*See also* Ordinary
differential equations)
double-walled membrane,
through, 577–581
membrane, through,
574–577
population problems,
581–585
Dirac delta function, 634–635,
663
Direction field, 28–43, 437.
See also First-order linear
equations
competing species system,
37–38, 37*f*
graphing in, 36
initial-value problems, 86
Interpolating Function,
95–96
numerical solution to, 95*f*
plot, 95*f*
primary purpose of, 32
Rayleigh's equation, 39*b*,
40*f*, 42*f*
various solutions, 34–35, 35*f*,
85*f*, 87*f*
Dirichlet problem, 809–811
Discontinuity, jump, 618
Disease
controlling the spread of,
514–525
modeling the spread of,
108–114
DisplayFunction, 136
Double pendulum, 709–714
Drug concentration, 88–89*b*
DSolve, 14, 24, 46, 54–55, 67, 78,
135, 138, 187, 331, 463,
710, 722
Dt, 10–11
duffingpoincareplot, 323–325
Duffing's equation, 316*b*,
554–555

E

Egyptian sarcophagus, 148*b*
Eigensystem, 429–430, 463
Eigenvalue problems and
fourier series, 721–731,
743–767

boundary-value problems,
721–724
Cosine Series, 731–742
even extensions, 754–760
odd extensions, 754–760
periodic extensions,
754–760
Sine Series, 731–742
Sturm-Liouville problems,
729–731
Eigenvalues, 429–433, 604
Eigenvectors, 429–433
Endemic, 515
Envelope, 359
Epidemic disease, 514
Equations
Bessel, 288*b*, 296*b*, 767
Cauchy-Euler, 262–274
characteristic, 215, 265
Duffing's, 316*b*
heat, 785–797
Indicial, 289, 293–294
integral, 670–672
integrodifferential, 670–672
Laplace, 809–816
Legendre, 282*b*
Lorenz, 559
nonlinear, 308
potential, 809–811
Rayleigh, 26*b*, 39*b*
Rössler, 448–449*b*
Van-der-Pol, 187–188*b*,
435–436
wave, 798–805
Equation Van-der-Pol,
187–188*b*
Equilibrium point, 552
Euler equation, 262–274
Euler's method, 114,
525–530
Evaluate, 496
Even extensions, 754–760
Exact differential equation
direction field
ContourPlot, 81–82, 81*f*
StreamPlot, 79*f*
Exclusions, 618, 653
Existence and uniqueness
theorem, 446
Existence and Uniqueness
theorem, 45, 194
Exponential growth, 98*b*
Exponential matrix, 499–507
Exponential order, 617–619
ExpToTrig, 629, 765

F
Factors
GaussianIntegers, 212, 218
integrating, 83
Field, direction, 28–43
FindRoot, 149, 179, 727
First-order equations
exact, 76–82
homogeneous, 66–76
initial-value problem
direction field, 46*f*, 49–50,
49*f*, 51*f*
DSolve, 47, 49–50
Existence theorem, 45, 50
NDSolve, 46
Plot, 49–50, 49*f*
StreamPlot, 46
Uniqueness theorem, 45,
48–50
linear, 82
separable, 52
First-order homogeneous
equations, 142–143
First-order initial-value
problem, 145–146
First-order linear equations,
642
application, antibiotic
production, 100–102
computational errors, 92–93
direction field
initial-value problems, 86
InterpolatingFunction,
95–96
numerical solution to, 95*f*
Plot, 95*f*
various solutions, 85*f*, 87*f*
DSolve, 83–84, 88
GraphicsGrid, 91, 91*f*
integrating factor approach,
83–96
numerical integration
techniques, 93
parameters, 96–100
solutions, numerical
approximations
application, disease
spreading, 103–108
built-in methods, 103–108
Euler's method, 115–120
The Runge-Kutta method,
124–125
standard form of, 82
undetermined coefficients,
method, 96–100

unit step function, 88
First-order linear systems,
565–585
L-R-C circuits, 565–572
spring-mass systems,
572–574
FitzHugh-Nagumo equation,
522
Flatten, 166–167, 463, 502
Floor function, 655–656
Forced motion, 352–369, 673
beats, 362
differential equations,
352–353
GraphicsGrid, 365
natural frequency, 354–355
resonance, 356
steady-state solution, 363*f*
three-dimensional surface,
365
trigonometric identity, 356
undetermined coefficients,
352–353
Forcing function, 666
Fourier, 319–320
coefficient, 799–800
Cosine series, 739–742
CosSeries, 739–742
series, 743–779
sin series, 733
transform, 319–320
Free-falling bodies
initial-value problem, 184,
186
velocity, 181
velocity function, 181–182
Free vibration
three-story building, 715–720
Frequency
natural, 354–355
FresnelS, 21
Frobenius, method of, 289–302
FullSimplify, 638
ExpToTrig, 629
Function
dirac delta, 663
even, 762–763
idealized unit impulse, 663
Jacobi elliptic, 444*b*
odd, 765, 766*f*
periodic, 652–662, 754
pure, 616, 627
unit impulse, 663
unit step, 643
weighting, 730

Functions
  Bessel, 297–298
  composing, 168
  envelope, 360, 361*f*
  spring, 332–333
  unit step, 88
Fundamental matrix, 459
Fundamental set of solutions,
    200–205, 456

G
Gain, 397–398
Gibbs phenomenon, 737–738
Graphics
  LogLogPlot, 400
GraphicsArray, 14, 302, 333–334
GraphicsGrid, 91, 169, 170*f*,
    190, 305–307, 333
GraphicsRow, 17, 30, 31*f*, 68–69,
    139
Graphing functions, 618

H
Hard springs, 372–373
  initial velocity, 374*f*
  physical system, 372
Harmonic motion
  aging springs, 374–376
  DSolve, 331
  forced, 352–369, 674*f*
  GraphicsArray, 333–334, 333*f*
  hard springs, 372–373
  simple, 329–339
  soft springs, 369–372
  zigzag, 332–333
Harvesting, 163–166
Heat equation, 4, 785–797
  insulated boundary,
    794–797
Heaviside Theta, 643, 646
Herd immunity, 516–517
Higher-order differential
    equations
  applications
    bodé plots, 397–402
    the catenary, 402
    long beam deflection,
      394–397
    L-R-C circuits, 390–393
  harmonic motion
    aging springs, 374–376
    forced motion, 352–369
    hard springs, 372–373
    simple, 329–339
    soft springs, 369–372

the pendulum problem
  DSolve, 381, 383
  initial-value problem, 381,
    383
  NDSolve, 381–382
  nonlinear equation,
    379–380
  numerical solution,
    381–382
  pen, 382
  Plot, 383, 384*f*
  PointSize, 384–385
  toshow2, 389, 389*f*
Higher-order equations,
    248–255, 259–262
  Cauchy-Euler equations,
    267–272
  constant coefficients
    parameters variation,
      nonhomogeneous
      equations, 259–262
    solving linear
      homogeneous
      equations, 215–223
    undetermined method,
      nonhomogeneous
      equations, 248–255
Homogeneous, 6
  linear system, 461–487
Homogeneous boundary
    conditions, 786–790
Homogeneous differential
    equation, 66
  direction field
    application, pursuit
      model, 72–76
    ContourPlot, 68–69
    DSolve, 67, 71
    graphs of, 71*f*
    Plot, 71, 71*f*, 75–76, 76*f*
    StreamPlot, 71, 71*f*
    various solutions of, 69*f*
Homogeneous equation
    corresponding, 82
Homogeneous first order
    equations, 66–76
Hooke's law, 329–330,
    572
HypergeometricU, 293

I
Idealized unit impulse
    function, 663
IdentityMatrix, 418–419
Implicit differentiation, 10

Implicit solution, 9
Improved Euler's method,
    120–124
Impulse, 662
  idealized, 663
Impulse functions, 662–667
Indicial equation, 289,
    293–294
Initial-boundary value
    problems (IBVP),
    786–787
Initial-value problems, 23,
    477–480, 635–642, 722.
    See also Separable
    differential equation
  ContourPlot, 57
  direction field, 58*f*, 59*f*, 62*f*
  DSolve, 57
  Integrate, 57
  kidney dialysis, application,
    62–66
  StreamPlot, 55
Integral
  convolution, 668
  equation, 670–672
  of a matrix, 434
Integrate, 52–53, 57, 614–615,
    656
Integrating factor, 83
Integrodifferential equations,
    670–672
Interpolating function, 95–96,
    106
Inverse, 423–424, 479
Inverse Laplace transform,
    626–635, 712
Irreducible quadratic factors,
    632–633
Irregular singular points, 287
Isotherms, 137*b*

J
Jacobian, 552
Jacobian matrix, 587, 604
Jacobi elliptic functions,
    444*b*
JacobiSN, 444
Jump discontinuity, 618

K
Kidney dialysis, 62–66
Kirchhoff's Law, 390–391
  current law, 565
  voltage law, 565

L

LaguerreL, 293
Laplace's equation, 5, 809–816
  circular region, 817–820
Laplace transform, 613–625,
    652–662, 711
  applications of
    coupled spring-mass
      systems, 704–709
    double pendulum, 709–714
  convolution theorem,
      668–670
  definition of, 613–617
  delta function, 662–667
  exponential order, 617–619
  integral and
      integrodifferential
      equations, 670–672
  inverse, 626–635
  inverse laplace transform
    integral, 633–635
    irreducible quadratic
      factors, 632–633
    linear factors, 629–631
    repeated linear factors,
      631–632
  jump discontinuities,
      617–619
  L-R-C circuits, 678–685
  methods, 690–704
  periodic functions, 652–653
  piecewise-continuous forcing
      functions, 648–652
  piecewise-continuous
      functions, 617–619
  population problems,
      685–686
  properties of, 620–625
  shifting property, 620
  solving initial-value
      problems, 635–642
  spring-mass systems,
      673–678
  unit step function, 643–648
Laplacian, in polar coordinates,
    304
LegendreP, 285–286
LegendreQ, 285–286
Legendre's equation, 282*b*
Limit cycle, 41, 437
Linear differential equation, 5,
    82, 193–199
  homogenous, 193
Linear equations systems,
    426–429

Linear homogeneous,
    differential equation, 193
Linearization, 535, 550–551
Linearly dependent functions,
    195
LinearSolve, 483, 486–487
Linear systems, 439, 451–461
  homogeneous, 461–487
  nonhomogeneous, 488–507
ListLinePlot, 166–167, 169
ListLogPlot, 321
ListPlot, 166–167
LogicalExpand, 278
Logistic difference equation,
    166–171
Logistic equations, 60
  applications, 166–171
  carrying capacity, 153–154
  direction field, 157*f*, 165*f*
  FindRoot, 166*f*
  growth rate, 153–154
  with predation, 156–157*b*
  solutions, 165*f*
  StreamPlot, 158
LogLogPlot, 400
Long food chain, chemostat,
    600–603
Lorenz equations, 559
Lorenzsol function, 559
Lotka-Volterra system, 586–603
L-R-C circuits, 390–393,
    565–566, 678–685
  One Loop, 566–569
  Two Loops, 569–572

M

Malthus model, 145–151
Manipulate function, 437–439
Map, 41, 90–91, 106–107, 202,
    605–606, 616, 625,
    713–714
Matrices, 415–421
Matrix
  calculus, 434–435
  derivative, 434
  determinant, 423
  exponential, 500
  fundamental, 459
  integral, 434
  inverse, 423–424
  Jacobian, 550–551, 587, 604
  transpose, 423
  variational, 550–551, 587
MatrixExp, 500
MatrixForm, 415–416

Membrane diffusion, 574–577
Method of Frobenius, 289–302
Modeling the Spread of a
    Disease, 108–114
Models of pursuit, 72–76
Motion
  damped, 339–352
  forced, 352–369, 673
  harmonic, 329–377
  simple harmonic, 329–339
Multiplicity, 215

N

N, 94
Natural frequency, 354–355
NDSolve, 20, 22, 26, 39–41,
    92–95, 103, 187, 441, 507,
    604*b*
Nest, 168
Nested lists, 415–421
Neumann problem, 809–811
Newton's Law of Cooling, 98*b*
  FindRoot, 173–174
  first-order initial-value
      problem, 171–172, 175
Newton's second law of
    motion, 177, 330
NIntegrate, 824
Node
  degenerate stable, 542
  degenerate unstable, 542
  stable, 551
  unstable, 536
Nonhomogeneous boundary
    conditions, 790–793
Nonhomogeneous equations
  constant coefficients,
      parameters variation
      higher-order equations,
      259–262
    second-order equations,
      255–258
  constant coefficients,
      undetermined method
      higher-order equations,
      248–255
    second-order equations,
      231–248
  general solution, 225
Nonhomogeneous equation,
    undetermined
    coefficients, 488–507
Nonhomogeneous first-order
    systems, 488–507

Nonhomogeneous linear
  system, 488–507
Nonlinear equations, 308
  ContourPlot, 314–315
  DisplayFunction, 317
  DSolve, 309
  duffingplot, 316–317
  duffingpower, 320–321
  Duffing's equation, 316b, 317f
  first-order homogeneous,
    313–314
  GraphicsGrid, 317
  initial-value problem, 313
  ListLogPlot, 321
  Manipulate function, 318,
    320f
  Plot, 313, 316
  second-order equation,
    321–322
  unique solution, 315f
Nonlinear systems, 535,
    550–551, 585–611.
    See also Ordinary
    differential equations
  Curvature, 609–611
  Long Food Chain in a
    Chemostat, 600–603
  Lotka-Volterra system,
    586–594
  Simple Food Chain in a
    Chemostat, 594–600
  variable damping, 603–609
Normal modes, 12b, 278
NRoots, 717–718
NSolve, 38, 149
Nuclides, 146–147b
Numerical
  Steady-state, 362
  transient, 362
Numerical methods, 507–534.
    See also Ordinary
    differential equations
  built-in, 507–514
  Euler's method, 114, 525–530
  improved Euler's method,
    120–124
  NDSolve, 103, 507
  Runge-Kutta method,
    124–125, 531–534

O
Oblique trajectories, 142–145
Odd extensions, 754–760
One-dimensional heat
  equation, 785–797

Options, assumptions, 614
Order
  exponential, 617
  reduction of, 207–210
Order of a differential
  equation, 4
Ordinary differential
  equation, 2
  Bessel's equations, 3
  diffusion and population
    problems
    double-walled membrane,
      through, 577–581
    membrane, through,
      574–577
    population problems,
      581–585
  first-order linear systems
    L-R-C circuits, 565–572
    spring-mass systems,
      572–574
  homogeneous linear systems
    complex conjugate
      eigenvalues, 468–477
    distinct real eigenvalues,
      462–467
    initial-value problems,
      477–480
    repeated eigenvalues,
      480–487
  matrix algebra and calculus
    computations matrices,
      423–426
    eigenvalues, 429–433
    eigenvectors, 429–433
    elements of matrices,
      421–423
    linear equations, 426–429
    matrices, 415–421
    matrix calculus, 434–435
    nested lists, 415–421
    vectors, 415–421
  nonhomogeneous first-order
    systems
    exponential matrix,
      499–507
    undetermined coefficients,
      489–493
    variation of parameters,
      493–499
  nonlinear systems, 6–7
    Curvature, 609–611
    Long Food Chain in a
      Chemostat, 600–603

Lotka-Volterra system,
    586–594
  Simple Food Chain in a
    Chemostat, 594–600
  variable damping, 603–609
numerical methods
  built-in, 507–514
  Euler's method, 525–530
  Runge-Kutta method,
    531–534
predator-prey equations, 3
systems of equations
  linear systems, 451–461
  preliminary theory,
    439–451
Ordinary differential
  equations, 249
Orthogonal curves, 134
Orthogonal trajectories
  application, 142–145
  ContourPlot, 136
  DSolve, 135
  family, 135, 139f

P
Palettes, 415–416
Parameters variation,
    nonhomogeneous
    equations. See also
    Constant coefficients
  higher-order equations,
    259–262
  second-order equations,
    255–258
ParametricPlot command, 41,
    437, 463, 611, 691–692,
    692f
ParametricPlot3D, 307, 476, 832
Parseval's Equality, 765–767
Part, 575
Part [[...]], 16
Part ([[...]]), 10–11, 423
Partial differential equations, 4,
    781–785
  D'Alembert's solution,
    805–809
  first-order derivative, 4b, 5
  heat, 785
  heat equation, 4
  highest-order derivatives,
    4b, 5
  Laplace's equation, 4,
    809–816
  one-dimensional heat
    equation

homogeneous boundary
conditions, 786–790
insulated boundary,
794–797
nonhomogeneous
boundary conditions,
790–793
potential equation, 809–811
separation of variables,
781–785
two dimensions problems
circular region, 816–833
Laplace's equation,
809–820
wave equation, 821–833
Particular solution, 224
Partition, 317, 333, 593
Pendulum, double, 709–714
Pendulum equation, 6
The Pendulum problem
DSolve, 381, 383
initial-value problem, 381,
383
NDSolve, 381–382
nonlinear equation, 379–380
numerical solution, 381–382
pen, 382
Plot, 383, 384*f*
PointSize, 384–385
toshow2, 389, 389*f*
Period, 652
Periodic extensions, 754–760
Periodic function, 652–662
Phase shift, 397–398
Picard-Lindelöf Existence
theorem, 45, 50
Piecewise-continuous forcing
functions, 648–652
Piecewise continuous function,
617–619, 626
Piecewise-defined function,
618, 619*f*
Piecewise smooth, 760
"Pitchfork Diagram,", 167*f*
Play, 376
Plot, 722
exclusions, 618, 653
Plot3D, 14
PlotField, 517
PlotLabel, 35–36
PlotRange, 32
Plots, 8, 331
AspectRatio, 32, 134
AxesStyle, 32
Bodé, 397–402

DisplayFunction, 136
PlotRange, 32
Points
equilibrium, 551
ordinary, 275–287
Poincaré, 323–325
rest, 552
saddle, 536
singular, 275
irregular, 287
regular, 287
Polar coordinates, Laplacian,
304
Population growth/decay
applications
harvesting, 163–166
logistic difference
equations, 166–171
carrying capacity, 153–154
first-order differential
equations, 145
growth rate, 153–154
logistic equation, 152–162
malthus model, 145–151
population decreases,
145–146
population increases,
145–146
radioactive decay, 146–147*b*
United States, 150
Population problems, 145, 150,
581–585, 685–686
Potential equation, 809–811
Power series solutions, 275–287
Power spectrum, 319–320
Predator-Prey, 586–603
chemostat, 594, 600–603
Principle of Superposition, 201,
783
Problem
boundary-value, 23, 721–724
eigenvalue, 721–731
initial-value, 23
Sturm-Liouville, 729–731
Product method, 783
p-series, 767
Pure function, 90, 616, 627
Pursuit models, 72–76

Q
Quasiperiod, 345

R
Radioactive decay, 146–147*b*
Radioactivity, 146–147*b*

RandomInteger, 418–419
RandomReal, 418–419
Rayleigh's equation, 26*b*, 39*b*
Real distinct eigenvalues,
535–541
Reduction of order, 207–210
Regular singular points, 287
Repeated linear factors,
631–632
ReplaceAll (/.), 22, 32
Replacement rule, 16
Resonance, 356, 376–377
Rest point, 552
Root, 717
Rössler attractor, 448–449*b*
Rössler equation, 448–449*b*
RowReduce, 426–429
Runge-Kutta method, 124–125,
128, 525, 531–534

S
Saddle point, 536
Scaling graphics, 134
Second law of motion,
Newton's, 330
Second-order equations,
211–215, 231–248,
255–258
Cauchy-Euler equations,
263–267
constant coefficients
parameters variation,
nonhomogeneous
equations, 255–258
solving linear
homogeneous
equations, 211–215
undetermined method,
nonhomogeneous
equations, 231–248
Second-order
nonhomogeneous
equations, 390
Second-order ordinary
differential equations,
799–800
Self-adjoint form, problem,
729–730
Separable differential
equation, 52
initial-value problem
ContourPlot, 57
direction field, 58*f*, 59*f*, 62*f*
DSolve, 57
Integrate, 57

Separable differential equation
(*Continued*)
kidney dialysis,
application, 62–66
StreamPlot, 55
Separation of variables,
783–785
Series solutions, 278
application
bessel function, 302–308,
303*f*
wave equation, 304–308
fourier, 743–767
sine, 731–739
frobenius method, 289–302
generalized fourier, 767–779
ordinary points
convergence of, 282
Legendre's equation,
282*b*
p, 767
singular points
irregular, 287
polynomial coefficients,
287
regular, 287–289
Shift, phase, 397–398
Short, 498
Show, 14, 32, 136
GraphicsRow, 17
Simple food chain, chemostat,
594
Simple harmonic motion,
329–339
Simplify, 8
ExpToTrig, 629
Sine Series, 731–742
Singular points, 275
SIR model, 514–515, 519*f*,
520*f*
SIS model, 108–109
Smooth, piecewise, 760
Soft springs, 369–372
NDSolve, 371–372
physical system, 369
three-dimensional plot, 370*f*
Solgraph, 189
Solution, 7–19
envelope, 359
fundamental set, 200–205
general, 19–20, 201–202
numerical
Euler's method, 114
Improved Euler's method,
120–124

NDSolve, 103
Runge-Kutta method,
124–125, 128
particular, 224
Plot, 331, 331*f*
Solve, 10–11, 38, 88, 428, 483,
515
SolveAlways, 231, 249, 630
Solving linear homogeneous
equations. *See also*
Constant coefficients
constant coefficients
higher-order equations,
215–223
second-order equations,
211–215
higher-order equations,
215–223
second-order equations,
211–215
Sound, 376
Spectrum, power, 319–320
Spiral
stable, 546
unstable, 546
Spring-mass systems, 572–574
coupled, 704–709
Stable
node, 551
spiral, 546
Standard form, 275
Steady-state charge, 393
Steadystate solution, 362
Steady-state temperature,
790–792
StreamDensityPlot, 28, 36
StreamPlot, 28, 36, 437,
442–443, 463
Sturm-Liouville problem,
729–731
self-adjoint form, 729–730
Sum, 653, 656
Superposition, principle of,
201, 783
System, 439
autonomous, 552
of differential equations, 439
linear, 439, 461–487
nonlinear, 551
undetermined coefficients,
488–507
Systems of equations. *See also*
Ordinary differential
equations
linear systems, 451–461

preliminary theory, 439–451
Systems of linear equations,
426–429

T
Table, 14, 32, 41, 47–48, 203,
307, 419, 605
TableForm, 100–101, 625
Tautochrone, 686–690
Temperature, 137*b*
steady-state, 790–792
transient, 790–792
Terminology, 275
Thread, 485, 497
Toplot, 19–20
Trajectories, oblique, 142–145
Transform, Laplace, 319–320,
613–625
Transient solution, 362
Transient temperature, 790–792
Transpose, 423
Two dimensions problems.
See also Partial
differential equations
circular region, 816–833
Laplace's equation, 809–820
wave equation, 821–833
Two-point boundary value
problem, 721–722

U
Uapprox, 804–805
Undetermined coefficients,
229–255, 488–507
Union, 517–518, 536, 605
Uniqueness and existence
theorem, 194, 446
Unit impulse function, 663
UnitStep, 88, 619, 643–648
Unstable
node, 536
spiral, 546

V
Vanderpol, 188
Van-der-Pol's equation,
187–188*b*, 435–436, 603,
604*b*
Variational matrix, 550–551, 587
Variation of parameters,
256–258, 260–262,
488–507
Vectors, 415–421
Verhulst equation, 60, 152

Vibration of a building, 715–720
Vital dynamics, 515

W
Wave equation, 12*b*, 798–805,
821–833

on a circular plate,
304–308
circular region, 821–833
D'Alembert's solution,
805–809
Weighting function, 730

Wolfram's CDF Player, 41
Wronskian, 195, 197, 260, 456
function, 198*f*

Z
Zigzag, 332

Printed in the United States
By Bookmasters